CULTURE OF ANIMAL CELLS

A Manual of Basic Technique

Third Edition

R. IAN FRESHNEY
Department of Medical Oncology
CRC Beatson Laboratories
University of Glasgow

WILEY-LISS
A JOHN WILEY & SONS, INC., PUBLICATION
New York • Chichester • Brisbane • Toronto • Singapore

Address All Inquiries to the Publisher
Wiley-Liss, Inc., 605 Third Avenue, New York, NY 10158-0012

Printed in the United States of America.

While the author and publisher believe that drug selection and dosage and the specification and usage of equipment and devices, as set forth in this book, are in accord with current recommendations and practice at the time of publication, they accept no legal responsibility for any errors or omissions, and make no warranty, express or implied, with respect to material contained herein. In view of ongoing research, equipment modifications, changes in governmental regulations and the constant flow of information relating to drug therapy, drug reactions, and the use of equipment and devices, the reader is urged to review and evaluate the information provided in the package insert or instructions for each drug, piece of equipment, or device for, among other things, any changes in the instructions or indication of dosage or usage and for added warnings and precautions.

Library of Congress Cataloging-in-Publication Data

Freshney, R. Ian.
Culture of animal cells : a manual of basic technique / R. Ian Freshney.—3rd ed.
p. cm.
Includes bibliographical references and index.
ISBN 0-471-58966-7
1. Tissue culture. 2. Cell culture. I. Title.
QH585.2.F74 1994
591′.0724-dc20 93-34771
CIP

The text of this book is printed on acid-free paper.

10 9 8 7 6 5 4 3 2 1

Contents

3 Design and Layout of the Laboratory

4 Equipping the Laboratory

5 Aseptic Technique

6 Laboratory Safety and Biohazards

7 The Culture Environment: Substrate, Gas Phase, Medium and Temperature

8 Preparation and Sterilization

9 Disaggregation of the Tissue and Primary Culture

10 Maintenance of the Culture: Cell Lines

11 Cloning and Selection of Specific Cell Types

12 Physical Methods of Cell Separation

13 Characterization

17 Instability, Variation, and Preservation

18 Quantitation and Experimental Design

19 Measurement of Viability and Cytotoxicity

20 Culture of Specific Cell Types

Preface to the Third Edition

In this, the second revision of *Culture of Animal Cells,* I have attempted a more extensive rewrite of some elements of the text, but have nevertheless felt restrained from eliminating much of the earlier work. This is not to imply that tissue culture has not progressed, for it has, quite significantly, in many areas; but, nevertheless, the basic start-up technology remains fundamentally the same. With this in mind, I have not attempted to alter sections where the basic technology is still relevant, such as simple primary culture, serial passage, characterization, cell preservation, and so forth, but I have updated sections such as specialized cell culture, chromosome painting, and DNA fingerprinting, where I have enlisted the help, as in the second edition, of those who are more familiar with the techniques than I am. Where I have found the greatest difficulty is in the interface between molecular biology and tissue culture. While the one is dependent on the other, it has been difficult to define a limit where tissue culture technology stops and molecular biology begins. I have not attempted to cover the vast expansion in cell culture-based gene technology that has emerged in the past decade; others have done this much better than I could, and it is not the remit of this text. It would, however, be remiss not to indicate where the bridges occur, so I have included previous and new contributions on gene technology where I have felt them particularly appropriate to the handling, characterization, and utilization of cell cultures.

While I have attempted to update many references in areas where there has been further development, I make no apology for retaining references that are now more than 20 or 30 years old, if they remain key references in an established technology that is now standard and, for the present needs at least, does not require further development. Puck and Marcus [1955] cloned cells by dilution and established a method now in routine use in many laboratories; Lovelock and Bishop [1959] demonstrated the superiority of dimethyl sulfoxide as a preservative in cell freezing. Neither of these need to be superseded by later references, as they established the technology now in current use without substantial modification. Having said this, however, I have not always cited the original reference in a particular area, but often a more recent development, which will in turn cite earlier literature.

Modern science and technology have become increasingly dependent on sophisticated technology. Procedures for antibody staining, ELISA, molecular probe analysis, cytotoxicity assay, and many others are now available in kit form, enabling easier and more rapid estimations of gene regulation and cell product formation, but at a considerable cost. The saving is in time and productivity, but, nevertheless, for a nonprofit laboratory with limited income, this presents problems in attracting sufficient running costs. Some of the recommendations that I have made will, I hope, still be operable on a restricted budget, even if more generous circumstances do make things easier.

One of the most exciting aspects about the current update is to see to what extent tissue culture has become an accepted technology in areas where it was previously an exploratory fringe. Industry has accepted the benefit, for the present at least, of producing biopharmaceuticals in cultured animal cells, *in vitro* assays for cytotoxicity and mutagenesis are a standard part of the repertoire in industrial toxicology, assays for inflammation are appearing, and the combination of gene technology and tissue replacement herald an era that we might have regarded as pure science fiction a few years ago.

At the mechanistic level, the fundamental understanding of the organization of the genome, the regulation of gene transcription, intracellular and extracellular mechanisms of growth control, signal transduction, and the biological basis of the specificity of cell interaction, both contact-mediated and by diffusible signals, have taken quantum steps forward. Transformation, seen initially as a major problem in cell line stability and associated with malignancy, is now seen in the light of controlled immortalization as a valuable tool at the disposal of the cell culturist presenting previously unsuspected possibilities for exploitation and resource management. Introduction of these genes into transgenic mice now makes it possible to isolate cell lines from several tissues that are already immortal but still phenotypically valid.

We are probably approaching a stage in cell and molecular biology where the prospects for the manipulation of the genome, and the regulation of product expression both *in vitro* and in transplants, are almost unlimited by *in vitro* technology, and the rising questions are becoming regulatory and ethical rather than scientific. Should neurons from one individual be transplanted into another, particularly after genetic manipulation? Is it ethical to use human fetal material for *in vitro* studies? Are genetically modified transformed cells appropriate vehicles for the transplantation of normal genes into genetically deficient individuals? I have little difficulty in satisfying my own conscience on these matters, but others may not feel the same, and there is an increasing need for debate.

Although the challenge of getting cells to grow *in vitro* has been met, and the diversity of cell types increases constantly, tissue culture is more in the public eye than ever. For some it presents an opportunity to reduce unnecessary animal experimentation, for others the ability to produce innovative pharmaceuticals at economically acceptable rates, while for yet others it still provides the only medium in which to explore the intricacies of cell regulation and the potential for medical intervention.

With each new edition, I acquire an extending list of people on whom I depend for advice and information on new developments. Some are well-recognized colleagues who have contributed directly with protocols or useful discussion, some contacts are transitory and the names not always recollected, but all are greatly appreciated and their advice incorporated whenever possible. In particular, I would like to recognize yet again the practical guidance of my wife Mary, as well as her many hours of proofreading. Others include Nicol Keith, Bob Brown, and Paul Workman, who have helped to nudge my rather traditionalist beliefs into the realm of modern molecular biology and molecular pharmacology. I would also like to thank Fiona Conway, Joanne Thomson and Liz Gordon for help in searching and creating databases for literature references and suppliers addresses and Fiona Cobban and David Talloch for additional illustrations.

Finally, I would like to acknowledge with sincere gratitude all those who have taken the trouble to write to me or to Wiley-Liss with advice and constructive criticism on the first and second editions. It is pleasant and satisfying to hear from those who have found the book beneficial, but even more important to hear from those who have found deficiencies, some of which I can attempt to rectify. I can only hope that those of you who use this book retain the same excitement that I feel about the future prospects emerging in this field.

Preface to the Second Edition

In revising *Culture of Animal Cells* I have tried to keep the emphasis on the practical aspects of cell culture and have discussed the theoretical background only when it seemed necessary to the understanding of the technique or the status of the culture. For example, cell transformation and some of its implications are dealt with more fully to help the reader to appreciate the phenotypic properties that these cells might be expected to express and the roles that they might usefully play in experimental studies and commercial exploitation.

Major changes have been introduced in the presentation of serum free medium formulations as these have gained more general acceptance, and some have become commercially available, since publication of the first edition. In parallel with this, and in many cases as a direct consequence, the culture of specific cell types such as epidermal keratinocytes, melanocytes, and breast epithelium have become more feasible, so a number of protocols are included for such specialized cultures.

To enable these areas to be covered more effectively, I have enlisted the help of experts in each respective field to present protocols from their own experience where I feel my own expertise is insufficient. These specialist protocols cover areas of new technology, such as somatic hybridization and production of hybridomas as well as the culture of specific cell types, and have been presented in the same style as the previous protocols. I am very grateful to these new contributors and feel that they have extended the scope of the text more than I could have hoped to do alone. In some cases these protocols will be sufficient for readers to fulfill their needs without further recourse to the literature, but to satisfy those whose demands are greater, or where the technique is more complex, the appropriate references are provided.

A more extensive treatment has also been given to cytotoxicity assay and the culture of tumor cells, particularly from human tumors, in line with the emphasis that these techniques are currently being given in hospitals, basic research laboratories, and the biotechnology and drug industry.

In addition to the contributors of specialized protocols referred to above, I am again indebted to my colleagues in the Department of Medical Oncology including Jane Plumb, Stephen Merry, Carol McCormick, Alison Mackie, and Ian Cunningham, and a succession of graduate and undergraduate students including John McLean, Alison Murray, Jim Miller, Iain Singer, Barbara Christie, and Alan Beveridge who have provided data and ideas. While trying to answer their questions, I was stimulated into thinking more about the potential needs of the reader.

My thanks are also due to Mrs. Rae Fergusson for typing new material faster than I could generate it and handling my poor handwriting and illegible corrections with unbelievable accuracy.

Most of all I would like to thank my wife and family for their continuing help and encouragement. They provided much practical help, advice, and moral support. In particular my wife's many hours collating, referencing, and proof reading, have spared me many hours of often tedious work.

Preface to the First Edition

Tissue culture is not a new technique. It has been in existence since the beginning of this century and has passed through its simple exploratory phase, a later expansive phase in the 1950's, and is now in a phase of specialization concerned with control mechanisms and differentiated function. Matching the current trends, recent additions to the range of available tissue culture books have been concerned with specialized techniques and the result of this is that the basic procedures have become a little neglected.

It has been my objective in preparing this book to provide the novice to tissue culture with sufficient information to perform the basic techniques. It is anticipated that the reader will have a fundamental grasp of elementary anatomy, histology, cell physiology, and the basic principles of biochemistry, but will have had little or no experience in tissue culture. This book should prove useful at the advanced undergraduate level for technicians in training, for graduate studies, and at the post-doctoral level. It is intended as an introduction to the theory of the technique, and biology of cultured cells as well as a practical, step-by-step guide to procedures, and should be of value to anyone without any, or with little, prior experience in tissue culture. Of necessity, some of the more exciting developments in recent years, e.g., production of monoclonal antibodies by hybridoma cultures, can only be described briefly and references provided to further reading.

A list of reagents and commercial suppliers is located at the end of the book. Occasionally, a supplier's name is incorporated in the text but in most cases reference should be made to the trade index. Other reference materials included at the rear of the book are a glossary, a list of cell banks, a subject index, and the literature references cited in the text.

It is inevitable when preparing a text such as this that, in addition to my own experience, I have called upon the help and advice of many others both during the preparation of the book and in the twenty years or so since I was first introduced to the field. As with many other similar techniques, there is much of tissue culture that is never documented, but passed on by word of mouth at meetings, or, more often, in moments of conviviality after meetings. Hence there may be occasions when I have reproduced advice or information as if it were my own, without due acknowledgment to published work, because I have been unable to trace a reference, or none exists. In all such cases I would like to thank those who have contributed consciously or unconsciously to my own accumulated experience in the field.

While it would be impossible to recall all of those with whom contact over the past two decades has influenced my current understanding of the field, there are those of whom I must make special mention. First among these is Dr. John Paul, who introduced me to the field and whose sound common sense and practicality were a good introduction to what can, in the correct hands, be a very precise discipline. I owe him my sincere gratitude, as his one-time student and now associate and friend.

In my years with the Beatson Institute I have had the privilege to work with many people, both resident and visitors, and share in their experience in the development of techniques to which I would otherwise not have been exposed. In some cases they are acknowledged in the text or figure legends, but I hope any who are not mentioned by name will still recognize my gratitude.

Among others who should be named are those who have worked most closely with me in recent years, helped in my own research activities, and generated some of the data that appear on these pages. They include Ms. Diana Morgan, Mrs. Elaine Hart, Mrs. Margaret Frame, Mr. Alistair McNab, Mrs. Irene Osprey, and Miss Sheila Brown. Although my wife and I do not work together usually, I have had the benefit of her skilled assistance at times, and, in addition, her experience in the field has added greatly to my own. Others who have worked with me for shorter periods, elements of whose work may be reported here in part, are Mohammad Hassanzadah, Peter Crilly, Fadik Akturk, Metyn Guner, Fahri Celik, Aileen Sherry, Bob Shaw, and Carolyn MacDonald.

I also have been indebted to many people in Glasgow and elsewhere for helpful advice and collaboration. Among many others, these include David G. T. Thomas, David I. Graham, Michael Stack-Dunne, Peter Vaughan, Brian McNamee, David Doyle, Rona MacKie, Kenneth C. Calman, and the late John Maxwell Anderson, with whom I had my first introduction to clinical collaboration.

I must also record my good fortune to have been able to spend time in other laboratories and learn from the approaches of others such as Robert Auerbach, Richard Ham, and Wally McKeehan.

I am also grateful to Flow Laboratories for their help and collaboration in running basic tissue culture courses and the resultant opportunity to broaden my knowledge of the field.

I would like to express my gratitude to Paul Chapple who first persuaded me that I should write a basic techniques book on tissue culture, and to numerous others, including Don Dougall, Wally and Kerstin McKeehan, Peter del Vecchio, John Ryan, Jim Smith, Rob Hay, Charity Waymouth, Sergey Federoff, Mike Gabridge, and Dan Lundin for help and advice during the preparation of the manuscript.

I would also like to thank Miss Donna Madore for converting my often illegible manuscript into typescript, Mrs. Marina LaDuke for expert photography, Miss Diane Leifheit for further help with the illustrations, and Ms. Jane Gillies for preparing the line drawings. These four ladies spent many hours on my behalf and their patience and skill is greatly appreciated. My thanks are also due to Mrs. Norma Wallace for completing the final retype quickly, efficiently, and at very short notice.

It would not be fitting for me to conclude this preface without further major acknowledgment to my wife, Mary, my daughter, Gillian, and son, Norman. Not only did I enjoy their sympathy and understanding at home, when I am sure, at times, I did not deserve it, but I also benefitted from the fruits of their labors during the day: drawing graphs, collecting references, researching and tabulating methods and information. My wife's experience in the field, plus countless hours of reading, revising, and collecting information, made her share in this work indispensable.

Abbreviations

ATCC	American Type Culture Collection
BPE	bovine pituitary extract
BSA	bovine serum albumin
BUdR, BrUdR, BrdU	bromodeoxyuridine
CAM	chorioallantoic membrane
CAM	cell adhesion molecule
cAMP	cyclic adenosine monophosphate
cDNA	complementary DNA
CE	cloning efficiency
CMF	calcium- and magnesium-free saline
CMRL	Connaught Medical Research Laboratories
CNTF	ciliary neurotropic factor
DMEM	Dulbecco's modification of Eagle's medium
DMSO	dimethyl sulfoxide
DNA	deoxyribonucleic acid
DT	population doubling time
EBSS	Earle's balanced salt solution
ECACC	European Collection of Animal Cell Cultures
ECGF	endothelial cell growth factor
EGF	epidermal growth factor
EM	electron microscope
FBS	fetal bovine serum
FCS	fetal calf serum
FGF	fibroblast growth factor
G_1	gap one (of the cell cycle)
G_2	gap two (of the cell cycle)
HAT	hypoxanthine, aminopterin, and thymidine
HBGF	heparin-binding growth factor
HBS	HEPES buffered saline
HBSS	Hanks' balanced salt solution
HC	hydrocortisone
hCG	human chorionic gonadotropin
HGF	hepatocyte growth factor (=scatter factor)
HGPRT	hypoxanthine guanosine phosphoribosyl transferase
HPRT	"
HITES	hydrocortisone, insulin, transferrin, estrodiol, and selenium
HMBA	hexamethylene-bis-acetamide
HS	horse serum
HSV	herpes simplex virus
HT	hypoxanthine/thymidine
IC_{50}	50% inhibitory concentration
IC_{10}, IC_{90}	10% and 90% as above
ID_{50}	50% inhibitory dose
ID_{10}, ID_{90}	10% and 90% as above
ITS	insulin, transferrin, selenium
KBM	keratinocyte basal medium
KGF	keratinocyte growth factor
KGM	keratinocyte growth medium
MEM	Eagle's Minimal Essential Medium
mRNA	messenger RNA
MSH	melanocyte stimulating hormone
NaBt	sodium butyrate
NBCS	new born calf serum

NCI	National Cancer Institute
PA	plasminogen activator
PBS	phosphate buffered saline
PBSA	phosphate buffered saline, solution A (Ca^{2+} and Mg^{2+} free)
PBSB	phosphate buffered saline, solution B (Ca^{2+} and Mg^{2+})
PCA	perchloric acid
PDGF	platelet-derived growth factor
PE	plating efficiency
PE	PBSA/EDTA
PEG	polyethylene glycol
PHA	phytohemaglutinin
PMA	phorbol myristate acetate (=TPA)
PWM	pokeweed mitogen
RNA	ribonucleic acid
RPMI	Roswell Park Memorial Institute
S	DNA synthetic phase of cell cycle
SD	saturation density
SIT	selenium, insulin, transferrin
SF	surviving fraction
SSC	sodium citrate/sodium chloride
TCA	trichloracetic acid
T_D	population doubling time
TEB	Tris/EDTA buffer
TGF	transforming growth factor
TK	thymidine kinase
tPA	tissue-type plasminogen activator
TPA	tetradecanoylphorbol acetate (=PMA)
uPA	urokinase-like plasminogen activator

CHAPTER 1

Introduction

BACKGROUND

Tissue culture was first devised at the beginning of this century [Harrison, 1907; Carrel, 1912] as a method for studying the behavior of animal cells free of systemic variations that might arise in the animal both during normal homeostasis and under the stress of an experiment. As the name implies, the technique was elaborated first with undisaggregated fragments of tissue, and growth was restricted to the migration of cells from the tissue fragment, with occasional mitoses in the outgrowth. Since culture of cells from such primary explants of tissue dominated the field for more than 50 years, it is not surprising that the name "tissue culture" has stuck in spite of the fact that most of the explosive expansion in this area since the 1950s has utilized dispersed cell cultures.

Throughout this book the term *tissue culture* is used as the generic term to include organ culture and cell culture. The term *organ culture* will always imply a three-dimensional culture of undisaggregated tissue retaining some or all of the histological features of the tissue *in vivo*. *Cell culture* refers to cultures derived from dispersed cells taken from the original tissue, from a primary culture, or from a cell line or cell strain by enzymatic, mechanical, or chemical disaggregation. The term *histotypic culture* will imply that cells have been reassociated in some way to recreate a three-dimensional tissue-like structure, e.g., by perfusion and overgrowth of a monolayer, reaggregation in suspension, or infiltration of a three-dimensional matrix such as collagen gel. *Organotypic* will imply the same procedures but recombining cells of different lineages, e.g., epidermal keratinocytes in combined reaggregated culture with dermal fibroblasts.

Harrison chose the frog as his source of tissue presumably because it was a cold-blooded animal, and consequently incubation was not required. Furthermore, since tissue regeneration is more common in lower vertebrates, he perhaps felt that growth was more likely to occur than with mammalian tissue. Although his technique may have sparked off a new wave of interest in cultivation of tissue *in vitro,* few later workers were to follow his example in the selection of species. The stimulus from medical science carried future interest into warm-blooded animals, where normal and pathological development are closer to human. The accessibility of different tissues, many of which grew well in culture, made the embryonated hen's egg a favorite choice; but the development of experimental animal husbandry, particularly with genetically pure strains of rodents, brought mammals to the forefront as favorite material. While chick embryo tissue could provide a diversity of cell types in primary culture, rodent tissue had the advantage of producing continuous cell lines [Earle et al., 1943] and a considerable repertoire of transplantable tumors. The development of transgenic mouse technology [Beddington, 1992; Peat et al., 1992], together with the well-established genetic background of the mouse, has added further impetus to the selection of mouse as a favorite species.

The demonstration that human tumors could also give rise to continuous cell lines [e.g., HeLa: Gey et al., 1952] encouraged interest in human tissue, helped later by Hayflick and Moorhead's [1961] classical studies with normal cells of a finite life-span.

For many years the lower vertebrates and the invertebrates have been largely ignored, though unique aspects of their development (tissue regeneration in amphibia, metamorphosis in insects) make them attractive systems for the study of the molecular basis of development. More recently the needs of agriculture and pest control have encouraged toxicity and virological studies in insects, and developments in gene technology have suggested that insect cell lines with baculovirus and other vectors may be useful producer cell lines with less requirement for temperature control. The rapidly developing area of fish farming has required more detailed knowledge of normal development and pathogenesis in fish.

In spite of this resurgence of interest, tissue culture of lower vertebrates and the invertebrates remains a specialized area, and the bulk of interest remains in avian and mammalian tissue. This has naturally influenced the development of the art and science of tissue culture, and much of what is described in the ensuing chapters of this book reflect this, as well as my own personal experience. Hence advice on incubation and the physical and biochemical properties of media refers to homiotherms and guidance on the appropriate modification for poikilothermic animals will require recourse to the literature. This is discussed in a little more detail in a later chapter. Many of the basic techniques of asepsis, preparation and sterilization, primary culture, selection and cell separation, quantitation, and so on, apply equally to poikilotherms and require only minor modification; on the whole the principles remain the same.

The types of investigation that lend themselves particularly to tissue culture are summarized in Figure 1.1: (1) intracellular activity, e.g., the replication and transcription of deoxyribonucleic acid (DNA), protein synthesis, energy metabolism, drug metabolism; (2) intracellular flux, e.g., RNA, translocation of hormone receptor complexes and resultant signal transduction processes, membrane trafficking; (3) environmental interaction, e.g., nutrition, infection, carcinogenesis, drug action, ligand receptor interactions; (4) cell–cell interaction, e.g., embryonic induction, cell population

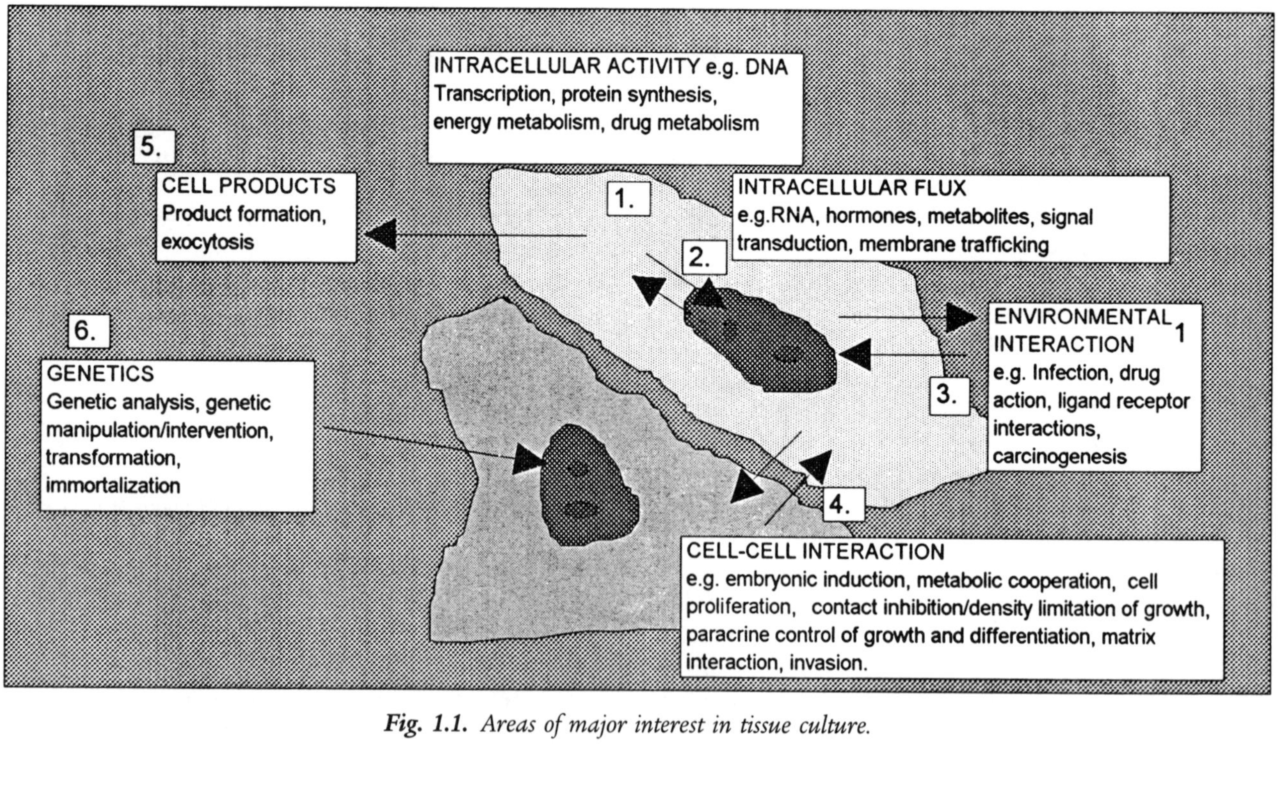

Fig. 1.1. Areas of major interest in tissue culture.

kinetics, cell–cell adhesion, and invasion; (5) cell products and secretion; and (6) genetics, including analyis, manipulation, transformation and immortalization.

The development of tissue culture as a modern, sophisticated technique owes much to the needs of two major branches of medical research: the production of antiviral vaccines and the understanding of neoplasia. The standardization of conditions and cell lines for the production and assay of viruses undoubtedly provided much impetus to the development of modern tissue culture technology, particularly the production of large numbers of cells suitable for biochemical analysis. This and other technical improvements made possible by the commercial supply of reliable media and sera, and by the greater control of contamination with antibiotics and clean-air equipment, has made tissue culture accessible to a wide range of interests.

An additional force of increasing weight from public opinion has been the expression of concern by many animal-rights groups over the unnecessary use of experimental animals. While most accept that some requirement for animals will continue for preclinical trials of new pharmaceuticals, there is widespread concern that extensive use of animals for cosmetics development and similar activities may not be morally justifiable. Hence there is an ever-increasing lobby for more *in vitro* assays, the adoption of which only requires their proper validation. While this seemed a distant prospect some years ago, the introduction of more sensitive and more readily performed *in vitro* assays, together with a very real prospect of assaying for inflammation *in vitro,* has promoted an unprecedented expansion in *in vitro* testing.

In addition to cancer research and virology, other areas of research have come to depend heavily on tissue culture techniques. The introduction of cell fusion techniques [Barski, et al., 1961; Sorieul and Ephrussi, 1961; Littlefield, 1964a; Harris and Watkins, 1965] and genetic manipulation [Merril, 1971; Horst et al., 1975; Maniatis et al., 1978; Sambrook et al., 1989; Frederick et al., 1993] established somatic cell genetics as a major component in the genetic analysis of higher animals, including humans, and contributed greatly, via the monoclonal antibody technique, to the study of immunology, already dependent on cell culture for assay techniques and production of hemopoietic cell lines.

The insight into the mechanism of action of antibodies, and the reciprocal information that this provided about the structure of the epitope, derived from monoclonal antibody techniques [Kohler and Milstein, 1975] was, like the technique of cell fusion itself, a prologue to a whole new field of studies in genetic manipulation. This has supplied much basic information on the control of gene transcription and a vast new technology has grown from the ability to insert exploitable genes into prokaryotic cells. Cell products such as human growth hormone, insulin, and interferon have been genetically engineered, but the absence of posttranscriptional modifications, such as glycosylation, in bacteria suggest that mammalian cells may provide more suitable vehicles. The insertion of the appropriate genes into normal human cells (1) to make them continuous cell lines (see Chapter 2) and (2) to make them produce pharmaceutically viable drugs will have profound effects on the drug industry, which can only be overshadowed by radical innovations in organic chemical synthesis that are, as yet, not apparent.

Other areas of major interest include the study of cell interactions and intracellular control mechanisms in cell differentiation and development [Auerbach and Grobstein, 1958; Cunha, 1984; Jessell and Melton, 1992] and attempts to analyze nervous function [Bornstein and Murray, 1958; Minna et al., 1972; Kingsbury et al., 1985; Snyder et al., 1992]. Progress in neurological research has not had the benefit, however, of working with propagated cell lines, as propagation of neurons has not been possible so far *in vitro* without resorting to the use of transformed cells (see Chapter 15).

Tissue culture technology has also been adopted into many routine applications in medicine and industry. Chromosomal analysis of cells derived from the womb by amniocentesis (see Chapter 23) can reveal genetic disorders in the unborn child, viral infections may be assayed qualitatively and quantitatively on monolayers of appropriate host cells (see Chapter 23), and the toxic effects of pharmaceutical compounds and potential environmental pollutants can be measured in colony-forming and other *in vitro* assays (see Chapter 19).

Further developments in the application of tissue culture to medical problems may follow from the demonstration that cultures of epidermal cells form functionally differentiated sheets in culture [Green et al., 1979] and endothelial cells may form capillaries [Folkman and Haudenschild, 1980], suggesting possibilities in homografting and reconstructive surgery using an individual's own cells [Burt et al., 1989; Gallico, 1990; Dennis, 1992]. It has now become accepted clinical practice in some burn units to biopsy a patient's skin, propagate the cells in culture, and graft the cultured cells back onto the areas of most severe burning [Boyce and Hansbrough, 1988].

With the ability to transfect normal genes into genetically deficient cells, it has become possible to graft such "corrected" cells back into the patient. Transfected cultures of rat bronchial epithelium, carrying the β-*gal* reporter gene, have been shown to become incorporated into the rat's bronchial lining when introduced as an aerosol into the respiratory tract [Rosenfeld et al., 1992]. Similarly, cultured satellite cells have been shown to be incorporated into wounded rat skeletal muscle, with nuclei from grafted cells appearing in mature, syncytial myotubes [Morgan et al., 1992].

The prospects for implantation of normal cells from adult or fetal, tissue-matched donors, or genetically reconstituted cells from the same patient, are now very real. The technical barriers are steadily being overcome, bringing the ethical questions to the fore. The technical feasibility of implanting normal fetal neurons into patients with Parkinson's disease has been demonstrated; society must now decide to what extent fetal material may be used for this purpose.

Where a patient's own cells can be grown and subjected to genetic reconstitution by transfection of the normal gene, e.g., transfecting the normal insulin gene into β-islet cells cultured from diabetics, or even transfecting other cell types, such as skeletal muscle satellite cells [Morgan et al., 1992], it would allow the cells to be incorporated into a low-turnover compartment and, potentially, give a long-lasting physiological benefit. The ethics of this type of approach seem less contentious.

It is clear that the study of cellular activity in tissue culture may have many advantages; but in summarizing these below considerable emphasis must also be placed on its limitations, in order to maintain some sense of perspective.

ADVANTAGES OF TISSUE CULTURE

Control of the Environment

The two major advantages, as implied above, are the control of the physiochemical environment (pH, temperature, osmotic pressure, O_2 and CO_2 tension), which may be controlled very precisely, and the physiological conditions, which may be kept relatively constant but cannot always be defined. Most cell lines still require supplementation of the medium with serum or other poorly defined constituents. These supplements are prone to batch variation [Olmsted, 1967; Honn et al., 1975] and contain undefined elements such as hormones and other regulatory substances. Gradually the essential components of serum are being identified, making replacement with defined constituents more practicable [Birch and Pirt, 1971; Ham and McKeehan, 1978; Barnes and Sato, 1980; Barnes et al., 1984a–d; Maurer, 1992] (see also Chapter 7).

Characterization and Homogeneity of Sample

Tissue samples are invariably heterogeneous. Replicates even from one tissue vary in their constituent cell types. After one or two passages, cultured cell lines assume a homogeneous (or at least uniform) constitution, as the cells are randomly mixed at each transfer and the selective pressure of the culture conditions tends to produce a homogeneous culture of the most vigorous cell type. Hence, at each subculture each replicate sample will be identical, and the characteristics of the line may be perpetuated over several generations, or indefinitely if the cell line is stored in liquid N_2. Since experimental replicates are virtually identical, the need for statistical analysis of variance is reduced.

Economy

Cultures may be exposed directly to a reagent at a lower and defined concentration, and with direct access to the cell. Consequently, less is required than for injection *in vivo* where 90% is lost by excretion and distribution to tissues other than those under study.

Screening tests with many variables and replicates are cheaper, and the legal, moral, and ethical questions of animal experimentation are avoided.

DISADVANTAGES

Expertise

Culture techniques must be carried out under strict aseptic conditions, because animal cells grow much less rapidly than many of the common contaminants such as bacteria, molds, and yeasts. Furthermore, unlike micro-organisms, cells from multicellular animals do not exist in isolation and, consequently, are not able to sustain independent existence without the provision of a complex environment, simulating blood plasma or interstitial fluid. This implies a level of skill and understanding to appreciate the requirements of the system and to diagnose problems as they arise. Tissue culture should not be undertaken casually to run one or two experiments.

Quantity

A major limitation of cell culture is the expenditure of effort and materials that goes into the production of relatively little tissue. A realistic maximum per batch for most small laboratories (two or three people doing tissue culture) might be 1–10 g of cells. With a little more effort and the facilities of a larger laboratory, 10–100 g is possible; above 100 g implies industrial pilot plant scale, beyond the reach of most laboratories but not impossible if special facilities are provided when kilogram quantities can be generated.

The cost of producing cells in culture is about ten times that of using animal tissue. Consequently, if large amounts of tissue (>10 g) are required, the reasons for providing them by tissue culture must be very compelling. For smaller amounts of tissue (≤10 g), the costs are more readily absorbed into routine expenditure, but it is always worth considering whether assays or preparative procedures can be scaled down. Semimicro- or micro-scale assays can often be quicker due to reduced manipulation times, volumes, centrifuge times, etc., and are often more readily automated (see under Microtitration, Chapter 19).

Dedifferentiation and Selection

When the first major advances in cell line propagation were achieved in the 1950s, many workers observed the loss of the phenotypic characteristics typical of the tissue from which the cells had been isolated. This was blamed on *dedifferentiation,* a process assumed to be the reversal of differentiation, but was later shown to be largely due to the overgrowth of undifferentiated cells of the same or a different lineage. The development of serum-free selective media (see Chapter 7) has now made the isolation of specific lineages quite possible, and it can be seen that, under the correct culture conditions, many of the differentiated properties of these cells may be restored (see Chapter 14).

Origin of Cells

If differentiated properties are lost, for whatever reason, it is difficult to relate the cultured cells to functional cells in the tissue from which they were derived. Stable markers are required for characterization (see Chapter 13); in addition, the culture conditions may need to be modified so that these markers are expressed (see Chapters 2 and 14).

Instability

Instability is a major problem with many continuous cell lines, resulting from their unstable aneuploid chromosomal constitution. Even with short-term cultures of untransformed cells, heterogeneity in growth rate and capacity to differentiate within the population can produce variability from one passage to the next. This is dealt with in more detail in Chapters 10 and 17.

MAJOR DIFFERENCES *IN VITRO*

Many of the differences in cell behavior between cultured cells and their counterparts *in vivo* stem from the dissociation of cells from a three-dimensional geometry and their propagation on a two-dimensional substrate. Specific cell interactions characteristic of the histology of the tissue are lost, and as the cells spread out, become mobile, and, in many cases, start to proliferate, so the growth fraction of the cell population increases. When a cell line forms it may represent only one or two cell types, and many heterotypic interactions are lost.

The culture environment also lacks the several systemic components involved in homeostatic regulation *in vivo,* principally those of the nervous and endocrine systems. Without this control, cellular metabolism may be more constant *in vitro* than *in vivo* but may not be truly representative of the tissue from which the cells were derived. Recognition of this fact has led to the inclusion of a number of different hormones in culture media (see Chapter 7) and it seems likely that this trend will continue.

Energy metabolism *in vitro* occurs largely by glycolysis, and although the citric acid cycle is still functional it plays a lesser role.

It is not difficult to find many more differences between the environmental conditions of a cell *in vitro* and *in vivo* (see also Chapter 19) and this has often led to tissue culture being regarded in a rather skeptical light. Although the existence of such differences cannot be denied, it must be emphasized that many specialized functions are expressed in culture, and as long as the limits of the model are appreciated, it can become a very valuable tool.

DEFINITION OF TYPES OF TISSUE CULTURE

There are three main methods of initiating a culture [Schaeffer, 1990] (see Glossary and Fig. 1.2): (1) *Organ culture* implies that the architecture characteristic of the tissue *in vivo* is retained, at least in part, in the culture (see Chapter 22). Toward this end, the tissue is cultured at the liquid–gas interface (on a raft, grid, or gel), which favors retention of a spherical or three-dimensional shape. (2) In *primary explant culture* a fragment of tissue is placed at a glass (or plastic)–liquid interface where, following attachment, migration is promoted in the plane of the solid substrate (see Chapter 9). (3) *Cell culture* implies that the tissue, or outgrowth from the primary explant, is dispersed (mechanically or enzymatically) into a cell suspension, which may then be cultured as an adherent monolayer on a solid substrate, or as a suspension in the culture medium (see Chapters 9 and 10).

Organ cultures, because of the retention of cell interactions as found in the tissue from which the culture was derived, tend to retain the differentiated properties of that tissue. They do not grow rapidly (cell proliferation is limited to the periphery of the explant and is restricted mainly to embryonic tissue) and hence cannot be propagated; each experiment requires fresh explantations and this implies greater effort and poorer sample reproducibility than with cell culture. Quantitation is, therefore, more difficult and the amount of material that may be cultured is limited by the dimensions of the explant ($\leq 1\ mm^3$) and the effort required

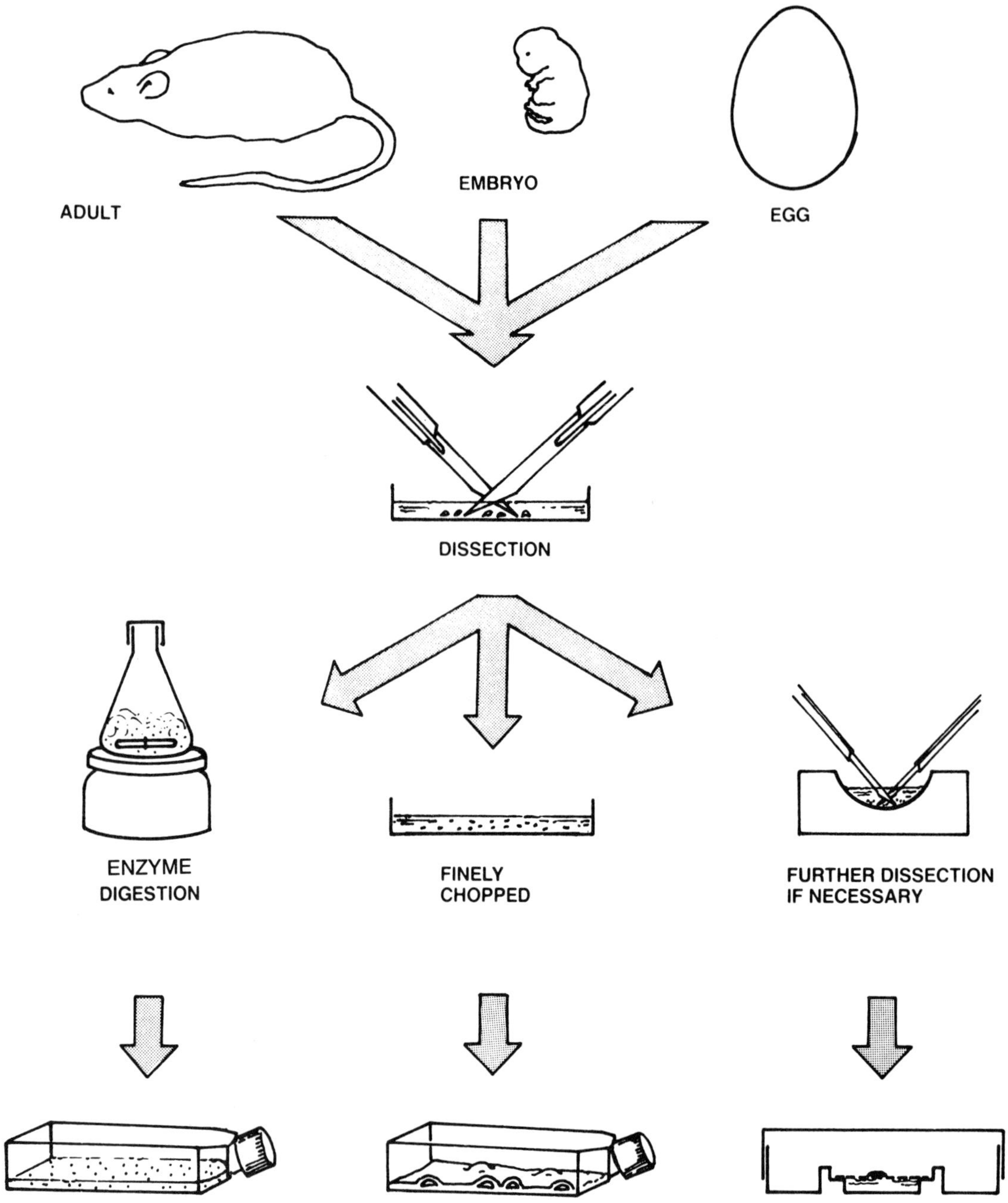

Fig. 1.2. *Types of tissue culture.*

for dissection and setting up the culture. However, it must be emphasized that organ cultures do retain specific histological interactions without which it may be difficult to reproduce the characteristics of the tissue.

Cell cultures may be derived from primary explants or dispersed cell suspensions. Because cell proliferation is often found in such cultures, propagation of cell lines becomes feasible. A monolayer or cell suspension, with a significant growth fraction (see Chapter 18), may be dispersed by enzymatic treatment or simple dilution and reseeded, or subcultured, into fresh vessels. This constitutes a *passage* and the daughter cultures so formed are the beginnings of a *cell line.*

The formation of a cell line from a primary culture implies (1) an increase in total cell number over several generations and (2) that cells or cell lineages with similar high growth capacity will predominate, resulting in (3) a degree of uniformity in the cell population. The line may be characterized, and those characteristics will apply for most of its finite life-span. The derivation of *continuous* (or "established," as they were once known) cell lines usually implies a phenotypic change or *transformation* and is dealt with in Chapters 2 and 15.

When cells are selected from a culture, by cloning or by some other method, the subline is known as a *cell strain*. Detailed characterization is then implied. Cell lines, or cell strains, may be propagated as an adherent monolayer or in suspension. *Monolayer* culture signifies that the cells will attach to the substrate given the opportunity and that normally the cells will be propagated in this mode. *Anchorage dependence* means that attachment to (and usually some degree of spreading on) the substrate is a prerequisite for cell proliferation. Monolayer culture is the mode of culture common to most normal cells with the exception of hemopoietic cells. *Suspension* cultures are derived from cells that can survive and proliferate without attachment (*anchorage-independent*); this ability is restricted to hemopoietic cells, transformed cell lines, or cells from malignant tumors. It can be shown, however, that a small proportion of cells that are capable of proliferation in suspension exists in many normal tissues (see Chapter 15). The identity of these cells remains unclear, but a relationship to the stem cell or uncommitted precursor cell compartment has been postulated. This concept implies that some cultured cells represent precursor pools within the tissue of origin; the generality of this observation is discussed more fully in the next chapter. Cultured cell lines are more representative of precursor cell compartments *in vivo* than of fully differentiated cells, as most differentiated cells do not normally divide.

Because they may be propagated as a uniform cell suspension or monolayer, cell cultures have many advantages in quantitation, characterization, and replicate sampling, but lack the potential for cell–cell interaction and cell–matrix interaction afforded by organ cultures. For this reason many workers have attempted to reconstitute three-dimensional cellular structures using aggregated cell suspension (*spheroids*) or perfused high-density cultures on microcapillary bundles or membranes (see Chapter 22). Such developments have required the introduction, or at least redefinition, of certain terms. *Histotypic* or *histiotypic* culture, or *histoculture* (I use *histotypic culture* here), has come to mean the high-density, or "tissue-like," culture of one cell type, while *organotypic* culture implies the presence of more than one cell type interacting as they might in the organ of origin or a simulation of it. This has given new prospects for the study of cell interaction among discrete, defined populations of homogeneous, and potentially genetically and phenotypically defined, cells.

In many ways some of the most exciting developments in tissue culture arise from recognizing the necessity of specific cell interaction in homogeneous or heterogeneous cell populations in culture. This may mark the transition from an era of fundamental molecular biology, where many of the regulatory processes have been worked out at the cellular level, to an era of cell or tissue biology, where this understanding is applied to integrated populations of cells, and to the ultimate definition of the signals transmitted among cells.

CHAPTER 2

Biology of the Cultured Cell

THE CULTURE ENVIRONMENT

The validity of the cultured cell as a model of physiological function *in vivo* has frequently been criticized. There are problems of characterization due to the alteration of the cellular environment; cells proliferate *in vitro* that would not normally *in vivo*, cell–cell and cell–matrix interactions are reduced because purified cell lines lack the heterogeneity and three-dimensional architecture found *in vivo*, and the hormonal and nutritional milieu is altered. This creates an environment that favors the spreading, migration, and proliferation of unspecialized cells rather than the expression of differentiated functions. The provision of the appropriate environment, nutrients, hormones, and substrate is fundamental to the expression of specialized functions (see Chapter 14). Before considering such specialized conditions, let us examine the events accompanying the formation of a primary cell culture and a cell line derived from it (Fig. 2.1).

CELL ADHESION

Most cells from solid tissues grow as adherent monolayers, and, unless they have transformed and become anchorage-independent (see Chapter 15), following tissue disaggregation or subculture they will need to attach and spread out on the substrate before they will start to proliferate (see also Chapter 10). It was found originally that cells would attach and spread on glass as long as it had a slight net negative charge, and subsequently it was found that cells will attach to some plastics, such as polystyrene, if they have been appropriately treated with electric arc discharge or high-energy ionizing radiation. We now know that cell adhesion is mediated by specific cell surface receptors to molecules in the extracellular matrix (see below), so it seems likely that cell spreading may be preceded by secretion of extracellular matrix proteins and proteoglycans by the cells. It is the matrix that adheres to the charged substrate and the cells then bind to the matrix via specific receptors. Hence, glass or plastic that has been conditioned by previous cell growth can often provide a better surface for attachment.

With fibroblast-like cells, the main requirement is for substrate attachment and spreading, as the cells are motile and independent at low densities, but it appears that epithelial cells must make the correct cell–cell contacts for optimum survival and growth, and consequently tend to grow as patches.

Three major classes of transmembrane proteins have

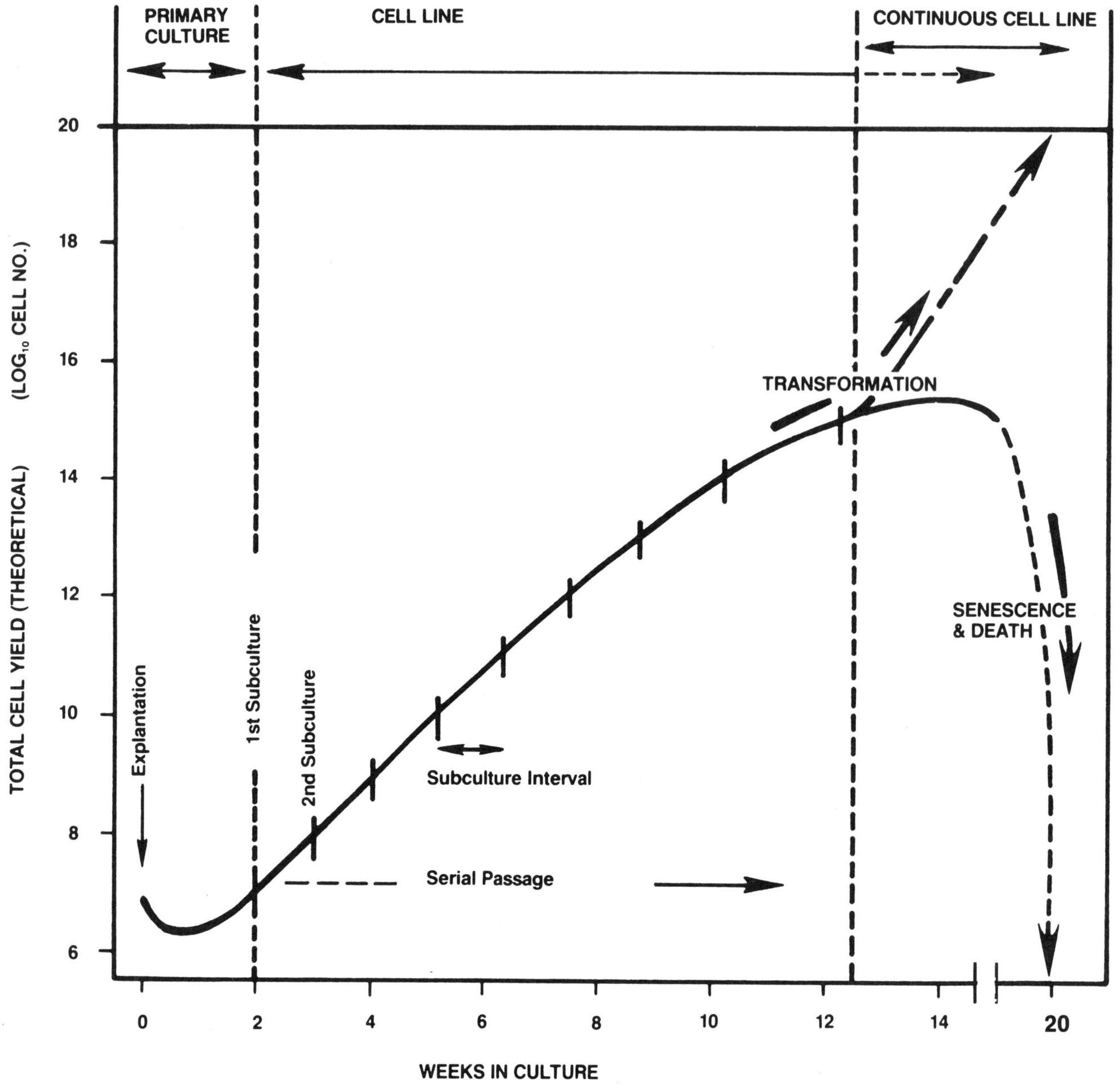

Fig. 2.1. *Evolution of a cell line. The vertical axis represents total cell growth (assuming no reduction at passage) for a hypothetical cell culture. Total cell number (cell yield) is represented on the Y-axis on a log scale and time in culture, on the X-axis on a linear scale. Although a continuous cell line is depicted as arising at 12½ wk it could, with different cells, arise at any time. Likewise, senescence may occur at any time, but for human diploid fibroblasts it is most likely to occur between 30 and 60 cell doublings or 10 to 20 wk, depending on the doubling time. Terms and definitions used are as in the Glossary. Transformation is explained in more detail in Chapter 15.*

been shown to be involved in cell–cell and cell–substrate adhesion. *Cell–cell adhesion molecules*, CAMs (Ca^{2+}-independent), and *cadherins* (Ca^{2+}-dependent) are involved primarily in interactions between homologous cells. They are self-interactive, i.e., like molecules in opposing cells interact with each other [Edelman, 1986, 1988; Rosenman and Gallatin, 1991]. Cell substrate interactions are mediated primarily by *integrins*, receptors for matrix molecules such as fibronectin, laminin, and collagen, which bind to them via a specific motif usually containing the RGD (arginine, glycine, aspartic acid) sequence [Yamada, 1991]. Each integrin comprises one α and one β subunit, both of which are highly polymorphic, generating considerable diversity among the integrins.

The third group of cell adhesion molecules is the

transmembrane proteoglycans, also interacting with matrix constituents such as other proteoglycans or collagen, but not via the RGD motif. Some transmembrane proteoglycans also act as low-affinity growth factor receptors [Klagsbrun and Baird, 1991].

Disaggregation of the tissue (or an attached monolayer culture) with protease will digest some of the extracellular matrix, and may even degrade some of the extracellular domains of transmembrane proteins, allowing cells to become dissociated from each other. Epithelial cells and endothelial cells are generally more resistant to disaggregation, as they tend to have tighter junctional complexes (desmosomes, tight junctions) holding them together, while mesenchymal cells, which are more dependent on matrix interactions for intercellular bonding, are more easily dissociated. In either case, the cells must resynthesize matrix proteins before they attach or must be provided with a matrix-coated substrate.

INITIATION OF THE CULTURE

Primary culture techniques are described in detail in Chapter 9. Briefly, a culture is derived either by outgrowth of migrating cells from a fragment of tissue or by enzymatic or mechanical dispersal of the tissue. Regardless of the method employed, this is the first in a series of selective processes (Table 2.1) that may ultimately give rise to a relatively uniform cell line. In primary explantation (see Chapter 9) selection occurs by virtue of the cells' capacity to migrate from the explant, while with dispersed cells, only those cells that (1) survive the disaggregation technique and (2) adhere to the substrate or survive in suspension will form the basis of a primary culture.

If the primary culture is maintained for more than a few hours, a further selection step will occur. Cells capable of proliferation will increase, some cell types will survive but not increase, and yet others will be unable to survive under the particular conditions used. Hence, the relative proportion of each cell type will change and continue to do so until, in the case of monolayer cultures, all the available culture substrate is occupied.

After confluence is reached (i.e., all the available growth area is utilized and the cells make close contact with one another), cells that are sensitive to density limitation of growth (see Chapters 9 and 15) will stop dividing, while any transformed cells, insensitive to density limitation, will tend to overgrow. Keeping the cell density low, e.g., by frequent subculture, helps to preserve the normal phenotype in cultures such as mouse fibroblasts, where spontaneous transformants tend to overgrow at high cell densities [Todaro and Green, 1963; Brouty-Boyé et al., 1979, 1980].

Some aspects of specialized function are expressed more strongly in primary culture, particularly when the culture becomes confluent. At this stage the culture will show its closest morphological resemblance to the parent tissue and retain some diversity in cell type.

EVOLUTION OF CELL LINES

After the first subculture—or passage (see Fig. 2.1)—the primary culture becomes known as a cell line (Chapter 10) and may be propagated and subcultured several times. With each successive subculture the component of the population with the ability to proliferate most rapidly will gradually predominate, and nonproliferating or slowly proliferating cells will be diluted out. This is most strikingly apparent after the first subculture, where differences in proliferative capacity are compounded with varying abilities to withstand the trauma of trypsinization and transfer (see Chapter 10).

Although some selection and phenotypic drift will continue, by the third passage the culture becomes

TABLE 2.1. Elements of Selection in the Evolution of Cell Lines

Stage	Factors influencing selection: Primary explant	Factors influencing selection: Enzymatic disaggregation
Isolation	Mechanical damage	Enzymatic damage
Primary culture	Adhesion of explant; outgrowth (migration)	Cell adhesion and spreading
First subculture	Trypsin sensitivity; nutrient, hormone, and substrate limitations	
Propagation as a cell line	Relative growth rates of different cells; selective overgrowth of one lineage; nutrient, hormone, and substrate limitations; effect of cell density on predominance of normal or transformed phenotype	
Senescence; transformation	Normal cells die out; transformed cells overgrow	

more stable, typified by a rather hardy, rapidly proliferating cell. In the presence of serum and without specific selection conditions (see Chapters 7 and 20), mesenchymal cells derived from connective tissue fibroblasts or vascular elements frequently overgrow the culture. While this has given rise to some very useful cell lines—e.g., WI38 human embryonic lung fibroblasts [Hayflick and Moorhead, 1961], BHK21 baby hamster kidney fibroblasts [Macpherson and Stoker, 1962] (see Table 10.2), and perhaps the most famous of all, the L-cell, a mouse subcutaneous fibroblast treated with methylcholanthrene [Earle et al., 1943; Sanford et al., 1948]—it has presented one of the major challenges of tissue culture since its inception: namely, how to prevent the overgrowth of the more fragile or slower-growing specialized cells such as hepatic parenchyma or epidermal keratinocytes. Inadequacy of the culture conditions is largely to blame for this problem and considerable progress has now been made in the use of selective media and substrates for the maintenance of many specialized cell lines (see Chapter 20).

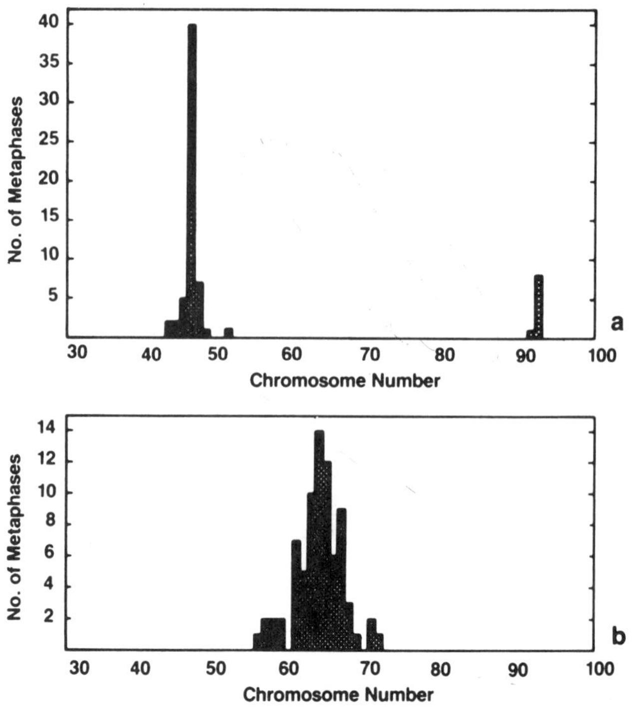

Fig. 2.2. Chromosome numbers of finite and continuous cell lines. a. A normal human glial cell line. b. A continuous cell line from human metastatic melanoma.

THE DEVELOPMENT OF CONTINUOUS CELL LINES

Most cell lines may be propagated in an unaltered form for a limited number of cell generations, beyond which they may either die out or give rise to continuous cell lines (Fig. 2.1). The ability of a cell line to grow continuously probably reflects its capacity for genetic variation allowing subsequent selection. Human fibroblasts remain predominantly euploid throughout their culture life-span and never give rise to continuous cell lines [Hayflick and Moorhead, 1961], while mouse fibroblasts and cell cultures from a variety of human and animal tumors often become aneuploid in culture and give rise to continuous cultures with fairly high frequency. The alteration in a culture giving rise to a continuous cell line is commonly called "*in vitro* transformation" (see Chapter 15) and may occur spontaneously or be chemically or virally induced. The word *transformation* is used rather loosely and can mean different things to different people. I use *immortalization* here to mean the acquisition of an infinite life-span and *transformation* to imply an alteration in growth characteristics (anchorage independence, loss of contact inhibition, and density limitation of growth) that will often, but not necessarily, correlate with tumorigenicity.

Continuous cell lines are usually *aneuploid* and often have a chromosome number between the diploid and tetraploid value (Fig. 2.2). There is also considerable variation in chromosome number and constitution among cells in the population (*heteroploidy*) (see also Chapter 16). It is not clear whether the cells that give rise to continuous lines are present at explantation in very small numbers or arise later as a result of transformation of one or more cells. The second would seem to be more probable on cell kinetic grounds, as continuous cell lines can appear quite late in a culture's life history, long after the time it would have taken for even one preexisting cell to overgrow. The possibility remains, however, that there is a subpopulation in such cultures with a predisposition to transform not shared by the rest of the cells.

The term *transformation* has been applied to the process of formation of a continuous cell line partly because the culture undergoes morphological and kinetic alterations, but also because the formation of a continuous cell line is often accompanied by an increase in tumorigenicity. A number of the properties of continuous cell lines are also associated with malignant transformations (see Chapter 10), such as reduced serum requirement, reduced density limitation of growth, growth in semisolid media, and aneuploidy (see also Table 10.3), and so forth. These are reviewed in more detail in Chapter 15. Similar morphological and behavioral changes can also be observed in cells that have undergone virally or chemically induced transformation.

Many (if not most) normal cells do not give rise to continuous cell lines. In the classic example [Hayflick and Moorhead, 1961] normal human fibroblasts remain euploid throughout their life-span and at crisis (usually around 50 generations) will stop dividing, though they may remain viable for up to 18 months thereafter. Human glia [Pontén and Westermark, 1980] and chick fibroblasts [Hay and Strehler, 1967] behave similarly. Epidermal cells, on the other hand, have shown gradually increasing life-spans with improvements in culture techniques [Green et al., 1979] and may yet be shown capable of giving rise to continuous growth. This may be related to the self-renewal capacity of the tissue *in vivo* (see below, this chapter). Continuous culture of lymphoblastoid cells is also possible [Gjerset et al., 1990] by transformation with Epstein-Barr virus.

It is possible that the condition that predisposes to the development of a continuous cell line is inherent genetic variation, so it is not surprising to find genetic instability perpetuated in continuous cell lines. A common feature of many human continuous cell lines is the development of a subtetraploid chromosome number (see Fig. 2.2).

For a further discussion of variation and instability, see Chapter 17.

DEDIFFERENTIATION

Dedifferentiation implies that differentiated cells lose their specialized properties *in vitro*, but it is often unclear whether (1) undifferentiated cells of the same lineage (Fig. 2.3) overgrow terminally differentiated cells of reduced proliferative capacity or (2) the absence of the appropriate inducers (hormones: cell or matrix interaction) causes deadaptation (see Chapter 14). In practice, both may occur. Continuous proliferation may select undifferentiated precursors, which, in the absence of the correct inductive environment, do not differentiate.

An important distinction should be made between dedifferentiation, deadaptation, and selection. Dedifferentiation implies that the specialized properties of the cell are lost irreversibly, e.g., a hepatocyte would lose its characteristic enzymes (arginase, aminotrans-

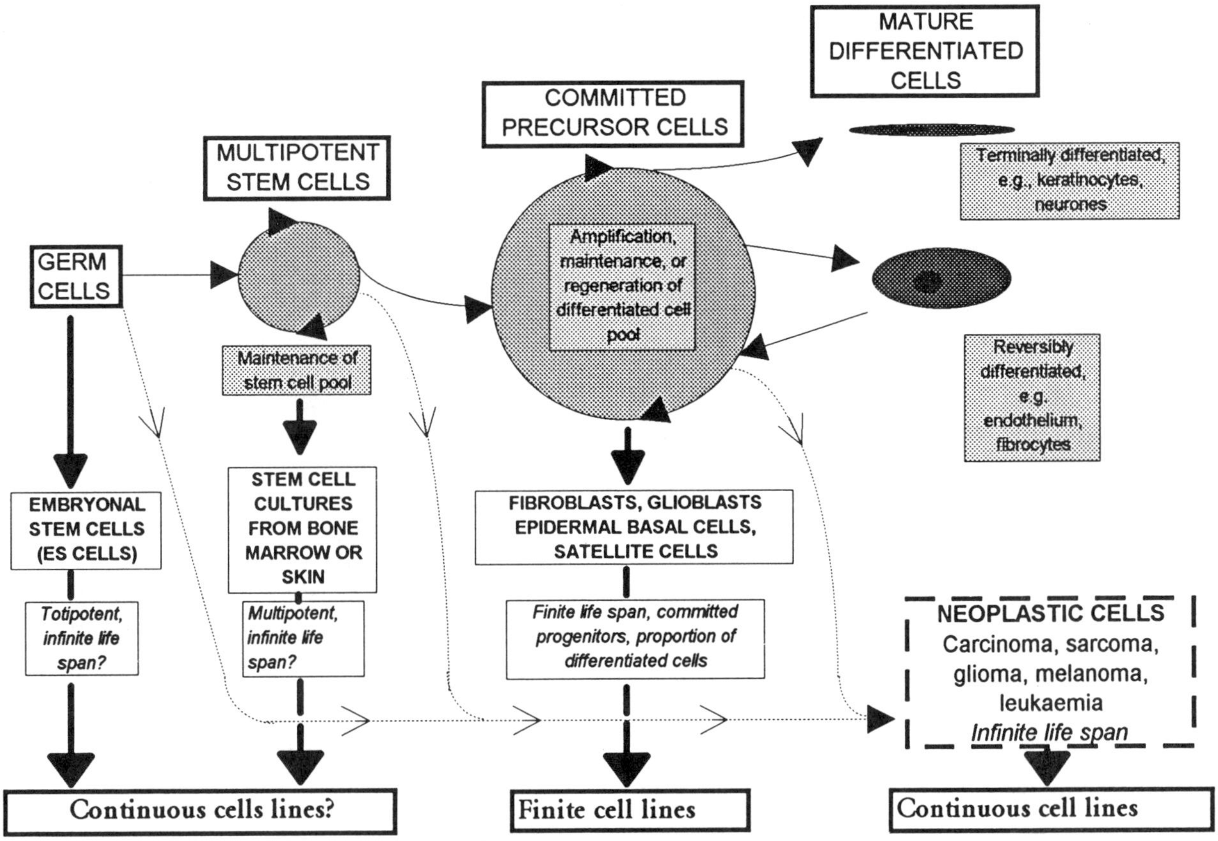

Fig. 2.3. Origin of cell lines. With a few exceptions (e.g., differentiated tumor cells) culture conditions select for the proliferating progenitor cell compartment of the tissue or induce cells that are partially differentiated to revert to a progenitor status. While neoplastic cells, and cell lines, may be derived from differentiated cells, it seems more likely that they arise from malignant progenitor cells, some of which retain the capability to divide, while others continue to differentiate.

ferases, etc.), could not store glycogen or secrete serum proteins, and these properties could not be reinduced once lost. Deadaptation, on the other hand, implies that synthesis of specific products, or other aspects of specialized function, are under regulatory control by hormones, cell–cell interaction, cell–matrix interaction, etc., and can be reinduced, given that the correct conditions can be recreated. The presence of matrix as a floating collagen raft [Michalopoulos and Pitot, 1975] allows induction of tyrosine aminotransferase in normal hepatocytes, and Matrigel has also been shown to stabilize the differentiated phenotype in hepatocytes [Bissell et al., 1987]. It is gradually becoming apparent that, given the correct culture conditions, differentiated functions can be expressed by a number of different cell types (Table 2.2), and the concept of dedifferentiation is now regarded as an unlikely explanation for the loss of specialized functions.

For correct inducing conditions to act, the appropriate cells must be present. In early attempts at liver cell culture, lack of expression of hepatocyte properties was due partly to overgrowth of the culture by connective tissue fibroblasts or endothelium from blood vessels or sinusoids. By using the correct disaggregation technique and the correct culture conditions [see Guguen-Guillouzo, 1992] (Chapter 20), hepatocytes can be selected preferentially. Similarly, epidermal cells can be grown either by using a confluent feeder layer [Rheinwald and Green, 1975] or selective medium [Peehl and Ham, 1980; Tsao et al., 1982]. Selective media have also been used for many other types of epithelium [e.g., Sundqvist et al., 1991; Freshney, 1992]. These and other examples, e.g., feeder selection for breast and colonic epithelium [Freshney et al., 1982b], D-valine for the isolation of kidney epithelium [Gilbert and Migeon, 1975], and the use of cytotoxic antibodies [Edwards et al., 1980] (selection procedures reviewed in Chapters 11, 12, and 20), clearly demonstrate that the selective culture of specialized cells is not the insuperable problem that it once appeared. New selective media, based mainly on supplemented Ham's F12:DMEM, or modifications of the MCDB series (see Chapters 7 and 11) are appearing all the time [Maurer, 1992].

WHAT IS A CULTURED CELL?

The question remains open, however, as to the exact nature of the cells that grow in each case. Expression of differentiated markers under the influence of inducing conditions may mean either that the cells being cultured are mature and only require induction to maintain synthesis of specialized proteins, or that the culture is composed of precursor or stem cells that are capable of proliferation but remain undifferentiated until the correct inducing conditions are applied, whereupon some or all of the cells mature to differentiated cells. It may be useful to think of a cell culture as being in equilibrium between multipotent stem cells, undifferentiated but committed precursor cells, and mature differentiated cells (see Fig. 2.3) and that the equilibrium may shift according to the environmental conditions. Routine serial passage at relatively low cell densities would promote cell proliferation and little differentiation, while high cell densities, low serum, and the appropriate hormones would promote differentiation and inhibit cell proliferation.

The source of the culture will also determine which cellular components may be present. Hence cell lines derived from the embryo may contain more stem cells and precursor cells and be capable of greater self-renewal than cultures from adults. In addition, cultures from tissues that are undergoing continuous renewal *in vivo* (epidermis, intestinal epithelium, hemopoietic cells) will still contain stem cells, which, under the appropriate culture conditions, may survive indefinitely, while cultures from tissues that renew only under stress (fibroblasts, muscle, glia) may only contain committed precursor cells with a limited culture life-span.

Thus, the identity of the cultured cell is not only defined by its lineage *in vivo* (hemopoietic, hepatocyte, glial, etc.) but also by its position in that lineage (stem cell, committed precursor cell, or mature differentiated cell). With the exception of mouse teratomas and one or two other examples from lower vertebrates, it seems unlikely that cells will change lineage (transdifferentiate), but they may well change position in the lineage, and may even do so reversibly in some cases.

When cells are cultured from a neoplasm, they need not adhere to these rules. Thus a hepatoma from rat may proliferate *in vitro* and still express some differentiated features, but the closer they are to the normal phenotype, the more induction of differentiation may inhibit proliferation. Although the relationship between position in the lineage and cell proliferation may become relaxed (though not lost; B16 melanoma cells still produce more pigment at high cell density and at a low rate of cell proliferation than at a low cell density and a high rate of cell proliferation), transfer between lineages has not been clearly established (see also Chapter 14).

FUNCTIONAL ENVIRONMENT

Since the inception of tissue culture as a viable technique, culture conditions have been adapted to suit two major requirements: (1) production of cells by continuous proliferation and (2) preservation of specialized functions. The upsurge of interest in cellular and molecular biology and virology in the 1950s and 1960s concentrated mainly on fundamental intracellu-

TABLE 2.2. Examples of Cultured Cell Lines and Strains That Express Differentiated Properties *In Vitro*

Origin	Cell line	Species	Marker[a]	Reference
Finite cell lines				
Epidermis	Keratinocytes	Mouse	Cornification	[Fusenig and Worst, 1975]
Epidermis	Keratinocytes	Human	Cornification	[Rheinwald and Green, 1975]
Buccal mucosa	Buccal epithelium	Human	Cytokeratin	[Sundqvist et al., 1991]
Gingiva	Gingival epithelium	Human	Cytokeratin	[Oda and Watson, 1990]
Epidermis	Melanocytes	Human	Melanin	[Gilchrest et al., 1984, 1985] (see Chapter 20)
Skeletal muscle	Myoblasts	Chick	Myogenesis, CK-MM	[Richler and Yaffe, 1970]
Skeletal muscle	Myoblasts	Rat	Myogenesis	[Yaffe, 1968]
Skeletal muscle	Myoblasts	Human	Myotubes, CK-MM	[Quax et al., 1992; see Barlovatz-Meimon, Chapter 20]
Optic nerve	Astrocytes	Rat	GFAP	[Raff et al, 1990 see Barnett, Chapter 20]
Hypothalamus	C7	Mouse	Neurophysin Vasopressin	[De Vitry et al., 1974]
Heart	Cardiac myoblasts	Human	Contraction	[Goldman and Wurzel, 1992]
Liver	Hepatocytes	Human	Albumin	[Li, A.P., et al., 1992]
Pancreas	β-islet cells	Human	Insulin	[Kinard et al., 1990]
Bone	Osteoblasts	Rat	Mineralization	[Bernier et al., 1990]
Continuous cell lines				
Spleen	Friend	Mouse	Hemoglobin	[Scher et al., 1971]
Hepatoma	H4-11-E-C3	Rat	Tyrosine aminotransferase	[Pitot et al., 1964]
Myeloid leukemia	K562	Human	Hemoglobin	[Andersson et al., 1979a,b]
Myeloid leukemia	HL60	Human	Phagocytosis NTB reduction	[Olsson and Ologsson, 1981]
Glioma	C6	Rat	GFAP, GPDH	[Benda et al., 1968]
Glioma	MOG-CCM	Human	Glutamyl synthetase	[Balmforth et al., 1986]
Pituitary tumor	GH2, GH3	Rat	Growth hormone	[Buonassisi et al., 1962]
Adrenal cortex tumor		Rat	Steroids	[Buonassisi et al., 1962]
Melanoma	B16	Mouse	Melanin	[Nilos and Makarski, 1978]
Neuroblastoma	C1300	Rat	Neurites	[Liebermann and Sachs, 1978]
Skeletal muscle	C2	Mouse	Myotubes	[Morgan et al., 1992]
Skeletal muscle	L6	Rat	Myotubes	[Yaffe, 1968, 1971]
Kidney	MDCK	Dog	Domes, transport	[Gaush et al., 1966; Rindler et al., 1979]
Kidney	LLC-PKI	Pig	Na^+-dependent glucose uptake	[Hull et al., 1976; Saier, 1984]
Lung carcinoma	A549	Human	Surfactant	[Giard et al., 1972]
Liver	Hepatocytes	Mouse	Aminotransferase	[Yeoh et al., 1990]
Placenta		Human	hCG	[Cou, 1978]
Teratocarcinoma	Various	Mouse	Various	[Martin, 1975]
Myeloma	Various	Mouse	IGG	[Horibata and Harris, 1970]
Pulmonary artery endothelium	CPAE	Cow	Factor VIII, ACE	[Del Vecchio and Smith, 1981]
Hepatoma, endothelium	SK HEP-1	Human	Factor VIII	[Heffelfinger et al., 1992]
Keratinocytes	HaCaT	Human	Cornification	[Boukamp et al., 1988]
Marrow	WEHI-3B D+	Mouse	Morphology	[Nicola, 1987]
Adrenal cortex		Cow	Steroids	[Simonian et al., 1987]
Breast	MCF-7	Human	Domes, α-lactalbumin	[Soule et al., 1973]

[a]CK-MM, creatine kinase, MM isoenzyme; NTB, neotarazolium blue; GFAP, glial fibrillary acidic protein; GPDH, glycerol phosphate dehydrogenase; hCG, human chorionic gonadotropin; IGG, immunogammaglobulin; ACE, angiotensin II converting enzyme; IL-2, interleukin 2.

lar processes such as the regulation of protein synthesis, often requiring large numbers of cells. Later, the development of such techniques as molecular hybridization and gene transfer allowed the emphasis to shift to the study of the regulation of specialized functions.

Cell Cycle Control

The entry of cells into cycle is regulated by signals from the environment, such as low cell density giving cells with free edges capable of spreading, and the presence of mitogenic growth factors, such as epidermal growth factor (EGF) or platelet-derived growth factor (PDGF) (see Table 7.8 and Chapter 15). Intracellular control is mediated by positive-acting factors, such as the cyclins [Pines, 1991, 1992], and negative-acting factors such as p53 [Finlay et al., 1989; Sager, 1992] or the Rb gene product [Huang et al., 1988; Sager, 1992]. The link between the extracellular control elements (positive-acting, e.g., PDGF, and negative-acting, e.g., TGF-β, growth factors) and intracellular effectors is made by cell membrane receptors and signal transduction pathways, often involving protein phosphorylation and second messengers such as cyclic adenosine monophosphate (cAMP) and diacylglycerol [Alberts et al., 1989]. Much of the evidence for the existence of these steps in cell proliferation control has emerged from studies of oncogene and suppressor gene expression in tumor cells, with the ultimate objective of therapeutic regulation of uncontrolled cell proliferation in cancer. The immediate benefit, however, has been a better understanding of the factors required in culture to regulate cell proliferation [Jenkins, 1992].

These studies have had other useful spinoffs in the identification of genes that will increase cell proliferation, some of which can be used to immortalize finite cell lines [Amsterdam et al., 1988; Su and Chang, 1989; Klein et al., 1990; Legrand et al., 1991; Madsen et al., 1991] (see also Chapter 20). The potential of this approach for cell production in biotechnology is under intense scrutiny.

It has been recognized for many years that specific functions are retained longer where the three-dimensional structure of the tissue is retained, as in organ culture (see Chapter 22). Unfortunately, organ cultures cannot be propagated, must be prepared *de novo* for each experiment, and are more difficult to quantify than cell cultures. For this reason there have been numerous attempts to recreate three-dimensional structures by perfusing monolayer cultures [e.g., Whittle and Kruse, 1973; Gullino and Knazek, 1979] and to reproduce elements of the environment *in vivo* by culturing cells on or in special matrices like collagen gel [Michalopoulos and Pitot, 1975; Yang et al., 1981; Reid, 1990; Jones et al., 1992; Chang et al., 1992], cellulose [Leighton, 1951], or gelatin sponge [Douglas et al., 1976, 1980], or matrices from other natural tissue matrix glycoproteins such as fibronectin, chondronectin, and laminin [Reid and Rojkind, 1979; Gospodarowicz et al., 1980; Kibbey et al., 1992; Kinsella et al., 1992; Blum and Wicha, 1988; Hay, 1991] (see Chapter 7). A number of commercial products, the best known of which is Matrigel (Becton-Dickinson), are available that reproduce the characteristics of extracellular matrix but are undefined.

These techniques present some limitations, but with their provision of homotypic cell interactions, cell–matrix interactions, and the possibility of introducing heterotypic cell interactions, they may hold considerable promise for the examination of tissue-specific functions.

The development of normal tissue functions in culture would facilitate investigation of pathological behavior such as demyelination and malignant invasion. But, from a fundamental viewpoint, it is only when cells *in vitro* express their normal functions that any attempt can be made to relate them to their tissue of origin. Expression of the differentiated phenotype need not be complete, since the demonstration of a single cell type–specific cell surface antigen may be sufficient to place a cell in the correct lineage. More complete functional expression may be required, however, to place a cell in its correct position in the lineage, and to reproduce a valid model of its function *in vivo*.

CHAPTER 3

Design and Layout of the Laboratory

PLANNING

The major requirement that distinguishes tissue culture from most other laboratory techniques is the need to maintain asepsis. This is accentuated by the much slower growth of cultured animal cells relative to most of the major potential contaminants. The introduction of laminar flow cabinets has greatly simplified the problem and allows the utilization of unspecialized laboratory accommodation (see below and Chapters 4 and 5).

There are several considerations to be taken into account when planning new accommodation. Is this a new building or a conversion? A conversion limits you to the structural confines of the building; new extracts and air-conditioning can be expensive and structural modifications that involve load-bearing walls can be difficult. Where a new building is contemplated there is more scope for integrated and innovative design, positioning facilities for ergonomic and energy-saving reasons, rather than structural ones.

(1) Consider where air extracts must be placed; should laminar flow units be ducted to the exterior to facilitate decontamination and improve air circulation, or would a recirculating system suffice and involve less heat loss? Laminar flow units and air-conditioning do not interact happily together, so distancing the air circulation from the laminar flow cabinets and arranging for total extract from the cabinets can simplify things considerably.

(2) Where will the wash-up and sterilization be located and will there be a height difference? Will an elevator be required or will a ramp suffice? If so, what is the gradient and what is the maximum that you can expect to be carried up this gradient without mechanical help?

(3) Ensure that tissue culture is reasonably accessible to, but not contiguous with, the animal facility.

(4) Give your preparation staff a reasonable outlook; they are usually performing fairly repetitive duties while the scientific and technical staff are going to be looking into a laminar flow cabinet and do not need a view. In fact, windows can be more of an encumbrance with heat gain, ultraviolet (UV) denaturation of medium, and incursion of micro-organisms if they are not properly sealed.

If a conversion is contemplated, then there will be significant structural limitations.

(1) Choose the location carefully, if you can, to avoid limitations in space and awkward projections into the room that will limit flexibility.

(2) Make sure that the doorways and ceilings allow

the installation of equipment such as laminar flow cabinets, incubators, and autoclaves, and allow for access for maintenance when in place.

Some questions are more general and apply to both situations:

(1) How many people will work in the facility, for how long each week, and what kinds of culture will they perform? What incubation do they require, how large, and how close? This determines how many laminar flow hoods will be required (whether people can share hoods or will require a hood for most of the day), and whether they need a large area to handle fermentors, animal-tissue dissections, or large numbers of cultures.

(2) What space is required for each facility? As a rough guide, the largest area should be given to culture, having to accommodate laminar flow cabinets, cell counter, centrifuge, incubators, microscopes and some stocks of reagents, media, glassware, and plastics. Second is wash-up, preparation, and sterilization, third is storage, and a fourth is incubation. A reasonable estimate is 4:2:1:1, in the order just presented.

(3) The space between hoods should also be considered carefully, as some people require space for apparatus and large media vessels, while others just need a place to put a few bottles and their notes.

(4) Will people require access to the animal house for animal tissue?

(5) What proportion will be cell line work with its requirement for liquid nitrogen storage?

(6) What quarantine facilities will you require? Newly introduced cell lines and biopsies need to be screened for mycoplasma before being handled in the same room as your general stocks, and some human and primate biopsies and cell lines may carry a biohazard risk that requires containment (see Chapter 6).

These questions will enable you to decide what size of facility you require and what type of accommodation: one small room, one large room, or a suite of rooms incorporating wash-up, sterilization, aseptic area(s), incubation room, microscope/dark room, refrigeration room, and storage.

CONSTRUCTION AND LAYOUT

The rooms should be supplied with air filtered to usual industrial or office standards and be designed for easy cleaning. Furniture should fit tight to the floor or be suspended from the bench, allowing space to clean underneath. Cover the floor with vinyl or other dustproof finish and allow a slight fall in the level toward a floor drain located toward the door side of the room, i.e., away from the sterile cabinets. This allows liberal use of water if the floor has to be washed, but, more important, it protects equipment from damaging floods if stills, autoclaves, or sinks overflow.

If the tissue culture lab can be separated from the preparation, wash-up, and sterilization areas, so much the better. Adequate floor drainage should still be provided in both areas, although clearly the wash-up and sterilization area will be most important. If you have a separate wash-up and sterilization facility, it will be convenient to have this on the same floor and adjacent, with no steps to negotiate, so that trolleys may be used. Across a corridor is probably ideal (see Fig. 3.5) (see next chapter for sinks, soaking baths, etc.).

Try to imagine the flow of traffic—people, reagents, wash-up, trolleys, etc.—and arrange for minimum conflict, easy and close access to stores, and easy withdrawal of soiled items. Make sure your doors are wide enough and high enough to allow entry of all the equipment you want, particularly laminar flow units, and allow space for maintenance alongside and above equipment.

Services that are required include power, combustible gas (domestic methane, propane, etc.), carbon dioxide, compressed air, and vacuum. Power is always underestimated, both in terms of the number of outlets and the amperage per outlet. Assess carefully the equipment that will be required and assume that both the number of appliances and their power consumption, will treble within the life of the building in its present form, and try to provide sufficient power, preferably at, or near, the outlets, but at least at the main distribution board.

Gas is more difficult to judge, as it requires some knowledge of the local provision of power. Electricity is cleaner and generally easier to manage from a safety standpoint, but gas may be cheaper and more reliable. Local conditions will generally determine the need for combustible gas.

If possible, carbon dioxide should be piped into the facility. The installation will pay for itself eventually in the cost of cylinders of mixed gases for gassing cultures, and it provides a better supply, which can be protected (see next chapter), for gassing incubators. Gas-mixing flow meters (Platon) can be provided at work stations to provide the correct gas mixture, and gas blendors, though expensive, are now available (Hotpak; Signal) to provide different gas mixtures at the flick of a switch or twist of a dial.

Compressed air is generally no longer required for incubators that regulate from pure CO_2 supplies only, but will be required if a gas mixture is provided at each work station, either via flow meters or a gas blendor. Compressed air is also used to expel cotton plugs from plugged glass pipettes before washing.

A vacuum line can be very useful for evacuation of culture flasks, but needs several precautions to run successfully. There must be a collection vessel and a trap with a filter between the operator and the line to avoid any fluid, vapor, or contamination entering the vacuum line. There must also be a system whereby the vacuum pump is protected against the line being left open inadvertently, usually via a pressure-activated foot switch that closes when no longer pressed.

Because of the potential problems of vacuum lines, many people choose a local pump (vacuum or peristaltic; see next chapter) for evacuation of culture flasks. There is much to recommend this, including the initial capital outlay and the ease of maintenance, provided spare pumps are held.

Inevitably, space will be the first problem and some degree of compromise will be inevitable, but a little thought ahead of time can save much space and ultimately people's tempers.

There are six main functions to be accommodated: sterile handling, incubation, preparation, wash-up, sterilization, and storage (Table 3.1). If a single room is used, the clean area for sterile handling should be located at one end of the room and wash-up and sterilization at the other, with preparation, storage, and incubation in between. The preparation area should be adjacent to the wash-up and sterilization areas, and storage and incubators should be readily accessible to the sterile working area.

STERILE HANDLING AREA

The sterile handling area should be located in a quiet part of the laboratory, its use should be restricted to tissue culture, and there should be no through traffic or other disturbance likely to cause dust or drafts. Use a separate room or cubicle if laminar flow cabinets are not available. The work area, in its simplest form, should be a plastic laminate-topped bench, preferably plain white or neutral gray, to facilitate observation of cultures, dissection, etc., and to enable accurate reading of pH when using phenol red as an indicator. Nothing should be stored on this bench and any shelving above should only be used in conjunction with sterile work, e.g., for holding pipette cans and instruments. The bench should either be free-standing (away from the wall) or sealed to the wall with a plastic sealing strip or mastic.

TABLE 3.1. Tissue Culture Facilities

Minimum requirements (essential)	Desirable features	Useful additions
Sterile area; clean and quiet area, no through traffic	Filtered air (air-conditioning)	Piped CO_2 and compressed air
Separate from animal house and microbiological labs	Service bench adjacent to culture area	Storeroom for bulk plastics
		Containment room for biohazard work
	Separate prep room	
Preparation area	Hot room with temperature recorder	Liquid N_2 storage tank (~500 l)
Wash-up area (not necessarily within tissue culture laboratory, but adjacent)	Separate sterilizing room	Microscope room
	Cylinder store	Dark room
		Vacuum line
Space for incubator(s)		
Storage areas:		
Liquids–ambient, 4°C, −20°C		
Glassware (shelving)		
Plastics (shelving)		
Small items (drawers)		
Specialized equipment (slow turnover), cupboard(s)		
Chemicals–ambient, 4°C, −20°C share with liquids but keep chemicals in sealed container over desiccant)		
Space for liquid N_2 freezer(s)		
Sink		

Laminar Flow

The introduction of laminar flow cabinets (or "hoods") with sterile air blown over the work surface (Fig. 3.1) (see Chapters 4 and 5) affords greater control of sterility at a lower cost than providing a separate sterile room. Individual free-standing cabinets are preferable, as they separate operators and can be moved around, but laminar flow wall or ceiling units in batteries can be used (Fig. 3.2). With cabinets, only the operator's arms enter the sterile area, while with laminar flow wall or ceiling units, there is no cabinet and the operator is part of the work area. While this may give more freedom of movement, particularly with large pieces of apparatus (roller bottles, fermentors), greater care must be taken by the operator not to disrupt the laminar flow, and it may prove necessary to wear caps and gowns to avoid contamination.

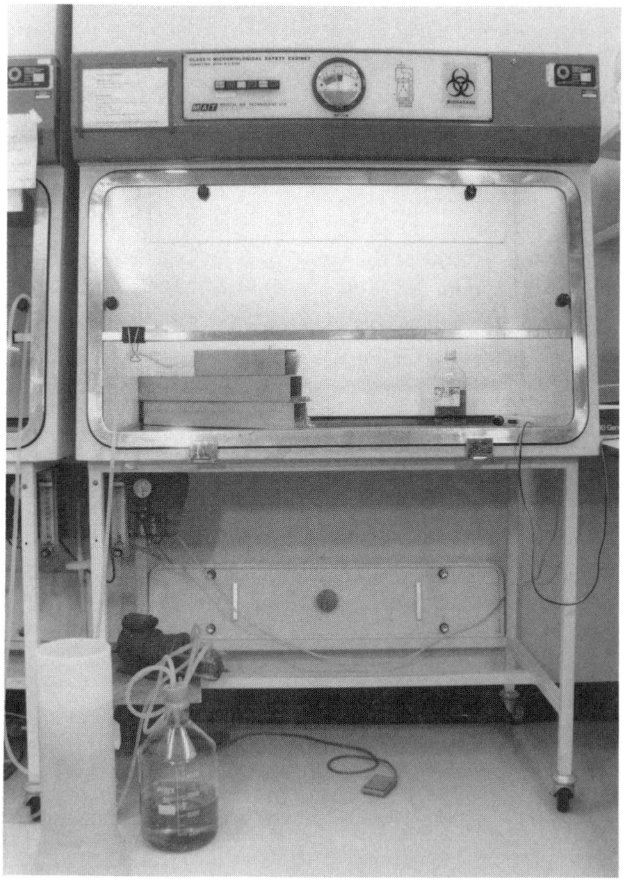

Fig. 3.1. Laminar flow cabinet, vertical flow (biohazard type). Work surface is laid out with minimal amount of apparatus, leaving maximal space for cultures and other reagents. Pipette cans are of the square variety, which prevent rolling, and an automatic pipetting device is placed at the right-hand side; the pipette hod is placed on the floor at the left. A peristaltic pump, connected to a receiver vessel, is shown on the left side below the hood, with a foot pedal to activate the pump. The suction line from the pump is attached to a clip at the left of the aperture to the work area, and a delivery tube travels across below the hood to provide a supply of mixed CO_2 in air. Orientation would be reversed for left-handed individuals. Below the hood, on the left-hand side, is a CO_2 and air supply with flow meters to regulate the flow rate and gas mixture (this may be replaced with electronic gas mixer [Signal]). The blanking screen, which may be used to seal off the hood when not in use, has been placed at the back below the hood, but can be hung alongside.

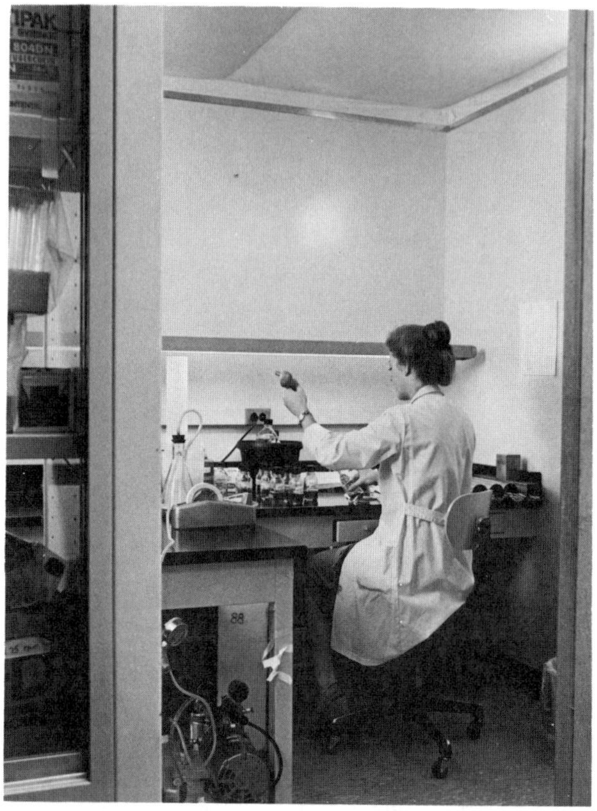

Fig. 3.2. Aseptic room. Filtered air is supplied from the ceiling, and the whole room is regarded as a sterile working area. Door has been left open for photograph but would normally be kept closed.

Select cabinets that suit your accommodation—free-standing or bench-top—and allow plenty of legroom underneath with space for pumps, aspirators, and so forth. Free-standing cabinets should be on castors so that they can be moved if necessary. As laminar flow cabinets contribute a significant amount of heat to the room (300–500 w/cabinet), and the airflow from the cabinets can cause turbulence if the ceiling height is low, it is preferable to have the exhaust ducted out of the room. This also facilitates fumigation of the hoods, required when the filters are changed.

Select chairs that are of a suitable height, preferably with adjustable seat height and back angle, and make sure that they can be drawn up close enough to the front edge to allow comfortable working well within the cabinet. A small section of bench, trolley, or folding flap (300–500 mm minimum) should be provided

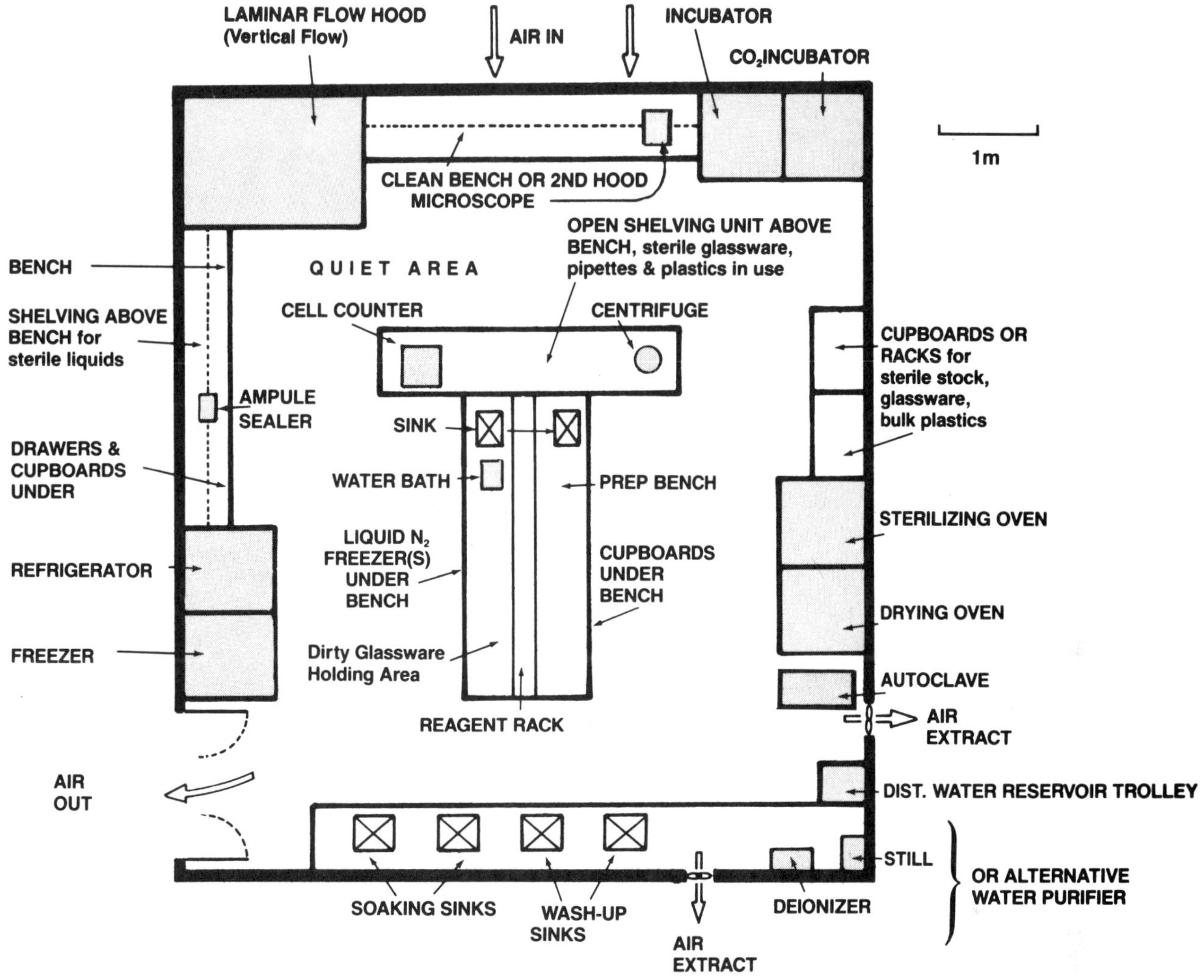

Fig. 3.3. Suggested layout for simple, self-contained tissue culture laboratory for use by two or three persons. Shaded areas represent movable equipment. Ampule sealer unnecessary if plastic ampules are used.

beside each hood for depositing apparatus or reagents not in immediate use.

It is a good principle, made easier by the introduction of laminar flow, to create a "sterility gradient" in the tissue culture laboratory. Hence, a single room housing all the necessary functions of a tissue culture laboratory should have its sterile cabinets located at one end, farthest from the door, while wash-up, preparation of glassware or reagents, centrifugation, etc., would be best performed at the opposite end of the room (Figs. 3.3, 3.4). This principle still applies where laminar flow is not available, in fact, even more so; but the introduction of laminar flow, particularly horizontal laminar flow, makes the gradient easier to maintain, although the use of horizontal laminar flow is now restricted by safety regulations (see Chapter 6).

In addition, you may wish to provide a small room or cubicle for use as a containment area (Fig. 3.5). This must be separated by a door or air lock from the rest of your suite and will need its own incubators, freezer, refrigerator, centrifuge, etc. It will also require a biohazard cabinet or pathogen hood with separate extract and pathogen trap (for fuller description of containment facilities, see Chapter 6).

INCUBATION

The requirement for cleanliness is not as stringent as with sterile handling, but clean air, low disturbance level, and no through traffic will endow your incubation area with a better chance of avoiding dust, spores, and drafts that carry them.

Incubation may be carried out in separate incubators or in a thermostatically controlled hot room. Incubators, bought singly, are inexpensive and economic

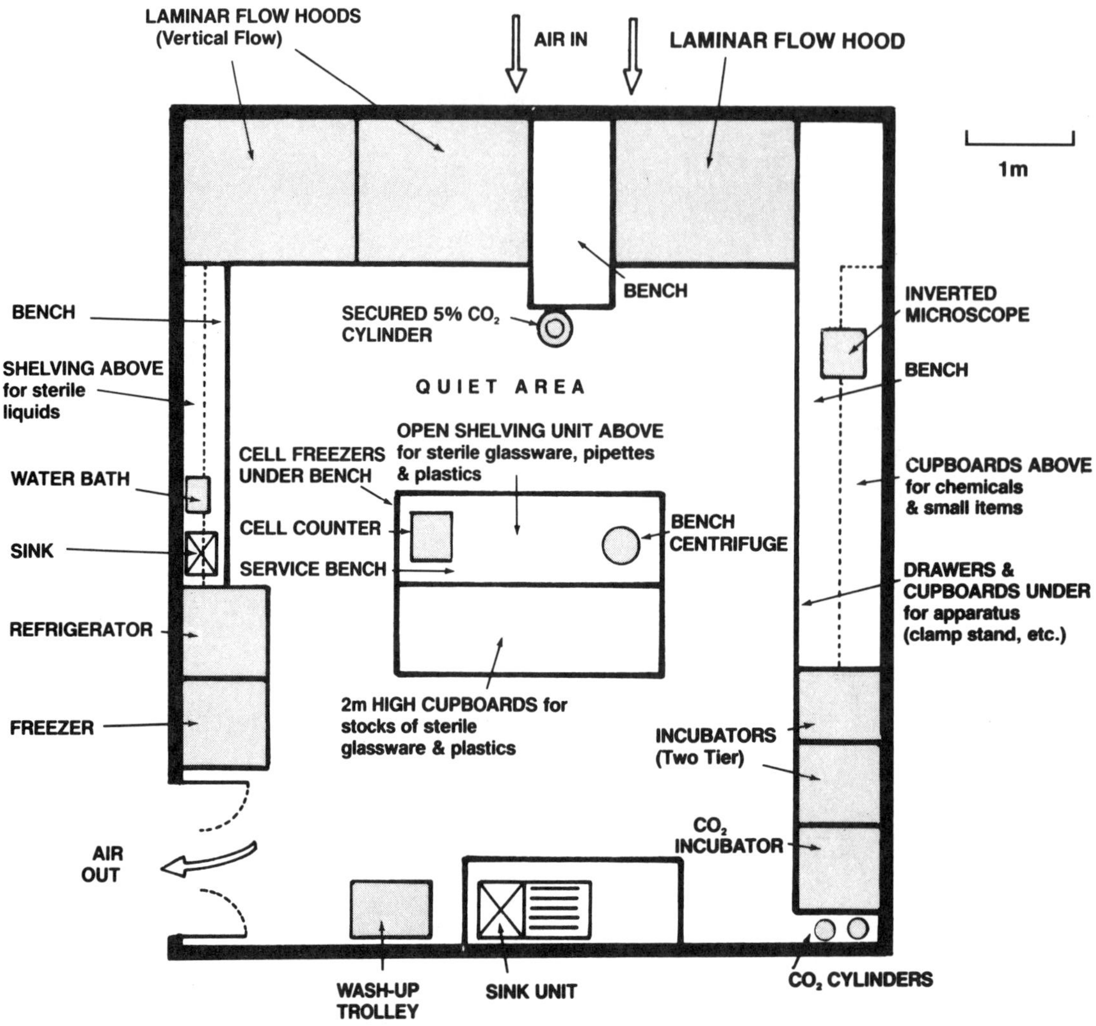

Fig. 3.4. *Tissue culture laboratory suitable for five or six persons with washing-up and preparation facility located elsewhere. Shaded areas represent movable equipment.*

in space; but as soon as you require more than two, their cost is more than a simple hot room and their use less convenient. Incubators also lose more heat when opened and are slower to recover than a hot room. As a rough guide, you will need 0.2 m^3 (200 l, 6 ft^3) of incubation space (0.5 m^2, 6 ft^2 shelf space) per person. Extra provision may need to be made for a humid incubator(s) with a controlled CO_2 level in the atmosphere (see Chapter 4).

Frequent cleaning of incubators, particularly humidified incubators, is essential, and flasks, dishes, or boxes containing them, taken from the incubator to the laminar flow or other sterile work station, should be swabbed with alcohol before opening (see also Chapter 5).

Hot Room

If you have the space within the laboratory area or have an adjacent room or walk-in cupboard readily available and accessible, it may be possible to convert this into a hot room (Fig. 3.6). It need not be specifically constructed but should be sufficiently well insulated not to allow "cold spots" to be generated on the walls. If insulation is required, line with plastic laminate-veneered board, separated from the wall by about 5 cm (2 in) fiberglass, mineral wool, or fire-retardant plastic foam. Mark the location of the straps or studs carrying the lining panel to identify anchorage points for shelving. Shelf supports should be spaced at 500–600 mm (21 in) to support shelving without sagging.

Do not underestimate the space that you will require in the lifetime of the hot room. It costs very little more to equip a large hot room than a small one. Calculate on the basis of the amount of shelf space you will require; if you have just started, multiply by five or ten; if you have been working for some time, by two or four. Allow 200–300 mm (9 in) between shelves and use wider shelves (450 mm, 18 in) at the bottom and

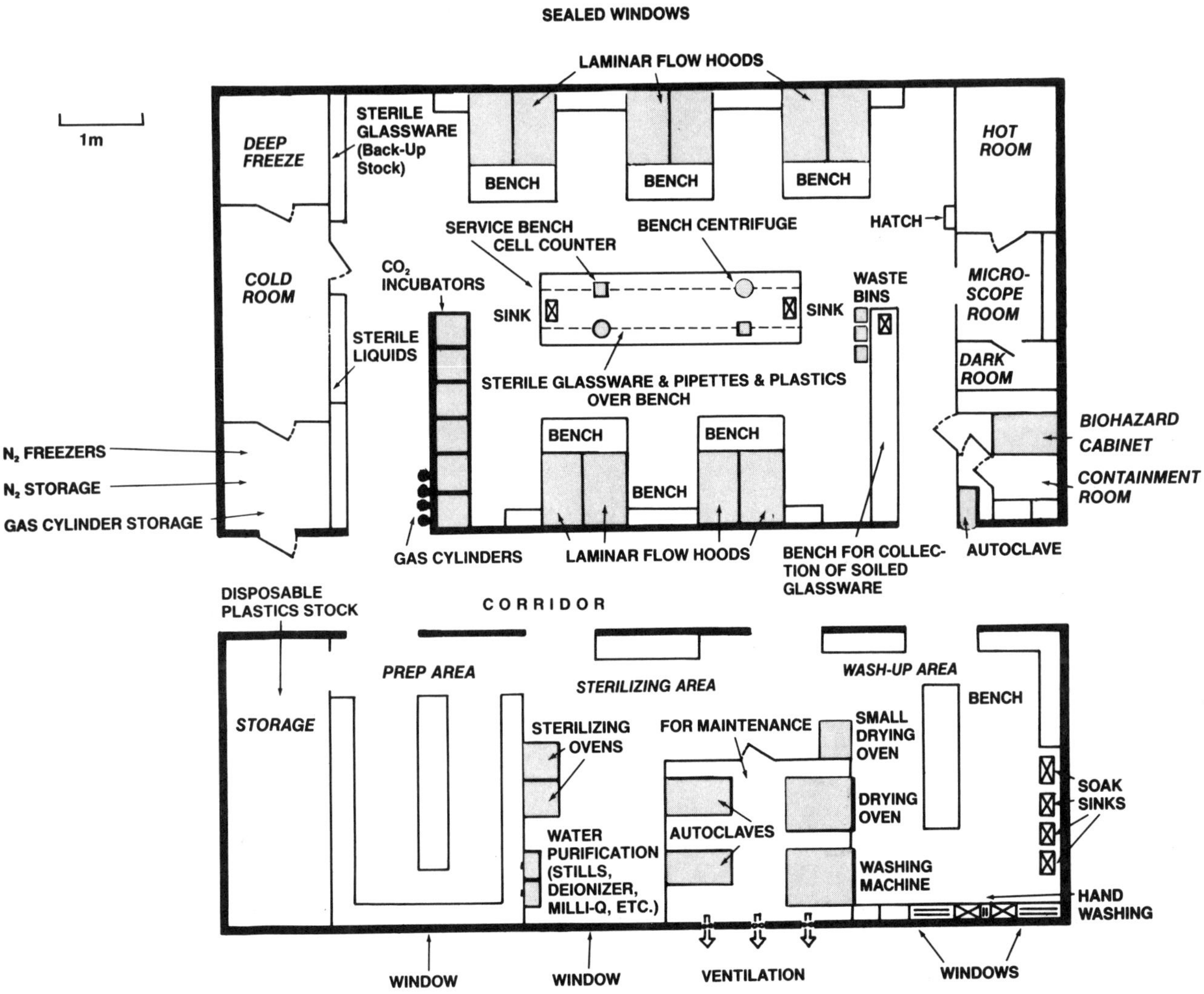

Fig. 3.5. Large-scale tissue culture laboratory with adjacent washing-up, sterilization, and preparation area. Suitable for 20 to 30 persons. Shaded areas represent equipment as distinct from furniture.

narrower (250–300 mm, 12 in) ones above eye level. Perforated shelving mounted on adjustable brackets should be used to allow for air circulation. It must be flat and perfectly horizontal with no bumps or irregularities.

Although perforated shelving gives the best air circulation, it should be realized that it can also lead to irregularities in cell distribution in monolayer cultures with variations in cell density following the pattern of spacing on the shelves. It is not clear what causes this, though it may be due to convection currents generated over points of contact relative to holes in the shelf, areas that may cool down quicker when the door is opened. Although this may not create a problem in routine maintenance, for experiments where uniform density is important, flasks and dishes should be placed on an insulated tile or metal tray.

Wooden furnishings should be avoided as much as possible, as they warp in the heat and can harbor infestations.

A small bench, preferably stainless steel or solid plastic laminate, should be provided at some part of the hot room. This should accommodate a microscope, its transformer, and the flasks that you wish to examine. If you contemplate doing cell synchrony experiments or having to make any sterile manipulations at 36.5°C, you should also allow space for a small laminar flow unit (300 × 300 or 450 × 450 mm, 12–18 in square, filter size) either wall mounted or on a stand over part of the bench. Alternatively a small laminar flow cabinet (not more than 1,000 mm, 3 ft, wide) could be located in the room. The fan motor should be tropically wound and should not run continuously. Apart from wear of the motor, it will generate heat in the room and the motor may burn out.

Once a hot room is provided, others may wish to use

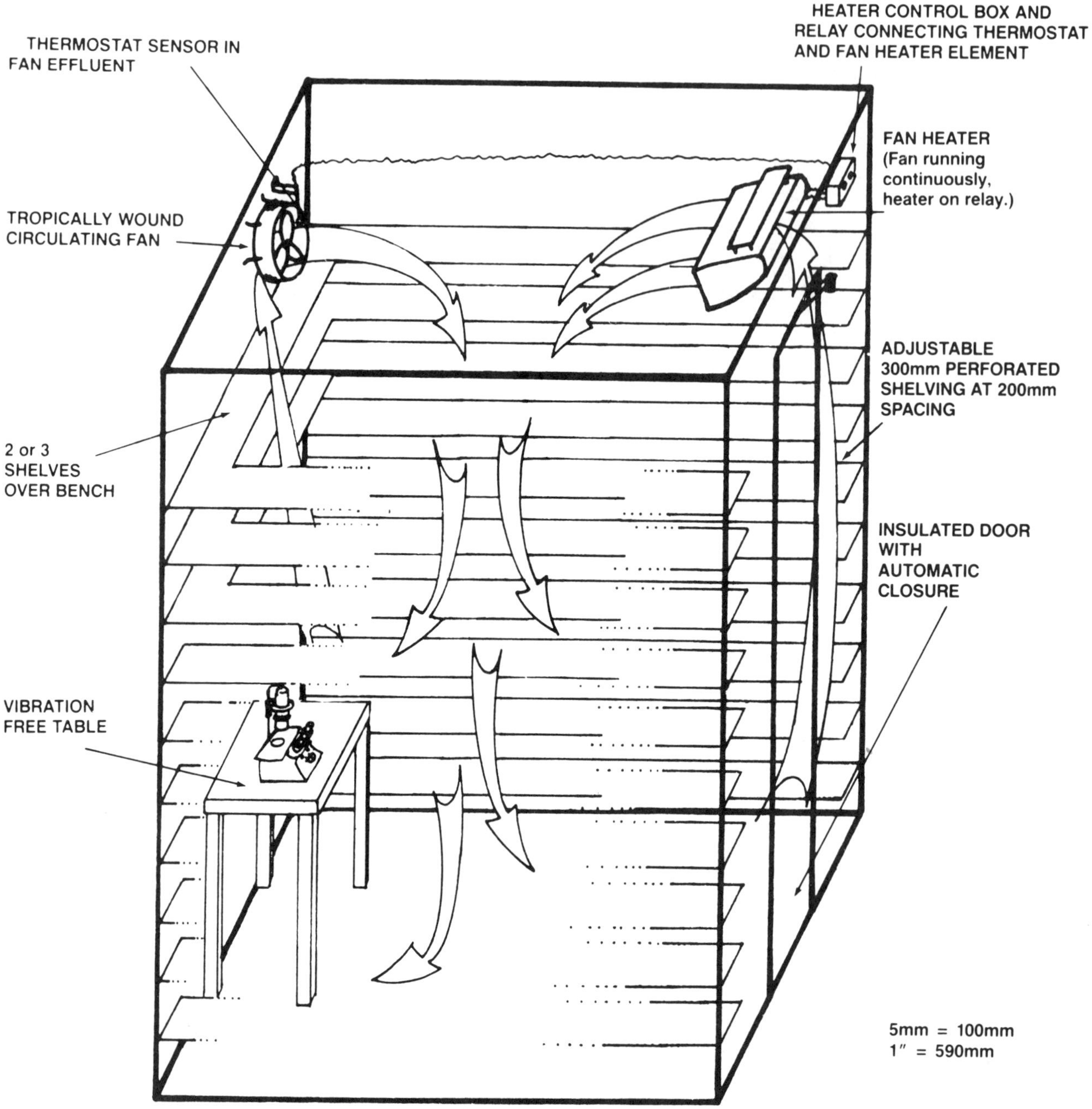

Fig. 3.6. Suggested design for a simple hot room. Arrows represent air circulation. (Based on an original design by Dr. John Paul.)

the space for non-tissue culture incubations, so the area of bench space provided should also take account of possible usage for incubation of tubes, shaker racks, etc.

Incandescent lighting is preferable to fluorescent, which can cause degradation of constituents of the medium. Furthermore, some fluorescent tubes have difficulty in striking in a hot room.

The temperature should be controlled within ±0.5°C at any point and at any time. This depends on: (1) the sensitivity and accuracy of the control gear; (2) the siting of the thermostat sensor; (3) the circulation of air in the room; (4) correct insulation; and (5) the evolution of heat by other apparatus (stirrers, etc.) in the room.

Heaters. Heat is best supplied via a fan heater, domestic or industrial, depending on the size of the room. Approximately 2–3 Kw per 20 m³ (700 ft³) will be required (or two 1.0–1.5 Kw), depending on the insulation. The fan on the fan heater should run continuously, and the power to the heating element should come from a proportional controller (see below).

Air circulation. A second fan, positioned on the opposite side of the room and with the airflow opposing that of the fan heater, will ensure maximum circulation. If the room is more than 2 × 2 m (6 × 6 ft) some form of ducting may be necessary. Blocking off the corners as in Figure 3.6 is often easiest and most economical in space in a square room. In a long, rectangular room, a false wall may be built at either end, but be sure to insulate it from the room and make it strong enough to carry shelving.

Thermostats. Thermostats should be of the "proportional controller" type, acting via a relay to supply heat at a rate proportional to the difference between the room temperature and the set point. So, when the door opens and the room temperature falls, recovery is rapid; but the temperature does not overshoot, as the closer the room temperature approaches the set point, the less heat is supplied.

If possible, dual thermostats in parallel, but preferably controlling separate heaters, should be installed so that if one fails it is overridden by the other (Fig. 3.7). One thermostat ("regulating") is set at the required temperature, 36.5°C, with a narrow fluctuation range, say 0.4°C (±0.2°C). The second, or safety, thermostat is set slightly below the first so that if the temperature falls below the range controlled by the first, the second will be activated. Pilot lights in series with each circuit will indicate when they are activated. Finally, there should be an overriding thermostat in series (on both heaters if two are installed) set at 38.5°C, so that if a thermostat locks on, the override thermostat will cut out at 38.5°C and illuminate a warning light (Fig. 3.8).

The thermostat sensors should be located in an area of rapid airflow close to the effluent from the second, circulating, fan for greatest sensitivity. A rapid-response, high thermal conductivity sensor (thermistor or thermocouple) should be used in preference to a pressure-bulb type.

Overheating. Since so much care is taken to provide heat and replenish its loss rapidly, another problem is often forgotten—namely, unwanted heat gain. This can arise (1) because of a rise in ambient temperature in the laboratory in hot weather or (2) due to heat produced from within the hot room by apparatus such as stirrer motors, roller racks, laminar flow units, etc.

Try to avoid heat-producing equipment in the hot room, and arrange for heat dissipation either by a thermostatically controlled fan extract (and inlet) or an air conditioner. In either case, set the thermostat well below (>2°C) the heater thermostats so that the latter will regulate the temperature.

Access. If a proportional controller, good circulation, and adequate heating are provided, an air lock will not be required. The door should still be well insulated (foam plastic or fiberglass filled), light, and easily closed, preferably self-closing. It is also useful to have a hatch leading into the tissue culture area, with a shelf on both sides, so that cultures may be transferred easily into the room. The hatch door should also have an insulated core. If the hatch is located above the bench, this will avoid any risk of a "cold spot" on the shelving.

A temperature recorder should be installed such that the chart is located in easy and obvious view of the people working in the tissue culture room. A weekly change of chart is convenient and still has sufficient resolution. If possible, one high-level and one low-level warning light should be placed beside the chart or at a different, but equally obvious, location.

SERVICE BENCH

It may be convenient to position a bench to carry cell counter, microscope, etc., close to the sterile handling area, either dividing the area or separating it from the other end of the lab (see Figs. 3.3–3.5). The service bench should also have provision for storage of sterile glassware, plastics, pipettes, screw caps, syringes, etc., in drawer units below and shelves above. This bench may also house other accessory equipment such as an ampule-sealing device and a small bench centrifuge. The bench should provide a close supply of all the immediate requirements.

PREPARATION

The need for extensive media preparation in small laboratories can be avoided if there is a proven source of reliable commercial culture media. While a large laboratory (approximately 50 people doing tissue culture) may still find it more economical to prepare their own media, most smaller enterprises may prefer to purchase ready-made media. This reduces preparation to reagents, such as salt solutions, ethylenediaminetetraacetic acid (EDTA), etc., bottling these and water, and packaging screw caps and other small items for sterilization. While this area should still be clean and quiet, sterile handling is not necessary, as all the items will be sterilized.

If there is difficulty in obtaining reliable commercial media, a larger area should be allocated for preparation to accommodate a coarse and fine balance, pH meter, and, preferably, an osmometer. Bench space will

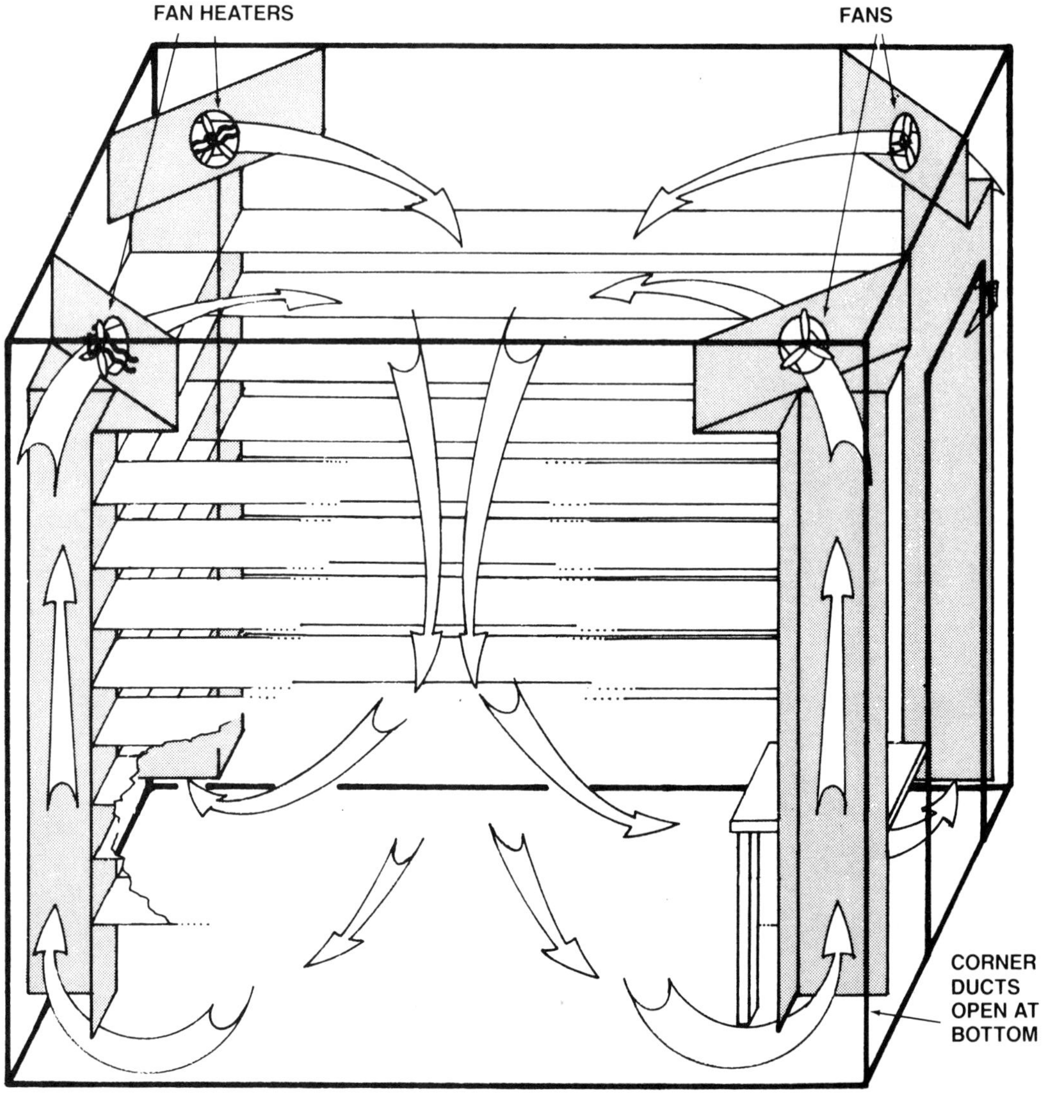
FAN HEATERS
FANS
CORNER DUCTS OPEN AT BOTTOM

a

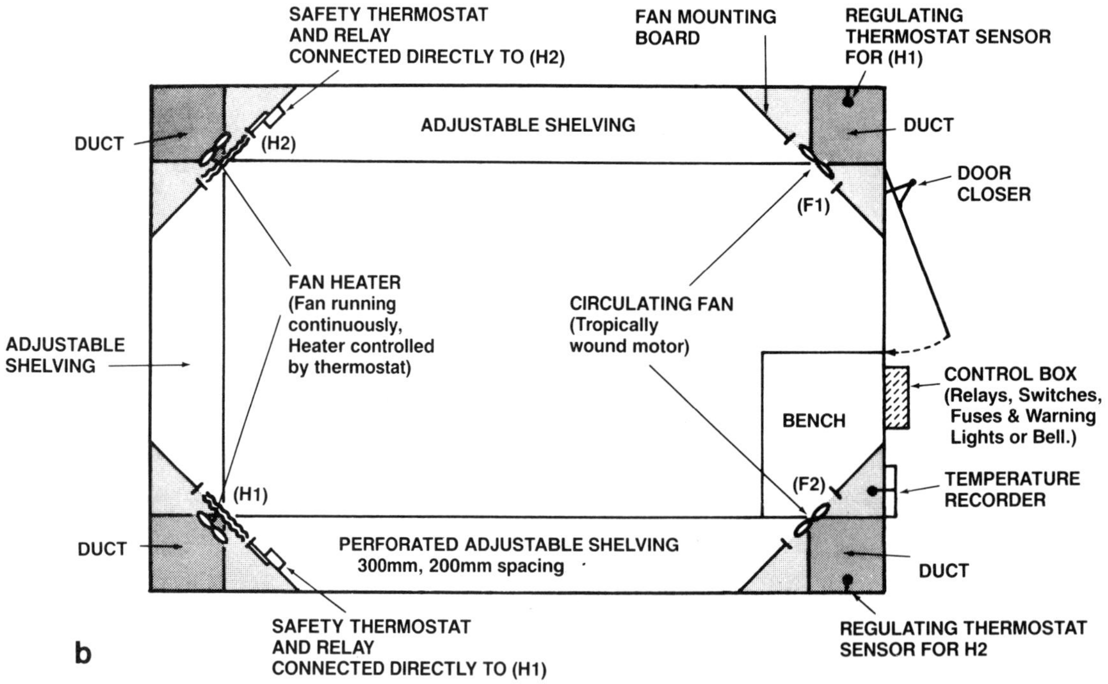
SAFETY THERMOSTAT AND RELAY CONNECTED DIRECTLY TO (H2)
FAN MOUNTING BOARD
REGULATING THERMOSTAT SENSOR FOR (H1)
DUCT
(H2)
ADJUSTABLE SHELVING
DUCT
DOOR CLOSER
(F1)
FAN HEATER (Fan running continuously, Heater controlled by thermostat)
CIRCULATING FAN (Tropically wound motor)
ADJUSTABLE SHELVING
BENCH
CONTROL BOX (Relays, Switches, Fuses & Warning Lights or Bell.)
(F2)
TEMPERATURE RECORDER
(H1)
DUCT
PERFORATED ADJUSTABLE SHELVING 300mm, 200mm spacing
DUCT
SAFETY THERMOSTAT AND RELAY CONNECTED DIRECTLY TO (H1)
REGULATING THERMOSTAT SENSOR FOR H2

b

be required for dissolving and stirring solutions and for bottling and packaging, and additional ambient and refrigerated shelf space will be required for reagents. If possible, an extra horizontal laminar flow cabinet should be provided in the sterile area for filtering and bottling sterile liquids, and incubator space must be allocated for quality control of sterility, i.e., incubation of samples of media in broth and after plating out.

Heat-stable solutions and equipment can be autoclaved or dry-heat sterilized at the nonsterile end of the area. Both streams then converge on the storage areas (see below).

WASH-UP

If possible, wash-up and sterilization facilities should be situated outside the tissue culture lab, as the humidity and heat that they produce may be difficult to dissipate without increasing airflow above desirable limits. Autoclaves, ovens, and distillation apparatus should be located in a separate room, if possible (see Fig. 3.5), with an efficient extraction fan. The wash-up area should have plenty of space for soaking glassware and space for an automatic washing machine, should you require one. There should also be plenty of bench space for handling baskets of glassware, sorting pipettes, packaging and sealing sterile packs, and a pipette washer and drier. If the sterilization facilities must be located in the tissue culture lab, site them where greatest ventilation is possible and farthest from the sterile handling area.

Trolleys are often useful for collecting dirty glassware and redistributing fresh sterile stocks, but remember to allocate parking space for them.

STORAGE

Storage must be provided for:

(1) Sterile liquids:
  (a) at room temperature (salt solutions, water, etc.)
  (b) at 4°C (media)
  (c) at −20°C or −70°C (serum, trypsin, glutamine, etc.)
(2) Sterile glassware:
  (a) media bottles, glass culture flasks
  (b) pipettes
(3) Sterile disposable plastics:
  (a) culture flasks and petri dishes
  (b) centrifuge tubes and vials
  (c) syringes
(4) Screw caps, filter tubes, stoppers, etc.
(5) Apparatus: filters and large receiver flasks
(6) Gloves, disposal bags, etc.
(7) Liquid nitrogen to replenish freezers:
  (a) as dewars (25–50 l) under the bench
  (b) a large storage vessel (100–150 l) on a trolley
  or
  (c) liquid nitrogen storage tanks (500–1,000 l) permanently sited in a room of their own

◇ If the last of these is chosen, adequate ventilation must be provided for the room where the nitrogen is stored and dispensed, preferably with an alarm to signify when the oxygen tension falls below safe levels.

(8) Cylinder storage for carbon dioxide
  (a) as separate cylinders for transferring to the laboratory as required (◇ tethered to the wall or bench in each case)
  (b) as a rack of cylinders from which a piped supply is taken to work stations
  or
  (c) a pressurized tank that is replenished regularly (and must therefore be accessible to delivery vehicles)

These considerations are based on scale of operation and unit cost, and can be calculated from quotations from suppliers. As a rough guide, 2–3 people will only require a few cylinders, 10–15 will probably benefit from a piped supply from a bank of cylinders, and for >15 it will probably pay to have a storage tank.

Storage areas 1–6 should be within easy reach of the sterile working area. Refrigerators and freezers should be located toward the nonsterile end of the lab, as the doors and compressor fans create dust and drafts and they may harbor fungal spores. Also, they require maintenance and periodic defrosting, which creates a level and kind of activity best separated from your sterile working area.

The keynote of storage areas is ready access both for withdrawal and replenishment. Double-sided units are useful because they may be restocked from one side

Fig. 3.7. *Hot room with dual heating circuits and safety thermometers. a. Oblique view. b. Plan view. Arrows represent air circulation. Layout and design were developed in collaboration with Malcolm McLean of Boswell, Mitchell & Johnson (architects) and Jimmy Lindsay of Kenneth Munro & Associates (consultant engineers).*

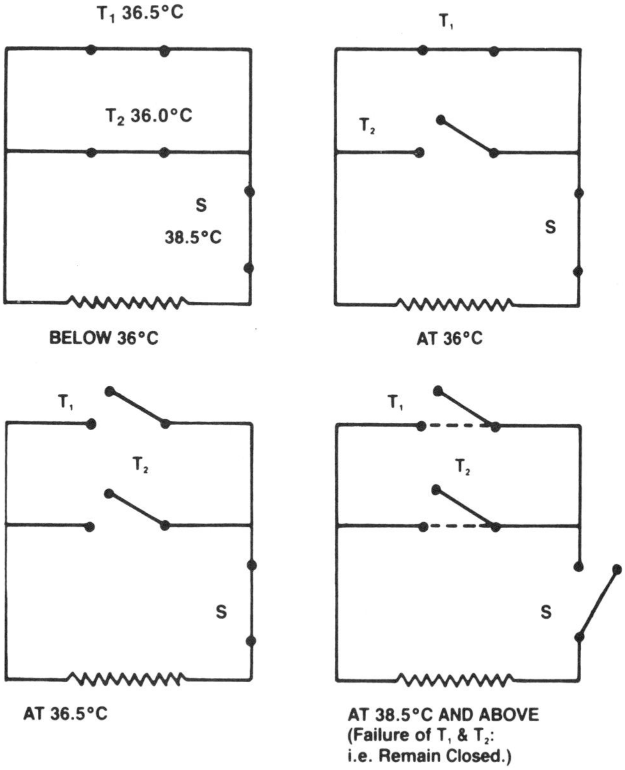

T_1 36.5°C
T_2 36.0°C
S
38.5°C
BELOW 36°C
T_1
T_2
S
AT 36°C
T_1
T_2
S
AT 36.5°C
T_1
T_2
S
AT 38.5°C AND ABOVE
(Failure of T_1 & T_2:
i.e. Remain Closed.)

a

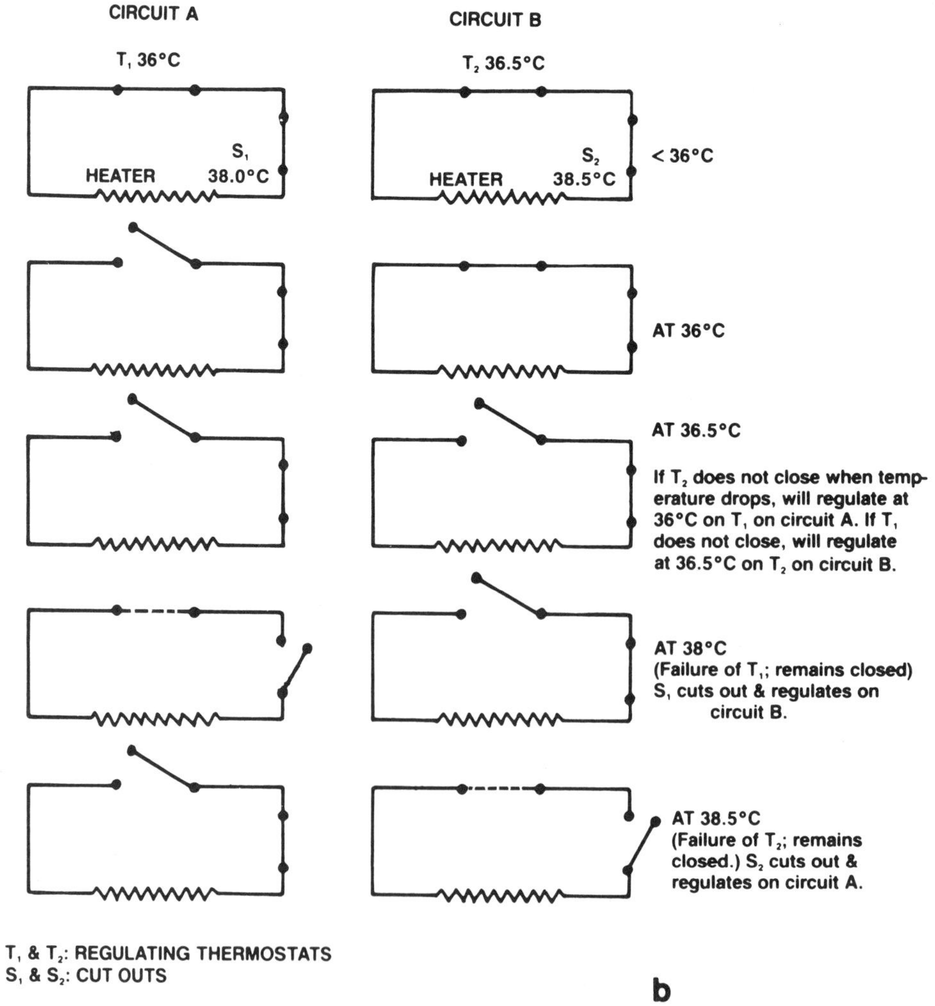

CIRCUIT A
T_1 36°C
HEATER
S_1
38.0°C
CIRCUIT B
T_2 36.5°C
HEATER
S_2
38.5°C
< 36°C
AT 36°C
AT 36.5°C
If T_2 does not close when temperature drops, will regulate at 36°C on T_1 on circuit A. If T_1 does not close, will regulate at 36.5°C on T_2 on circuit B.
AT 38°C
(Failure of T_1; remains closed)
S_1 cuts out & regulates on circuit B.
AT 38.5°C
(Failure of T_2; remains closed.) S_2 cuts out & regulates on circuit A.
T_1 & T_2: REGULATING THERMOSTATS
S_1 & S_2: CUT OUTS

b

and used from the other. Storage boxes or trays that can be taken away and filled and replaced when full are also useful.

Remember to allocate sufficient space for storage, as this will allow you to make bulk purchases and thereby save money, and at the same time reduce the risk of running out of valuable stocks at times when they cannot be replaced. As a rough guide, you will need 200 l (~8 ft^3) of 4°C storage and 100 l (~4 ft^3) of −20°C storage per person. The volume per person increases with fewer people. Thus, one person may need a 250-l (10-ft^3) fridge and a 150-l (6-ft^3) freezer. This refers to storage space only, and allowance must be made for access for working in the cold room where walk-in cold rooms and deep freezers are planned.

In general, separate −20°C freezers are better than a walk-in cold room. They are easier to clean out and maintain, and they provide better backup if one unit fails. You may also wish to consider whether a cold room has any advantage over refrigerators. There is no doubt that a cold room will give more storage per cubic meter, but the utilization of that space is important; how easy is it to clean and defrost, and how well can space be allocated to individual users? Several independent refrigerators will occupy more space than the equivalent volume of cold room but may be easier to manage and maintain in the event of failure.

It is also well worth considering budgeting for additional freezer and refrigerator space to allow for routine maintenance and unpredicted breakdown.

Fig. 3.8. *Circuit diagrams for regulating and safety thermostats for hot room illustrated in Figures 3.6 and 3.7: The switches indicated are electronic relays activated by thermostats or thermistors.* T_1 *and* T_2*, regulating thermostats,* S_1 *and* S_2*, safety thermostats. a. Single heater circuit. b. Dual heater circuit.*

C H A P T E R 4

Equipping the Laboratory

The specific needs of a tissue culture laboratory, like most labs, can be divided into three categories: (1) essential—you cannot perform a job without them; (2) beneficial—the work would be done better, more efficiently, quicker, or with less labor; and (3) useful—it would make life easier, improve working conditions, reduce fatigue, enable more sophisticated analyses to be made, or generally make your working environment more attractive (Table 4.1).

The need for a particular piece of equipment is often very subjective, a product of personal aspirations, high-pressure salesmanship, technical innovation, and peer pressure. The real need is harder to define but is determined, objectively, by the type of work, the saving in time that the equipment would produce, the greater technical efficiency in terms of asepsis, quality of data, analytical capability, and sample requirements, the saving in time and/or personnel, the number of people who would use it, the available budget and potential cost benefit, and your own factor as determined by the special requirements of your own procedures.

ESSENTIAL EQUIPMENT

Incubator

The incubator should be large enough, probably 50–200 l (1.5–6 ft^3) per person, have forced air circulation, temperature control ±0.5°C, and a safety thermostat that cuts out if the incubator overheats or, better, that regulates it if the first fails. It should be corrosion-resistant, e.g., stainless steel (anodized aluminum is acceptable for a dry incubator), and easily cleaned. A double cabinet, one above the other, independently regulated, is preferable to one large cabinet. This gives more accommodation but with better temperature control, and with the added protection that if one half fails or needs to be cleaned, the other can still be used. This is also useful when you need to clean out one compartment.

Many incubators are provided with a heated water jacket as a method for distributing the heat evenly around the cabinet and avoiding the risk of cold spots. They also hold their temperature in the event of a heater failure or power cut. However, the advent of high-efficiency insulation and diffuse surface heater elements has all but eliminated the need for a water jacket and makes moving the incubator much simpler. A water jacket generally needs to be emptied if the incubator is to be moved.

Sterilizer

The simplest and cheapest sterilizer is a domestic pressure cooker that will generate 100 kPa (1 atm, 15 lb/in^2) above ambient. More complex autoclaves exist,

TABLE 4.1. Tissue Culture Equipment

Minimum requirements (essential)	Desirable features (beneficial)	Useful additions
Incubator	Laminar flow hood(s), vertical, horizontal, biohazard	−70°C freezer
Sterilizer (autoclave, pressure cooker, oven)	Cell counter	Glassware washing machine
Refrigerator	Vacuum pump	CCTV for inverted microscope(s)
Freezer (for −20°C storage)	CO_2 incubator	Filing for freezer records
Inverted microscope	Coarse and fine balance	Colony counter
Soaking bath or sink	pH meter	High-capacity centrifuge (6 × 1 l)
Deep washing sink	Osmometer	Cell sizer (e.g., Coulter ZB series)
Pipette cylinder(s)	Phase-contrast and fluorescence microscope(s)	Time-lapse cinemicrographic equipment
Pipette washer	Portable temperature recorder	Interference-contrast microscope
Still or water purifier	Permanent temperature recorders on sterilizing oven and autoclave	Polythene bag sealer (for packaging
sterile items	Roller racks for roller bottle culture	Controlled-rate cooler (for cellfreezing)
for long-term storage)	Pipette drier	Density meter (for density gradient cell separation)
Bench centrifuge	Pipette plugger	MT plate scintillation counter
Liquid N_2 freezer (~35 l, 1,500–3,000 ampules)	Trolleys for collecting soiled glassware and redistributing fresh supplies	Fluorescence-activated cell sorter
Magnetic stirrer racks for suspensioncultures	Pipette aid	Confocal microscope
Liquid N_2 storage flask (~25 l)	Autopipette or other form of automatic dispensor, dilutor	Centrifugal elutriator centrifuge and rotor
	Separate sterilizing oven and drying oven	

but the main consideration is the capacity: Will it accommodate all you want to do? A simple bench-top autoclave (Fig. 4.1a) may be sufficient, but a larger model with a timer and a choice of presterilization and poststerilization evacuation (Fig. 4.1b) will give more capacity and greater flexibility in use. A "wet" cycle (water, salt solutions, etc.) is performed without evacuation before or after sterilization. Dry items (instruments, swabs, screw caps, etc.) require that the chamber be evacuated before sterilization, to allow efficient access of hot steam. It should be evacuated after sterilization to remove steam and promote subsequent drying; otherwise the articles will emerge wet, leaving a trace of contamination from the condensate on drying. To minimize this risk, always use deionized or reverse-osmosis water to supply the autoclave. If you require a high sterilization capacity (300 l, 9 ft^3, or more), buy two smaller autoclaves rather than one large one, so that during routine maintenance and accidental breakdowns you still have one functioning machine. Furthermore, a smaller machine will heat up and cool more quickly and can be used more economically for small loads. Leave sufficient space around them for maintenance and ventilation and provide adequate air extraction to remove heat and steam.

Most small autoclaves will come with their own steam generator (calorifier), but larger machines may have the option of a self-contained steam generator, a separate steam generator, or the facility to use a steam line. If high-pressure steam is available on line, then this will be the cheapest and simplest method of heating and pressurizing the autoclave; if not, it is best to purchase a sterilizer complete with its own self-contained steam generator. It will be cheaper to install and easier to move. With the largest machines, you may not have a choice, as they are frequently offered only with a separate generator. In this case, you will need to allow space for it at the planning stage.

Refrigerators and Freezers

Usually a domestic item will be found to be quite efficient and cheaper than special laboratory equipment. Domestic refrigerators are available with no icebox ("larder refrigerators"), giving more space and eliminating the need for defrosting. However, if you require a lot of accommodation (400 l, 12 ft^3, or more; see

Fig. 4.1. Autoclaves. a. Bench-top model. b. Large, recessed model. Both types now require that the pressure vessel cannot be opened until a safety pressure-release valve releases a safety lock, indicating a fall in pressure and temperature to a safe level.

Chapter 3), a large hospital (blood bank) or catering refrigerator may be better.

If the accommodation is available, and the number of people using tissue culture is more than three or four, it is worth considering installation of a cold room. It is more economical in space than several separate refrigerators and easier to access (see also Chapter 3). The walls should be smooth and easily cleaned and the racking on castors to facilitate cleaning. Cold rooms should be cleaned out regularly to eliminate old stock, and so that the walls and shelves may be washed to minimize fungal contamination.

Similar advice applies to freezers; several inexpensive domestic freezers will be cheaper and just as effective as a special laboratory freezer. Most tissue culture reagents will store satisfactorily at −20°C, so an ultra-deep freeze is not necessary. A deep-freeze room is not recommended. They are very difficult and unpleasant to clear out and create severe problems in relocation if extensive maintenance is required.

While autodefrost freezers may be bad for some reagents (enzymes, antibiotics, etc.), they are very useful for most tissue culture stocks where their bulk and nature precludes severe cryogenic damage. Conceivably, serum could deteriorate during oscillations in the temperature of an autodefrost freezer, but in practice it does not seem to. Many of the essential constituents of serum are small proteins, polypeptides, and simpler organic and inorganic compounds that may be insensitive to cryogenic damage.

Inverted Microscope

It cannot be overstressed that, in spite of considerable and highly desirable progress toward quantitative analysis of cultured cells and remote sensing techniques, it is still vital to look at them regularly. A morphological change is often the first sign of deterioration in a culture (see Chapter 10) and the characteristic pattern of microbiological infection (see Chapter 16) is easily recognized.

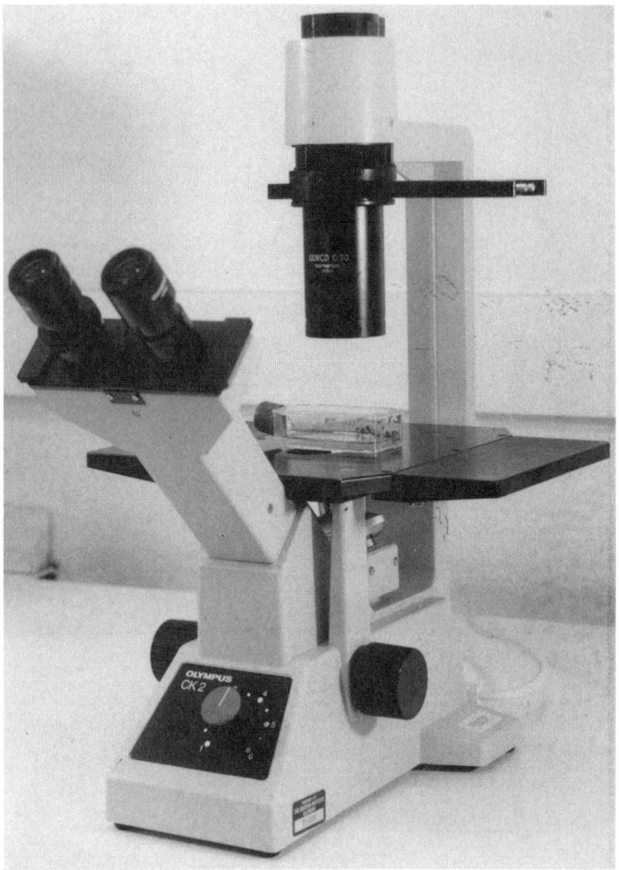

Fig. 4.2. *Olympus CK2 routine inverted microscope. Also available with a trinocular head allowing photomicrography, a useful addition if the budget will allow. This model is fitted with phase-contrast optics, a major advantage for assessing the status of a culture by cell morphology.*

A simple inverted microscope is essential (Fig. 4.2). Make certain that the stage is large enough to accommodate large roller bottles (see Chapter 23) in case you should require them. Large bottles will also require an adequate space between stage and condenser. There are many simple and inexpensive inverted microscopes on the market, but if you foresee the need for photography of living cultures, then you should invest in one with high-quality optics, a long working distance phase-contrast condenser and objectives, with provision to take a camera (e.g., Olympus CM, Leitz Diavert). The Nikon marking ring is a useful accessory to the inverted microscope. This device is placed on the objective and can be used to mark the underside of a dish where an interesting colony or patch of cells is located. The colony can then be picked (see Chapter 11, Fig. 11.7), or the development of a particularly interesting area in a culture followed.

Washing Up

Soaking Baths or Sinks. Soaking baths or sinks should be deep enough so that all your glassware (except pipettes and the largest bottles) can be totally immersed in detergent during soaking, but not so deep that the weight of glass is sufficient to break smaller items at the bottom, e.g., 400 mm (15 in) wide × 600 mm (24 in) long × 300 mm (12 in) deep.

If you are designing a lab from scratch, then you can get sinks built in of the size that you want. Stainless steel or polypropylene are best, the former if you plan to use radioisotopes and the latter for hypochlorite disinfectants.

Washing sinks should be deep enough (450 mm, 18 in) to allow manual washing and rinsing of your largest items without having to stoop too far to reach into them, and about 900 mm (3 ft) from floor to rim (Fig. 4.3). It is better to be too high than too low; a short person can always stand on a raised step to reach a high sink but a tall person will always have to bend down if the sink is too low. There should be a raised edge around the top of the sink to contain spillage and prevent the operator from getting wet when bending over the sink. The raised edge should go around behind the taps at the back.

Each washing sink will require four taps: a single cold water, combined hot/cold mixer, a cold hose connection for a rinsing device, and a nonmetallic or stainless steel tap for deionized water, from a reservoir above the sink (see Fig. 4.3). A centralized supply for deionized water should be avoided, as the pipework can build up dirt and algae and is difficult to clean.

Pipette cylinders (or "hods"). Pipette cylinders should be made from polypropylene and be free-standing, distributed around the lab, one per work station, with sufficient in reserve to allow full cylinders to stand for 2 hr in disinfectant before washing.

Pipette washer. Following an overnight soak in detergent, reusable pipettes are easily washed in a standard siphon-type washer (see Chapter 8 and Fig. 8.4). This should be placed at floor rather than bench level to avoid awkward lifting of the pipettes and connected to the deionized water supply so that the final rinse can be done in deionized water. If possible, a simple changeover valve should be incorporated into the deionized water feed line (see Fig. 4.3).

Pipette drier. If a stainless steel basket is used in the washer, this may then be transferred directly to an electric drier. Alternatively, pipettes can be dried on a rack or in a regular drying oven.

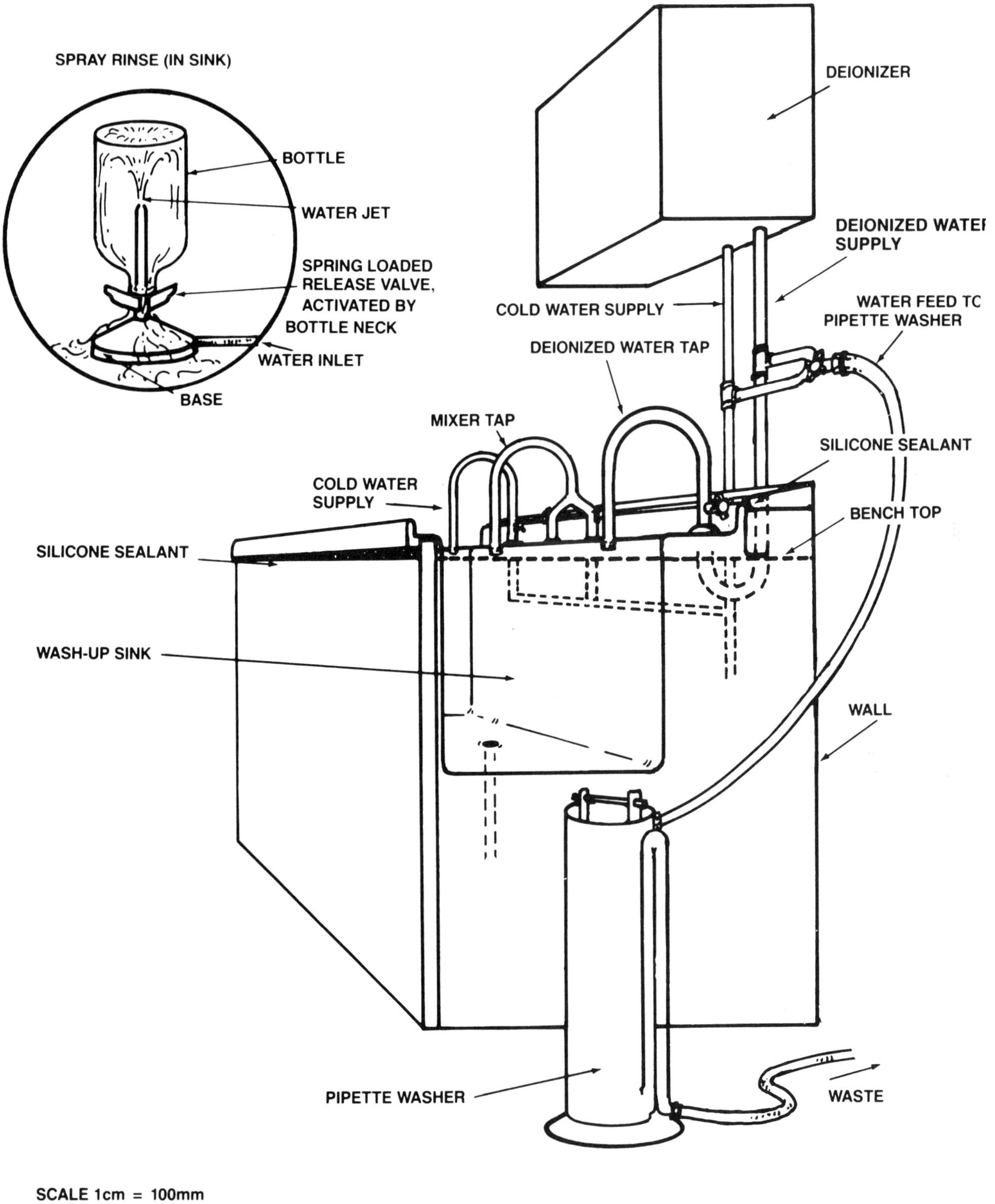

Fig. 4.3. *Washing-up sink and pipette washer, drawn to scale (bench height, 900 mm). Inset: bottle-rinsing device, located in sink.*

Sterilizing and Drying

Although all sterilizing can be done in an autoclave, it is preferable to sterilize pipettes and other glassware by dry heat, avoiding the possibility of chemical contamination from steam condensate or corrosion of pipette cans. This will require a high-temperature (160–180°C) fan-powered oven to ensure even heating throughout the load. As with autoclaves, do not get an oven that is too big for the size of glassware that you use. It is better to use two small ovens than one big one; heating is easier, more uniform, quicker, and more economical when only a little glassware is being used. You are also better protected during breakdowns.

Water Purification

Pure water is required for rinsing glassware, dissolving powdered media, or diluting concentrates. The first of these is usually satisfied by deionized water, but the second and third require a higher degree of purity, demanding a three- or four-stage process (Fig. 4.4). The important principle is that each stage is qualitatively different; reverse osmosis may be followed by charcoal filtration, deionization, and micropore filtration (e.g., Millipore), or distillation (with a silica-sheathed element) may be substituted for the first stage. Reverse osmosis is cheaper if you pay the fuel bills, but if not, distillation is better and more likely to give a sterile product. If reverse osmosis is used, the type of cartridge should be chosen to suit the pH of the water supply.

The deionizer should have a conductivity meter monitoring the effluent to indicate when the cartridge must be changed. Other cartridges should be dated and replaced according to the manufacturer's instructions.

Purified water should not be stored but recycled through the apparatus continually to minimize algal infection. Any tubing or reservoirs in the system should be checked regularly (every 3 months or so) for algal infection, cleaned out with hypochlorite and detergent (e.g., Chloros), and thoroughly rinsed in purified water before reuse.

Water is the simplest but probably the most critical constituent of all media and reagents, particularly with serum-free media (see Chapter 7). It is a good quality-control measure to check the cloning efficiency of a sensitive cell line (see Chapter 11) at regular intervals.

Centrifuge

Periodically, cell suspensions require centrifugation to increase the concentration of cells or to wash off a reagent. A small bench-top centrifuge, preferably with proportionally controlled braking, is sufficient for most purposes. Cells sediment satisfactorily at 80–100 *g*; higher gravity may cause damage and promote agglutination of the pellet. A large-capacity refrigerated centrifuge, 4 × 1 l or 6 × 1 l, will be required if large-scale suspension cultures (see Chapter 23) are contemplated.

Cell Freezing

The procedures for cell freezing are dealt with in detail elsewhere (Chapter 17), but the basic facilities should be considered here. The freezing process can be carried out satisfactorily without sophisticated equipment, but storage requires a properly constructed liquid N_2 freezer and storage dewar (see Fig. 17.2). Freezers range in size from around 25 l to 500 l, i.e., 250–15,000 1-ml ampules. It is best to freeze a minimum of 5 ampules for each cell strain, 20 for a commonly used strain, and 100 for one in continuous use. A capacity of 1,200–1,500 ampules is appropriate for most small laboratories.

The choice of freezer is determined by three factors: (1) capacity (number of ampules); (2) economy and static holding time (the time taken for all the liquid N_2 to evaporate)—both of which are governed by the evaporation rate, which in turn is dependent on the frequency of access; and (3) convenience of access. Generally speaking, number 2 is inversely proportional to 1 and 3.

There are two main types of freezer: narrow-necked with slow evaporation but with more difficult access, and wide-necked with easier access but three times the evaporation rate (see Fig. 17.8). If the cost of liquid nitrogen and its supply presents no problem, then a wide-necked freezer may be more convenient (e.g., Taylor Wharton 3K or equivalent, 3,000 ampules capacity, 3–5 l/d evaporation) although the holding time will only be about 1 week to 10 days. A narrow-necked freezer, on the other hand, will be more economical and last up to 2 months if N_2 supplies run out (e.g., L'aire Liquide 35-1, 1,500 ampules capacity, 0.5 l/d evaporation).

It is also possible to compromise between the two by purchasing a narrow-necked freezer with a rack inventory system (e.g., Thermolyne, Cryomed, and many of the major suppliers; see Trade Index). These freezers have the advantage of drawer storage and have rectangular drawers rather than triangular, but still have a relatively narrow neck with consequent savings in evaporation of liquid N_2. The total storage per unit volume is still not as great as a wide-necked freezer, but the ease of access is equivalent.

If you require bulk storage (~10,000 ampules), then you will need to consider a vessel of around 300 l capacity. Wide-necked freezers are most common in this size because of the mechanical difficulties in operating narrow-necked freezers of high capacity, but the latter are available and will save a considerable amount in

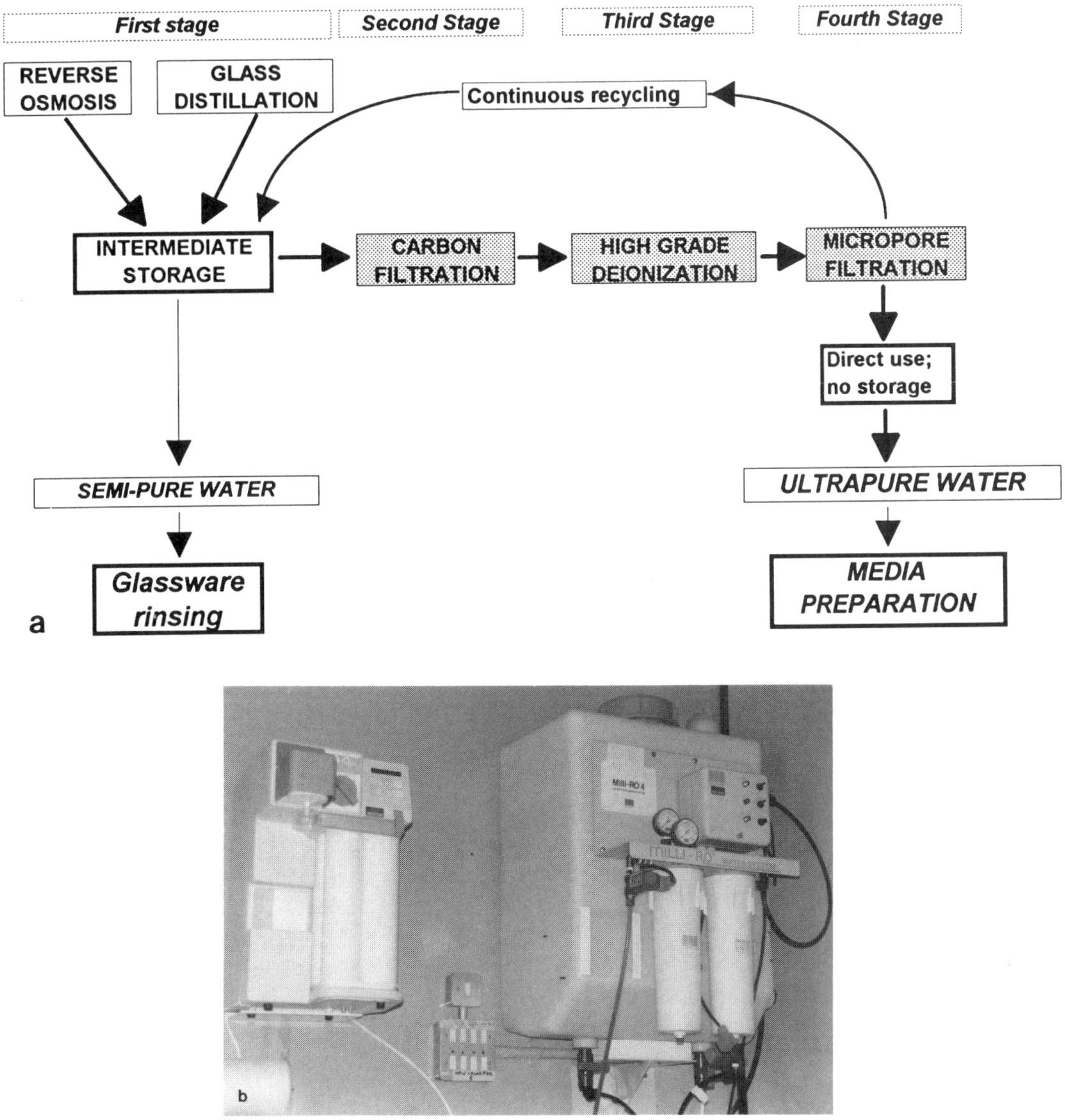

Fig. 4.4. *a. Preparation of ultrapure water. First stage can be either reverse osmosis or distillation. Ultrapure water is not stored but continuously recycled to intermediate store. Water collected first thing in the morning will be of maximum purity. b. Suggested layout for high-purity water supply. The water is fed to right-hand unit, reverse osmosis (glass distillation may be substituted for this stage), to supply a storage container. This semipurified water is then passed through carbon filtration, deionization, and micropore filtration and recycled back to the storage container. Reagent-quality water is available at all times from storage bottom of right-hand container; media-quality water is available from the micropore filter supply (bell-shrouded delivery, top left-hand corner of photograph). If the apparatus recycles continuously (left-hand unit returning to right-hand unit; not shown) then the highest purity water will be collected first thing in the morning for the preparation of medium.*

expenditure on liquid N_2. At 300 l the evaporation rate is approximately 10 l/d in a wide-necked freezer.

The advantages of gas-phase and liquid-phase storage are discussed elsewhere (Chapter 17), but one major implication of storing in the gas phase is that the liquid phase is necessarily reduced to the space below your ampule storage area, usually 20–30% of the full volume. Hence, the static holding time is reduced to one-third or one-fifth that of the filled freezer, filling must be carried out more regularly, and the chances of accidental thawing are increased. Where the investment is higher (many ampules or rare cell strains), automatic alarm systems should be fitted and, for the high-capacity freezers, an automatic filling system is recommended. However, automatic systems can fail and a twice-weekly check of liquid levels with a dipstick should be maintained and a record kept.

An appropriate storage vessel should also be purchased to enable a backup supply of liquid N_2 to be held. The size of this depends on (1) the size of the freezer; (2) the frequency and reliability of delivery of liquid N_2; and (3) the rate of evaporation. A 40-l, wide-necked freezer will require about 20–30 l/wk, so a 50-l dewar flask (or two 25-l flasks, which are easier to handle) is advisable. A 35-l, narrow-necked freezer, on the other hand, using 5–10 l/wk, will only require a 25-l dewar. Larger freezers are best supplied on line from a dedicated storage tank, e.g., a 160-l storage vessel linked to a 320-l freezer with automatic filling and alarm, or a 500-l tank for a larger freezer or for more smaller freezers.

BENEFICIAL EQUIPMENT

The above describes the essential equipment for a modest tissue culture facility; but there are several items of equipment that will make your laboratory easier to use and more efficient and better controlled. If the facility is to be used frequently for more than one or two people, then many of the following items will probably join the "essential" list.

Laminar Flow

Usually one hood is sufficient for two to three people (see below and Chapter 5). A horizontal flow hood is cheaper and gives best sterile protection to your cultures, but for potentially hazardous materials (radioisotopes, carcinogenic or toxic drugs, virus-producing cultures, or any primate [including human] cell lines), a Class II biohazard cabinet should be used (Figs. 3.1, 5.3). It is important to consult local and national biohazard regulations before equipping, as legal requirements and recommendations vary (see Chapter 6).

Choose a hood that is (1) large enough (usually 1,200 mm [4 ft] wide × 600 mm [2 ft] deep) for one person to use at a time; (2) quiet (noisy hoods are more fatiguing); (3) easily cleaned both inside the working area and below the work surface in the event of spillage; and (4) comfortable to sit at (some cabinets have awkward ducting below the work surface, which leaves no room for your knees, have lights or other accessories above that strike your head, or have screens that obscure your vision). The front screen should be able to be raised, lowered, or removed completely to facilitate handling bulky culture apparatus and cleaning. Remember, however, that a biohazard cabinet will not give you, the operator, the required protection if you remove the front screen.

Insist that you be allowed to examine and sit at a hood as if using it before committing yourself to purchase. Check the following points:

(1) Can you get your knees under it while sitting comfortably and close enough to work, with your hands at least halfway into the hood?

(2) Is there a footrest in the correct place?

(3) Is your head conveniently placed to see what you are doing without placing strain on your neck?

(4) Is the work surface perforated and will this give you trouble with spillage (a solid work surface vented at front and back is preferable)?

(5) Is the work surface easy to remove for cleaning?

(6) If the work surfaces are lifted, are the edges sharp, or are they rounded so that you will not cut yourself when cleaning out the hood?

(7) Are there crevices in the work surface that might accumulate spillage and contamination? Some cabinets have sectional work surfaces that are easier to remove for cleaning but leave capillary spaces when in place.

(8) Is the lighting convenient and adequate?

There are also some more general questions about locating hoods in a laboratory, already referred to (Chapter 3).

(1) Will you be able to get the hood into the tissue culture laboratory?

(2) When in place, can it be serviced easily (ask the service engineer, not the salesman!) and will it have sufficient headroom for venting to the room or for ducting (as required)?

(3) Will the airflow from the room ventilation or air-conditioning interfere with the integrity of the work space of the hood; i.e., will air spill in or aerosols leak out due to turbulence? This will probably require correct testing by an engineer with experience in laminar flow units.

Cell Counter

A cell counter (see Fig. 18.2) is a great advantage when more than two or three cell lines are carried and is essential for precise quantitative growth kinetics. Several companies now market models ranging in sophistication from simple particle counting up to automated cell counting and size analysis. For routine counting, the Coulter "D Industrial" is more than adequate and much less expensive than equipment with cell-sizing facilities (see also Cell Counting, Chapter 18).

Aspiration Pump

A vacuum pump or simple tap siphon saves a lot of time and effort when handling large numbers of cultures or large fluid volumes. Tap siphons require a minimum of 6 m (20 ft) head of water to create sufficient suction, but are by far the cheapest, simplest, and

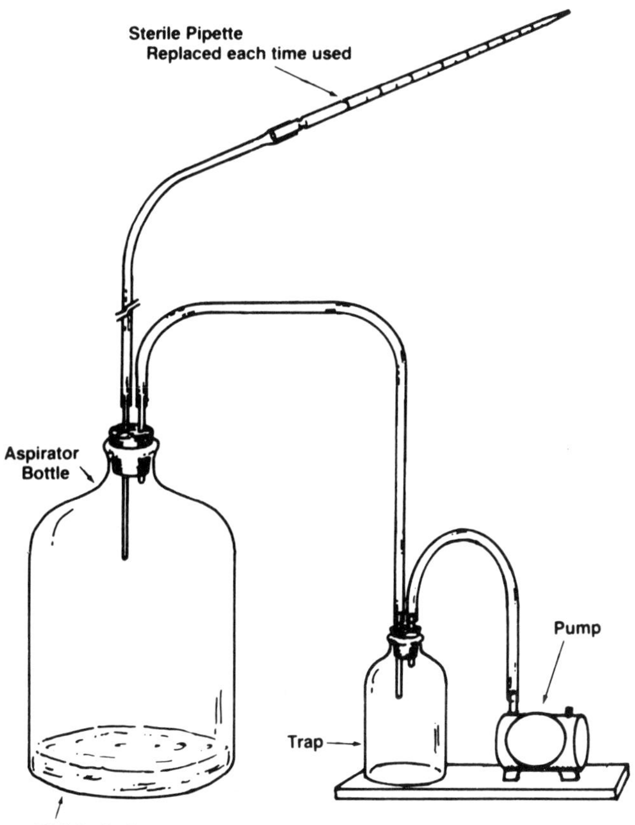

Fig. 4.5. Vacuum-pump assembly for withdrawing spent medium, etc. Alternatively, a peristaltic pump can be inserted between the pipette and the aspirator, in which case a trap is not required and the problem of carry-over of chlorine from the disinfectant into the pump is eliminated (see Fig. 3.1).

most efficient way to dispose of nonhazardous tissue culture effluent. If you do not have sufficient water pressure, have a water shortage, or are handling potentially hazardous material, use a vacuum pump similar to that supplied for sterile filtration. If necessary, the same pump could serve both duties. The effluent should be collected in a reservoir into which a sterilizing agent such as glutaraldehdye or hypochlorite has been added when work is finished and at least 30 min before the reservoir is emptied (Fig. 4.5). A drying agent, hydrophobic filter (Gelman), or second trap placed in the line to the pump prevents fluid being carried over. Do not draw air through a pump from a reservoir containing hypochlorite, as the free chlorine will corrode the pump and could be toxic. Avoid vacuum lines; if they become contaminated with fluids, they can be very difficult to clean out.

A peristaltic pump may be used instead of a vacuum pump. No trap is required, effluent can be collected directly into disinfectant, and there is less chance of discharging aerosol into the atmosphere. However, the pump tubing should be checked regularly for wear and the pump operated by a self-canceling foot switch.

Always switch on the pump before inserting a pipette in the tubing to avoid effluent running back.

CO_2 Incubator

Although incubations can be performed in sealed flasks in a regular dry incubator or hot room, some vessels, e.g., petri dishes or multiwell plates, require a controlled atmosphere with high humidity and elevated CO_2 tension. The cheapest way of controlling the gas phase is to place the cultures in a plastic box, a desiccator, or an anaerobic jar (Fig. 4.6). Gas the container with the correct CO_2 mixture and then seal. If the container is not full, include an open dish of water to increase the humidity. Making the culture medium about 10% hypotonic will also help to counteract evaporation.

CO_2 incubators (Fig. 4.7) are more expensive, but their ease of use and the superior control of CO_2 tension and temperature (anaerobic jars and desiccators take longer to warm up) justify the expenditure. A controlled atmosphere is achieved by blowing air over a humidifying tray (Fig. 4.8) and controlling the CO_2 tension with a CO_2-monitoring device. Alternatively, CO_2 tension may be controlled by mixing air and CO_2 in the correct ratio. CO_2 controllers, although they add to the capital cost of the incubator, reduce CO_2 consumption considerably and give better control and recovery after opening the incubator. They function by

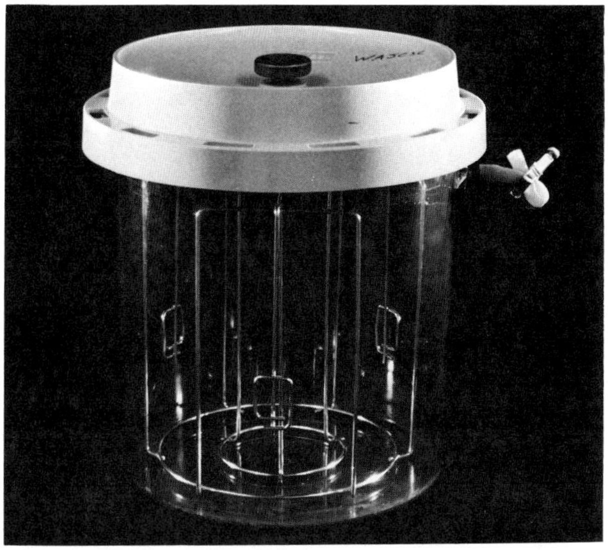

Fig. 4.6. Becton Dickinson anaerobic jar. This type of jar or a desiccator (preferably plastic) can be used to maintain a regulated atmosphere in the absence of a CO_2 incubator. It is purged from a cylinder and humidified by its contents, so if the number of vessels is low, compensate with dishes of phosphate-buffered saline (PBS) or saline.

Fig. 4.7. Automatic CO_2 incubator (Heraeus). This is a dual-chamber model with the top chamber open. Dual controls for temperature regulation and the CO_2 controller are located in the side panels (see also Fig. 4.8).

drawing air from the incubator into the sample chamber, determining the concentration of CO_2, and injecting pure CO_2 into the incubator to make up any deficiency. Air is circulated around the incubator by a fan to keep both the CO_2 level and the temperature uniform. Most inexpensive CO_2 controllers require calibration every few months, but most "top-of-the-line" models have "auto-zero" capability (e.g., Heraeus).

Since humid incubators require regular cleaning, the interior should dismantle readily without leaving inaccessible crevices or corners.

Balances, pH Meter, and Osmometer

A coarse and a fine balance and a simple pH meter are useful additions to the tissue culture area for the preparation of media and special reagents. Although a phenol-red indicator is sufficient for monitoring pH in most solutions, a pH meter will be required when phenol red cannot be used, e.g., in preparation of cultures for fluorescence assays and in the preparation of stock solutions.

One of the most important physical properties of culture medium, and one that is often difficult to predict, is the osmolality. An osmometer (see Fig. 7.10) is a useful accessory to check solutions as they are made up, to adjust new formulations, or to compensate for the addition of reagents to the medium. They usually work by freezing-point depression or elevation of vapor pressure. Choose one with a low sample volume ($\leq$1 ml), since on occasion you may want to measure a valuable or scarce reagent and the accuracy ($\pm$10 mOsmol) may be less important than the value or scarcity of the reagent.

Upright Microscope

An upright microscope may be required, in addition to an inverted microscope, for chromosome analysis, mycoplasma detection, and autoradiography. Select a high-grade research microscope (Leitz, Zeiss, or Nikon) with regular brightfield optics up to 100× objective magnification, phase contrast up to at least 40× objective magnification, and preferably 100×, and fluorescence optics with epi-illumination and 40× and 100× objectives, for mycoplasma testing by fluorescence (see Chapter 16) and fluorescent antibody observation. Leitz supplies a 50× water-immersion objective, which is particularly useful for observation of routine mycoplasma preparations with Hoechst stain (see Chapter 16). An automatic camera should also be fitted for photographic records of permanent preparations.

Dissecting Microscope

There are a number of activities in a tissue culture laboratory that will require a dissecting microscope (Nikon, Olympus, Leica [Wild]). The first and most obvious is for the dissection of small pieces of tissue, e.g., embryonic organs or tissue from smaller invertebrates. It is also essential for counting colonies, particularly small colonies in agar, and for setting the lower size threshold for counting monolayer colonies. It will also be found to be essential for picking colonies from agar.

Temperature Recording

Ovens, incubators, and hot rooms should be monitored regularly for uniformity and stability of temperature control. A recording thermometer with ranges from below −50°C to about +200°C will enable you to monitor frozen storage, cell freezing, incubators, and sterilizing ovens with one instrument fitted

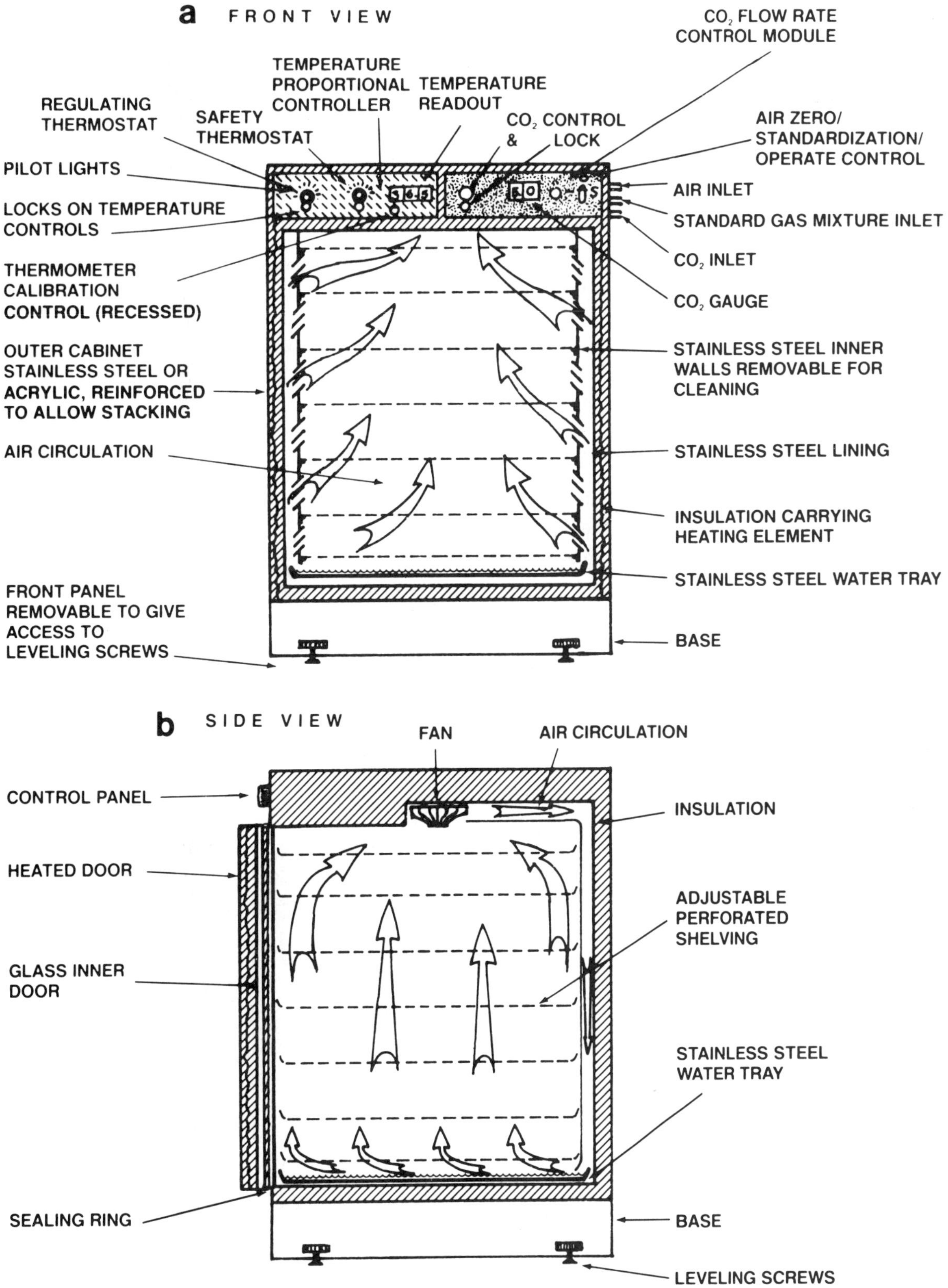
a FRONT VIEW
CO2 FLOW RATE CONTROL MODULE
TEMPERATURE PROPORTIONAL CONTROLLER
TEMPERATURE READOUT
REGULATING THERMOSTAT
SAFETY THERMOSTAT
CO2 CONTROL & LOCK
AIR ZERO/ STANDARDIZATION/ OPERATE CONTROL
PILOT LIGHTS
AIR INLET
LOCKS ON TEMPERATURE CONTROLS
STANDARD GAS MIXTURE INLET
CO2 INLET
THERMOMETER CALIBRATION CONTROL (RECESSED)
CO2 GAUGE
STAINLESS STEEL INNER WALLS REMOVABLE FOR CLEANING
OUTER CABINET STAINLESS STEEL OR ACRYLIC, REINFORCED TO ALLOW STACKING
AIR CIRCULATION
STAINLESS STEEL LINING
INSULATION CARRYING HEATING ELEMENT
STAINLESS STEEL WATER TRAY
FRONT PANEL REMOVABLE TO GIVE ACCESS TO LEVELING SCREWS
BASE
b SIDE VIEW
FAN
AIR CIRCULATION
CONTROL PANEL
INSULATION
HEATED DOOR
ADJUSTABLE PERFORATED SHELVING
GLASS INNER DOOR
STAINLESS STEEL WATER TRAY
SEALING RING
BASE
LEVELING SCREWS

Fig. 4.8. Components of a typical CO_2 incubator. a. Front view. b. Side view.

with a resistance thermometer or thermocouple with a long Teflon-coated lead.

Recording thermometers should be permanently fixed into the hot room, sterilizing oven, and autoclave, and dated records kept to allow a regular check for abnormal behavior, particularly in the event of a problem arising.

Magnetic Stirrer

There are certain specific requirements for magnetic stirrers. A rapid stirring action for dissolving chemicals is available with any stirrer, but to be used for disaggregation (Chapter 9) or suspension culture (Chapters 10 and 22), (1) the stirrer motor should not heat the culture (use the rotating field type of drive or belt drive from an external motor); (2) the speed must be controlled down to 50 rpm; (3) the torque at low rpm should still be capable of stirring up to 10 l of fluid; (4) it should have more than one place if several cultures are to be maintained simultaneously; (5) each stirrer position should be individually controlled; and (6) there should be a readout of rpm at each position.

Roller Racks

Roller racks are used to scale up monolayer culture (see Chapter 23). The choice of apparatus is determined by the scale, i.e., the size and number of bottles to be rolled. This may be calculated from the number of cells required, the maximum attainable cell density, and the surface area of the bottles. A large number of small bottles gives the highest surface area but tends to be more labor-intensive in handling, so a usual compromise is around 125 mm (5 in) diameter and various lengths from 150 to 500 mm (6–20 in) long. The length of the bottle will determine the maximum yield but is limited by the size of the rack; the height of the rack will determine the number of tiers, i.e., rows of bottles. Although it is cheaper to buy a larger rack than several small ones, the latter alternative (1) allows you to build up gradually (having confirmed that the system works); (2) can be easier to locate in a hot room; and (3) will still provide accommodation if one rack requires maintenance. Bellco or Luckhams bench-top models may be satisfactory for smaller scale activities and Bellco and New Brunswick Scientific make larger racks.

Fluid Handling

An increase in the caution applied to the handling of tissue cultures and reagents, combined with an increase in the scale of operation in many laboratories, has eliminated traditional mouth pipetting and transformed the use of pipettes and other devices for the handling of reagents.

There are four main tasks required for fluid transfer devices: (1) nonsterile loading of containers with water, salt solutions, and other reagents, (2) simple addition of sterile liquids to and removal from small numbers of flasks, etc., (3) repeated additions to and removals from a large number of vessels, and (4) simultaneous addition of reagents to multiple replicate cultures. The choice of equipment is also determined by the volume of liquid being transferred, the frequency of the operation, the relative cost of labor or equipment, the ergonomic efficiency, and the safety of the procedure.

Removal of fluids. Removal of fluids can be achieved simply and rapidly by use of a vacuum pump or vacuum line, with suitable reservoirs to collect the effluent and prevent contamination of the pump, or by use of a simple peristaltic pump discharging into a reservoir with disinfectant (see above).

Nonsterile loading. Nonsterile loading applies to the preparation of reagents for sterilization and the dispensing of nonsterile reagents such as counting fluid. Bottletop dispensers (Boehringer, Cole Parmer) are suitable for volumes up to about 50 ml; above this, gravity dispensing from a reservoir (see Fig. 4.1), which may be a graduated bottle or plastic bag, is acceptable where accuracy is not critical. Where the volume dispensed is more critical, then a peristaltic pump (Fig. 4.9)(Cole Parmer, Zinsser, Jencons) will be preferable. The duration of dispensing and diameter of tube determine the volume and accuracy. A long dispense cycle with a narrow delivery tube will be more accurate, but a wide-bore tube will be faster.

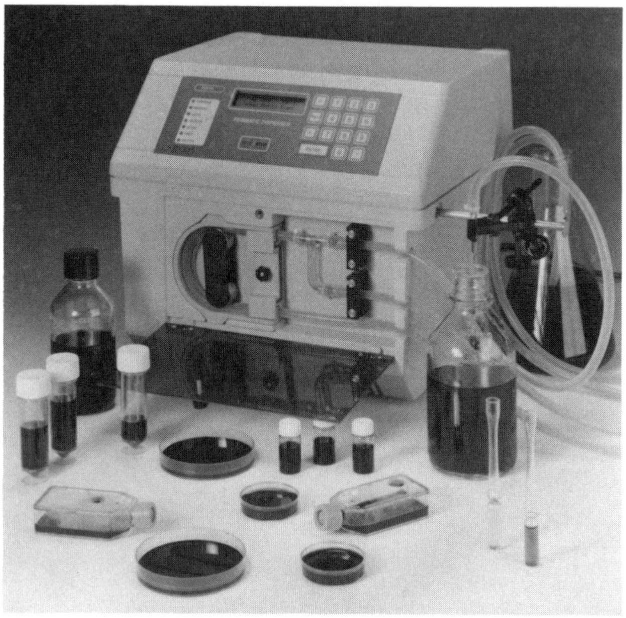

Fig. 4.9. *Automated pipetting device. The Perimatic Premier, suitable for repetitive dispensing and dilution in the 1–100 ml range. Only the delivery tube requires sterilization, and it is available in different diameters.*

When using bottletop dispensers for counting fluid for cell counting, leave the fluid to settle for a few minutes before counting, as there are often small bubbles generated by rapid dispensing that would be counted by the cell counter.

Simple pipetting. Simple pipetting is one of the most frequent tasks required in the normal handling of routine culture. While a small-scale, low-budget operation can be quite satisfactorily achieved with a rubber bulb or other proprietary pipetting device (Fig. 4.10), speed, accuracy, and reproducibility are greatly enhanced by using a motorized pipette aid (see Fig. 4.10). These are obtainable with separate or built-in pump, and can be main operated or rechargeable (Bellco, ICN-Flow, Costar). The major determinant is the weight and feel of the instrument during continuous use, and it is best to try one out before purchase.

See Figure 6.1 for the proper method of inserting a pipette into a pipetting device.

These pipette aids usually have a filter at the pipette insert to minimize transfer of infection. Some are disposable, and some reusable after resterilization. If unplugged pipettes are being used, these filters must be changed when you change to a different cell line. However, it is preferable to use plugged pipettes.

Micropipettes. Micropipettes are used extensively for dispensing small volumes (<1 ml). Only the tip needs to be sterile, but this limits the size of vessels used. If a sterile fluid is withdrawn from a container with a micropipette, the nonsterile shank of the pipette must not touch the sides of the container. Reagents of 10–20 ml volume may be sampled in 5 μl–1 ml volumes from a universal container, or 5–200 μl from a bijou bottle. Eppendorf-type tubes may also be used for volumes of 100 μl–1 ml but should be of the shrouded-cap variety.

It is assumed that the inside of a micropipette is sterile, or does not displace enough air for this to matter. There are situations where it clearly does matter, however. If you are performing serial subculture of a stock cell line (as opposed to a short-term experiment with cells that ultimately will be sampled or discarded but *not* propagated), the security of the line is paramount and you must either use a plugged regular glass or disposable pipette with a sterile length sufficient to reach into the vessel that you are sampling. If you are using a small enough container to preclude contact from nonsterile parts of a micropipette, then it is permissible to use a micropipette, provided that the tip is of the type that contains a filter. Otherwise, you run the risk of microbial contamination from nonsterile parts of the pipette or, more subtle and potentially more serious, cross-contamination from aerosol or fluid drawn up into the shank of the micropipette.

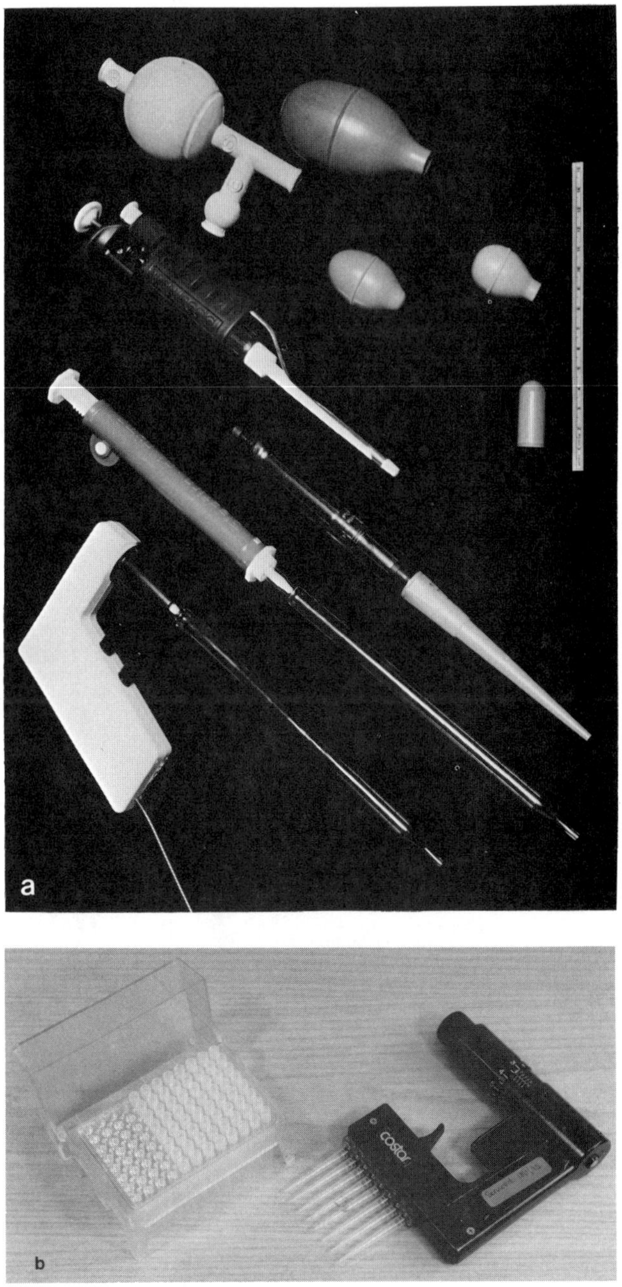

Fig. 4.10. *a. Pipetting aids. Top to bottom: various plain bulbs and bulb with inlet and outlet valves; Gilson and Finn-pipette micropipettes (take special plastic tips); Bellco Pi-pump (standard pipettes); (see also Figs. 4.12, 5.3, and 5.6). b. Costar multipoint pipettor for loading microtitration plates.*

Routine subculture, which should be rapid, secure from microbial and cross-contamination, but need not be very accurate, is best performed with conventional glass or disposable plastic pipettes. Experimental work, which must be accurate but should not involve stock propagation of the cells used, may benefit from using micropipettes.

Micropipette tips are available with a filter near the top of the tip. These prevent cross-contamination and minimize microbial contamination but add considerably to the unit cost of the tip.

Tips can be bought loose and packaged and sterilized in the laboratory, or can be bought already sterile and mounted in racks ready for use. Loose tips are cheaper, though more labor-intensive. Prepacked tips are much more convenient, but considerably more expensive. Some racks can be refilled and resterilized, which presents a reasonable compromise.

Large-volume dispensing. Where culture vessels exceed 100 ml medium volume, there is a need for a different approach to fluid delivery. If only a few flasks are involved, a gravity dispenser such as in Figure 4.11 may be quite adequate, but if very large volumes or a large number of high-volume replicates are required, then a pressurized dispenser will probably be necessary. These may be metered by a horizontally displaced float and two-way valve, capable of withstanding the increased pressure, where accuracy is important.

Single-fluid transfers of large volumes (10–10,000 l) are usually achieved by preparing the medium in a sealed pressure vessel and then displacing it by positive pressure into the culture vessel (Alpha Laval).

Repetitive dispensing. The traditional repetitive dispenser was the type known as a Cornwall Syringe (see Fig. 8.9b), where liquid is alternately taken into a syringe via one tube and expelled via another, using a simple two-way valve. The syringe plunger is spring-loaded so the whole procedure is semiautomatic and repetitive. There are many variants of this type of dispenser, many of which are still in regular use. The major problems arise from the valves sticking, but this can be minimized by avoiding the drying cycle after autoclaving and flushing these out with medium or salt solution before and after use.

Similar repetitive dispensers have been used in conjunction with micropipettes (Gilson), and small-volume repetitive dispensing can also be achieved by incremental movement of the piston in a syringe (Hamilton, ICN Flow, Boehringer).

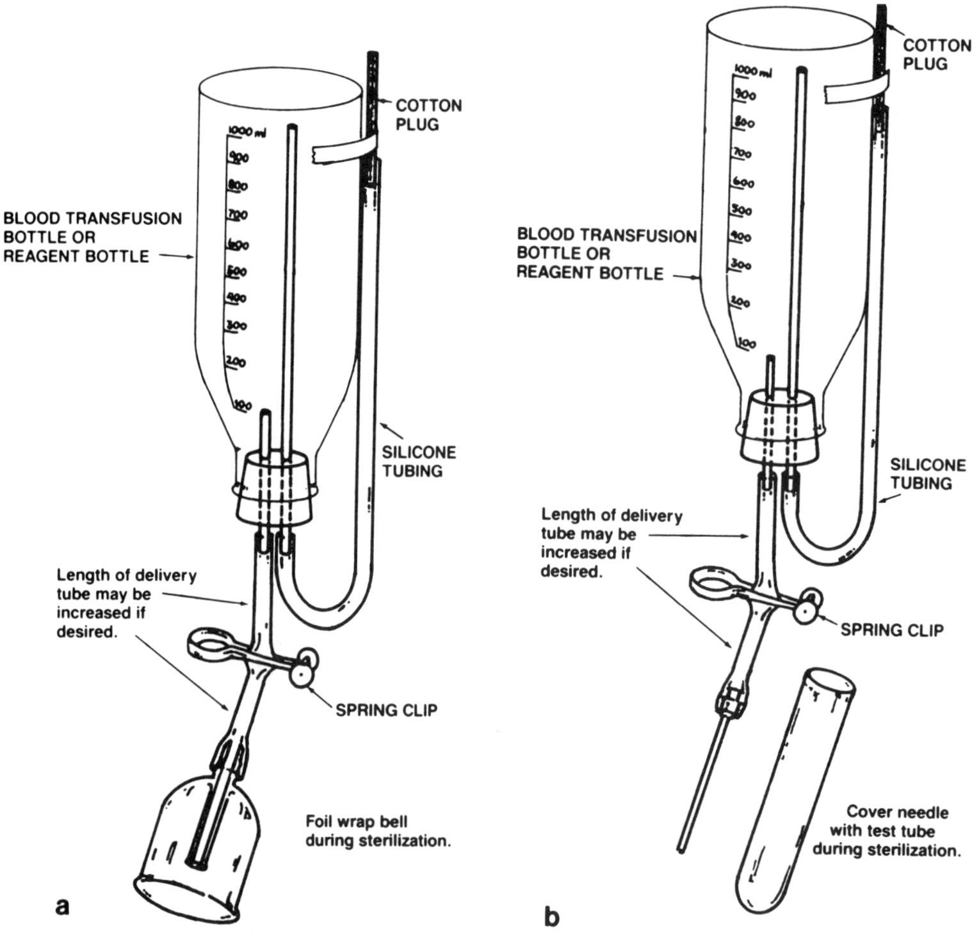

Fig. 4.11. *Simple dispensing devices for use with a graduated bottle. a. With bell used in conjunction with open bottle. b. With needle for slower delivery via a skirted cap or membrane type closure. (From a design by Dr. John Paul.)*

The bottletop dispenser, referred to above for nonsterile dispensing, can also be used in a sterile environment. Here the flask will need to be taken to the dispenser, rather than vice versa, and a protective bell will be required on the delivery end of the dispenser. All of these repeating pipettes have problems in use resulting from the necessity to autoclave glass syringes, two-way valves, etc. The valves tend to stick (though making these of Teflon helps), syringe pistons deform, or, if Teflon, may contract due to compression if autoclaved in a glass syringe barrel.

It is preferable to have a nonsterile metering and repeating mechanism so that only the dispensing element need be sterile. The Bellco automated pipette handle, though it has no facility for repetitive pipetting, conforms to this requirement, as does the Tridak Stepper (Bellco) syringe dispenser.

A peristaltic pump (Perimatic) can also be used for repetitive serial deliveries, and has the advantage that it may be activated via a foot switch, leaving the hands free. Care must be taken setting up such devices to avoid contaminating the tubing at the reservoir and delivery ends. In general, they are only worthwhile if a very large number of flasks is being handled. Automated pipetting provided by a peristaltic pump can be controlled in small increments (e.g., see Fig. 4.9). In addition, only the delivery tube is autoclaved, and accuracy and reproducibility can be maintained to high levels over ranges from 10 μl up to 10 ml. A number of delivery tubes may be sterilized and held in stock, allowing a quick changeover in the event of accidental contamination or change in cell type or reagent. Larger pumps (10–50 ml) are also available (e.g., Watson-Marlow, Zinsser Perifill).

Automation. Many attempts have been made at automating fluid changing in tissue culture, but few devices or systems have the flexibility required for general use. Where a standard production system is in use, automatic feeding may be useful, but the time taken in setting it up, the changes that may be needed if the production system changes, and the overriding importance of complete sterility have deterred most laboratories from investing the necessary time and funds. The introduction of microtitration trays (see Fig. 7.4) has brought with it many automated dispensers, diluters (Fig. 4.12), and other accessories. Transfer devices using perforated trays or multipoint pipettes make it easier to seed from one plate to another, and there are also plate mixers and centrifuge carriers available. The range of equipment is so extensive that it cannot be covered here and the appropriate trade catalogues should be consulted (Flow, Microbiological Associates, Dynatech, GIBCO, Millipore). Two items worthy of note, however, are the Rainin programmable single or

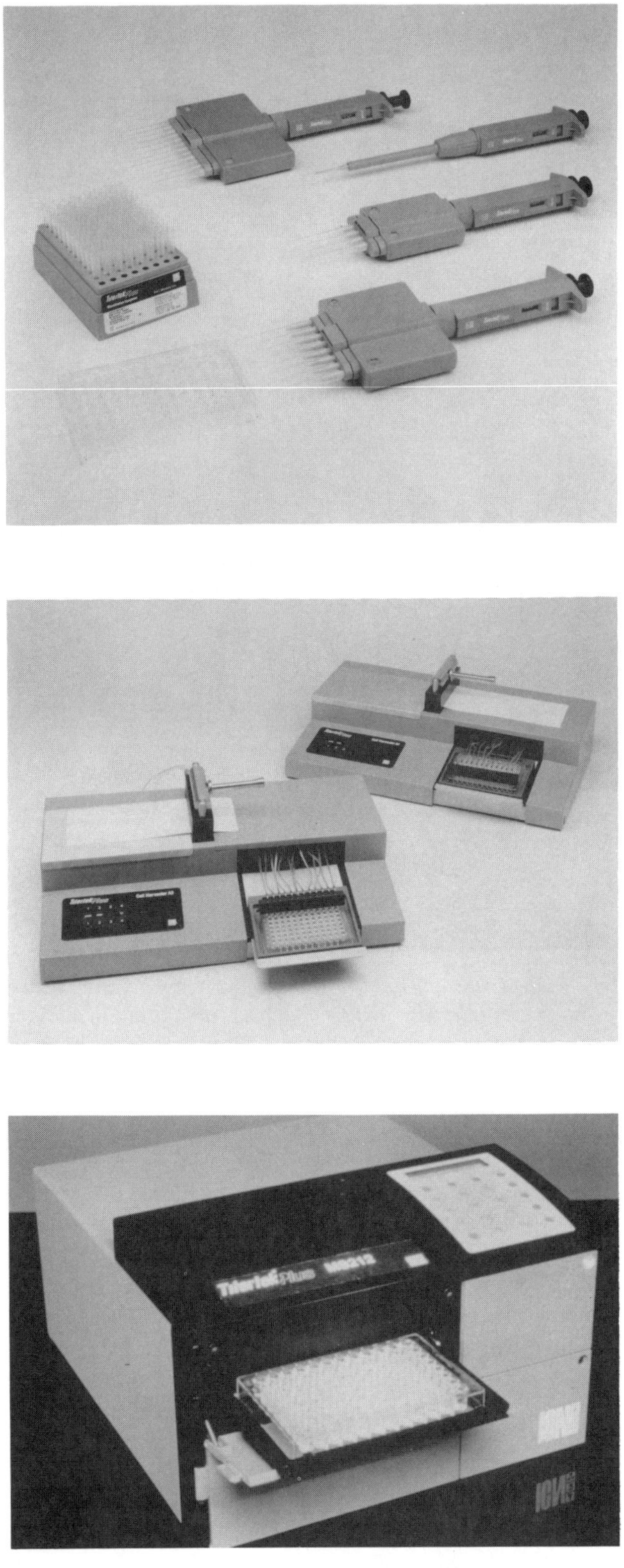

Fig. 4.12. Microtitration instruments. Top. Single and multipoint micropipettes. Center. Transfer device for transferring whole cells, nuclei, or protein precipitates from microtitration wells to glass-fiber sheet. Bottom. Densitometer (plate reader) for measuring absorbance of each well; some models also measure fluorescence. Photographs reproduced by permission of ICN-Flow Ltd.

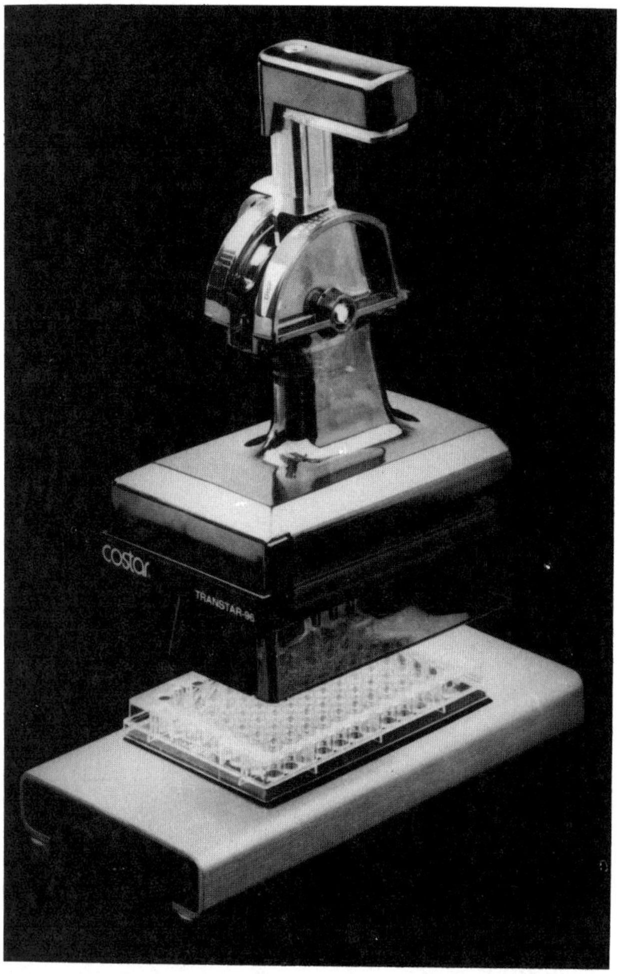

Fig. 4.13. *Transtar (Costar) transfer device for seeding, transferring medium, replica plating, and other similar manipulations with microtitration plates, enabling simultaneous handling of all 96 wells. (Reproduced by permission of Northumbria Biologicals Ltd.)*

multitip micropipette and the Costar Transtar media transfer and replica plating device (Fig. 4.13).

Robotic systems (e.g., Canberra Packard) are now being introduced into tissue culture assays as a natural extension of the microtitration system. These provide totally automated procedures to be used, but also allow for reprogramming if the analytical approach changes.

Choice of system. Whether a simple manual system or complex automated one is chosen, the choice is governed by five main criteria:

(1) Ease of use and ergonomic efficiency

(2) Cost relative to time saved and increased efficiency

(3) Accuracy and reproducibility in serial or parallel delivery

(4) Ease of sterilization and effect on accuracy and reproducibility

(5) Safety–mechanical, electrical, chemical, biological, and radiological

◇ Most pipetting devices tend to expel fluid at a higher rate than normal manual operation and have, in consequence, a greater ability to generate aerosols. This must be kept in mind when using substances that are potentially hazardous.

USEFUL ADDITIONAL EQUIPMENT

Low-Temperature Freezer

Most tissue culture reagents can be stored at 4°C or −20°C, but occasionally some drugs, reagents, or products from cultures may require a temperature of −70°C to −90°C, where most if not all the water is frozen and most chemical and radiolytic reactions are severely limited. Such a freezer is also a useful accessory for cell freezing (see Chapter 17). The chest type is more efficient at maintaining a low temperature for minimum power consumption, but vertical cabinets are much less extravagant in floor space. If you do choose a vertical cabinet type, make sure that it has individual compartments (six to eight in a 400-l [15-ft^3] freezer) with separate close-fitting doors, and expect to pay at least 20% more than for a chest type.

Low-temperature freezers generate a lot of heat and this must be dissipated for them to work efficiently (or at all). They should be located in a well-ventilated room or one with air-conditioning, such that the ambient temperature does not rise above 23°C. If this is not possible, invest in a freezer designed for tropical use; otherwise you will be faced with constant maintenance problems and a short working life for the freezer, with all the attendant problems of relocatng valuable stocks. One or two failures costing $1,000 or more in repairs and loss of valuable material soon cancels any saving in buying a cheap freezer.

Glassware Washing Machine

A reliable person doing your washing up is probably the best way of producing clean glassware, but when the amount gets to be too great, or reliable help is not readily available, it may be worth considering an automatic washing machine (Fig. 4.14). There are several of these currently available that are quite satisfactory. You should look for the following principles of operation:

(1) Choice of racks with individual spigots over which you can place bottles, flasks, etc. Open vessels such as petri dishes and beakers will wash satisfactorily in a whirling arm spray, but narrow-necked vessels need individual jets. Each jet should have a cushion at

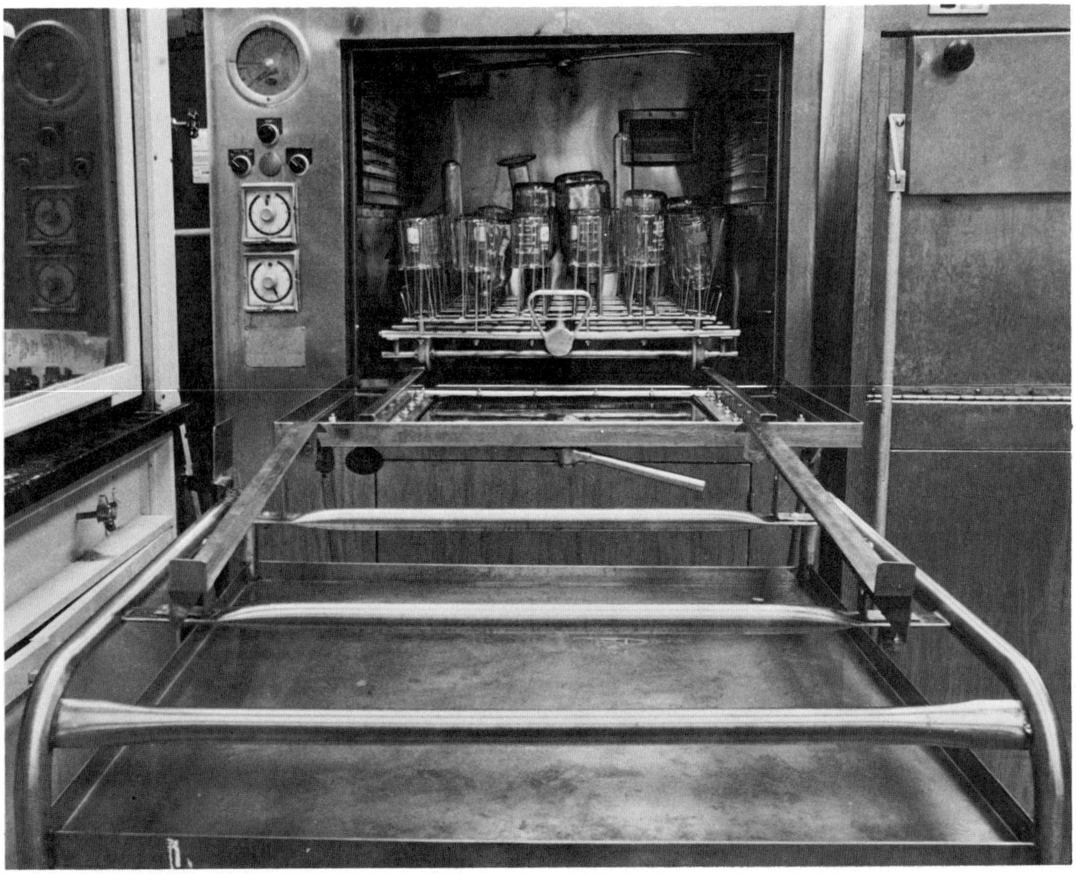

Fig. 4.14. Automatic glassware washing machine. Glassware is placed on individual jets, which ensures thorough washing and rinsing. After washing, glassware is withdrawn on the rack onto the trolley (front) and transferred to drier (right); fitted with same rails as washing machine and drier (Betterbuilt).

its base to protect the neck of the bottle from chipping.

(2) The water pump that pumps the water through the jets should have a high delivering pressure, requiring around 2–5 hp, depending on the size of machine.

(3) Washing water should be heated to 90°C.

(4) There should be a facility for a deionized water rinse at the end of the cycle. This should be heated to 50–60°C; otherwise the glassware may crack after the hot wash and rinse, and should be delivered as a continuous flush, discarded, and not recycled. If recycling is unavoidable, a minimum of three separate deionized rinses will be required.

(5) Preferably, rinse water from the end of the previous wash cycle should be discarded and not retained for the prerinse of your next wash. This reduces the risk of cross-contamination when the machine is used for chemical and radioisotope wash-up.

(6) The machine should be lined with stainless steel and plumbed in stainless steel or nylon pipework.

(7) If possible, a glassware drier should be chosen that will accept the same racks (see Fig. 4.14), so that they may be transferred directly via a suitably designed trolley without unloading.

Betterbuilt makes such machines of different sizes with compatible drying ovens.

Video Camera and Monitor

Since the advent of cheap microcircuits, television cameras and monitors have become a valuable aid to the discussion of cultures and the training of new staff or students (Fig. 4.15). Choose a high-resolution but not high-sensitivity camera, as the standard camera sensitivity is usually sufficient, and high sensitivity may lead to problems of overillumination. Black and white usually gives better resolution and is quite adequate for phase-contrast observation of living cultures. Color is preferable for fixed and stained specimens. If you will be discussing cultures with a technician or one or two associates, a 12- or 15-in monitor is adequate and gives better definition, but if you are teaching a group of 10 or more students, then go for a 19- or 21-in monitor.

High-resolution charge-coupled device (CCD) video

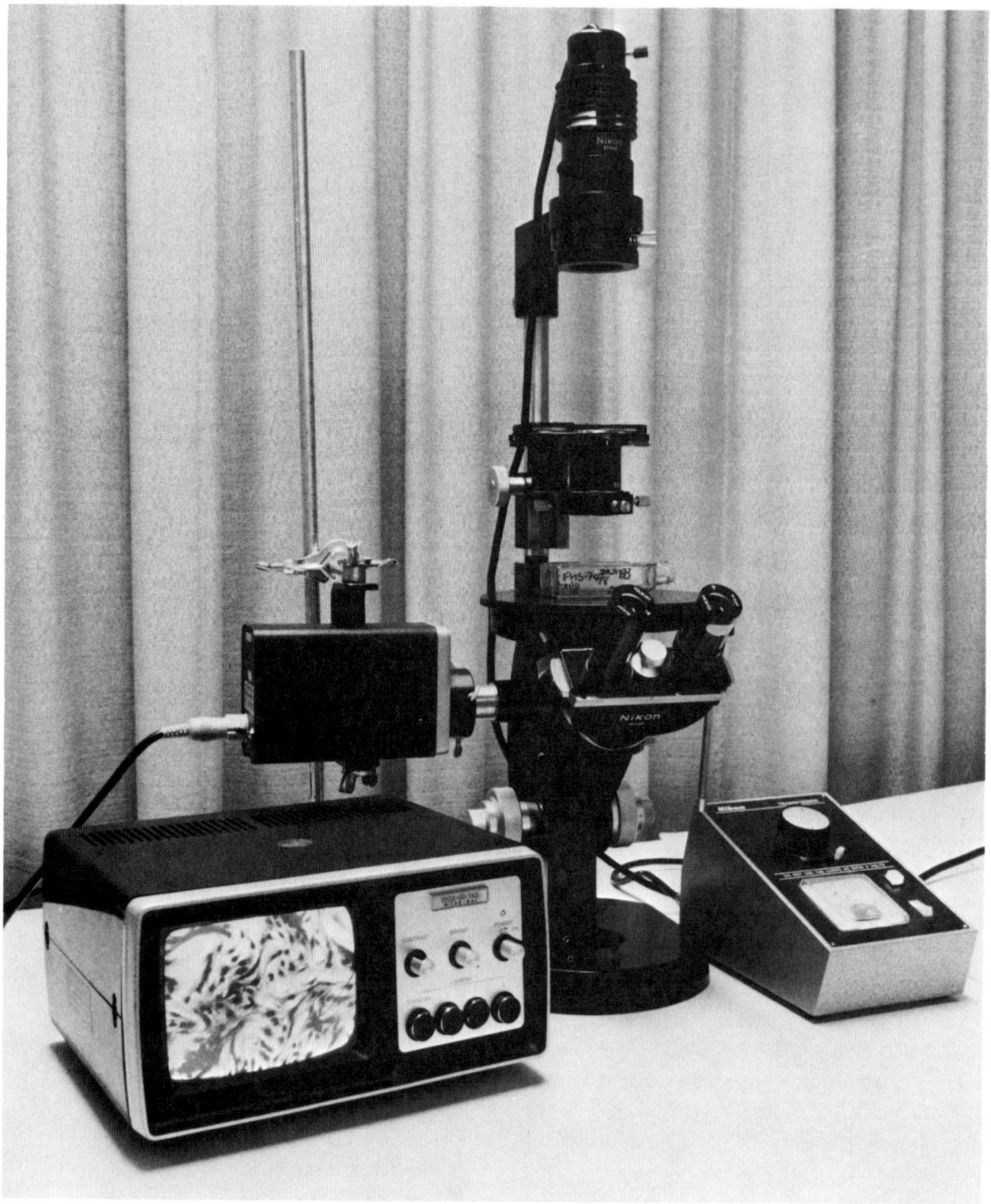

Fig. 4.15. Closed-circuit television attached to Nikon inverted microscope. Linked to a video recorder, it can be used for time-lapse studies (see Chapter 23).

cameras can be used to record and digitize images for subsequent analysis, and addition of a video recorder will enable real-time or time-lapse recordings to be made (see Chapter 23). Electronic printers are available (Polaroid, Kodak) that provide permanent records of video images. These can be color or monochrome, are of reasonable quality, and even publishable in some cases.

Colony Counter

Monolayer colonies are easily counted by eye or on a dissecting microscope with a felt-tip pen to mark off the colonies, but if many plates are to be counted, then an automated counter will help.

There are three levels of sophistication in colony counters. The simplest use an electrode-tipped marker pen, which counts when you touch down on a colony. They often have a magnifying lens to help visualize the colonies (see Fig. 18.6). From there, a large increase in sophistication and cost takes you to an electronic counter employing a fixed program, which counts colonies using a hard-wired program. These counters are very rapid and can discriminate between colonies of different diameters (though this is not necessarily proportional to cell number per colony; see also Chapter 19).

At the highest level of sophistication, image-analysis equipment may be used for colony counting but will need more skill and experience in programming. Because of this programmable feature, however, image analysis will cope with almost any size or shape of colony and will perform other complex tasks such as measuring area of outgrowth around an explant.

Cell Sizing

A dual-threshold cell counter (e.g., the Coulter) (Fig. 18.2) with the facility for pulse-height analysis scans a cell population at a range of threshold settings simultaneously and prints out cell-size distributions automatically.

Controlled-Rate Freezer

While cells may be frozen by simply placing them in an insulated box at −70°C, some cells may require different cooling rates or complex programmed cooling curves [Mazur et al., 1970; Leibo and Mazur, 1971]. A programmable freezer (Planer, Cryo-med) (see Chapter 17) enables the cooling rate to be varied by controlling the rate of injecting liquid nitrogen into the freezing chamber, under the control of a preset program (see Fig. 17.6).

Cheaper alternatives for controlling the cooling rate during cell freezing are the variable neck plug (Taylor Wharton) and specialized cooling box (Nalgene), a simple expanded polystyrene packing container, or foam insulation for water pipes (see Chapter 17).

Centrifugal Elutriator

The centrifugal elutriator is a specially adapted centrifuge suitable for separating cells of different sizes (see Cell Separation, Chapter 12). They are costly but very effective.

Flow Cytometer

This instrument, also known as a fluorescence-activated cell sorter (impulse cytophotometer or cytofluorimeter), can analyze cell populations and separate them according to a variety of criteria (see Chapters 12 and 18). It has almost unlimited potential but is too expensive to come within most tissue culture laboratory equipment budgets. Its major advantages include sophisticated cell population analyses and isolation of minority constituents of a cell population by single or multiple criteria.

It is always very tempting to purchase new pieces of equipment as they appear on the market, but weigh the advantages that they may offer against the space they will occupy and what they will cost. Try to be sure also that (1) they will be of lasting benefit, (2) you and others will want to use them, and (3) you can get them into the laboratory.

CONSUMABLE ITEMS

This category includes general items such as pipettes and pipette canisters, culture flasks, ampules for freezing, centrifuge tubes (10–15 ml, 50 ml, 250 ml; Sterilin, Nunc, Corning), universal containers (Sterilin, Nunclon), disposable syringes and needles (21–23 g for withdrawing fluid from vials, 19 g for dispensing cells), filters of various sizes (see Chapter 8) for sterilization of fluids, surgical gloves, and paper towels.

Pipettes

These should be "blow out" and wide tipped for fast delivery, graduated to the tip, with the maximum point of the scale at the top rather than at the tip. Disposable pipettes can be used but are expensive and may need to be reserved for holidays or crises in washup or sterilization. Pipettes to be reused are collected in pipette cylinders or hods, one per work station.

Pasteur pipettes are best regarded as disposable and should not be discarded into pipette cylinders but into secure glassware waste.

Pipettes are usually sterilized in aluminum or nickel-plated steel cans. Square section cans are preferable, as they stack more easily and will not roll about the work surface. Versions are available with silicone rubber-lined top and bottom ends to avoid chipping the pipettes in handling (Denley, Bellco).

Many laboratories have now adopted disposable plastic pipettes. They have the advantage of being prepacked and presterilized, and do not have the safety problems associated with chipped or broken glass pipettes. They also avoid the relatively difficult problem of washing and the tedious job of plugging. On the down side, they are very expensive and slower to use if singly packed. If bulk packed, there is a high wastage rate unless packs are shared, which is not recommended (see Chapter 5).

As a rough guide, plastic pipettes will cost about $2,000 per person per annum. The number of people will therefore determine whether to hire a washer/sterilizer or to buy plastic, remembering that glass pipettes also must be purchased at around $200 per person per annum, and require energy for washing and sterilizing.

Culture Vessels

Choice of culture vessels is determined by (1) the yield (cell number) required (see Table 7.1); (2) whether the cell is grown in monolayer or suspension; and (3) the sampling regime, i.e., are the samples to be collected

TABLE 4.2. Glass Versus Plastic Flasks

	Glass	Plastic
Attachment and growth	Good	Usually better than glass, but needs testing
Handling	Heavy; chip in wash-up	Easy, but can have inaccessible corners
Preparation	Intensive washing, rinsing, and sterilizing required	Come already clean and sterilized
Optical clarity	Poor	Excellent
Solvent resistance	Excellent	Poor
Convenience	Moderate	High
Cost in handling and preparation time	High	Low
Robustness	Strong unless dropped	Resilient but can crack if knocked or during heat expansion
Flatness of growth surface	Uneven	Flat
Definition of growth area	Approximate	Accurate
Stacking	Difficult	Good to five or six high
Storage	Based on weekly usage; limited shelf life, high turnover desirable	Unlimited shelf life if in sealed package; very bulky to store; risk of delays in delivery mean holding about 2 months' worth in reserve
Cost	More economical in large-scale operations	More economical in small-scale operations

simultaneously or at intervals over a period of time (see Chapter 18)?

"Shopping around" will often result in a cheaper price, but do not be tempted to change too often and always test a new supplier's product before committing yourself.

Care should be taken to label "sterile," "nonsterile," "tissue culture" grade, and "nontissue culture" grade plastics clearly, and preferably they should be stored separately. Glass bottles with flat sides can be used instead of plastic, provided a suitable wash-up and sterilization service is available (see Chapter 8). A similar rationale to the use of disposable pipettes applies, but tends to be overridden by the optical superiority, sterility, quality assurance, and general convenience of plastic flasks (Table 4.2). Nevertheless, they account for approximately 40% of the tissue culture budget, even more than serum.

Petri dishes are much less expensive, though more prone to contamination and to spillage. Depending on the pattern of work and the sterility of the environment, they are worth considering, at least for use in experiments if not for routine propagation of cell lines. They are particularly useful for colony-formation assays, where colonies have to be stained and counted or isolated at the end of an experiment.

Sterile Containers

Petri dishes (9 cm) are required for dissection, 5-ml bijoux bottles, 30-ml universal containers, or 50-ml sample pots for storage, 10- and 50-ml centrifuge tubes for centrifugation, and 1.2-ml plastic vials (Nunc, Nalgene) for freezing in liquid nitrogen (see Chapter 17).

Syringes and Needles

While it is not recommended that syringes and needles be used extensively in normal handling (for reasons of safety, sterility, and problems with shear stress in the needle when handling cells), syringes are required for filtration in conjunction with syringe filter adapters (see below). They may also be required for extraction of reagents (drugs, antibiotics, or radioisotopes) from sealed vials.

Sterilization Filters

While permanent apparatus is available for sterile filtration, most laboratories now tend to use disposable filters. It is useful to hold some of the commoner sizes in stock, such as 25-mm syringe adapters (Gelman, Millipore), 47-mm bottletop adapters or filter flasks (Falcon, Nalgene), and a small selection of larger sizes. This is discussed in more detail in Chapter 8.

CHAPTER 5

Aseptic Technique

In spite of the introduction of antibiotics, contamination by micro-organisms remains a major problem in tissue culture. Bacteria, mycoplasma, yeasts, and fungal spores may be introduced via the operator, atmosphere, work surfaces, solutions, and many other sources (see Table 16.1). Contaminations can be minor and confined to one or two cultures, can spread among several and infect a whole experiment, or can be widespread and wipe out your (or even the whole laboratory's) entire stock. Catastrophes can be minimized if (1) cultures are checked on the microscope, preferably by phase contrast, every time that they are handled; (2) they are kept antibiotic-free for at least part of the time to reveal cryptic contaminations (see Chapters 10 and 16); (3) reagents are checked for sterility before use (by yourself or the supplier); (4) bottles of media, etc., are not shared with other people or used for different cell lines; and (5) the standard of sterile technique is kept high at all times.

Mycoplasmal infection, invisible under regular microscopy, presents one of the major threats. Undetected, it can spread to other cultures around the laboratory. It is therefore essential to back up visual checks with a mycoplasma test, particularly if cell growth appears abnormal. (For a more detailed account of contamination, see Chapter 16.)

OBJECTIVES OF ASEPTIC TECHNIQUE

Correct aseptic technique should provide a barrier between micro-organisms in the environment outside the culture and the pure uncontaminated culture within its flask or dish. Hence, all materials that will come into direct contact with the culture must be sterile and manipulations designed such that there is no direct link between the culture and its nonsterile surroundings.

It is recognized that the sterility barrier cannot be absolute without working under conditions that would severely hamper most routine manipulations. Since testing the need for individual precautions would be an extensive and lengthy controlled trial, procedures are adopted largely on the basis of common sense and experience. Aseptic technique is a combination of procedures designed to reduce the probability of infection, and the correlation between the omission of a step and subsequent contamination is not always absolute. The operator may abandon several precautions before the probability rises sufficiently that a contamination occurs (Fig. 5.1). By then, the cause is often multifactorial and consequently no simple solution is obvious. If, once established, all precautions are maintained consistently, breakdown will be rarer and more easily detected.

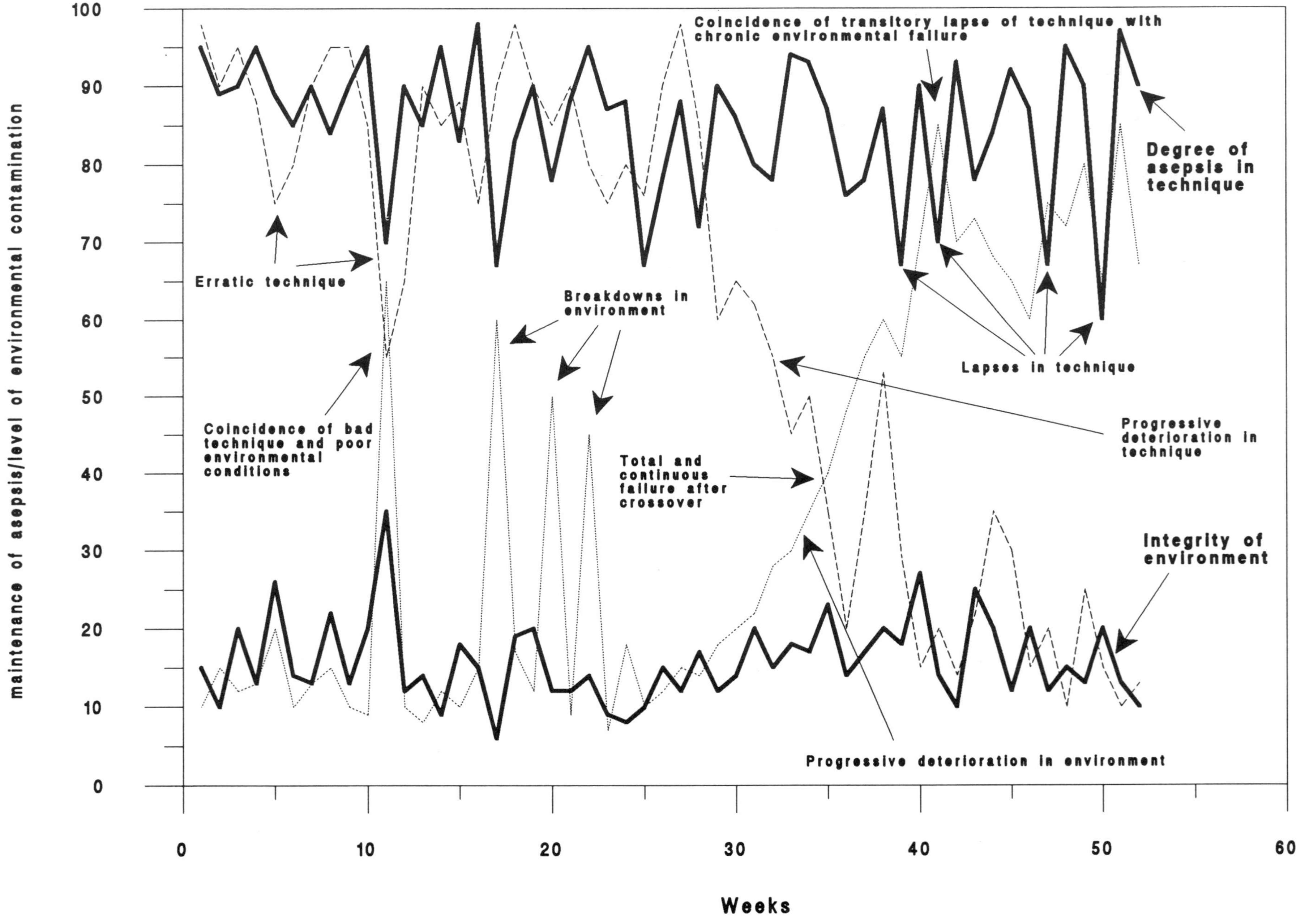
Coincidence of transitory lapse of technique with chronic environmental failure
Degree of asepsis in technique
Erratic technique
Breakdowns in environment
Lapses in technique
Coincidence of bad technique and poor environmental conditions
Progressive deterioration in technique
Total and continuous failure after crossover
Integrity of environment
Progressive deterioration in environment
maintenance of asepsis/level of environmental contamination
Weeks
100
90
80
70
60
50
40
30
20
10
0
0
10
20
30
40
50
60

Although laboratory conditions have improved in some respects (air-conditioning and filtration, laminar flow facilities, etc.), the modern laboratory is often more crowded and accommodation may have to be shared. However, with reasonable precautions, maintenance of sterility is not difficult.

QUIET AREA

In the absence of laminar flow, a separate sterile room should be used if possible (see Fig. 3.2). If not, pick a quiet corner of the laboratory with little or no traffic and no other activity (see Chapter 3). With laminar flow, an area should be selected that is free from drafts and traffic should still be kept to a minimum. Animals and microbiological culture should be excluded from the tissue culture area. It should be kept clean and dust-free and should not contain equipment other than that connected with tissue culture. Nonsterile activities, such as sample processing, staining, or extractions, should be carried out elsewhere.

WORK SURFACE

One of the more frequent examples of bad technique is the failure to keep the work surface clean and tidy. The following rules should be observed:

(1) Start with a completely clear surface.

(2) Swab down liberally with 70% alcohol.

(3) Bring onto the surface only those items you require for a particular procedure; swab bottles, cans, etc., with 70% alcohol beforehand.

(4) Remove everything that is no longer required, and swab down before the next procedure.

(5) Arrange your apparatus (a) to have easy access to all of it without having to reach over one item to get at another and (b) to leave a wide, clear space in the center of the bench (not just the front edge!) to work on (Fig. 5.2). If you have too much equipment too close to you, you will inevitably brush the tip of a sterile pipette against a nonsterile surface.

(6) Work within your range of vision, e.g., insert a pipette in a bulb or pipetting device with the tip of the pipette pointing away from you so that it is in your line of sight continuously and not hidden by your arm.

(7) Mop up any spillage immediately and swab with 70% alcohol.

(8) Remove everything when you have finished and swab down again.

PERSONAL HYGIENE

There has been much discussion about whether hand washing encourages or reduces the bacterial count on the skin. Regardless of this debate, washing will moisten the hands and remove dry skin likely to blow onto your culture and will reduce loosely adherent microorganisms, which are the greatest risk to your cultures. Surgical gloves may be worn and swabbed frequently, but it may be preferable to work without them (where no hazard is involved) and retain the extra sensitivity that this allows.

Caps, gowns, and face masks are often worn but are not always strictly necessary, particularly when working with laminar flow. However, if you have long hair, tie it back. When working on the open bench, do not talk while working aseptically; and if you have a cold, wear a face mask, or, better still, do not do any tissue culture during the height of the infection. Talking is permissible when working in vertical laminar flow with a barrier between you and the culture but should still be kept to a minimum.

PIPETTING

Standard glass or disposable plastic pipettes are still the easiest form of manipulating liquids. Syringes are often used, but regular needles are too short to reach

Fig. 5.1. *Probability of infection. The top graph (solid line) represents variability in technique against a scale of 100, where 100 would represent perfect aseptic technique. The bottom graph (solid line) represents fluctuations in environmental contamination, where zero would be perfect asepsis. Both show fluctuations representing lapses in technique (forgetting to swab the work surface, handling a pipette too far down, touching nonsterile surfaces with pipette, etc.) or crises in environmental contamination (high spore count, contaminated incubator, contaminated reagents, etc.). As long as these lapses or crises are minimal in degree and duration, the two graphs do not overlap. Where particularly bad lapses in technique coincide with severe environmental crises (left-hand side of chart, dashed and dotted lines overlap briefly), there is an increased probability of infection. If the breakdown in technique is progressive (dashed line, center) and the deterioration in the environment is progressive (dotted line, center), when the two cross, the probability of infection is high, resulting in frequent, multispecific, and multifactorial contamination.*

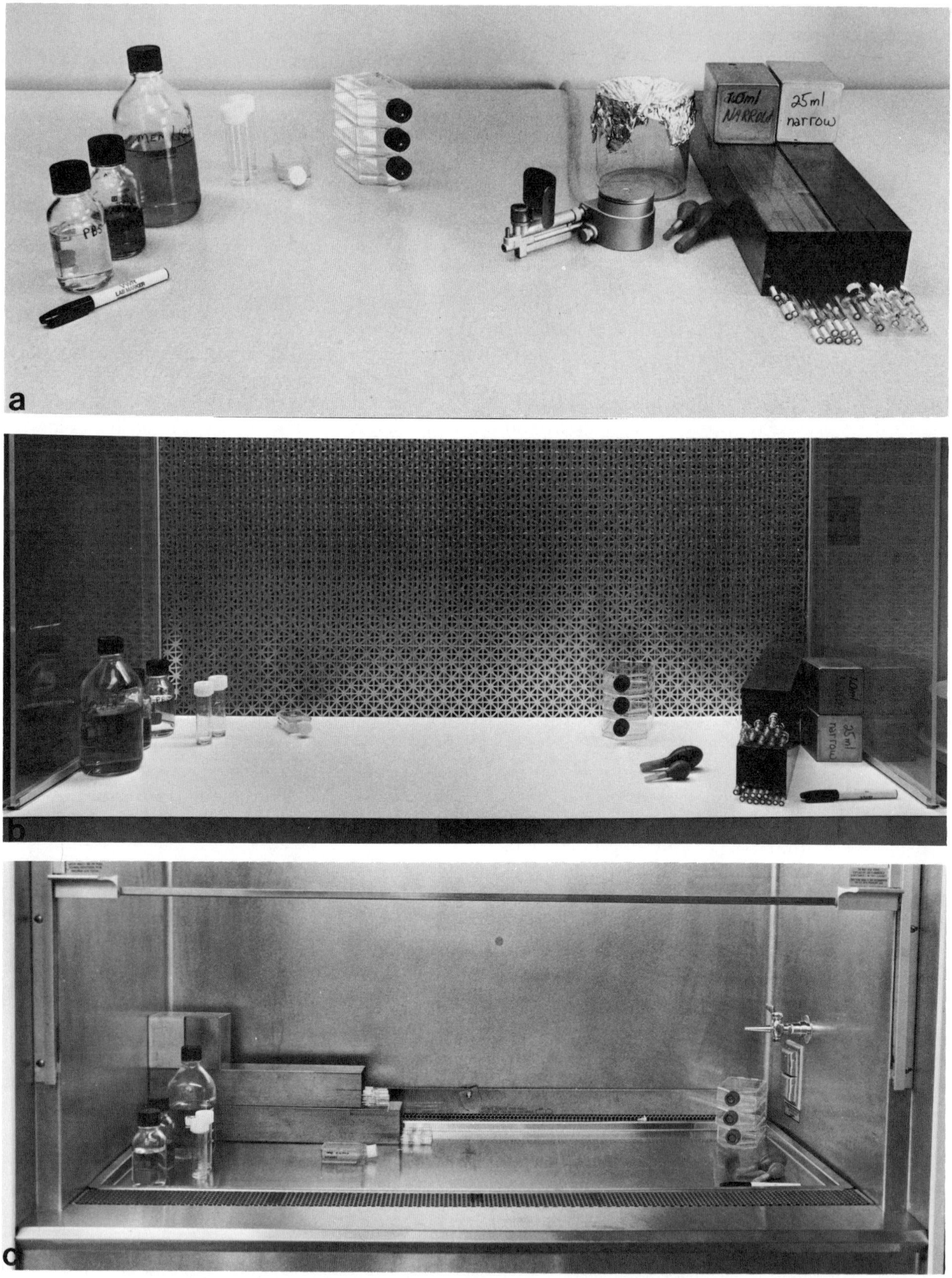

Fig. 5.2. Suggested layout of work area. a. Open bench. b. Horizontal laminar flow. c. Vertical laminar flow.

into most bottles. Syringing may produce high shearing forces when dispensing cells and increases the risk of self-inoculation. Syringes with blunt, wide-bore canulae are preferable but still not as rapid to use as pipettes, except where multiple-stepping dispensers (see Chapter 4) are used.

Pipettes of a convenient size range should be selected—1 ml, 2 ml, 5 ml, 10 ml, and 25 ml cover most requirements. If you only require a few of each, make up mixed cans and save space. Mouth pipetting, even with plugged pipettes or a filter tube/mouthpiece should be avoided, as it has been shown to be a contributory factor in mycoplasmal infection and may introduce an element of hazard to the operator, e.g., with virus-infected cell lines and human biopsy or autopsy specimens or other potential biohazards (see Chapter 6). Inexpensive bulbs and pipetting devices are available; try a selection of these to find one that suits you (see Fig. 4.10). They should accept securely all the sizes of pipette that you use without forcing them and without the pipette falling out. The regulation of flow should be easy and rapid but at the same time capable of fine adjustment. You should be able to draw liquid up and down repeatedly (e.g., to disperse cells) and there should be no fear of carry-over. The device should fit comfortably in your hand and should be easy to operate with one hand.

The Marburg-type pipette (see Fig. 4.10) (Gilson, Oxford, Eppendorf, etc.) is particularly useful for small volumes (1 ml and less), though there can be some difficulty in reaching down into larger vessels with most of them. They are best used in conjunction with a shallow vial or bottle and are particularly useful when dealing with microtitration assays and other multiwell dishes but should not be used for serial propagation unless filter tips are used. Multipoint pipettors (4, 8, or 12 point) are available for microtitration dishes (see Fig. 4.10).

It is necessary to insert a cotton plug in the top of a glass pipette before sterilization to maintain sterility in the pipette during use. If this becomes wet in use, discard the pipette into the wash-up. Plugging pipettes for sterile use is a very tedious job, as is the removal of plugs before washing. Automatic pipette pluggers are available, and, although expensive, speed up the process and reduce the tedium (see Fig. 8.5). Alternatively, a sterile filter tube (Fig. 5.3) may be attached to the bulb, eliminating the need to plug pipettes. It is vital that the filter tube be changed between handling of different cell lines to avoid the risk of cross-contamination.

These problems are avoided with plastic pipettes (see Chapter 4), which come already plugged. However, they are slower to use if individually wrapped, wasteful if bulk wrapped (some pipettes are always lost, as it is

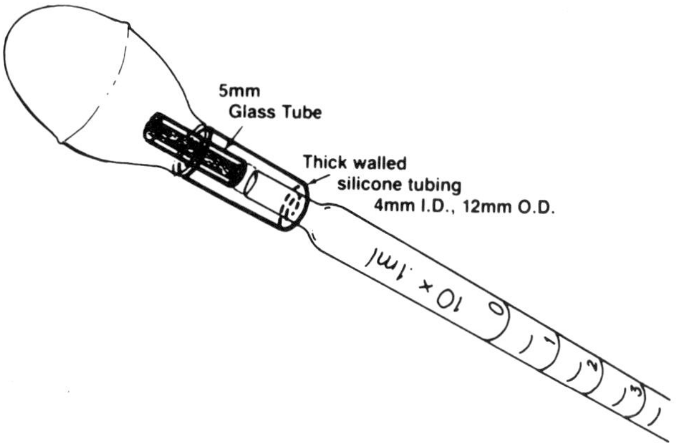

Fig. 5.3. *Filter tube. Interposed between bulb and pipettes, this avoids the necessity to plug pipettes. Filter tube must be renewed between cell lines or if wetted. (Developed at the Beatson Institute from an original idea by Dr. John Paul.)*

not advisable to share or reuse a pack once opened), and carry a slightly higher risk of contamination in use, as removing them from a plastic wrapper is not as clean as withdrawal from a can. They are, however, more likely to be free of chemical and microbial contamination than recycled glass pipettes.

Automatic pipetting devices and repeating dispensers are discussed in Chapter 4.

STERILE HANDLING

Swabbing

Swab down the work surface before and during work, particularly following any spillage, and swab down again when you have finished. Swab bottles, particularly from the cold room, before using for the first time each day.

Capping

Deep screw caps should be used in preference to stoppers, although care must be taken when washing caps to ensure that all detergent is rinsed from behind rubber liners. Wadless polypropylene caps should be used if possible. The screw cap should be covered with aluminum foil to protect the neck of the bottle from sedimentary dust. Introduction of deep polypropylene caps (e.g., Duran) has made foil shrouding less necessary.

Flaming

When working on the open bench, the necks of bottles and screw caps should be flamed before and after opening a bottle and before and after closing. Pipettes should be flamed before use. Work close to the flame where there is an up-current due to convection, and do not leave bottles open. Screw caps should be placed open-side down on a clean surface and flamed before replacing on the bottle. Alternatively, screw caps may be held in the hand during pipetting, avoiding the need to flame or lay down (see Fig. 5.6).

Flaming is not necessary when working in a laminar flow hood, and is best avoided. The presence of an open flame within the hood disrupts the laminar airflow and can be a fire hazard.

Handling Bottles and Flasks

When working on the open bench or in vertical laminar flow, bottles should not be vertical when open, but should be kept at an angle as shallow as possible without risking spillage (see Fig. 5.7). Culture flasks should be laid down horizontally when open and held at an angle, as for bottles, during manipulations. When working in horizontal laminar flow, do not let your hands or any other items come between an open vessel or sterile pipette and the air filter.

Pouring

Whenever possible, do not pour from one sterile container into another unless the bottle you are pouring from is to be used once only and, preferably, is to deliver all its contents (premeasured) in one single delivery. The major risk in pouring lies in the generation of a bridge of liquid between the outside of the bottle and the inside, which may permit infection to enter the bottle, so bottles or flasks that are stored or incubated after pouring are at a significantly higher risk.

LAMINAR FLOW

The major advantage of working in laminar flow is that the working environment is protected from dust and contamination by a constant, stable flow of filtered air passing over the work surface (Fig. 5.4) (see also Fig. 3.1). There are two main types: (1) *horizontal*, where the airflow blows from the side facing you, parallel to the work surface, and is not recirculated; and (2) *vertical*, where the air blows down from the top of the cabinet onto the work surface and is drawn through the work surface and either recirculated or vented. In *recirculating* hoods, 20% is vented and made up by drawing in air at the front of the work surface. This is designed to minimize overspill from the work area of the cabinet. Horizontal flow hoods give the most stable airflow and best sterile protection to the culture and reagents; vertical flow gives more protection to the operator. If potentially hazardous material (radioisotopes, mutagens, human- or primate-derived cultures, virally infected cultures, etc.) is being handled, a Class II vertical flow biohazard hood should be used (see Fig. 6.4a and Chapter 6).

If known human pathogens are handled, a Class III pathogen cabinet with a pathogen trap on the vent is obligatory (see Fig. 6.4b and Chapter 6).

Laminar flow hoods depend for their efficiency on a minimum pressure drop across the filter. When filter resistance builds up, the pressure drop increases and the flow rate of air in the cabinet falls. Below 0.4 m/s (80 ft/min), the stability of the laminar airflow is lost and sterility can no longer be maintained. The pressure drop can be monitored with a manometer fitted to the cabinet, but direct measurement of airflow with an anemometer is preferable.

Routine maintenance checks of the primary filters are required (every 3–6 months). They may be removed (after switching off the fan) and discarded or washed in soap and water, as they are usually made of polyurethane foam. Every 6 months the main high efficiency particulate air (HEPA) filter above the work surface should be checked for airflow and holes (detectable by locally increased airflow and an increased particulate count). This is best done by professional engineers on a contract basis. Class II biohazard cabinets will have HEPA filters on the exhaust, which will also need to be changed periodically. Again, this should be done by a professional engineer, with proper precautions taken for bagging and disposing of the filters by incineration. Ideally, cabinets should be sealed and fumigated before the filters are changed.

Regular weekly checks should be made below the work surface and any spillage mopped up, the tray washed with hot detergent, and the area sterilized with 5% phenolic disinfectant in 70% alcohol. Spillages should, of course, be mopped up when they occur; but occasionally they go unnoticed, so a regular check is imperative.

Laminar flow hoods are best left running continuously because this keeps the working area clean. Should any spillage occur, either on the filter or below the work surface, it dries fairly rapidly in sterile air, reducing the chance of growth of micro-organisms.

Ultraviolet lights are used to sterilize the air and exposed work surfaces in laminar flow cabinets between use. The effectiveness of this is doubtful because crevices are not reached, and these are treated more effectively with alcohol or other sterilizing agents, which will run in by capillarity. Ultraviolet irradiation is hazardous to use and will also lead to crazing of some clear plastic panels (e.g., Perspex) after 6 months to 1 year.

◇ Where UV is used, protective goggles must be worn and all exposed skin covered.

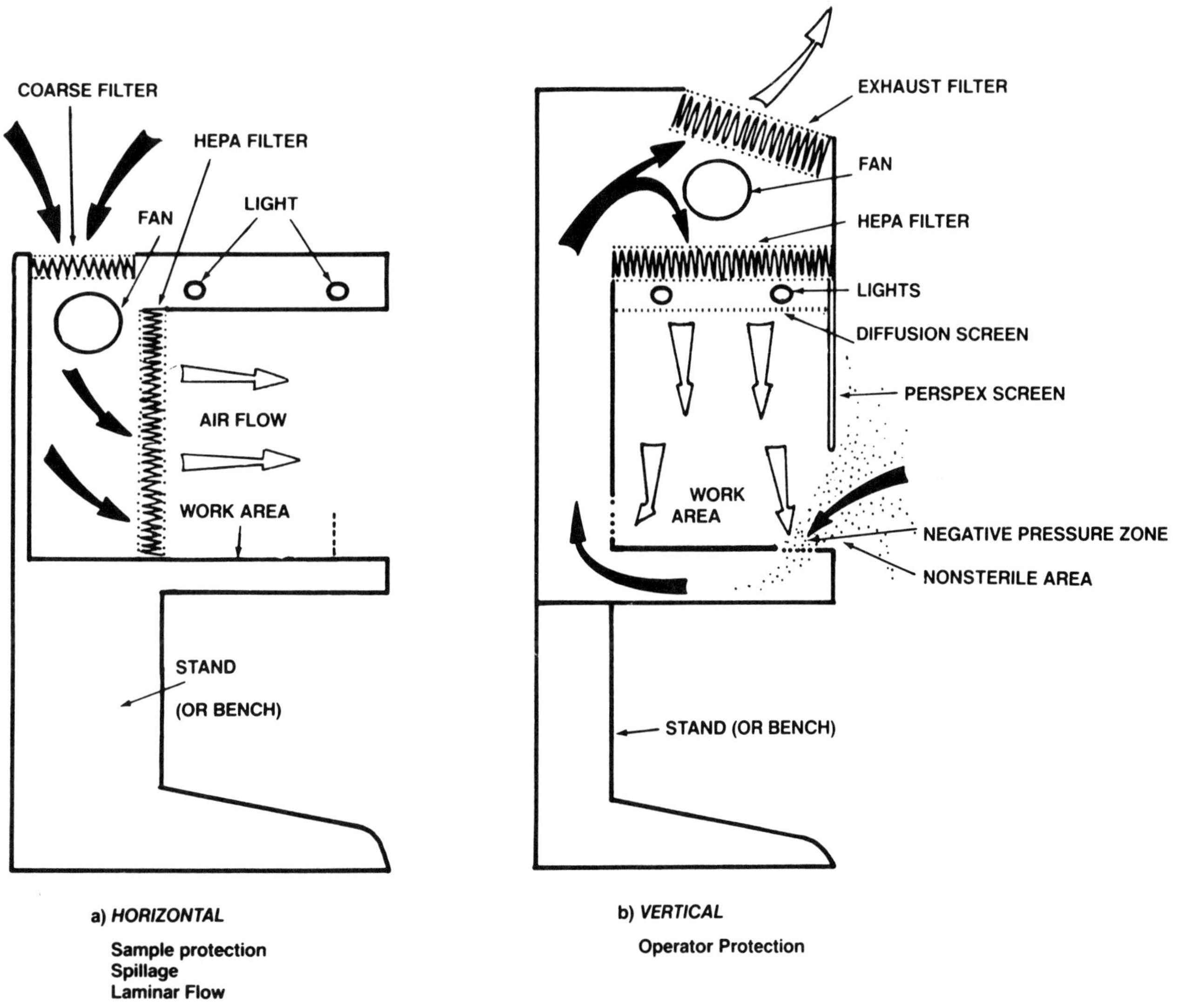

Fig. 5.4. *Horizontal (a) and vertical (b) laminar flow hoods. Filled arrows, nonsterile air; open arrows, sterile air. For vertical flow biohazard cabinet, see next chapter.*

STANDARD PROCEDURE

Emphasis is being placed here on aseptic technique. Media preparation and other manipulations are discussed in more detail under the appropriate headings.

Outline

Clean and prepare work area, with bottles, pipettes, etc. Carry out preparative procedures first. Flame articles as necessary and keep the work surface clean and clear. Finally, tidy up and wipe over surface with 70% alcohol.

Materials

70% alcohol
Swabs
Bunsen (not in laminar flow) and lighter
Pipette aid or bulb (see Fig. 4.10)
Waste beaker (Fig. 5.5) or aspiration pump (see Fig. 4.5)
Scissors
Marker pen
Media, stocks, etc.

Protocol

1. Swab down bench surface or all inside surfaces of laminar flow hood with 70% alcohol.
2. Bring media, etc., from cold store and freezer, swab bottles with alcohol, and place those that you will need first on the bench or in the hood.
3. Collect pipettes and place at the rear or side of the work surface in an accessible position (see Fig. 5.1). Open pipette cans and place lids out

of the way but still in sterile work area (within the hood).

4. Collect any other glassware, plastics, instruments, etc., that you will need and place them close by.
5. (a) Flame necks of bottles, rotating neck in flame, and slacken caps–they should be flamed outside the hood, if one is used–and flame again after slackening.

   (b) *On open bench*: (i) take pipette from can, touching other pipettes as little as possible, particularly at the tops; (ii) flame top to burn off any cotton protruding from the pipette; (iii) insert in bulb, pointing pipette away from you and holding it well above the graduations.

   ◇ Note. Take care not to exert too much pressure, as pipettes can break when being forced into a bulb (see Chapter 6 [Fig. 6.1]).

   (iv) Flame pipette by pushing lengthwise through flame, rotate 180°, and pull back through flame. This should only take 2–3 s or the pipette will get too hot. You are not attempting to sterilize the pipette, merely to fix any dust that may have settled on it. If you have touched anything or contaminated the pipette in any other way, discard it into the wash-up; do not attempt to resterilize it by flaming; (v) holding the pipette still pointing away from you, remove the cap of your first bottle into the crook formed between your little finger and the heel of your hand (Fig. 5.6); (vi) flame the neck of the bottle; (vii) withdraw the requisite amount of fluid and hold; (viii) flame bottle-neck and recap; (ix) remove caps of receiving bottle, flame neck, insert fluid, reflame, and replace cap; (x) when finished, tighten caps, flame thoroughly, and replace foil. Work with the bottles tilted so that your hand does not come over the open neck. If you have difficulty in holding the cap in your hand while you pipette, place the cap on the bench open-side down. If bottles are to be left open, they should be sloped as close to horizontal as possible, laying them on the bench or on a bottle rest (Fig. 5.7).

   (c) *In laminar flow*, proceed as for open bench but omit flaming during manipulations. Bottles may be left open more safely but should still be closed if you leave the hood for more than a few minutes. In vertical laminar flow, do not work immediately above an open bottle or dish. In horizontal laminar flow, do not work behind an open bottle or dish.
6. On completion of the operation, remove stock solutions from work surface, keeping only the bottles that you will require.
7. Check cultures, decide what they require, and bring to sterile work area.
8. Swab bottles, flame necks, and place on work surface together with cultures, preferably one cell strain at a time with its own bottle of medium and other solutions.
9. For fluid change, proceed as follows:

   (a) Take sterile pipette, flame (if not in hood), and insert bulb.

   (b) Flame neck of culture bottle, open, withdraw medium, and discard into waste beaker (see Fig. 5.5) or, preferably, via a suction pump with a collection trap, or a tap siphon, with a suction line into the hood (see Fig. 4.5). Discard pipette.

   (c) With a fresh pipette, transfer fresh medium to culture flask as in (5) above. Discard pipette.

   (d) Tighten caps, flame necks, and replace foil.
10. Return culture flasks to incubator and media to cold room.
11. Clear away all pipettes, glassware, etc., and swab down the work surface.

All *apparatus* used in the tissue culture area should be cleaned regularly to avoid the accumulation of dust and microbial growth in accidental spillages. Items of equipment, such as gas cylinders, must be cleaned before being introduced, and any major movement of equipment should not take place while people are working aseptically.

Humidified incubators are a major source of contamination. They should be cleaned out at regular intervals (weekly or monthly, depending on the level of atmospheric contamination) by removing the contents, including all the racks or trays, and washing down the interior and the racks or shelves with a nontoxic detergent such as Decon or Roccal. Traces of de-

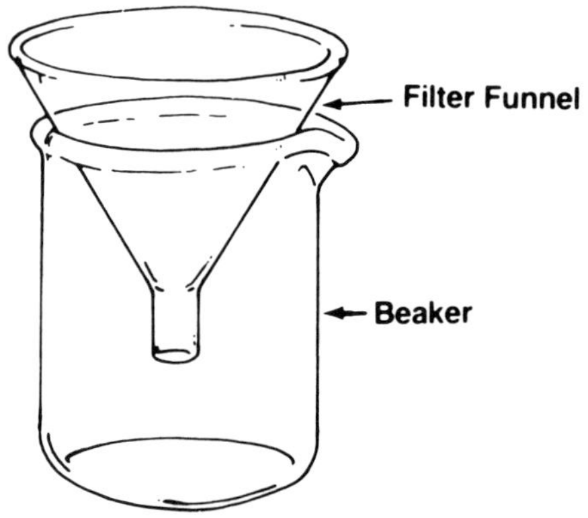

Fig. 5.5. *Waste beaker. Filter funnel prevents splashback from beaker.*

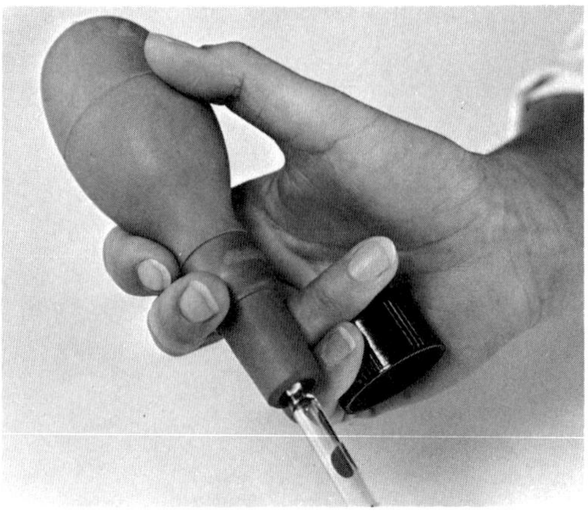

Fig. 5.6. *Uncapping bottle and holding cap. Hand may need to be moved up or down bulb between uncapping and pipetting. With this particular bulb (Aspirette), the forefinger is used to seal the top of the bulb when pipetting.*

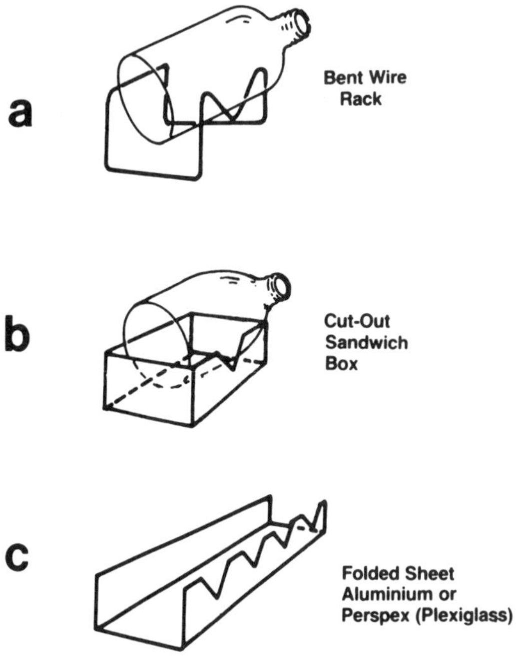

Fig. 5.7. *Suggested designs for bottle rest for use during pipetting. a. Wire rack, suggested by M. Stack-Dunne. b. V-cut in a plastic storage box. c. Folded aluminum or Plexiglas suggested by A.C. McKirdy.*

tergent should then be removed with 70% alcohol, which should be allowed to evaporate completely before returning the shelves and restocking the incubator.

A fungicide may be placed in the humidifier tray at the bottom of the incubator to retard fungal growth, but the success of these is limited, and there is no real substitute for regular cleaning. Some incubators have high-temperature sterilization cycles, but these are seldom able to generate sufficient heat for long enough to be effective. More recently, incubators have been introduced with micropore filtration and laminar airflow to inhibit the circulation of micro-organisms (Forma, Jencons). The author has no experience of the use of this type of incubator, but it would seem a reasonable step forward; time will determine the efficacy.

Where there are recurrent problems with contamination in humidified incubators, it is advantageous to enclose dishes and plates in plastic sandwich boxes. The box should be swabbed before use, inside and outside, and allowed to dry in sterile air. When the box is subsequently removed from the incubator, it should be swabbed with 70% alcohol before being opened or introduced into your work area. The dishes are then carefully removed, and the box interior swabbed prior to reuse.

One major source of contamination in plates and petri dishes is the formation of a capillary film between plate or dish and its lid. Most plastic dishes and plates now have raised or "vented" lids (see Chapter 7 [Fig. 7.8]) to facilitate gas exchange; this reduces, but does not eliminate, the risk of trapping medium. Trapped medium forms a bridge with the nonsterile outer air and may cross-contaminate wells in a multiwell plate, so it must be prevented, or, having happened, eliminated. Liquid films may be removed by the careful application of a tissue moistened with 70% alcohol; the lid should be replaced with a fresh one.

It is common practice to place flasks, with the caps slackened, in a humid CO_2 incubator to allow for gaseous equilibration, but this does increase the risk of infection. It is preferable to purge the flasks from a sterile, premixed gas supply and then seal them. This avoids the need for a gassed incubator for flasks, and gives the most uniform and rapid equilibration. Several manufacturers (Corning, Falcon, Nunc, Costar) now provide flasks with permeable caps to allow rapid equilibration in a CO_2 atmosphere but without the risk of contamination.

The essence of good sterile technique embodies many of the principles of standard good laboratory practice. Keep a clean, clear space to work and have on it only what you require at one time. Prepare as much as possible in advance so that cultures are out of the incubator for the shortest possible time and the various manipulations can be carried out quickly, easily, and smoothly. Keep everything in direct line of sight and develop an awareness of accidental contacts between sterile and nonsterile surfaces. Leave the area clean and tidy when you finish.

It is vital that new staff introduced into the sterile area receive adequate training in techniques and procedures. Before they are allowed to work independently, they should be apprenticed to a skilled operator for a period of up to 3 months.

CHAPTER 6

Laboratory Safety and Biohazards

Accidents happen in the laboratory as in other work places, and a greater understanding of the biological or medical consequences offers no greater protection—perhaps the converse, as familiarity often leads to a more casual approach in dealing with regular, biological, and radiological hazards just as it does in the factory for those dealing with equally hazardous engineering tools.

To draw attention to items in the text that refer specifically to safety, they are identified with this diamond symbol: ◊.

A valuable principle that has become incorporated into modern safety legislation is that of risk assessment. Determining the degree and nature of hazard of a particular chemical or biological is only part of the process; the way in which the material is used is equally important in determining risk. Such considerations as amount, degree and frequency of exposure, and handling conditions and ancillary hazards like heat, frost, electric current, the type of protective clothing worn, and the type of training and experience of the operator will all contribute to the risk of a given procedure though the nature of the chief hazard may remain constant. Hence hazardous substances, equipment, and conditions should not be thought of in isolation but as part of a procedure that should be assessed for all its constituents (Table 6.1). Try to think in terms of the stages of handling: procurement, storage, operating procedure, and disposal, and be aware of how individual components of a procedure will interact and alter the level of risk.

A major problem that arises constantly in establishing safe practices in a biology laboratory is the disproportionate concern given to the more esoteric and poorly understood risks, such as those arising from genetic manipulation, relative to the known proven hazards of chemicals, toxins, fire, ionizing radiation, electrical shock, and broken glass. No one should ignore potential biohazards [Barkley, 1979; Wells et al., 1989; Aldhous, 1991], but they should not displace the recognition of everyday safety problems.

The following should not be interpreted as a code of practice, but rather as advice that might help in compiling safety regulations. The information that is provided here is designed to provide the reader with some guidelines and suggestions to help construct a local code of practice, in conjunction with regional and national legislation, and in full consultation with your local safety committee. It has no legal standing and should not be quoted as such.

General safety regulations are usually available from the safety office in the institution or company in which you work. In addition, these are available from the Occupational Health and Safety Administration (OHSA) in the United States. From January 1, 1993, Europe, including the United Kingdom, will come under new joint regulations [Management of Health and Safety at Work Regulations, 1992; Provision and Use of Work Equipment, 1992]. This will cover all matters of general safety. The relevant regulations and recommendations for biological safety for the United States are contained in Biosafety in Microbiological and Biomedical Laboratories [1984], a joint document prepared by the Centers for Disease Control, Atlanta, Georgia, and the National Institutes of Health, Bethesda, Maryland. For the United Kingdom, the Health Services Advisory Committee of the Health and Safety Commission has published two booklets, Safe Working and the Prevention of Infection in Clinical Laboratories [1991], and Safe Working and the Prevention of Infection in Clinical Laboratories–Model Rules for Staff and Visitors [1991], available from Her Majesty's Stationery Office. Genetically modified cells are dealt with in A Guide to the Genetically Modified Organisms (Contained Use) Regulations [1992] also from Her Majesty's Stationery Office.

Table 6.1 provides a checklist highlighting some of the factors to be considered when elaborating safety policy in a tissue culture laboratory. In most cases the headings and subheadings are self-explanatory, but a few need a little explanation.

GENERAL SAFETY (Table 6.2)

Operator

It is the responsibility of the institution to provide the correct training or to determine that the individual is already trained in appropriate laboratory procedures. It is the supervisor's responsibility to ensure that procedures are carried out correctly and the correct protective clothing is worn at the appropriate times.

Equipment

A general supervisor should be appointed in charge of all equipment maintenance, electrical safety, and mechanical reliability, and a curator should be appointed in charge of each specific piece of equipment to ensure that the day-to-day operation is satisfactory and to train others in its use. Particular risks include the generation of toxic fumes or aerosols from centrifuges and homogenizers. These must either be contained by the design of the equipment or by placing it in a fume cupboard.

The electrical safety of equipment is dealt with in the Provision and Use of Work Equipment Regulations [1992], and will shortly be incorporated in new European Community (EC) guidelines (see above). In the United States this is covered by OHSA.

TABLE 6.1. Elements of Risk Assessment

Operator	Experience:
	Level
	Appropriateness
	Background
	Training
	Protective clothing
Equipment	Age
	Condition
	Suitability for task
	Mechanical stability
	Electrical safety
	Containment (aerosols, leakage)
	Generation of heat
	Generation of toxic fumes
Physical Risks	Heat
	Intense cold
	Electric shock
	Fire:
	General precautions
	Equipment wiring, installation, and maintenance
	Incursion of water near electrics
	Fire drills, procedures, escape routes
	Solvent usage and storage
	Flammable mixtures
	Identification of stored biohazards and radiochemicals
Chemicals	Toxicity
	Carcinogenicity
	Teratogenicity
	Mutagenicity
	Corrosiveness
	Reaction with water
	Reaction with solvents
	Volatility
	Asphyxiation
	Generation of powders and aerosols
Biological	Pathogenicity
	Infectivity
	Host specificity
	Stability
Radioisotopes	Energy
	Nature of emission:
	Penetration
	Interaction/ionization
	Half-life
	Shielding
	Volatility
	Chemical toxicity
	Localization
General	Amount
	Location
Special Circumstances	Pregnancy
	Illness
	Cuts and abrasions
	Immunosuppressant drugs

(*continued*)

TABLE 6.1. *(Continued)*

Procedures	Scale
	Complexity
	Duration
	Number of persons involved (increase or diminish risk?)
Precautions	Protective clothing
	Containerization (physical)
	Containment (chemical, radiological, biological)
	Observation
	Automatic monitoring
	Inspection
	Record keeping

Try to think in terms of the stages of handling: procurement, storage, operating procedure, and disposal, and be aware of how individual components of a procedure will interact and alter the level of risk.

Glassware and Sharp Items

◇ The most common form of injury in tissue culture results from accidental handling of broken glass and syringes, e.g., broken pipettes in a wash-up cylinder when too many pipettes, particularly Pasteur pipettes, are forced into too small a container. Pasteur pipettes should be discarded or if reused handled separately and with great care. Avoid syringes unless they are needed for loading ampules or withdrawing fluid from a vial. When disposable needles are discarded, use a rigid plastic or metal container. Do not attempt to bend or manipulate the needle. Provide separate receptacles for the disposal of sharp items and broken glass and do not use them for general waste.

◇ Take care when fitting a bulb or pipetting device onto a pipette. Choose the correct size to guard against the risk of the pipette breaking at the neck and lacerating your hand. Check that the neck is sound, hold the pipette as near the end as possible, and apply gentle pressure with the pipette pointing away from your knuckles (Fig. 6.1).

TABLE 6.2. Typical General Safety Hazards in a Tissue Culture Laboratory

Broken glass
Pipettes
Cables
Tubing
Cylinders
Radioisotopes in sterile cabinet
Bunsen burners
Sharp instruments
Pasteur pipettes
Syringe needles

Chemical Toxicity

Relatively few major toxic substances are used in tissue culture, but when they are, the conventional precautions should be taken, paying particular attention to the distribution of aerosols by laminar flow cabinets (see Biohazards, this chapter). Detergents, particularly those used in automatic machines, are usually caustic; even when they are not they can cause irritation to the skin, eyes, and lungs. Use dosing devices where possible, wear gloves, and avoid procedures that cause the detergent to spread as dust. Liquid detergent concentrates are more easily handled but are often more expensive.

◇ Chemical disinfectants such as hypochlorite should also be used cautiously and with a dispenser. Hypochlorite disinfectants will bleach clothing and cause skin irritations and will even corrode stainless steel.

◇ Specific chemicals used in tissue culture requiring special attention are (1) dimethyl sulfoxide (DMSO), which is a powerful solvent and skin penetrant and can, therefore, carry many substances through the skin [Horita and Weber, 1964] and even through protective gloves, and (2) mutagens and carcinogens, which should be handled only in a safety cabinet (see below) and are sometimes dissolved in DMSO.

The handling of chemicals in the United States is regulated by OHSA and in the United Kingdom is covered by the Control of Substances Hazardous to Health Regulations [1988], shortly to be incorporated in EC legislation and guidelines (see above).

Gases

◇ Most gases used in tissue culture (CO_2, O_2, N_2) are not harmful in small amounts but are nevertheless dangerous if handled improperly. They are contained in pressurized cylinders that must be properly secured (Fig. 6.2). When a major leak occurs, there is a risk of asphyxiation from CO_2 and N_2 and of fire from O_2. Evacuation and maximum ventilation are necessary in each case; for O_2, call the fire department.

◇ Glass ampule sealing is usually performed in a gas oxygen flame, so great care must be taken both to guard the flame and to prevent unscheduled mixing of the gas and oxygen. A one-way valve should be incorporated in the gas line so that oxygen cannot blow back.

Liquid Nitrogen

◇ There are three major risks associated with liquid N_2: frostbite, asphyxiation, and explosion. Since the temperature of liquid N_2 is $-196°C$, direct contact with it (splashes, etc.) or with anything, particularly metallic, submerged in it presents a serious hazard. Gloves thick enough to act as insulation but flexible enough to allow manipulation of ampules should be

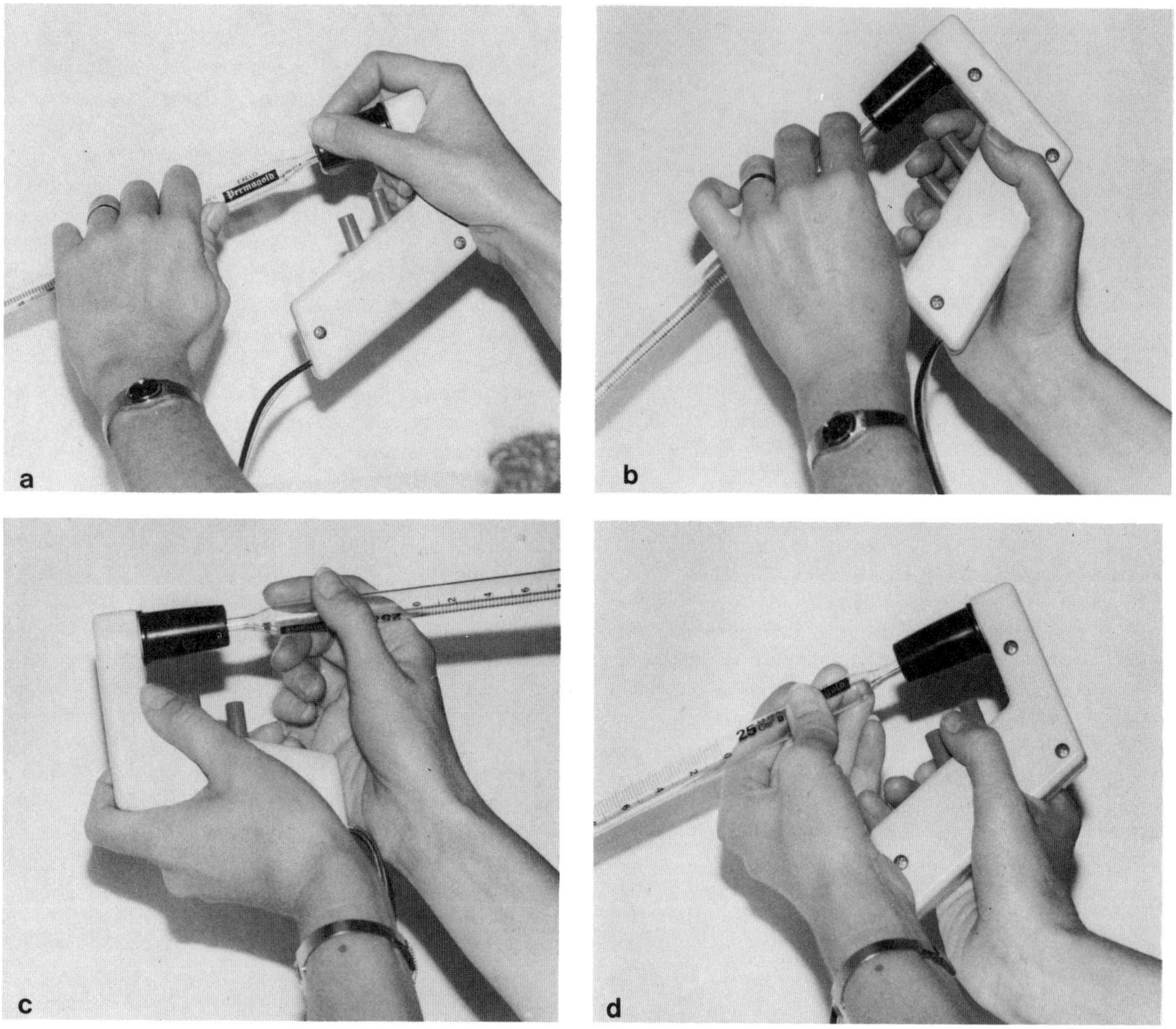

Fig. 6.1. *Inserting a pipette into a pipetting device. a. Wrong position. Left hand too far down pipette, risking contamination of the pipette and exerting too much leverage, which might break the pipette; right hand too far over and exposed to end of pipette or splinters, should the pipette break at the neck during insertion. b–d. Correct positions. Left hand farther up pipette, right hand clear of top of pipette.*

worn. When liquid N_2 boils off during routine use of the freezer, regular ventilation is sufficient to remove excess nitrogen; but when nitrogen is being dispensed, or a lot of material is being inserted in the freezer, extra ventilation will be necessary.

◇ When ampules are submerged in liquid N_2, a high pressure difference results between the outside and the inside of the ampule. If they are not perfectly sealed, this results in inspiration of liquid N_2, which will cause the ampule to explode violently when thawed. This can be avoided by storing in the gas phase (see Chapter 17) or by ensuring that the ampules are perfectly sealed. Thawing from storage under liquid N_2 should always be performed in a container with a lid, such as a plastic bucket (see Chapter 17), and a face shield or goggles must be worn.

FIRE

◇ Particular fire risks associated with tissue culture stem from the use of bunsen burners for flaming, together with alcohol for swabbing or sterilization. Keep the two separate; always ensure that alcohol for sterilizing instruments is kept in a narrow-necked bottle or flask that is not easily upset, and with the minimum

Fig. 6.2. Cylinder clamp. Clamps onto edge of bench and secures gas cylinder with fabric strap. Fits different sizes of cylinder and can be moved from one position to another if necessary. Available from most laboratory suppliers.

volume of alcohol (Fig. 6.3). Alcohol for swabbing should be kept in a plastic wash bottle. When instruments are sterilized in alcohol and the alcohol is subsequently burnt off, care must be taken not to return the instruments to the alcohol while they are still alight.

RADIATION

◇ There are three main types of radiation hazard from tissue culture–associated procedures. The first is from ingestion of radiolabeled compounds. Tritiated nucleotides, if accidentally ingested, will become incorporated into DNA, and, due to the short path length of the low-energy β-emission from ^{3}H, will cause radiolysis within the DNA. Radioactive isotopes of iodine will concentrate in the thyroid and may also cause local damage.

◇ The second type of risk is from irradiation from higher energy β and γ emitters such as ^{32}P, ^{125}I, ^{131}I, and ^{51}Cr. Protection can be obtained by working behind a 2-mm-thick lead shield and storing the concentrated isotope in a lead pot. Perspex screens (5 mm) can be used with ^{32}P at low concentrations for short periods.

◇ In both cases, work in a biohazard cabinet to contain aerosols and wear gloves. The items that you are working with should be held in a shallow tray lined with paper tissue or Benchcote to contain any accidental spillage. Clean up carefully when finished and monitor regularly for any spillage.

◇ The third type of irradiation risk is from x-ray machines, high-energy sources such as ^{60}Co, or UV sources used for sterilizing apparatus or stopping cell proliferation in feeder layers (see Chapters 11 and 23). Since the energy, particularly from x-rays or ^{60}Co, is high, these sources are usually located in specially designed accommodation and subject to strict control. UV sources can cause burns to the skin and can damage the eyes. They should be carefully screened to prevent direct irradiation of the operator and barrier filter goggles should be worn.

Consult your local radiological officer and code of practice before embarking on radioisotopic experiments. Local rules vary but most places have strict controls on the amount of radioisotopes that can be used, stored, and discarded. The advice given above is gener-

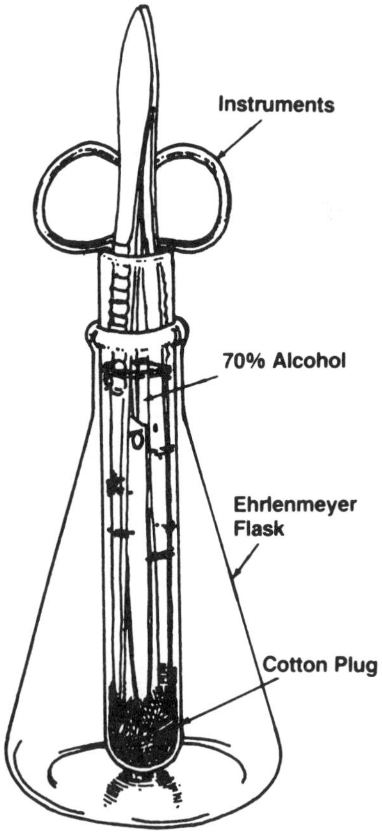

Fig. 6.3. Flask for alcohol sterilization of instruments. Wide base prevents tipping and center tube reduces the amount of alcohol required so that spillage, if it occurs, is minimized. (From an original idea by M.G. Freshney.)

al and should not be construed as satisfying any legal requirement.

BIOHAZARDS

The need for protection against biological hazards [see Barkley, 1979] is defined (1) by the source of the material and (2) by the nature of the operation being carried out. It is also governed by the conditions under which culture is performed. Standard microbiological technique on the open bench has the advantage that the techniques in current use have been established as a result of many years of accumulated experience. Problems arise when new techniques are introduced or when the number of people sharing the same area increases. With the introduction of horizontal laminar flow cabinets, the sterility of the culture was protected more effectively, but the exposure of the operator to aerosols was increased. This led to the development of vertical laminar hoods with an air curtain at the front (see Chapters 4 and 5) to minimize overspill from within the cabinet.

Levels of Containment

We can define three levels of handling:

(1) Sealed pathogen cabinet with filtered air entering and leaving via a pathogen trap filter (Class III). This will generally be housed in a separate room with restricted access and with showering facilities and protection for solid and liquid waste, depending on the nature of the hazard.

(2) Vertical laminar flow cabinet with front protection in the form of an air curtain and a filtered exhaust (Class II) (Fig. 6.4, National Sanitation Foundation Standard 49, NIH specification NIH 03-112, British Standard BS5726). If recognized pathogens are being handled, these cabinets will be housed either in separate rooms, at containment levels II, III, or IV, depending on the nature of the pathogen. If there is no reason to suppose that the material is infected, other than by adventitious infections, then these cabinets will generally be housed in the main tissue culture accommodation, which may be categorized as containment level I, with restricted access, control of waste disposal, protective clothing worn, and no food or drink in the area.

◇ All biohazard cabinets must be subject to a strict maintenance program, the filters tested at regular intervals, proper arrangements made for fumigation of the cabinets before changing filters, and disposal of old filters made safe (see Chapter 4).

(3) Open bench, depending on good microbiological technique. Again this will normally be conducted in a specially defined area, which may simply be defined as the "tissue culture laboratory," but which will have the level I conditions applied to it.

Table 6.3 lists common procedures with suggested levels of containment. These are suggestions only, however, and you should seek the advice of your local safety committee and the appropriate biohazard guidelines (see above) for legal requirements.

Human Biopsy Material

◇ There is often less doubt when known classified pathogens are being used, since the regulations were laid down by the Howie Report [1978], the Advisory Committee on Dangerous Pathogens (ACDP) [1984] (U.K.), and the National Institutes of Health (NIH) and the Centers for Disease Control (U.S.), and are now summarized in the documents referred to above [Biosafety in Microbiological and Biomedical Laboratories, 1984], or when harmless, sterile solutions are being prepared. It is the "gray area" in the middle that causes concern, as culture of primate cell lines [Wells et al., 1989] and the development of new techniques such as interspecific cell hybridization and facilities such as laminar flow introduce putative risks for which there are no epidemiological data available for assessment. Transforming viruses, amphitropic viruses, transformed human cell lines, human-mouse hybrids, and cell lines derived from xenografts in immune-deficient mice, for example, should be treated cautiously until data accumulates that they carry no risk.

◇ Risks that are more easily recognized are those associated with biopsy and autopsy specimens from human and primate tissue. When infection has been confirmed, the type of organism will determine the degree of containment, but where there is no known infection, the possibility remains that the sample may yet carry hepatitis B, human immunodeficiency virus (HIV), tuberculosis, or other pathogens as yet undiagnosed. Confidentiality frequently prevents HIV testing without the patient's consent, and, for most of the adventitious infections, the appropriate information will not be available. If possible, biopsy material should be tested for potential adventitious infections before handling, but the need to get samples into culture quickly will often mean that you must proceed without this information.

Such samples should be handled with caution:

(1) Transport specimens in a double-wrapped container, e.g., a universal container or screw-cap vial within a second screw-top vessel, such as a polypropylene sample jar. This in turn should be enclosed in an opaque, plastic or waterproof paper envelope and transported by a nominated carrier to the lab.

(2) Enter all specimens into a log book on receipt, and place in a secure refrigerator marked with a biohazard label.

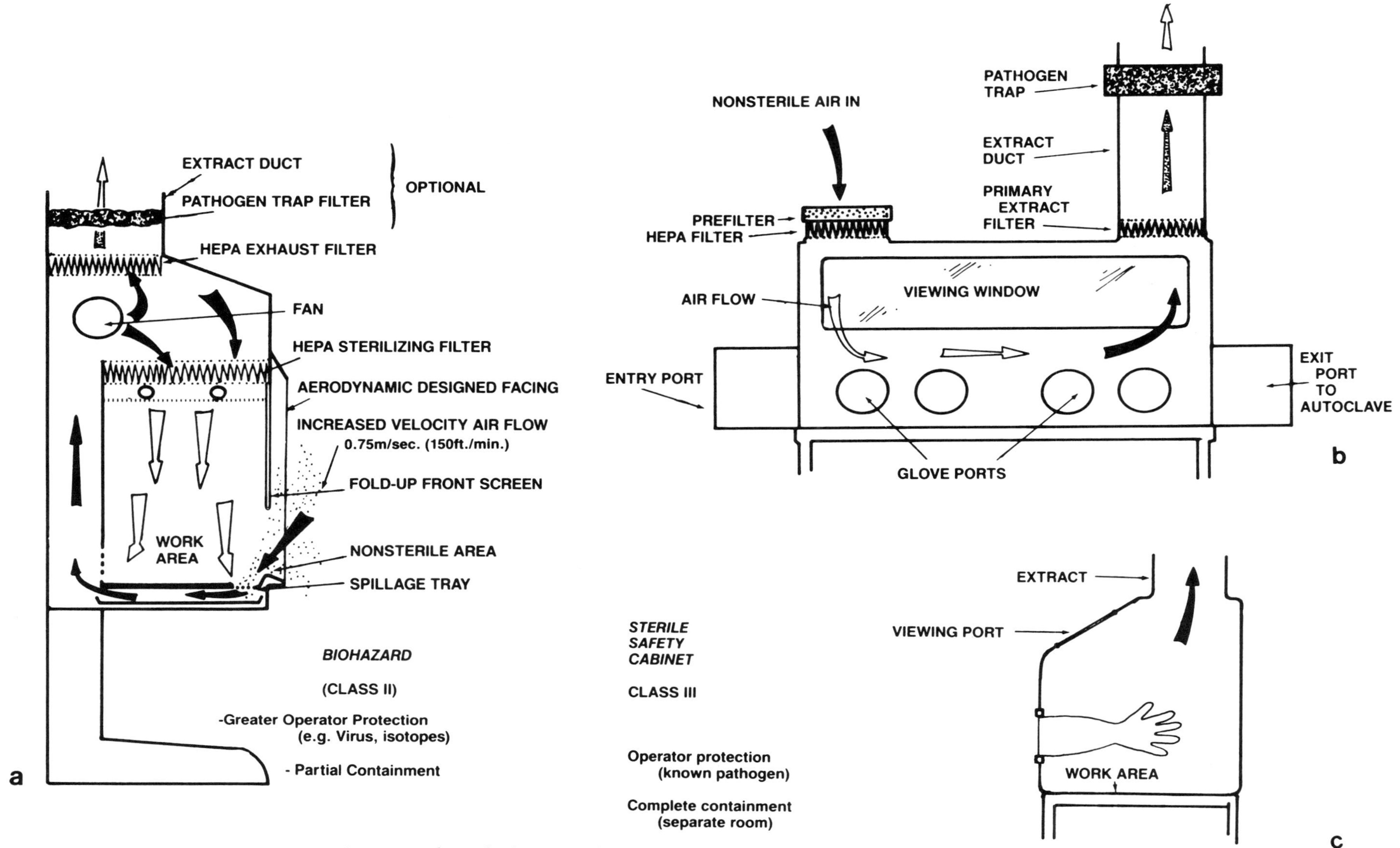

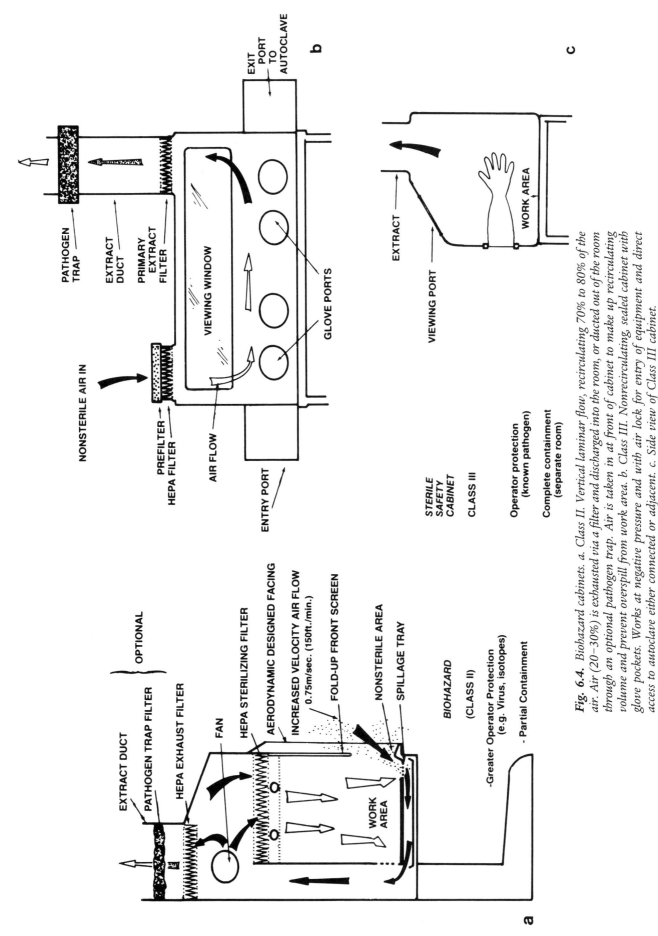

Fig. 6.4. *Biohazard cabinets. a. Class II. Vertical laminar flow, recirculating 70% to 80% of the air. Air (20–30%) is exhausted via a filter and discharged into the room, or ducted out of the room through an optional pathogen trap. Air is taken in at front of cabinet to make up recirculating volume and prevent overspill from work area. b. Class III. Nonrecirculating, sealed cabinet with glove pockets. Works at negative pressure and with air lock for entry of equipment and direct access to autoclave either connected or adjacent. c. Side view of Class III cabinet.*

TABLE 6.3. Biohazard Procedures and Suggested Levels of Containment

Procedure	Level of protection
Media preparation	Open bench, standard microbiological practice, or horizontal laminar flow
Cell lines other than human and other primates	Open bench, standard microbiological practice, or horizontal or vertical laminar flow
Primary culture and serial passage of human, and other primate cells	Vertical laminar flow cabinet with air curtain and protection at front and filtered extract (Class II)
Interspecific hybrids or other recombinants, transfected cells, human cells and animal tumor cells	Vertical laminar flow cabinet with air curtain between protection at front and filtered extract (Class II)
Human cells infected with retroviral constructs	Vertical laminar flow cabinet with air curtain between protection at front and filtered extract (Class II). Separate room with separate facilities, incubator, cell counting, centrifuge, etc., separate autoclaved or incinerated waste or chemical decontamination
Virus-producing human cell lines Cell lines infected with amphitrophic virus	Pathogen cabinets with glove pockets, filtered air entering and pathogen trap on vented air (Class III) Located in a separate room with separate provision for incubation, centrifugation, cell counting, etc. No access except to designated personnel All waste, soiled glassware, etc., to be sterilized and extracted air to be filtered as it leaves the room
Tissue samples and cultures carrying known human pathogens	Pathogen cabinet with glove pockets, filtered air, and pathogen trap on vented air (Class III). Located in separate room with separate provision for incubation, centrifugation, cell counting, etc. No access except to designated personnel. All waste, soiled glassware, etc., to be sterilized as it leaves the room and extracted air to be filtered. Shower facilities and change of clothing on entering and leaving.

(3) All dissection and subsequent culture work should be carried out in a designated Class II biohazard cabinet, preferably located in a separate room. This will minimize the risk of spreading infections, such as mycoplasma, to other cultures, but will also reduce the number of people associated with the specimen, should it eventually be found to be infected.

(4) No sharp instruments, e.g., syringes, scalpels, Pasteur pipettes, should be used in handling. Clearly, this rule may need to be compromised when a dissection is required, but this should proceed with extra caution.

(5) Discard all glassware, pipettes, and instruments, etc., into disinfectant or autoclave.

If the material can be shown to be uninfected by the appropriate clinical diagnostic tests, and when it has been shown to be sterile and mycoplasma-free, it may then be cultured with other stocks. However, if $>10^9$ cells are to be generated, or if pure DNA is to be prepared, the advice of the local safety committee should be sought.

If a specimen is found to be infected, it should be discarded into a biohazard bag with all instruments and reagents used with it, double wrapped, and autoclaved or incinerated. If it is necessary to carry on working with the material, the level of containment must increase, according to the category of the pathogen (see CDC/NIH or ACDP regulations).

Disposal

◇ Potentially biohazardous materials must be sterilized before disposal. They may be placed in autoclavable sacks (unsealed) and autoclaved, or immersed in a sterilizing agent such as hypochlorite or glutaraldehyde. Various proprietary preparations are available, e.g., Clorox (Biomedical Products; Polysciences), Chloros liquid concentrate (I.C.I.), or Precept Tablets (Johnson and Johnson). Recommended concentrations vary according to local rules, but a rough guide can be obtained from the manufacturer's instructions. Hypochlorite is often used at 300 ppm available chlorine but some authorities demand 2,500 ppm, as recommended in the Howie Report [1978]. Hypochlorite is effective and easily washed off items to be reused but is highly corrosive, particularly in alkaline solutions. It will bleach clothing and even corrodes stainless steel, so gloves and a lab coat or apron should be worn when handling hypochlorite, and soaking baths and cylinders should be made of polypropylene.

There is also evidence accumulating that glutaraldehyde may generate allergic reactions in some people, so it should be handled with care, concentrated solutions handled in a fume cupboard, and pouring carried out slowly to avoid excessive generation of aerosols.

CHAPTER 7

The Culture Environment: Substrate, Gas Phase, Medium, and Temperature

The influence of the environment on the culture is expressed via four routes: (1) the nature of the substrate or phase on or in which the cells grow—this may be solid, as in monolayer growth on plastic, semisolid, as in a gel such as collagen or agar, or liquid, as in suspension culture; (2) the physicochemical and physiological constitution of the medium; (3) the constitution of the gas phase; and (4) the incubation temperature. It is perhaps useful to think of the four elements of the ancient alchemists in remembering these routes: "air"—the gas phase, "earth"—the substrate, "fire"—the temperature, and "water"—the medium.

THE SUBSTRATE

The majority of vertebrate cells cultured *in vitro* have been grown as monolayers on an artificial substrate. Spontaneous growth in suspension is restricted to hemopoietic cell lines, rodent ascites tumors, and a few other selected cell lines, such as human small cell lung cancer [Carney et al., 1981]. From the earliest attempts, glass has been used as the substrate, initially because of its optical properties, but subsequently because it appears to carry the correct charge for cells to attach and grow. With the exception of the above-mentioned cells and other transformed cell lines, most cells need to spread out on a substrate in order to proliferate [Folkman and Moscona, 1978; Ireland et al., 1989; Shiba and Kanno, 1989]. Inadequate spreading due to poor adhesion or overcrowding will inhibit proliferation of most normal cells. Cells shown to require attachment for growth are said to be "anchorage-dependent" (see also Chapter 15). Cells that have undergone transformation frequently become anchorage-independent and can grow in suspension when stirred or held in suspension with semisolid media such as agar.

This assumes, however, that cell proliferation is the principal objective. It may not be; cells that are anchored only to each other as spheroids in suspension or that are growing as a secondary layer on top of a confluent monolayer may proliferate more slowly but may still reflect, more accurately, behavior *in vivo*.

Glass

Glass is now rarely used as a substrate although it is cheap, easily washed without losing its growth-supporting properties, can be sterilized readily by dry or moist heat, and is optically clear. Treatment with strong alkali (e.g., NaOH or caustic detergents) renders glass unsatisfactory for culture until it is neutralized by an acid wash (see Chapter 8).

Disposable Plastic

Single-use polystyrene flasks provide a simple, reproducible substrate for culture. They are usually of good optical quality and the growth surface is flat, providing uniform and reproducible cultures. Polystyrene, as manufactured, is hydrophobic and does not provide a suitable surface for cell growth, so tissue culture plastics are treated by γ-irradiation, chemically, or with an electric ion discharge to produce a charged surface that is then wettable. As the resulting product varies in quality from one manufacturer to another, samples from a number of sources should be tested by determining the plating efficiency and growth rate of your cells (see Chapters 11 and 19) in medium containing the optimal and half-optimal concentration of serum (high serum concentrations may mask imperfections in the plastic).

While polystyrene is by far the most common and cheapest plastic substrate, cells may also be grown on polyvinylchloride (PVC), polycarbonate, polytetrafluorethylene (PTFE), melinex, thermanox (TPX), and a number of other plastics. If you need to use a different plastic, it is worth trying to grow a regular monolayer and then attempting to clone cells on it (see Chapters 11 and 19), with and without pretreatment of the surface (see below). Teflon (PTFE) is available in a charged (hydrophilic) and uncharged (hydrophobic) form; the charged form can be used for regular monolayer cells and the uncharged for macrophages and some transformed cell lines. PTFE films are available as disposable petri dishes ("Petriperm," Heraeus), or as membranes to be incorporated in an autoclavable reusable culture vessel ("Chamber/Dish," Bionique) (Fig. 13.2). These dishes have two other advantages: (1) the substrate is permeable to O_2 and CO_2 and (2) the plastic is thin and, therefore, well suited to histological sectioning for light or electron microscopy.

Permeable substrates have been in use for many years. In 1965, Sandström suggested that hepatocytes survived better in the higher oxygen tension provided by growth in a cellophane sandwich. Growth of cells on floating collagen [Michalopoulos and Pitot, 1975; Lillie et al., 1980] and cellulose nitrate membranes [Savage and Bonney, 1978] have been used to improve the survival of epithelial cells and promote terminal differentiation (see Chapters 14 and 20). There are now several manufacturers providing permeable supports in the form of disposable filter wells of many different sizes, materials, and membrane porosities (Costar, Falcon, Millipore, Nunc). These can be used as tissue culture-treated plastic, or coated with collagen (Costar), or can be coated before use with collagen, laminin, or other matrix material (e.g., Matrigel [Falcon]) as required.

It is possible that growth of cells on a permeable substrate contributes more than increased diffusion of oxygen, CO_2, and nutrients. Attachment to a natural substrate such as collagen may exert some biological control of phenotypic expression due to the interaction of receptor sites on the cell surface with specific sites in the extracellular matrix (see Chapters 2 and 14). Permeability of the surface to which the cell is anchored may, in itself, signify polarity to the cell by simulating the basement membrane underlying an epithelial cell layer or between tissue cells and endothelium surrounding the vascular space. Such polarity may be vital to full functional expression in secretory epithelia and many other cell types. This prompted Reid and Rojkind [1979], Gospodarowicz [Vlodavsky et al., 1980], and others to explore the growth of cells on natural substrates related to basement membrane (see below and Chapter 14), and to determine the functional role of polarity [Chambard et al., 1987]. Synthetic matrices and defined matrix macromolecules are now available for controlled studies on matrix interaction (Matrigel, Natrigel, Collagen, laminin, vitronection [Becton Dickinson, Gibco/Life Technologies]) (see below).

Microcarriers

Polystyrene (Nunclon, GIBCO), Sephadex (Flow Laboratories and Pharmacia), polyacrylamide (Biorad), and collagen (Pharmacia) or gelatin (Ventrex) are available in bead form for propagation of anchorage-dependent cells in suspension (see Chapter 23).

Sterilization of Plastics

Disposable plasticware is usually supplied sterile and cannot be reused, as washing in detergents renders the surface unsuitable for monolayer culture. For the sterilization of other plastics, see Chapter 8.

Alternative Artificial Substrates

Although glass and plastic are employed for more than 90% of all cell propagation, there are alternative substrates that can be used for specialized applications. Westermark [1978] developed a method for the growth of fibroblasts and glia on palladium. Using electron microscopy shadowing equipment, he produced islands of palladium on agarose, which does not allow cell attachment in fluid media. The size and shape of the islands was determined by masks made in the manner of electronic printed circuits, and the palladium was applied by "shadowing" under vacuum, as used in electron microscopy.

Cells may be grown on stainless steel discs [Birnie and Simons, 1967] or other metallic surfaces [Litwin, 1973]. Observation of the cells on an opaque substrate requires surface interference microscopy, unless very thin metallic films are used, as with Westermark's palladium islands.

Treated Surfaces

Cell attachment and growth can be improved by pretreating the substrate in a variety of ways [Barnes et al., 1984a]. It is a well-established piece of tissue culture lore that used glassware supports growth better than new. This may be due to etching of the surface or minute traces of residue left after culture. Growth of cells in a flask also improves the surface for a second seeding, and this type of conditioning may be due to collagen [Hauschka and Konigsberg, 1966] or fibronectin or other matrix products [Crouch et al., 1987] released by the cells. The substrate can be conditioned by treatment with spent medium for another culture [Stampfer et al., 1980], or by purified fibronectin (1 ng/ml) added to the medium [Gilchrest et al., 1980] or collagen [Elsdale and Bard, 1972; Kibbey et al., 1992; Kinsella et al., 1992]. Treatment with denatured collagen improves the attachment of many cells such as epithelial cells [Freeman et al., 1976; Lillie et al., 1980] and muscle cells [Hauschka and Konigsberg, 1966], and it may be necessary for the expression of differentiated functions by these cells (see Chapters 14 and 20).

Coating with denatured collagen may be achieved using rat-tail collagen or commercially supplied alternatives (Vitrogen, Flow) and simply pouring the collagen solution over the surface of the dish, draining off the excess, and allowing the residue to dry. As this sometimes leads to detachment of the collagen layer during culture, a protocol was devised by Macklis et al. [1985] to ensure the collagen remains firmly anchored to the substrate, by cross-linking to the plastic with carbodiimide. This is presented in detail in Chapter 14.

Collagen may also be applied as an undenatured gel (see Chapter 14), and this type of substrate has been shown to support neurite outgrowth from chick spinal ganglia [Ebendal, 1976], morphological differentiation of breast [Yang et al., 1981] and other epithelia [Sattler et al., 1978], and to promote expression of tissue-specific functions of a number of other cells *in vitro* [Meier and Hay, 1974, 1975; Kosher and Church, 1975]. In this case the collagen is diluted 1:10 with culture medium and neutralized to pH 7.4. This causes the collagen to gel, so the dilution and dispensing must be rapid. It is best to add the growth medium to the gel for a further 4–24 hr to ensure the gel equilibrates with the medium before adding cells. At this stage fibronectin (25–50 μg/ml) and/or laminin (1–5 μg/ml) may be added to the medium.

Evidence is gradually accumulating that specific treatment of the substrate with biologically significant compounds can induce specific alterations in attachment or behavior of specific cell types. For example, chondronectin enhances chondrocyte adherence [Varner et al., 1984] and laminin epithelial cells [Kleinman et al., 1981]. Reid and Rojkind [1979] described methods for preparing reconstituted "basement membrane rafts" from tissue extracts for optimization of culture conditions for cell differentiation.

Commercially available matrices, such as Matrigel (Becton Dickinson) from the Engelbreth Holm Swarm (EHS) sarcoma, contain laminin, fibronectin, and proteoglycans, with laminin predominating (see Chapter 14). Numerous studies have evaluated Matrigel in studies of differentiation and malignant invasion [Repesh, 1989; Schlechte et al., 1990]. Other matrix products include Pronectin F (Protein Polymer Technologies), laminin, fibronectin, entactin (UBI), heparan sulfate, EHS Natrix (Becton Dickinson), ECM (IBT), and Cell-tak (Becton Dickinson). Some of these are purified, if not completely chemically defined; others are a mixture of matrix products that have been poorly characterized and may also contain bound growth factors. If cell adhesion for cell survival is the main objective, and defined substrates are inadequate, the use of these matrices is acceptable, but if mechanistic studies are being carried out, they can only be an intermediate stage on the road to a completely defined substrate.

Gelatin coating has been found to be beneficial for the culture of muscle [Richler and Yaffe, 1970] and endothelial cells [Folkman et al., 1979] (see Chapter 20), and it is necessary for some mouse teratomas. McKeehan and Ham [1976] found that it was necessary to coat the surface of plastic dishes with 1 mg/ml poly-D-lysine before cloning in the absence of serum (see Chapter 11).

This raises the interesting question of whether the cell requires at least two components of interaction with the substrate: (1) adhesion to allow the attachment and spreading necessary for cell proliferation [Folkman and Moscona, 1978] and (2) specific interactions, reminiscent of the interaction of an epithelial cell with basement membrane, with other extracellular matrix constituents, or with adjacent tissue cells. The second type of interaction may be less critical to sustained proliferation of undifferentiated cells, but may be required for the expression of some specialized functions (see Chapters 2 and 14).

While inert coating of the surface may suffice, it may yet prove necessary to use a monolayer of an appropriate cell type to provide the correct matrix for maintenance of some specialized cells. Gospodarowicz et al. [1980] were able to grow endothelium on confluent monolayers of 3T3 cells that had been extracted with Triton X100, leaving a residue on the surface of the substrate. This so-called extracellular matrix (ECM) has also been used to promote differentiation in ovarian granulosa cells [Gospodarowicz et al., 1980] and in studying tumor cell behavior [Vlodavsky et al., 1980].

Outline

Remove a postconfluent monolayer of matrix-forming cells with detergent, wash, and seed required cells onto residual matrix so produced.

Materials

3T3 mouse fibroblasts
MRC-5 human fibroblasts
or CPAE bovine pulmonary arterial endothelial cells (or any other cell line shown to be suitable for producing extracellular matrix)
1% Triton X100 in sterile distilled deionized water
Sterile distilled deionized water

Protocol

1. Set up matrix-producing cultures, and grow to confluence.
2. After 3–5 days at confluence, remove the medium and add an equal volume of sterile 1% Triton X100 in distilled water to the cell monolayer.
3. Incubate for 30 min at 37°C.
4. Remove Triton X solution and wash residue three times with the same volume of sterile distilled water.
5. Flasks may be used directly or stored at 4°C for up to 3 weeks.

Feeder Layers

Cultures of mouse embryo fibroblasts, or other cells, have been used for many years to enhance growth, particularly at low cell densities (see Chapter 11) [Puck and Marcus, 1955]. This action is due partly to supplementation of the medium by leakage or secretion from the fibroblasts, but may also be due to conditioning of the substrate by cell products. Feeder layers grown as a confluent monolayer may make the surface suitable for attachment for other cells (see Chapters 20 and 21). We have shown selective growth of breast and colonic epithelium, and of glioma, on confluent feeder layers of normal fetal intestine [Freshney et al., 1982b].

The survival and neurite extension by central and peripheral neurons can be enhanced by culturing the neurons on a monolayer of glial cells, although in this case the effect is due to a diffusible factor rather than direct cell contact [Lindsay, 1979; Seifert and Müller, 1984].

After a monolayer culture reaches confluence, subsequent proliferation causes cells to detach from the artificial substrate and migrate over the surface of the monolayers. Their morphology may change (Fig. 7.1), and the cells are less well spread, more densely staining, and may be more highly differentiated. Apparently, and not too surprisingly, the interaction of a cell with a cellular underlay is different from the interaction with a synthetic substrate. This can cause the change in morphology and reduce proliferative potential.

Three-Dimensional Matrices

It has long been realized that while growth in two dimensions is a convenient way of preparing and observing a culture and allows a high rate of cell proliferation, it lacks the cell–cell and cell–matrix interaction characteristic of whole tissue *in vivo*. The very first attempts to culture animal tissues [Harrison, 1907; Carrel, 1912] were performed with gels formed of clotted lymph or plasma on glass. In these cases, however, the cells migrated along the glass/clot interface rather than within the gel, and tissue architecture and cell–cell interaction was gradually lost. Migration was often accompanied by proliferation of cells in the outgrowth, leading, in later studies, to the development of propagated cell lines. It gradually became apparent that many functional and morphological characteristics were lost during serial subculture, as discussed in Chapter 2.

These deficiencies encouraged the exploration of three-dimensional matrices such as collagen gel [Douglas et al., 1980]; cellulose sponge, alone [Leighton et al., 1968] or collagen-coated [Leighton et al., 1968]; or Gelfoam (see Chapter 22). Fibrin clots were one of the first media to be used for primary culture, and are still used either as crude plasma clots (see Chapter 9) or as purified fibrinogen mixed with thrombin. Both systems generate a three-dimensional gel in which cells may migrate and grow, either on the solid/gel interface or within the gel [Leighton, 1991].

Many different cell types can be shown to penetrate such matrices and establish a tissue-like histology. Breast epithelium, seeded within collagen gel, displays a tubular morphology, while breast carcinoma grows in a more disorganized fashion, confirming the correlation between this mode of growth and the condition *in vivo* [Yang et al., 1981]. The kidney epithelial cell line, MDCK, responds to paracrine stimulation from fibroblasts by producing tubular structures in collagen gel [Kenworthy et al., 1992].

Neurite outgrowth from sympathetic ganglia neurons growing on collagen gels follows the orientation of the collagen fibers in the gel [Ebendal, 1976] (see further discussion of three-dimensional cultures, Chapter 22).

Nonadhesive Substrates

There are situations where attachment of the cells is undesirable. The selection of virally transformed colonies, for example, can be achieved by plating cells in

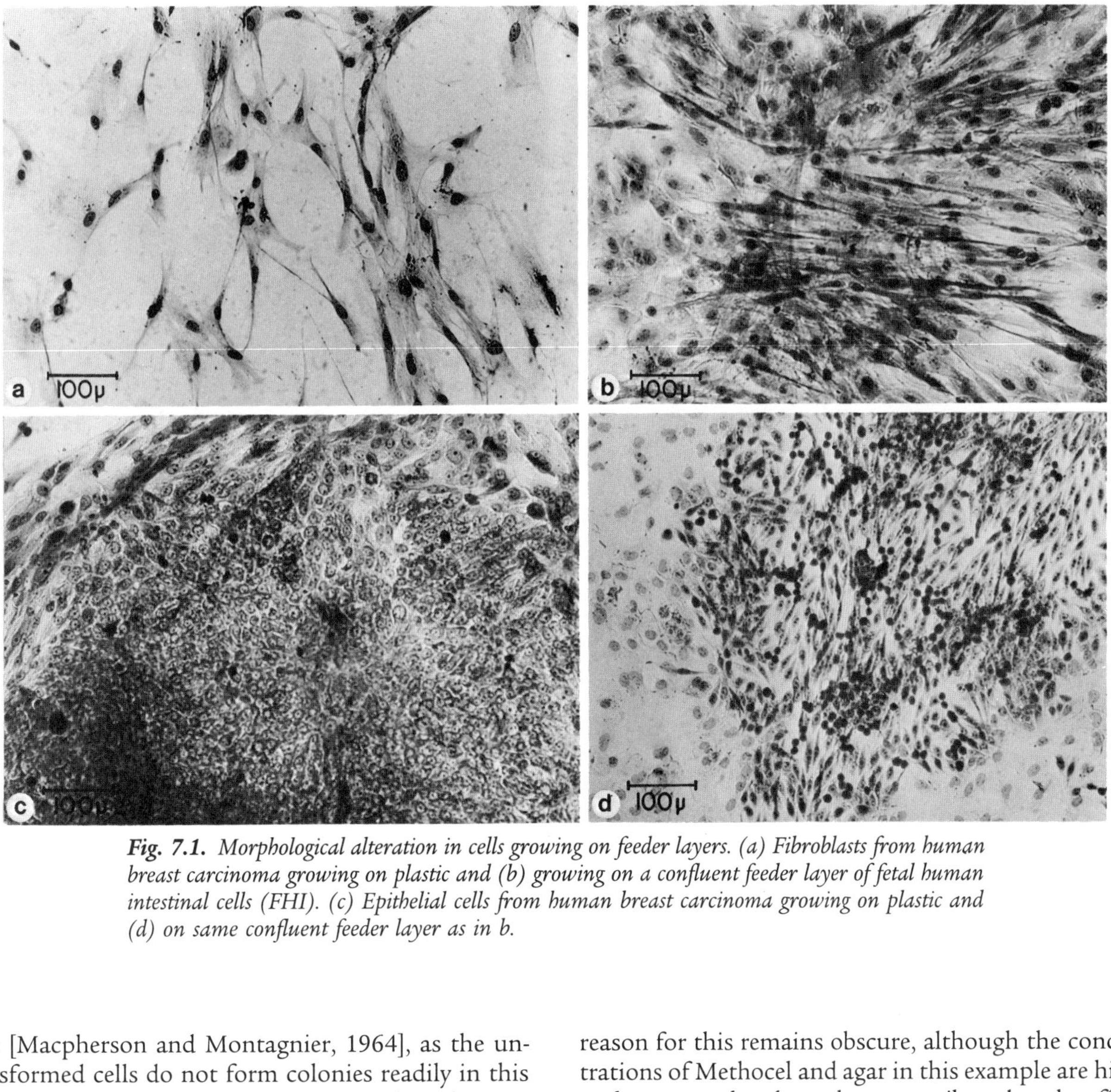

Fig. 7.1. Morphological alteration in cells growing on feeder layers. (a) Fibroblasts from human breast carcinoma growing on plastic and (b) growing on a confluent feeder layer of fetal human intestinal cells (FHI). (c) Epithelial cells from human breast carcinoma growing on plastic and (d) on same confluent feeder layer as in b.

agar [Macpherson and Montagnier, 1964], as the untransformed cells do not form colonies readily in this matrix. There are two principles involved in this system: (1) prevention of attachment at the base of the dish where spreading would occur and (2) immobilization of the cells such that daughter cells remain associated with the colony, even if nonadhesive. The usual agents employed are agar, agarose, or Methocel (methylcellulose viscosity 4,000 cps). The first two are gels and the third is a high-viscosity sol. Because Methocel is a sol, cells will sediment slowly through it. It is, therefore, commonly used with an underlay of agar (see Chapter 11). Nontissue culture-grade dishes can be used without an agar underlay, but some attachment and spreading may occur.

Liquid-Gel or Liquid-Liquid Interfaces

While the Methocel-over-agar system usually gives rise to discrete colonies at the interface of the agar and the Methocel, some cells can migrate across the gel surface and form monolayers or cords of cells (Fig. 7.2). The reason for this remains obscure, although the concentrations of Methocel and agar in this example are higher than normal and may have contributed to the effect. Cell spreading and monolayer formation at the liquid-liquid interface between various fluorinated hydrocarbons (FC43, FC73) and aqueous culture media have been observed [Nagaoka et al., 1990]. The occurrence of spreading and locomotion on nonrigid substrates conflicts somewhat with current concepts of cell adhesion and locomotion unless denatured serum protein or some other substance forms a layer at the interface sufficient to permit anchorage. Methocel, particularly, often contains particulate debris that may help to promote this.

Hollow Fibers

Knazek et al. [1972] developed a technique for the growth of cells on the outer surface of bundles of plastic microcapillaries (Fig. 7.3) (see Chapter 22). The plastic allows the diffusion of nutrients and dissolved gases from medium perfused through the capillaries.

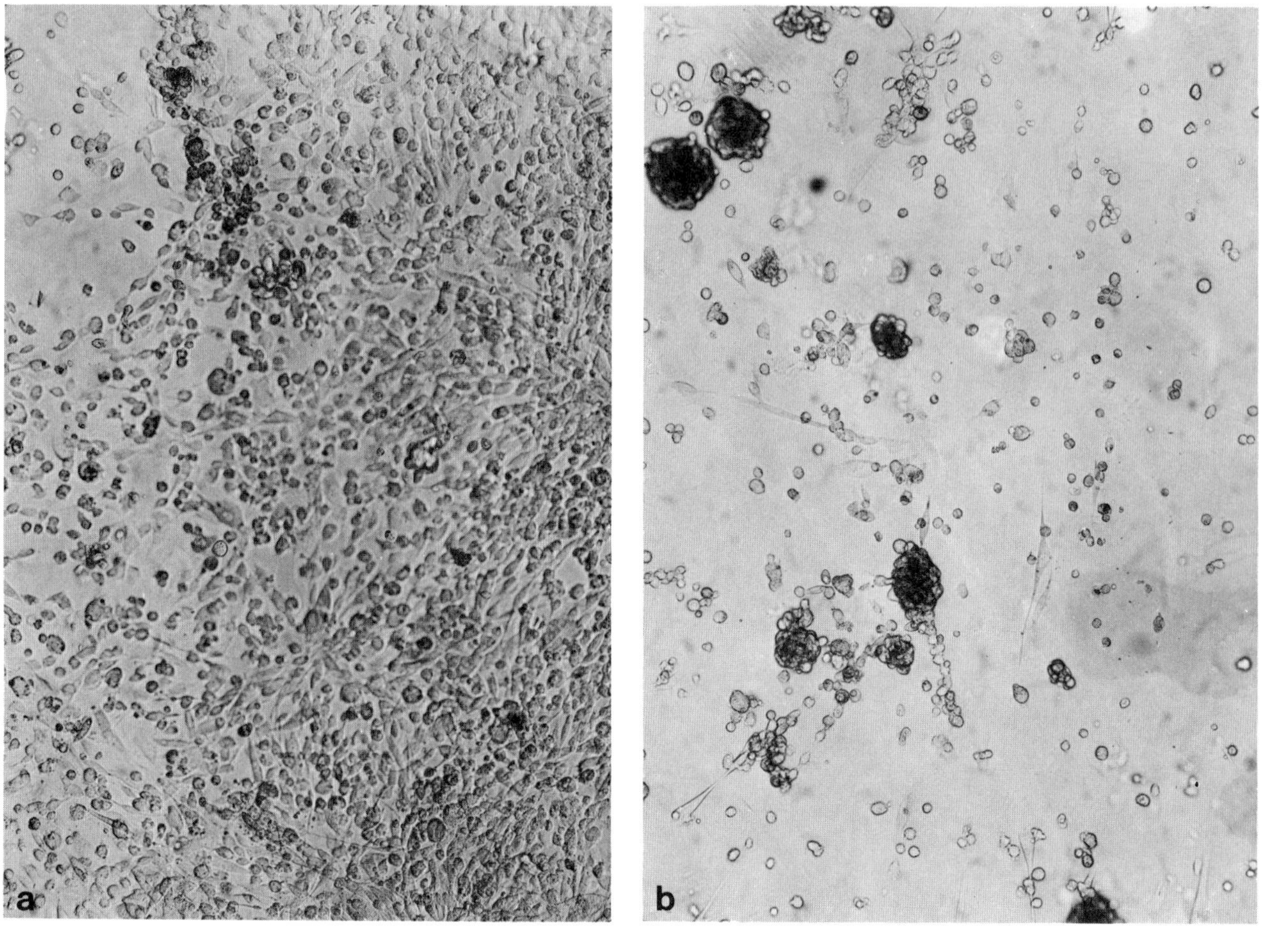

Fig. 7.2. Cell growth at interface between Methocel-containing medium and agar gel. Methocel concentration, 1.5%; agar, 1.25%. Human metastatic melanoma. In a, 2.5 × 10^5 cells/ml, cloned alone. In b, 5 × 10^4 cells/ml cloned with 2 × 10^5 homologous feeder cells/ml.

Cells will grow up to several cells deep on the outside of the capillaries, and an analogy with whole tissue is suggested.

Filter Well Inserts

Filter wells have also provided an environment for the generation of polarized, high-density cultures, with adequate medium supply and the potential for establishing histotypic cell interactions (see Chapter 22).

Culture Vessels

Some typical culture vessels are listed in Table 7.1. The anticipated yield of HeLa cells is quoted for each vessel; the yield from a finite cell line, e.g., diploid fibroblasts, would be about one-fifth of the HeLa figure. Several factors govern the choice of culture vessel, including: (1) the cell yield; (2) whether the cells grow in suspension or as a monolayer; (3) whether the culture should be vented to the atmosphere or sealed; (4) what form of sampling and analysis is to be performed; and (5) the anticipated cost.

Cell yield. For monolayer cultures, the cell yield is proportional to the available surface area of the flask. Small volumes and multiple replicates are best performed in multiwell dishes (Fig. 7.4), which range from Terasaki plates (60–72 wells, 10-μl culture volume) up to four wells, 50 mm in diameter, 5-ml culture volume. The most popular are microtitration dishes (96 or 144 wells, 0.1–0.2 ml, 0.25-cm^2 growth area) and 24-well "cluster dishes" (1–2 ml each well, 1.75 cm^2) (see Table 7.1). The middle of the size range embraces both petri dishes (Fig. 7.5) and flasks ranging from 25 cm^2 to 175 cm^2 (Fig. 7.6). Flasks are usually designated by their surface area, e.g., No. 25 or No. 175. Glass bottles are more variable since they are usually drawn from standard pharmaceutical supplies (Fig. 7.7). They should have: (1) one reasonably flat surface; (2) a deep screw cap with a good seal and nontoxic liner; and (3) shallow sloping shoulders to facilitate harvesting monolayer cells after trypsinization and to improve the efficiency of washing.

If you require large cell yields (e.g., ~10^9 HeLa cervi-

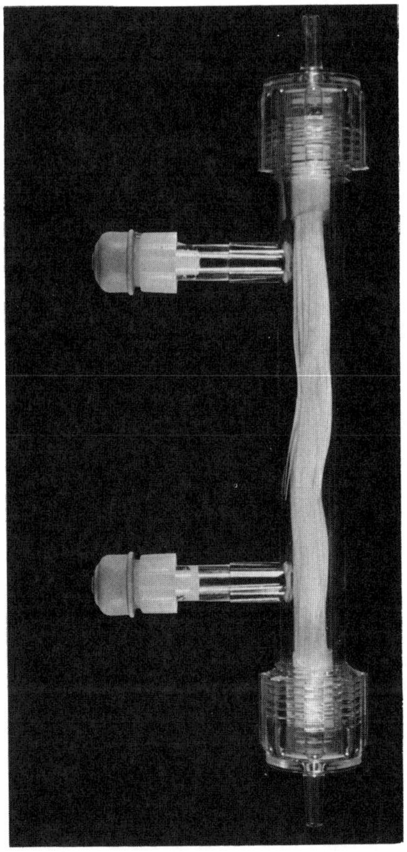

Fig. 7.3. Vitafiber Chamber (Amicon). A bundle of hollow fibers of permeable plastic is enclosed in a transparent plastic outer chamber, accessible via either of the two side arms for seeding cells. During culture, the chamber is perfused down the center of the hollow fibers through connections attached to either end of the chamber (see also Fig. 22.3).

cal carcinoma cells or 2×10^8 MCR-5 diploid human fibroblast), then increasing the size and number of conventional bottles becomes cumbersome and special vessels are required. Flasks with corrugated surfaces (Corning, Becton Dickinson) or multilayered flasks (Nunc) offer an intermediate step in increasing the surface area. Beyond that requires roller bottles on special racks, or multisurface propagators. These are described in Chapter 23. Increasing the yield of cells growing in suspension requires only that the medium volume be increased, as long as cells in deep culture are kept agitated and sparged with 5% CO_2 in air (see Chapter 23).

Venting. Multiwell dishes and petri dishes chosen for replicate sampling or cloning have loose-fitting lids to give easy access to the dish. Consequently, they are not sealed and will require a humid atmosphere with control of the CO_2 tension (see below and Chapter 4). Because a thin film of liquid may form around the inside of the lid, partially sealing some dishes, vented lids with molded plastic supports inside the lid should be used (Fig. 7.8a). If a perfect seal is required, some multiwell dishes can be sealed with self-adhesive Mylar film (Conway). Flasks may be vented by slackening the caps. Again, because of variable sealing due to liquid inside the cap, the cap must be slackened one full turn. Flasks are vented in this way to allow CO_2 to enter (in a CO_2 incubator) or to allow excess CO_2 to escape in excessive acid-producing cell lines. Caps are now available with permeable filters to allow equilibration with the gas phase (Fig. 7.8b).

Sampling and analysis. Multiwell plates are ideal for replicate culture if all samples are to be removed simultaneously and processed in the same way. If, on the other hand, samples need to be withdrawn at different time intervals and processed immediately, it may be preferable to use separate vessels (flasks, test tubes, etc.) (Fig. 7.9). Individual wells in microtitration plates can be sampled by cutting and removing only that part of the adhesive plate sealer overlying the wells to be sampled. Alternatively, microtitration plates are available with removable wells for individual processing, although you should ensure that these are tissue culture–treated if you wish to use adherent cells.

If processing of the sample involves extraction in acetone, toluene, ethyl acetate, or certain other organic solvents, then a problem will arise with polystyrene. Since this problem is often associated with histological procedures, Lux supplies Thermanox (TPX) plastic coverslips, suitable for histology, to fit into regular multiwell dishes (which need not be tissue culture grade). However, they are of poor optical quality and should be mounted cells uppermost with a conventional glass coverslip on top.

Glass vessels are required for procedures such as hot perchloric acid extractions of DNA. Plain-sided test tubes or Erlenmeyer flasks (no lip) used in conjunction with sealing tape or Oxoid caps are quick to use and are best kept in a humid CO_2-controlled atmosphere. Regular glass scintillation vials, or "minivials," are also good culture vessels, as they are flat-bottomed and have a screw closure. Once used with scintillant, however, they should not be reused for culture.

Cost. Cost always has to be balanced against convenience—e.g., petri dishes are always cheaper than flasks of an equivalent surface area but require humid CO_2-controlled conditions and are more prone to infection. Cheap soda glass bottles, though not always of good optical quality, are often better for culture than higher grade Pyrex or optically clear glass, which usually contains lead.

A major disadvantage of glass is that it is labor-

TABLE 7.1. Culture Vessel Characteristics[a]

Culture vessel	Plastic or glass	No. of replicates	Vol. (ml)	Surface area	Approx. cell yield	Supplier
Microtest (Terasaki)	P	60, 72	0.01	0.78 mm^2	2.5×10^3	F, N
Microtitration plate	P	96, 144	0.1	32 mm^2	10^5	C, F, N, L
Multiwell plate[b]	P	4 round	1.0	2 cm^2	5×10^5	C, F, N
Multiwell plate	P	12 round	2.0	4.5 cm^2	10^6	F
Multiwell plate	P	24 round	1.0	2 cm^2	5×10^5	C, F, L
Multiwell plate	P	8 rectangular	2.0	7.8 cm^2	2×10^6	Lu
Multiwell plate	P	4 rectangular	3.0	16.1 cm^2	4×10^6	Lu
Multiwell plate	P	6 round	2.5	9.6 cm^2	2.5×10^6	C, L
Multiwell plate	P	4 round	5.0	28 cm^2	7×10^6	F, N, L
Petri dishes[b]						
30 mm	P		2.0	6.9 cm^2	1.7×10^6	S
35 mm	P		3.0	8.0 cm^2	2.0×10^6	C, Cg
50 mm	P		4.0	17.5 cm^2	4.4×10^6	S, F, N
60 mm	P		5.0	21 cm^2	5.2×10^6	C, Cg, S, N
90 mm	P		10.0	49 cm^2	12.2×10^6	S, F
100 mm	P		10.0	55 cm^2	13.7×10^6	C, Cg, F, N
100 mm, square	P		15.0	100 cm^2	20×10^6	S
Tissue culture tubes						
Leighton[b]	P & G		1.0	4.00 cm^2	10^6	Be
One side flattened	P		2.0	5.50 cm^2	10^6	N
Round with screw cap	P		2.0			N
Flasks						
25	P		5.0	25 cm^2	5×10^6	C, Cg, F, N
50	G		10–20	50 cm^2	10^7	
75	P & G		15–30	75 cm^2	2×10^7	C, Cg, F, N
120	G		40–100	120 cm^2	5×10^7	
150	P		75	150 cm^2	6×10^7	C, Cg
175	P		50–100	175 cm^2	7×10^7	F, N
500, triple layer	P		150–300	500 cm^2	2×10^8	N
175, corrugated	P		50–250	350 cm^2	1.5×10^8	Cg, F
Roller bottles						
2,500 ml	G		100–250	700 cm^2	2.5×10^8 (~1 g)	NB
Roller disposable	P		100–250	850 cm^2	3.0×10^8	F, Cg
Large	G		100–500	1,585 cm^2	6.0×10^8	NB
Nunc cell factory	P		1,800	6,000 cm^2	2.0×10^9	N
Microcarriers	See Chapter 21			See Stirrer bottles, below		Ph, B, N, F
Stirrer bottles						
Reagent bottle, round (500 ml)	G		200		3×10^8	Cg, Be
Reagent bottle, round (1,000 ml)	G		400		8×10^8	Cg, Be
Flask (2,000 ml)	G		600		10^9	T, Be
Flask (5,000 ml)	G		4,000	Gas with 5% CO_2	6×10^9	T, Be
Flask (10,000 ml)	G		8,000	Gas with 5% CO_2	8×10^9	T, Be

[a]Abbreviations: B, Bio Rad; Be, Bellco; Cg, Corning; C, Costar; F, Falcon; L, Linbro (Flow); Lu, Lux; N, Nunc (GIBCO); NB, New Brunswick; P, Pyrex; Ph, Pharmacia; S, Sterilin; T, Techne.

[b]Dishes and Leighton tubes can be used on their own or with a coverslip, e.g., glass, TPX (Lux), Polystyrene (Lux), Melinex (I.C.I.). Nontissue culture-grade dishes may be used with coverslips. Petri-dish sizes often refer to the outside diameter of the base or lid. Surface area must be calculated from the inside diameter of the base.

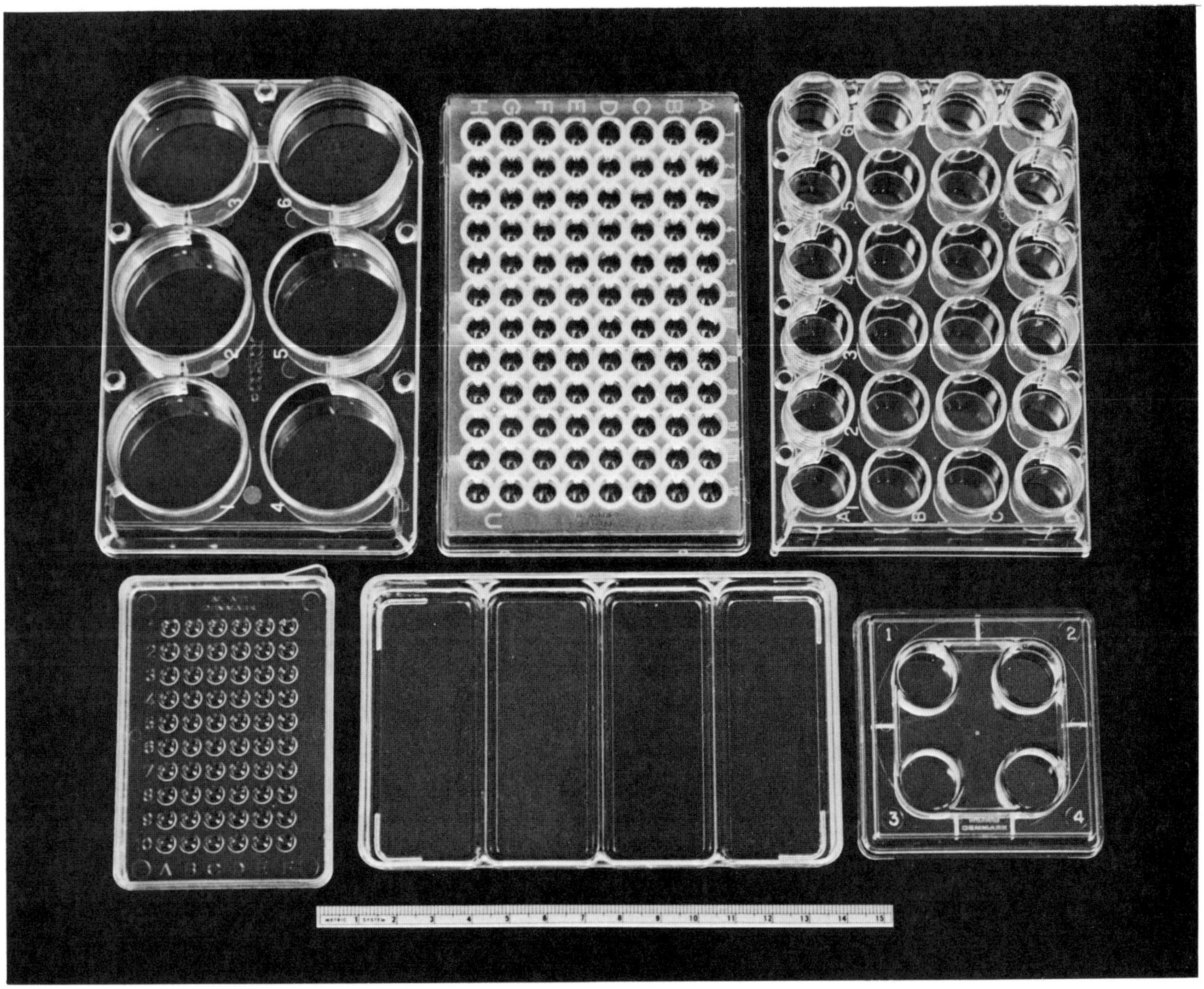

Fig. 7.4. Multiwell plates (see Table 7.1 for sizes and capacities).

intensive in preparation, as it must be carefully washed and resterilized before it can be reused. The cost of this will depend on your existing staff and the number of flasks used. To employ a new member of staff to wash and sterilize glassware will cost about half the amount per flask, relative to disposable plastic, for an annual output of 10,000 flasks. If your usage is substantially less than this, it will be better to use disposable plastic, particularly for smaller flasks (25 cm^2). If you do not need to meet the cost of employing washing-up staff, glass will be found to be much cheaper than plastic. Because of the convenience, optical clarity, and quality assurance, most laboratories now tend to use plastic.

THE GAS PHASE

Oxygen

The significant constituents of the gas phase are oxygen and carbon dioxide. Cultures vary in their oxygen requirement, the major distinction lying between organ and cell cultures. While atmospheric, or lower, oxygen tensions [Cooper et al., 1958; Balin et al., 1976] are preferable for most cell cultures, some organ cultures, particularly from late-stage embryo, newborn, or adult, require up to 95% O_2 in the gas phase [Trowell, 1959; De Ridder and Mareel, 1978]. This may be a problem of diffusion related to the geometry of organ cultures (see Chapter 22) rather than a distinct cellular requirement, since most dispersed cells prefer lower oxygen tensions, and some systems, e.g., human tumor cells in clonogenic assay [Courtenay et al., 1978] and human embryonic lung fibroblasts [Balin et al., 1976], do better in less than the normal atmospheric oxygen tension. It has been suggested [McKeehan et al., 1976] that the requirement for selenium in medium is related to oxygen tension and that this element helps to remove free radicals of oxygen. This requirement may only arise in the absence of serum proteins and, as

Fig. 7.5. Some common sizes of disposable plastic petri dishes. Sizes range from 35 mm to 90 mm diameter, circular, and 9 × 9 cm, square. Larger dishes are available but are seldom used for cell culture. A grid pattern can be provided to help in scanning the dish, to count colonies, for example.

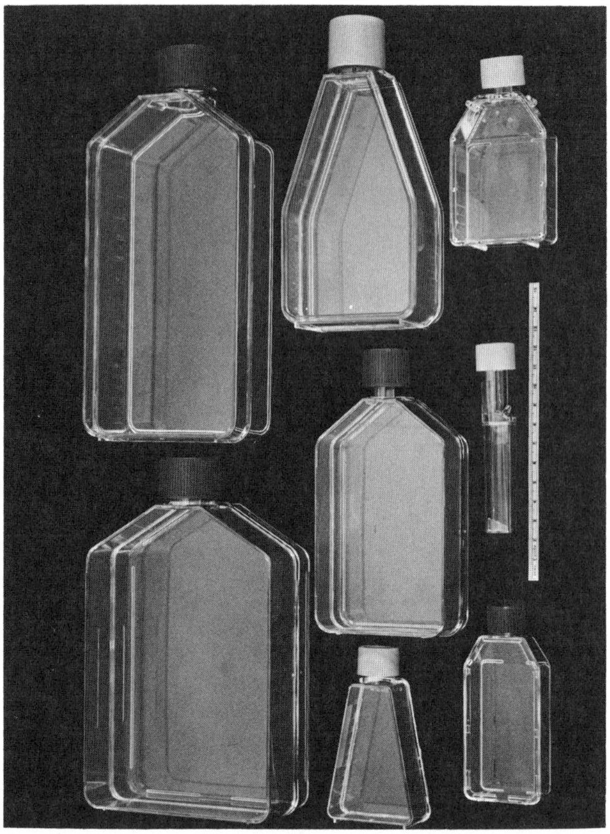

Fig. 7.6. Disposable plastic culture vessels (Falcon, Costar, and Corning). The triangular bottles (Costar) are designed to improve access to all of the growth surface when dispersing a monolayer (see Table 7.1 for sizes and capacities).

there is a trend toward serum-free media, the role of dissolved O_2 may become more important in the future, requiring, perhaps, controlled O_2 tension during incubation. As the depth of the culture medium can influence the rate of oxygen diffusion to the cells, it is advisable to keep the depth of medium within the range 2–5 mm (0.2–0.5 ml/cm^2) in static culture.

Carbon Dioxide

Carbon dioxide has a rather complex role to play, and because many of its actions are interrelated, e.g., dissolved CO_2, pH, and HCO_3^- concentration, it is difficult to determine its major direct effect. The atmospheric CO_2 tension will regulate the concentration of dissolved CO_2 directly, as a function of temperature. This in turn produces H_2CO_3, which dissociates:

$$H_2O + CO_2 \rightleftharpoons H_2CO_3 \rightleftharpoons H^+ + HCO_3^- \quad (1)$$

As HCO_3^- has a fairly low dissociation constant with most of the available cations, it tends to reassociate, leaving the medium acid. The net result of increasing atmospheric CO_2 is to depress the pH, so the effect of elevated CO_2 tension is neutralized by increasing the bicarbonate concentration:

$$NaHCO_3 \rightleftharpoons Na^+ + HCO_3^- \quad (2)$$

The increased HCO_3^- concentration pushes equation 1 to the left until equilibrium is reached at pH 7.4. If another alkali, e.g., NaOH, is used instead, the net result is the same.

$$NaOH + H_2CO_3 \rightleftharpoons NaHCO_3 + H_2O \rightleftharpoons Na^+ + HC0_3^- + H_2O \quad (3)$$

The equivalent $NaHCO_3$ concentrations commonly used with different CO_2 tensions are listed in Tables 7.2 and 7.4. Intermediate values of CO_2 and HCO_3^- may be used, provided the concentration of both is varied simultaneously. As many media are made up in acid solution and may incorporate a buffer, it is difficult to predict how much bicarbonate to use when other alkali may also end up as bicarbonate, as in equation 3 above. When preparing a new medium for the

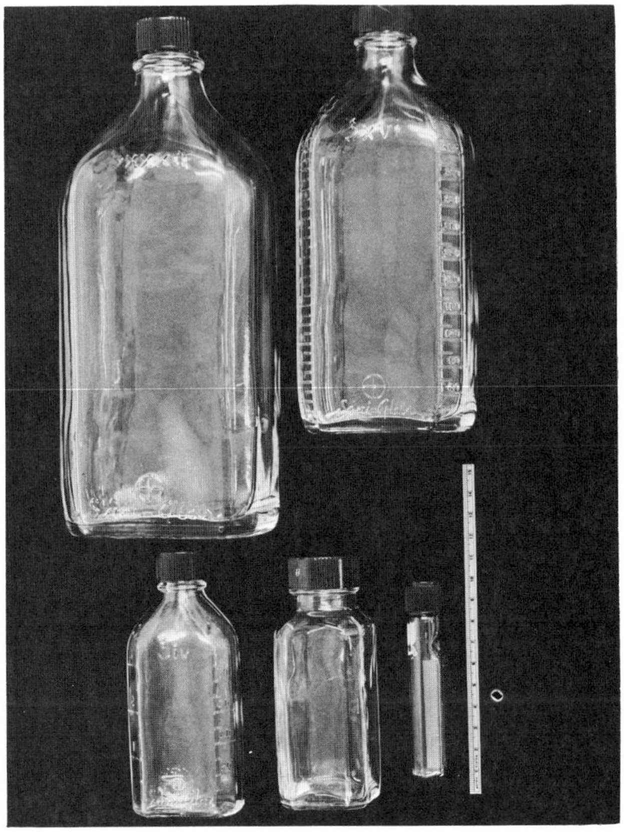

Fig. 7.7. Examples of standard glass bottles that may be used as culture flasks.

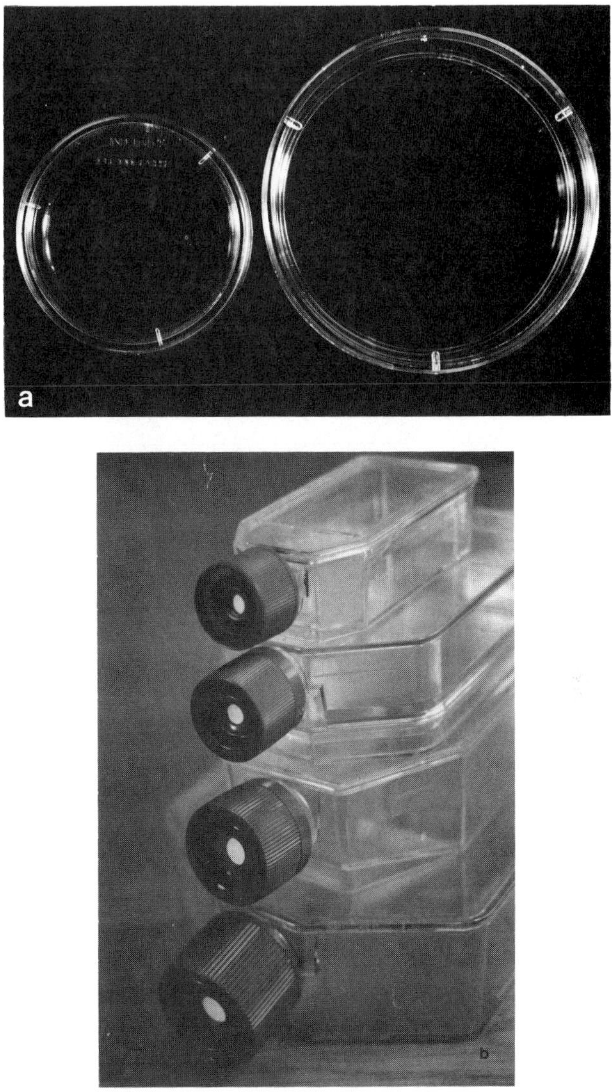

Fig. 7.8. Venting to the atmosphere. a. "Vented" dishes (9 cm and 6 cm diameter). Small pegs, 120° apart, raise the lid from the base and prevent a thin film of liquid from sealing the lid and reducing the rate of gas exchange. b. Gas-permeable caps (Falcon, Becton Dickinson).

first time, add the specified amount of bicarbonate and then sufficient 1 *N* NaOH such that the medium equilibrates to the desired pH after incubation at 36.5°C overnight. When dealing with media already at working strength, vary the amount of HCO_3^- to suit the gas phase (Table 7.2) and leave overnight to equilibrate at 36.5°C. Each medium has a recommended bicarbonate concentration and CO_2 tension to achieve the correct pH and osmolality, but minor variations will occur in different methods of preparation.

With the introduction of Good's buffers (e.g., HEPES, Tricine) [Good et al., 1966] into tissue culture,

TABLE 7.2. Relationship Between CO_2, Bicarbonate Concentration, and HEPES Buffer in Various Media

	Eagle's MEM Hanks' salts	Low HCO_3^- + buffer	Eagle's MEM Earle's salts	Dulbecco's Mod. MEM
$NaHCO_3$ (m*M*)	4	8	26	44
CO_2	Atmospheric and evolved from culture	2%	5%	10%
HEPES (if used; m*M*)	10	20	50	–

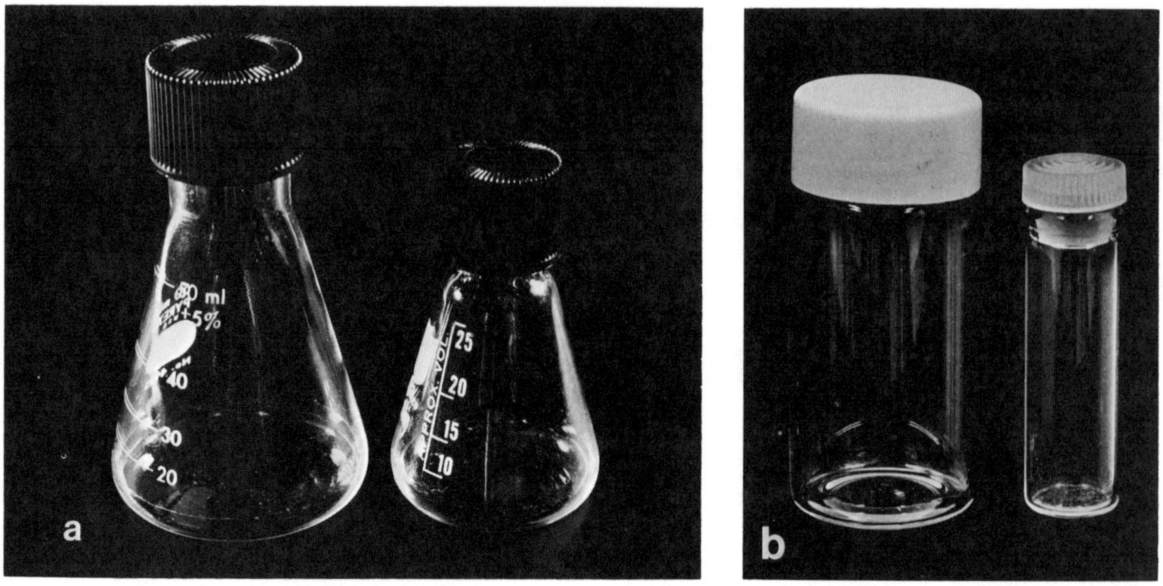

Fig. 7.9. Screw-cap vials and conical flasks suitable for replicate cultures or sample storage. a. Screw caps are preferable to stoppers, as they are less likely to leak and they protect the neck of the flask from contamination. b. Scintillation vials are particularly useful for isotope incorporation studies but should not be reused for culture after containing scintillation fluid.

there was some speculation that since CO_2 was no longer necessary to stabilize the pH, it could be omitted. This has since proved to be untrue [Itagaki and Kimura, 1974], at least for a large number of cell types, particularly at low cell concentrations. Although 20 m*M* HEPES buffer can control pH within the physiological range, the absence of atmospheric CO_2 allows equation 1 to move to the left, eventually eliminating dissolved CO_2, and ultimately HCO_3^-, from the medium. This appears to limit cell growth, although whether the cells require the dissolved CO_2 or the HCO_3^- (or both) is not clear. Recommended HCO_3^-, CO_2, and HEPES concentrations are given in Table 7.2. The inclusion of pyruvate in the medium enables cells to increase their endogenous production of CO_2, making them independent of exogenous CO_2, and HCO_3^-. Leibovitz L15 medium [Leibovitz, 1963] contains a higher concentration of sodium pyruvate (550 mg/l) but no $NaHCO_3$ and does not require CO_2 in the gas phase. Sodium β-glycerophosphate can also be used to buffer autoclavable media lacking CO_2 and HCO_3^- [Waymouth, 1979] and Life Technologies (Gibco BRL) markets a CO_2-independent medium. If elimination of CO_2 is important for cost saving, convenience, or other reasons, it might be worth considering one of these formulations, but only after rigorous testing.

In summary, cultures at low cell concentration in an open vessel need to be incubated in an atmosphere of CO_2, the concentration of which is in equilibrium with the sodium bicarbonate in the medium. At very low cell concentrations (e.g., during cloning), it is necessary to add CO_2 to the gas phase of sealed flasks for most cultures. At high cell concentrations, it will not be necessary to add CO_2 to the gas phase in sealed flasks, but it may yet be necessary in open dishes. Where the culture produces a lot of acid, and the endogenous production of CO_2 is high, it may be desirable to slacken the cap of a culture flask and allow excess CO_2 to escape. In these cases it is advisable to incorporate HEPES (20 m*M*) in the medium to stabilize the pH.

MEDIA AND SUPPLEMENTS

The discovery that cells from explants could be subcultured and propagated *in vitro* led to attempts to provide more defined media to sustain continuous cell growth and replace the "natural" media like embryo extract, protein hydrolysates, lymph, etc. Basal media of Eagle [1955a, 1959] and the more complex media 199 of Morgan et al. [1950] and CMRL 1066 of Parker et al. [1957], although "defined," are usually supplemented with 5–20% serum. It was the desire to eliminate this remaining undefined constituent that led to the evolution of such complex media as NCTC 109 [Evans et al., 1956] and 135 [Evans and Bryant, 1965], Waymouth's MB 572/1 [1959], Ham's F10 [1963] and F12 [1965], Birch and Pirt [1971], the MCDB series [Ham and McKeehan, 1978], Sato's hormone-supplemented media [Barnes and Sato, 1980], and many others (see below, under "Serum-Free Medium").

One approach to developing a medium is to start with a rich medium such as Ham's F12 [1965] or medium 199 supplemented with a high concentration of serum (say 20%) and gradually attempt to reduce the serum concentration by manipulating the concentrations of existing constituents and by adding new ones. This is a very laborious procedure but it has resulted in a number of different formulations for the culture of many diverse cell types, either in low serum concentrations or in its complete absence (see below).

Even after many years of exhaustive research into matching particular media to specific cell types and culture conditions, the choice of medium is not obvious and is often empirical. No all-purpose medium has been developed for the more demanding requirements of specialized cells, and even transformed cells, cultured from spontaneous tumors, have highly specific requirements, differing among tumors, even of one type, and often differing from the normal cells of the same tissue. Hence serum has been retained for many cell types and is only gradually being eliminated after many years of careful and painstaking work.

PHYSICAL PROPERTIES

pH

Most cell lines will grow well at pH 7.4. Although the optimum pH for cell growth varies relatively little among different cell strains, some normal fibroblast lines perform best at pH 7.4–7.7, and transformed cells may do better at pH 7.0–7.4 [Eagle, 1973]. There have been reports that epidermal cells may be maintained at pH 5.5 [Eisinger et al., 1979]. In special cases it may prove advantageous to do a brief growth experiment (see Chapters 11 and 18) or special function analysis (e.g., Chapter 14) to determine the optimum pH.

Phenol red is commonly used as an indicator. It is red at pH 7.4, becoming orange at pH 7.0, yellow at pH 6.5, lemon yellow below pH 6.5, more pink at pH 7.6, and purple at pH 7.8. Since the assessment of color is highly subjective, it is useful to make up a set of standards using sterile balanced salt solution (BSS) and phenol red at the correct concentration and in the same type of bottle, with the same headspace for air, that you normally use for preparing medium.

Preparations of pH Standards

Materials

Hanks' Balanced Salt Solution (HBSS), 10× concentrate or powder, without bicarbonate or glucose, with 20 m*M* HEPES

Nine bottles of a size closest to your standard medium bottles or culture flasks

Distilled deionized water

Sterile 0.1 *N* NaOH (make up in distilled deionized water and filter sterilize; see Chapter 8)

pH meter

Protocol

1. Make up the BSS at pH 6.5, dispense into nine bottles (two extra to allow for breakage) of the appropriate size, and autoclave.
2. Adjust the pH to 6.5, 6.8, 7.0, 7.2, 7.4, 7.6, 7.8, with sterile 0.1 *N* NaOH, checking the pH of a sample of each bottle on a pH meter after allowing each bottle to equilibrate with the atmosphere.
3. Keep sterile and sealed.

Buffering

Culture media must be buffered under two sets of conditions: (1) open dishes, where evolution of CO_2 causes the pH to rise, and (2) overproduction of CO_2 and lactic acid in transformed cell lines at high cell concentrations, when the pH will fall. A buffer may be incorporated in the medium to stabilize the pH, but in (1) exogenous CO_2 may still be required by some cell lines, particularly at low cell concentrations, to prevent total loss of dissolved CO_2 and bicarbonate from the medium (see above). In (2) it is usually preferable to leave the cap slack (shrouded in aluminum foil) or to use a CO_2-permeable cap (Camlab, Corning, Costar, Falcon, Nunclon) to promote the release of CO_2.

A bicarbonate buffer is still used more frequently than any other, in spite of its poor buffering capacity at physiological pH, because of its low toxicity, low cost, and nutritional benefit to the culture. HEPES is a much stronger buffer in the pH 7.2–7.6 range and is now used extensively at 10 or 20 m*M*. When HEPES is used with exogenous CO_2, it has been found that the HEPES concentration must be more than double that of the bicarbonate for adequate buffering (see Table 7.2). A variation of Ham's F12 with 20 m*M* HEPES, 8 m*M* bicarbonate, and 2% CO_2 has been used successfully in the author's laboratory for the culture of a number of different cell lines. It allows the handling of microtitration, and other multiwell plates, out of the incubator without excessive rise in pH, and minimizes the amount of HEPES, which is both toxic and expensive.

Osmolality

Most cultured cells have a fairly wide tolerance for osmotic pressure [see also Waymouth, 1970]. Since the osmolality of human plasma is about 290 mOsm/kg, it is reasonable to assume that this is the optimum for human cells *in vitro*, although it may be different for other species (e.g., around 310 mOsm/kg for mice [Waymouth, 1970]). In practice, osmolalities between 260 mOsm/kg and 320 mOsm/kg are quite acceptable for most cells, but once selected, should be kept consistent at ±10 mOsm/kg. Slightly hypotonic medium may be better for petri-dish or open-plate culture to

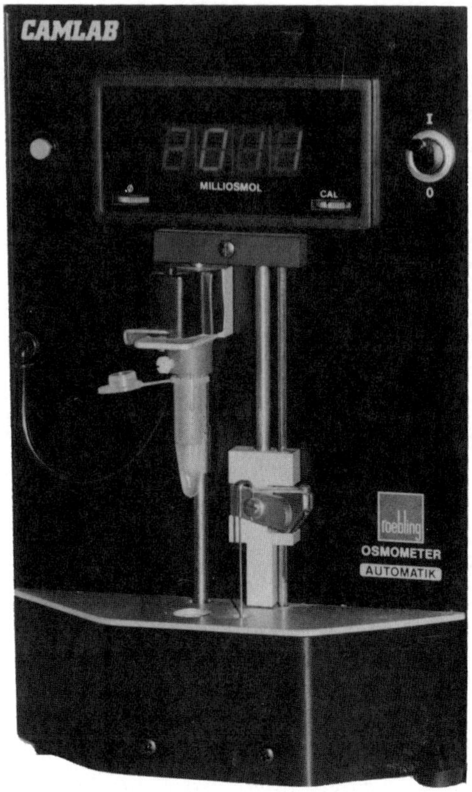

Fig. 7.10. *Roebling osmometer (Camlab, Cambridge, England). This model accepts samples of 50 μl.*

compensate for evaporation during incubation. Osmolality is usually measured by freezing-point depression (Fig. 7.10) or elevation of vapor pressure. The measurement of osmolality is a useful quality-control step if you are making up medium yourself, as it helps to guard against errors in weighing and dilution, etc. It is particularly important to monitor osmolality if alterations are made in the constitution of the medium. Addition of HEPES, drugs dissolved in strong acids and bases, and subsequent neutralization can all markedly affect the osmolality.

Temperature

Apart from the direct effect of temperature on cell growth (see below), it will also influence pH due to the increased solubility of CO_2 at lower temperatures and, possibly, due to changes in ionization and the pK_a of the buffer. The pH should be adjusted to 0.2 units lower at room temperature than at 36.5°C. It is best to make up the medium complete with serum, if used, and incubate a sample overnight at 36.5°C, under the correct gas tension, to check the pH when making up a medium for the first time.

Viscosity

The viscosity of culture medium is influenced mainly by the serum content, and in most cases will have little effect on cell growth. It becomes important, however, whenever a cell suspension is agitated, e.g., when a suspension culture is stirred, or when cells are dissociated after trypsinization. If there is cell damage under these conditions, then this may be reduced by increasing the viscosity of the medium with carboxy methyl cellulose or polyvinyl pyrolidone (see Reagent Appendix) [Birch and Pirt, 1971]. This becomes particularly important in low serum concentrations, in the absence of serum, and in fermentor cultures.

Surface Tension and Foaming

The surface tension of medium may be used to promote adherence of primary explants to the substrate (see Chapter 9) but is seldom controlled in any way. In suspension cultures, where 5% CO_2 in air is bubbled through medium containing serum, foaming may result. The addition of a silicone antifoam (Dow Chemical) or Pluronic F68 (Wyandotte) helps to prevent this by reducing surface tension.

The effects of foaming have not been clearly defined. As the rate of protein denaturation may increase, and the risk of contamination increases if the foam reaches the neck of the culture vessel, it is best avoided. It will also limit gaseous diffusion as an overlay and can limit aqueous permeation through a loose cap if a film from a foam or spillage gets into the capillary space between cap and bottle.

CONSTITUENTS OF MEDIA

Balanced Salt Solutions

It is useful for the present discussion to distinguish balanced salt solutions from media. A balanced salt solution (BSS) is composed of inorganic salts, usually including sodium bicarbonate and, by most recipes, glucose, although glucose and bicarbonate are often omitted. The compositions of some common examples are given in Table 7.3. HEPES buffer (5–20 m*M*) may be added to these if necessary and the equivalent weight of NaCl omitted to maintain the correct osmolality. BSS is used as a diluent for more complete media, as a washing or dissection medium, for short incubations up to about 4 hr, and for a variety of other purposes that require an isotonic solution that is not necessarily nutritionally complete.

The choice of BSS is dependent on: (1) the CO_2 tension (see above and Tables 7.2 and 7.3). The bicarbonate concentration must be such that equilibrium is reached at pH 7.4 at 36.5°C; e.g., Earle's BSS (EBSS) is commonly used as a diluent for Eagle's media for equilibration with 5% CO_2, while Hanks' BSS [Hanks and Wallace, 1949] (HBSS) is used with air; (2) its use

TABLE 7.3. Balanced Salt Solutions

Component	Earle's balanced salt solution (g/l)	Dulbecco's phosphate-buffered saline (solution A) (PBSA) (g/l)	Hanks' balanced salt solution (g/l)	Spinner salt solution (Eagle) (g/l)
Inorganic salts				
$CaCl_2$ (anhyd.)	0.02		0.14	
KCl	0.40	0.20	0.40	0.40
KH^2PO_4		0.20	0.06	
$MgCl_2 \cdot 6H_2O$			0.10[a]	
$MgSO_4 \cdot 7H_2O$	0.20		0.10	0.20
NaCl	6.68	8.00	8.00	6.80
$NaHCO_3$	2.20		0.35	2.20
$Na_22HPO_4 \cdot 7H_2O$		2.16	0.09[c]	
$NaH_2PO_4 \cdot H_2O$	0.14[b]			1.40
Other components				
D-glucose	1.00		1.00	1.00
Phenol red	0.01		0.01[d]	0.01
Gas phase	5% CO_2	10% CO_2	Air	5% CO_2

[a]$MgCl_2 \cdot 6H_20$ added to original formula.
[b]Original formulation calls for 0.15 g/l $NaH_2PO_4 \cdot 2H_20$, Earle (1934); Morton, [197].
[c]Original formulation calls for 0.06 g/l $Na_2HPO_4 \cdot H_20$, Hanks and Wallace [1949].
[d]Original formulation calls for 0.02 g/l, Hanks and Wallace [1949].

for tissue disaggregation, or monolayer dispersal. In these cases Ca^{2+} and Mg^{2+} are usually omitted, as in Moscona's [1952] calcium- and magnesium-free saline (CMF) or Dulbecco and Vogt's [1954] phosphate-buffered saline, solution A (PBSA) (see Table 7.3); and (3) its use for suspension culture of adherent cells. MEM(S), based on Eagle's Spinner salt solution, is a variant of Eagle's [1959] minimum essential medium, deficient in Ca^{2+}, to reduce cell aggregation and attachment (see Table 7.3).

HBSS, EBSS, and PBS rely on the relatively weak buffering of phosphate, which is not at its most effective at physiological pH. Paul [1975] constructed a tris-buffered BSS that is more effective, but for which the cells sometimes require a period of adaptation. HEPES (10–20 m*M*) is currently the most effective buffer in the pH 7.2–7.8 range and TRICINE in the pH 7.4–8.0 range, although these tend to be expensive if used in large quantities. Sodium β-glycerophosphate (pK_a 6.6) also works well (Waymouth, personal communication).

DEFINED MEDIA

Defined media vary in complexity from Eagle's MEM [Eagle, 1959], which contains essential amino acids, vitamins, and salts, to complex media such as 199 [Morgan et al., 1950], CMRL 1066 [Parker et al., 1957], RPMI 1640 [Moore et al., 1967], and F12 [Ham, 1965] (Table 7.4), and a wide range of serum-free formulations (Table 7.5) (see below). The complex media contain a larger number of different amino acids, including nonessential amino acids and vitamins, and are often supplemented with extra metabolites (e.g., nucleosides, tricarboxylic acid intermediates, and lipids) and minerals. Nutrient concentrations are, on the whole, low in F12 (which was optimized by cloning) and high in Dulbecco's modification of Eagle's MEM (DMEM) [Dulbecco and Freeman, 1959; Morton, 1970], optimized at higher cell densities for viral propagation, although the latter has fewer constituents. Barnes and Sato [1980] employed a 1:1 mixture of DMEM and F12 as the basis for their serum-free formulations to combine the richness of F12 and the higher nutrient concentration of DMEM.

Although not always entirely rational, this has provided an empirical formula, suitable as a basic medium for supplementation with special additives for many different cell types.

The common constituents of medium may be grouped as follows (Table 7.4).

Amino Acids

The essential amino acids, i.e., those which are not synthesized in the body, are required by cultured cells with, in addition, cysteine and tyrosine, although individual requirements for amino acids will vary from one cell to another. Other nonessential amino acids are often added to compensate either for a particular cell type's incapacity to make them or because they are

TABLE 7.4. Media[a]

Component	Eagle's MEM (mg/l)	Dulbecco's modification (mg/l)	Ham's F12 (mg/l)	CMRL 1066 (mg/l)	RPMI 1640 (mg/l)	199 (mg/l)	L15 (mg/l)	Fischer's (mg/l)	Waymouth's MB 752/1 (mg/l)
Amino acids									
L-alanine	—	—	8.90	25.0	—	25.0	225	—	—
L-arginine (free base)	—	—	—	—	200	—	500	—	—
L-arginine HCl	126	84.0	211	70.0	—	70.0	—	15.0	75.0
L-asparagine	—	—	—	—	50.0	—	—	—	—
L-asparagine H_2O	—	—	15.0	—	—	—	260	11.4	—
L-aspartic acid	—	—	13.3	30.0	20.0	30.0	—	—	60.0
L-cysteine (free base)	—	—	—	—	—	—	120	—	61.0
L-cystine	24	48.0	—	20.0	50.0	—	—	—	15.0
L-cystine, 2Na	—	—	—	—	—	23.7	—	23.7	—
L-cysteine HCl · H_2O	—	—	35.1	260	—	0.0987	—	—	—
L-glutamic acid	—	—	14.7	75.0	20.0	66.8	—	—	150
L-glutamine	292	584	146	100	300	100	300	200	350
Glycine	—	30.0	7.50	50.0	10.0	50.0	200	—	50.0
L-histidine (free base)	—	—	—	—	15.0	—	250	—	128
L-histidine HCl · H_2O	42.0	42.0	21.0	20.0	—	21.9	—	81.1	—
L-hydroxy-proline	—	—	—	10.0	20.0	10.0	—	—	—
L-isoleucine	52.0	105	3.94	20.0	50.0	20.0	125	75.0	25.0
L-leucine	52.0	105	13.1	60.0	50.0	60.0	125	30.0	50.0
L-lysine HCl	73.1	146	36.5	70.0	40.0	70.0	93	50.0	240
L-methionine	15.0	30.0	4.48	15.0	15.0	15.0	75.0	100	50.0
L-phenylalanine	33.0	66.0	4.96	25.0	15.0	25.0	125	67.0	50.0
L-proline	—	—	34.5	40.0	20.0	40.0	—	—	50.0
L-serine	—	42.0	10.5	25.0	30.0	25.0	200	15.0	—
L-threonine	48.0	95.0	11.9	30.0	20.0	30.0	300	40.0	75.0
L-tryptophan	10.0	16.0	2.04	10.0	5.00	10.0	20.0	10.0	40.0
L-tyrosine	36.0	72.0	5.40	40.0	20.0	—	—	—	40.0
L-tyrosine 2Na	—	—	—	—	—	49.7	373	74.6	—
L-valine	47.0	94.0	11.7	25.0	20.0	25.0	100	70.0	65.0
Vitamins									
L-ascorbic acid	—	—	—	50.0	—	0.050	—	—	17.5
Biotin	—	—	0.0073	0.010	0.200	0.010	—	0.010	0.02
D-Ca pantothenate	1.00	4.00	0.480	0.010	0.250	0.010	1.00	0.500	1.00
Calciferol	—	—	—	—	—	0.100	—	—	—
Choline chloride	1.00	4.00	14.0	0.500	3.00	0.500	1.00	1.50	250
Folic acid	1.00	4.00	1.30	0.010	1.00	0.010	1.00	10.0	0.40
i-inositol	2.00	7.20	18.0	0.050	35.0	0.050	2.00	1.50	1.00
Nicotinamide	1.00	4.00	0.04	0.025	1.00	0.025	1.00	0.50	1.00
Pyridoxal HCl	1.00	4.00	0.062	0.025	—	0.025	—	0.50	—
Riboflavin	0.10	0.40	0.038	0.010	0.20	0.010	—	0.50	1.00
Thiamin HCl	1.00	4.00	0.34	0.010	1.00	0.010	—	1.00	10.0
Vitamin B_{12}	—	—	1.36	—	0.005	—	—	—	0.20
Pyridoxine HCl	—	—	0.062	0.025	1.00	0.025	—	—	1.00
Cholesterol	—	—	—	0.200	—	—	—	—	—
Para-aminobenzoic acid	—	—	—	0.050	1.00	0.050	—	—	—

(*continued*)

TABLE 7.4. *(Continued)*

Component	Eagle's MEM (mg/l)	Dulbecco's modification (mg/l)	Ham's F12 (mg/l)	CMRL 1066 (mg/l)	RPMI 1640 (mg/l)	199 (mg/l)	L15 (mg/l)	Fischer's (mg/l)	Waymouth's MB 752/1 (mg/l)
Nicotinic acid	—	—	—	—	—	0.025	—	—	—
Menaphthone sodium bisulphite $3H_2O$	—	—	—	—	—	0.019	—	—	—
Dl-α tocopherol $PO_4 \cdot 2Na$	—	—	—	—	—	0.01	—	—	—
Vitamin A acetate	—	—	—	—	—	0.115	—	—	—
Riboflavin PO_4, 2Na	—	—	—	—	—	—	0.10	—	—
Thiamin mono PO_4, $2H_2O$	—	—	—	—	—	—	1.00	—	—
Inorganic salts									
$CaCl_2$ (anhyd.)	200	200	—	200	—	—	—	—	—
$CaCl_2 \cdot 2H_2O$	—	—	44.0	—	—	186.00	186	91.0	120
$Fe(NO_3)_3 \cdot 9H_2O$	—	0.10	—	—	—	—	—	—	—
KCl	400	400	224	400	400	400	400	400	150
$KH2PO_4$	—	—	—	—	—	60.0	60.0	—	80.0
$MgCl_2 \cdot 6H_2O$	—	—	122	—	—	—	—	—	240
$MgSO_4 \cdot 7H_2O$	200	200	—	200	100	200	400	121	200
NaCl	6,800	6,400	7,599	6,799	6,000	8,000	8,000	8,000	6,000
$NaHCO_3$	2,200	3,700	1,176	2,200	2,200	350	—	1,125	2,240
$NaH_2PO_4 \cdot H_20$	140	125	—	140	—	—	—	78.0	—
Na_2HPO_4 (anhyd.)	—	—	—	—	—	47.5	190	60.0	—
$Na_2HPO_4 \cdot 7H_2O$	—	—	268	—	1,512	—	—	—	566
$CuSO_4 \cdot 5H_2O$	—	—	0.00249	—	—	—	—	—	—
$FeSO_4 \cdot 7H_2O$	—	—	0.834	—	—	—	—	—	—
$ZnSO_4 \cdot 7H_2O$	—	—	0.863	—	—	—	—	—	—
Other components									
D-glucose	1,000	4,500	1,802	1,000	2,000	1,000	—	1,000	5,000
D-galactose	—	—	—	—	—	—	9,000	—	—
Lipoic acid	—	—	0.21	—	—	—	—	—	—
—									
Phenol red	10.0	15.0	12.0	20.0	5.00	17.0	10.0	5.00	10.0
Sodium pyruvate	—	110	110	—	—	—	550	—	—
Hypoxanthine	—	—	4.10	—	—	0.30	—	—	—
Linoleic acid	—	0.084	—	—	—	—	—	—	25.0
Putrescine 2HCl	—	—	0.161	—	—	—	—	—	—
Thymidine	—	—	0.73	10.0	—	—	—	—	—
Cocarboxylase	—	—	—	1.00	—	—	—	—	—
Coenzyme A	—	—	—	2.50	—	—	—	—	—
Deoxyadenosine	—	—	—	10.0	—	—	—	—	—
Deoxycytidine HCl	—	—	—	10.0	—	—	—	—	—
Deoxyguanosine	—	—	—	10.0	—	—	—	—	—
Diphosphopyridine nucleotide $\cdot 4H_2O$	—	—	—	7.00	—	—	—	—	—
Ethanol for solubilizing lipid components	—	—	—	16.0	—	—	—	—	—

(continued)

TABLE 7.4. (*Continued*)

Component	Eagle's MEM (mg/l)	Dulbecco's modification (mg/l)	Ham's F12 (mg/l)	CMRL 1066 (mg/l)	RPMI 1640 (mg/l)	199 (mg/l)	L15 (mg/l)	Fischer's (mg/l)	Waymouth's MB 752/1 (mg/l)
Flavine adenine dinucleotide	—	—	—	1.00	—	—	—	—	—
Glutathione (reduced)	—	—	—	10.0	1.00	0.05	—	—	15.0
5-methyl-deoxycytidine	—	—	—	0.10	—	—	—	—	—
Sodium acetate · $3H_2O$	—	—	—	83.0	—	—	—	—	—
Sodium glucuronate · H_2O	—	—	—	4.20	—	—	—	—	—
Tri-phosphopyridine nucleotide	—	—	—	1.00	—	—	—	—	—
Tween 8	—	—	—	5.00	—	5.00	—	—	—
Uridine triphosphate · $4H_2O$	—	—	—	1.00	—	—	—	—	—
Adenine SO_4	—	—	—	—	—	10.0	—	—	—
5'AMP	—	—	—	—	—	0.20	—	—	—
ATP-2Na	—	—	—	—	—	10.0	—	—	—
Cholesterol	—	—	—	—	—	0.20	—	—	—
2-Deoxyribose	—	—	—	—	—	0.50	—	—	—
Guanine HCl	—	—	—	—	—	0.30	—	—	—
D-ribose	—	—	—	—	—	0.50	—	—	—
Na acetate	—	—	—	—	—	36.7	—	—	—
Thymine	—	—	—	—	—	0.30	—	—	—
Uracil	—	—	—	—	—	0.30	—	—	—
Xanthine	—	—	—	—	—	0.30	—	—	—
CO_2 (gas phase)	5%	10%	5% (pH 7.0) 2% (pH 7.4)	5%	5%	Ambient	Ambient	5% (pH 7.0) 2% (pH 7.4)	5%

[a]Note: For use in a gas phase of air in sealed containers, Hanks' salts may be substituted in Eagle's MEM. Likewise, to use 199 in 5% CO_2, substitute Earle's salts (see Table 7.3).

made but lost into the medium. The concentration of amino acids usually limits the maximum cell concentration attainable, and the balance may influence cell survival and growth rate. Glutamine is required by most cells, although some cell lines will utilize glutamate, but evidence suggests that glutamine is also used by cultured cells as an energy and carbon source [Reitzer et al., 1979].

Vitamins

Eagle's MEM contains only the B-group vitamins (see Table 7.4), other requirements presumably being derived from the serum. The requirement for extra vitamins is most apparent where the serum concentration is reduced, but there are other cases (e.g., low cell densities for cloning) where they may be essential even in the presence of serum. Vitamin limitation is usually expressed in terms of cell survival and growth rate rather than maximum cell density.

Salts

The salts are chiefly those of Na^+, K^+ Mg^{2+}, Ca^{2+} Cl^-, SO_4^{2-}, PO_4^{3-}, and HCO_3^- and are the major components contributing to the osmolality of the medium. Divalent cations, particularly Ca^{2+}, are Na^+, K^+, and Cl^-, regulate membrane potential, while SO_4^{2-}, PO_4^{3-}, and HCO_3^- have roles as matrix requirements and nutritional precursors for macromolecules, as well as regulators of intracellular charge.

Calcium is reduced for suspension cultures to minimize cell aggregation and attachment (see above). The sodium bicarbonate concentration is determined by the concentration of CO_2 in the gas phase (see above) and has a significant nutritional role in addition to its buffering capacity.

Glucose

Glucose is included in most media as an energy source. It is metabolized principally by glycolysis to form pyru-

TABLE 7.5. Examples of Serum-Free Media Basal Formulas[a]

Component	MCDB 110	MCDB 202	MCDB 402	MCDB 153	Iscove's	LHC
Amino acids (all as L-enantiomers)						
Alanine	1.0 E-4	1.0 E-4	—	1.0 E-4	2.8 E-4	1.0 E-4
Arginine HCl	1.0 E-3	3.0 E-4	3.0 E-4	1.0 E-3	4.0 E-4	2.0 E-3
Asparagine	1.0 E-4	1.0 E-3	1.0 E-4	1.0 E-4	1.9 E-4	1.0 E-4
Aspartic acid	1.0 E-4	1.0 E-4	1.0 E-5	3.0 E-5	2.3 E-4	3.0 E-5
Cysteine · HCl	5.0 E-5	2.0 E-4	—	2.4 E-4	—	2.4 E-4
Cystine	—	2.0 E-4	4.0 E-4	—	2.9 E-4	
Glutamic acid	1.0 E-4	1.0 E-4	1.0 E-5	1.0 E-4	5.1 E-4	1.0 E-4
Glutamine	2.5 E-3	1.0 E-3	5.0 E-3	6.0 E-3	4.0 E-3	6.0 E-3
Glycine	3.0 E-4	1.0 E-4	1.0 E-4	1.0 E-4	4.0 E-4	1.0 E-4
Histidine · HCl	1.0 E-4	1.0 E-4	2.0 E-3	8.0 E-5	2.0 E-4	1.6 E-4
Isoleucine	3.0 E-5	1.0 E-4	1.0 E-3	1.5 E-5	8.0 E-4	3.0 E-5
Leucine	1.0 E-4	3.0 E-4	2.0 E-3	5.0 E-4	8.0 E-4	1.0 E-3
Lysine HCl	2.0 E-4	2.0 E-4	8.0 E-4	1.0 E-4	8.0 E-4	2.0 E-4
Methionine	3.0 E-5	3.0 E-5	2.0 E-4	3.0 E-5	2.0 E-4	6.0 E-5
Phenylalanine	3.0 E-5	3.0 E-5	3.0 E-4	3.0 E-5	4.0 E-4	6.0 E-5
Proline	3.0 E-4	5.0 E-5	—	3.0 E-4	3.5 E-4	3.0 E-4
Serine	1.0 E-4	3.0 E-4	1.0 E-4	6.0 E-4	4.0 E-4	1.2 E-3
Threonine	1.0 E-4	3.0 E-4	5.0 E-4	1.0 E-4	8.0 E-4	2.0 E-4
Tryptophan	1.0 E-5	3.0 E-5	1.0 E-5	1.5 E-5	7.8 E-5	3.0 E-5
Tyrosine	3.0 E-5	5.0 E-5	2.0 E-4	1.5 E-5	4.6 E-4	3.0 E-5
Valine	1.0 E-4	3.0 E-4	2.0 E-3	3.0 E-4	8.0 E-4	6.0 E-4
Vitamins						
d-Biotin	3.0 E-8	3.0 E-8	3.0 E-8	6.0 E-8	5.3 E-8	6.0 E-8
Folic acid	—	—	—	1.8 E-6	9.1 E-6	1.8 E-6
Folinic acid	1.0 E-9	1.0 E-8	1.0 E-6	—	—	—
DL-a-lipoic acid	1.0 E-8	1.0 E-8	1.0 E-8	1.0 E-6	—	1.0 E-6
Niacinamide	5.0 E-5	5.0 E-5	5.0 E-5	3.0 E-7	3.3 E-5	3.0 E-7
D-pantothenate ½Ca	1.0 E-6	1.0 E-6	5.0 E-5	1.0 E-6	1.7 E-5	1.0 E-6
Pyridoxal	—	—	—	—	2.0 E-S	—
Pyridoxine HCl	3.0 E-7	3.0 E-7	1.0 E-4	3.0 E-7	—	3.0 E-7
Riboflavin	3.0 E-7	3.0 E-7	1.0 E-6	1.0 E-7	1.1 E-6	1.0 E-7
Thiamin · HCl	1.0 E-6	1.0 E-6	1.0 E-4	1.0 E-6	1.2 E-5	1.0 E-6
Vitamin B_{12}	1.0 E-7	1.0 E-7	1.0 E-8	3.0 E-7	9.6 E-9	3.0 E-7
Other organic constituents						
Acetate	—	—	—	3.7 E-3	—	3.7 E-3
Adenine	1.0 E-5	1.0 E-6	1.0 E-6	1.8 E-4	—	1.8 E-4
Choline chloride	1.0 E-4	1.0 E-4	1.0 E-4	1.0 E-4	2.9 E-5	2.0 E-4
D-glucose	4.0 E-3	8.0 E-3	5.5 E-3	6.0 E-3	2.5 E-2	6.0 E-3
i-Inositol	1.0 E-4	1.0 E-4	4.0 E-5	1.0 E-4	4.0 E-5	1.0 E-4
Linoleic acid	—	2.0 E-7	3.0 E-7	—	—	—
Putrescine · 2HCl	1.0 E-9	1.0 E-9	1.0 E-9	1.0 E-6	—	1.0 E-6
Na Pyruvate	1.0 E-3	5.0 E-4	1.0 E-3	5.0 E-4	1.0 E-3	5.0 E-4
Thymidine	3.0 E-7	3.0 E-7	1.0 E-6	3.0 E-6	—	3.0 E-6
Major inorganic salts						
$CaCl_2$	1.0 E-3	2.0 E-3	1.6 E-3	3.0 E-5	1.5 E-3	1.1 E-4
KCl	5.0 E-3	3.0 E-3	4.0 E-3	1.5 E-3	4.4 E-3	1.5 E-3
KNO_3	—	—	—	—	7.5 E-7	—
$MgCl_2$	—	—	—	6.0 E-4	—	2.2 E-2
$MgSO_4$	1.0 E-3	1.5 E-3	8.0 E-4	—	8.1 E-4	—
NaCl	1.1 E-1	1.2 E-1	1.2 E-1	1.2 E-1	7.7 E-2	1.0 E-1
Na_2HPO_4	3.0 E-3	5.0 E-4	5.0 E-4	2.0 E-3	1.0 E-3	2.0 E-3
Trace elements						
$CuSO_4$	1.0 E-9	1.0 E-9	5.0 E-9	1.1 E-8	—	1.0 E-8
$FeSO_4$	5.0 E-6	5.0 E-6	1.0 E-6	5.0 E-6	—	5.4 E-4

(continued)

TABLE 7.5. *(Continued)*

Component	MCDB 110	MCDB 202	MCDB 402	MCDB 153	Iscove's	LHC
H_2SeO_3	3.0 E-8	3.0 E-8	1.0 E-8	3.0 E-8	1.0 E-7	3.0 E-8
$MnSO_4$	1.0 E-9	5.0 E-10	1.0 E-9	1.0 E-9	—	1.0 E-9
Na_2SiO_3	5.0 E-7	5.0 E-7	1.0 E-5	5.0 E-7	—	5.0 E-7
$(NH_4)_6Mo_7O_{24}$	1.0 E-9	1.0 E-9	3.0 E-9	1.0 E-9	—	1.0 E-9
NH_4VO_3	5.0 E-9	5.0 E-9	5.0 E-9	5.0 E-9	—	5.0 E-9
$NiCl_2$	5.0 E-10	5.0 E-12	3.0 E-10	5.0 E-10	—	5.0 E-10
$SnCl_2$	5.0 E-10	5.0 E-12	—	5.0 E-10	—	5.0 E-10
$ZnSO_4$	5.0 E-7	1.0 E-7	1.0 E-6	5.0 E-7	—	4.8 E-7
Buffers and indicators						
HEPES	3.0 E-2	3.0 E-2	—	2.8 E-2	2.5 E-2	2.3 E-2
$NaHCO_3$	—	—	1.4 E-2	1.4 E-2	3.6 E-2	1.2 E-2
Phenol red	3.3 E-6	3.3 E-6	3.3 E-5	3.3 E-5	4.0 E-5	3.3 E-6
CO_2	2%	2%	5%	5%	10%	5%

[a]Computer-style notation used for concentrations, e.g., 3.0 E-2 = 30 m*M*. Sufficient NaOH is added to give pH 7.3–7.4 at 37°C in correct gas phase.

vate, which may be converted to lactate or to acetoacetate and enter the citric acid cycle to form CO_2. The accumulation of lactic acid in the medium, particularly evident in embryonic and transformed cells, implies that the citric acid cycle may not function entirely as *in vivo*, and recent data have shown that much of its carbon is derived from glutamine rather than glucose. This may explain the exceptionally high requirement of some cultured cells for glutamine or glutamate.

Organic Supplements

A variety of other compounds including proteins, peptides, nucleosides, citric acid cycle intermediates, pyruvate, and lipids appear in complex media. Again these constituents have been found to be necessary when the serum concentration is reduced (see below) and may help in cloning and in maintaining certain specialized cells, even in the presence of serum.

Hormones and Growth Factors

See below for discussion.

SERUM

The sera used most in tissue culture are calf, fetal bovine, horse, and human. The choice of serum is discussed below in more detail. Calf serum is the most widely used, fetal bovine second, usually for more demanding cell lines and for cloning, and human serum is sometimes used in conjunction with some human cell lines. Horse serum is preferred to calf serum by some workers, as it can be obtained from a closed herd and is often more consistent from batch to batch. It may also be less likely to metabolize polyamines, due to lower polyamine oxidase; polyamines are mitogenic for some cells [Hyvonen et al., 1988; Kaminska et al., 1990].

Although most cell lines still require the supplementation of the medium with serum, there are now many instances where cultures may be maintained and may proliferate serum-free (see Table 7.5). Continuous cell strains such as the L929 and HeLa were among the first to be grown serum-free [Evans et al., 1956; Waymouth, 1959; Birch and Pirt, 1971; Higuchi, 1977] and a degree of selection may have been involved. However, results from the laboratories of Ham [Ham and McKeehan, 1978], Sato [Barnes and Sato, 1980], and others [Carney et al., 1981] have demonstrated that serum may be reduced or omitted without cellular adaptation if nutritional and hormonal modifications are made to the media appropriate to the cell line being studied [Barnes et al., 1984a–d]. This has provided indirect evidence for the constitution of serum, which is considered before further discussion of serum-free media.

Protein

Although proteins are a major component of serum, the functions of many of these *in vitro* remain obscure; it may be that relatively few proteins are required other than as carriers for minerals, fatty acids, and hormones, or as hormones themselves. Those proteins that have been found beneficial are albumin [Iscove and Melchers, 1978; Barnes and Sato, 1980] and globulins [Tozer and Pirt, 1964]. Fibronectin (cold-insoluble globulin) promotes cell attachment [Yamada, 1991; Hynes, 1992] and α2-macroglobulin inhibits trypsin [De Vonne and Mouray, 1978]. Fetuin in fetal serum enhances cell attachment [Fisher et al., 1958], and transferrin [Guilbert and Iscove, 1976] binds iron, making it less toxic but bioavailable. There may be other proteins, as yet uncharacterized, essential for cell attachment and growth.

Polypeptides

Natural clot serum stimulates cell proliferation more than serum from which the cells have been removed physically, e.g., by centrifugation. This appears to be due to the release of a polypeptide from the platelets during clotting. This polypeptide, platelet-derived growth factor (PDGF) [Antoniades et al., 1979; Heldin et al., 1979], is one of a family of polypeptides with mitogenic activity and is probably the major growth factor in serum. PDGF stimulates growth in fibroblasts and glia, but other platelet-derived factors, such as TGF-β, may be inhibitory to growth or promote differentiation in epithelial cells [Lechner et al., 1981]. Other growth factors, such as fibroblast growth factor (FGF) [Gospodarowicz, 1974], epidermal growth factor (EGF) [Cohen, 1962; Carpenter and Cohen, 1977; Gospodarowicz et al., 1978a], endothelial growth factor [Folkman et al., 1979; Maciag et al., 1979], and insulin-like growth factors IGF-1, IGF-2 [Le Roith and Raizada, 1989], which have been isolated from whole tissue or released into the medium by cells in culture, have varying degrees of specificity [Hollenberg and Cuatrecasas, 1973] and are probably present in serum in small amounts [Gospodarowicz and Moran, 1974]. Many of these growth factors are available commercially (see "Growth Factors," Trade Index) in pure form.

Hormones

Hormones may exhibit a variety of different effects on cells, and it is often difficult to recognize the key pathway. Insulin promotes the uptake of glucose and amino acids [Kelley et al., 1978; Lammers et al., 1989; Traxinger and Marshall, 1989] and may owe its mitogenic effect to this property or to activity via the IGF-1 receptor. Some growth factors (IGF-1, IGF-2) bind to the insulin receptor but also have their own specific receptors, to which insulin may bind with lower affinity. IGF-2 also stimulates glucose uptake [Sinha et al., 1990]. Growth hormone may be present in sera, particularly fetal sera, and, in conjunction with the somatomedins (IGFs), may have a mitogenic effect. Hydrocortisone is also present in serum, particularly fetal bovine serum, in varying amounts. It can promote cell attachment [Ballard and Tomkins, 1969; Fredin et al., 1979] and cell proliferation [Guner et al., 1977; McLean et al., 1986] (see also Chapter 20) but under certain conditions (e.g., high cell density) may be cytostatic [Freshney et al., 1980a,b] and can induce cell differentiation [Moscona and Piddington, 1966; Ballard, 1979; McLean et al., 1986; Speirs et al., 1991].

Serum replacement experiments suggest that other hormones required for culture may be present in serum (see below), making it necessary at present to retain serum for culture, but underlining the desirability for its replacement.

Nutrients and Metabolites

Serum also contains amino acids, glucose, ketoacids, lipids, such as oleic acid, ethanolamine, and phosphoethanolamine, and a number of other nutrients and intermediary metabolites. These may be important in simple media but less so in complex media, particularly those with higher amino acid concentrations and other defined supplements.

Minerals

Serum replacement experiments have also suggested that trace elements and iron, copper, and zinc may be provided by serum [Ham and McKeehan, 1978] bound to serum protein. A requirement for selenium has also been demonstrated in the same way [McKeehan et al., 1976] and probably helps to detoxify free radicles as a cofactor for GSH synthetase.

Inhibitors

Serum may also contain substances inhibiting cell proliferation [Harrington and Godman, 1980]. Some of these may be artifacts of preparation, e.g., bacterial toxins from contamination prior to filtration or the γ-globulin fraction may contain antibodies cross-reacting with the culture, but others may be physiological negative growth regulators, such as TGF-β [Massague, 1990; Massague et al., 1992]. Heat inactivation removes complement from the serum and reduces the cytotoxic action of immunoglobulins without damaging polypeptide growth factors, but it may remove some more labile constituents and is not always as satisfactory as untreated serum.

SERUM-FREE MEDIA

Ever since the observation in the 1950s that natural media could be replaced in part by synthetic media, attempts have been made to culture cells without serum. NCTC 109 [Evans et al., 1956], 135 [Evans and Bryant, 1965], Waymouth's MB752/1 [Waymouth, 1959], MB705/1 [Kitos et al., 1962], and the media of Birch and Pirt [1970, 1971] and Higuchi [1977] were all able to sustain the growth of L-cells without serum. Pirt and co-workers further modified their formulation by adding insulin and, with other minor modifications, were able to culture HeLa cells without serum [Blaker et al., 1971]. Ham [1963, 1965] was able to clone CHO cells serum-free and, more recently, specific

TABLE 7.6. Examples of Serum-Free Media Supplements and Modifications[a]

	MCDB 110	MCDB 170MDS	MCDB 153KDS	WAJC 404	Iscove	HITES	Masui	Lechner LHC-9
	F	*ME*	*K*	*PE*	*L*	*SCLC*	*AL*	*LE*
Basal medium	MCDB110	MCDB202	MCDB153	MCDB151	Iscove	RPMI1640	F12/DME	MCDB153
Modifications								
$CaCl_2$				1.3 E-4				
Cysteine		7.0 E-S		Delete				
Cystine				2.4 E-4				
Glutamine		2.0 E-3						
Pyruvate		1.0 E-3	1.0 E-4					
Trace elements				As in 110				
CuS4				1.0 E-9				
$ZnSO_4$								
Supplements								
Na_2Se0_3						3.0 E-8	2.5 E-8	
Dithiothreitol	6.5 E-6							
Glutathione	6.5 E-7							
Phosphoenolpyruvate	1.0 E-5							
Phosphoethanolamine		1.0 E-4						5.0 E-7
Ethanolamine		1.0 E-4					5.0 E-7	
PGE1	2.5 E-8		2.5 E-8					
Cholesterol	7.6 E-6							
Soya lecithin	6 μg/ml							
Soybean lipid					50 μg/ml			
Sphingomyelin	1 μg/ml							
Retinoic acid								0.1 μg/ml
Vitamin E	1.4 E-7							
Vitamin E acetate	4.2 E-7							
Dexamethasone	5.0 E-7			5.0 E-7				
Hydrocortisone		1.4 E-7	1.4 E-7			1.0 E-8		2.0 E-7
Estradiol						1.0 E-8		
EGF (ng/ml)	30	10	25	25				5.0
Insulin (μg/ml)	1	5	5	10		5	5	5.0
Glucagon							0.2 μg/ml	
Prolactin		5 μg/ml						
Triiodothyronine							5 E-10	6.5 ng/ml
Epinephrine								0.5 μg/ml
Transferrin, Fe^{3+} saturated (μg/ml)		5			30–300	100		10
BSA					0.5–10 mg/ml			
Bovine pituitary extract		70 μgP/ml[b]		25 μgP/ml				35 μg/ml
Ovine prolactin		1.0 μg/ml						
Dialysed FBS			1 mgP/ml					
Cholera toxin				2.0 E-10				

[a]*F,* fibroblasts; *ME,* mammary epithelium; *K,* keratinocytes; *PE,* prostatic epithelium; *L,* Lymphoblasts; *SCLC,* small lung cell cancer; *AL,* adenocarcinoma of the lung; *LE,* lung epithelium; *PG,* prostaglandin; *BSA,* bovine serum albumin. For preparation of lipid and peptides, see Barnes et al. [1984a]..See footnote a in Table 7.5.
[b]Ovine prolactin may be substituted for bovine pituitary extract. Bovine pituitary extract is prepared according to Tsao et al. [1982].

formulations (e.g., MCDB 110) have been derived to culture human fibroblasts [Ham, 1984], many normal and neoplastic murine and human cells [Barnes and Sato, 1980], lymphoblasts [Iscove and Melchers, 1978], and several different primary cultures [Mather and Sato, 1979a,b; Sundqvist et al., 1991] in the absence of serum [Maurer, 1992], with, in some cases, some protein added [Tsao et al., 1982; Benders et al., 1991]. Some examples of serum-free media are given in Tables 7.5 and 7.6.

Advantages of Serum-Free Medium

1. The constitution of serum is well known in terms of its major constituents like albumin, transferrin, etc.,

but it also contains a wide range of minor components that may have a considerable effect on cell growth. These include nutrients, such as amino acids, nucleosides, and sugars, peptide growth factors, hormones, minerals, and lipids, the presence and action of which have not been fully determined. While substances such as PDGF may be mitogenic to fibroblasts, there are other constituents of serum that can be cytostatic. Hydrocortisone, present at around 10^{-8} M in fetal serum, is cytostatic to many cell types such as glia and lung epithelium at high cell densities (though it may be mitogenic at low cell densities), and TGF-β, released from platelets, is cytostatic to many epithelial cells.

2. Serum varies from batch to batch and at best a batch will last 1 year, perhaps deteriorating during that time. It then must be replaced with another batch that may be selected as similar but will never be identical.

3. Changing serum batches requires extensive testing to ensure that the replacement is as close as possible to the previous batch. This can involve several tests (growth, plating efficiency, special functions) (see above) and may involve several different cell lines.

4. If more than one cell type is used, each type may require a different batch of serum, requiring several batches to be held on reserve simultaneously. Coculture of different cell types will present an even greater problem.

5. Periodically the supply of serum is restricted due to drought in the cattle-rearing areas, spread of disease among the cattle, or economic or political reasons. This can create problems at any time, restricting the amount available and number of batches to choose from, but can be particularly acute at times of high demand. Demand is currently increasing and will probably exceed supply unless the majority of commercial users are able to adopt serum-free media. While an average research laboratory may reserve 100–200 l of serum per year, a commercial biotechnology laboratory can use that or more in a week.

6. For anyone interested in downstream processing of culture medium to recover cell products, the presence of serum creates a major obstacle to purification and may even limit the pharmaceutical acceptance of the product.

7. Serum is frequently contaminated with viruses, many of which may be harmless to cell culture but represent an additional unknown factor outside the operator's control. Fortunately, improvements in serum sterilization techniques have virtually eliminated the risk of mycoplasma infection from sera from most reputable suppliers, but this cannot be guaranteed for viral infection, in spite of claims that some filters may remove virus (Pall). Because of the risk of spreading bovine spongiform encephalitis among cattle, sample of cell cultures and/or serum shipped to the United States or Australia requires information on the country of origin and the batch number of the serum. Serum derived from cattle in New Zealand has probably the lowest endogenous viral contamination, as many of the viruses found in European and North American cattle are not found in New Zealand.

8. Cost is often cited as a disadvantage of serum supplementation. Certainly, serum constitutes the major part of the cost of a bottle of medium (more than ten times the cost of the chemical constituents), but if it is replaced by defined constituents the cost of these may be as high as the serum. However, it is to be hoped that as demand for such items as transferrin, selenium, insulin, etc., increases the cost will come down with increasing market size, and serum-free media will become relatively cheaper. The availability of recombinant growth factors, coupled to market demand, particularly as pharmaceuticals, may help to reduce their intrinsic cost.

9. As well as its growth-promoting activity, serum contains growth-inhibiting activity and the net effect of the serum is an unpredictable combination of both inhibition and stimulation of growth.

10. Serum growth factors such as PDGF and TGF-β tend to stimulate fibroblastic overgrowth in mixed cultures. They may also induce differentiation in some epithelia, removing them from the proliferative state.

11. Standardization of experimental and production protocols is difficult.

Selective Media

One of the major advantages of the control over growth-promoting activity afforded by serum-free media is in the ability to make a medium selective for a particular cell type (Table 7.7). The long-standing problem of overgrowth by stromal fibroblasts can now be tackled effectively in breast and skin cultures using MCDB 170 and 153, melanocytes can be cultivated in the absence of fibroblasts, and keratinocytes and separate lineages and even stages of development may be selected in hemopoietic cells by selecting the correct growth factor or group of growth factors (see Chapter 20). Add to this the possibility of switching from a growth factor, after necessary amplification of the culture, to a differentiation factor or set of factors, and the amplified culture may then be made to perform one or more specialized functions.

Disadvantages of Serum-Free Media

Unfortunately the transition to serum-free conditions, however desirable, is not as straightforward as it seems. Each cell type appears to require a different recipe and cultures from malignant tumors may vary in requirements from tumor to tumor even within one class of tumors. Removal of serum also requires that the degree of purity of reagents and water and the degree of

TABLE 7.7. Examples of Selective Media

Cells or cell line	Medium	Reference
Fibroblasts	MCDB 202	[McKeehan et al., 1977]
Fibroblasts	MCDB 110	[Bettger et al., 1981]
Keratinocytes	MCDB 153	[Tsao et al., 1982]
Bronchial epithelium	LHC	[Lechner and LaVeck, 1985]
Mammary epithelium	MCDB 170	[Hammond et al., 1984]
Prostate epithelium (rat)	WAJC 401	[McKeehan et al., 1982]
Prostate epithelium (human)	WAJC 404	[McKeehan et al., 1984]
Glial cells		[Michler-Stuke and Bottenstein, 1982]
Melanocytes		[Gilchrest (see Chapter 20)]
Small cell lung cancer	HITES	[Carney et al., 1981]
Adenocarcinoma of lung		[Brower et al., 1986]
Colon carcinoma		[Van der Bosch et al., 1981]
Endothelium	MCDB 130	[Knedler and Ham, 1987]

cleanliness of all apparatus must be extremely high, as the removal of serum also removes the protective, detoxifying action that some serum proteins may have.

Some of the constituents of serum-free media are not commercially available as yet (see Chapter 20) and may require preparation in the lab.

Growth is often slower in serum-free media, fewer generations are achieved with finite cell lines, and some degree of selection may be involved in finite and continuous lines adapting to serum-free medium.

Finally, and often important, availability of serum-free media, subject to proper quality control, is quite limited and the products are often expensive.

Replacement of Serum

The essential factors in serum have been described above and include: (1) adhesion factors such as fibronectin; (2) peptides regulating growth and differentiation such as insulin, PDGF, and TGF-β (a tumor-derived growth factor also extractable from platelets); (3) essential nutrients such as minerals, vitamins, fatty acids, and intermediary metabolites; and (4) hormones regulating membrane transport, phenotypic status, and cell surface constitution such as insulin, hydrocortisone, estrogen, and triiodotyrosine.

Adhesion factors. When serum is removed it may be necessary to treat the plastic growth surface with fibronectin (25–50 μg/ml) or laminin (1–5 μg/ml) added directly to the medium [Barnes et al., 1984a]. Pretreat the plastic with polylysine, 1 mg/ml, and wash off [McKeehan and Ham, 1976]. (See also "Treated Surfaces" above and Barnes et al., 1984a.)

Protease inhibitors. Following trypsin-mediated subculture, the addition of serum inhibits any residual proteolytic activity. Consequently, protease inhibitors such as soya bean trypsin inhibitor or aprotinin (Sigma) must be added to serum-free media after subculture [Rockwell et al., 1980]. Furthermore, because crude trypsin is a complex mixture of proteases, some of which may require different inhibitors, it is preferable to use pure trypsin (e.g., Sigma Gr. III). Alternatively, cells may be washed by centrifugation to remove trypsin.

Special care may be required when trypsinizing cells from serum-free media, as they are more fragile and may need to be chilled to reduce damage [McKeehan, 1977].

Hormones. Hormones that have been used include growth hormone (somatotropin), 50 ng/ml, insulin at 1–10 U/ml, which improves plating efficiency in a number of different cell types, and hydrocortisone, which improves the cloning efficiency of glia and fibroblasts (see Table 7.6 and Chapter 11) and has been found necessary for the maintenance of epidermal keratinocytes and some other epithelial cells (see Chapter 20). Barnes and Sato [1980] described the use of 10 pM 5-tri-iodo tyrosine (T_3) as a necessary supplement for MDCK (dog kidney) cells, and various combinations of estrogen, androgen, or progesterone with hydrocortisone and prolactin at around 10 nm can be shown to be necessary for the maintenance of mammary epithelium (see Chapter 20).

Other hormones with functions not usually associated with the cells they were tested on were found to be effective in replacing serum, e.g., follicle-stimulating hormone (FSH) with B 16 murine melanoma [Barnes and Sato, 1980]. It is possible that sequence homologies exist between some growth-stimulating polypeptides and well-established peptide hormones. Alternatively, processing of some of the large proteins or polypeptides may release active peptide sequences with quite different functions.

Peptide growth factors. The family of polypeptides that has been found to be mitogenic *in vitro* is now quite extensive (Table 7.8) and includes the heparin-

TABLE 7.8. Growth Factors and Mitogens

Systematic abbreviation	Name and synonyms	Mol. Wt. (KDa)	Source[a]	Function
EGF	Epidermal growth factor Urogastrone	6.0	Submaxillary salivary gland (mouse); human urine; guinea pig prostate	Active transport DNA, RNA, protein, synthesis; mitogen for epithelial and fibroblastic cells; synergizes with IGF-1 and TGF-β
FGF-1	Acidic fibroblast gf; aFGF; heparin binding gf 1, HBGF-1; endothelial cell gf (ECGF); myoblast gf (MGF)	13 *h*	Bovine brain; pituitary	Mitogen for endothelial cells
FGF-2	Basic fibroblast gf; bFGF; HBGF-2; Prostatropin;	13 *h*	Bovine brain; pituitary	Mitogen for many mesodermal and neurectodermal cells; adipocyte and ovarian granulosa cell diff.
FGF-3		*h*		
FGF-4		*h*		
FGF-5		*h*		
FGF-6		*h*		
FGF-7	Keratinocyte gf, KGF	*h*	Fibroblasts	Keratinocyte proliferation and diff.; prostate epithelial proliferation and diff.
HGF	Hepatocyte gf, HBGF-8; Scatter factor	*h*	Fibroblasts	Epithelial morphogenesis; hepatocyte proliferation
IFN-α1	Interferon-α1; Leukocyte interferon	18–20		
IFN-α2	Interferon-α2; Leukocyte interferon	18–20		
IFN-β1	Fibroblast interferon	22–27 *g*		
IFN-β2	Fibroblast interferon; IL-6, BSF-2 (see IL-6, below)	22–27 *g*	Activated T-cells; fibroblasts; tumor cells	Keratinocyte diff.; PC12 diff. (see IL-6 below)
IFN-α	Immune interferon; macrophage activating factor (MAF)	20–25		
Ins	Insulin	5.7		Glucose uptake and oxidation; amino acid uptake; glyconeogenesis
IGF-1	Insulin-like gf 1; Somatomedin-C; NSILA-1			Mediates effect of growth hormone on cartilage sulphation; insulin-like activity
IGF-2	Insulin-like gf 2; MSA in rat	7.1	BRL-3A cell-conditioned medium	Mediates effect of growth hormone on cartilage sulphation; insulin-like activity
PDGF	Platelet-derived gf	30	Blood platelets	Mitogen for mesodermal and neurectodermal cells; wound repair; synergizes with EGF and IGF-1
NGF-β	Nerve gf, β subunit	13.5	♂ mouse submaxillary salivary gland	Trophic factor; chemotactic factor; diff. factor
NGF-α	Nerve gf, α-subunit		♂ mouse submaxillary salivary gland	No known function
NGF-γ	Nerve gf, γ-subunit		♂ mouse submaxillary salivary gland	Arginine-specific serine protease
TGF-α	Transforming gf α	5.6		Induces anchorage-independent growth and loss of contact inhibition

(*continued*)

TABLE 7.8. (*Continued*)

Systematic abbreviation	Name and synonyms	Mol. Wt. (KDa)	Source[a]	Function
TGF-β1-6	Transforming gf β (six species)	23–25 *dimer*	Blood platelets	Proliferation inhibitor; squamous diff. inducer
TNF-α	Cachectin	17	Monocytes	Catabolic; cachexia; shock
TNF-β	Lymphotoxin	20–25	Lymphocytes	
Tfn	Transferrin	78	Liver	Iron transport; mitogenic?
Inhibin			Ovary	Pituitary FSH
Activin				
Mullerian Inhibition factor			Testis	Inhibition Mullerian; Inhibition ovarian ca.
G-CSF	Granulocyte colony-stimulating factor; pluripoietin; CSF-β	18–22		Granulocyte prognosis proliferation and diff.
M-CSF	Monocyte/macrophage colony-stimulating factor CSF-1	47–74		Macrophage progenitor proliferation and diff.
GM-CSF	Granulocyte/macrophage colony-stimulating factor CSA; human CSFα	14–35 *g*		Granulocyte/Macrophage progenitor proliferation
EPO	Erythropoietin	34–39 *g*	Juxta-glomerular cells of kidney	Erythroid progenitor proliferation and diff.
	Thrombopoietin			
IL-1α, IL-1β	Lymphocyte activating factor, LAF; hematopoietin-1	31,17.5 *g*		T-cell activation IL-2
IL-2	T-cell gf (TCGF)	15	CD4+ve lymphocytes (NK); murine LBRM-5A4 and human Jurkat FHCRC cell lines	Supports growth of activated T-cells; stimulates LAK cells
IL-3	Multi-CSF; mast cell gf	14–28 *g*	Activated T-cells WEHI-3b phyelomonocytic cell lines	Growth and diff. of early hematopoietic precursors
IL-4	B-cell gf; BCGF-1; BSF-1	15-20	Activated CD4+ve lymphos	Competence factor for resulting B-cells; mast cell maturation (with IL-3)
IL-5	T-cell replacing factor (TRF); eosinophil differentiating factor (EDF) BCGF-2	12-18 *g*	T-lymphocytes	Eosinophil diff.; progression factor for competent B-cells
IL-6	Interferon β2; B-cell stimulating factor (BSF-2); hepatocyte stimulating factor; hybridoma-plasmacytoma gf	22–27 *g*	Activated T-cells macrophage/monocytes; fibroblasts; tumor cells	Acute phase response B-cell diff.; keratinocyte diff.; PC12 diff.
IL-7	Lymphopoietin 1	15–17 *g*	Bone marrow stroma	Pre- and pro-B-cell gf
IL-8	Neutrophil activating protein (NAP-1); NCF; T-lymphocyte chemotactic factor (TCF); monocyte-derived neutrophil chemotactic factor (MDNCF)	8–10 *h*	LPS monocytes PHA lymphocytes; endothelial cells; IL-1 and TNF stimulated fibroblasts and keratinocytes	Chemotactic factor of lymphocytes and polymorphs (heparin binding)
IL-9	Human P-40; mouse T-helper gf; mast cell enhancing activity (MEA)	30–40	CD4+ve T-cells; stimulated by anti-CD4 antibody PHA or PMA	Growth factor for T-helper, megakaryocytes, most cells (with IL-3)
IL-10				
IL-11				
SCF	Stem cell factor			Promotes first maturation division of pluripotent hemopoietic stem cell

(*continued*)

TABLE 7.8. (*Continued*)

Systematic abbreviation	Name and synonyms	Mol. Wt. (KDa)	Source[a]	Function
LIF	Leukemia inhibitory factor; HILDA	24	SCO cells	Inhibits diff. in embryonal stem cells
LPS	Lipopolysaccharide Endotoxin		Bacteria	Stimulates TNF production
PHA	Phytohemagglutinin			Lymphocytes activation
PMA	Pokeweed mitogen			Lymphocytes activation
CT	Cholera toxin		Cholera bacillus	Mitogen for some normal epithelia
TPA/PMA	Phorbol meristate acetate		Croton oil	Mitogen for some epithelia and melanocytes; diff. factor for HL-60; tumor promoter

[a]Sources described are some of the original tissues from which the natural product was isolated. In many cases the natural product has been replaced by cloned recombinant material, available commercially (Life Technologies, Upstate Biochemicals, ICN-Flow, Becton Dickinson).
Abbreviations: diff. = differentiation; gf = growth factor; *g* = glycosylated; *h* = heparin binding.
Information in this table was taken from Barnes et al. [1984a], Lange et al. [1991], and Jenkins [1992].

binding growth factors (the FGF family), EGF, PDGF [Barnes et al., 1984a,c], IGF-1 and −2, and the interleukins [Thomson, 1991] active in the 1–10 ng/ml range. Until recently it was felt that the bulk of these peptide growth factors had very low specificity, but recently a number of new factors have been isolated, mostly active in the hemopoietic system [Barnes et al., 1984d] (see Chapter 20), but also acting on non-hemopoietic cells such as keratinocytes (KGF) [Aaronson et al., 1991] and hepatocytes (HGF) [Kenworthy et al., 1992] (see Chapter 20). Both the hemopoietic factors and solid tissue factors are often highly purified, many are genetically engineered, and several are commercially available (Becton Dickinson, Life Technologies, UBI).

The range of commercially available growth factors has extended rapidly due to the development of recombinant genetic techniques at the biotechnological level. While these are still often very expensive ($1,000/μg!), it is to be hoped that increasing demand together with improved production techniques will make these products more readily available to the more modest laboratories. Many of the growth factors that are required in culture (interferons, interleukins, heparin-binding growth factors) are acquiring pharmaceutical potential and this may create a market that would bring their research availability within the reach of noncommercial laboratories.

Growth factors may act synergistically or additively with each other or prostaglandin $F_2\alpha$ and hydrocortisone [Westermark and Wasteson, 1975; Gospodarowicz, 1974]. Some growth factors are dependent on the activity of a second growth factor before they will act [Phillips and Christofalo, 1988]. Bombesin is not mitogenic alone in normal cells but requires the simultaneous or prior action of insulin or one of the IGFs [Aaronson et al., 1991].

Nutrients. Iron, copper, and a number of minerals have been included in serum-free recipes, although the evidence for a positive requirement for some of the rarer minerals is still lacking. Selenium (Na_2SeO_3), at around 20 n*M*, is found in most formulas and there appears to be some requirement for lipids or lipid precursors such as choline, linoleic acid, ethanolamine, or phosphoethanolamine.

Proteins and polyamines. The inclusion of proteins such as bovine serum albumin (BSA) or tissue extracts often increases growth and survival but adds undefined constituents to the medium. BSA, fatty acid free, is used at 1–10 mg/ml. Transferrin, at around 10 ng/ml, is required as a carrier for iron and may also have a mitogenic role. Putrescine is used at 100 n*M*.

Selection and Development of Serum-Free Medium

In some cases recipes may already exist in the literature and may be used with only minor modification. Otherwise it will be necessary to develop a new formulation.

There are two general approaches to the development of serum-free medium for a particular cell line or primary culture [Maurer, 1992]. The first is to take a known recipe for a related cell type and alter the constituents individually until the medium is optimized for your own particular requirement. This has been the approach adopted by Ham and co-workers [Ham, 1984] and generally will provide the optimal conditions. However, it is a very time-consuming and laborious process, involving growth curves and clonal growth assays at each stage, and it is not unreasonable to expect to spend at least 3 years developing a new medium for a new cell type.

This has led to the second approach, using existing media or combinations such as RPMI 1640 [Carney et al., 1981] or Ham's F12 and DMEM [Barnes and Sato, 1980] and restricting the manipulation of the constituents to a shorter list of such substances as selenium, transferrin, albumin, insulin, hydrocortisone, estrogen, triiodotyrosine, ethanolamine, phosphoethanolamine, growth factors (EGF, FGF, PDGF, endothelial growth supplement, etc.), prostaglandins (PGE_1, $PGF_{2\alpha}$), and any others that may have special relevance to your own system. Among these, selenium, transferrin, and insulin will usually be found to be essential, the requirement for the others more variable.

So far there seem to be no clear guidelines to indicate which supplements may be required. Some may be fairly universal, like insulin, transferrin, and selenium, while others such as estrogens, androgens, and T_3 may be more specific for individual cell types (though not necessarily those cell types that would be traditionally associated with those hormones). As with medium and serum selection, trial and error may be the only method to select the correct supplements. If a group of compounds is found to be effective in reducing serum supplementation, the active constituents may be identified by systematic omission of single components, and then their concentrations optimized [Ham, 1984].

Preparation of Serum-Free Media

A number of recipes are now available for particular cell types [Maurer, 1992] (Tables 7.6 and 7.7; see Chapter 20). Some of these are available commercially (see Trade Index); others will have to be made up. The procedure for making up serum-free recipes is similar to regular media (see above) [Waymouth, 1984]. Ultrapure reagents and water should be used and care taken with solutions of Ca^{2+} and Fe^{2+} or Fe^{3+} to avoid precipitation. Metal salts tend to precipitate in alkaline pH in the presence of phosphate, particularly when autoclaved, so cations in stock solutions should be kept at a low pH (below 6.5) and phosphate-free. They should be sterilized by autoclaving or filtration (see Chapter 8). It is often recommended that these be added last, immediately before use. Otherwise the constituents are generally made up as a series of stock solutions, minerals and vitamins 1,000×, tyrosine, tryptophan, and phenylalanine in 0.1 *N* HCl at 50×, essential amino acids 100× in water, salts 10× in water, and any other special cofactors, lipids, etc., 1,000× in the appropriate solvents. These are combined in the correct proportions, diluted to the final concentration, and the pH and osmolality checked.

Growth factors, hormones, and cell adhesion factors are best added separately just before use, as these may need to be adjusted to suit particular experimental conditions.

Commercially Available Serum-Free Media

Several suppliers (ICN/Flow, GIBCO, Clonetics [see Trade Index]) now make serum-free media. Most of these are designed primarily for hybridoma culture, but others are applicable to other cell types. While evidence exists in the literature for the use of media with specific cell types (see Table 7.9), commercial recipes are often a trade secret and you can only rely on the supplier's advice, or, better, screen a number of media over several subcultures with your own cells. This can be an extensive exercise but is justified if you are planning a long program of work with the cells.

Serum Substitutes

A number of products have been developed commercially to replace all or part of the serum in conventional media. These include SerXtend (NEN), Ventrex (Ventrex Laboratories Ltd.), Nutricyte (Brooks Laboratories), Nu-serum (Collaborative Research), and CLEX (Dextran Products Ltd.). While these may offer a degree of consistency not obtainable with regular sera, batch variations can still occur and their constitution is not fully defined. They may be useful as an *ad hoc* measure or to increase economy but are not a replacement for serum-free media. Nutridoma (BCL), ITS (Flow), SIT (Sigma), Selectakit (GIBCO), and Ultraser-G (LKB) are defined supplements to replace serum, partially or completely [Maurer, 1992].

Conclusions

However desirable serum-free conditions may be, there is no doubt that the relative simplicity of retaining serum, the lack of a reliable source of most serum-free media, the specialization in technique required for their use, and the considerable investment in time, effort, and resources that must go into preparing new recipes or even existing ones all act as considerable deterrents to most laboratories to enter the serum-free arena. There is no doubt, however, that the need for consistent and defined conditions for the investigation of regulatory processes governing growth and differentiation, the pressure from biotechnology for easier product purification, and the gradually worsening situation worldwide in the supply of serum will eventually force the adoption of serum-free media on a more general scale. But first, recipes must be found that are less temperamental than some in current use and that can be used with equal facility and effectiveness in different laboratories.

TABLE 7.9. Selecting a Suitable Medium[a]

Cells or cell line	With serum: Medium	With serum: Serum	Serum-free medium
Chick embryo fibroblasts	Eagle's MEM	CS	MCDB 202
Chick embryo pigmented retina and cartilage	Ham's F10		
Chinese hamster ovary (CHO)	Eagle's MEM	CS	MCDB 302
Chondrocytes	F12	FB	[Adolphe, 1984]
Continuous cell lines	Eagle's MEM	CS	199 [Waymouth, 1984] MB752/1, MD7505/1 [Kitos et al., 1962], CMRL1066
Endothelium	DME, 199, MEM	FB	MCDB 130
Fibroblasts	Eagle's MEM	CS	MCDB 110, 202, 402
Glial cells	MEM, SF12	FB	[Michler-Stuke and Bottenstein, 1982]
Glioma	MEM, SF12	FB	
HeLa cells	Eagle's MEM	CS	[Blaker et al., 1971]
Hemopoietic cells	RPMI 1640, Fischer's	FB	[Iscove, 1984] (see also Chapter 20)
Human diploid fibroblasts	Eagle's MEM	CS	MCDB 110, 202
Human leukemia	RPMI 1640	FB	[Breitman et al., 1984; Iscove, 1984]
Human tumors	SF12, L15, RPMI 1640, DME	FB	See Table 7.6
L cells (L929, LS)	Eagle's MEM	CS	[Birch and Pirt, 1970, 1971; Higuchi, 1977]
Lymphoblastoid cell lines (human)	RPMI 1640	FB	[Iscove, 1984]
MDCK dog kidney epithelium	DME, F10/DME	FB	[Taub, 1984]
Melanocytes	See Chapter 20		
Melanoma	MEM, SF12	FB	[Barnes and Sato, 1980]
Mouse embryo fibroblasts	Eagle's MEM	CS	MCDB 402
Mouse leukemia	Fischer's RPMI 1640	FB, HoS	[Murakami, 1984]
Mouse erythroleukemia	SF12, RPMI 1640	FB, HoS	[Iscove, 1984]
Mouse myeloma	DME, RPMI 1640	FB	[Murakami, 1984], HB101 (Hana Biologics)
Mouse neuroblastoma	DME, F12/DME	FB	MCDB411
NRK rat kidney fibroblasts	MEM, DME	CS	
Rat minimal deviation hepatoma (HTC, MDH)	Swim S77, SF12	FB	
Skeletal muscle	DME, F12	FB, HoS	F12
Syrian hamster fibroblasts, e.g., BHK 21	MEM, GMEM, DME	CS	[Pardee et al., 1984]
3T3 cells	MEM, DME	CS	MCDB402

[a]Abbreviations: CS, calf serum; FB fetal bovine serum; HoS, horse serum. SF12 is Ham's F12 plus Eagle's essential amino acids and nonessential amino acids as in DME (available as 100× stock from Flow Labs (GIBCO, etc.). Further recommendations on the choice of media can be found in McKeehan [1977], Barnes et al. [1984(a–d)], and Maurer [1992].

SELECTION OF MEDIUM AND SERUM

Information for the selection of the appropriate medium for a given cell type is usually available in literature articles on the origin of the cell line or the culture of similar cells, or from the source of the cells, for cell lines currently available (Table 7.9; see Chapter 20). Failing this, the choice is either empirical or by comparative testing of several media (see below). Many continuous cell lines (e.g., HeLa, L-cells, BHK21), primary cultures of human, rodent, and avian fibroblasts, and cell lines derived from them can be maintained on a relatively simple medium such as Eagle's MEM, supplemented with calf serum. More complex media may be required where a specialized function is being expressed (see also Chapter 14) or when cells are passaged at low seeding density ($<10^3$/ml), as for cloning (see Chapter 11). Frequently the more demanding culture conditions that require complex media also require fetal bovine serum rather than calf or horse serum, unless the formulation specifically allows for the omission of serum.

Some examples of cell types and the media used for them are given in Table 7.9, but this list is neither exhaustive nor binding. For a more complete list, see Maurer [1992], Barnes et al. [1984a–d], Ham and McKeehan [1979], and Morton [1970]. If a clear indication of the correct culture conditions is not available, a simple cell growth experiment with commercially available media and multiwell plates (see Chapters 18 and 19) can be carried out in about 2 weeks. Assaying for clonal growth (see Chapters 11 and 19) and measuring the expression of specialized functions may narrow the choice further. You may be surprised to find that your best conditions do not agree with the literature, and you will have to decide between the optimal growth and behavior of the cells as you find them. Reproducing the conditions found in another laboratory may be difficult due to variations in the impurities present in reagents and water and in batches of serum, if present. It is to be hoped that as serum requirements are reduced and reagent purity increases, medium standardization will improve and the need for such a choice will not arise.

Finally, you may have to compromise in your choice of medium or serum because of cost. Autoclavable media are available from commercial suppliers (ICN Flow, GIBCO). They are simple to prepare from powder and are suitable for many continuous cell strains. They may need to be supplemented with glutamine for most cells and will usually require serum. The cost of serum should be calculated on the basis of medium volume where cell yield is not important, but where the objective is to produce large quantities of cells, calculate serum costs on a per cell basis. If a culture grows to 10^6/ml in serum A and 2×10^6 ml in serum B, serum B becomes the less expensive by a factor of two, given that product formation or other specialized function is as efficient.

If fetal bovine serum seems essential, try mixing it with calf serum. This may allow you to reduce the concentration of the more expensive fetal serum. If you can, leave out serum altogether, or reduce the concentration and use one of the serum-free formulations suggested above.

Batch Reservation

Serum standardization is difficult, as batches vary considerably and one batch will only last about 6 months to 1 year, stored at −20°C. Select the type of serum that is most appropriate and request batches to test from a number of suppliers. Most serum suppliers will normally reserve a batch until a customer can select the most suitable one (provided this does not take longer than 3 weeks or so). When a suitable batch has been selected, the supplier is requested to hold it for up to 1 year for regular dispatch. Other suppliers should also be informed so that they may return theirs to stock.

Testing Serum

The quality of a given serum is assured by the supplier, but their quality control is usually performed with one of a number of continuous cell lines. If your requirements are more discriminating, then you will need to do your own testing. There are four main parameters for testing serum, as follows.

Cloning efficiency. During cloning, the cells are at a low density and hence are at their most sensitive, making this a very stringent test. Plate the cells out at 10 to 100 cells/ml and look for colonies after 10 days to 2 weeks. Stain and count the colonies (see Chapters 11 and 19) and look for differences in cloning efficiency (survival) and colony size (cell proliferation). Each serum should be tested at a range of concentrations from 2% to 20%. This will reveal whether one serum is equally effective at a lower concentration, thereby saving money and prolonging the life of the batch, and will show up any toxicity at high serum concentration.

Growth curve. A growth curve should be performed in each serum (see Chapter 18) determining the lag period, doubling time, and saturation density (density at "plateau"). A long lag implies that the culture is having to adapt; short doubling times are preferable if you want a lot of cells quickly; and a high saturation density will provide more cells for a given amount of serum and will be more economical.

Preservation of cell culture characteristics. Clearly the cells must do what you require of them in the new serum, whether they are acting as host to a given virus, producing a certain cell product, or expressing a characteristic sensitivity to a given drug.

Sterility. Serum from a reputable supplier will have been tested and shown to be free of micro-organisms, but occasionally odd bottles or parts of a batch may slip through even the most stringent sampling procedures. To be certain, you should grow cells in antibiotic-free medium supplemented with the serum, look for any microbiological contamination, and stain the cells for mycoplasma (see Chapter 16).

The previous precautions and tests are advisable in selecting a new batch of serum partly because of the investment both in terms of cash and cell lines, but also because of the need for stringent control of your culture conditions. However, it is not always within the capacity of a small laboratory to cover all these requirements, and shortcuts may have to be taken. If this is the case, you may be obliged to accept the assurances of the supplier regarding sterility and growth-promoting capacity, but you must still test the serum for any special functional requirements.

OTHER SUPPLEMENTS

Tissue extracts and digests have been used since tissue culture was first developed as supplements to tissue culture media in addition to serum. Many are derived from microbiological culture techniques and autoclavable broths, e.g., bactopeptone, tryptose, and lactalbumin hydrolysate, which are proteolytic digests of beef heart or lactalbumin, and contain mainly amino acids and small peptides (Difco). Bactopeptone and tryptose may also contain nucleosides and other heat-stable tissue constituents such as fatty acids and carbohydrates.

Embryo Extract

Embryo extract is a crude homogenate of 10-day chick embryo clarified by centrifugation (see Reagent Appendix). The crude extract was fractionated by Coon and Cahn [1966] to give high and low molecular weight fractions. The low molecular weight fraction promoted cell proliferation, while the high molecular weight fraction promoted pigment and cartilage cell differentiation. Although these fractions were not fully characterized, more recent evidence would suggest that the low molecular weight fraction probably contained peptide growth factors and the high molecular weight fraction proteoglycans and other matrix constituents.

Embryo extract was originally used as a component of plasma clots (see Chapters 1 and 9) to promote cell migration from the explant. It has been retained in some organ culture techniques (Chapter 22) and can still be used in nerve and muscle culture (Chapter 20), although this is gradually being replaced with defined growth factors and matrix components. It has been shown that embryo extract can be replaced by hemin in the induction of skeletal muscle differentiation [Smith and Schroedl, 1992].

Conditioned Medium

Puck and Marcus [1955] found that the survival of low-density cultures could be improved by growing the cells in the presence of feeder layers (see Chapter 11). While part of this effect may have been due to conditioning of the substrate, the main effect was presumed to be conditioning of the medium by release of small molecular metabolites and macromolecules into the medium [Takahashi and Okada, 1970]. Hauschka and Konigsberg [1966] showed that the conditioning of culture medium necessary for the growth and differentiation of myoblasts was due to collagen released by the feeder cells. Feeder layers and conditioning of the medium by embryonic fibroblasts or other cell lines remains a useful method of culturing difficult cells [e.g., Stampfer et al., 1980] (see Chapter 11).

Attempts have been made to isolate active fractions from conditioned medium, and the original supposition is still probably close to the correct interpretation. Conditioned medium will contain both substrate-modifying matrix constituents like collagen, fibronectin, and proteoglycans, and growth factors such as those of the heparin-binding group (FGF, etc.), insulin-like growth factors (IGF-1 and −2), PDGF, and several others (see Table 7.8), in addition to the intermediary metabolites previously proposed.

They represent undefined components of medium that should be eliminated by determination of the active constituents but, nevertheless, often represent an easy alternative to many months of tedious medium optimization and may be the only economical route to successful culture, given that you do not wish to invest the time on development of culture conditions rather than on the specific problem of interest.

INCUBATION TEMPERATURE

The optimal temperature for cell culture is dependent on (1) the body temperature of the animal from which the cells were obtained, (2) any regional variation in temperature (e.g., skin and testis may be lower), and (3) the incorporation of a safety factor to allow for minor errors in incubator regulation. Thus, the temperature recommended for most human and warm-blooded an-

Fig. 7.11. *Examples of ridges seen in cultured monolayers in dishes and flasks, probably due to resonance in the incubator from fan motors or to opening and closing the incubator. (Courtesy of Nunc.)*

imal cell lines is 36.5°C, close to body heat but set a little lower for safety, as overheating is a more serious problem than underheating.

Avian cells, because of the higher body temperature in birds, should be maintained at 38.5°C for maximum growth but will grow quite satisfactorily, if more slowly, at 36.5°C.

Cultured cells will tolerate considerable drops in temperature, can survive several days at 4°C, and can be frozen and cooled to −196°C (see Chapter 17), but they cannot tolerate more than about 2°C above normal (39.5°C) for more than a few hours, and will die quite rapidly at 40°C and over.

Attention must be paid to consistency of temperature (with time) to ensure reproducible performance. Doors of incubators or hot rooms must not be left open longer than necessary, and large items or volumes of liquid, placed in the warm room to heat, should not be placed near any cultures. The spatial distribution of temperature within the incubator or hot room must also be uniform (see Chapter 3); there should be no "cold spots," and air should circulate freely. This means that a large number of flasks should not be stacked together when first placed in the incubator or hot room; space must be allowed between them for air circulation.

Another problem is in the design of perforated shelving, which sometimes causes nonuniform distribution of cells (see Chapter 3). Vibration, caused by opening and closing of the incubator, a faulty fan motor, or adjacent people or equipment, can cause perturbation of the medium, which can result in resonance or standing waves in the flask, which, in turn, results in a wave pattern in the monolayer (Fig. 7.11), implying variations in cell density. Elimination of vibration and minimizing entry into the incubator will help to reduce this.

Nonrandom distribution of cells during incubation can be most problematical when associated with cloned cultures, when colonies form preferentially at the center of the plate. This can be due to incorrect seeding, either from seeding the cells into the center of a plate already containing medium or due to swirling the plate such that cells tend to focus in the center, but it can also be due to resonance in the incubator. Placing a heavy weight in the tray or box with the plates and separating it from the shelf with plastic foam may help to minimize this problem [Nielsen, 1989], but great care must be taken to wash and sterilize such foam pads, as they will tend to harbor contamination.

Much of the preceding discussion has been based on observations with warm-blooded animals but is, nevertheless, applicable in principle to lower vertebrates and perhaps, to a lesser extent, to invertebrates. Temperature must be considered separately, however. In general the cells of poikilothermic animals have a wide temperature tolerance but should be maintained at a constant level within the normal range of the donor species. This requires incubators with cooling as well as heating, as the incubator temperature may need to be below ambient (e.g., for fish). As for a hot room, cooling capacity should be sufficient to lower the temperature by about 2°C, or more, below ambient so that regulation is performed by the heater circuit that is more sensitive.

If necessary, poikilothermic animal cells can be maintained at room temperature, but the variability of the ambient temperature in laboratories makes this undesirable.

Regulation of temperature should be kept within ±0.5°C, as consistency is more important than accuracy. As cells vary in growth rate and metabolism, dependent on the temperature, the incubation temperature should be kept constant both in time and at different parts of the incubator. Water baths give the most accurate control of temperature but present problems of contamination, particularly since the flasks need to be immersed for proper temperature control. They are, therefore, seldom used and incubators are preferable. The air should be circulated by a fan to give even temperature distribution, and cultures should be placed on perforated shelves and not on the floor or touching the sides of the incubator. Further discussion of temperature control in hot rooms is given in Chapter 3.

Otherwise, temperature will be determined by the origin of the cells. Poikilotherms (cold-blooded animals that do not regulate blood heat within narrow limits) will tolerate a wide range between 15°C and 30°C, while homiotherms (which regulate blood temperature within narrow limits) will require 36–38°C for mammals and 37–39°C for birds. Generally reducing the temperature will reduce growth but not viability, unless hypothermy is prolonged (>3 days), while increasing the temperature will initially increase growth but reduce viability quite rapidy as the temperature reaches, and rises above, 39°C.

A number of temperature-sensitive mutant cell lines have been developed, which allow expression of specific genes below a set temperature but not above it [Su et al., 1991; Foster and Martin, 1992; Wyllie et al., 1992]. These mutants facilitate studies on cell regulation, but also emphasize the narrow range within which one can operate, as the two discriminating temperatures are usually only about 2–3°C apart.

Superficial tissues, such as epidermis or testis, may survive better at lower temperatures, but, in general, unless there is a selective advantage for a particular cell type, most laboratories will not distinguish among cell types and select an average temperature of 36.5°C, below most mammalian blood temperatures but sufficient for most cells to survive and grow adequately, while allowing for some that prefer a lower temperature and protecting against override in incubator thermostat regulation, which could soon become harmful if cells were exposed to hyperthermic conditions for more than an hour or so (see also Chapter 4, Incubators).

CHAPTER 8

Preparation and Sterilization

All stocks of chemicals and glassware used in tissue culture should be reserved for that purpose alone. Traces of heavy metals or other toxic substances can be difficult to detect other than by a gradual deterioration of your cultures. It also follows that separate stocks imply separate glassware washing. The requirements of tissue culture washing are higher than for general glassware; a special detergent may be necessary (see below) and cross-contamination from chemical glassware must be avoided. As most laboratories propagate cells in disposable plastic flasks, the sensitivity of monolayer attachment is less of a problem. Nevertheless, whenever glass is used, whether for storage or culture, there remains a problem of chemical contaminants leaching out into media or reagents, and absolute cleanliness is therefore essential.

All apparatus and liquids that come in contact with cultures or other reagents must be sterile. A summary of the procedures used is given in Tables 8.1 and 8.2.

APPARATUS

Glassware

Items of glassware used for dispensing and storage of media and for cell culture must be cleaned very carefully to avoid traces of toxic materials, contaminating the inner surfaces, becoming incorporated into the medium (Fig. 8.1). Where the glass surface is to be used for cell propagation it must not only be clean but also carry the correct charge. Caustic alkaline detergents render the surface of the glass unsuitable for cell attachment and require subsequent neutralization with HCl or H_2SO_4, but many modern detergents do not alter the glass surface and can be removed completely. For the most effective washing procedure: (1) Do not let soiled glassware dry out. A sterilizing agent, such as sodium hypochlorite, should be included in the water used to collect soiled glassware (a) to remove any potential biohazard and (b) to prevent microbial contamination growing up in the water; (2) select a detergent that is effective in the water of your area, rinses off easily, and is nontoxic (see below); (3) before drying, ensure that the glassware is thoroughly rinsed in tap water and deionized or distilled water; (4) dry inverted; and (5) sterilize by dry heat to minimize the risk of depositing toxic residues from steam sterilization.

Plastic culture flasks are, on the whole, meant for single use, as washing detergent renders them unsuitable for cell propagation (in monolayer) and resterilization is difficult. Cells may be reseeded back into the same flask after subculture but this tends to increase

TABLE 8.1. Sterilization of Equipment and Apparatus

Item	Sterilization
Ampules for freezer, glass	Dry heat[a]
Ampules for freezer, plastic	Autoclave[b]
Apparatus containing glass and silicone tubing	Autoclave
Disposable tips for micro-pipettes	Autoclave, in autoclavable trays or nylon bags
Filters: Millipore, Sartorius	Autoclave-do not prevac. or postvac.
Glassware	Dry heat
Glass bottles with screw caps	Autoclave, with cap slack
Glass coverslips	Dry heat
Glass slides	Dry heat
Glass syringes	Autoclave (remove piston if PTFE)
Instruments	Dry heat
Magnetic stirrer bars	Autoclave
Pasteur pipettes, glass	Dry heat
Pipettes, glass	Dry heat
Plexiglas, Perspex, Lucite	70% EtOH (see text)
Polycarbonate	Autoclave
Repeating pipettes/syringes	Autoclave (remove PTFE pistons from glass barrels)
Screw caps	Autoclave
Silicone tubing	Autoclave
Stoppers: rubber, silicone	Autoclave
Test tubes	Dry heat

[a]Dry heat, 160°C/l hr.
[b]Autoclave, 100 kPa (1 bar/15 1b/in^2) 121°C for 20 min.

the risk of contamination. However, with care, cells that grow in suspension can be subcultured in the same flask for many generations.

Sterilization procedures are designed not just to kill the bulk of micro-organisms but to eliminate spores that may be particularly resistant. Moist heat is more effective than dry heat but does carry a risk of leaving a residue. Dry heat is preferable but at a minimum of 160°C for 1 hr. Moist heat (for fluids and perishable items) need only be maintained at 121°C for 15–20 min (Fig. 8.1). For moist heat to be effective, steam penetration must be assured and for this the sterilization chamber must be evacuated prior to steam injection. Insertion of Thermalog indicators (Johnsen and Jorgensen) monitors both temperature and humidity during sterilization.

Materials

Pipette cylinders (to collect used pipettes)
Disinfectant (if required)
Detergents
Soaking baths
Bottle brushes
Stainless steel baskets (to collect washed and rinsed glassware for drying)
Aluminum foil
Sterility indicators (Browne's tubes, Thermalog indicators)
Glass petri dishes (for screw caps)
Autoclavable plastic film (Portex, Cedanco) or paper sterilization bags
Sterile-indicating autoclave tape

Collection and Washing

1. Collect immediately after use into detergent containing a disinfectant such as sodium hypochlorite. It is important that apparatus should not dry before soaking, or cleaning will be much more difficult.
2. Soak overnight in detergent (see below).
3. Machine wash, or brush by hand or machine (Fig. 8.2), the following morning and rinse thoroughly in four complete changes of tap water followed by three changes of deionized water. Machine rinses, if done on a spigot header (see Chapter 4), can be reduced to two tap water and one deionized or reverse osmosis water. If rinsing is done by hand, a sink spray (Fig. 4.3) is a useful accessory; otherwise bottles must be emptied and filled completely each time. Clipping bottles in a basket will help to speed up this stage.
4. After rinsing thoroughly, invert bottles, etc., in stainless steel wire baskets and dry upside down.
5. Cap with aluminum foil when cool and store.

Sterilization

1. Place in an oven with fan-circulated air at 160°C.
2. Check that the temperature has returned to 160°C, seal the oven with a strip of tape with the time recorded on it, or use automatic locking, and leave for 1 hr; ensure that the center of the load achieves 160°C by using a sterility indicator or recording thermometer with the sensor in the middle of the load. Do not pack the load too tightly; leave room for hot air circulation.
3. After 1 hr, switch off the oven and allow to cool with the door closed. It is convenient to put the oven on an automatic timer so that it can be left to switch off on its own overnight and be accessed in the morning. This allows for cooling in a sterile environment, and also minimizes the heat generated during the day, when it is hardest to deal with.
4. Use within 24–48 hr. Alternatively, bottles may be loosely capped with screw caps, autoclaved for 20 min at 121°C with prevacuum and post-vacuum cycle (see Chapter 4), and the caps

tightened when cool. Caps must be very slack (one complete turn) during autoclaving to allow entry of steam and to prevent the liner being sucked out of the cap and sealing the bottle. If the bottle seals during sterilization in an autoclave, sterilization will not be complete. Unfortunately, misting often occurs when bottles are autoclaved and a slight residue may be left when this evaporates. There is also a risk of the bottles becoming contaminated as they cool by drawing in nonsterile air before they are sealed. Dry-heat sterilization is better, allowing the bottles to cool down within the oven before removal.

Pipettes

1. Place in water with detergent (e.g., 1% Decon) and a sterilizing agent (hypochlorite or glutaraldehyde), tip first, immediately after use (Fig. 8.3). Do not put pipettes that have been used with agar or silicones (water repellents, antifoams, etc.) in the same cylinder as regular pipettes. Use disposable pipettes for silicones and either rinse agar pipettes after use in hot tap water or use disposable pipettes.
2. Soak overnight and remove plugs with compressed air the following morning.
3. Transfer to pipette washer (Fig. 8.4; see also Fig. 4.3), tips uppermost.
4. Rinse by siphoning action of pipette washer for a minimum of 4 hr or in an automatic washing machine with a pipette adaptor.
5. Turn over valve to deionized water (see Fig. 4.3), or wait until last tap water finally runs out, turn off tap water, and empty and fill three times with deionized water (automatic deionized rinse cycle in machine).
6. Transfer to pipette drier or drying oven and dry with tips uppermost.
7. Plug with cotton (Fig. 8.5). Alternatively, pipette plugs may be dispensed with and a short sterile filter tube placed on the pipette bulb during use (see Fig. 5.2). To avoid cross-contamination, this *must* be replaced before starting work and *every time* that you change to a new cell strain.
8. Sort pipettes by size and store dust-free.

Sterilization

1. Place pipettes in pipette cans (square aluminum or stainless steel with silicone cushions at either end; square cans do not roll on the bench) and label both ends of the cans. Fill each can with one pipette size with a few cans containing an assortment of 1-ml, 2-ml, 10-ml, and 25-ml pipettes in the ratio 1:1:3:2.
2. Seal with sterile-indicating tape and sterilize by dry heat for 1 hr at 160°C. This is different from the sterile-indicating tape used in autoclaves, as the sterilizing temperature is higher. Most tend to char and release traces of volatiles from the adhesive, which can leave a deposit on the oven, and potentially on the pipettes. Use the smallest amount of tape possible or replace with temperature indicator tabs (Bennett), which are small and have less volatile material. The temperature should be measured in the center of the load, to ensure that this, the most difficult part to reach, attains the minimum sterilizing conditions. Leave spaces between cans when loading the oven, to allow for circulation of hot air.
3. Remove from oven, allow to cool, and transfer cans to tissue culture laboratory. If you anticipate that pipettes will lie for more than 48 hr before use, seal cans around the cap with adhesive tape.

Screw Caps

There are two main types of caps for glass bottles in common use: (1) aluminum or phenolic plastic caps with synthetic rubber or silicone liners and (2) wadless polypropylene caps, reusable (Duran), which are deeply shrouded and have ring inserts to improve the seal and improve pouring (although pouring is not recommended in sterile work), or disposable (Johnsen and Jorgensen). Disposable polypropylene caps will only seal if screwed down very tightly on a bottle with no chips or imperfections on the lip of the opening. Do not leave aluminum caps or any other aluminum items in alkaline detergents for more than 30 min, as they will corrode. Do not have glassware in the same detergent bath or the aluminum may contaminate the glass. Avoid machine washing detergents as they are very caustic.

Reusable caps. Soak 30 min in detergent and rinse thoroughly for 2 hr (make sure all caps are submerged). Liners should be removed and replaced after rinsing. Rinsing may be done in a beaker (or pail) with running tap water led by a tube to the bottom. Stir the caps by hand every 15 min. Alternatively, place in a basket or, better, in a pipette washing attachment, and rinse in an automatic washing machine, but do not use detergent in the machine.

Disposable caps. Disposable caps should not need to be washed unless reused. They may be washed and rinsed by hand as above (extending the detergent soak if necessary). Because these caps may float, they must be weighted down during soaking and rinsing. For au-

TABLE 8.2. Sterilization of Liquids

Solution	Sterilization	Storage
Agar	Autoclave[a]	Room temperature
Amino acids	Filter[b]	4°C
Antibiotics	Filter	−20°C
Bacto-peptone	Autoclave	Room temperature
Bovine serum albumin	Filter–use stacked filters (see text)	4°C
Carboxylmethyl cellulose	Steam, 30 min[c]	4°C
Collagenase	Filter	−20°C
DMSO	Self-sterilizing, aliquot into sterile tubes	Room temperature; keep dark, avoid contact with rubber or plastics (except polypropylene)
Drugs	Filter (check for binding; use low binding filter, e.g., Millex-GV, if necessary)	−20°C
EDTA	Autoclave	Room temperature
Glucose, 20%	Autoclave	Room temperature
Glucose, 1–2%	Filter (low concentrations caramelize if autoclaved)	Room temperature
Glutamine	Filter	−20°C
Glycerol	Autoclave	Room temperature
Growth factors	Filter (low protein binding, e.g. Millex-GV)	−20°C
HEPES	Autoclave	Room temperature
HCl 1 *N*	Filter	Room temperature
Lactalbumin hydrolysate	Autoclave	Room temperature
Methocel	Autoclave	4°C
$NaHCO_3$	Filter	Room temperature
NaOH 1 *N*	Filter	Room temperature
Phenol red	Autoclave	Room temperature
Salt solutions (without glucose)	Autoclave	Room temperature
Serum	Filter; use stacked filters (see text)	−20°C
Sodium pyruvate, 100 m*M*	Filter	−20°C
Transferrin	Filter	−20°C
Tryptose	Autoclave	Room temperature
Trypsin	Filter	−20°C
Vitamins	Filter	−20°C
Water	Autoclave	Room temperature

[a]Autoclave, 100 kPa (15 lb/in^2) 121°C for 20 min.
[b]Filter, 0.2-μm pore size.
[c]Steam, 100°C for 30 min.

tomatic washers, use pipette washing attachment and normal cycle with machine detergent.

Stoppers. Use silicone or heavy metal–free white rubber stoppers in preference to natural rubber. Wash and sterilize as for disposable caps. (There will be no problem with flotation in washing and rinsing.)

Sterilization. Place caps in a glass petri dish with the open side down. Wrap in cartridge paper or steam-permeable nylon film (Portex) and seal with autoclave tape (Fig. 8.6). Autoclave for 20 min at 121°C, 100 kPa (1 bar, 15 lb/in^2) (see Fig. 8.8).

Keep organic matter out of the oven. Do not use paper tape or packaging unless you are sure that it will

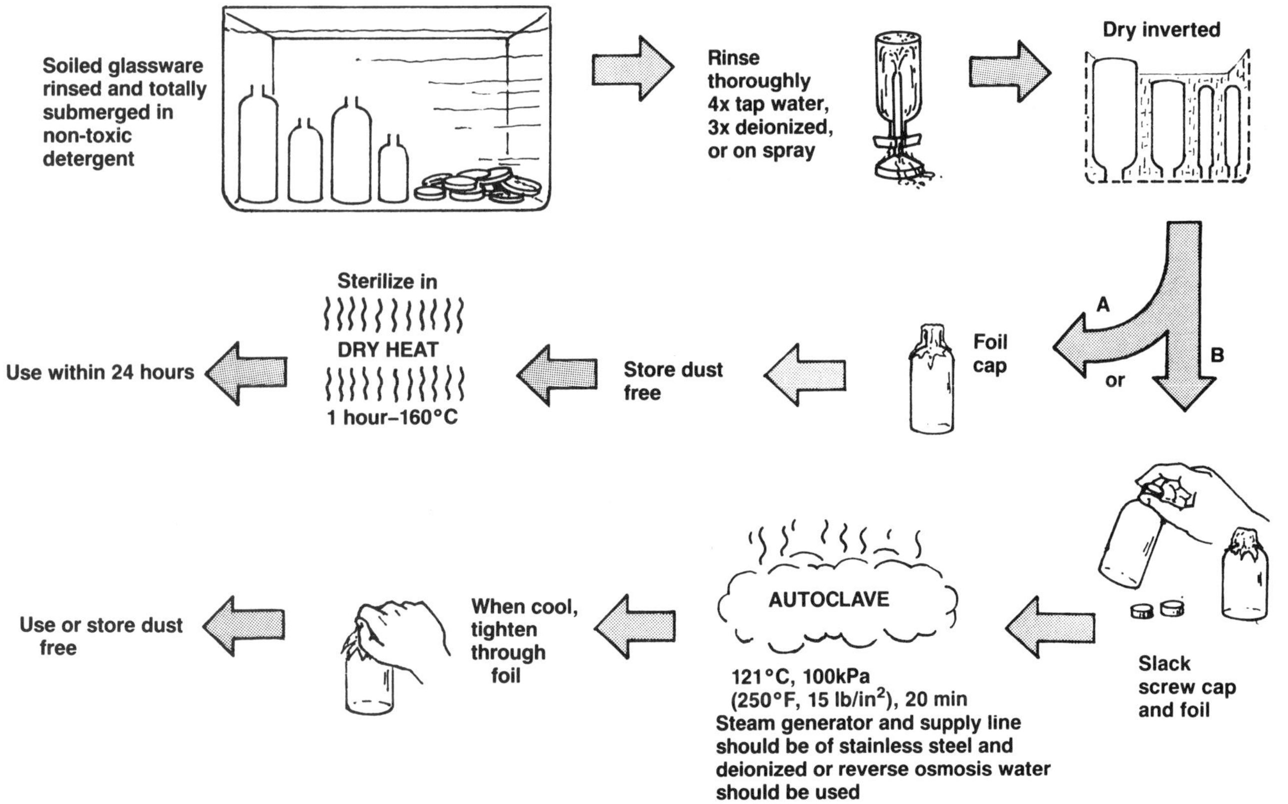

Fig. 8.1. Wash-up and sterilization of glassware.

not release volatile products on heating. Such products will eventually build up on the inside of the oven, making it smell when hot, and some deposition may occur inside the glassware being sterilized.

Fig. 8.2. Motorized bottle brushes. Care must be taken not to press down too hard on the brush lest the bottle break.

Selection of Detergent

Solicit samples from local suppliers and test them (1) for their ability to wash heavily soiled glassware; (2) for the quality of the growth surface afterwards; and (3) for the toxicity of the detergent (see also Chapters 11 and 19).

(a) Washing efficiency

Materials

Standard 75-cm² or 120-cm² glass culture vessels
Samples of detergents made up to working-strength HBSS

Protocol

1. Autoclave glass flasks carrying cell monolayers, standing vertically, for 20 min at 120°C.
2. Soak overnight in detergent, three flasks per detergent.
3. Rinse out detergent with water, note flasks with residue, and brush as necessary to get them clean.
4. Rinse completely four times in water and three times in deionized water.

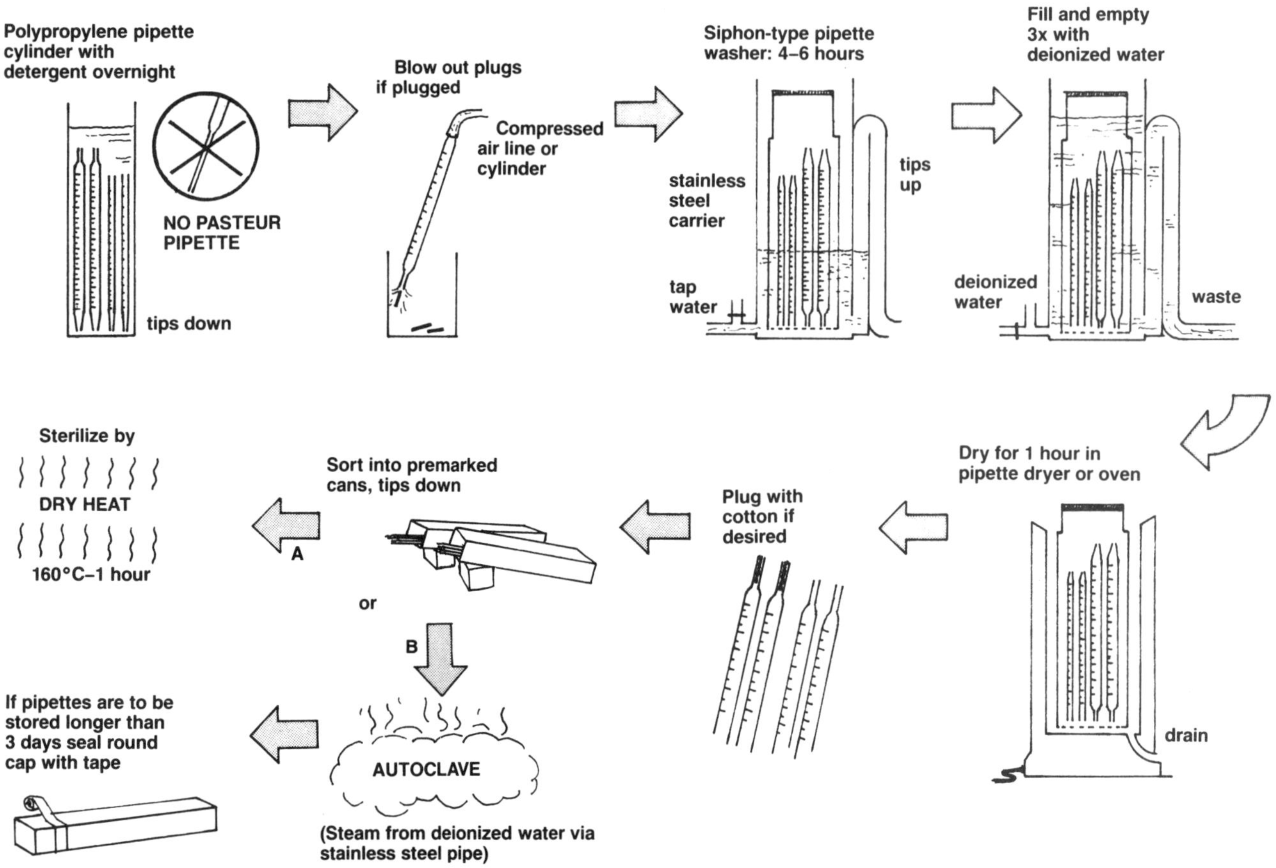

Fig. 8.3. *Wash-up and sterilization of pipettes.*

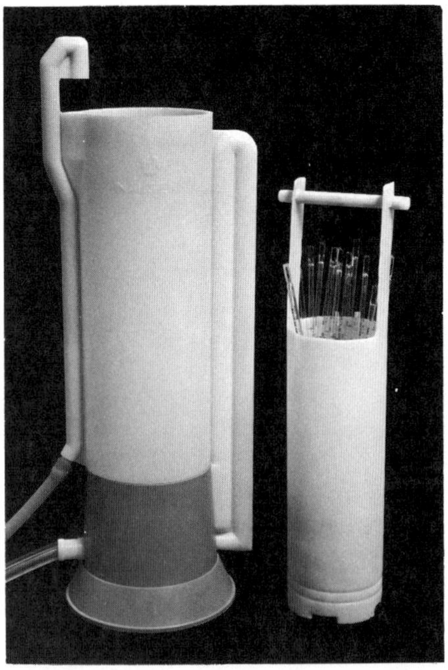

Fig. 8.4. *Siphon pipette washer. The model illustrated is made of polypropylene. Versions are available made of stainless steel, allowing direct transfer to a drier.*

5. Dry neck down, as above.
6. Check flasks again for apparent cleanliness and any sign of residue and record your findings.
7. Add a little BSS containing phenol red to each flask (approximately 1 ml/100 cm^2), rinse over the inside of the bottle, and look for any color change. If it becomes pink (alkaline), detergent has not been completely rinsed out of the bottle. Record result.
8. When you are satisfied that the flasks are clean, sterilize them by dry heat (see above).

(b) Quality of growth surface

Materials

Monolayer cells with good cloning efficiency (10% or more)
Medium
Serum, if required
(For materials for fixing and staining cells, see Chapter 13)

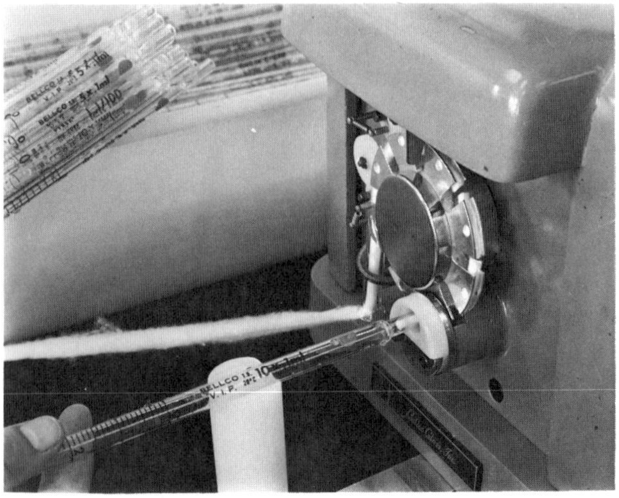

Fig. 8.5. Semiautomatic pipette plugger (Bellco).

Protocol

1. Taking flasks from (a).6 (above), add cell suspension to each flask and to three control flasks (e.g., disposable tissue culture-grade plastic of same surface area), using a cell concentration suitable for cloning (e.g., 20 cells/ml, 5 cells/cm^2, CHO-K1; 100 cells/ml, 25 cells/cm^2, MRC-5; see Chapter 11). Use the minimum concentration of serum that will allow the cells to clone (see under "Testing Serum," Chapter 7).
2. Incubate for 10–20 d, depending on the growth rate of the cells (see Chapters 11 and 19), fix, stain, and count colonies (see Chapters 11 and 13).
3. Determine relative plating efficiency (see Chapter 19) using disposable plastic flask as control, and record.

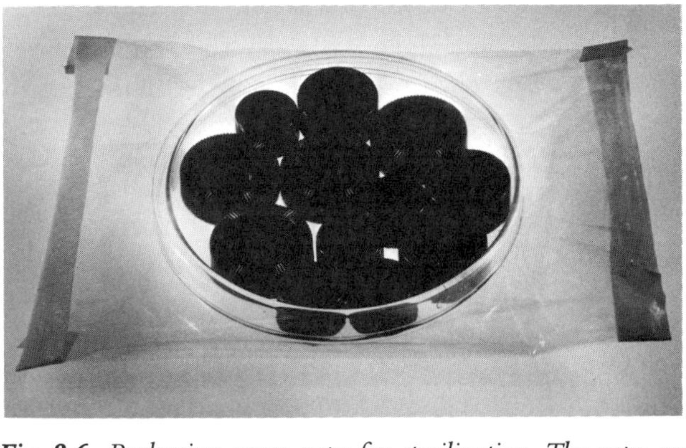

Fig. 8.6. Packaging screw caps for sterilization. The caps are enclosed in a glass petri dish, which is then sealed in an autoclavable nylon bag (Portex).

(c) Cytotoxicity (see also Chapter 19)

Materials

Microtitration plate
Medium
Serum
Detergent samples diluted to working strength and filter sterilized (see below)
(For fixing and staining cells, see Chapter 13)

Protocol

1. Set up microtitration plate with suitable target cells 1,000/well, incubate to 20% confluence, and change to fresh medium, 100 μl/well.
2. Add 100 μl of detergent, diluted to usual working strength in medium and filter sterilized, to the first well in each row and dilute serially, 1:2, across plate in that row, leaving the last two wells in each row as untouched controls.
3. Continue growth for 3–5 d but *not beyond* the time that the control wells reach confluence. Wash, fix, and stain plate or run MTT assay (see Chapter 19).
4. Determine titration point (point at which cell number per well is reduced by approximately 50%) and record. A high detergent concentration at this point means low cytotoxicity.

Miscellaneous Equipment

Cleaning. All new apparatus and materials (silicone tubing, filter holders, instruments, etc.) should be soaked in detergent overnight, thoroughly rinsed, and dried. Anything that will corrode in the detergent—mild steel, aluminum, copper, or brass, etc.—should be washed directly by hand, without soaking, or soaking for 30 min only, using detergent if necessary, then rinsed and dried.

Used items should be rinsed in tap water and immersed in detergent immediately after use. Allow to soak overnight, rinse, and dry. Again, do not expose materials that might corrode to detergent for longer than 30 min. Aluminum centrifuge buckets and rotors must never be allowed to soak in detergent.

Particular care must be taken with items treated with silicone grease or silicone fluids. They must be treated separately and the silicone removed, if necessary, with carbon tetrachloride. Silicones are very difficult to remove if allowed to spread to other apparatus, particularly glassware.

Packaging. Ideally, all apparatus for sterilization should be wrapped in a covering that will allow steam penetration but be impermeable to dust, micro-organisms, and mites. Proprietary bags are available bearing sterile-indicating marks that show up after sterilization. Semipermeable transparent nylon film (Portex Plastics, Cedanco) is sold in rolls of flat tube of different diameters and can be made up into bags with sterile-indicating tape. Although expensive, it can be reused several times before becoming brittle.

Tubes and orifices should be covered with tape and paper or nylon film before packaging, and needles or other sharp points should be shrouded with a glass test tube or other appropriate guard.

Sterilization. The type of sterilization used will depend on the material (see Table 8.1). Metallic items are best sterilized by dry heat. Silicone rubber (which should be used in preference to natural rubber), Teflon, polycarbonate, cellulose acetate, and cellulose nitrate filters (see below for filters in holders), etc., should be autoclaved for 20 min at 121°C, 100 kPa (1 bar, 15 lb/in^2) with preevacuation and postevacuation steps in the cycle except when used for sterilizing filters in a filter assembly (see below). In small bench-top autoclaves and pressure cookers, make sure that the autoclave boils vigorously for 10–15 min before pressurizing to displace all the air. (Take care that enough water is put in at the start to allow for this.) After sterilization, the steam is released and the items removed to dry off in an oven or rack.

◇ Take care releasing steam and handling hot items to avoid burns. Wear elbow-length insulated gloves and keep face well clear of escaping steam when opening doors, lids, etc. Use safety locks on autoclaves.

Sterilizing Filters

Reusable filter holders should be made up and sterilized as follows:

1. After thorough washing in detergent (see above) rinse in water, then deionized water, and dry.
2. Insert support grid in filter and place filter membrane on grid. If polycarbonate, apply wet to counteract static electricity.
3. Place prefilters (glass fiber and others as required; see below) on top of filter.
4. Reassemble filter holder, but do not tighten up completely (leave about one-half turn on collars, one whole turn on bolts).
5. Cover inlet and outlet of filter with aluminum foil.
6. Pack filter assembly in sterilizing paper or steam-permeable nylon film and close with sterile-indicating tape.
7. Autoclave at 121°C, 100 kPa (1 bar, 15 lb/in^2) with no preevacuation or postevacuation ("liquids cycle" in automatic autoclaves).
8. Remove and allow to cool.
9. Do not tighten filter holder completely until the filter is wetted at the beginning of filtration (see below).

Alternative Methods of Sterilization

Many plastics cannot be exposed to the temperature required for autoclaving or dry-heat sterilization. To sterilize such items, immerse in 70% alcohol for 30 min and dry off under UV light in a laminar flow cabinet. Care must be taken with Plexiglas (Perspex, Lucite), as it may crack in alcohol or UV treatment due to release of stresses built in during manufacture.

Ethylene oxide may be used to sterilize plastics; 2–3 wk are required for the Et_20 to clear from plastic surface.

γ-irradiation, 2,000–3,000 Gy (0.2–1.0 mrad), is the best method for plastics. Items should be packaged and sealed. Polythene may be used and sealed by heat welding.

REAGENTS AND MEDIA

The ultimate objective in preparing reagents and media is to produce them in a pure form (1) to avoid the accidental inclusion of inhibitors and substances toxic to cell survival, growth, and expression of specialized functions, and (2) to enable the reagent to be totally defined and the functions of its constituents to be fully understood.

Most reagents or media can be sterilized either by autoclaving if they are heat stable (water, salt solutions, amino acid hydrolysates) or by membrane filtration if heat labile. During autoclaving the container should be kept sealed in borosilicate glass or polycarbonate. Soda-glass bottles are better left with the caps slack to minimize breakage. Evolution of vapor will help to prevent ingress of steam from the autoclave, but the liquid level will need to be restored with sterile distilled water later.

Media and reagents supplied on line to large-scale culture vessels and industrial or semi-industrial fermentors can be sterilized on line by ultrahigh-temperature treatment for a short time (Alfa-Laval). Adaptation of this process to media production might allow increased automation and ultimately reduce costs.

Water (see also Chapter 4)

Water must be of a very high purity for use in tissue culture, particularly if serum-free media are required. As water supplies vary greatly, the degree of purification required may vary. Hard water will need a conventional, ion-exchange water softener on the supply before entering the purification system, but this will not be necessary with soft water.

There are four main approaches to water purification: distillation, deionization, carbon filtration, and ultrafiltration. Simple systems depend on the first two alone or combined, while the more efficient systems employ multiple stages, each operating on a different principle.

Double glass distillation. Double glass distillation was the first system to be used widely and can still be quite effective in some areas where there is a low solute concentration in the water. The still should be electric, automatically controlled if possible, and feed the first distillate into the second boiler. The heating elements should be borosilicate glass- or silica-sheathed.

Deionization and glass distillation. Deionization is used to replace the first-stage distillation, and again feeds directly, with a level controller, into the distillation boiler (see Fig. 4.4a). The quality of the deionized water should be monitored by conductivity at regular intervals and the cartridge changed when an increase in conductivity is observed. Distillation can precede deionization, providing sterile water to the deionization stage. This has significant advantages, provided care is taken to exclude the escape of resin from the deionization cartridge.

Multistage purification (Fig. 8.7). With a universal increase in the demand for water, and a decrease in quality-control measures, acceptable to public health control where they are controlled by chlorination, the amount of solute contamination, particularly colloid contamination, often increases. This means that for many laboratories the need for water purification has increased in the last decade. This is compounded by the increased use of low-serum or serum-free media, where the loss of the protective effect of serum protein has meant an increased requirement in water purity.

For ultrapure water, the first stage is usually reverse osmosis but can be replaced by distillation. Distillation has the advantage that the water is heat sterilized, but it is more expensive and the still needs to be cleaned out regularly. Reverse osmosis (RO) depends on the integrity of the filtration membrane, and hence the effluent must be monitored. If the costs of both are deducted directly from your budget, RO will probably work out cheaper, but if power is supplied free or costed independently of usage, then distillation will be

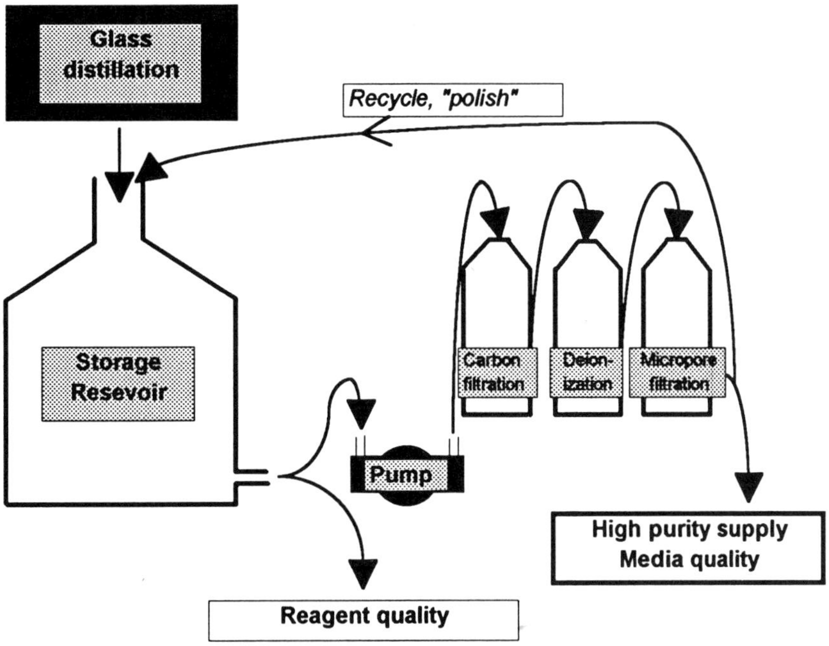

Fig. 8.7. Alternative layout for high-purity water supply. Tap water is fed through glass distillation to a storage container. This semipurified water is then recycled via carbon filtration, deionization, and micropore filtration back to the storage container. Reagent-quality water is available at all times from storage; media-quality water is available from the micropore filter supply (right of diagram). If the apparatus recycles continuously, then the highest purity water will be collected first thing in the morning for the preparation of medium. (Based on Elga system; see also Fig. 4.4a.)

cheaper. The type of RO cartridge used is determined by the pH of the water supply (see manufacturer for details).

The second stage is carbon filtration, which will remove both organic and inorganic material. The third stage is high-grade deionization and the final stage micropore filtration to eliminate microbial contamination. In the Millipore system, the water is collected directly from the final-stage micropore filter without storage to minimize pollution during storage. Water is only stored after the first stage, and can be used from this point as rinsing water. In the Elga system, water is recycled continuously from the micropore filter to the store, and if the supply from the first stage is turned off (e.g., overnight), then the stored water gradually "polishes," i.e., increases in purity. Hence, in this system, water should be used first thing in the morning for preparation of media.

In the Millipore system, an ultrafiltration stage can be inserted between deionization and micropore filtration to produce pyrogen-free water. Millipore does not recommend storage, so a system should be selected to give the rate of supply that you require on line.

Water is sterilized by autoclaving at 121°C, 100 kPa (15 lb/in^2, 1 bar) for 10–15 min. It should be dispensed in aliquots suitable to use, e.g., for media preparation from concentrates. The bottles should be sealed during sterilization. This will require borosilicate glass (Pyrex) or polycarbonate (Nalgene). If bottles are unsealed (e.g., if soda glass is used it may break if sealed), allow 10% extra volume per bottle to allow for evaporation.

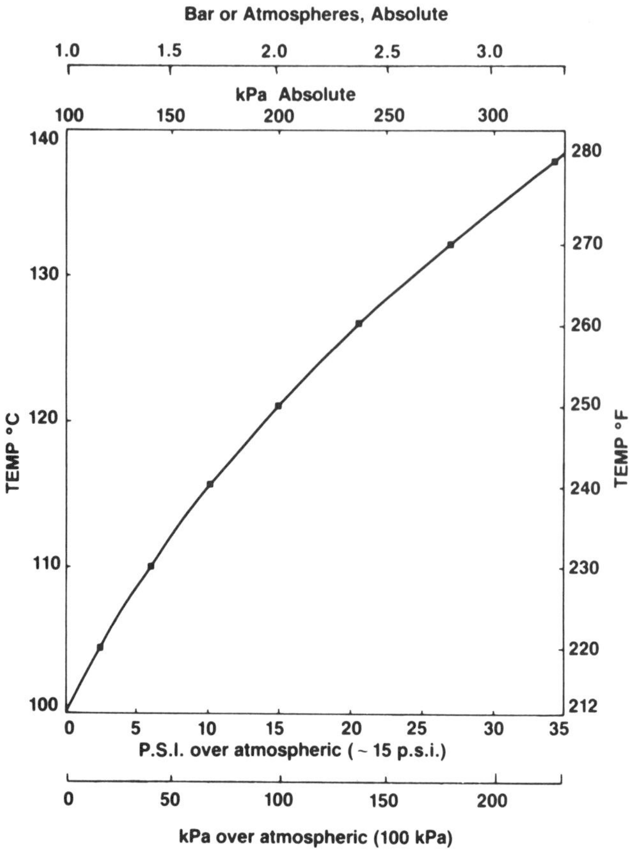

Fig. 8.8. *Relationship between pressure and temperature: 121°C and 100 kPa or 1 bar (250°F, 15 lb/in^2) for 15–20 min are the conditions usually recommended.*

Balanced Salt Solutions

A selection of formulations is given in Chapter 7. The formula for Hanks' BSS [after Paul, 1975] contains magnesium chloride in place of some of the sulfate originally recommended and should be autoclaved below pH 6.5 and neutralized before use. Similarly, Dulbecco's PBS is made up without calcium and magnesium (PBSA), which are made up separately (PPSB) and added just before use. These precautions are designed to minimize precipitation of calcium and magnesium phosphate during autoclaving and storage.

Most balanced salt solutions contain glucose. Since glucose may caramelize on autoclaving, it is best omitted and added later. If prepared as a 10× concentrate (20% W/V), caramelization during autoclaving is reduced.

Salt solutions may be sterilized by autoclaving for 10–15 min at 121°C, 100 kPa (1 bar, 15 lb/in^2), in sealed bottles as for water (above) (Fig. 8.8).

Media

During the preparation of complex solutions, care must be taken to ensure that all the constituents dissolve and do not get filtered out during sterilization and that they remain in solution after autoclaving or storage. Concentrated media are often prepared at a low pH (between 3.5 and 5.0) to keep all the constituents in solution, but even then some precipitation may occur. If properly resuspended this will usually redissolve on dilution; but if the precipitate has been formed by degradation of some of the constituents of the medium, then the quality of the medium may be reduced. If a precipitate forms, the medium performance should be checked by cell growth, cloning, and assay of special functions (see below).

The preparation of media is rather complex. It is convenient to make up a number of concentrated stocks, essential amino acids at 50× or 100×, vitamins at 1,000×, tyrosine and tryptophan at 50× in 0.1 *N* HCl, glucose, 200 g/l, and single-strength BSS. The requisite amount of each concentrate is then mixed and filtered through a 0.2-μm porosity cellulose acetate, cellulose nitrate (Millipore, Gelman, Sartorius), or polycarbonate (Nuclepore) filter and diluted with the BSS (see below and Fig. 8.9, Table 8.2), e.g.:

Amino acid concentrate, 100× (in water) 100 ml
Tyrosine and tryptophan, 50× (in 0.1 *N* HCl) 200 ml

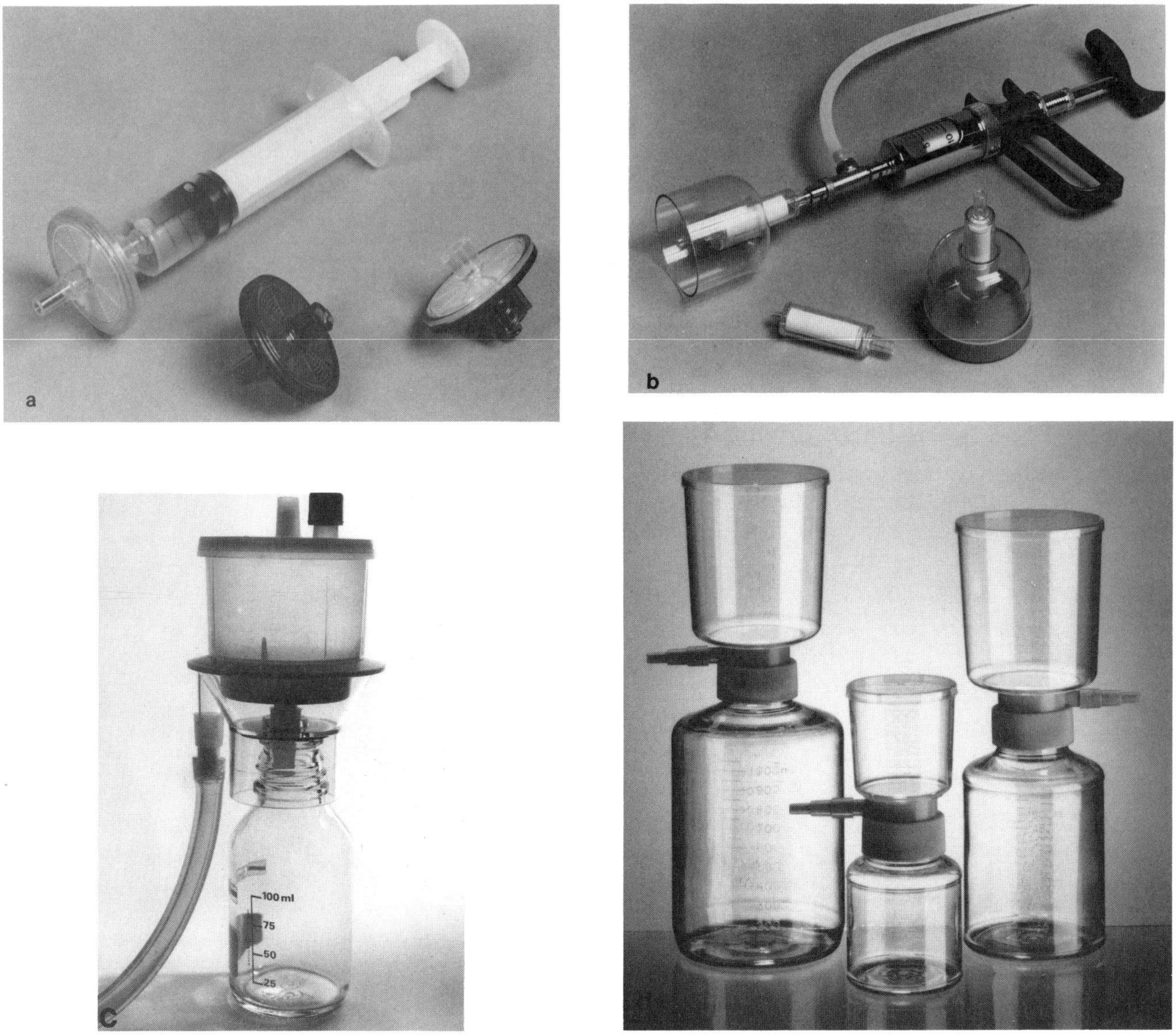

Fig. 8.9. Disposable sterilizing filters. a. Millex 25-mm disc syringe filter (Millipore). b. Sterivex in use with repeating syringe. c. Bottle-top fitting (Becton-Dickinson). d. Filter cup and storage vessels (Stericup, Millipore). (a,b) positive pressure, (c,d) negative pressure. See also Figure 8.15. (a,b,d) courtesy of Millipore (UK) Ltd.

Vitamins, 1,000× (in water) 10 ml
Glucose, 100× (in 200 g/l BSS) 100 ml

Mix and sterilize by filtration. Store frozen. For use, dilute 41 ml of concentrate mixture with 959 ml sterile × 1 BSS.

The advantage of this type of recipe is that it can be varied; extra nutrients (keto-acids, nucleosides, minerals, etc.) can be added or the major stock solutions altered to suit requirements, but, in practice, this procedure is so laborious and time-consuming that few laboratories make up their own media from basic constituents unless they wish to alter individual constituents regularly. The reliability of commercial media depends entirely on the application of appropriate quality-control measures. There are now several reputable suppliers (see Trade Index) of standard formulations, and many of them will supply specialized, serum-free formulations and will also prepare media to your own formulation, but it is important to ensure that the quality control is relevant to the medium and the cells you wish to propagate. You might buy MCDB 153, which is tested on HeLa cell colony formation, but this is of little relevance if you wish to grow primary keratinocytes.

Commercial media are supplied as: (1) working-strength solutions, complete with sodium bicarbonate and glutamine; (2) 10× concentrates without $NaHCO_3$ and glutamine, which are available as separate concentrates; or (3) powdered media, complete or without glutamine. Powdered media are the cheapest and not a great deal more expensive than making up your own, if you include time for preparation, sterilization, and quality control, cost of raw materials of high purity, and overheads such as power and wages. Powdered media are quality controlled by the manufacturer for their growth-promoting properties but not, of course, for sterility. Tenfold concentrates cost about twice as much per liter of working-strength medium but are purchased sterile. Buying media at working strength is the most expensive (about five times the cost of a 10× concentrate) but is the most convenient, as no further preparation is required other than the addition of serum, if required.

Preparation of Medium from 10× Concentrate

Sterilize aliquots of deionized distilled water (see above) of such a size that one aliquot, when made up to full-strength medium, will last from 1 to 3 wk. Add concentrated medium and other constituents as follows:

1. For sealed culture flask; gas-phase air, low buffering capacity, and low CO_2/HCO_3^- concentration:

Sterile water	885.5
10× concentrate	100
200 m*M* glutamine	10
7.5% $NaHCO_3$	4.5
	1,000 ml

   Add 1 *N* NaOH to give pH 7.2 at 20°C. When incubated, this will rise to pH 7.4 at 36.5°C, but this may need to be checked by a trial titration the first time the recipe is used.

   If extra constituents are added, e.g., HEPES buffer, extra amino acids, this becomes as in (2) below:

2. For sealed culture flask; high buffering capacity, gas-phase air; may be vented to atmosphere by slackening cap for some cell lines at a high cell density if a lot of acid is produced; atmospheric CO_2/HCO_3^-:

Sterile water	855.5
10× concentrate	100
200 m*M* glutamine	10
100× nonessential amino acids	10
1.0 M HEPES	20
7.5% $NaHCO_3$	4.5
	1,000 ml

   Correct pH as in (1) above.

3. For cultures in open vessels in a CO_2 incubator or under CO_2 in sealed flasks, two suggested formulations are as follows:

	(A) 2% CO_2	(B) 5% CO_2
Sterile water	861	863
10× concentrate	100	100
200 m*M* glutamine	10	10
1.0 M HEPES	20	–
7.5% $NaHCO_3$	9	27
	1,000 ml	1,000 ml

   Add 1.0 *N* NaOH to give pH 7.2 at 20°C and pH 7.4 after incubation at 36.5°C, as for recipes above in (1) and (2).

   (A) will give good buffering capacity, at a moderate CO_2/HCO_3 concentration, while (B) will give moderate buffering capacity, with a high CO_2/HCO_3 concentration.

   Always equilibrate and check pH at 36.5°C, as the solubility of CO_2 decreases with increased temperature and the pK_a of the HEPES will change.

The amount of alkali needed to neutralize a 10× concentrated medium (which is made up in acid to maintain solubility of the constituents) may vary from batch to batch and from one medium to another, and, in practice, titrating medium to pH 7.4 at 36.5°C can sometimes be a little difficult. When making up a new medium for the first time, add the stipulated amount of $NaHCO_3$ and allow samples with varying amounts of alkali to equilibrate overnight at 36.5°C in the appropriate gas phase. Check the pH the following morning, select the correct amount of alkali, and prepare the rest of the medium accordingly.

Some media are designed for use with a high bicarbonate concentration and elevated CO_2 in the atmosphere, e.g., Ham's F12, while others have a low bicarbonate concentration for use with a gas phase of air, e.g., Eagle's MEM with Hanks' salts. If the use of a medium is changed and the bicarbonate concentration altered, it is important to check that the osmolality is still within an acceptable range.

The bicarbonate concentration is important in establishing a stable equilibrium with atmospheric CO_2, but regardless of the amount of bicarbonate used, if the medium is at pH 7.4 and 36.5°C, the bicarbonate concentration at each concentration of CO_2 will be as in Table 7.2 (see Chapter 7).

The osmolality should be checked (see Chapter 7) where alterations are made to a medium that are not in the original formulation.

If your consumption of medium is fairly high (>200 l/yr) and you are buying medium ready-made, then it may be better to get extra constituents included in the formulation, as this will work out to be cheaper.

HEPES particularly is very expensive to buy separately. Glutamine is supplied separately, as it is unstable and should be kept frozen. The half-life in medium at 4°C is about 3 wk, and at 36.5°C about 1 wk. Some dipeptides of glutamine have increased stability, while retaining the bioavailability of the glutamine. One such is Glutamax, available from GIBCO/Life Technologies.

Care should be taken with 10× concentrates to ensure that all of the constituents are in solution, or at least evenly suspended before dilution. Some constituents, e.g., folic acid or tyrosine, can precipitate and be missed at dilution. Incubation at 36.5°C for several hours may overcome this.

The final step is the addition of serum, which, since it is close to isotonic, is added to the final volume, e.g.:

Complete medium	1,000 ml
Serum	111 ml (10% final)
	1,111 ml

Remember to allow space for all additions when choosing the container and the volume of H_2O to be sterilized.

Having once tested a batch of medium for growth promotion, etc., and found it to be satisfactory, then it need not be tested each time it is made up to working strength. The sterility should be checked, however, by incubating the medium at 36.5°C for 48 hr before adding glutamine, serum, or antibiotic, or incubating an aliquot of the complete medium.

Powdered Media

Select a formulation lacking glutamine. If there are other unstable constituents, they also should be omitted and added later as a sterile concentrate just before use. Dissolve the powder in the recommended amount of water (choose a pack size that you can make up all at once and use within 3 months), taking care that all the constituents are completely dissolved. (Follow the manufacturer's instructions.) Do not store, but filter sterilize immediately. Precipitation may occur on storage, and microbiological contamination may also appear.

Alternatively, for people using smaller amounts (<1.0 l/wk) or several different types of medium, smaller volumes may be prepared, complete with glutamine, and filtered directly into storage bottles using a bottle-top filter sterilizer (Becton Dickinson, Nalgene, Costar) (Fig. 8.9c). With this and other negative-pressure filtration systems, some dissolved CO_2 may be lost during filtration and the pH may rise. Provided the correct amount of $NaHCO_3$ is in the medium to suit the gas phase (see Chapter 7), the medium will reequilibrate in the incubator, but this should be confirmed the first time used to make sure.

Sterilization. Filter through 0.2-μm polycarbonate, cellulose acetate, or cellulose nitrate filter (see below).

Autoclavable Media

Some commercial suppliers offer autoclavable versions of Eagle's MEM and other media. Autoclaving is much less labor-intensive, less expensive, and has a much lower failure rate than filtration. The procedure to follow is supplied in the manufacturer's instructions and is similar to that described above for BSS. The medium is buffered to pH 4.25 with succinate to stabilize the B vitamins during autoclaving and is subsequently neutralized. Glutamine is replaced by glutamate or added sterile after autoclaving. As with BSS, care should be taken with evaporation and any deficit made up with sterile deionized distilled water.

Filter Sterilization

This method is suitable for filtering heat-labile solutions (Figs. 8.9–8.12). Reusable filters are made up and sterilized by autoclaving, or presterilized disposable filters may be used. The latter are more expensive but less time-consuming to use and give fewer failures.

Preparation and sterilization of filter–see above.

Materials for Filtration

Pressure vessel
Pump
Clamp to secure filter
Sterile container to receive filtrate with outlet at the base
Tubing
Spring clip and glass bell (see Figs. 8.12, 8.13)
Sterile bottles with caps and foil

Protocol

1. Choose an appropriate filter size from the range available (Table 8.3), secure the filter holder in position, assemble the sterile and nonsterile components, and make the necessary connections (Figs. 8.12 and 8.13).
2. Decant the medium into the pressure vessel.
3. Position the sterile receiver under the outlet of the filter.
4. For reusable filters, turn on the pump just long enough to wet the filter. Stop the pump and tighten up the filter holder.
5. Switch on the pump to deliver 100 kPa (15 lb/in^2). When the receiver starts to fill, draw off aliquots into medium stock bottles of the desired volume, cap, and store.

When sterilizing small volumes (100–200 ml), the solution may fit within the filter holder (Fig. 8.11) and a pressure vessel and receiver will not be required. Collect directly into storage bottle.

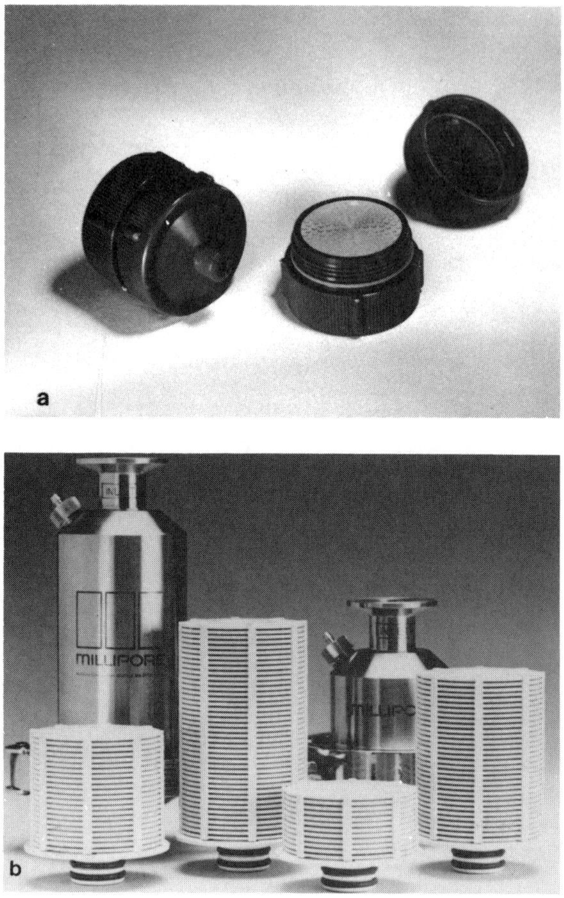

Fig. 8.10. Reusable filter holders (Millipore). a. 47-mm in-line polyproplene. b. Millidisk range; stainless steel housings, high-capacity cartridge-type filters. (Courtesy of Millipore [UK] Ltd.)

Positive pressure is recommended for optimum filter performance and to avoid removal of CO_2, which results from negative-pressure filtration. However, the latter method is often used for filtering small volumes, as the equipment is simple. It is important to incorporate a filter and trap on the outlet from the filter flask (Figs. 8.9a,c and 8.14).

Sterility Testing

Positive pressure may also be applied using a peristaltic pump (Fig. 8.15) in line between a nonsterile reservoir and a disposable in-line filter such as the Millipak (Millipore). No receiver is required and only the filter, which is bought sterile and is disposable, and the media bottles need be sterile. This is a simple, effective, and inexpensive method for batch filtration applicable to small, medium, and even large laboratories.

(1) When filtration is complete and all the liquid has passed through the filter, disconnect the outlet and raise the pump pressure until bubbles form in the effluent from the filter. This is the "bubble point" and should occur at more than twice the pressure used for filtration (see manufacturer's instructions). If the filter bubbles at the sterilizing pressure (100 kPa) or lower, then it is perforated and should be discarded. Any filtrate that has been collected should then be regarded as nonsterile and refiltered.

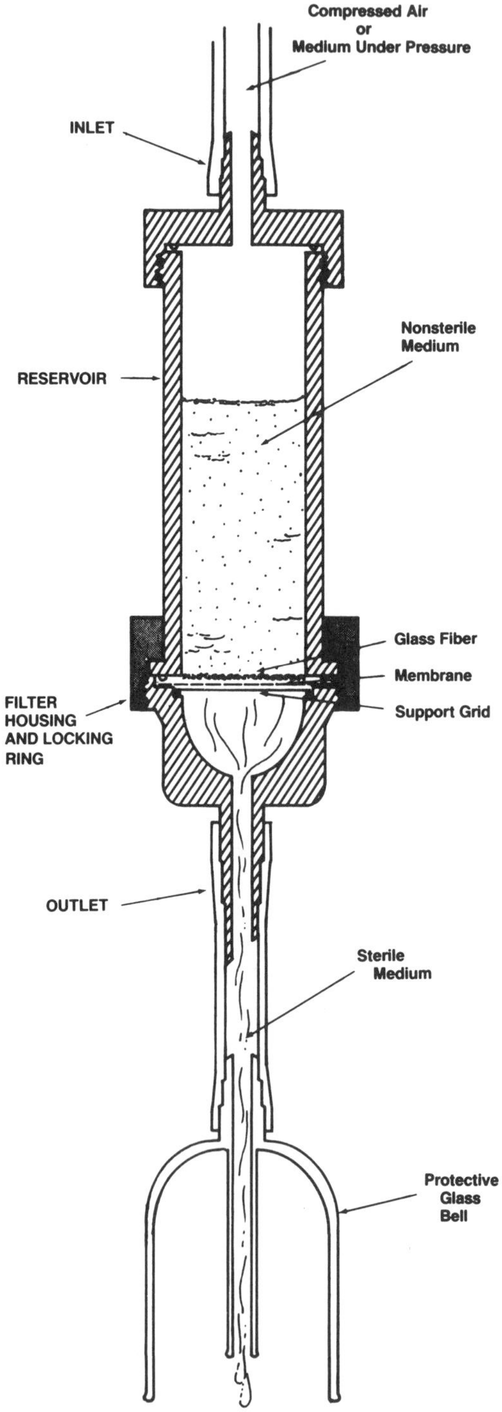

Fig. 8.11. Longitudinal section through 47-mm reusable filter holder. The whole apparatus is sterilized. Fluid may be added directly to the reservoir or may be delivered from a pressurized tank (see Figs. 8.12 and 8.13).

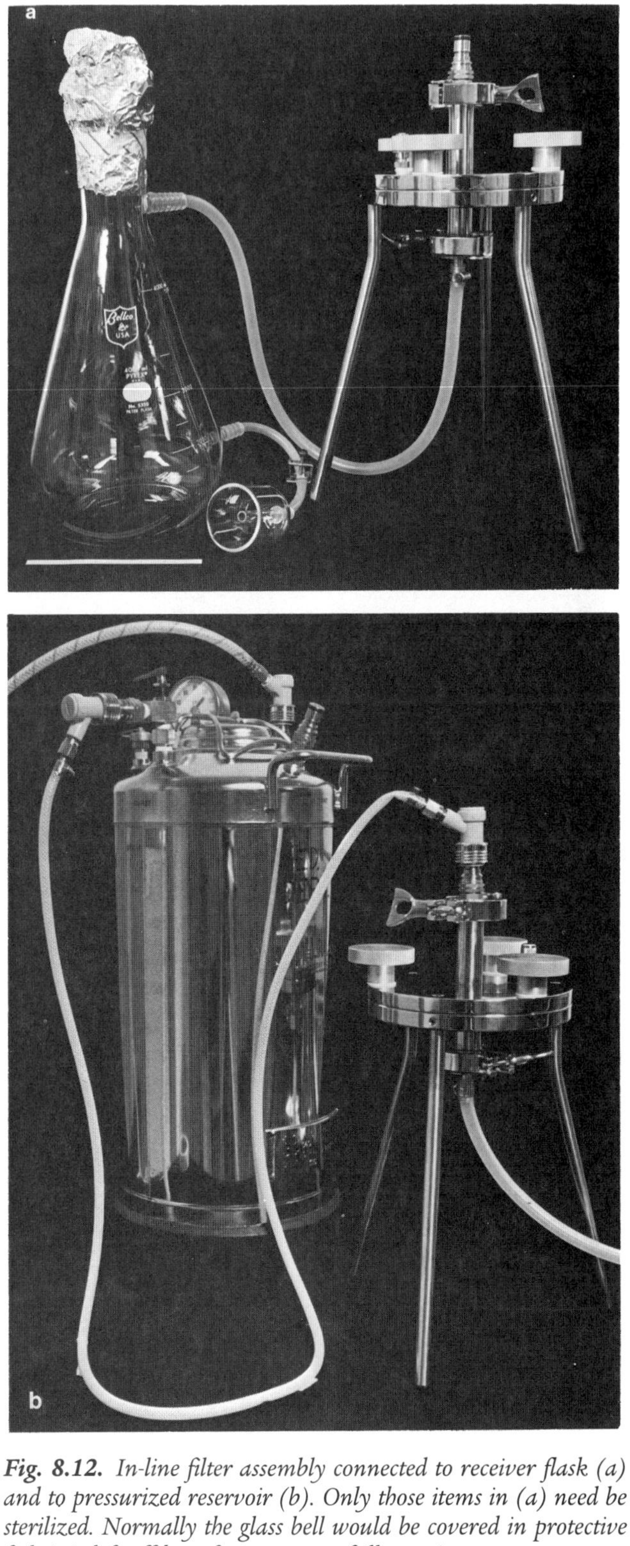

Fig. 8.12. In-line filter assembly connected to receiver flask (a) and to pressurized reservoir (b). Only those items in (a) need be sterilized. Normally the glass bell would be covered in protective foil; it is left off here for purposes of illustration.

(2) If the filter passes the "bubble-point" test, withdraw aliquots from the beginning, middle, and end of the run and test for sterility by incubating at 36.5°C for 72 hr. If the samples become cloudy, discard them and resterilize the batch. If there are signs of contamination in the other stored bottles, the whole batch should be discarded.

TABLE 8.3. Filter Size and Fluid Volume

Filter size or designation	Disposable (D) or reusable (R)	Approximate volume that may be filtered	
		Crystalloid	Colloid
25 mm	D	1–100 ml	1–20 ml
47 mm or Sterivex cartridge	R,D	0.1–1 l	100–250 ml
90 mm	R	1–10 l	0.2–2 l
Millipak-20	D	2–10 l	200 ml–2 l
Millipak 40	D	10–20 l	2–5 l
Millipak-60	D	20–30 l	5–7 l
Millipak-100	D	30–75 l	7–10 l
Millipak-200	D	75–150 l	10–30 l
Millidisk	D	30–300 l	5–50 l
142 mm	R	10–50 l	1–5 l
293 mm	R	50–500 l	5–20 l

The above examples are quoted from Millipore catalogues. Similar products are available from Gelman and Sartorius.

For a more thorough test, take samples, as in (2), dilute one-third of each into nutrient broths, e.g., beef-heart hydrolysate and thioglycollate. Divide each in two and incubate one at 36.5°C and one at 20°C for 10 days, with uninoculated controls. If there is any doubt after this incubation, mix and plate out aliquots on nutrient agar.

Alternatively, place a demountable sterile filter in the effluent line from the main sterilizing filter. Any contamination that passes due to failure in the first filter will be trapped in the second. At the end of the run, remove the second filter and place on nutrient agar. If colonies grow, discard or refilter. This method has the advantage that it monitors the entire filtrate and not just a small fraction, but it does not cover risks of contamination during bottling and capping.

Sterility testing of autoclaved stocks is much less essential, provided proper monitoring (temperature and time at the sterilizing temperature) of the autoclave is carried out.

Culture Testing (see also "Selection of Detergent," above)

Media that have been produced commercially will have been tested for their ability to sustain growth of one or more cell lines. (If they have not, then you should change your supplier!) However, there are certain circumstances when you may wish to test your own media for quality, e.g.: (1) if it has been made up in the laboratory from basic constituents; (2) if any additions

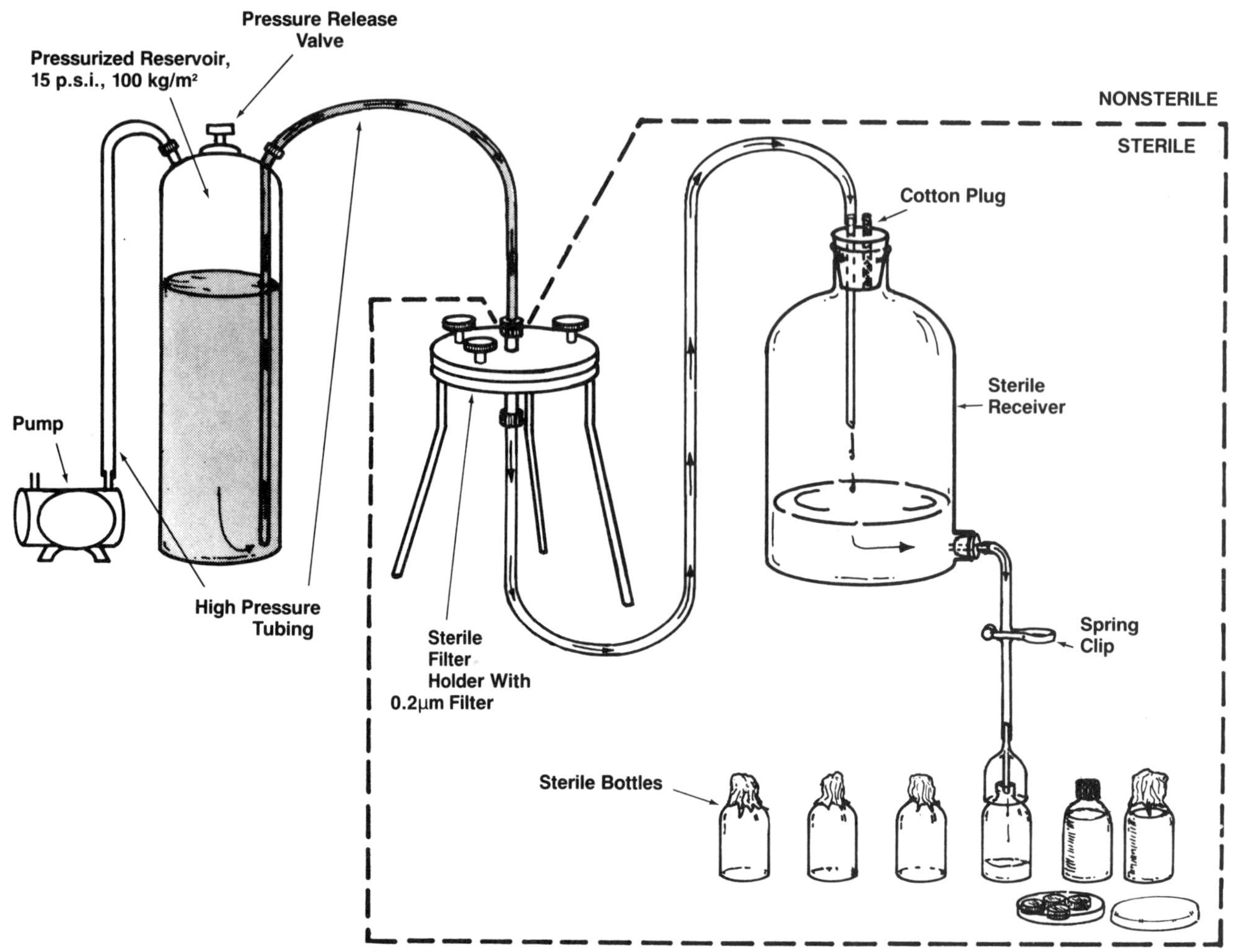

Fig. 8.13. Diagram of complete system illustrated in Figure 8.12. The broken line encloses those items that are sterilized. A disposable filter unit of equivalent surface area may be substituted for the metal, reusable filter holder and stand illustrated here.

are made to the medium; (3) if the medium is for a special purpose that the commercial supplier is not able to test; and (4) if the medium is made up from powder and there is a risk of losing constituents during filtration.

Contamination of medium with toxic substances can arise during filtration. Some filters are treated with traces of detergent to facilitate wetting, and this may leach out into the medium during filtration. Such filters should be washed by passing PBS or BSS through before use, or by discarding the first aliquot of filtrate. Polycarbonate filters, e.g., Nuclepore, are wettable without detergents and are preferred by some workers, particularly where the serum concentration in the medium is low.

There are three main types of culture test: (1) plating efficiency; (2) growth curve at regular passage densities and up to saturation density; and (3) expression of a special function, e.g., differentiation in the presence of an inducer, virus propagation, specific product formation, or expression of a specific antigen. All of these should be performed on the new batch of medium with your regular medium as a control.

Plating efficiency (see Chapter 19). Plating efficiency is the most sensitive test, as it will detect minor deficiencies and low concentrations of toxins not apparent at higher cell densities. Ideally it should be performed at a range of serum concentrations from 0% to 20%, as serum may mask deficiencies in the medium.

Growth curve. Clonal assay will not always detect insufficiencies in the amount of particular constituents. For example, if the concentration of one or more amino acids is low, it may not affect clonal growth but could influence the maximum cell concentration attainable.

A growth curve (see Chapter 18) gives three parame-

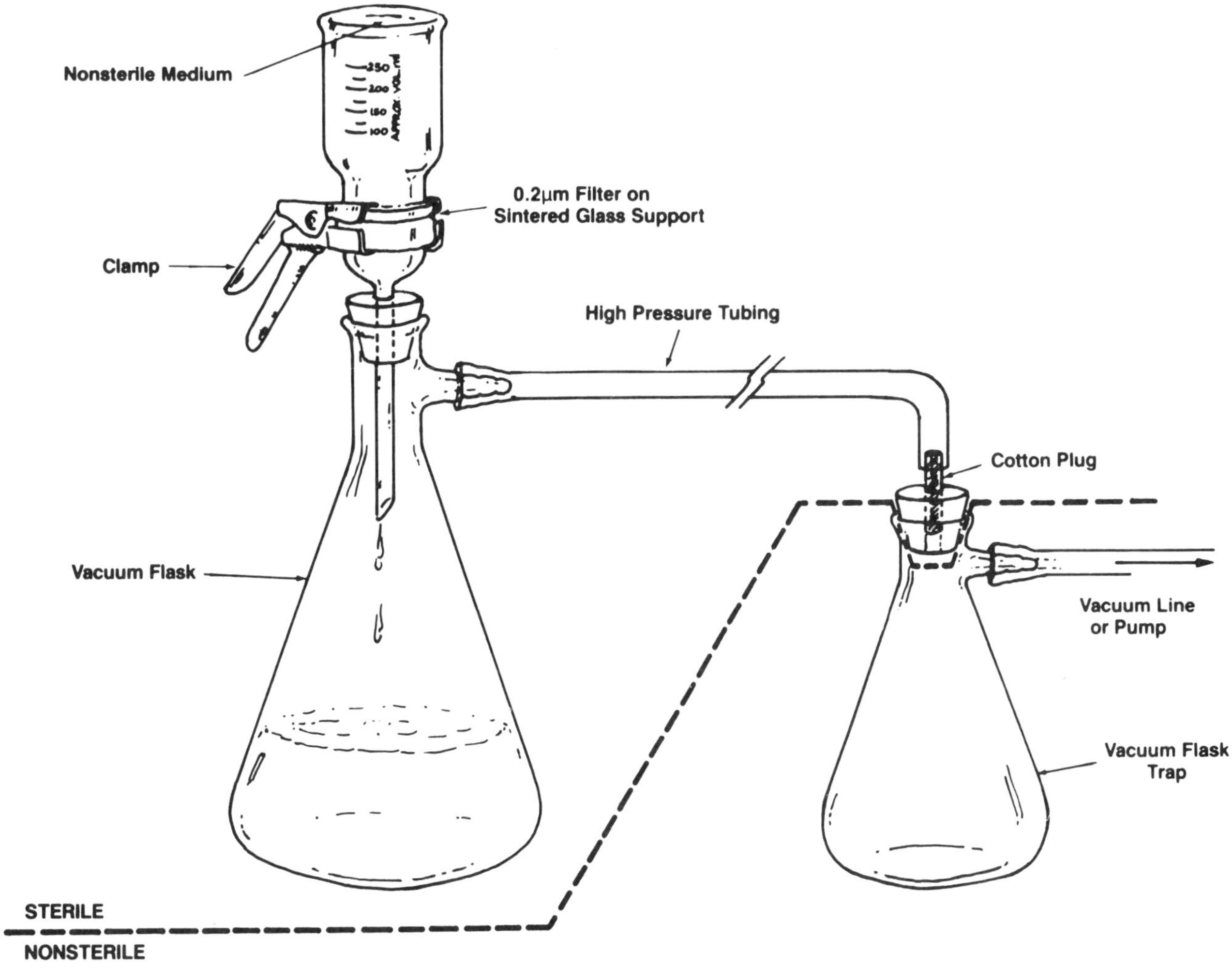

Fig. 8.14. Apparatus for vacuum filtration. Only those items above the broken line need be sterilized. Vacuum filtration is also possible with some disposable filters (Nalgene and Falcon; see Fig. 8.8).

ters of measurement: (1) the lag phase before cell proliferation is initiated after subculture, indicating whether the cells are having to adapt to different conditions; (2) the doubling time in the middle of the exponential growth phase, indicating the growth-promoting capacity; and (3) the terminal cell density. In cell lines that are not sensitive to density limitation of growth, e.g., continuous cell lines (see Chapters 2 and 15), this indicates the total yield possible and usually reflects the total amino acid or glucose concentration. Remember that a medium that gives half the terminal cell density is costing twice as much per cell produced.

Special functions. In this case a standard test from the experimental system you are using, e.g., virus titer in medium after a set number of days, should be performed on the new medium alongside the old.

A major implication of these tests is that they should be initiated well in advance of the exhaustion of the current stock of medium so that (1) proper comparisons may be made and (2) there is time to have fresh medium prepared if the medium fails any of these tests.

Storage

Opinions differ as to the shelf-life of different media. As a rough guide, media made up without glutamine should last 6–9 months at 4°C. Once glutamine, serum, or antibiotics are added, storage time is reduced to 2–3 weeks. Hence media that contain labile constituents should either be used within 2–3 weeks of preparation or stored at −20°C.

Some forms of room fluorescent lighting will cause deterioration of riboflavin and tryptophan into toxic by-products [Wang, 1976]. Incandescent lighting

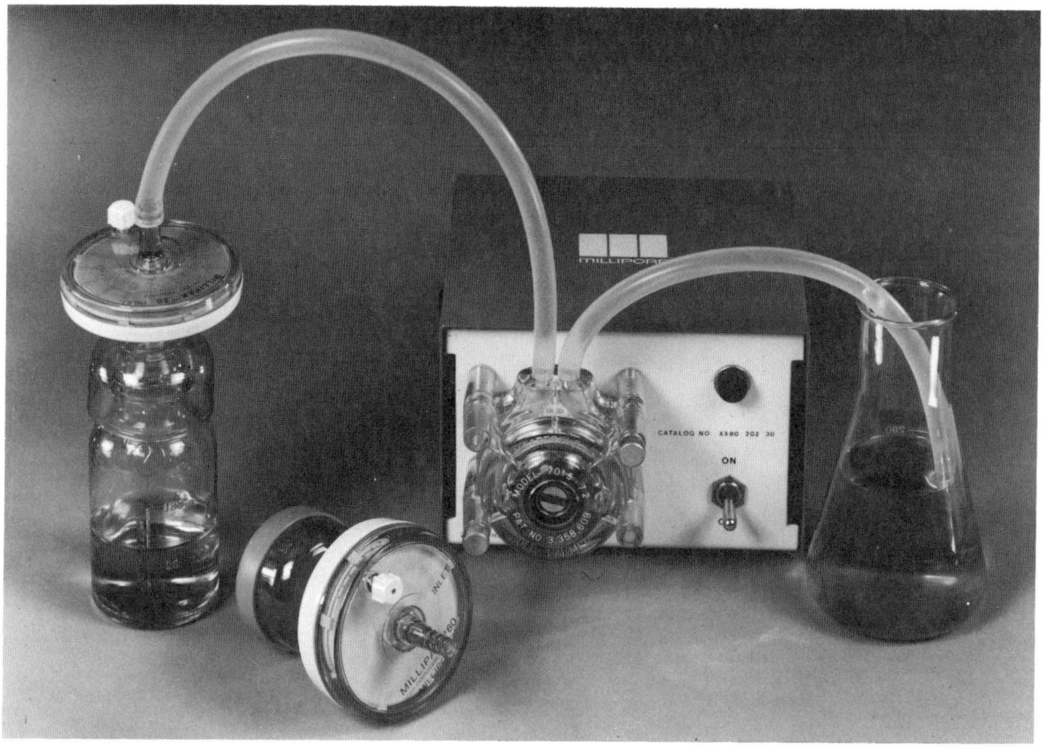

Fig. 8.15. Sterile filtration with peristaltic pump between nonsterile reservoir and sterilizing filter Millipak, Millipore. (Courtesy of Millipore [UK] Ltd.)

should be used in cold rooms where media is stored and in hot rooms where cells are cultured, and the light extinguished when the room is not occupied. Bottles of media should not be exposed to fluorescent lighting for longer than a few hours; a dark freezer is recommended for long-term storage.

Serum

This is one of the more difficult preparative procedures in tissue culture because of variations in the quality and consistency of the raw material and because of the difficulties encountered in sterile filtration. However, it is also the highest constituent of the cost of doing tissue culture, accounting for 30–40% of the total budget if bought from a commercial supplier. Buying sterile serum is certainly the best approach from the point of view of consistency and simplicity (see Chapter 7), but it may be prepared as follows.

Outline

Collect blood, allow to clot, and separate the serum. Filter serum through gradually reducing porosity of filters, bottle, and freeze.

Protocol

Large scale, 20–100 l per batch.

Collection

Arrangements may be made to collect whole blood from a slaughterhouse. It should be collected directly from the bleeding carcass and not allowed to lie around after collection. Alternatively, blood may be withdrawn from live animals under proper supervision. The second routine, if performed consistently on the same group of animals, gives a more reproducible serum but a lower volume for greater expenditure of effort. If done carefully, it may be collected aseptically.

Clotting

Allow the blood to clot by standing overnight in a covered container at 4°C. This so-called "natural clot" serum is superior to serum physically separated from the blood cells by centrifugation and defibrination, as platelets release growth factor into the serum during clotting. Separate the serum from the clot and centrifuge at 2,000 *g* for 1 hr to remove sediment.

Sterilization. Serum is usually sterilized by filtration through a 0.2-μm-porosity sterilizing filter, but because of its viscosity and high particulate content, it should be passed through a graded series of cartridge-type glass-fiber prefilters before passing through the final sterilizing filter. Only the last filter, a 350-mm in-line disc filter or equivalent, need be sterilized.

The prefilter assemblies may be stainless steel with replaceable cartridges (Pall) or may be a single bonded unit (Gelman). The latter are easier to use but more difficult to clean and reuse. Reuse is possible, however.

For smaller volumes (less than 1 liter) graded filters may be stacked in a single unit (Fig. 8.16).

Materials

Pump
Clamp
142-mm filter holder with series of nonsterile stacked filters, e.g., glass fiber, 5-, 1.2-, 0.45-μm
Millidisk disposable filter (Millipore) or equivalent
Sterile receiving vessel with outlet at base
Sterile bottles with caps and foil

For Volumes of 5–20 l

1. Connect a nonsterile reusable filter holder (142 mm) or cartridge-type filter (Pall, Millipore) in line with a sterile Millipak 200 or Millidisk 500 (disposable) or sterile 142-mm (reusable) filter holder (Fig. 8.17) with a 0.2 μm–porosity filter and connected to a sterile receiver.
2. Place a 0.2 μm–porosity filter on the support screen of the nonsterile holder, a 1.2-μm filter on top of it, and a 5-μm filter on top of that. Finally place a glass-fiber filter on the top of the 5-μm filter, wet the filters with sterile water, and close up the holder.
3. Connect a pressure reservoir (see Figs. 8.12, 8.13) to the top of the nonsterile filter.
4. Add serum to the pressure vessel, close, and apply pressure (15 kPa, 2 lb/in^2) until the first sign

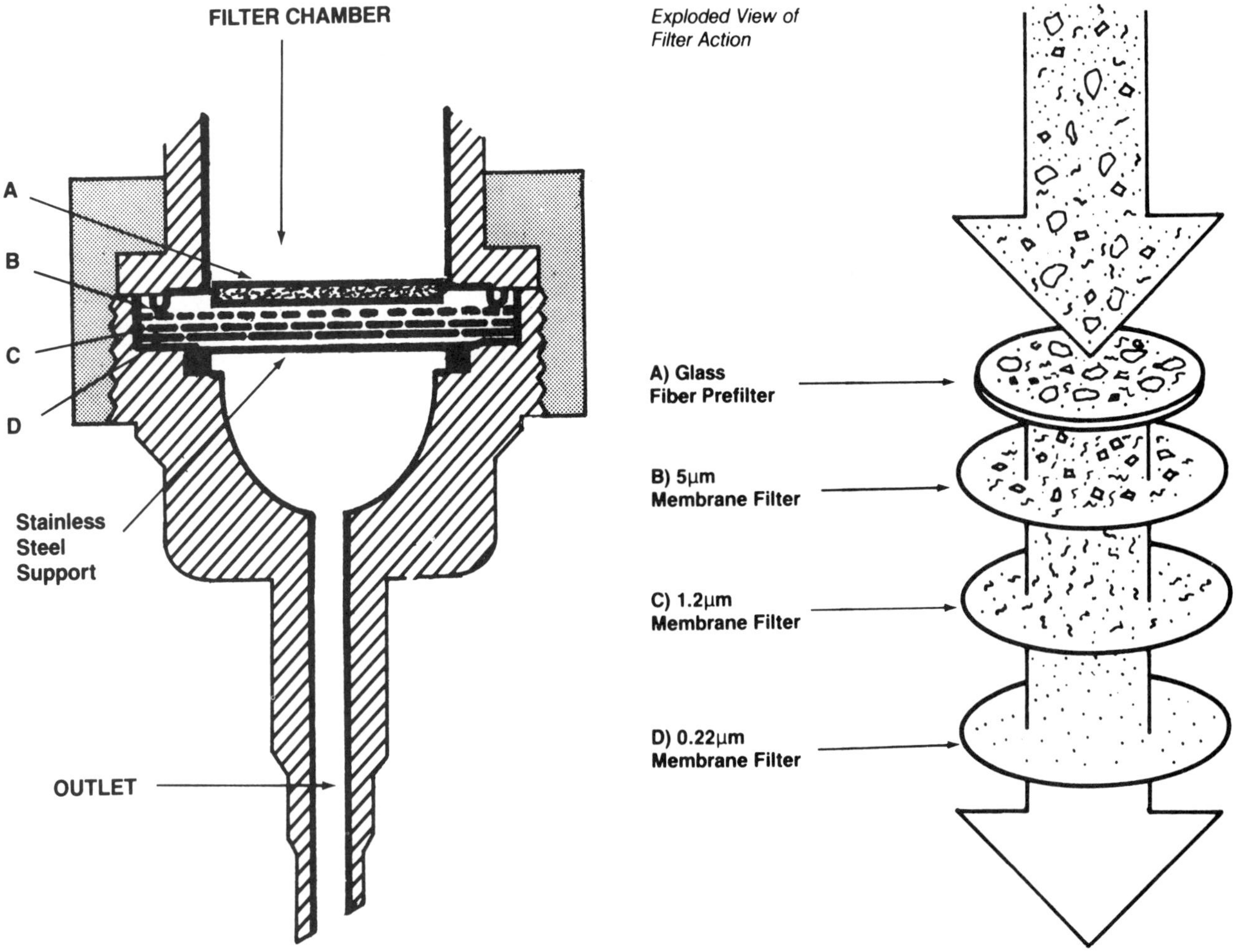

Fig. 8.16. Stacking filters for filtering colloidal solutions (e.g., serum) or solutions with high particulate content.

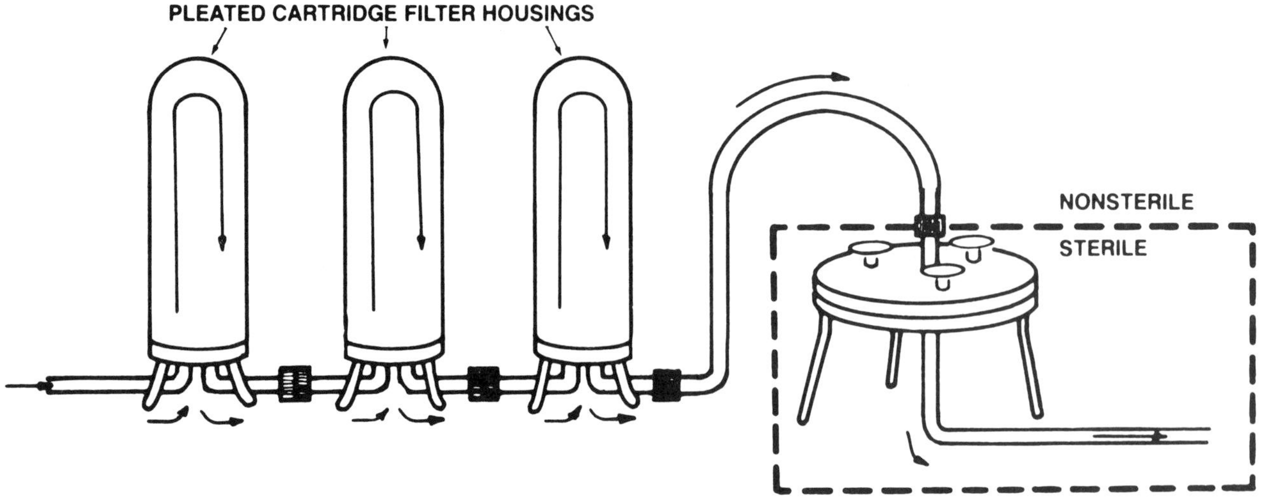

Fig. 8.17. Series of pleated cartridge filters for large-scale filtration of colloidal solutions. Sterile filtration is completed by a reusable or disposable in-line filter.

of liquid appears leaving sterile filter. Stop and tighten sterile filter.

5. Reapply pressure and continue filtration, checking for leaks or blockages. Increasing the filtration pressure will increase the rate of filtration but may cause packing or clogging of the filters.
6. When all serum is through, bubble-point test the filter (see above), then bottle and freeze the serum, taking samples out previously to test (see above).

Small-scale serum processing. If small amounts (<1 liter) of serum are required, then the process is similar but can be scaled down. After clot retraction (see above) small volumes of serum may be centrifuged (5–10,000 *g*) and then filtered through a reusable in-line filter assembly (47 or 90 mm) containing glass-fiber, 5-μm, 1.2-μm, and 0.45-μm filters as described above and finally through a 47-mm, 0.2 μm–porosity sterile disposable filter.

With very small volumes (10–20 ml), centrifuge at 10,000 *g* and filter directly through one or more disposable 25-mm, 0.2-μm filters. A graded series of the syringe-type filters (e.g., Millex, Millipore) is now available.

Storage. Bottle the serum in sizes that will be used up within 2–3 weeks after thawing. Freeze the serum as rapidly as possible and if thawed, do not refreeze unless further prolonged storage is required.

Serum is best used within 6–12 months of preparation if stored at −20°C, but more prolonged storage may be possible at −70°C. The bulk of serum stocks usually makes this impractical. Polycarbonate or high-density polypropylene bottles will eliminate the risk of breakage if storage at −70°C is desired. Regardless of the temperature of the freezer or the nature of the bottles, do not fill them completely. Allow for the anomalous expansion of water during freezing.

Human serum. Outdated blood-bank human blood or plasma can be used, pooled instead of or in addition to bovine or horse. It will be sterile and free from major infections.

◇ Care must be taken with donor serum to ensure that it is screened for hepatitis, AIDS virus, tuberculosis, and so forth.

Titrate out the heparin or citrate anticoagulant with Ca^{2+}, allow to clot overnight, then separate the serum and freeze.

Quality control. Use same procedures as for medium.

A major problem emerging with the use of serum is the possibility of viral infection. When the possibility of bacterial infection was first appreciated, it was relatively easy to devise filtration procedures to filter out anything above 1.0 μm and a porosity of 0.45 μm became standard. Subsequently, it was appreciated that mycoplasma would pass through filters as low as 0.2 μm, and commercial suppliers of serum dropped the exclusion limits of their filters to 0.1 μm. This appears to have virtually eliminated mycoplasma from serum batches used in culture, but has accentuated the remaining problem of viral contamination. Filtering out virus would seem to be a much more significant task, but some filtration companies (Pall) claim that this is possible.

Dialysis. For certain studies, the presence of small molecular weight constituents (amino acids, glucose, nucleosides, etc.) may be undesirable. These may be removed by dialysis through conventional dialysis tubing.

Materials

Dialysis tubing
Beaker with distilled water
Bunsen
Tripod with wire gauze
Serum to be dialyzed
HBSS at 4°C
Measuring cylinder
Sterile stacked filters: 0.22, 0.45, 1.2, 5.0 μm, glass fiber
Sterile bottles with caps

Protocol

1. Boil five pieces, 30-mm × 500-mm dialysis tubing, in three changes of distilled water.
2. Transfer to Hanks' Balanced Salt Solution (HBSS) and allow to cool.
3. Tie double knots at one end of each tube.
4. Half-fill each dialysis tube with serum (20 ml).
5. Express air and knot other end, leaving a space between the serum and the knot of about half the tube.
6. Place in 5 l of HBSS and stir on a magnetic stirrer overnight at 4°C.
7. Change HBSS and repeat twice.
8. Collect serum into measuring cylinder and note volume. (If volume is reduced, add HBSS to make up to starting volume. If increased, make due allowance when adding to medium later.)
9. Sterilize through graded series of filters (see above).
10. Bottle and freeze.

Preparation and Sterilization of Other Reagents

Individual recipes and procedures are given in the Reagent Appendix. On the whole, most reagents are sterilized by filtration if heat-labile and by autoclaving if stable (see summary, Table 8.2).

Filters with low binding properties (e.g., Millex-GV) are available for filter sterilization of proteins and peptides.

CHAPTER 9

Disaggregation of the Tissue and Primary Culture

A primary cell culture may be obtained either by allowing cells to migrate out of fragments of tissue adhering to a suitable substrate or by disaggregating the tissue mechanically or enzymatically to produce a suspension of cells, some of which will ultimately attach to the substrate. It appears to be essential for most normal untransformed cells (with the exception of hemopoietic cells) to attach to a flat surface in order to survive and proliferate with maximum efficiency. Transformed cells (see Chapter 15), on the other hand, particularly cells from transplantable animal tumors, are often able to proliferate in suspension.

The enzymes used most frequently are crude preparations of trypsin, collagenase, elastase, hyaluronidase, DNase, pronase, dispase, or various combinations. Trypsin and pronase give the most complete disaggregation but may damage the cells. Collagenase and dispase give incomplete disaggregation but are less harmful. Hyaluronidase can be used in conjunction with collagenase to digest intracellular matrix, and DNase is employed to disperse DNA released from lysed cells, as it tends to impair proteolysis and promote reaggregation.

Although each tissue may require a different set of conditions, certain common requirements are shared by most primary cultures.

(1) Fat and necrotic tissue are best removed during dissection.

(2) The tissue should be chopped finely with minimum damage.

(3) Enzymes used for disaggregation should be removed subsequently by gentle centrifugation.

(4) The concentration of cells in the primary culture should be much higher than that normally used for subculture, since the proportion of cells from the tissue that survives primary culture may be quite low.

(5) A rich medium, such as Ham's F10 or F12, should be used in preference to a simple, basal medium, such as Eagle's BME; and, if serum is required, fetal bovine often gives better survival than calf or horse. Isolation of specific cell types will probably require selective media (see Chapter 20).

(6) Embryonic tissue disaggregates more readily, yields more viable cells, and proliferates more rapidly in primary culture than adult.

ISOLATION OF THE TISSUE

Before attempting to work with human or animal tissue, make sure that your work fits within medical ethical rules or current animal experiment legislation. Recent changes in the law in the United Kingdom mean that the use of embryos or fetuses beyond 50% gestation or incubation comes under the Animal Experiments (Scientific Procedures) Act, 1986. Work with human biopsies or fetal material will usually require the consent of the local ethical committee and the patient and/or relatives [Warnock, 1985; Winterton, 1989; Royal College of Physicians guidelines, 1990].

An attempt should be made to sterilize the site of the dissection with 70% alcohol if likely to be contaminated. Remove tissue aseptically and transfer to tissue culture laboratory in BSS or medium as soon as possible. If a delay is unavoidable, refrigerate the tissue. (Viable cells can be recovered from chilled tissue several days after explantation.)

Mouse Embryos

Outline

Remove uterus aseptically from timed pregnant mouse and dissect out embryos.

Materials

70% alcohol in wash bottle
Bunsen burner
70% alcohol to sterilize instruments sterile BSS in 50-ml sterile beaker to cool instruments after flaming
HBSS (with antibiotics if required) in 25–50 ml screw-capped vial or tube
Timed pregnant mice (see below)
◇ When sterilizing instruments, by dipping in alcohol and flaming, take care not to return instruments to alcohol while they are still alight!

Protocol

1. If males and females are housed separately, when they are put together for mating, estrus will be induced in the female 3 d later, when the maximum number of successful matings will occur. This enables the planned production of embryos at the appropriate time. The timing of successful matings may be determined by examining the vaginas each morning for a hard mucous plug. The day of detection of the plug (the "plug date") is noted as day zero, and the development of the embryos timed from this date. Full term is about 19–21 d. The optimal age for preparing cultures from whole disaggregated embryo is around 13 d, when the embryo is relatively large (Figs. 9.1, 9.2) but still contains a high proportion of undifferentiated mesenchyme. It is from this mesenchyme that most of the culture will be derived.

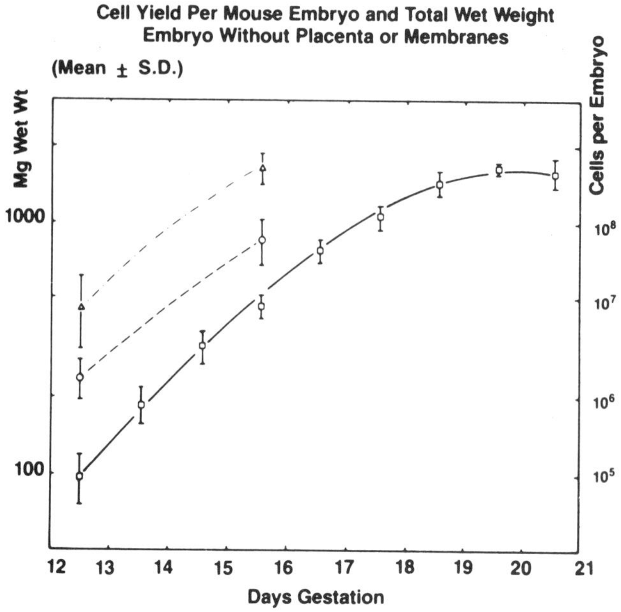

Fig. 9.1. Total wet weight and yield of cells per mouse embryo. Total wet weight of embryo without placenta or membranes, mean ± standard deviation (squares) [From Paul et al., 1969]. Cell yield per embryo after incubation in 0.25% trypsin at 36.5°C for 4 hr (circles). Cell yield per embryo after soaking in 0.25% trypsin at 4°C for 5 hr and incubation at 36.5°C for 30 min (triangles; see text).

Most individual organs, with the exception of brain and heart, begin to form about the 9th day of gestation but are difficult to isolate until about the 11th day. Dissection is easier by 13–14 d, and most of the organs are completely formed by the 18th day.

2. Kill the mouse by cervical dislocation (UK Schedule I procedure) and swab the ventral surface liberally with 70% alcohol (Fig. 9.3a). Tear the ventral skin transversely at the median line just over the diaphragm (Fig 9.3b) and, grasping the skin on both sides of the tear, pull in opposite directions to expose the untouched ventral surface of the abdominal wall (Fig. 9.3c).
3. Cut longitudinally along the median line with sterile scissors, revealing the viscera. At this stage, the uteri filled with embryos will be obvious posteriorly and may be dissected out into a 25-ml or 50-ml screw-capped tube containing 10 or 20 ml BSS (Fig. 9.3d–f). Antibiotics may be added to the BSS where there is a high risk of infection (see Reagent Appendix, Dissection BSS [DBSS]).

All of the preceding steps should be done outside the tissue culture laboratory; a small laminar flow hood and rapid technique will help to maintain sterility. Do not take live animals into

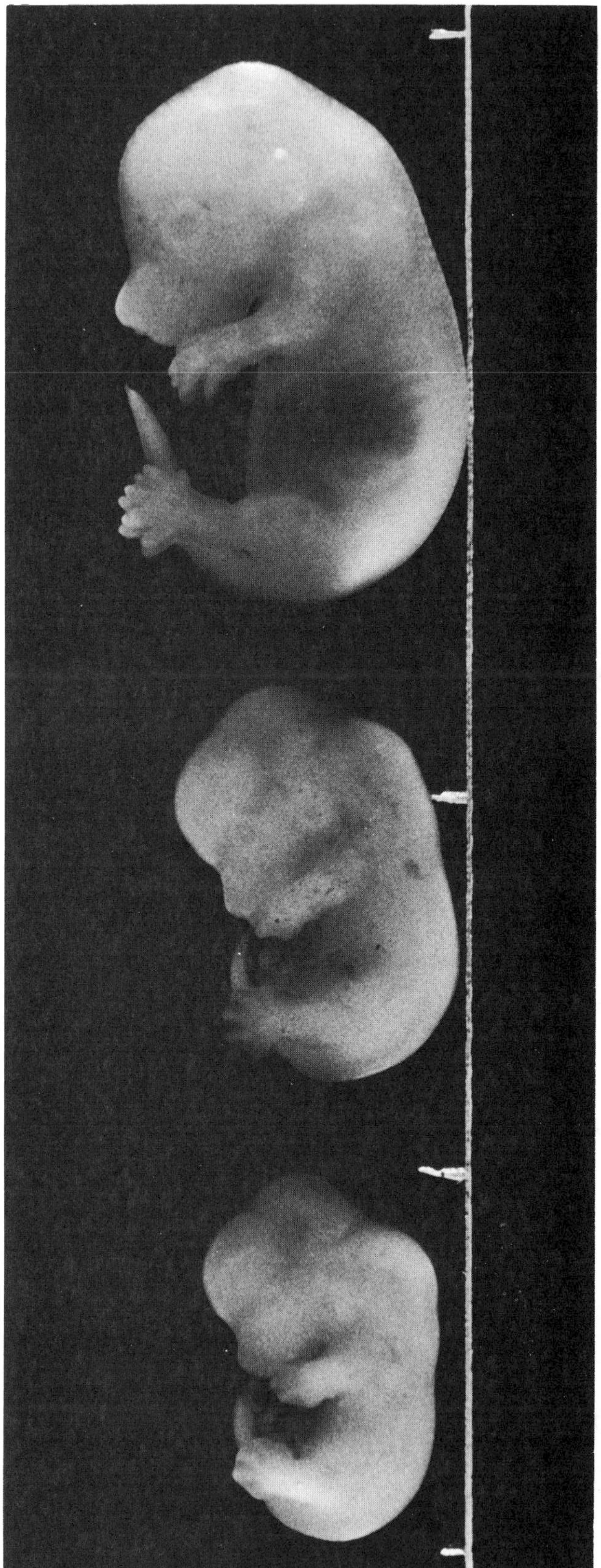

Fig. 9.2. Mouse embryos from the 12th, 13th, and 14th day of gestation. The 12-d embryo (bottom) came from a small litter (three) and is larger than would normally be found at this stage. Scale, 10 mm between marks.

the tissue culture laboratory; they may carry contamination. If the carcass must be handled in the tissue culture area, make sure it is immersed in alcohol briefly, or thoroughly swabbed, and disposed of quickly after use.

4. Take the intact uteri to the tissue culture laboratory and transfer to a fresh dish of sterile BSS.
5. Dissect out the embryos (Fig. 9.3g,h). Tear the uterus with two pairs of sterile forceps, keeping their points close together to avoid distorting the uterus and bringing too much pressure to bear on the embryos. As the uterus is torn apart, the embryos may be freed from the membranes and placenta and placed to one side of the dish to bleed, then transferred to a fresh dish. If a large number of embryos is required (more than four or five litters), it may be helpful to place the last dish on ice (for subsequent dissection and culture, see below).

Hen's Egg

Outline

Remove embryo from egg and transfer to dish.

Materials

70% alcohol
Swabs
Small beaker 20–50 ml
Forceps, straight and curved
9-cm petri dishes
BSS 11-day embryonated eggs
Humid incubator (no additional CO_2 above atmospheric)

Protocol

1. Incubate the eggs at 38.5°C in a humid atmosphere and turn through 180° daily. Although hen's eggs hatch at around 20 to 21 d, the lengths of the developmental stages are different from the mouse. For culture of dispersed cells from the whole embryo, the egg should be taken at about 8 d, for isolated organ rudiments, at about 10 to 13 d.
2. Swab the egg with 70% alcohol and place with blunt end uppermost in a small beaker (Fig. 9.4a).
3. Crack the top of the shell and peel off to the edge of the air sac with sterile forceps (Fig. 9.4b).
4. Resterilize the forceps (dip in a beaker of alcohol, burn off alcohol, and cool in sterile BSS) and peel off the white shell membrane to reveal the chorioallantoic membrane (CAM) below, with its blood vessels (Fig. 9.4c,d).
5. Pierce the CAM with sterile curved forceps and

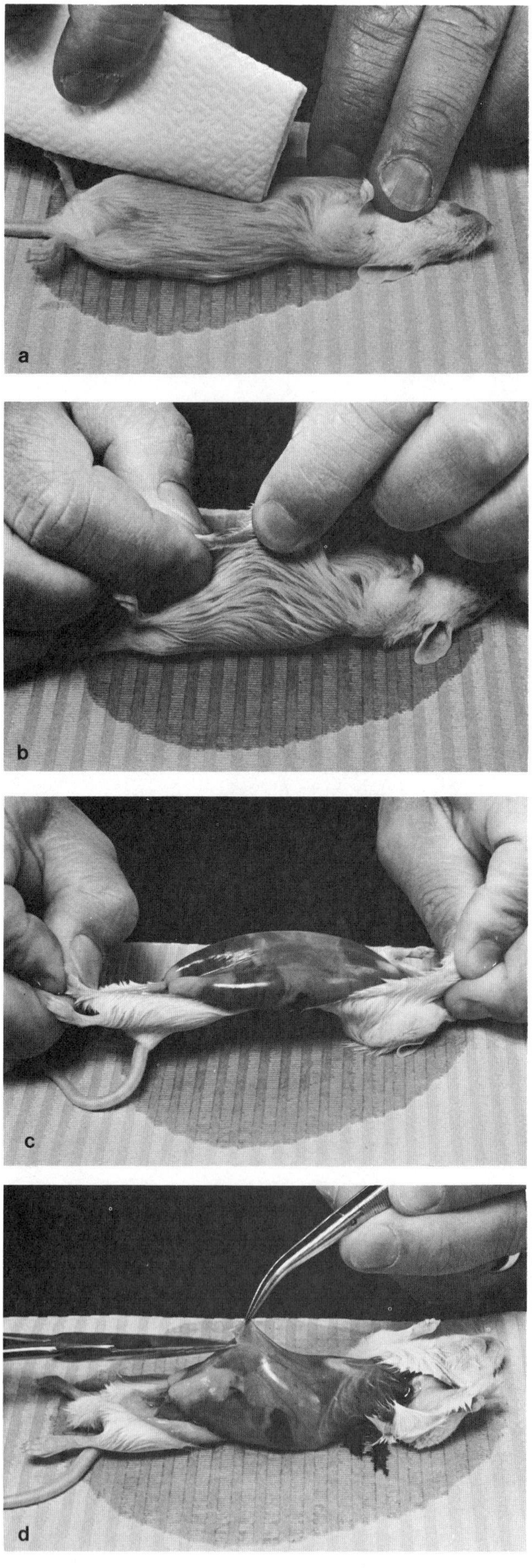

Fig. 9.3. Stages in the aseptic removal of mouse embryos for primary culture (see text). a. Swabbing the abdomen. b,c. Tearing the skin to expose the abdominal wall. d. Opening the

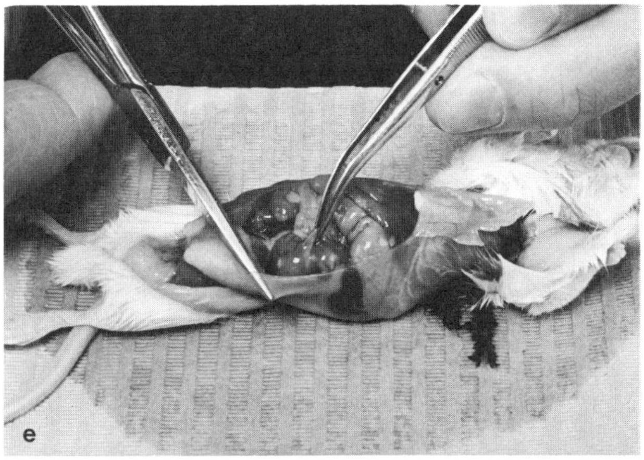

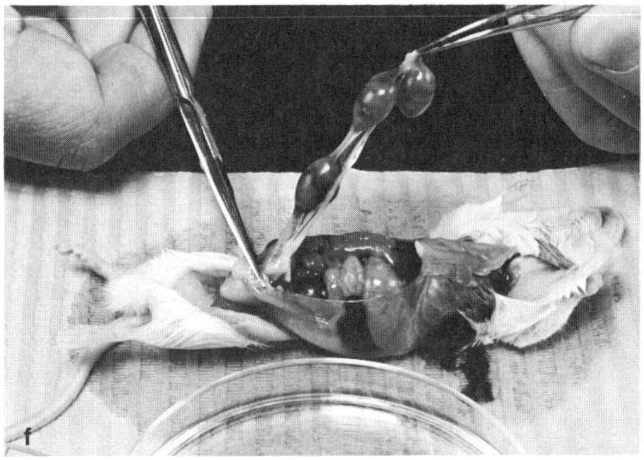

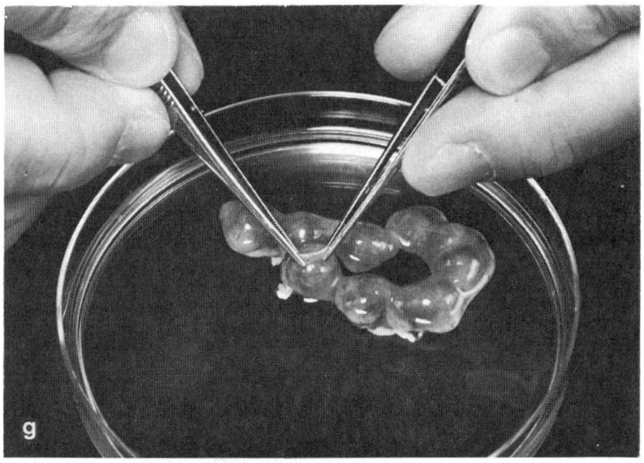

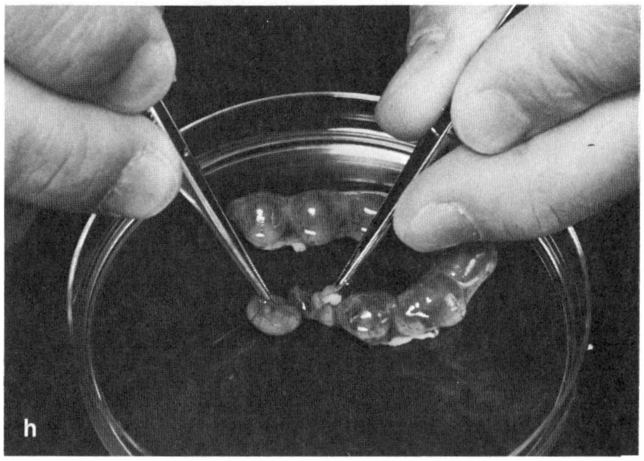

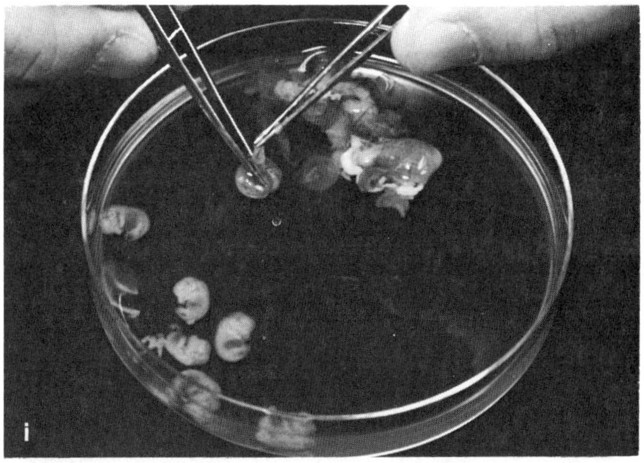

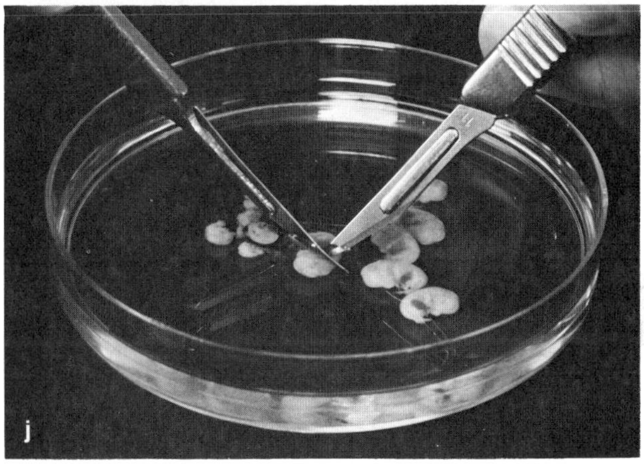

abdomen. e. Uterus in situ. *f. Removing the uterus. g,h. Dissecting embryos from uterus. i. Removing membranes. j. Chopping embryos.*

Fig. 9.4. Stages in the aseptic removal of chick embryo from the egg (see text).

lift out the embryo by grasping gently under the head. Do not close the forceps completely or the neck will sever (Fig. 9.4e–g).

6. Transfer embryo to a 9-cm petri dish containing 20 ml DBSS. (For subsequent dissection and culture, see below.)

Human Biopsy Material

Handling human biopsy material presents certain problems not encountered with animal tissue. It will usually be necessary to obtain consent (1) from the hospital ethical committee; (2) from the attending physician or surgeon; and (3) from the patient or the patient's relatives. Furthermore, biopsy sampling is usually performed for diagnostic purposes and hence the needs of the pathologist must be met first. This is less of a problem if extensive surgical resection or nonpathological tissue (e.g., placenta or umbilical cord) is involved.

The operation will be performed by one of the resident staff at a time that is not always convenient to the tissue culture laboratory, so some formal collection or storage system must be initiated for times when you or someone on your staff cannot be there. If delivery to your lab is arranged, there must be a system for receipt of specimens, recording details, and alerting the person who will perform the culture, otherwise valuable material may be lost or spoiled.

◇ Biopsy material carries a risk of infection (see Chapter 6), so it should be handled in a Class II biohazard cabinet, and all media and apparatus disinfected after use by autoclaving or immersion in a suitable disinfectant (see Chapter 6). If possible, the tissue should be screened for infections such as hepatitis, AIDS, tuberculosis, and so forth, unless the patient has already been tested.

Outline

Provide labeled container(s) of medium, consult with hospital staff, and collect sample from operating room or pathologist.

Materials

Specimen tubes (15–30 ml) with leakproof caps about one-half full with culture medium containing antibiotics (see Reagent Appendix, Collection Medium), and labeled with your name, address, and telephone number

Protocol

1. Provide a container of collection medium clearly labeled, and either arrange to be there to collect it after surgery or have on it your name, address, and, preferably, telephone number, so that it can be sent to you immediately and you can be informed easily when it has been dispatched.
2. Transfer sample to tissue culture laboratory. Usually, if kept at 4°C, biopsy samples will survive for at least 24 hr and even up to 3 or 4 d, although the longer the time from surgery to culture, the more deterioration that may be expected.
3. Decontamination. A disinfectant wash is given before skin biopsy, and an oral antibiotic before gut surgery. Most surgical specimens, however, from the needs of surgery, are sterile when removed, though problems may arise with subsequent handling. Superficial (skin biopsies, melanomas, etc.) and gastrointestinal tract (colon and rectal samples) are particularly prone to infection. It may be advantageous to consult a medical microbiologist to determine what flora to expect in a given tissue and choose your antibiotics accordingly. If the surgical sample is large enough (200 mg or more), a brief dip (30 s–1 min) in 70% alcohol will help to reduce superficial contamination without causing much harm to the center of the tissue sample.

PRIMARY CULTURE

Primary Explant Technique

The primary explant technique was the original method developed by Harrison [1907], Carrel [1912], and others for initiating a tissue culture. A fragment of tissue was embedded in blood plasma or lymph, mixed with embryo extract and serum, and placed on a slide or coverslip. The plasma clotted and held the tissue in place. The embryo extract plus serum both supplied nutrients and stimulated migration out of the explant across the solid substrate. Heterologous serum was used to promote clotting of the plasma. This technique is still used but has been largely replaced by the simplified method below.

Outline

The tissue is chopped finely, rinsed, and the pieces seeded onto the culture surface in a small volume of medium with a high concentration (40–50%) of serum, such that surface tension holds the pieces in place until they adhere spontaneously to the surface (Fig. 9.5). Once this is achieved, outgrowth of cells usually follows (Fig. 9.6).

Materials

Petri dishes
100 ml BSS
Forceps

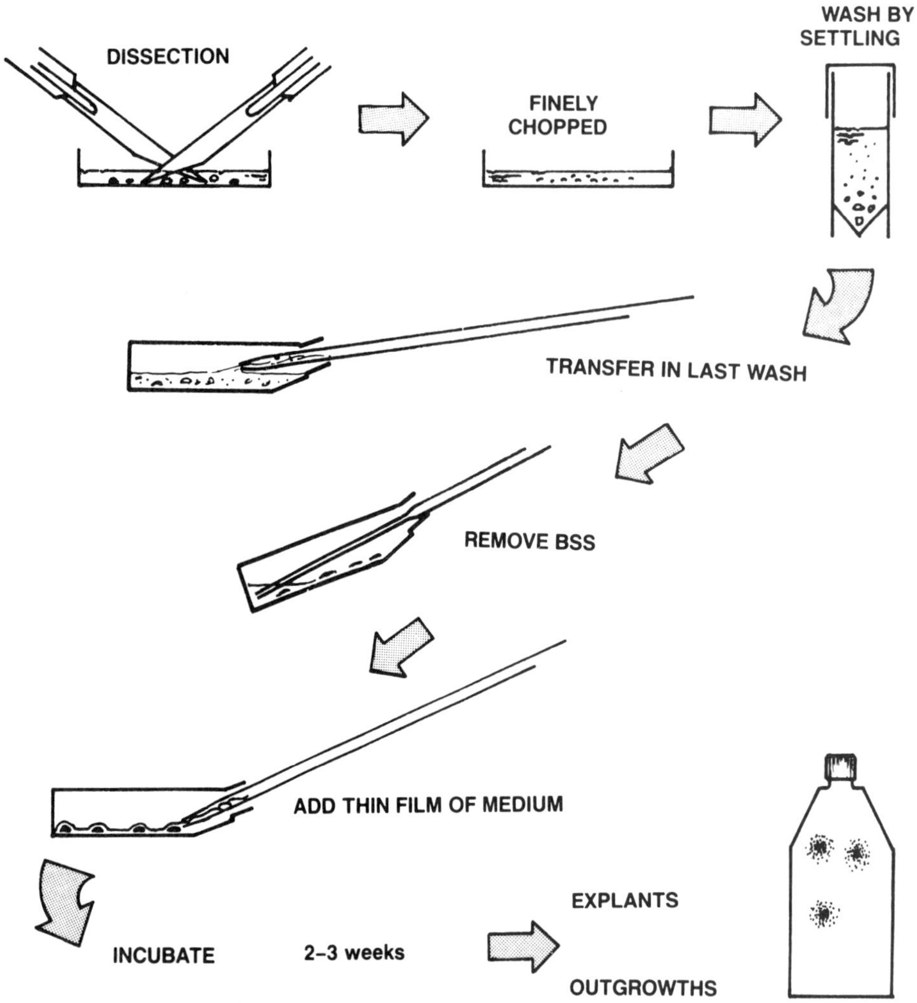

Fig. 9.5. Primary explant technique.

Scalpels
10-ml pipettes
15- or 20-ml centrifuge tubes or universal containers
Culture flasks
Growth medium

The size of flasks and volume of growth medium will depend on the amount of tissue–roughly five 25-cm² flasks per 100 mg tissue, and initially 1 ml medium per flask, building up to 5 ml per flask over the first 3–5 d.

Protocol

1. Transfer tissue to fresh sterile BSS and rinse.
2. Transfer to a second dish, dissect off unwanted tissue such as fat or necrotic material, and chop finely with crossed scalpels (Fig. 9.5) to about 1-mm cubes.
3. Transfer by pipette (10–20 ml with wide tip) to a 15- or 50-ml sterile centrifuge tube or universal container (wet the inside of the pipette first with BSS or the pieces will stick). Allow the pieces to settle.
4. Wash by resuspending the pieces in BSS, allowing the pieces to settle, and removing the supernatant fluid, two or three times.
5. Transfer the pieces (remember to wet the pipette) to a culture flask, about 20–30 pieces per 25-cm² flask.
6. Remove most of the fluid and add about 1 ml growth medium per 25-cm² growth surface. Tilt the flask gently to spread the pieces evenly over the growth surface.
7. Cap the flask and place in incubator or hot room at 36.5°C for 18–24 hr.
8. If the pieces have adhered, the medium volume may be made up gradually over the next 3–5 d

to 5 ml per 25 cm^2 and then changed weekly until a substantial outgrowth of cells is observed (see Fig. 9.6).

9. The explants may then be picked off from the center of the outgrowth with a scalpel and transferred by prewetted pipette to a fresh culture vessel. (Return to step 7 above.)
10. Replace medium in the first flask until the outgrowth has spread to cover at least 50% of the growth surface, at which point the cells may be passaged (see Chapter 10).

This technique is particularly useful for small amounts of tissue such as skin biopsies where there is a risk of losing cells during mechanical or enzymatic disaggregation. Its disadvantages lie in the poor adhesiveness of some tissues and the selection of cells in the outgrowth. In practice, however, most cells, fibroblasts, myoblasts, glia, epithelium, particularly from the embryo, will migrate out successfully.

Both adherence and migration may be stimulated by a plasma clot:

1. Place the tissue pieces in position in the culture flask as in step 7 above.
2. Mix 2 parts of chicken plasma with 1 part chicken embryo extract and 1 part fetal bovine serum. Immediately pipette gently over the tissue pieces, spacing the pieces evenly on the surface of the dish as you do so.
3. Allow to clot and place at 36.5°C [see Paul, 1975, for further description].

Alternatively, a glass coverslip may be placed on top of the explant, with the explant near the edge of the coverslip, or the plastic dish may be scratched through the explant to attach the tissue to the flask [Elliget and Lechner, 1992] (see also LaVeck and Lechner protocol, Chapter 20). Attachment may be promoted by treating the plastic with polylysine or fibronectin (see Chapter 11), or extracellular matrix or feeder layers (see Chapter 7).

Enzymatic Disaggregation

Cell–cell adhesion in tissues is mediated by a variety of homotypic interacting glycopeptides (cell adhesion molecules or CAMs), some of which are calcium dependent (cadherins), and hence sensitive to chelating agents such as EDTA or EGTA. Intercellular matrix and basement membranes also contain other glycoproteins, such as fibronectin and laminin, which are protease sensitive, and proteoglycans, which are less so, and can sometimes be degraded by glycanases such as hyaluronidase or heparanase. The simple approach is to proceed from a simple disaggregation solution to a more complex (as in Table 10.4) with trypsin alone or trypsin/EDTA as a starting point, adding other proteases to improve disaggregation, and deleting trypsin if necessary to increase viability. In general, increasing the purity of an enzyme will give better control and less toxicity with increased specificity, but may result in less disaggregating activity.

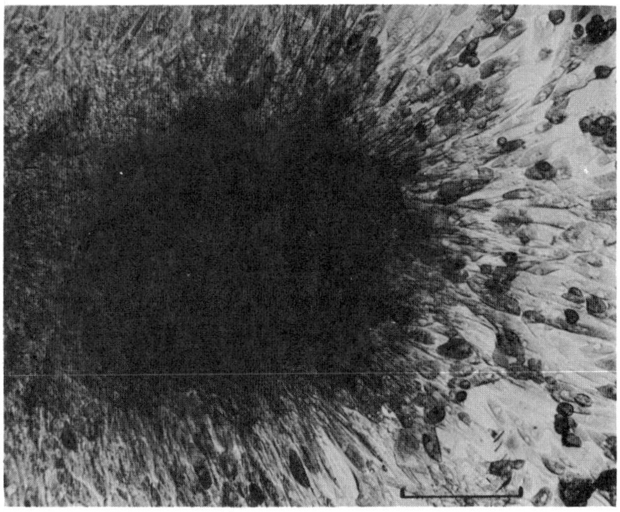

Fig. 9.6. Primary explant from human mammary carcinoma. Dense area in center is an undisaggregated tissue fragment and cells are seen migrating radially from the explant. This is a good size for an explant, promoting maximal radial outgrowth, but routinely it is difficult to dissect below 0.5 mm. Scale bar, 100 μm.

Mechanical and enzymatic disaggregation of the tissue avoids problems of selection by migration, but more important, perhaps, yields a higher number of cells more representative of the whole tissue in a shorter time. However, just as the primary explant technique selects on the basis of cell migration, dissociation techniques select cells resistant to the method of disaggregation and still capable of attachment.

Embryonic tissue disperses more readily and gives a higher yield of proliferating cells than newborn or adult. The increasing difficulty in obtaining viable proliferating cells with increasing age is due to several factors, including the onset of differentiation, an increase in fibrous connective tissue and extracellular matrix, and a reduction of the undifferentiated proliferating cell pool. Where procedures of greater severity are required to disaggregate the tissue, e.g., longer trypsinization or increased agitation, the more fragile components of the tissue may be destroyed. In fibrous tumors, for example, it is very difficult to obtain complete dissociation with trypsin while still retaining viable carcinoma cells.

The choice of trypsin grade has always been difficult,

as there are two opposing trends: (1) the purer the trypsin the less toxic it becomes, and the more predictable its action; (2) the cruder the trypsin, the more effective it may be due to other proteases. In practice, a preliminary test experiment may be necessary to determine the optimum grade for viable cell yield, as the balance between sensitivity to toxic effects and disaggregation ability may be difficult to predict.

Crude trypsin is by far the most common enzyme used in tissue disaggregation [Waymouth, 1974], as it is tolerated quite well by many cells, is effective for many tissues, and any residual activity left after washing is neutralized by the serum of the culture medium. A trypsin inhibitor (e.g., soya bean trypsin inhibitor, Sigma) must be included when serum-free medium is used.

It is important to minimize the exposure of cells to active trypsin to preserve maximum viability. Hence when trypsinizing whole tissue at 36.5°C, dissociated cells should be collected every half-hour and the trypsin removed by centrifugation and neutralized with serum in medium. Soaking the tissue for 6–18 hr in trypsin at 4°C (see below) allows penetration with minimal tryptic activity, and digestion may then proceed for a much shorter time (20–30 min) at 37°C [Cole and Paul, 1966]. Although the cold-trypsin method gives a higher yield of viable cells and is less effort, the warm-trypsin method is still used extensively and is presented here for comparison.

Disaggregation in Warm Trypsin

Outline

The tissue is chopped and stirred in trypsin for a few hours, collecting dissociated cells every half-hour, and the dissociated cells are then centrifuged and pooled in medium containing serum (Fig. 9.7).

Materials

250-ml Erlenmeyer flask (preferably indented as in Fig. 9.8 [Bellco])
Magnetic bar
Magnetic stirrer
100 ml 0.25% trypsin (crude: Difco 1:250, Flow or GIBCO) in CMF or PBSA (see Reagent Appendix)
Two 50-ml centrifuge tubes
Hemocytometer or cell counter
Growth medium with serum (e.g., Ham's F12 with 10% fetal bovine serum)
Culture flasks, 5–10 g tissue (will vary depending on cellularity of tissue)

Protocol

1. Proceed as for primary explant, although the pieces need only to be chopped to about 3-mm diameter. As described here, this method requires about 20 times as much tissue as the primary explant technique, although it can be scaled down if desired.
2. After washing as in step 4 for primary explantation, transfer the chopped pieces to the trypsinization flask (Bellco makes one specially designed for the purpose, Fig. 9.8, but a 250-ml conical Erlenmeyer flask will do), and add 100 ml trypsin.
3. Stir at about 200 rpm for 30 min at 36.5°C.
4. Allow pieces to settle, collect supernatant, centrifuge at approximately 500 g for 5 min, resuspend pellet in 10 ml medium with serum, and store cells on ice.
5. Add fresh trypsin to pieces and continue to stir and incubate for a further 30 min. Repeat steps 3 to 5 until complete disaggregation occurs or until no further disaggregation is apparent.
6. Collect and pool chilled cell suspensions, and count by hemocytometer or electronic cell counter (see Chapter 18).

   Remember you are dealing with a very heterogeneous population of cells; electronic cell counting will initially require confirmation with a hemocytometer, as a calibration "plateau" (see Chapter 18) is rather difficult to obtain.
7. Dilute to 10^6 per ml in growth medium and seed as many flasks as are required with approximately 2×10^5 cells per cm^2. Where the survival rate is unknown or unpredictable, a cell count is of little value (e.g., tumor biopsies where the proportion of necrotic cells may be quite high). In this case, set up a range of concentrations from about 5 to 25 mg tissue per ml. Change the medium at regular intervals (2 to 4 d as dictated by depression of pH).

This technique is useful for the disaggregation of large amounts of tissue in a relatively short time, particularly whole mouse embryo or chick embryo. It does not work as well with adult tissue where there is a lot of fibrous connective tissue, and mechanical agitation can be damaging to some of the more sensitive cell types such as epithelium.

Trypsin at 4°C

One of the disadvantages of using trypsin to disaggregate tissue is the damage that may result from prolonged exposure to trypsin at 36.5°C–hence the need to harvest cells after 30-min incubations in the warm-trypsin method rather than have them exposed for the full time (3 to 4 hr) required to disaggregate the whole tissue. A simple method of minimizing damage to the cells during disaggregation is to soak the tissue in tryp-

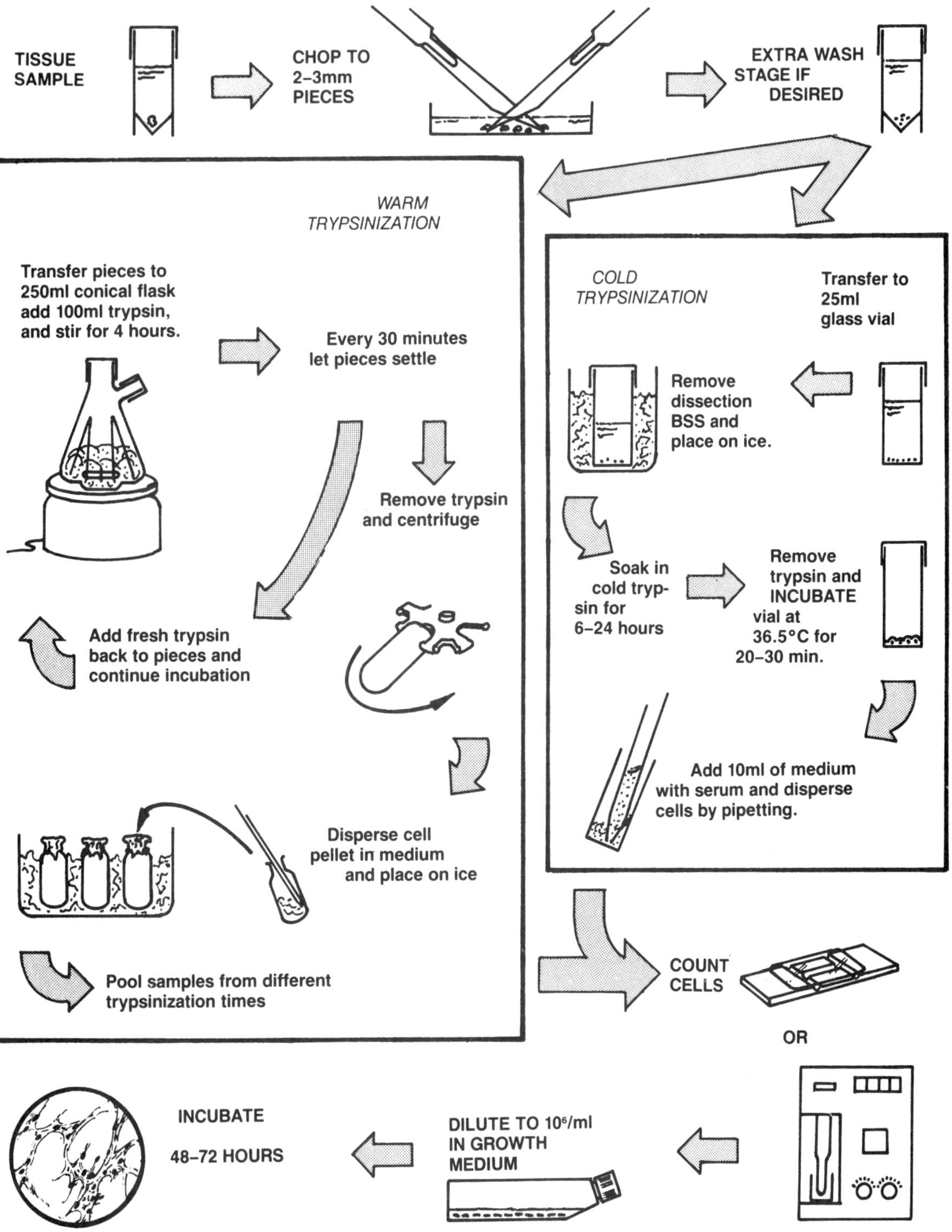

Fig. 9.7. *Preparation of primary culture by disaggregation in trypsin; warm trypsin method on the left, cold trypsin method on the right.*

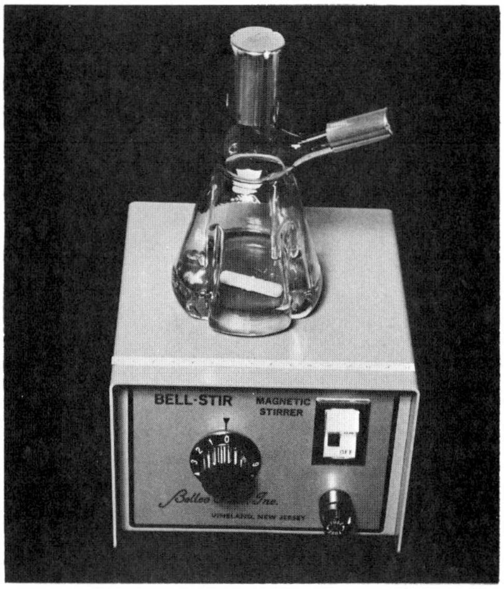

Fig. 9.8. Trypsinization flask (Bellco). The indentations in the side of the flask improve mixing, and the rim around the neck, below the side arm, allows the cell suspension to be poured off while leaving the stirrer bar and any larger fragments behind.

sin at 4°C to allow penetration of the enzyme with little tryptic activity (Table 9.1). Following this, the tissue will require much shorter incubation at 36.5°C for disaggregation [Cole and Paul, 1966].

Outline

Chop tissue and place in trypsin at 4°C for 6–18 hr. Incubate after removing the trypsin, and disperse the cells in warm medium (Fig. 9.7).

Materials

Petri dish
BSS
Forceps, straight and curved
0.25% crude trypsin
Scalpels
25-ml screw-capped Erlenmeyer flask(s)
25-cm^2 or 75-cm^2 culture flasks
Pipettes
Culture medium

Protocol

1. Follow steps 1–4 as for primary explants, but collect tissue in glass tube or vial to facilitate chilling (see below). Tissue need only be chopped to 3–4-mm pieces. Embryonic organs, if they do not exceed this size, are better left whole.
2. After washing, place the container on ice, remove the last BSS wash, and replace with 0.25% trypsin in PBSA at 4°C (approximately 1 ml for every 100 mg of tissue).
3. Place at 4°C for 6–18 hr.
4. Remove and discard the trypsin carefully, leaving the tissue with only the residual trypsin.
5. Place tube at 36.5°C for 20–30 min.
6. Add warm medium, approximately 1 ml for every 100 mg, and gently pipette up and down until the tissue is completely dispersed.
7. If some tissue is left undispersed, the cell suspension may be filtered through sterile muslin or stainless steel mesh (100–200 μm), or Falcon 70-μm "Cell Strainer" (Becton Dickinson; Fig. 9.12b), or the larger pieces may simply be allowed to settle. Where there is a lot of tissue, increasing the volume of suspending medium to 20 ml for each gram of tissue will facilitate settling and subsequent collection of supernatant fluid. Two to three minutes should be sufficient to get rid of most of the larger pieces.
8. Determine the concentration of the cell suspension and seed the vessels at 10^6 cells/ml (2×10^5 cells/cm^2).

The cold-trypsin method usually gives a higher yield of cells (see Fig. 9.1 and Table 9.1) and preserves more

TABLE 9.1. Relative Cell Yield From 12-d Mouse Embryos by Warm or Cold Trypsin Methods

Trypsin temperature and duration of exposure (hr)			Viable cells per embryo			
4°C	36.5°C	Total cell number recovered per embryo ($\times 10^{-7}$)	No. ($\times 10^{-7}$)	% Viability by dye exclusion (Trypan blue)	Percentage of total recovered after 24 hr in culture	Percentage of viable cells recovered
–	4[a]	1.69	1.45	86	47.2	54.9
5.5	0.5[b]	3.32	1.99	60	74.5	124
24	0.5[b]	3.40	2.55	75	60.3	80.2

[a]Stirred continuously at 36.5°C; fractions were *not* collected at 30-min intervals.
[b]Incubated without agitation.

different cell types than the warm method. Cultures from mouse embryos contain more epithelial cells when prepared by the cold method, and erythroid cultures from 13-d fetal mouse liver respond to erythropoietin after this treatment but not after warm trypsin or mechanical disaggregation [Cole and Paul, 1966; Conkie, personal communication]. The method is also convenient, as no stirring or centrifugation is required, and the incubation at 4°C may be done overnight. This method does take longer, however, and is not as convenient where large amounts of tissue (greater than 10 g) are being handled. A particular advantage in the cold-trypsin method is the handling of small amounts of tissue, such as embryonic organs. Taking 10–13-d chick embryo as a starting point, the following procedure gives good reproducible cultures with evidence of several different cell types characteristic of the tissue of origin.

Chick Embryo Organ Rudiments

Outline

Dissect out individual organs or tissues, and place, preferably whole, in cold trypsin overnight. Remove the trypsin, incubate briefly, and disperse in culture medium. Dilute and seed cultures.

Materials

Petri dish
BSS
Scalpels (No. 11 blade for most steps)
Iridectomy knives (Beaver, blade 21) for fine dissection
Curved and straight fine forceps
Pipettes (Pasteur, 2 ml, 10 ml)
0.25% crude trypsin (Difco 1:250 or equivalent) in CMF or PBS on ice; lower concentrations may be used with purer grades of trypsin, e.g., 0.05–0.1%
Sigma crystalline or Worthington Grade IV
10–15-ml test tubes with screw caps
25-cm² culture flasks (two per tissue)
Culture medium (e.g., Ham's F12 + 10% fetal bovine serum) 12 ml per tissue, minimum

Protocol

1. Remove the embryo from the egg as described above and place in sterile BSS.
2. Remove the head (Fig. 9.9a,b).
3. Remove an eye and open carefully, releasing the lens and aqueous and vitreous humors (Fig. 9.9c,d).
4. Grasp the retina in two pairs of fine forceps and gently peel the pigmented retina off the neural retina and connective tissue (Fig. 9.9e). (A brief exposure to 0.25% trypsin in 1 m*M* EDTA will separate the two tissues more easily.) Put tissue to one side.
5. Pierce the top of the head with curved forceps and scoop out the brain (Fig. 9.9f). Place this, as for the retina above, and for other tissues below, at the side of the dish.
6. Halve the trunk transversely where the pink color of the liver shows through the ventral skin (Fig. 9.9g). If the incision is made on the line of the diaphragm, it will pass between the heart and the liver; but sometimes the liver will go to the anterior instead of the posterior half.
7. Gently probe into the cut surface of the anterior half and draw out the heart and lungs (Fig. 9.9h; tease the organs out and do not cut until you have identified them). Separate.
8. Probe the posterior half, and draw out the liver with the folds of the gut enclosed in between the lobes (Fig. 9.9i). Separate.
9. Fold back the body wall to expose the inside of the dorsal surface of the body cavity in the posterior half. The elongated lobulated kidneys should be visible parallel to and on either side of the midline.
10. Gently slide the tip of the scalpel under each kidney and tease away from the dorsal body wall (Fig. 9.9j). Carefully cut free and place on one side.
11. Place the tips of the scalpels together on the midline at the posterior end and, advancing the tips forward, one over the other, express the spinal cord like toothpaste from a tube (Fig. 9.9k).
12. Turn the posterior trunk of the embryo over and strip the skin off the back and upper part of the legs (Fig. 9.9l). Collect and place on one side.
13. Dissect off muscle from each thigh and collect together (Fig. 9.9m).
14. Transfer all of these tissues, and any others you may want, to separate test tubes containing 1 ml of 0.25% trypsin in PBSA and place on ice. Make sure the tissue slides right down the tube into the trypsin.
15. Leave 6–18 hr at 4°C.
16. Remove trypsin carefully; tilting and rolling the tube slowly will help.
17. Incubate the tissue in the residual trypsin for 15–20 min at 36.5°C.
18. Add 4 ml medium to each of two 25-cm² flasks for each tissue to be cultured.
19. Add 2 ml medium to tubes after step 18, and pipette up and down gently to disperse the tissue.

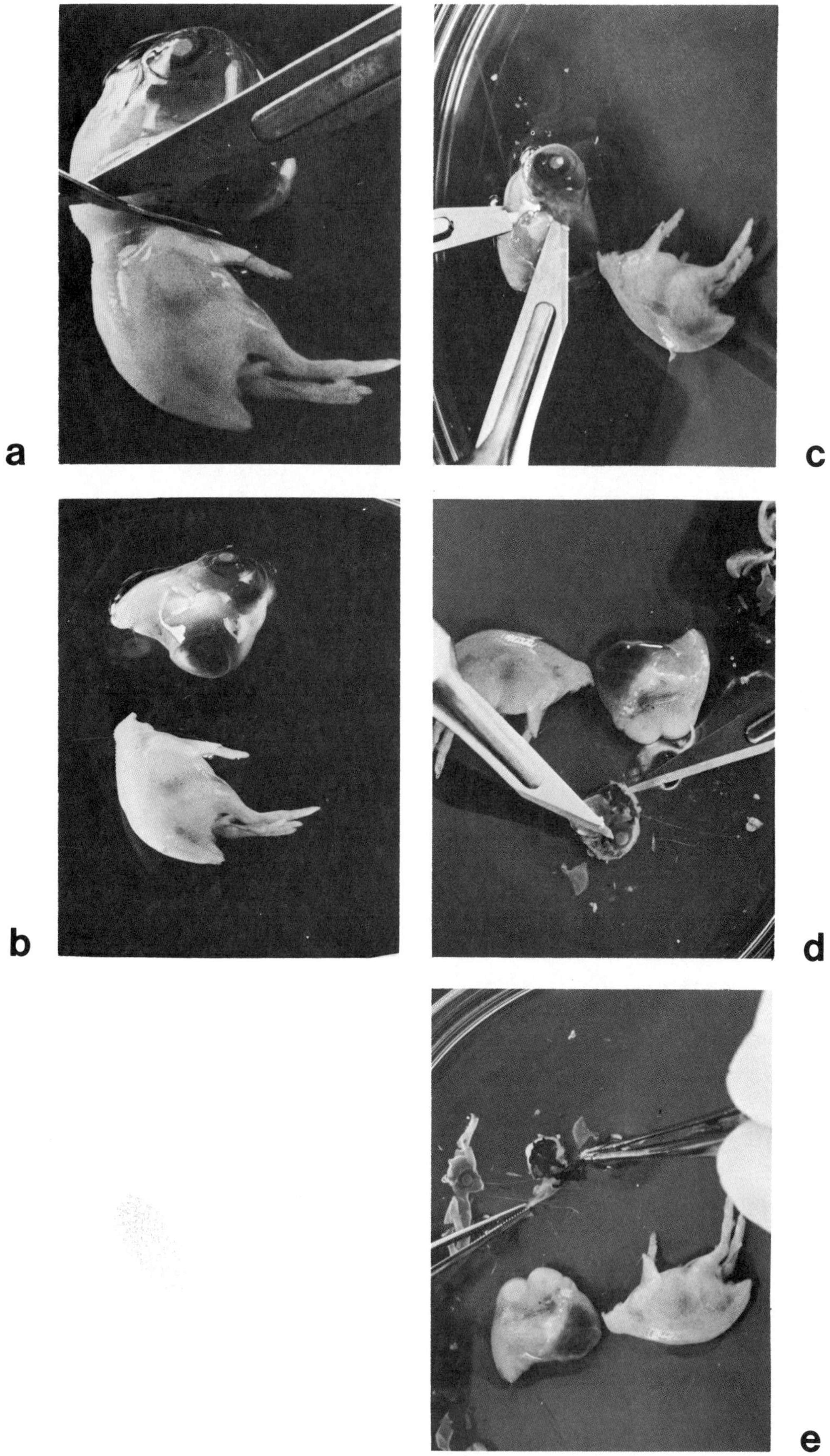

Fig. 9.9. Dissection of chick embryo. a,b. Removing the head. c. Removing the eye. d. Dissecting out the lens. e. Peeling off the retina. f. Scooping out the brain. g. Halving the trunk. h. Teasing out the heart and lungs from the anterior half. i. Teasing out the liver and gut from the posterior half. j. Inserting the tip of the scalpel between the left kidney and the dorsal body wall. k. Squeezing out

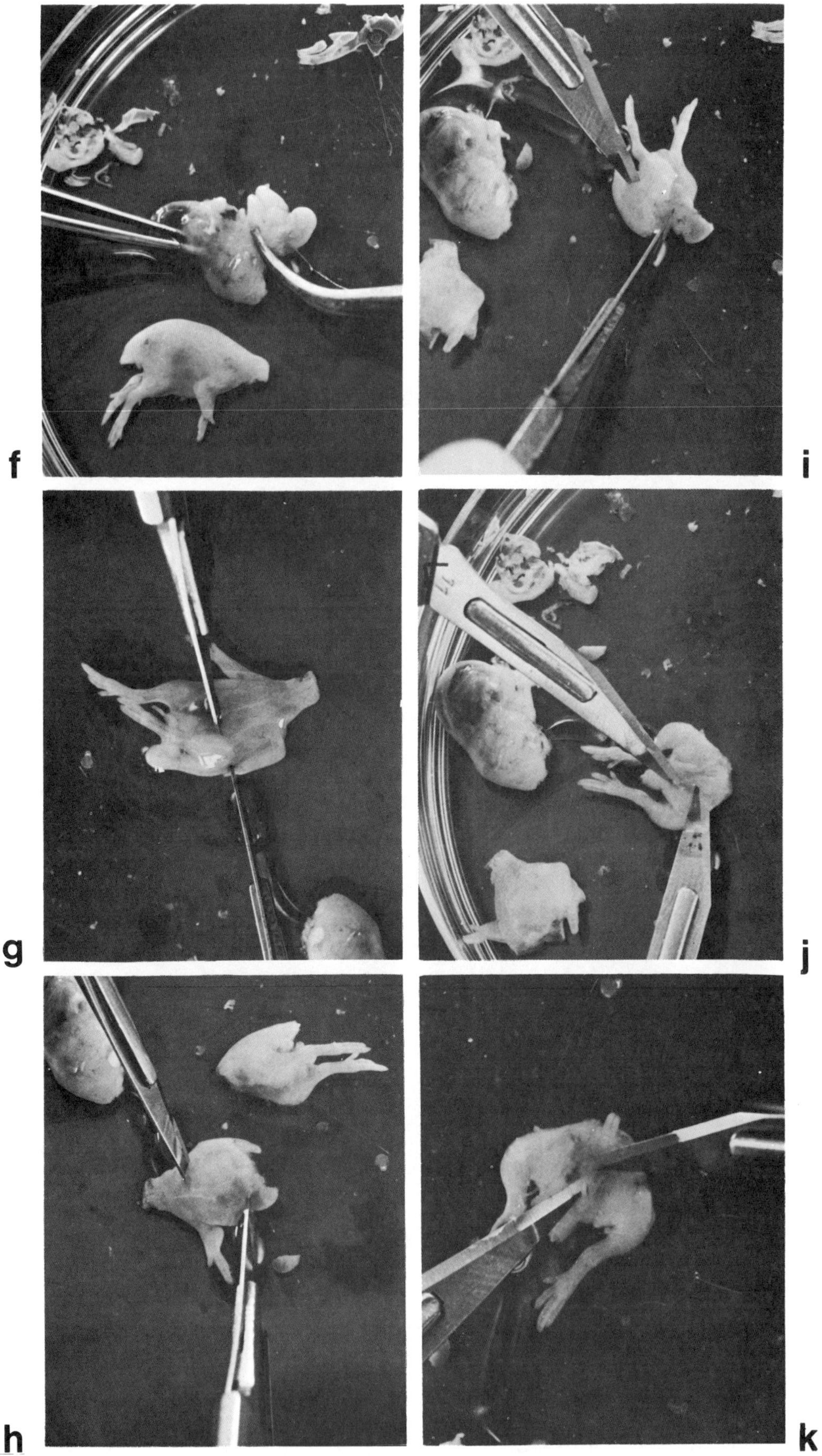

the spinal cord. l. Peeling skin off the back of the trunk and hind leg. m. Slicing muscle from the thigh. n. Organ rudiments arranged around periphery of dish; from the right, clockwise: brain, heart, lungs, liver, gizzard, kidneys, spinal cord, skin, and muscle.

20. Allow any large pieces to settle, pipette off supernatant fluid into the first flask, mix, and transfer 1 ml of diluted suspension to second flask. This gives two flasks at different cell concentrations and avoids the need to count the cells. Experience will determine the appropriate cell concentration to use in subsequent attempts.
21. Change the medium as required (e.g., with brain it may need to be changed after 24 hr, but pigmented retina will probably last 5–7 d), and check for characteristic morphology and function. After 3 to 5 d, contracting cells may be seen in the heart cultures, colonies of pigmented cells in the pigmented retina culture, and the beginning of myotubes in skeletal muscle cultures.

Other Enzymatic Procedures

Disaggregation in trypsin can be damaging (e.g., to some epithelial cells) or ineffective (e.g., for very fibrous tissue such as fibrous connective tissue), so attempts have been made to utilize other enzymes. Since the extracellular matrix often contains collagen, particularly in connective tissue and muscle, collagenase has been the obvious choice [Freshney, 1972 (colon carcinoma); Lasfargues, 1973 (breast carcinoma); Chen et al., 1989 (kidney); Kralovanszky et al., 1990 (gut); Heald et al., 1991 (pancreatic islet cells)] (see also Chapter 20). Other bacterial proteases such as pronase [Wiepjes and Prop, 1970; Gwatkin, 1973; Prop and Wiepjes, 1973] and dispase [Matsamura et al., 1975] (Boehringer-Mannheim Biochemicals) have also been used with varying degrees of success. The participation of carbohydrate in intracellular adhesion has led to the use of hyaluronidase [Berry and Friend, 1969] and neuraminidase in conjunction with collagenase. Other proteases continue to appear on the market (Sigma; Boehringer; Worthington). With the selection now available, screening available samples would seem to be the only option if trypsin, collagenase, dispase, pronase, hyaluronidase, or DNAse, alone and in combinations, have not proved to be successful.

It is not possible to describe here all the primary disaggregation techniques that have been used, but the following method has been found to be effective in several normal and malignant tissues.

Collagenase

This technique is very simple and effective for embryonic and normal and malignant adult tissue. It is of greatest benefit where the tissue is either too fibrous or

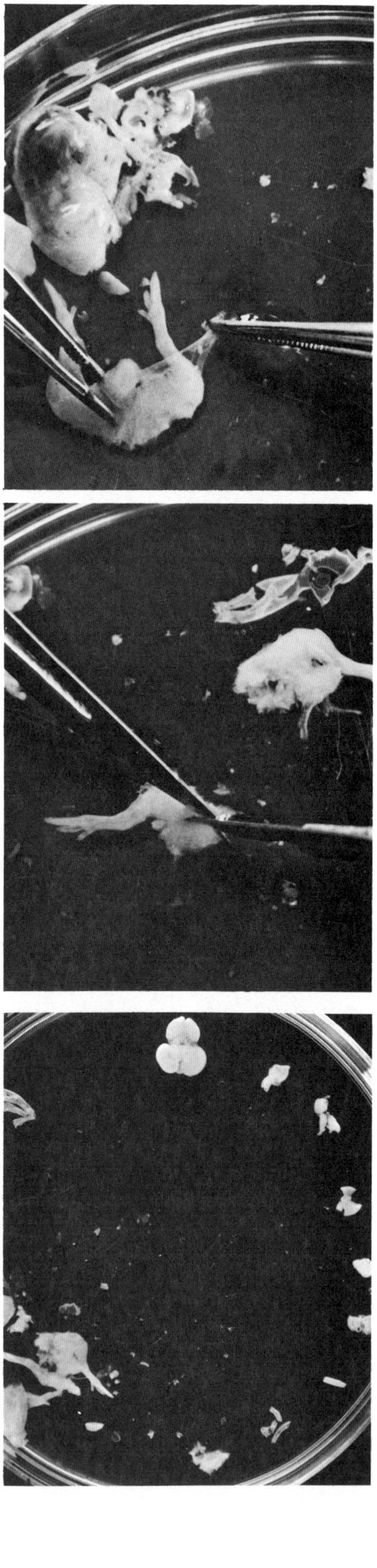

l

m

n

Fig. 9.9. (Continued)

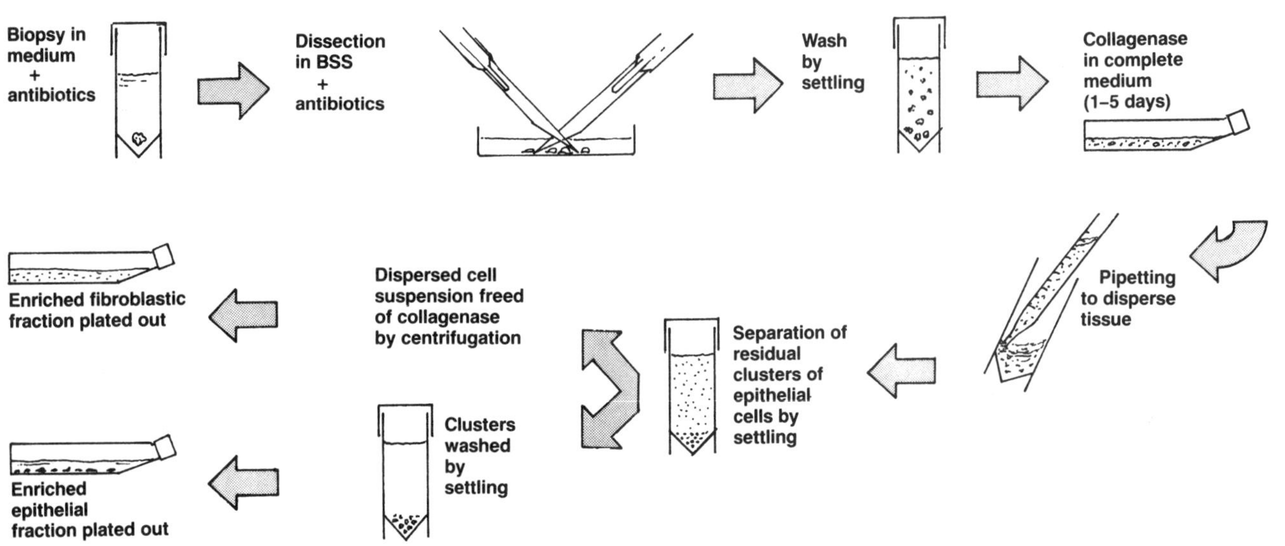

Fig. 9.10. *Stages in disaggregation of tissue for primary culture by collagenase.*

too sensitive to allow the successful use of trypsin [Freshney, 1972]. Crude collagenase is often used and may depend for some of its action on contamination with other nonspecific proteases. More highly purified grades are available if nonspecific proteolytic activity is undesirable but may not be as effective.

Outline

Place finely chopped tissue in complete medium containing collagenase and incubate. When tissue is disaggregated, remove collagenase by centrifugation, seed cells at a high concentration, and culture (Fig. 9.10).

Materials

Pipettes
25-cm^2 culture flasks
Culture medium
Collagenase (2,000 units/ml) Worthington CLS or Sigma 1A
Centrifuge tubes, 15–50 ml depending on amount of tissue being processed
Centrifuge

Protocol

1. Proceed as for primary explant up to step 5, but transfer 20–30 pieces to one 25-cm^2 flask and 100–200 pieces to a second.
2. Drain off BSS and add 4.5 ml growth medium with serum to each flask.
3. Add 0.5 ml crude collagenase, 2,000 units/ml, to give a final concentration of 200 units/ml collagenase.
4. Incubate at 36.5°C for 4–48 hr without agitation. Tumor tissue may be left up to 5 d or more if disaggregation is slow, e.g., in scirrhous carcinomas of breast or colon, although it may be necessary to centrifuge the tissue and resuspend in fresh medium and collagenase before then if an excessive drop in pH is observed (to less than pH 6.5).
5. Check for effective disaggregation by gently moving the flask; the pieces of tissue will "smear" on the bottom of the flask and, with moderate agitation, will break up into single cells and small clusters (Fig. 9.11). With some tissues (e.g., lung, kidney, and colon or breast carcinoma) small clusters of epithelial cells can be seen to resist the collagenase and may be separated from the rest by allowing them to settle for about 2 min. If these clusters are further washed with BSS by resuspension and settling and the sediment seeded in medium, they will form healthy islands of epithelial cells. Epithelial cells generally survive better if not completely dissociated.
6. Where complete disaggregation has occurred, or when the supernatant cells are collected after allowing clusters to settle, centrifuge at 50–100 g for 3 min. Discard supernatant DBSS, resuspend, combine pellets in 5 ml medium, and seed in a 25-cm^2 flask. If the pH fell during collagenase treatment (to pH 6.5 or less by 48 hr), dilute twofold to threefold in medium after removing the collagenase.
7. Replace medium after 48 hr.

Some cells, particularly macrophages, may adhere to the first flask during the collagenase incubation.

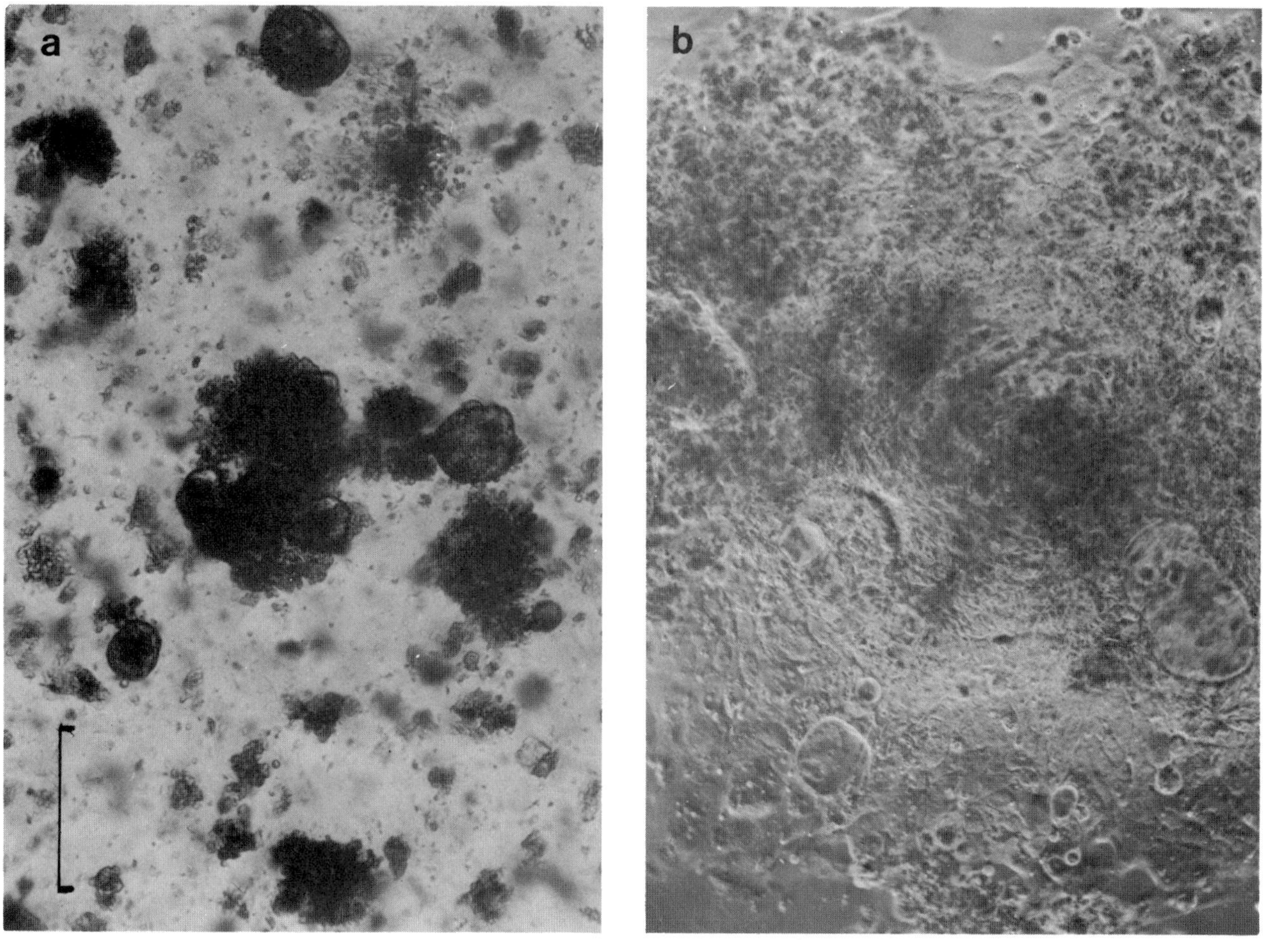

Fig. 9.11. *Cells and cell clusters from human colonic carcinoma after 48 hr dissociation in crude collagenase (Worthington CLS grade). a. Before removal of collagenase. b. After removal of collagenase, further disaggregation by pipetting, and culture for 48 hr (scale bar 250 μm). The clearly defined rounded clusters in (a) form epithelial-like sheets in (b) and the more irregularly shaped clusters produce fibroblasts.*

Transferring the cells to a fresh flask after collagenase treatment (and removal) removes many of the macrophages from the culture. The first flask may be cultured as well if required. Light trypsinization will remove any adherent cells other than macrophages.

Disaggregation in collagenase has proved particularly suitable for the culture of human tumors, mouse kidney, human adult and fetal brain, lung, and many other tissues, particularly epithelium. It is gentle and requires no mechanical agitation or special equipment. With more than 1 g of tissue, however, it becomes tedious at the dissection stage and can be expensive due to the amount of collagenase required. It will also release most of the connective tissue cells, accentuating the problem of fibroblastic outgrowth, so it may require being followed by selective culture or cell separation (see Chapters 7, 12, and 20).

Many epithelial tissues (e.g., kidney tubules and glomeruli, clusters of carcinoma cells of breast and gastrointestinal tract, and lung alveoli) are not disaggregated by collagenase and may be separated from connective tissue cells by allowing the epithelial clusters or tubules to sediment for 5–10 min. Connective tissue cells are completely dissociated by the collagenase and remain in suspension. If the sediment is resuspended in BSS and allowed to settle twice more, the final sediment is enriched for epithelial cells. The survival of this epithelium is probably enhanced by culturing the cells as undissociated clusters and tubule fragments.

The discrete clusters of epithelial cells produced by disaggregation in collagenase can be selected under a dissection microscope and transferred to individual wells in a microtitration plate, alone or with irradiated or mitomycin C–treated feeder cells (see Chapter 11).

The addition of hyaluronidase aids disaggregation

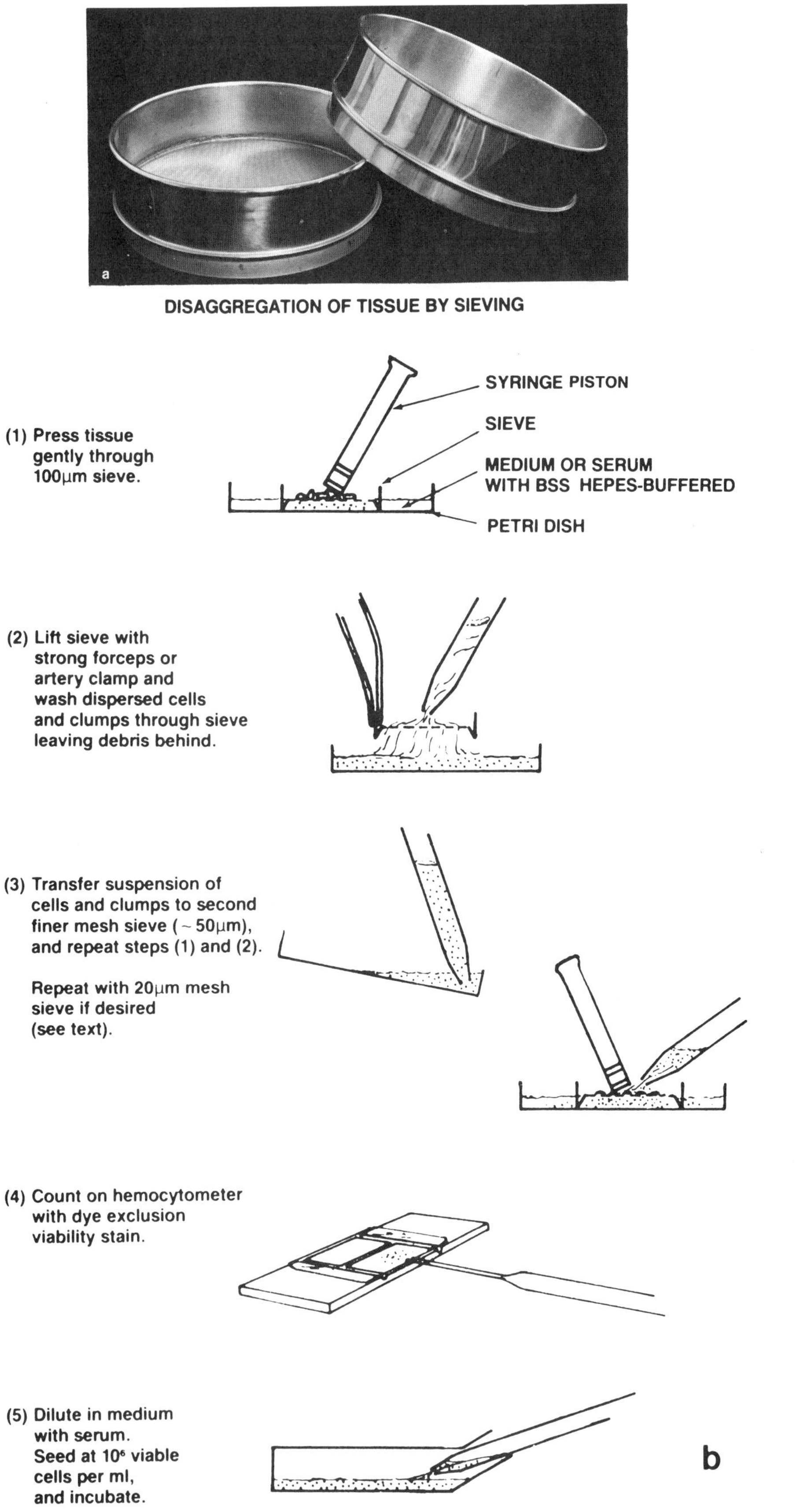

a
DISAGGREGATION OF TISSUE BY SIEVING
SYRINGE PISTON
SIEVE
MEDIUM OR SERUM
WITH BSS HEPES-BUFFERED
PETRI DISH
(1) Press tissue gently through 100μm sieve.
(2) Lift sieve with strong forceps or artery clamp and wash dispersed cells and clumps through sieve leaving debris behind.
(3) Transfer suspension of cells and clumps to second finer mesh sieve (~50μm), and repeat steps (1) and (2).
Repeat with 20μm mesh sieve if desired (see text).
(4) Count on hemocytometer with dye exclusion viability stain.
(5) Dilute in medium with serum. Seed at 10⁶ viable cells per ml, and incubate.
b

Fig. 9.12. *a. Stainless steel sieves suitable for disaggregating tissue. b. Disaggregation of tissue by sieving. c. Falcon disposable sieve; nests in top of 50-ml centrifuge tube. Can be used for mechanical disaggregation, as in (b), or for filtering aggregates out of a disaggregated suspension. (Part c appears on following page.)*

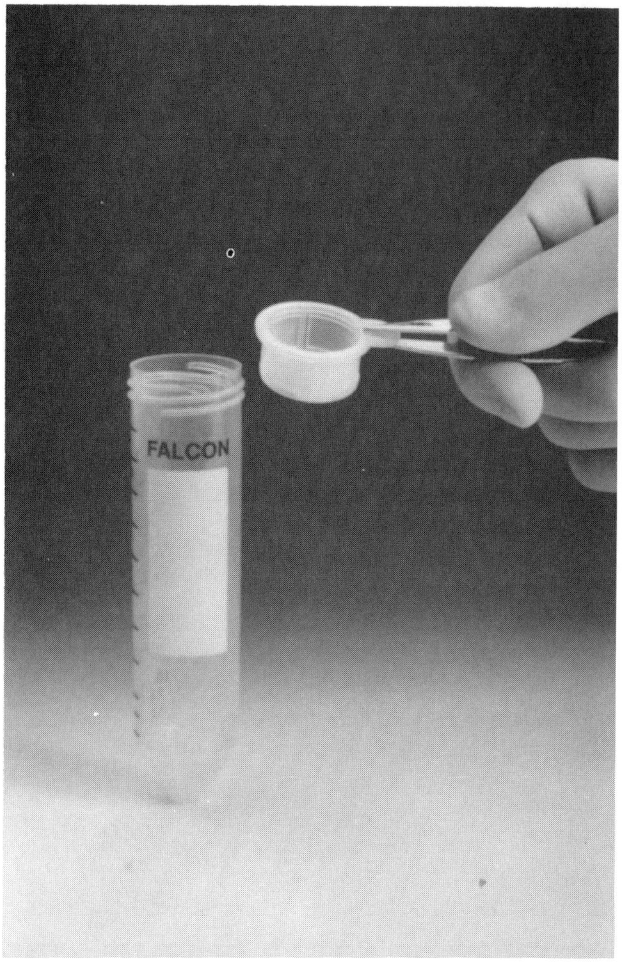

c

Fig. 9.12. (Continued)

by attacking terminal carbohydrate residues on the surface of the cells. This combination has been found to be particularly effective for dissociating rat or rabbit liver, by perfusing the whole organ *in situ* [Berry and Friend, 1969; Seglen, 1975; Kralovansky et al., 1990] (see also Chapter 20: liver) and completing the disaggregation by stirring the partially digested tissue in the same enzyme solution for a further 10–15 min, if necessary.

This technique gives a good yield of viable hepatocytes and is a good starting point for further culture (see Chapter 20).

Collagenase may also be used in conjunction with trypsin. Cahn and others [1967] developed a formulation including chick serum, collagenase, and trypsin. Chick serum has a moderating effect on the activity of the trypsin but does not inhibit it to the extent expected of other sera.

Mechanical Disaggregation

The outgrowth of cells from primary explants is a relatively slow process and can be highly selective. Enzymatic digestion (see above) is rather more labor-intensive, though potentially it gives a culture that is more representative of the tissue. As there is a risk of damaging cells during enzymatic digestion, many people have chosen the alternative of mechanical disaggregation, e.g., collecting the cells that spill out when the tissue is carefully sliced [Lasfargues, 1973], pressing the cells through sieves of gradually reduced mesh (Fig. 9.12), forcing cells through a syringe and needle [Zaroff et al., 1961], or simply pipetting repeatedly. This gives a cell suspension more quickly than with enzymatic digestion but may cause mechanical damage. The following is one method of mechanical disaggregation found to be moderately successful with soft tissues such as brain.

Outline

The tissue in culture medium is forced through sieves of gradually reduced mesh until a reasonable suspension of single cells and small aggregates is obtained. The suspension is then diluted and cultured directly.

Materials

Forceps
Sieves (1 mm, 100 μm, 20 μm, Fig. 9.12a; or Falcon "Cell Strainer," Fig. 9.12b)
9-cm petri dishes
Scalpels
Culture medium
Disposable plastic syringes (2 ml or 5 ml)
Culture flasks

Protocol

1. After washing and preliminary dissection (see above), chop tissue into pieces about 5–10 mm across, and place a few at a time into a stainless steel sieve of 1-mm mesh in a 9-cm petri dish (Fig. 9.12b).
2. Force the tissue through the mesh into medium by applying gentle pressure with the piston of a disposable plastic syringe. Pipette more medium into the sieve to wash the cells through.
3. Pipette the partially disaggregated tissue from the petri dish into a sieve of finer porosity, perhaps 100-μm mesh, and repeat step 2.
4. The suspension may be diluted and cultured at this stage or sieved further through 20-μm mesh if it is important to produce a single cell suspension. In general, the more highly dispersed, the higher the sheer stress required, and the lower the resulting viability.
5. Seed culture flasks at 10^6 cells/ml and 2×10^6 cells/ml by dilution of cell suspension in medium.

Only such soft tissues as spleen, embryonic liver, embryonic and adult brain, and some human and animal soft tumors respond at all well to this technique. Even with brain, where fairly complete disaggregation can be obtained easily, the viability of the resulting suspension is lower than that achieved with enzymatic digestion, although the time taken may be very much less. Where the availability of tissue is no limitation and the efficiency of the yield unimportant, it may be possible to produce as many viable cells as with enzymatic digestion in a shorter time but at the expense of very much more tissue.

Separation of Viable and Nonviable Cells

When an adherent primary culture is prepared from dissociated cells, nonviable cells will be removed at the first medium change. With primary cultures maintained in suspension, nonviable cells are gradually diluted out when cell proliferation starts. If necessary, however, nonviable cells may be removed from the primary disaggregate by centrifuging the cells on a mixture of Ficoll and sodium metrizoate (e.g., Hypaque or Triosil) [Vries et al., 1973]. This technique is similar to the preparation of lymphocytes from peripheral blood described in Chapter 23. Up to 2 x 10^7 cells in 9 ml medium may be layered on top of 6 ml Ficoll/Hypaque ("Lymphoprep," Flow Laboratories) in a 25-ml screw-capped centrifuge bottle, centrifuged, and viable cells collected from the interface.

The disaggregation of tissue and preparation of the primary culture is the first, and perhaps most vital, stage in the culture of cells with specific functions. If the required cells are lost at this stage, then the loss is irrevocable. Many different cell types may be cultured by choosing the correct techniques (see Chapters 7, 12, and 20). In general, trypsin is more severe than collagenase but sometimes more effective in creating a single cell suspension. Collagenase does not dissociate epithelial cells readily, but this can be an advantage in separating them from stromal cells. Mechanical disaggregation is much quicker but will damage more cells. The best approach is to try out the techniques described above and select the method that works best in your system. If none of these is successful, try additional enzymes such as pronase, dispase, and DNase and consult the literature.

CHAPTER 10

Maintenance of the Culture: Cell Lines

The first subculture represents an important transition for the culture. The need to subculture implies that the primary culture has increased to occupy all of the available substrate. Hence cell proliferation has become an important feature. While the primary culture may have a variable growth fraction (see Chapter 18) depending on the type of cells present in culture, after the first subculture the growth fraction is usually high (80% or more). From a very heterogeneous primary culture, containing many of the cell types present in the original tissue, a more homogeneous cell line emerges. In addition to its biological significance, this also has considerable practical importance, as the culture can now be propagated, characterized, and stored, and the potential increase in cell number and uniformity of the cells opens up a much wider range of experimental possibilities (Table 10.1). Once a primary culture is subcultured (or "passaged," or "transferred"), it becomes known as a "cell line." This term implies the presence of several cell lineages either of similar or distinct phenotypes. If one cell lineage is selected, by cloning (see Chapter 11), by physical cell separation (see Chapter 12), or by any other selection technique, to have certain specific properties that have been identified in the bulk of the cells in the culture, this becomes known as "cell strain" (see Glossary). Some commonly used cell lines and cell strains are listed in Table 10.2. If a cell line transforms *in vitro*, this gives rise to a continuous cell line (see Chapters 2 and 15), and if selected or cloned and characterized it is known as a continuous cell strain. The relative advantages and disadvantages of finite cell lines or continuous cell lines are listed in Table 10.3.

NOMENCLATURE

The first subculture gives rise to a "secondary" culture, the secondary to a "tertiary," and so on, although in practice this nomenclature is seldom used beyond tertiary. Since the importance of culture lifetime was highlighted by Hayflick and others with diploid fibroblasts [Hayflick and Moorhead, 1961], where each subculture divided the culture in half ("split ratio"–1:2), passage number has often been confused with "generation number."

The passage number refers to the number of times that the culture has been subcultured, while the generation number implies the number of doublings that the cell population has undergone. Where the split ratio is 1:2, then, as in Hayflick's experiments above, passage number is approximately equal to generation

TABLE 10.1. Subculture

Advantages	Disadvantages
Propagation	Selection, overgrowth
More cells	Loss of differentiated properties (may be indicable)
Cloning	Genetic instability
Homogeneity	Trauma of disaggregation, enzymatic and mechanical damage, eventually
Characterization of replicate samples	
Frozen storage	

number. Many laboratories will subculture at split ratios greater than 1:2, and if this is the case, it is useful to use a power of two, e.g., 1:2, 1:4, 1:8, 1:16, to give the equivalent of one, two, three, or four doublings. None of these approximations takes account of cell loss through necrosis, apoptosis, or differentiation or premature aging and withdrawal from cycle, which probably takes place at every growth cycle between each subculture.

Cell lines with limited culture life-spans ("finite" cell lines) behave in a fairly reproducible fashion (see Chapter 2). They will grow through a limited number of cell generations, usually between 20 and 80 population doublings, before extinction. The actual number depends on strain differences, clonal variation, and culture conditions but is consistent for one cell line grown under the same conditions. It is, therefore, important that reference to a cell line should express the approximate generation number or number of doublings since explantation, "approximate" because the number of generations that have elapsed in the primary culture is difficult to assess.

The cell line should also be given a code or designation (e.g., normal human brain, NHB), a cell strain or cell line number (if several cell lines were derived from the same source), NHB1, NHB2, etc., and if cloned, a clone number, NHB2-1, NHB2-2, etc. It is useful to keep a log book or computer data file where the receipt of biopsies or specimens is recorded prior to initiation of a culture. In this case the accession number in the log book, perhaps linked to an identifier letter code, can be used to establish the cell line designation, e.g., LT156 would be *l*ung *t*umor biopsy number 156. This is less likely to generate ambiguities, where the same letter code is used for two different cell lines, and gives automatic reference to the record of accession of the line.

For finite cell lines the number of population doublings should be estimated and indicated by a slash number, e.g., NHB2/2, and will increase by one for a split ratio of 1:2 (NHB2/2, NHB2/3, etc.), by two for a split ratio of 1:4 (NHB2/2, NHB2/4, etc.), and so on. This is more difficult when dealing with a continuous cell line, where the generation number is very high and often unknown. If the line is immortal, the generation number is less informative, and the slash number may be used to indicate the number of generations or, more simply, the number of passages since the last thaw from the freezer (see Chapter 17).

For publication, each cell line should be prefixed with a code designating the laboratory in which it was derived, e.g., WI, Wistar Institute; NCI, National Cancer Institute; SK, Sloan Kettering [Federoff, 1975]. In publications or reports, the cell line should be given its full designation the first time it is mentioned, and in descriptive tables and thereafter may be abbreviated to its trivial name.

ROUTINE MAINTENANCE

Once a culture is initiated, whether it be a primary culture or a subculture of a cell line, it will need a periodic medium change ("feeding") followed eventually by subculture if the cells are proliferating. In nonproliferating cultures, the medium will still need to be changed periodically, as the cells will still metabolize and some constituents of the medium will become exhausted or will degrade spontaneously. Intervals between medium changes and between subcultures vary from one cell line to another, depending on the rate of growth and metabolism; rapidly growing cell lines such as HeLa are usually subcultured once per week and the medium changed 4 days later. More slowly growing cell lines may only need to be subcultured every 2, 3, or even 4 weeks, and the medium changed weekly between subcultures. (For a more detailed discussion of the growth cycle, see below.)

Replacement of Medium

Four factors indicate the need for the replacement of culture medium (see also Chapter 7), as follows.

A drop in pH. The rate of fall and absolute level should be considered. Most cells will stop growing as the pH falls from pH 7.0 to pH 6.5 and will start to lose viability between pH 6.5 and pH 6.0, so if the medium goes from red through orange to yellow, the medium should be changed. Try to estimate the rate of fall; a culture at pH 7.0 that falls 0.1 pH units in 1 day will not come to harm if left a day or two longer before feeding, but a culture that falls 0.4 pH units in 1 day will need to be fed within 24–48 hr and cannot be left over a weekend.

TABLE 10.2. Cell Lines and Cell Strains in Regular Use

Name	Morphology[a]	Origin	Age[b]	Tissue[c]	Ploidy	Characteristic	Reference
Finite cell lines							
MRC5	Fibroblast	Human lung	Emb	Nor	Diploid	Susceptible to human viral infection	[Jacobs, 1970]
MRC9	Fibroblast	Human lung	Emb	Nor	Diploid	Susceptible to human viral infection	[Jacobs, 1979]
WI38	Fibroblast	Human lung	Emb	Nor	Diploid	Susceptible to human viral infection	[Hayflick and Moorhead, 1961]
IMR90	Fibroblast	Human lung	Emb	Nor	Diploid	Susceptible to human viral infection	[Nichols et al., 1977]
Continuous cell lines							
A2780	Epithelial	Human ovary	Ad	Neo	Aneuploid	Chemosensitive with resistant variants	[Tsuruo et al., 1986]
A549	Epithelial	Human lung	Ad	Neo	Aneuploid	Synthesizes pulmonary surfactant	[Giard et al., 1972]
A9	Fibroblast	Mouse subcutaneous	Ad	Neo	Aneuploid	HGPRT-ve: deriv. L929	[Littlefield, 1964b]
BHK21, C13	Fibroblast	Syrian hamster kidney	NB	Nor	Aneuploid	Transformable by polyoma virus	[Macpherson and Stoker, 1962]
BRL3A	Epithelial	Rat liver	NB	Nor		Produce IGF-II	[Coon, 1968]
Caco-2	Epithelial	Human colon	Ad	Neo	Aneuploid	Transports ions and amino acids	[Fogh, 1977]
CHANG liver	Epithelial	Human liver	Emb	Nor?	Aneuploid	HeLa contaminated[d]	[Chang, 1954]
CHOK1	Fibroblast	Chinese hamster ovary	Ad	Nor	Diploid	Simple karyotype	[Puck et al., 1958]
EB-3	Lymphocytic	Human	Ju	Neo	Diploid	EB virus + ve	[Epstein and Barr, 1964]
GH1, GH3	Epithelial	Rat	Ad	Neo	Aneuploid	Produce growth hormone	[Yasumura et al., 1966]
HeLa	Epithelial	Human cervix	Ad	Neo	Aneuploid	G6PD Type A	[Gey et al., 1952]
HeLa-S_3	Epithelial	Human cervix	Ad	Neo	Aneuploid	High plating efficiency; will grow well in suspension	[Puck and Marcus, 1955]
Hep-2	Epithelial	Human larynx	Ad	Neo	Aneuploid	HeLa contaminated[d]	[Moore et al., 1955]
HT-29	Epithelial	Human colon	Ad	Neo	Aneuploid	Differentiation inducible with NaBt	[Fogh and Trempe, 1975]
KB	Epithelial	Human oral	Ad	Neo	Aneuploid	HeLa contaminated[d]	[Eagle, 1955b]
L1210	Lymphocytic	Mouse	Ad	Neo	Aneuploid	Rapidly growing; suspension	[Law et al., 1949]
L5178Y	Lymphocytic	Mouse	Ad	Neo	Aneuploid	Rapidly growing suspension	
L929	Fibroblast	Mouse	Ad	Nor	Aneuploid	Clone of L-cell	[Sanford et al., 1948]
LS	Fibroblast	Mouse	Ad	Neo	Aneuploid	Grow in suspension: deriv. L929	(Paul and Struthers, unpublished)
MCF7	Epithelial	Human breast pleural effusion	Ad	Neo	Aneuploid	Estrogen recep +ve	[Soule et al., 1973]
P388D_1	Lymphocytic	Mouse	Ad	Neo	Aneuploid	Grow in suspension	[Dawe and Potter, 1957; Koren et al., 1975]
S180	Fibroblast	Mouse	Ad	Neo	Aneuploid	Canccr chemotherapy screening	[Dunham & Stewart, 1953]
STO	Fibroblast	Mouse	Emb	Nor	Aneuploid	Used as feeder layer for embryonal stem cells	[Bernstein, 1975]
3T3-L1	Fibroblast	Mouse Swiss	Emb	Nor	Aneuploid	Adipose diff.	[Green and Kehinde, 1974]

(continued)

TABLE 10.2. *(Continued)*

Name	Morphology[a]	Origin	Age[b]	Tissue[c]	Ploidy	Characteristic	Reference
3T3 A31	Fibroblast	Mouse BALB/c	Emb	Nor	Aneuploid	Contact inhibited; readily transformed	[Aaronson and Todaro, 1968]
NRK49F	Fibroblast	Rat kidney	Ad	Nor	Aneuploid	Induction of suspension growth by transforming growth factors	[DeLarco and Todaro, 1978]
Vero	Fibroblast	Monkey kidney	Ad	Nor	Aneuploid	Viral substrate and assay	[Hopps et al., 1963]
WISH	Epithelial	Human amnion	NB	Nor?	Aneuploid	Hela contaminated	[Hayflick, 1961]
ZR-75-1	Epithelial	Human breast, ascites fluid	Ad	Neo	Aneuploid	ER-ve, EGFr+ve	[Engel, 1978]

[a]Fibroblast, epithelial, lymphocytic refer to appearance and not lineage.
[b]Emb, embryonic; Ad, adult; NB, newborn; Ju, juvenile.
[c]Nor, from normal tissue; neo, from neoplastic tissue.
[d]HeLa contamination refers to cross-contamination with HeLa cells at some indeterminate period in the life history of the cell line before acquired by the American Type Culture Collection (ATCC). It does not imply cross-contamination by the originator or by ATCC.

Cell concentration. Cultures at a high cell concentration will exhaust the medium faster than at a low concentration. This is usually evident in the rate of pH change but not always.

Cell type. Normal cells (e.g., diploid fibroblasts) will usually stop dividing at a high cell density (density limitation of growth; see Chapter 15) due to growth factor depletion and other factors. The cells block in the G_1 phase of the cell cycle and deteriorate very little even if left for 2–3 weeks or longer. Transformed cells, continuous cell lines, and some embryonic cells, however, will deteriorate rapidly at high cell densities unless the medium is changed daily or they are subcultured.

Cell morphology. When checking a culture for routine maintenance, be alert to signs of morphological deterioration: granularity around the nucleus, cytoplasmic vacuolation, and rounding up of the cells with

TABLE 10.3. Advantages and Disadvantages of Finite and Continuous Cell Lines

	Finite	Continuous
Ploidy	Diploid Euploid	Heteroploid Aneuploid
Transformation	Normal	Transformed
Tumorigenicity	Nontumorigenic	Tumorigenic
Anchorage dependence	Yes	No
Contact inhibition	Yes	No
Density limitation of growth	Yes	No (or less so)
Mode of growth	Monolayer	Monolayer or suspension
Maintenance	Cyclic	Steady state possible
Serum requirement (in simple media)	High	Low
Cloning efficiency	Low	High
Markers	May be tissue specific	Chromosomal, enzymic
Special functions (e.g., virus susceptibility, differentiation)	May be retained	Often lost
Growth rate	Slow (24–96 hr doubling time)	Rapid (12–24 hr doubling time)
Yield	Low (6 cells/ml, 5 cells/cm²)	High (>10^6 cells/ml, 10^5 cells/cm²)
Control features	Generation number *In vivo* markers	Time from last thawing Strain characteristics

detachment from the substrate (Fig. 10.1). This may imply that the culture requires a medium change, or may indicate a more serious problem, e.g., inadequate or toxic medium or serum, microbiological contamination, or senescence of the cell line. During routine maintenance, the medium change or subculture frequency should prevent such deterioration.

Volume, Depth, and Surface Area

The usual ratio of medium volume to surface area is 0.2–0.5 ml/cm². The upper limit is set by gaseous diffusion through the liquid layer and the optimum will depend on the oxygen requirement of the cells. Cells with a high O_2 requirement will be better in shallow medium (2 mm) and those with a low requirement may do better in deep medium (5 mm). If the depth is greater than 5 mm, then gaseous diffusion may become limiting. With monolayer cultures, this can be overcome by rolling the bottle or perfusing the culture with medium and arranging for gas exchange in an intermediate reservoir (see Chapter 23). When the depth of suspension culture is increased, it should be stirred with a bar magnet (see Chapter 23). To prevent frothing, the depth of stirrer cultures must be a minimum of 5 cm. For intermediate depths of medium between 5 mm and 5 cm, use a roller bottle (see Table 7.1).

Changing the Medium or "Feeding" a Culture

Outline

Examine the culture by eye and on an inverted microscope. If indicated, remove the old medium and add fresh medium. Return the culture to the incubator.

Materials

Pipettes
Medium (both sterile)

Protocol

1. Examine culture carefully for signs of contamination or deterioration (see Figs. 10.1 and 16.1).
2. Check the criteria described above: pH and cell density or concentration, and, based on your knowledge of the behavior of the culture, decide whether or not to replace the medium. If feeding is required, proceed as follows.
3. Take to sterile work area and remove and discard medium (see Chapter 5).
4. Add same volume of fresh medium, prewarmed to 36.5°C if it is important that there be no check in cell growth.
5. Return culture to incubator.

   Note. Where a culture is at a low density and growing slowly, it may be preferable to "half-feed." In this case, remove only half the medium at step 3 and replace it in step 4 with the same volume as was removed.

Maintenance Medium

A maintenance medium, or "holding medium," may be used where stimulation of mitosis, which usually accompanies a medium change, even at high cell densities, is undesirable. Maintenance media are usually regular media with the serum concentration reduced to 0.5 or 2% or eliminated completely. This will not stimulate mitosis in most untransformed cells unless a special serum-free formulation is used (see Chapter 7). Transformed cell lines are unsuitable for this procedure, as they may either continue to divide successfully or the culture may deteriorate, as transformed cells do not block in a regulated fashion in G_1.

Holding medium is also used to maintain cell lines with a finite life-span without using up the limited number of cell generations available to them (see Chapter 2). Reduction of serum and cessation of cell proliferation also promotes expression of the differentiated phenotype in some cells [Maltese and Volpe, 1979; Schousboe et al., 1979]. Medium used for the collection of biopsy samples can also be referred to as "holding medium."

Subculture

The growth of cells in culture usually follows the pattern depicted in Figure 10.3. A lag following seeding is followed by a period of exponential growth (log phase). When all the available substrate is occupied or when the cell concentration exceeds the capacity of the medium, growth ceases or is greatly reduced. Then either the frequency of medium changing must increase or the culture must be divided. The usual practice in subculturing an adherent cell line involves removal of the medium and dissociation of the cells in the monolayer with trypsin, although some loosely adherent cells (e.g., HeLa-S_3) may be subcultured by shaking the bottle and collecting the cells in the medium, and diluting as appropriate in fresh medium in new bottles. Exceptionally, some cell monolayers cannot be dissociated in trypsin and require the action of alternative proteases such as pronase, dispase, or collagenase (Table 10.4) [Foley and Aftonomos, 1973]. The attachment of cells to each other and to the culture substrate is mediated by cell surface glycoproteins and Ca^{2+} and ions. Other proteins, and proteoglycans, derived from the cells and from the serum, become associated with the cell surface and the surface of the substrate and facilitate cell adhesion.

Subculture usually requires chelation of Ca^{2+} and degradation of cell adhesion and extracelluar matrix molecules. The severity of the treatment will depend

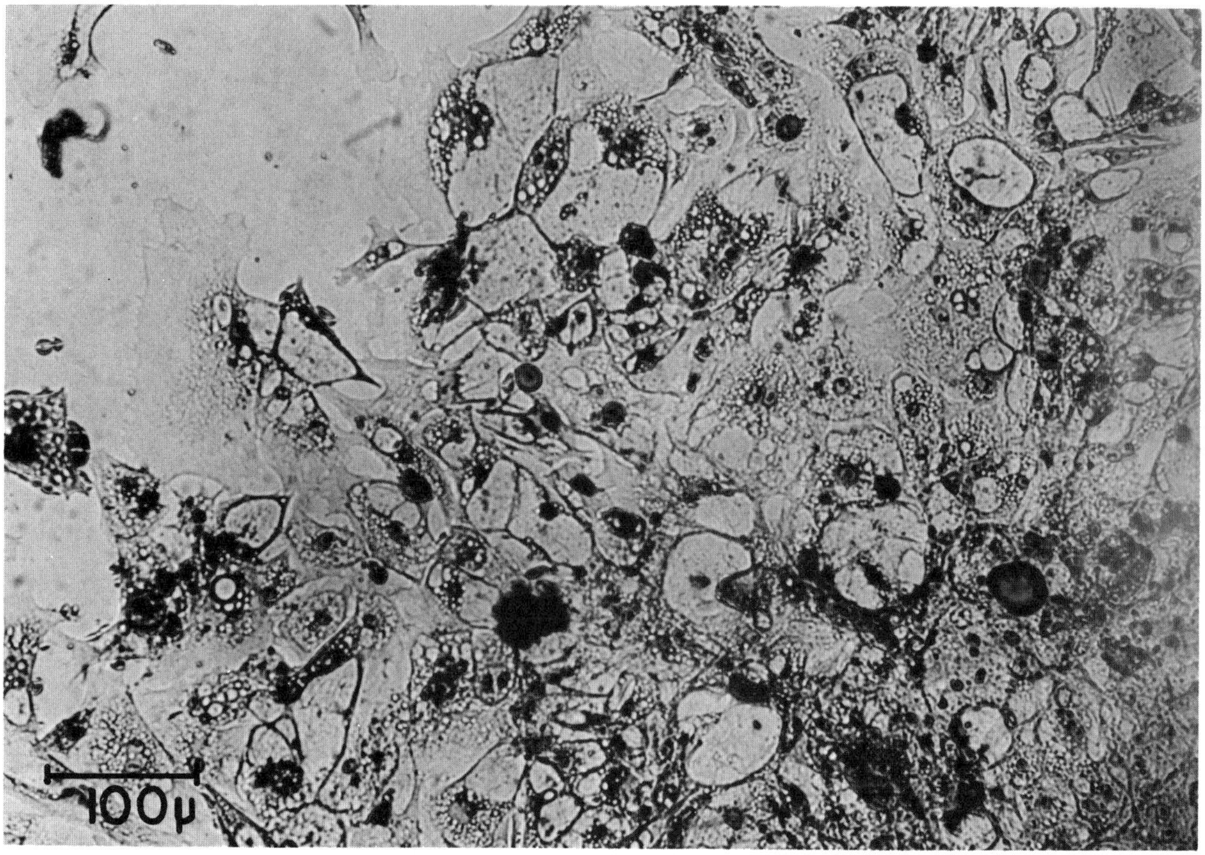

Fig. 10.1. Signs of deterioration of the culture. Cytoplasm of cells becomes granular, particularly around the nucleus, and vacuolation occurs. Cells may become more refractile at the edge if cell spreading is impaired.

on the cell type, and a protocol should be selected with the least severity compatible with the generation of a single cell suspension of high viability (Table 10.4).

Outline

Remove medium, expose cells briefly to trypsin, incubate, and disperse cells in medium.

Materials

Pipettes (sterile)
Medium (sterile)
PBSA (sterile)
0.25% trypsin in PBSA, saline citrate, or EDTA (sterile) (see Table 10.4)
Culture flasks
Hemocytometer or cell counter

Protocol

1. Withdraw medium and discard.
2. Add PBSA prewash (5 ml/25 cm^2) to the side of the flask opposite the cells, so as to avoid dislodging cells, rinse the cells, and discard rinse. This step is designed to remove traces of serum that would inhibit the action of the trypsin.
3. Add trypsin (3 ml/25 cm^2) to the side of the flask opposite the cells. Turn the flask over to cover the monolayer completely. Leave 15–30 s and withdraw the trypsin, making sure that the monolayer has not detached. Using trypsin at 4°C helps to prevent this.
4. Incubate until cells round up; when the bottle is tilted, the monolayer should slide down the surface (this usually occurs after 5–15 min). Do not leave longer than necessary, but do not force the cells to detach before they are ready to do so, or clumping may result.

   Note. In each case the main dissociating agent, be it trypsin or EDTA, is present only briefly and the incubation is performed in the residue after most of the dissociating agent has been removed. If difficulty is encountered in getting cells to detach and, subsequently, in preparing a single cell suspension, alternative procedures as described in Table 10.4 may be employed.
5. Add medium (0.1–0.2 ml/cm^2) and disperse

cells by repeated pipetting over the surface bearing the monolayer. Finally, pipette the cell suspension up and down a few times, with the tip of the pipette resting on the bottom corner of bottle, taking care not to create a foam. The degree of pipetting required will vary from one cell line to another; some disperse easily, others require vigorous pipetting. Almost all will incur mechanical damage from shearing forces if pipetted too vigorously; primary suspensions and early passage cultures are particularly prone to damage due partly to their greater fragility and partly to their larger size, but continuous cell lines are usually more resilient and require vigorous pipetting for complete disaggregation. Pipette up and down sufficiently to disperse the cells into a single cell suspension. If this is difficult, apply a more aggressive dissociating agent (see Table 10.4).

A single cell suspension is desirable at subculture to ensure an accurate cell count and uniform growth on reseeding. It is essential where quantitative estimates of cell proliferation or of plating efficiency are being made and where cells are to be isolated as clones.

6. Count cells by hemocytometer or electronic particle counter (see Chapter 18).

7. Dilute to the appropriate seeding concentration (a) by adding the appropriate volume of cells to a premeasured volume of medium in a culture flask or (b) by diluting the cells to the total volume required and distributing that among several flasks. Procedure (a) is useful for routine subculture when only a few flasks are used and precise cell counts and reproducibility are not critical, but (b) is preferable when setting up several experimental replicate samples, as the total number of manipulations is reduced and the concentration of cells in each flask will be identical.

8. If the cells are grown in elevated CO_2, gas the flask by blowing the correct gas mixture from a premixed cylinder, or gas blendor unit, through a filtered line into the flask above the medium, but not bubbled into it. If the normal gas phase is air, as with Eagle's medium with Hanks' salts, this stage may be omitted.

9. Cap the flask(s) and return to the incubator. Check after about 1 hr for pH change. If the pH rises in a medium with a gas phase of air, return to aseptic area and gas culture briefly (1–2 s) 5% CO_2. Since each culture will behave predictably in the same medium, you will know eventually which cells to gas when they are passaged, without having to incubate them first. If the medium already has a 5% CO_2 gas phase, either increase to 7% or 10% or add sterile 0.1 *N* HCl.

Expansion of air inside plastic flasks causes larger plastic flasks to swell and prevents them from lying flat. Release the pressure by slackening the cap briefly, or, alternatively, this may be prevented by compressing the top and bottom

TABLE 10.4. Cell Dissociation for Transfer or Counting; Procedures of Gradually Increasing Severity

1. Shake-off	Mitotic or other loosely adherent cells
2. Trypsin[a] in PBS (0.01–0.5% as required, usually 0.25%, 5–15 min)	Most continuous cell lines
3. Prewash with PBS or CMF, then 0.25% trypsin[a] in PBS or saline-citrate	Some strongly adherent continuous cell lines and many cell lines at early passage stages
4. Prewash with 1 m*M* EDTA in PBS or CMF, then 0.25% trypsin[a] in citrate	Some strongly adherent early passage cell lines
5. Prewash with 1 m*M* EDTA, then EDTA 2nd rinse, and leave on, 1 ml/5 cm	Epithelial cells, although some may be sensitive to EDTA
6. EDTA prewash, then 0.25% trypsin[a] with 1 m*M* EDTA	Strongly adherent cells, particularly epithelial and some tumor cells (note: EDTA can be toxic to some cells)
7. 1 m*M* EDTA prewash, then 0.25% trypsin[a] and collagenase,[a] 200 units/ml PBS or saline-citrate or EDTA/PBS	Thick cultures, multilayers, particularly collagen-producing dense cultures
8. Scraping	All cultures, but may cause mechanical damage and usually will not give a single cell suspension
9. Add dispase (0.1–1.0 mg/ml) or pronase (0.1–1.0 mg/ml) to medium and incubate until cells detach	Will dislodge most cells, but requires centrifugation step to remove enzyme not inactivated by serum. May be harmful to some cells

[a]Digestive enzymes are available (Difco, Worthington, Boehringer Mannheim, Sigma) in varying degrees of purity. Crude preparations, e.g., Difco trypsin 1:250 or Worthington CLS grade collagenase, contain other proteases that may be helpful in dissociating some cells but may be toxic to others. Start with a crude preparation and progress to purer grades if necessary. Purer grades are often used at a lower concentration (mg/ml) as their specific activities (enzyme units/g) are higher. Purified trypsin at 4°C has been recommended for cells grown in low serum concentrations or in the absence of serum [McKeehan, 1977], and will generally be found to be more consistent. Batch testing and reservation, as for serum, may be necessary for some applications.

of large flasks before sealing them. Incubation then restores the correct shape. Care must be taken not to crack the flasks.

For finite cell lines, it is convenient to reduce the cell concentration at subculture by two-, four-, eight-, or 16-fold, making the calculation of the number of population doublings easier (see above; 2≡1, 4≡2, 8≡3, 16≡4), e.g., a culture divided eightfold will require three doublings to achieve the same cell density. With continuous cell lines, where generation number is not usually recorded, the cell concentration is more conveniently reduced to a round figure, e.g., 5×10^4 cells/ml. In both cases, the cell number should be recorded so that growth rate can be estimated at each subculture and consistency monitored (see below under "Growth Cycle").

Cell Concentration and Density at Subculture

When handling a cell line for the first time, or with an early passage culture with which you have little experience, it is good practice to subculture to a split ratio of 2 or 4 at the first attempt. As experience is gained and the cell line seems established in the laboratory, it may be possible to increase the split ratio, i.e., reduce the cell concentration after subculture, but always keep a flask at a low split ratio as well when attempting to increase the split ratio.

The ideal method of determining the correct seeding density is to perform a growth curve at different seeding densities (Fig. 10.3; see Chapter 18) and determine the minimum cell concentration that will give a short lag period and early entry in rapid logarithmic growth (short population doubling time), and will reach the top of the exponential phase at a time convenient for the next subculture.

As a general rule, most continuous cells lines will subculture satisfactorily at a seeding concentration of between 10^4 and 5×10^4 cells/ml, finite fibroblasts cell lines about the same, and more fragile cultures such as endothelium and some early passage epithelia around 10^5/ml. For a new culture, trial and error will be the only reasonable approach.

Propagation in Suspension

The preceding instructions refer to subculture of monolayers, as most primary cultures or continuous lines grow in this way. Cells that grow continuously in suspension, either because they are nonadhesive (e.g., many leukemias and murine ascites tumors) or because they have been kept in suspension mechanically, or selected (see also Chapter 23), may be subcultured like micro-organisms. Trypsin treatment is not required and the whole process is quicker and less traumatic for the cells. Medium replacement is not usually carried out with suspension cultures, as this would require centrifugation of the cells. Routine maintenance is, therefore, reduced to one of two alternative procedures, i.e., subculture by dilution or increase of the volume without subculture.

Outline

Count cells, withdraw cell suspension, and add fresh medium to restore cell concentration to starting level.

Materials

Culture flasks (sterile)
Medium (sterile)
Pipettes (sterile)
Bar magnet (sterile)
Magnetic stirrer
Hemocytometer or cell counter

Protocol

1. Mix cell suspension and disperse any clumps by pipetting.
2. Remove sample and count.
3. Add medium to fresh flask.
   Note. Any culture flask with a reasonably flat surface may be used for cells that grow spontaneously in suspension. Where stirring is required, e.g., for larger cultures or cells that would normally attach, use standard round reagent bottles or flasks, siliconized if necessary (see Reagent Appendix), and insert a magnetic stirrer bar, Teflon coated and with a ridge around the middle (Fig. 10.2) or suspended from the top of the bottle. Select the appropriate size of bottle to give between 5 and 8 cm depth with the volume of medium that you require (see also Chapter 23).
4. Add sufficient cells to give a final concentration of 10^5 cells/ml for slow-growing cells (24–48 hr doubling time) or 2×10^4/ml for rapidly growing cells (12–18 hr doubling time).
5. Cap and return culture to incubator.
6. Culture flasks should be laid flat as for monolayer culture. Stirrer bottles should be placed on a magnetic stirrer and stirred at 60–100 rpm. Take care that the stirrer motor does not overheat the culture. Insert a polystyrene foam mat under the bottle if necessary. Induction-drive stirrers generate less heat and have no moving parts.

Suspension cultures have a number of advantages (see Table 10.5). The production and harvesting of

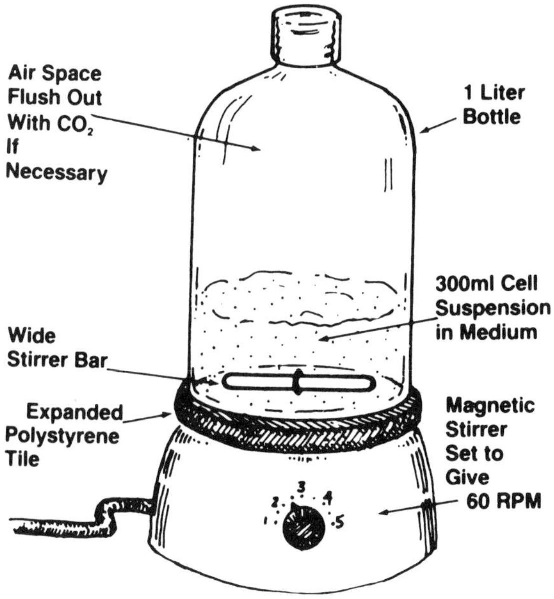

Fig. 10.2. Simple stirrer culture for cells growing in suspension. An expanded polystyrene mat (shaded area below bottle) should be interposed between the bottle and the magnetic stirrer to avoid heat transfer from the stirrer motor. Alternatively, a moving field stirrer, which generates substantially less heat, may be used.

large quantities of cells may be achieved without increasing the surface area of the substrate (see Chapter 23). Furthermore, if dilution of the culture is continuous and the cell concentration kept constant, a steady state can be achieved; this is not readily achieved in monolayer culture. Maintenance of monolayer cultures is essentially cyclic, with the result that growth rate and metabolism varies depending on the phase of the growth cycle.

Monolayers are convenient for cytological and immunological observations, cloning, mitotic "shake-off" (for cell synchronization of chromosome preparation), and *in situ* extractions without centrifugation.

SLOW CELL GROWTH

Even in the best-run laboratories, problems may arise in routine cell maintenance. Some may be attributed to microbiological contamination (see Chapter 16), but often the cause lies in one or more alterations in culture conditions. The following checklist may help to track these down:

TABLE 10.5. Properties of Monolayer and Suspension Cultures

	Monolayer	Suspension
Maintenance	Cyclic pattern of propagation (see text)	Can be maintained at "steady state"
	Require dissociation	Simple dilution at passage
	Dependent on availability of substrate	Dependent on medium volume only (with adequate gas exchange)
Results of differences in geometry	Cell interaction: metabolic cooperation, junctional communication; contact inhibition of movement and membrane activity, density limitation of growth	Homogeneous suspension
		Cell density limited by nutrient and hormonal concentration of the medium only
		Shearing effects in stirred cultures may damage some cells
	Diffusion boundary of effects (see text)	
	Establishment of polarity, differentiation	
	Cell shape and cytoskeleton–spreading, motility, overlapping, underlapping	
Sampling and analysis	Good cytological preparation, chromosomes, immunofluorescence, histochemistry	Bulk production of cells
		Ease of harvesting (no trypsinization required)
	Enrichment of mitoses by "shake-off" (see Chapter 23)	
	Serial extractions *in situ* possible without centrifugation	
Which cells?	Most cell types except some hemopoietic cells and ascites tumors	Transformed cells and lymphoblastoid cell lines

1. Any change in procedure or equipment?
2. Medium:
   a. Medium adequate?–check against other media (see Chapter 8).
   b. Frequency of changing correct?
   c. pH: check that it is within 7.0–7.4 during culture.
   d. Osmolality: check on osmometer.
   e. Component missed out: make up fresh batch.
   f. New batch of stock medium that is faulty?
   g. If BSS-based, is BSS satisfactory? (Check with other users as in 5, below.)
   h. If water-based, is water satisfactory (check with other users; or against fresh 1× medium, bought in).
   i. Check still, deionizer, conductivity, contamination, glass, boiler, and residue.
   j. Storage vessel, for algal or fungal contamination; chemical traces in plastic.
   k. HCO_3^-.
   l. Antibiotics.
3. Serum:
   a. New batch? Check supplier's quality control.
   b. Check concentration. Too low or too high?
   c. Reconfirm lack of toxicity, growth promotion, and plating efficiency.
4. Glassware or plastics:
   a. If new, check against previous stock.
   b. Wash-up; other cells showing symptoms? Other users have trouble?
   c. Trace contamination on glass? Check growth on plastic.
5. Cells (if other people's cells are all right):
   a. Contamination (see also Chapter 16).
      i. Bacterial, fungal–grow up without antibiotics.
      ii. Mycoplasma:
         A. Stain culture with Hoechst 33258, Chapter 16.
         B. Check for cytoplasmic DNA (incorporation of radioactive thymidine) by autoradiography.
         C. Get commercial test done (see Trade Index).
      iii. Viral; difficult to detect; try electron microscopy or fluorescent antibody.

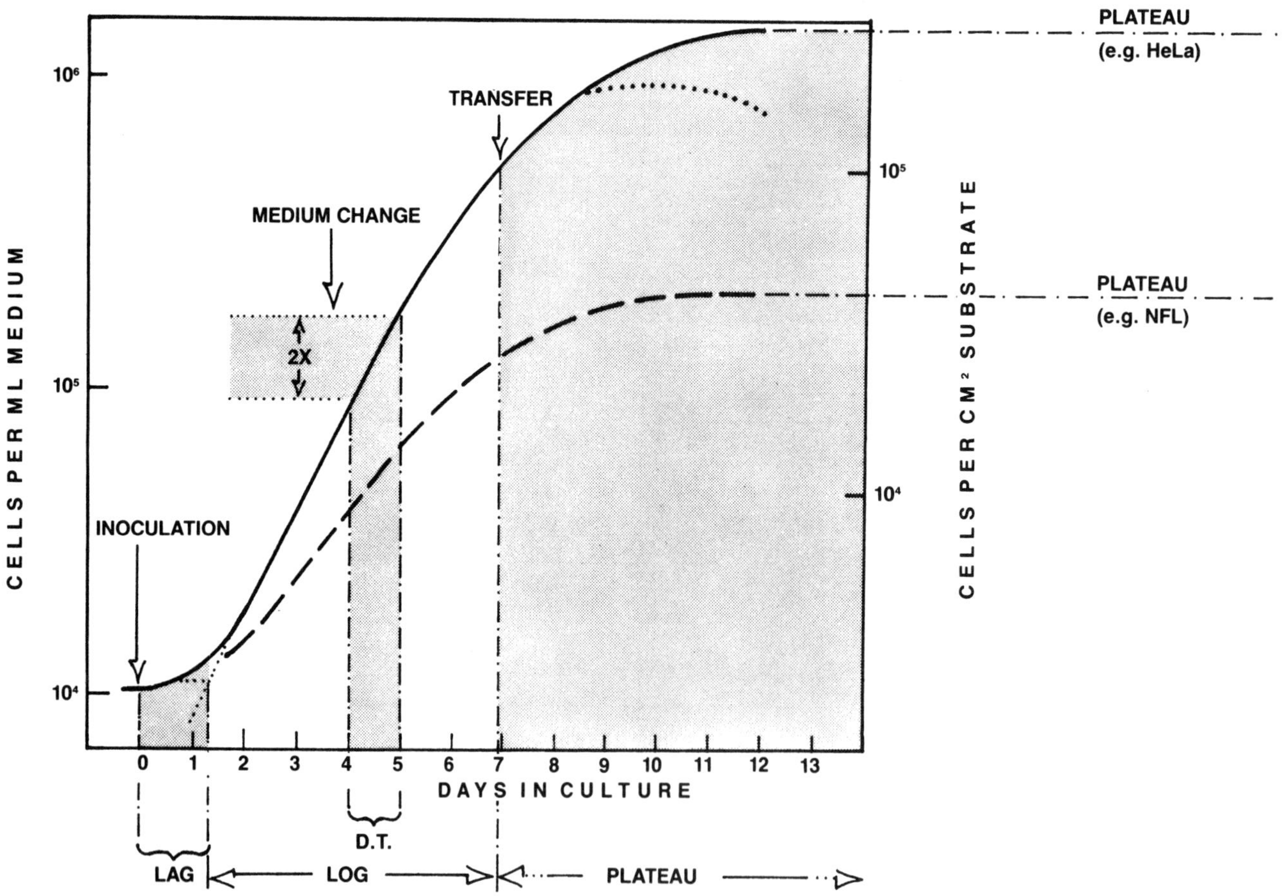

Fig. 10.3. Diagram of growth curve of a continuous cell line such as HeLa and a finite cell line, NFL (normal human fetal lung fibroblasts). The solid line represents growth of HeLa; the dashed line illustrates the lower plateau obtained with NFL (see text).

   b. Seeding density too low at transfer.
   c. Transferred too frequently.
   d. Allowed to remain for too long in plateau before transfer.
6. Subculture routine:
   a. Change in batch of trypsin or other dissociation agent.
   b. Severity of dissociation: too long, agent too concentrated or too high specific activity.
   c. Pipetting during dissociation too vigorous.
   d. Sensitive to EDTA (if EDTA used).
7. Hot room and incubators: check temperature and stability.
   a. Faulty thermostats.
   b. Access too frequent.
   c. Humidity of humid $C0_2$ incubators.
   d. CO_2 concentration (check pH *in situ*).

At subculture a fragile or slowly growing line should be split 1:2, and a vigorous, rapidly growing line, 1:8 or 1:16. Once a cell line becomes continuous (usually taken as beyond 150 or 200 generations) the generation number is disregarded and the culture should simply be cut back to between 10^4 and 10^5 cells/ml. The split ratio or dilution is also chosen to establish a convenient subculture interval (perhaps 1 or 2 wk) and to ensure that the cells (1) are not diluted below that concentration that permits them to re-enter the growth cycle with a lag period of 24 hr or less and (2) do not enter plateau before the next subculture.

Even when a standard split ratio is employed, cell counts should still be performed to ensure a consistent growth rate is maintained. Otherwise minor alterations are not detected for several passages.

Routine passage leads to the repetition of a standard growth cycle (see also Chapter 18). It is essential to become familiar with this cycle for each cell line that is handled, as this controls the seeding concentration, the duration of growth before subculture, the duration of experiments, and the appropriate times for sampling to give greatest consistency. Cells at different phases of the growth cycle behave differently with respect to proliferation, enzyme activity, glycolysis and respiration, synthesis of specialized products, and many other properties.

CHAPTER 11

Cloning and Selection of Specific Cell Types

It can be seen from the preceding two chapters that a major recurrent problem in tissue culture is the preservation of a specific cell type and its specialized properties. While environmental conditions undoubtedly play a significant role in maintaining the differentiated properties of specialized cells in a culture (see Chapter 14), the selective overgrowth of unspecialized cells is still a major problem.

CLONING

The traditional microbiological approach to the problem of culture heterogeneity is to isolate pure cell strains by cloning, but the success of this technique in animal cell culture is limited by the poor cloning efficiencies of most primary cultures.

A further problem of cultures derived from normal tissue is that they may only survive for a limited number of generations (see Chapter 2), and by the time a clone has produced a usable number of cells, it may already be near to senescence (Fig. 11.1). Cloning is most successful in isolating variants from continuous cell lines, but even then considerable heterogeneity may arise within the clone as it is grown up for use (see Chapter 17).

Clark and Pateman [1978] isolated a Kupffer cell line from Chinese hamster liver by cloning the primary culture and Sertoli cells [Zwain et al., 1991], and juxtaglomerular cells [Muirhead et al., 1990] and glomerular cells [Troyer and Kreisberg, 1990] from kidney have also been isolated by cloning.

Cloning has also been used to isolate specific biochemical mutants and cell strains with marker chromosomes and may help to reduce the heterogeneity of a culture (see below).

Dilution Cloning [Puck and Marcus, 1955]

Outline

Seed cells at low density, incubate until colonies form, isolate, and propagate into cell strain (Fig. 11.2).

Materials

Pipettes
Medium (as normally used for the cell line selected; serum concentration may need to be increased, if serum-dependent, and fetal bovine is usually better than calf)
Trypsin (as normally used)
Culture flasks or dishes

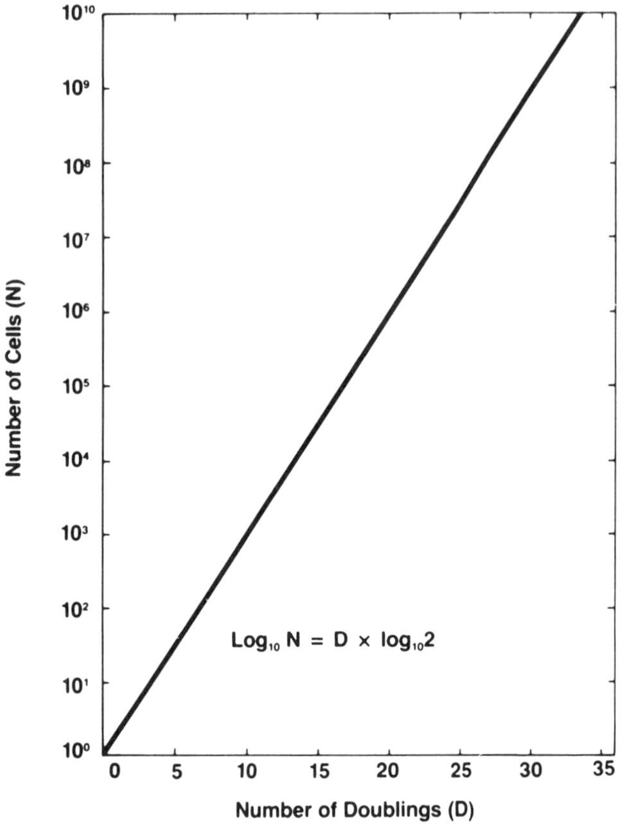

Fig. 11.1. Relationship of cell yield in a clone to the number of population doublings; e.g., 20 doublings are required to produce 10^6 cells.

Tubes, or universal containers, for dilution
Hemocytometer or cell counter

Protocol

1. Trypsinize cells (see Chapter 10) to produce a single cell suspension. Under-trypsinizing will produce clumps and over-trypsinizing will reduce viability, but it is fundamental to the concept of cloning that the cells be singly suspended. It may be necessary when cloning a new cell line for the first time to try different lengths of trypsinization and different recipes (see Table 10.4), to give the optimum plating efficiency from a good single cell suspension.
2. While cells are trypsinizing, number flasks or dishes and measure out medium for dilution steps. (Up to four dilution steps may be necessary to reduce a regular monolayer to a concentration suitable for cloning.)
3. When cells round up and start to detach, disperse the monolayer in medium containing serum or trypsin inhibitor, count, and dilute to desired seeding concentration. If cloning the cells for the first time, choose a range of 10, 50, 100, and 200 cells/ml (Table 11.1).
4. Seed petri dishes or flasks with the requisite amount of medium (see Chapter 10), place petri dishes, in a transparent plastic sandwich or cake box, in a humid CO_2 incubator or gassed sealed container (2–10% CO_2, see Chapter 7), or gas flasks with CO_2, seal with cap, and place in dry incubator.
5. Leave untouched for 1 wk. If colonies have formed, isolate (see below); if not, replace medium and continue to culture for a further week, or feed again and culture for 3 wk if necessary. If no colonies have appeared by 3 wk, it is unlikely that they will do so.

Stimulation of Plating Efficiency

When cells are plated at low densities, the survival falls in all but a few cell lines. This does not usually present a severe problem with continuous cell lines where the plating efficiency seldom drops below 10%, but with primary cultures and finite cell lines, the plating efficiency may be quite low—0.5–5% or even zero. Numerous attempts have been made to improve plating efficiencies, based on the assumption either that cells require a greater range of nutrients at low densities, because of loss by leakage, or that cell-derived diffusible signals or conditioning factors are present in high-density cultures and absent or too dilute at low densities. The intracellular metabolic pool of a leaky cell in a dense population will soon reach equilibrium with the surrounding medium, while that of an isolated cell never will. This was the basis of the capillary technique of Sanford et al. [1948], when the L929 clone of L-cells was first produced. The confines of the capillary tube allowed the cell to create a locally enriched environment mimicking the higher cell density state. In microdrop techniques developed later, the cells were seeded as a microdrop under liquid paraffin. Keeping one colony separate from another, as in the capillary techniques, colonies could be isolated subsequently. As media improved, however, plating efficiencies increased,

TABLE 11.1. Relationship of Seeding Density to Plating Efficiency

Expected plating efficiency (%)	Optimal cell number to be seeded	
	Per ml	Per cm²
0.1	10^4	2×10^3
1.0	10^3	200
10	100	20
50	20	4
100	10	2

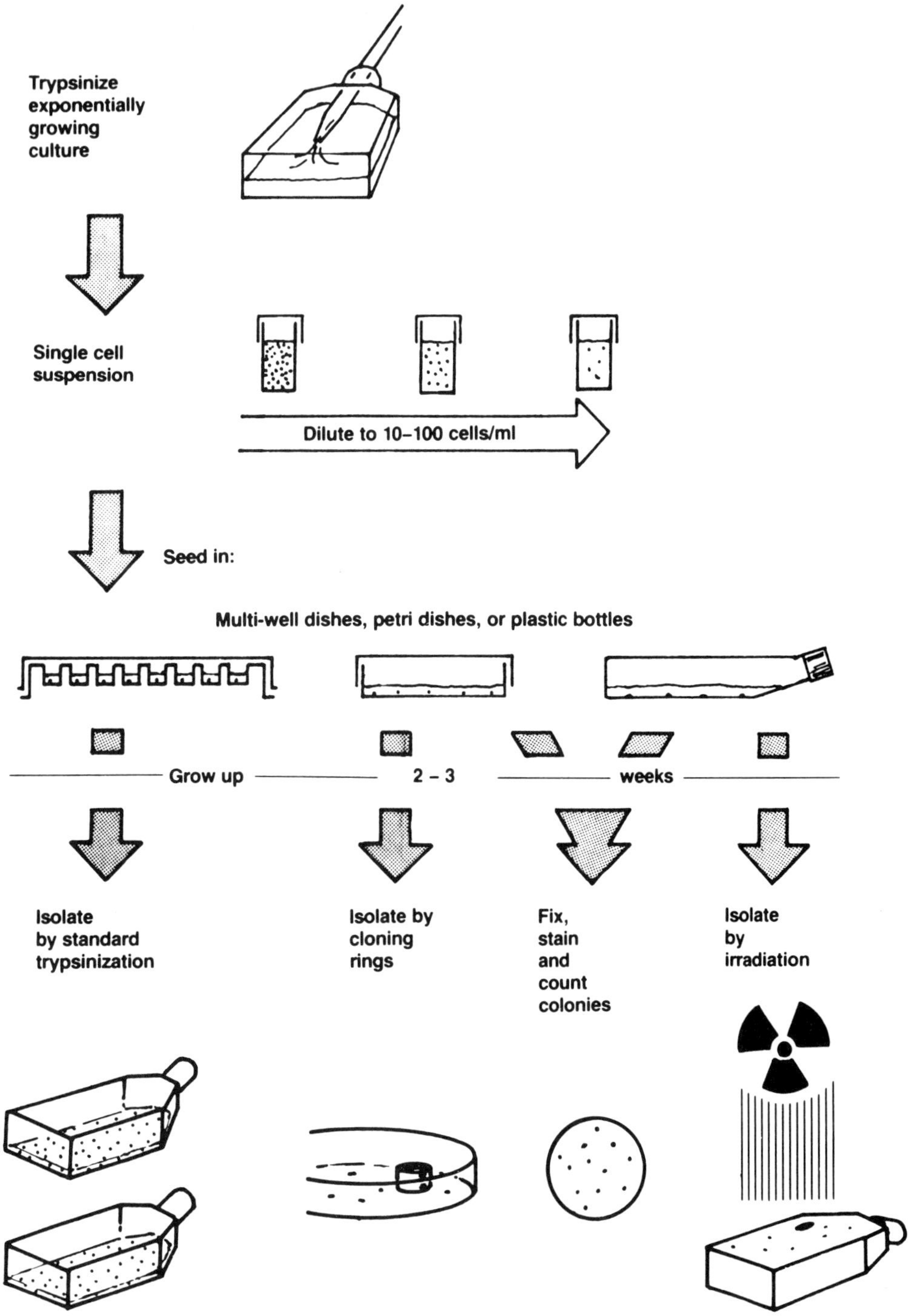

Fig. 11.2. *Cloning cells in monolayer culture. When clones form, they may be isolated directly from multiwell dishes (center left and lower left of figure), by the cloning ring technique (center and lower center), or by irradiating the flask while shielding one colony (center right and lower right). If isolation is not required and the cloning is being performed for quantitative assay, the colonies are fixed, stained, and counted (right of center and lower right of center).*

and Puck and Marcus [1955] were able to show that cloning cells by simple dilution (as described above) in association with a feeder layer of irradiated mouse embryo fibroblasts (see below) gave acceptable cloning efficiencies, although subsequent isolation required trypsinization from within a collar placed over each colony.

Some modifications that may improve clonal growth are listed below.

Medium. Choose a rich medium such as Ham's F12 or one that has been optimized for the cell type in use, e.g., MCDB 110 [Ham, 1984] for human fibroblasts, Ham's F12 or MCDB 302 for CHO [Ham, 1963; Hamilton and Ham, 1977] (see Chapters 7 and 20).

Serum. Where serum is required, fetal bovine is generally better than calf or horse. Select a batch for cloning experiments that gives a high plating efficiency during tests.

Conditioning. (1) Grow homologous cells, embryo fibroblasts, or another cell line to 50% of confluence, change to fresh medium, incubate for a further 48 hr, and collect the medium. (2) Filter through a 0.2-μm sterilizing filter (the medium may need to be clarified first by centrifugation 10,000 *g*, 20 min, or filtration through 5-μm and 1.2-μm filters) (see Chapter 9, section on sterilization of serum). (3) Add to cloning medium 1 part conditioned medium to 2 parts cloning medium.

Feeder layers (Fig. 11.3, regular feeder layer).

Outline

Homologous or heterologous cells, e.g., from mouse embryo, are rendered nonproliferative by irradiation or drug treatment and plated at medium density before cloning of test cells.

Materials

Secondary culture of 13-d mouse embryo fibroblasts (see Chapters 9 and 10)
x-ray or ^{60}Co source capable of delivering 30 Gy (3,000 rad) in 30 min or less
or
Mitomycin C, 5 μg/ml, in BSS or serum-free medium
Culture medium for cells to be cloned

Protocol

1. Trypsinize embryo fibroblasts from primary culture (see Chapters 9 and 10) and reseed at 10^5 cells/ml.
2. At 50% confluence, add mitomycin C, 2 μg/10^6 cells, 0.25 μg/ml, overnight [Macpherson and Bryden, 1971], or irradiate culture with 30 Gy (3,000 rad).
3. Change the medium after treatment, and after a further 24 hr, trypsinize the cells and reseed in fresh medium at 5×10^4 cells/ml (10^4 cells/cm^2).
4. Incubate for a further 24–48 hr and then seed cells for cloning.

The feeder cells will remain viable for up to 3 weeks but will eventually die out and are not carried over if the colonies are isolated. Other cell lines or homologous cells may be used to improve the plating efficiency, but heterologous cells have the advantage that if clones are to be isolated later, chromosome analysis will rule out accidental contamination from the feeder layer.

Hormones. Insulin, 1–10 IU/ml, has been found to increase the plating efficiency of several cell types [Hamilton and Ham, 1977]. Dexamethasone, 2.5×10^{-5} M, 10 μg/ml (a soluble synthetic hydrocortisone analogue), improves the plating efficiency of human normal glia, glioma, fibroblasts, and melanoma, and chick myoblasts, and will give increased clonal growth (colony size) if removed 5 days after plating [Freshney et al., 1980a,b]. Lower concentrations (10^{-7} M) have been found preferable for epithelial cells (see Chapter 20).

Intermediary metabolites. Keto acids, e.g., pyruvate or α-ketoglutarate [Griffiths and Pirt, 1967; McKeehan and McKeehan, 1979] and nucleosides [α-medium, Stanners et al., 1971], have been used to supplement media and are already included in the formulation of a rich medium like Ham's F12. Pyruvate is also added to Dulbecco's modification of Eagle's MEM [Dulbecco and Freeman, 1959; Morton, 1970].

Carbon dioxide. CO_2 is essential to obtain maximum cloning efficiency for most cells. While 5% is most usual, 2% is sufficient for many cells, and may even be slightly better for human glia and fibroblasts. HEPES (20 m*M*) may be used with 2% CO_2, protecting the cells against pH fluctuations during feeding and in the event of failure of the CO_2 supply. (Using 2% CO_2 also cuts down on the consumption of CO_2.) At the other extreme, Dulbecco's modification of Eagle's MEM is normally equilibrated with 10% CO_2 and is frequently used for cloning myeloma hybrids for monoclonal antibody production. The concentration of bicarbonate must be adjusted if the CO_2 tension is

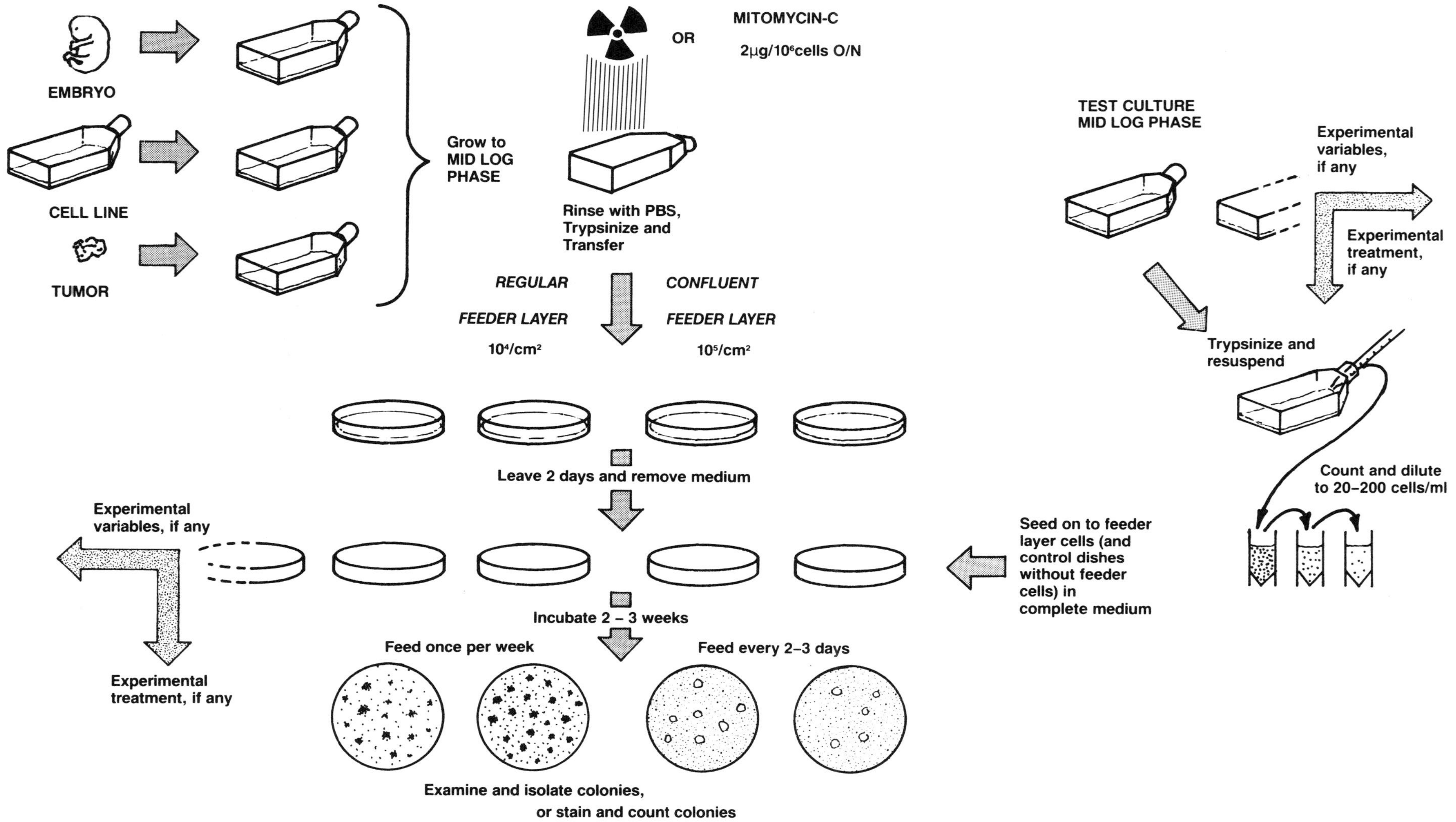

Fig. 11.3. *Cloning cells in the presence of a feeder layer. A low-density feeder layer (~10^4 cells/cm^2, 3T3 cells or secondary mouse or chick embryo fibroblasts; left-hand side of figure) is used to enhance cloning efficiency and clonal growth, while a high-density, aconfluent feeder layer (e.g., 2 × 10^5 cells/cm^2, normal human diploid fibroblasts, glia, or 3T3 cells) can be used as a selective substrate to minimize fibroblastic overgrowth (see also Fig. 11.8 and text).*

altered so that equilibrium is reached at pH 7.4 (see Table 7.2).

Treatment of substrate. Polylysine improves the plating efficiency of human fibroblasts in low serum concentrations [McKeehan and Ham, 1976] (see Chapter 7). (1) Add 1 mg/ml poly-D-lysine in water to plates (~5 ml/25 cm^2). (2) Remove and wash plates with 5 ml PBSA per 25 cm^2. The plates may be used immediately or stored for several weeks before used.

Fibronectin also improves the plating of many cells [Barnes and Sato, 1980]. The plates should be pretreated with 5 μg/ml fibronectin incorporated in the medium.

Trypsin. Purified (twice recrystallized) trypsin used at 0.05 μg/ml may be preferable to crude trypsin, but there are conflicting reports on this. McKeehan [1977] noted a marked improvement in plating efficiency when trypsinization (pure trypsin) was carried out at 4°C.

Multiwell Dishes

If clones are to be isolated, cloning by dilution directly into microwells (microtitration dishes or 24-well plates, see Fig. 7.4) makes subsequent harvesting easier. The plates must be checked regularly after seeding, however, to confirm that either only one cell is present per well at the start or, if there is more than one cell per well, that they are not clumped and only one cell gives rise to a colony, i.e., that the colonies which form are truly clonal in origin, and only one colony forms in the well.

Semisolid Media

Some cells, particularly hemopoietic stem cells and virally transformed fibroblasts, will clone readily in suspension. To hold the colony together and prevent mixing, the cells are suspended in agar or Methocel and plated out over an agar underlay or into nontissue culture-grade dishes.

Cloning in agar. See Figure 11.4 and Chapters 15, 20, and 21. The following protocol has been submitted by Mary Freshney, Beatson Institute for Cancer Research, Garscube Estate, Switchback Road, Bearsden, Glasgow, G61 1BD, Scotland.

Outline

Agar is liquid at high temperatures but is a gel at 36.5°C. Cells are suspended in warm agar medium, and, when incubated after the agar gels, will form discrete colonies that may be isolated easily.

Materials

Difco Noble agar
Medium at double strength, i.e., Ham's F12, RPMI 1640, Dulbecco's MEM, or CMRL 1066. Prepare from 10× concentrate to half the recommended final volume and add twice the normal concentration of serum.
Fetal bovine serum
Pipettes, including sterile plastic disposable pipettes for agar solutions
35-mm petri dishes, nontissue culture grade
Bunsen and tripod
Water bath at 45°C
Water bath at 36.5°C
Sterile tissue culture-grade water
Sterile conical flask
Universal containers, bijoux bottles or tubes
Cell counter or hemocytometer
1× medium for cell dilutions
Tray

Note. Before preparing medium and cells, work out cell dilutions and label petri dishes. For an assay to measure the cloning efficiency of a cell line, prepare to set up three dishes for each cell dilution. Convenient cell numbers per 35-mm dish are 1,000, 333, 111, 37, i.e., one-third dilution of the cell suspension. If any growth factors, etc., are to be added to the dishes they should be added to the 0.6% agar underlayer.

Protocol

1. Number or label petri dishes; putting them on a tray is convenient.
2. Prepare 2× medium containing 40% FBS, keep at 36.5°C.
3. Weigh out agar 1.2 g.
4. Measure 100 ml sterile tissue culture-grade water into a sterile conical flask, and a second 100 ml into a sterile bottle. Add 1.2 g agar to the flask, cover, and boil for 2 min. Alternatively, the agar may be sterilized in the autoclave in advance.
5. Transfer the boiled agar and the bottle of sterile water to a water bath at 45°C.
6. Prepare 0.6% agar underlayer by combining an equal volume of 2× medium and 1.2% agar. Keep at 36.5°C.
7. If any growth factors, etc., are being used they should be added to the underlayer medium now.

   Note. If a titration of growth factors is being carried out or a selection of different factors used, add the required amount to the petri dishes before the underlay is added.
8. Add 1 ml of 0.6% agar medium to the dishes, mix, and let it cover the base of the dish. Leave to set.

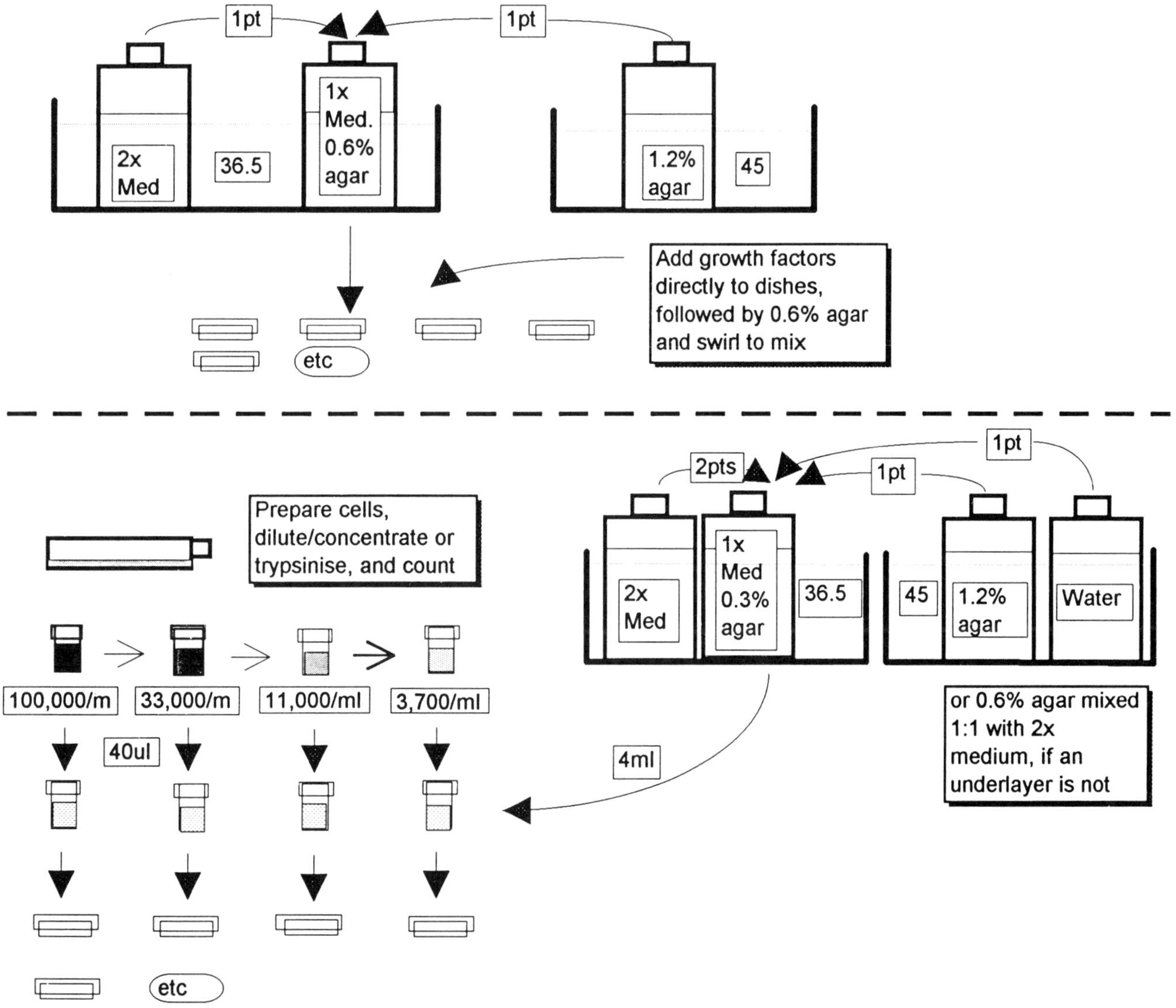

Fig. 11.4. Cloning cells in suspension in agar.

9. Prepare cell suspension and count.
10. Prepare 0.3% agar/medium and keep at 36.5°C. This may be done by diluting 2× medium at 36.5°C with 1.2% agar at 45°C and water at 45°C in the proportions of 2:1:1.
11. Prepare cell dilutions by making the first concentration of cells 10^5/ml, then:
    a. 10^5/ml dilute ⅓ to give
    b. 3.3×10^4/ml dilute ⅓ to give
    c. 1.1×10^4/ml dilute ⅓ to give
    d. 3.7×10^3/ml
12. Label four bijoux bottles or tubes and pipette 40 μl of each cell dilution into them. Add 4 ml of 0.3% agar medium at 36.5°C to them, mix, and pipette 1 ml onto each of three petri dishes.
13. Allow to gel at room temperature.
    Note. Always be sure that the top agar medium has had adequate time to cool to 36.5°C before adding the cells to it.
14. Put petri dishes into a clean plastic box with a lid and incubate at 36.5°C in a humid incubator for 10 days.

Cloning in Methocel over agar base [Buick et al., 1979].

Outline

Suspend cells in medium containing Methocel and seed into dishes containing gelled agar medium (Fig. 11.5).

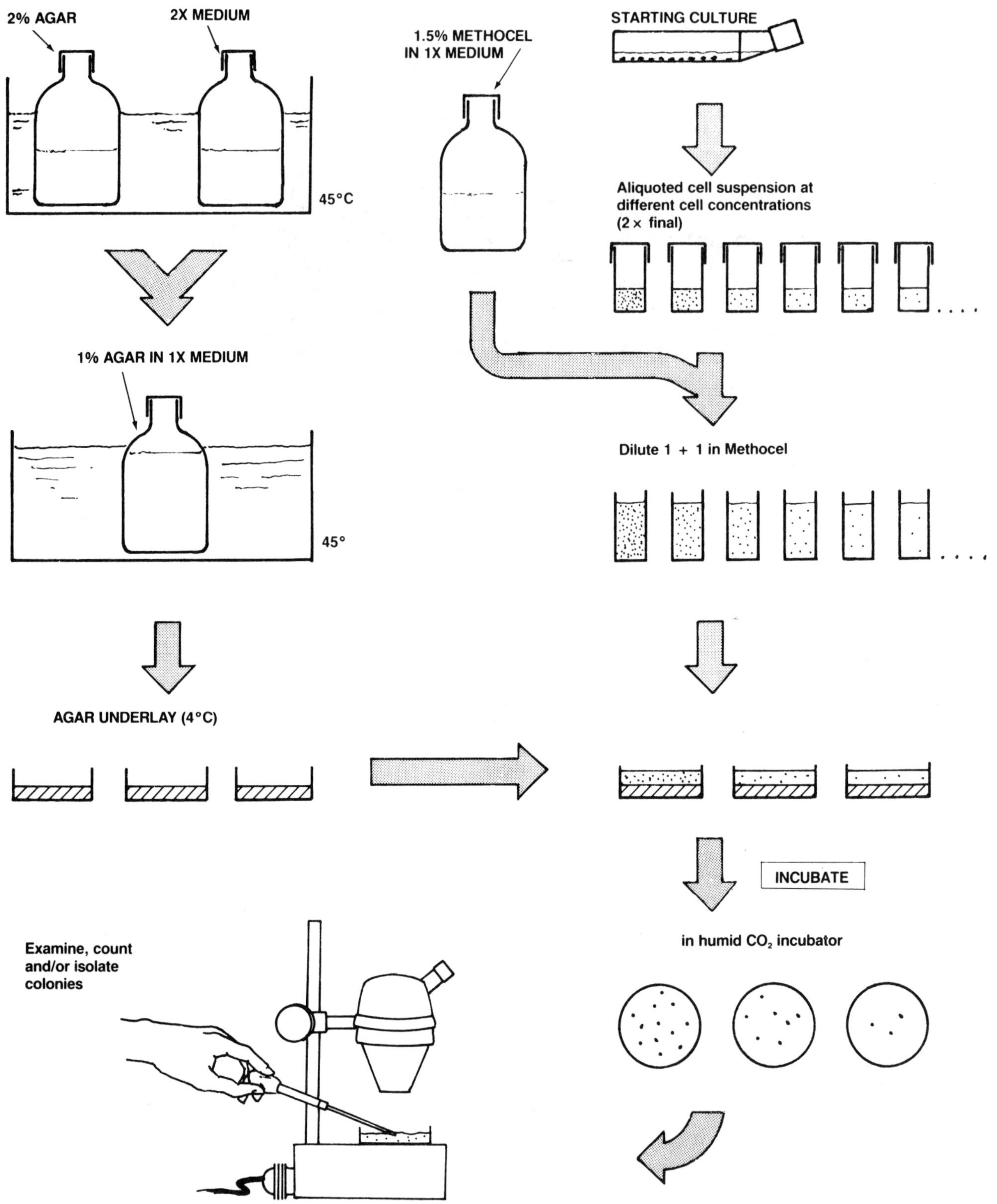

Fig. 11.5. Cloning cells in suspension in Methocel over an agar underlay.

Materials

As for agar cloning; 1.36% Methocel 4 Nsm^{-2} (4,000 cps) in deionized distilled water

Protocol

1. Prepare agar underlay by heating sterile 1.0–2.0% agar to 100°C, bring to 45°C, and dilute with an equal volume of double-strength medium at 45°C (prepare from 10× concentrate to half the recommended final volume and add twice the normal concentration of serum). Plate out 1 ml immediately into 35-mm dishes or 6 × 35-mm multiwell plates, and allow to gel at 4°C for 10 min.
2. Trypsinize or collect cells from suspension and dilute to double the required final concentration.
3. Dilute the cells with an equal volume of Methocel and plate out 1 ml over the agar underlay (10–1,000 cells per dish for continuous cell strains but up to 5×10^5 per dish may be needed for primary cultures).
4. Incubate until colonies form. Since the colonies form at the interface between the agar and the Methocel, fresh medium may be added, 1 ml per dish or well, after 1 wk and removed and replaced with more fresh medium after 2 wk without disturbing the colonies.

Many of the recommendations applying to medium supplementation for monolayer cloning also apply to suspension cloning. In addition, sulfhydryl compounds such as mercaptoethanol (5×10^{-5} M), glutathione (1 m*M*), or α-thioglycerol (7.5×10^{-5} M) [Iscove et al., 1980] are sometimes used. Macpherson [1973] found the inclusion of DEAE dextran was beneficial for cloning, later confirmed by Hamburger et al. [1978], who also found macrophages enhanced cloning of tumor cells, although others have found them detrimental. Courtenay et al. [1978] incorporated rat red blood cells into the medium and demonstrated that a low oxygen tension enhanced cloning.

Most cell types clone in suspension with a lower efficiency than in monolayer, some cells by two or three orders of magnitude. Isolation of colonies is, however, much easier.

Isolation of Clones

Monolayer clones—multiwell plates. If cells are cloned directly into multiwell plates (see above), colonies may be isolated by trypsinizing individual wells. It is necessary to confirm the clonal origin of the colony during its formation by regular microscopic observation.

Cloning rings. If cloning is performed in petri dishes, there is no physical separation between colonies. This must be created by removing the medium and placing a stainless steel or ceramic ring around the colony to be isolated (Fig. 11.6).

Outline

The colony is trypsinized from within a porcelain, glass, Teflon, or stainless steel ring and transferred to one of the wells of a 24- or 12-well plate, or directly to a 25-cm^2 flask (see step 3 above) (Fig. 11.7).

Materials

Cloning rings (Bellco, Fisher)
Silicone grease
Pasteur pipettes with bent end, or 50–100 μl micropipette with yellow tip
0.25% trypsin
Medium
24-well plate and/or 25 cm^2 flasks
Sterile forceps

Protocol

1. Sterilize cloning rings and silicone grease separately in glass petri dishes, by dry heat. Prepare about 20 bent Pasteur pipettes by heating briefly in a bunsen flame and allowing about 12

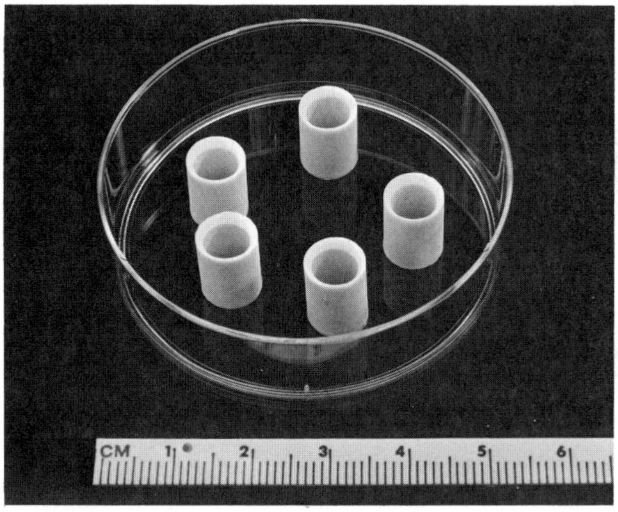

Fig. 11.6. *Cloning rings. Porcelain rings (Fisher) are illustrated, but thick-walled stainless steel rings (e.g., roller bearings) or plastic (e.g., cut from nylon, silicone, or Teflon thick-walled tubing) can be used. Whatever the material, the base must be smooth to seal with silicone grease onto the base of the petri dish and the internal diameter just wide enough to enclose one whole clone without the external diameter overlapping adjacent clones.*

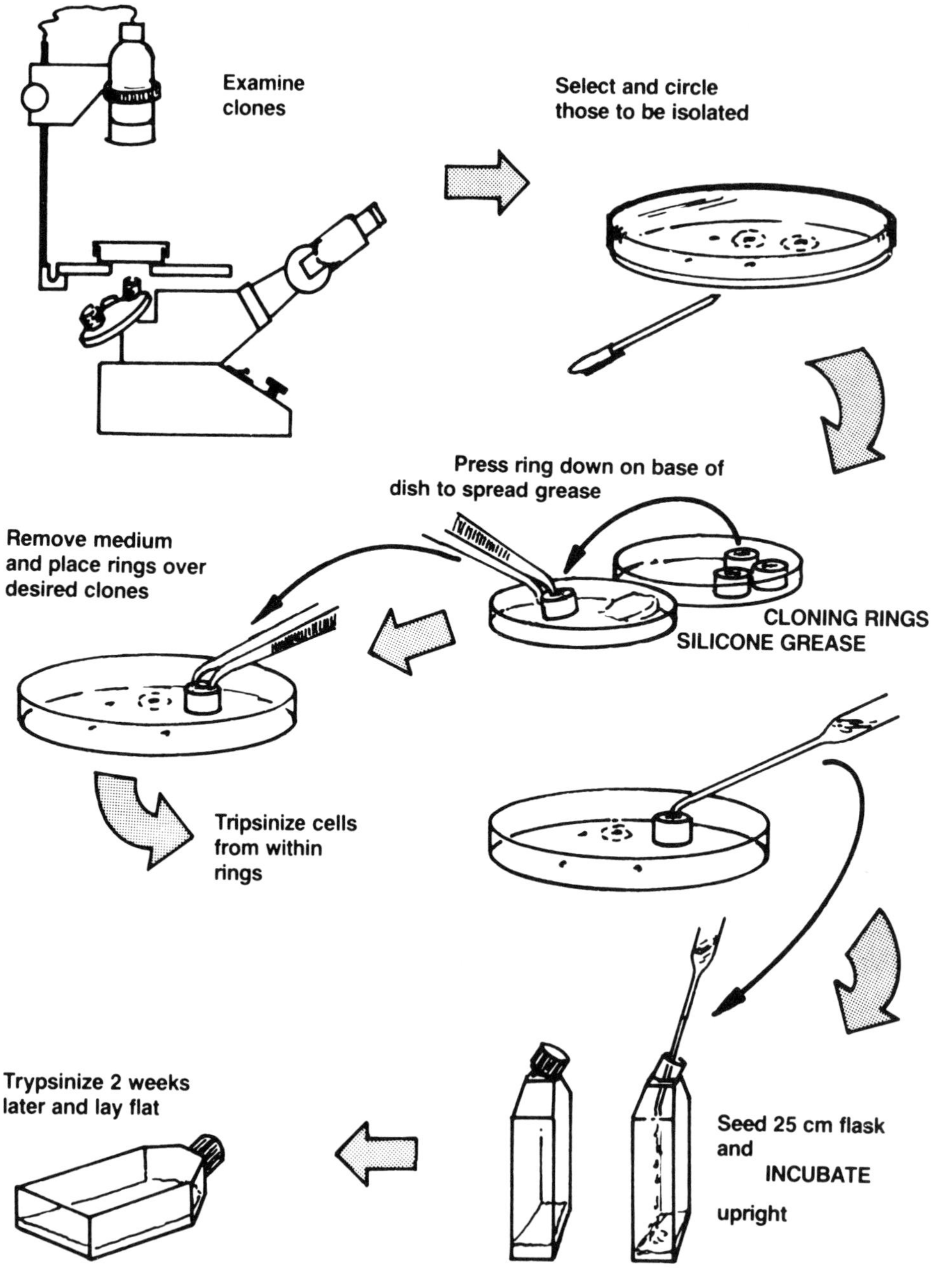

Fig. 11.7. Isolation of clones with cloning rings.

mm of the tip to drop under gravity. If the pipette is held at 30° above horizontal, the bend will be 120°. Place pipettes in sterile test tubes and allow to cool before use.

2. Examine clones and mark those that you wish to isolate with a felt-tip marker on the underside of the dish or use objective fitted with ring marker (Nikon).
3. Remove medium from dish and rinse clones gently with PBS.
4. Using sterile forceps, take one cloning ring, dip in silicone grease, and press down on dish alongside silicone grease to spread the grease around the base of the ring.
5. Place ring around desired colony.
6. Repeat steps 4 and 5 for two or three other colonies in same dish.
7. Add sufficient 0.25% trypsin to fill the hole in ring (~0.4 ml), leave 20 s, and remove.
8. Close dish and incubate for 15 min.
9. Add 0.1–0.4 ml medium to each ring (depending on pipette being used to harvest clones).
10. Taking each clone in turn, pipette medium up and down to disperse cells and transfer to a

well of a 24-well plate or to a 25-cm^2 flask standing on end. Use a separate pipette, or micropipette tip, for each clone.

11. Wash out ring with a second 0.1–0.4 ml medium and transfer to same well.
12. Make up medium in wells to 1.0 ml, close plate, and incubate. If using flasks, add 1 ml medium and incubate standing on end.
13. When clone grows to fill well, transfer up to 25-cm^2 flask, incubated conventionally with 5 ml medium. If using upended flask technique, remove medium when end of flask is confluent, trypsinize cells, resuspend in 5 ml medium, and lay flask down flat. Continue incubation.

The cloning-ring technique may be applied when cells are cloned in a plastic flask by swabbing the flask with alcohol and slicing the top off with a heated sterile scalpel or hot wire. Thereafter proceed as for petri dishes. Flasks are available with a removable top film (Nunc) that may be peeled off to allow harvesting of clones.

Irradiation. Alternatively, where an irradiation source is available, clones may be isolated by shielding one and irradiating the rest of the monolayer (30 Gy, 3,000 rad).

Outline

Invert the flask under an x-ray machine or ^{60}Co source, screening the desired colony with lead.

Materials

x-ray or cobalt source
Piece of lead 2 mm thick
PBSA
0.25% trypsin
Medium

Protocol

1. Select desired colony.
2. Invert flask under x-ray or cobalt source.
3. Cover colony with a piece of lead 2 mm thick.
4. Irradiate with 30 Gy (3,000 rad).
5. Return to sterile area and remove medium, trypsinize, and allow cells to re-establish in the same bottle, using the irradiated cells as a feeder layer.

If irradiation and trypsinization is carried out when the colony is about 100 cells in size, then the trypsinized cells will reclone. Three serial clonings may be performed within 6 wk by this method.

Other isolation techniques include:

(1) Distributing small coverslips or broken fragments of coverslips on the bottom of a petri dish. When plated out at the correct density, some colonies are found singly distributed on a piece of glass and may be transferred to a fresh dish or multiwell plate.

(2) Capillary technique of Sanford et al. [1948]. A dilute cell suspension is drawn into a glass capillary tube (e.g., 50-μl Drummond Microcap), allowing colonies to form inside the tube. The tube is then carefully broken on either side of a colony and transferred to a fresh plate. This technique was exploited by Maurer [1988] for clonogenic assay of hemopoietic cells and tumor cells, where the colony-forming efficiency can be quantified by scanning the capillary in a densitometer.

(3) Petri-perm dish. This is a petri dish with a thin gas-permeable base (see Chapter 7), which may be cut with scissors or a scalpel to isolate colonies. Since this means keeping the outside of the dish sterile, it needs to be handled aseptically and kept inside a larger sterile petri dish.

Suspension Clones

Outline

Draw colony into micropipette and transfer to flask or the well of a multiwell plate.

Materials

24-well plates
Medium
100-μl micropipette with yellow tips
Dissecting microscope
25-cm^2 culture flask

Protocol

Picking colonies is best done on a dissecting microscope.

1. Pipette 1 ml of medium into each well of a 24-well plate.
2. Using a separate tip for each clone, set the micropipette to 100 μl, draw approximately 50 μl into the pipette tip, place the tip of the pipette against the colony to be isolated, and gently draw in the colony.
3. Transfer to a 24-well dish and flush out colony with medium. If from Methocel, the colony will settle, adhere, and grow out. If from agar, you may need to pipette the colony up and down a few times in the well to remove the agar.
4. Clones may also be seeded directly into a 25-cm^2 plastic flask standing on end (see above).

SELECTIVE MEDIA

Manipulating the culture conditions by using a selective medium is a standard method for selecting microorganisms. Its application to animal cells in culture is limited, however, by the basic metabolic similarities of most cells isolated from one animal in terms of their nutritional requirements. The problem is accentuated by the effect of serum, which tends to mask the selective properties of different media. Peehl and Ham [1980] were able to demonstrate that by using two different media, MCDB 105 and MCDB 151, with minimal amounts of dialyzed serum, either fibroblasts or epithelial cells could be grown preferentially from human foreskin (see also Chapter 7, Table 7.7).

Gilbert and Migeon [1975, 1977] replaced the L-valine in the culture medium with D-valine and demonstrated that cells possessing D-amino acid oxidase would grow preferentially. Kidney tubular epithelium and epithelial cells from fetal lung and umbilical cord may be selected this way, as have Schwann cell cultures [Armati and Bonner, 1990] and bovine mammary epithelium [Sordillo et al., 1988].

Much of the effort in developing selective conditions has been aimed at suppressing fibroblastic overgrowth. Kao and Prockop [1977] used cis-OH-proline for this purpose, although this substance can prove toxic to other cells. Fry and Bridges [1979] found phenobarbitone inhibited fibroblastic overgrowth in cultures of hepatocytes, and Braaten et al. [1974] were able to reduce the fibroblastic contamination of neonatal pancreas by treating the culture with sodium ethylmercurithiosalicylate. One of the more successful approaches was the development of a monoclonal antibody to the stromal cells of a human breast carcinoma [Edwards et al., 1980]. Used with complement, this antibody proved cytotoxic to fibroblasts from several tumors and helped to purify a number of malignant cell lines (see also Chapter 21). Attempts have also been made to kill cells selectively with drug- or toxin-conjugated antibodies, but this approach is complicated by the need to get stably bound drug or toxin that will yet become dissociated at the target cell. This has been difficult to achieve. Selective antibodies are now more successfully used in "panning" or magnet and bead separation techniques (see Chapter 12).

Selective media are also commonly used to isolate hybrid clones from somatic hybridization experiments. HAT medium, a combination of hypoxanthine, aminopterin, and thymidine, selects hybrids with both hypoxanthine guanine phosphoribosyltransferase and thymidine kinase from parental cells deficient in one or the other enzyme (see Chapter 23) [Littlefield, 1964a].

Where a mixture of cells shows different responses to growth factors, it is possible to stimulate one cell type with the appropriate growth factor and then, taking advantage of the increased sensitivity of the more rapidly growing cells, kill them selectively with irradiation or cytosine arabinoside (see glial protocol, Chapter 20). Alternatively, if an inhibitor is known, or a growth factor removed, which will take one population out of cycle, then remaining cycling cells can be killed with ara-c or irradiation.

Isolation of Genetic Variants

The following protocol for the development of mutant cell lines that amplify the dihydrofolate reductase (DHFR) gene has been contributed by June Biedler, Memorial Sloan-Kettering Cancer Center, New York, New York.

Principle

Cells exposed to gradually increasing concentrations of folic acid antagonists such as methotrexate (MTX) over a prolonged period of time will develop resistance to the toxic effects of the drug [Biedler et al., 1972]. Resistance resulting from amplification of the DHFR gene generally develops the most rapidly, although other mechanisms, e.g., alteration in antifolate transport and/or mutation affecting enzyme structure or affinity, may confer part or all of the resistant phenotype.

Outline

Expose cells to a graded series of concentrations of MTX for 1–4 wk, periodically replacing the medium with fresh medium containing the same drug concentration. Select for subculturing those flasks in which a small percentage of cells survive and form colonies; repeatedly subculture such cells in the same and in two- to tenfold higher MTX concentrations until cells acquire the desired degree of resistance.

Materials

Sterile:

Chinese hamster cells or rapidly growing human or mouse cell lines

Methotrexate Sodium Parenteral (Lederle Laboratories)

0.15 M NaCl

Tissue culture flasks

Pipettes

Culture medium that does not contain thymidine and hypoxanthine (e.g., Eagle's MEM with 10% fetal bovine serum)

Nonsterile:

Inverted microscope

Ultralow temperature cabinet or liquid nitrogen freezer

Protocol

1. Clone parental cell line to obtain a rapidly growing, genotypically uniform population to be used for selection.
2. Dilute MTX with sterile 0.15 M (0.85%) NaCl. Drug packaged for use in the clinic is in solution at 2.5 mg/ml.
3. Inoculate 2.5×10^5 cells into replicate 25-cm^2 flasks containing no drug or 0.01, 0.02, 0.05, and 0.1 μg/ml of MTX in complete tissue culture medium. Adjust pH and incubate at 37°C for 5–7 d.
4. Observe cultures with an inverted microscope. Replace medium with fresh medium containing the same amount of MTX in cultures showing clonal growth of a small proportion of cells amid a background of enlarged, substrate-adherent, and probably dying cells and reincubate.
5. Allow cells to grow for another 5–7 d, changing growth medium as necessary but continuously exposing the cells to methotrexate. When cell density has attained $2-10 \times 10^6$ cells/flask, subculture cells at 2.5×10^5 cells/flask into new flasks containing the same and two- to tenfold higher drug concentrations.
6. After another 5–7 d, observe new passage flasks as well as cultures from the previous passage that had been exposed to higher drug concentrations; change medium and select for viable cells as before.
7. Continue selection with progressively higher drug concentrations at each subcultivation step until the desired level of resistance is obtained: 2–3 months for Chinese hamster cells with low to moderate levels of resistance, increase in DHFR activity, and/or transport alteration; 4–6 months or more for high levels of resistance and enzyme overproduction, when Chinese hamster, mouse, or fast-growing human cells are used.
8. Periodically freeze samples of developing lines at −70°C or in liquid nitrogen (see Chapter 17).

Analysis

Characterize resistant cells for levels of resistance to drug in a clonal or cell growth assay (see Chapter 19), for increase in activity or amount of DHFR by biochemical or gel electrophoresis techniques [Albrecht et al., 1972; Melera et al., 1980], and/or for increase in mRNA and copy number of the reductase gene by Northern or Southern or dot blots [Scotto et al., 1986] with DHFR-specific probes to determine the mechanism(s) of resistance.

Variations

Cell culture media other than Eagle's MEM can be used; medium composition, e.g., folic acid, content, can be expected to influence rate and type of MTX resistance development. Media containing thymidine, hypoxanthine, and glycine will prevent development of antifolate resistance and should be avoided. Cells can be treated with chemical mutagens prior to selection [Thompson and Baker, 1973]; this treatment may also alter rate and type of mutant selection.

Selection can also be done using cells plated in drug at low density in 100-mm tissue culture dishes (with isolation of individual colonies using cloning cylinders; see above), as single cells in 96-well cluster dishes, or in soft agar, to enable isolation of one or multiple clonal populations at each or any step during resistance development.

Cells can be made resistant to a number of other agents, such as antibiotics, other antimetabolites, toxic metals, etc., by similar techniques; differences in the mechanism of action or degree of toxicity of the agents, however, may require that treatment with the agent be intermittent rather than continuous and may increase the time necessary for selection.

Cell lines of different species or with slower growth rates, such as some human tumor cell lines, may require different (usually lower) initial drug concentrations, longer exposure times at each concentration, and smaller increases in concentration between selection steps.

Solubilization of MTX other than the Lederle product will require addition of equimolar amounts of NaOH and sterilization through a 0.2-μm filter.

INTERACTION WITH SUBSTRATE

Selective Adhesion

Different cell types have different affinities for the culture substrate and will attach at different rates. If a primary cell suspension is seeded into one flask and transferred to a second after 30 min, a third after 1 hr, and so on for up to 24 hr, the most adhesive cells will be found in the first flask and the least adhesive in the last. Macrophages will tend to remain in the first flask, fibroblasts in the next few flasks, then epithelial, and finally hemopoietic cells in the last flask. Polinger [1970] used a similar procedure for the separation of embryonic heart muscle cells from fibroblasts.

If collagenase in complete medium is used for primary disaggregation of the tissue (see Chapter 9), most of the cells released will not attach within 48 hr unless the collagenase is removed. However, macrophages migrate out of the fragments of tissue and attach during this period and can be removed from other cells by transferring the disaggregate to a fresh flask after 48–72 hr treatment with collagenase. This technique works well during disaggregation of biopsy specimens from human tumors.

Selective Detachment

Treatment of a heterogeneous monolayer with trypsin or collagenase will remove some cells more rapidly than others. Periodic brief exposure to trypsin removed fibroblasts from cultures of fetal human intestine [Owens et al., 1974] and skin [Milo et al., 1980]. Lasfargues [1973] found exposure of cultures of breast tissue to collagenase for a few days at a time removed fibroblasts and left the epithelial cells. EDTA, on the other hand, may release epithelial cells more readily than fibroblasts [Paul, 1975].

Dispase II (Boehringer, Mannheim) selectively dislodges sheets of epithelium from human cervical cultures grown on feeder layers of 3T3 cells (see below) without dislodging the 3T3 cells (see Chapter 20). This technique may be effective in subculturing epithelial cells from other sources, excluding stromal fibroblasts.

Nature of Substrate

The hydrophilic nature of most culture substrates (see also Chapter 7) appears to be necessary for cell attachment, but little is known about variations in charge

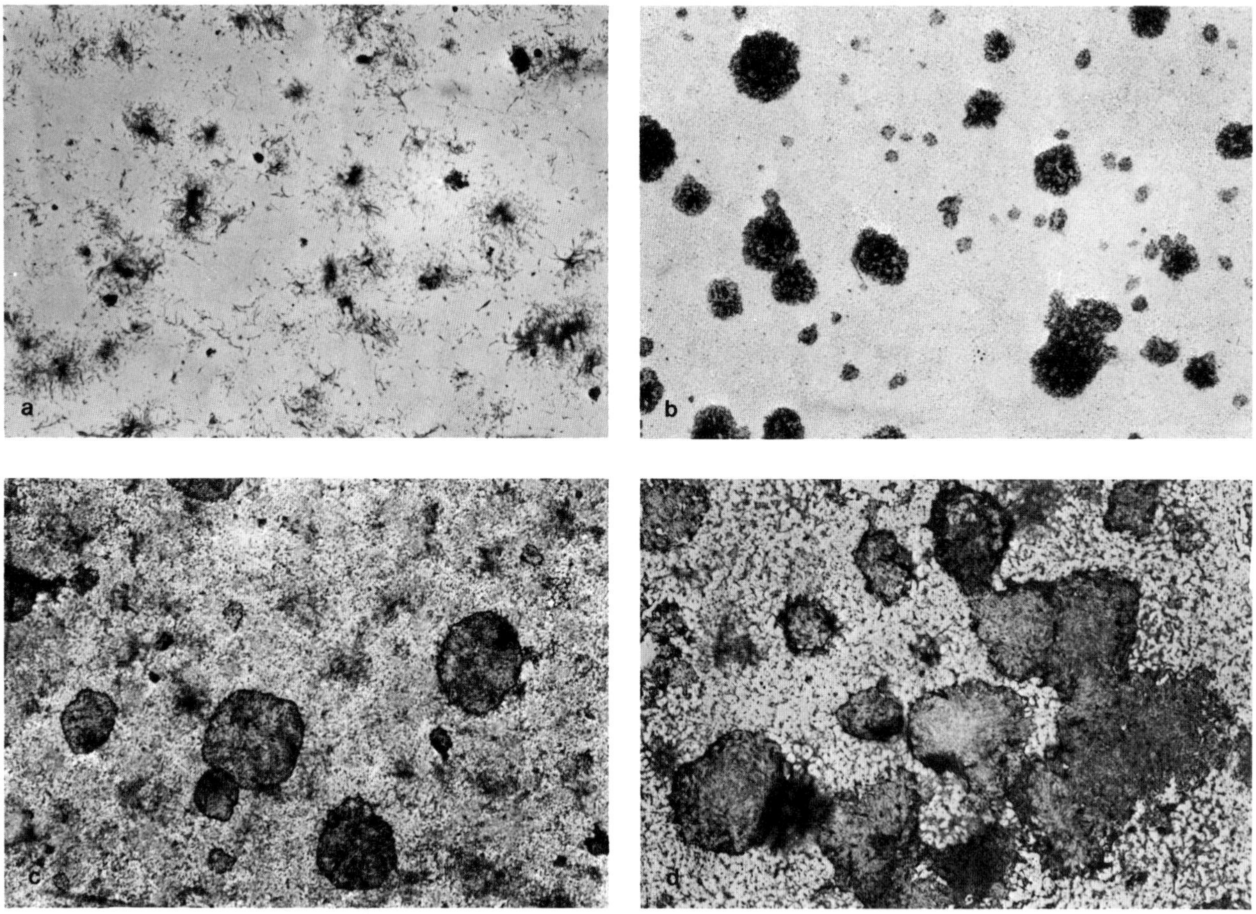

Fig. 11.8. *Selective cloning of breast epithelium on a confluent feeder layer. a. Colonies forming on plastic alone after seeding 4,000 cells/cm² (2 × 10⁴ cells/ml) from a breast carcinoma culture. Small dense colonies are epithelial cells, larger stellate colonies are fibroblasts. b. Colonies of cells from the same culture, seeded at 400 cells/cm² (2,000 cells/ml) on a confluent feeder layer of FHS74Int cells [Owens et al., 1974]. The epithelial colonies are much larger than in (a), the plating efficiency is higher, and there are no fibroblastic colonies. c. Colonies from a different breast carcinoma culture plated onto the same feeder layer. Note different colony morphology with lighter stained center and ring at point of interaction with feeder layer. d. Colonies from normal breast culture seeded onto FHI cells (fetal human intestine similar to FHS74Int). There are a few small fibroblastic colonies present in (c) and (d). (After a technique described by Dr. A. J. Hackett, personal communication.)*

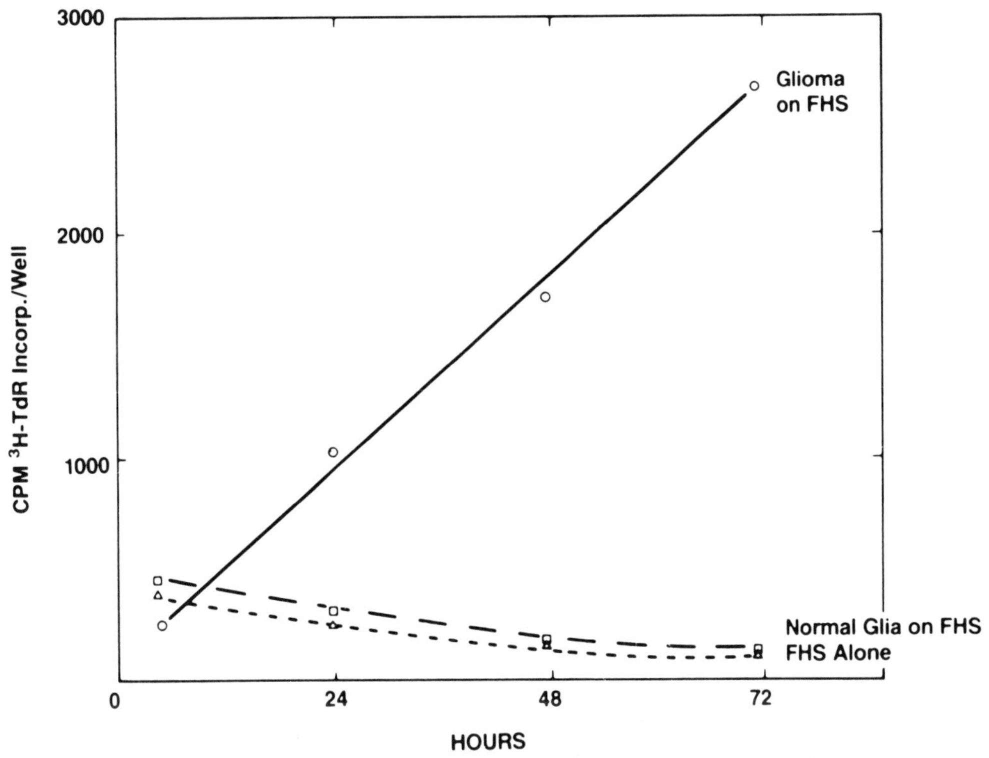

Fig. 11.9. *Selective growth of glioma on confluent feeder layer. Cells were seeded at* 2×10^4*/ml (*4×10^3*/cm*^{2}*) onto confluent, mitomycin-C-treated feeder layers (see text) of FHS74Int cells [Owens et al., 1974] and labeled at intervals thereafter with [*^3H*]-thymidine (see text), extracted, and counted.*

distribution on the cell surface and how different mosaic patterns may interact with different substrates. Since cell sorting in the embryo is a highly selective process and probably relates to differences in the distribution of charged molecules and specific receptor sites on the cell surface, qualitative and quantitative variations in substrate affinity should be anticipated in cultured cells. The relative infrequency with which this is actually found probably illustrates our ignorance of the subtlety of cell–cell and cell–substrate interactions (see also Chapter 14).

Cell–cell adhesion is mediated mainly by cell adhesion molecules (CAMs), some of which, the cadherins, are Ca^{2+} dependent. Hence the action of trypsin is to cleave peptide regions of these molecules, while chelating agents such as EDTA and EGTA sequester the Ca^{2+}. Substrate adhesion, on the other hand, is mainly controlled by integrins, receptors that bind to extracellular matrix (ECM) constituents such as fibronectin and laminin. Proteoglycans also occur in the ECM and act as low-affinity receptors in the plasma membrane [Casillas et al., 1991; Klagbrun & Baird, 1991]. While several sources of ECM are now available (Matrigel, Collagen [Becton Dickinson], Biomatrix [IBT]), the emphasis so far has been on promoting cell survival or differentiation, and little has been made of the potential for selectivity in "designer" matrices, although collagen has been reported to favor epithelial proliferation [Kibbey et al., 1992; Kinsella et al., 1992] and Matrigel also favors epithelial survival and differentiation [Bissell et al., 1987; Ghosh et al., 1991; Kibbey et al., 1992]. As the constituents are now better understood, mixing various collagens with proteoglycans, laminin, and/or fibronectin could be used to create more selective substrates.

The selective effect of substrates on growth may depend on both differential rates of attachment and growth, although in practice the two are indistinguishable. Polyacrylamide layers allow the cloning of tumor cells but not normal fibroblasts [Jones and Haskill, 1973, 1976]. Transformed cells proliferate on Teflon, while most other cells will not [Parenjpe et al., 1975]. Macrophages will also attach to Teflon but do not proliferate. The dermal surface of freeze-dried pig skin was shown to allow growth of epidermal cells but not fibroblasts [Freeman et al., 1976]. Collagen, presumably the basis of the selection, has also been used in gel form to favor epithelial cell growth [Lillie et al., 1980] and in its denatured form to support endothelial outgrowth from aorta into a fibrin clot [Nicosia and Leighton, 1981].

Becton Dickinson has introduced a range of plastics

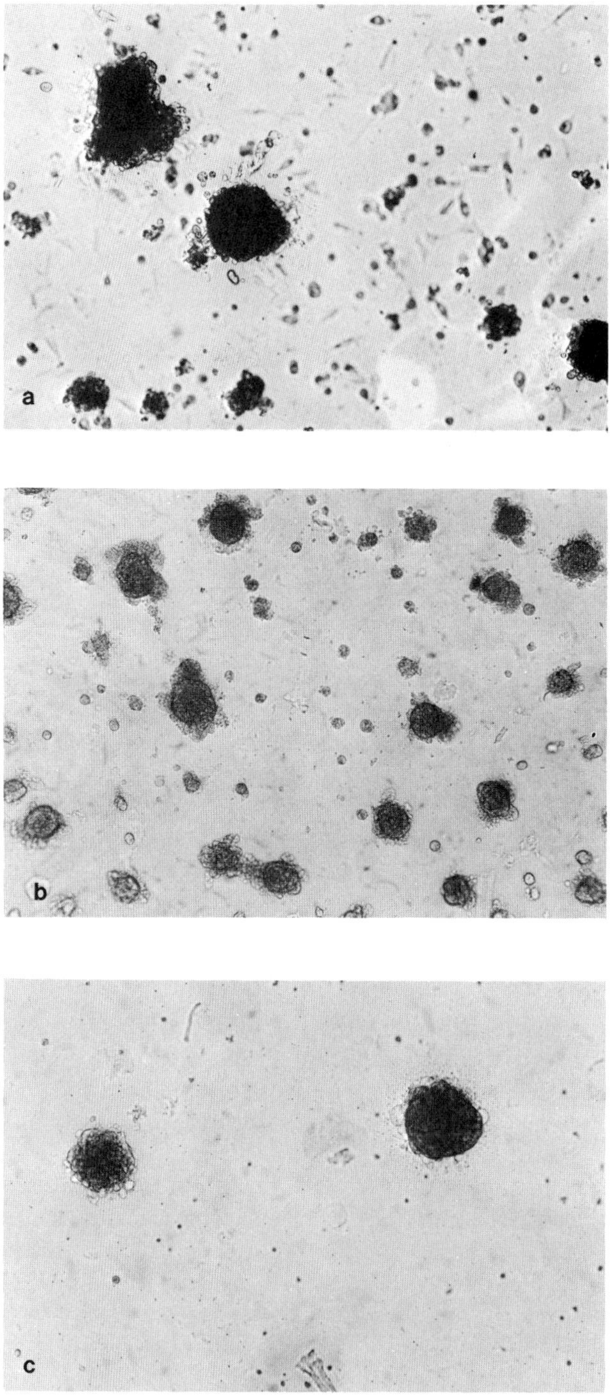

Fig. 11.10. *Growth of melanoma, fibroblasts, and glia in suspension. Cells were plated out at 5 × 10⁵ per 35-mm dish (2.5 × 10⁵ cells/ml) in 1.5% Methocel over a 1.25% agar underlay. Colonies were photographed after 3 wk. a. Melanoma. b. Human normal embryonic skin fibroblasts. c. Human normal adult glia. d. Colony-forming efficiency of normal and malignant*

d 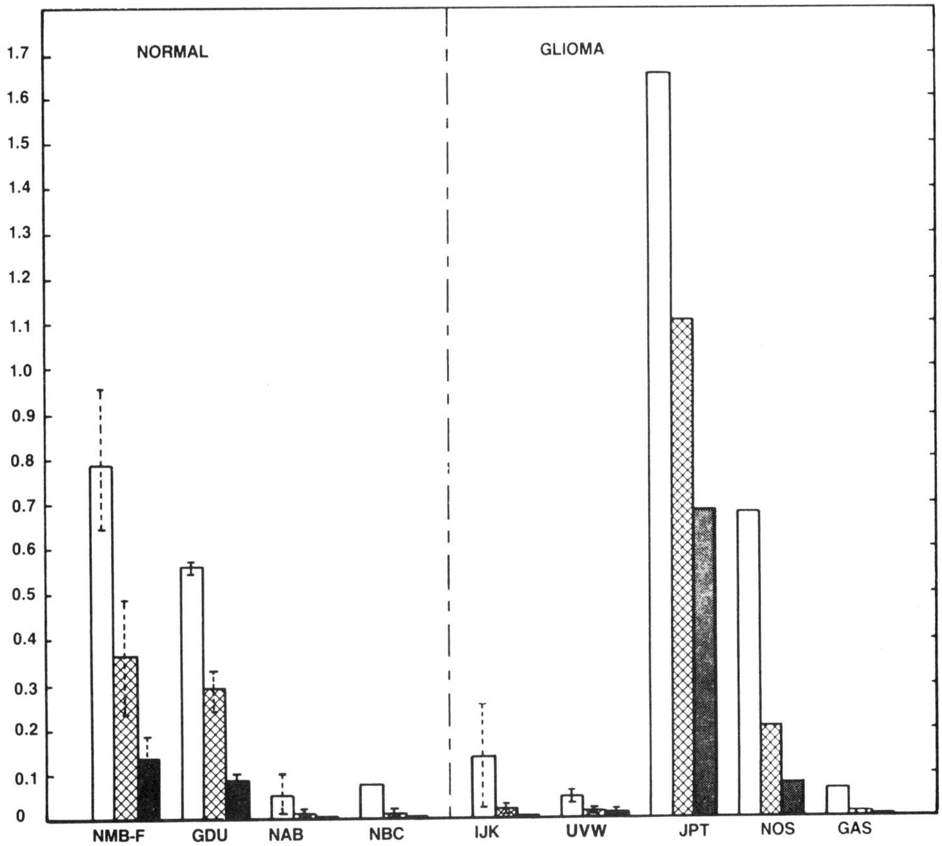

Fig. 11.10. (Continued)
glial cells in suspension. Unshaded bars, colonies of >8 cells; cross-hatched bars, colonies >16 cells; stippled bars, colonies of >32 cells (approximate). Colony counts were done on an Artek Colony Counter (see Fig. 18.6c) at different threshold settings.

in the Falcon series called Primaria. These have a different charge on the plastic surface from conventional tissue culture plastics and are designed to enhance epithelial growth relative to fibroblasts. A number of companies are also supplying plastics coated in natural or synthetic matrices that may facilitate growth of more fastidious cell types but are probably not selective.

Feeder layers. The conditioning of the substrate by feeder layers is discussed in Chapter 14. Feeder layers can also be used for the selective growth of epidermal cells [Rheinwald and Green, 1975] and for repressing stromal overgrowth in cultures of breast (Fig. 11.8) and colon carcinoma (see Chapter 21) [Freshney et al., 1982b]. The author has also been able to demonstrate that human glioma will grow on confluent feeder layers of normal glia, while cells derived from normal brain will not [MacDonald et al., 1985] (Fig. 11.9; see Chapter 15).

Semisolid supports. Transformation of many fibroblast cultures reduces anchorage dependence of cell proliferation (see Chapter 15) [Macpherson and Montagnier, 1964]. By culturing the cells in agar (see above) after viral transformation, it is possible to isolate colonies of transformed cells and exclude most of the normal cells. Most normal cells will not form colonies in suspension with the same high efficiency as virally transformed cells, although they will often do so with low plating efficiencies. The difference between virally transformed fibroblasts and untransformed cells is not seen as clearly in attempts at selective culture of spontaneously arising tumors. Experiments in the author's laboratory have shown that normal glia and fetal skin fibroblasts will form colonies in suspension just as readily as glioma and melanoma (Fig. 11.10).

Cell cloning and the use of selective conditions have a significant advantage over physical cell separation techniques (next chapter), in that contaminating cells are either eliminated entirely by clonal selection or re-

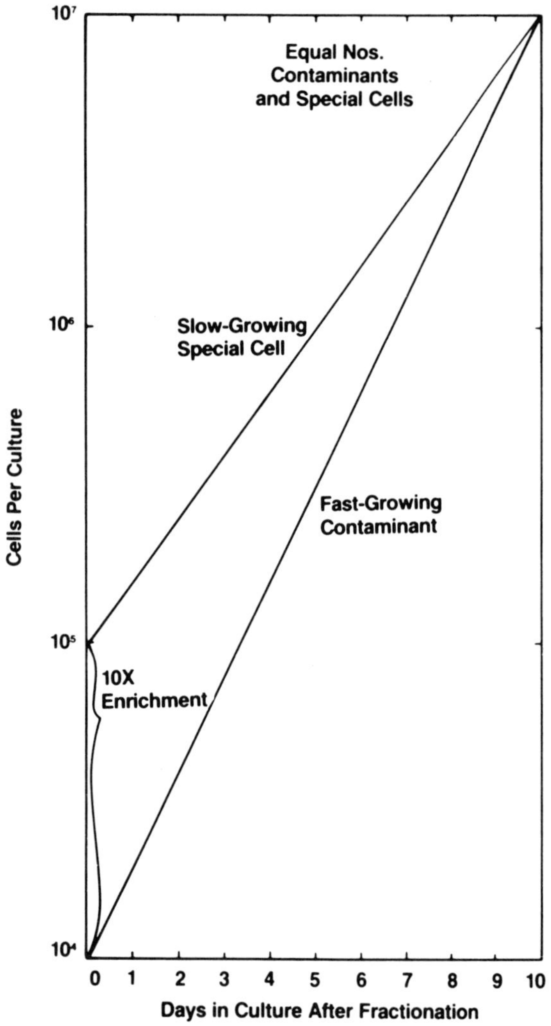

Fig. 11.11. *Overgrowth of a slow-growing cell line by a rapidly growing contaminant. This is a hypothetical example, but it demonstrates that a 10% contamination with a cell population that doubles every 24 hr will reach equal proportions with a cell population that doubles every 36 hr after only 10 d growth.*

pressed by constant or repeated application of selective conditions. Even the best physical cell separation techniques will still allow some overlap between cell populations such that overgrowth will recur. As long as this situation exists, a steady state cannot be achieved and the constitution of the culture is altering continuously. From Figure 11.11 it can be seen that a 90% pure culture of line A will be 50% overgrown by a 10% contamination with line B in 10 days, given that B grows 50% faster than A. For continued culture, therefore, selective conditions are required in addition to, or in place of, physical separation techniques.

CHAPTER 12

Physical Methods of Cell Separation

While cloning or selective culture conditions are the preferred methods for purifying a culture (see Chapter 11), there are occasions when cells do not grow with a high enough plating efficiency to make cloning possible or when appropriate selection conditions are not available. It may then be necessary to resort to a physical separation technique such as density sedimentation or flow cytometry. Physical separation techniques have the advantage that they give a high yield more quickly than cloning, although not with the same purity.

The more successful separation techniques (Fig. 12.1) depend on differences in (1) cell size; (2) cell density (specific gravity); (3) cell surface charge; (4) cell surface chemistry (affinity for lectins, antibodies, or chromatographic media); (5) total light scatter per cell; and (6) fluorescence emission of one or more cellular constituents or adsorbed antibody. The apparatus required ranges from about \$10 worth of glassware to \$200,000 worth of complex laser and computer technology; the choice depends on the parameter that you are obliged to use (see Table 12.1) and on your budget.

CELL SIZE AND SEDIMENTATION VELOCITY

The relationship between particle size and sedimentation rate at 1 *g*, though complex for submicron particles, is fairly simple for cells, and can be expressed approximately as

$$v \approx \frac{r^2}{4} \qquad (4)$$

[Miller and Phillips, 1969], where v = sedimentation rate in mm/hr and r = radius of the cell in μm (see Table 18.1).

Unit Gravity Sedimentation

The apparatus required is illustrated in Figure 12.2 and can be assembled from routine laboratory glassware. To ensure stability of the column of liquid supporting the cells in the sedimentation chamber, it is formed from a serum, Ficoll, Percoll, or bovine serum albumen gradient, and run into the chamber through a baffle (Fig. 12.3) to prevent turbulence.

The height of the separation chamber determines how long the sedimentation may run. Since this is usually 2–4 hr, in a low-viscosity medium, 10 cm is approximately correct. A longer sedimentation time may give better resolution but may cause deterioration of the cells.

The width of the chamber controls the number of cells that may be loaded onto the gradient, as the cell layer should be kept thin (~5 mm) and the cell concentration low (~10^6/ml).

TABLE 12.1. Cell Separation Methods

Method	Basis for separation	Comments	Reference
Sedimentation velocity at 1 g	Cell size	Simple technique	[Miller and Phillips, 1969]
Sedimentation by centrifugation "isokinetic gradient"	Cell size, density, and surface configuration	Computer-designed gradient in zonal rotor	[Pretlow, 1971]
Centrifugal elutriation	Cell size, density, and surface configuration	Rapid; high cell yield	[Meistrich et al., 1977a,b]
Isopyknic sedimentation	Cell density	Simple and rapid	[Pertoft and Laurent, 1977]
Flow cytophotometry (fluorescence-activated cell sorting)	Cell surface area fluorescent markers, fluorogenic enzyme substrates, multiparameter	Complex technology and expensive; very effective; high resolution but low yield	[Kreth and Herzenberg, 1974]
Affinity chromatography	Cell surface antigens, cell surface carbohydrate	Elution of cells from columns difficult, better in free suspension	[Edelman, 1973]
Counter current distribution	Affinity of cell surface constituents for solvent phase	Some cells may suffer loss of viability but quite successful for others	[Walter, 1977]
Electrophoresis in gradient or curtain	Surface charge		[Kreisberg et al., 1977; Platsoucas et al., 1979]
Magnetic separation with coated beads	Surface antibody	Specific given highly specific surface antibody	[Gaudernack et al., 1986]
Panning	Surface antibody	Simple, low technology, with precoated plates available, but also depends on specific surface antibody	[Wysocki and Sata, 1978; see also Holyoake protocol]

The height and width (i.e., volume) also affect the filling rate. The chamber must not be filled too rapidly or turbulence will result, and it cannot be filled too slowly or the cells will sediment faster than the liquid level rises. The dimensions given in Figure 12.3 are optimal for separating about 2×10^7 cells of 15–18 μm diameter or up to 10^8 cells of 10–12 μm diameter.

The procedure for fractionating a typical cell suspension of average cell diameter 15 μm is as follows.

Outline

Float cells on top of a gradient of serum in medium, allow cells to sediment through the gradient for about 3 hr, and run off gradient into culture vessels (see Fig. 12.2).

Materials

300 ml Eagle's MEMS (suspension salts, see Table 7.4) + 30% serum
300 ml MEMS + 15% serum
20 ml MEMS + 5% serum
10 ml 0.25% trypsin-citrate (see Reagent Appendix)
30 ml PBS
20 ml MEMS + 3% serum hemocytometer or cell counter
Flotation medium (1 M sucrose or 20% Ficoll in PBS)
25-cm² flasks
Culture medium

Protocol

1. Prepare apparatus as in Figure 12.2. Incorporate Luer connections to allow for disassembly for sterilization. Package and autoclave.
2. Assemble and check that valves V_1, V_2, V_3, and V_4 are closed.
3. Add 300 ml 30% serum in medium to mixer vessel M_1, and 300 ml 15% serum in medium to mixer vessel M_2.
4. Check that stirrer S_1 is functioning.
5. Add 20 ml 5% serum in medium to mixer M_3 and check that stirrer S2 is functioning.
6. Open V_4 to connect syringe to separation chamber and insert 20 ml PBS into separation chamber.
7. Open V_4 to M_3 line, open V_3, and draw a little 5% serum into the syringe (just enough to fill line). Close V_3 and V_4.
8. Prepare cell suspension, e.g., by trypsinizing primary culture for 15 min in 0.25% trypsin-citrate. Disperse cells carefully in 3% serum in medium and check that a single cell suspension is formed.

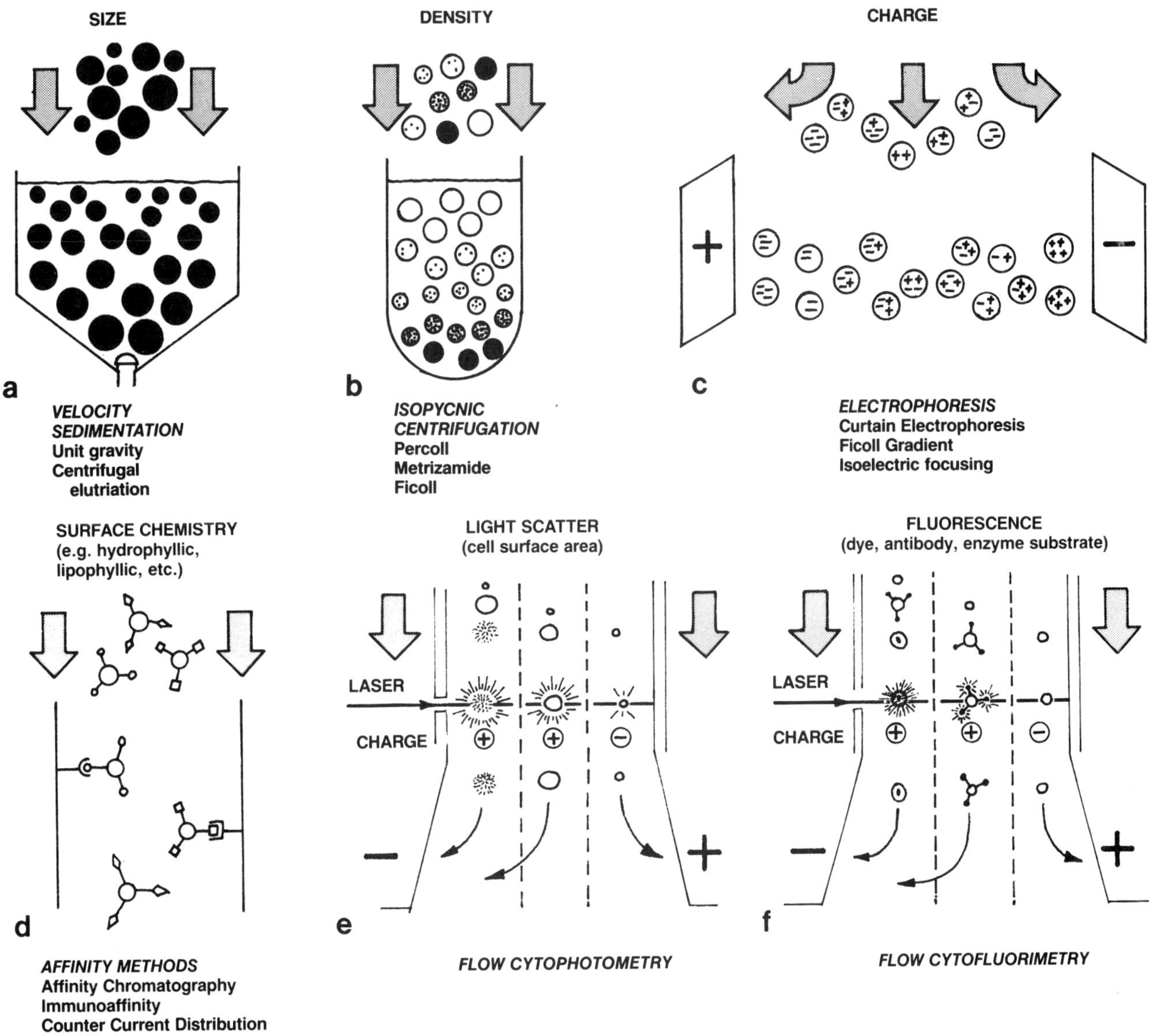

Fig. 12.1. *Physical cell separation techniques. a. Velocity sedimentation, influenced by cell size mainly but also by cell density and surface area at elevated g. b. Isopyknic sedimentation. Cells sediment to a point in a density gradient equivalent to their own density. c. Electrophoresis. Cells migrate to either polarized plate according to net surface charge. d. Affinity methods. Cells are separated by their differential affinities for (1) chromatographic media, (2) antibodies or lectins bound to chromatographic media, or (3) two-phase aqueous polymer systems. e. Flow cytophotometry. Cells are diverted to either of two charged plates according to their light-scattering potential (proportional to surface area). f. Flow cytofluorimetry. Specific fluorochromes are used to label cells and electrophoretic separation is based on fluorescence emission. Both (e) and (f) are carried out on a single cell stream passing through the flow chamber of an instrument such as the Fluorescence-Activated Cell Sorter (FACS) (Becton Dickinson), the Cytofluorograph (Ortho), or the Coulter Cell Sorter.*

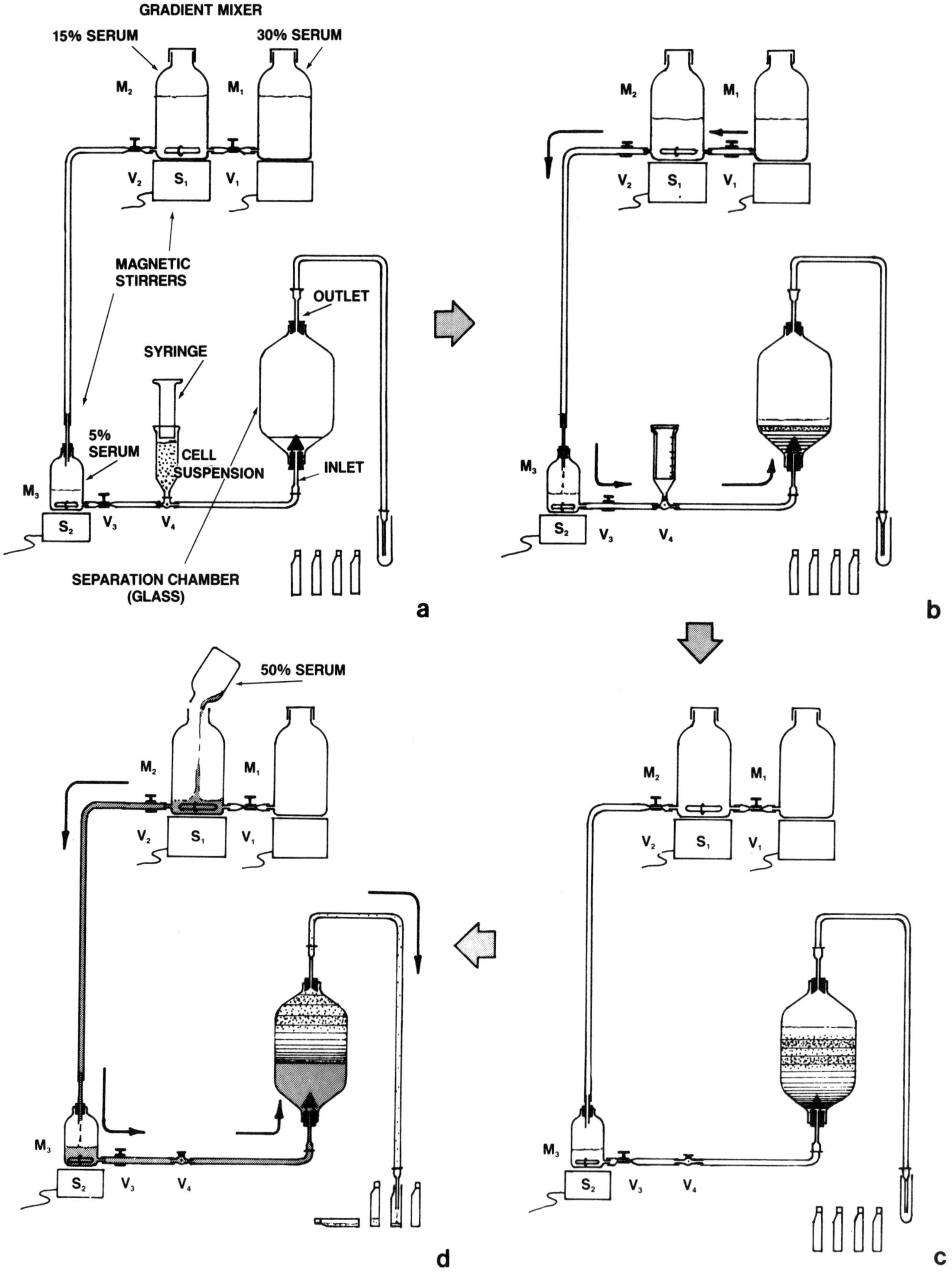

Fig. 12.2. Cell separation by velocity sedimentation at unit gravity. Apparatus and position of valves at different stages of the procedure. a. At start, loading cells. b. Running in gradient. c. After cell sedimentation. d. Harvesting [Miller and Phillips, 1969]. V, valve; M, mixer vessel; S, stirrer.

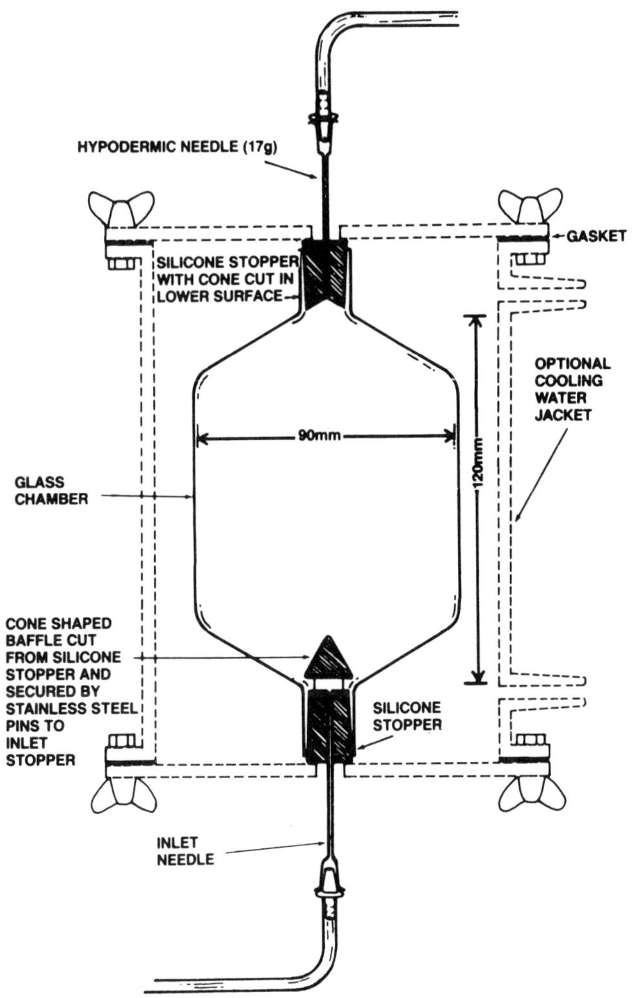

Fig. 12.3. Separation chamber for unit gravity sedimentation. The dotted outline is a cooling jacket for carrying out sedimentations at 4°C. [Modified from Miller and Phillips, 1969].

9. Take up 20 ml at 10^6/ml (maximum) into syringe and connect to V_4 inlet.
10. With syringe held vertically, open V_4 to M_3 line, open V_3, and draw a little 5% serum into syringe to clear any bubbles from M_3 line and V_4.
11. Turn valve V_4 to separation chamber line and draw a little PBS into syringe to clear any bubbles from this line.
12. Insert cell suspension slowly into chamber; avoid mixing cell suspension with the overlying PBS layer. Take care to stop while a little fluid is left in the syringe to avoid injecting any air bubbles back into the line. If difficulty is encountered injecting cells smoothly, without turbulence, remove piston from syringe and allow cells to run in under gravity alone by raising V_4.
13. Start stirrers S_1 and S_2.
14. Open V_4 to connect M_3 line to separation chamber.
15. Open V_1 and adjust flow rate by opening V_2 to give 15 ml/min (~5 drops/s) at M_3. Cell suspension will now float up into separation chamber on gradient of serum. Check for turbulence at baffle as suspension and gradient run in. If there is any, reduce flow rate at M_3 by closing V_2.
16. When gradient mixers M_1 and M_2 are empty, but before M_3 empties, close V_3 and V_2.
17. It should be possible to see the cell layer in the sedimentation chamber and to follow the cells as they sediment. As they do, the cell band will become wider and more diffuse. Check for signs of "streaming" in the early stages of sedimentation (tails of cells that sediment ahead of the main band). This occurs when the cell concentration is too high or the step between the cells and the gradient is too steep.
18. After about 20 min, close V_1 and add 90 ml 50% serum to M_2. Open V_3 and adjust flow rate at M_3 to 15 ml/min. Stop when M_2 is empty but before M_3 empties, by closing V_3 and V_2.
19. Add 500 ml flotation medium (1 M sucrose or 20% Ficoll) to M_1 and M_2, open V_1 and V_2, and let some of the flotation medium run into M_3. Close V_2.
20. When sedimentation is complete, i.e., cell band midway down separation chamber, open V_3, and adjust V_2 to give a flow rate of 15 ml/min in M_3.
21. Collect eluate from top of chamber via elution line and run into graduated culture vessels, e.g., 25-cm^2 flasks, 10 ml per flask. Mix the contents of each flask and take sample for cell counting.
22. Seal and incubate flasks for 24 hr, replace medium with fresh medium at standard serum concentration and volume.

Variations

Gradient medium. If serum and regular culture medium are used, then it is possible to culture cells directly from the eluate. Fetal bovine serum causes less reaggregation than calf or horse serum. If serum is found to be unsuitable, gradients can be formed from bovine serum albumen [Catsimpoolas et al., 1978], Ficoll, or Percoll (Pharmacia).

Aggregation. Aggregation can be reduced by enclosing the separation chamber in a cooling jacket and running the whole process at 4°C. Mixers M_1, M_2, and M_3 must all be kept cold also. Water-driven magnetic stirrers (Calbiochem) may be used at S_1 and S_2 to minimize overheating of gradient.

Pump. Gravity is used in this example as the cheapest and simplest method for generating the flow of gradient medium, but if desired, a peristaltic pump (pulse free) may be inserted at V_3.

Sedimentation of cells at unit gravity is a simple low-technology method of separating cells. It works well for many cell types, e.g., brain [Cohen et al., 1978], hemopoietic cells [McCool et al., 1970; Petersen and Evans, 1967], and HeLa/fibroblast mixtures (Fig. 12.4), and can be performed in regular physiological media. The cells must be singly suspended, however, and there is a practical limit of about 10^8 cells that may be separated.

Pretlow and others [Pretlow, 1971; Pretlow et al., 1978; Hemstreet et al., 1980; Pretlow and Pretlow, 1989] used specially formed gradients of Ficoll (isokinetic gradients) to separate cells by sedimentation velocity at higher *g* forces on a zonal centrifuge rotor. The gradients are shallow and of relatively low density to minimize the effect of cell density on sedimentation rate. Cells of many different types were separated by this method, and it appeared to have a wide application, though perhaps it was overtaken by centrifugal elutriation techniques, where the technology, though no less complex, is perhaps more available.

Centrifugal Elutriation

The centrifugal elutriator (Beckman) is a device for increasing the sedimentation rate and improving the yield and resolution by performing the separation in a specially designed centrifuge and rotor (Fig. 12.5) [Conkie, 1992; Lutz et al., 1992]. Cells in the suspending medium are pumped into the separation chamber in the rotor while the rotor is turning. While the cells are in the chamber, centrifugal force will tend to force the cells to the outer edge of the rotor (Fig. 12.6). Meanwhile the suspending medium is pumped through the chamber such that the centripetal flow rate approximates to the sedimentation rate of the cells. If the cells were uniform, they would remain stationary, but since they vary in size, density, and cell surface configuration, they tend to sediment at different rates. As the sedimentation chamber is tapered, the flow rate increases toward the edge of the rotor and a continuous range of flow rates is generated. Cells of differing sedimentation rates will, therefore, reach equilibrium at different positions in the chamber. The sedimentation chamber is illuminated by a stroboscopic light and can be observed through a viewing

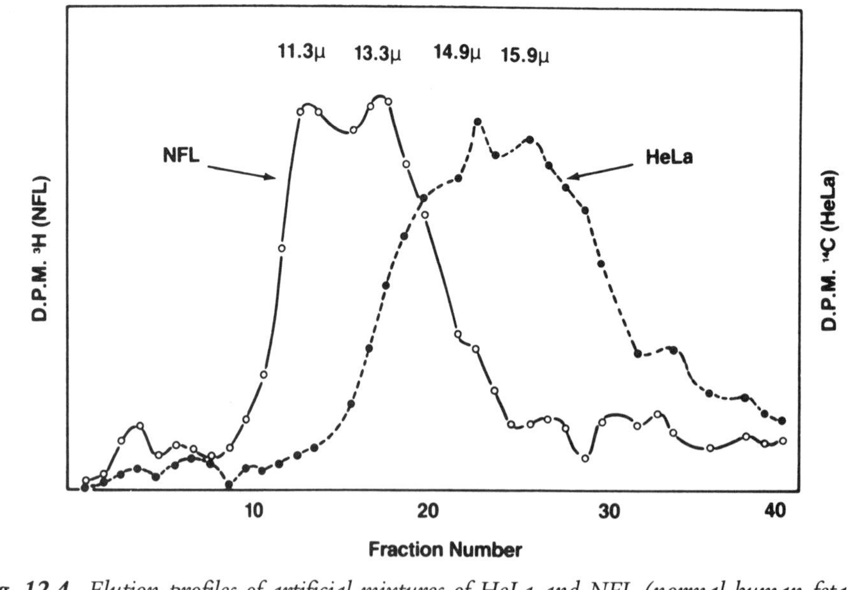

Fig. 12.4. Elution profiles of artificial mixtures of HeLa and NFL (normal human fetal lung fibroblasts) after sedimentation at 1 g for 3 hr. Cultures were prelabeled with ^{3}H-leucine or ^{14}C-thymidine and the distribution of cells in the eluate was determined by dual-isotope scintillation counting. Open circles and solid line, NFL; solid circles and broken line, HeLa. The numbers above each peak are the calculated cell diameters derived from the sedimentation velocity by the equation in the opening paragraphs (see text). The values obtained for HeLa were confirmed by micrometry, but the values for NFL did not agree with the micrometer readings, which were around 16–18 μm. [After Freshney et al., 1982b, reproduced with permission from the publisher].

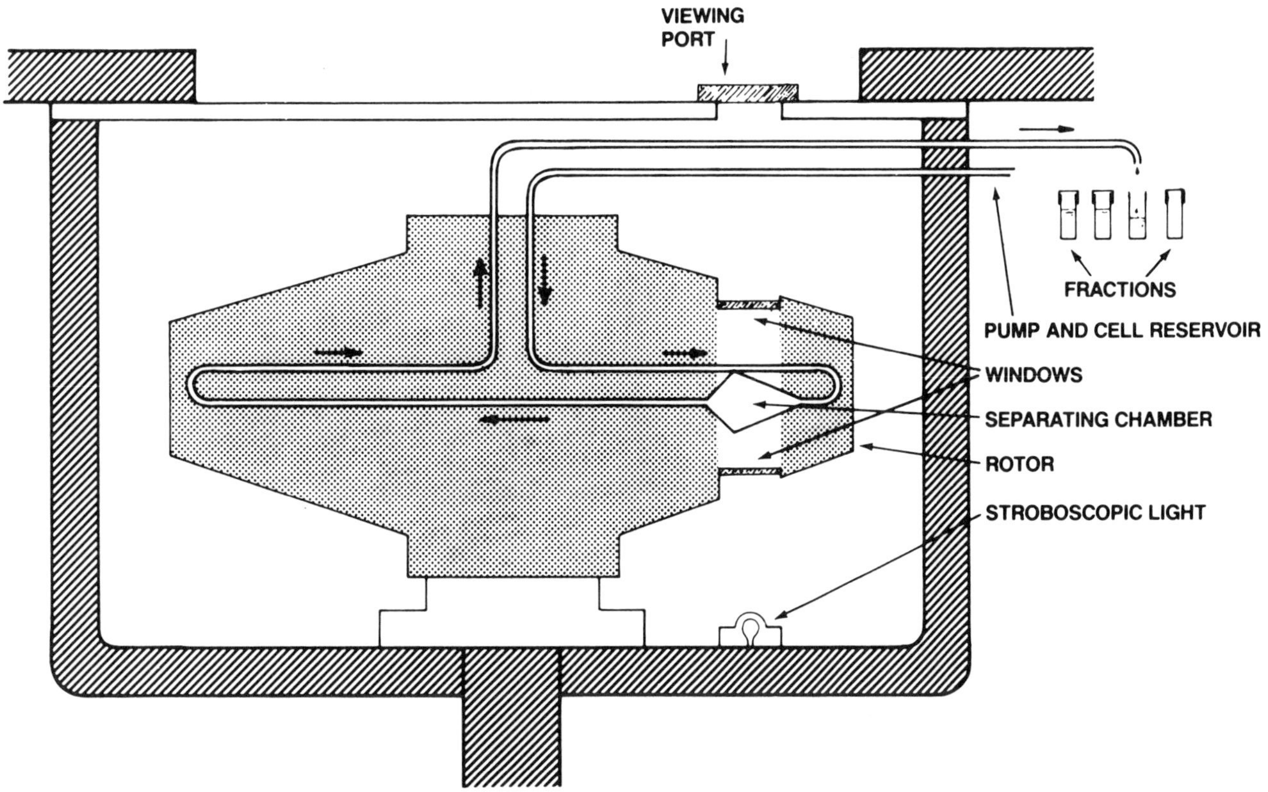

Fig. 12.5. Centrifugal elutriator rotor (Beckman). Cell suspension and carrier liquid enter at the center of the rotor and are pumped to the periphery to enter the outer end of the separating chamber. The return loop is via the opposite side of the rotor to maintain balance.

port. When the cells are seen to reach equilibrium, the flow rate is increased and the cells are pumped out into receiving vessels. The separation can be performed in complete medium and the cells cultured directly afterward.

Equilibrium is reached in a few minutes and the whole run may take 30 min. On each run 10^8 cells may be separated and the run may be repeated as often as necessary. The apparatus is, however, fairly expensive and a considerable amount of experience is required before effective separations may be made. A number of cell types have been separated by this method [Meistrich et al., 1977b; Greenleaf et al., 1979; Pretlow and Pretlow, 1989; Hering et al., 1990; Conkie, 1992], as have cells of different phases of the cell cycle [Meistrich et al., 1977a].

CELL DENSITY AND ISOPYKNIC SEDIMENTATION

Separation of cells by density can be performed at low or high *g* using conventional equipment [Sharpe, 1988]. The cells sediment in a density gradient to an equilibrium position equivalent to their own density (isopyknic sedimentation). Physiological media must be used and the osmotic strength carefully monitored. The density medium should be nontoxic, nonviscous at high densities (1.10 g/ml), and exert little osmotic pressure in solution. Serum albumen [Turner et al., 1967], dextran [Schulman, 1968], Ficoll (Pharmacia) [Sykes et al., 1970], metrizamide (Nygaard) [Munthe-Kaas and Seglen, 1974], and Percoll (Pharmacia) [Pertoft and Laurent, 1977; Wolff and Pertoft, 1972] have all been used successfully; Percoll (colloidal silica) is one of the more effective media currently available. For isolation of lymphocytes on Ficoll/Metrizoate, see Chapter 23.

Outline

Form gradient (1) by layering different densities of Percoll; (2) by high-speed spin; or (3) with special gradient former. Centrifuge cells through Percoll gradient (or allow to sediment at unit gravity), collect fractions, and culture directly (Fig. 12.7).

Materials

Culture medium (sterile)
Medium + 20% Percoll (sterile)
25-ml centrifuge tubes (sterile)

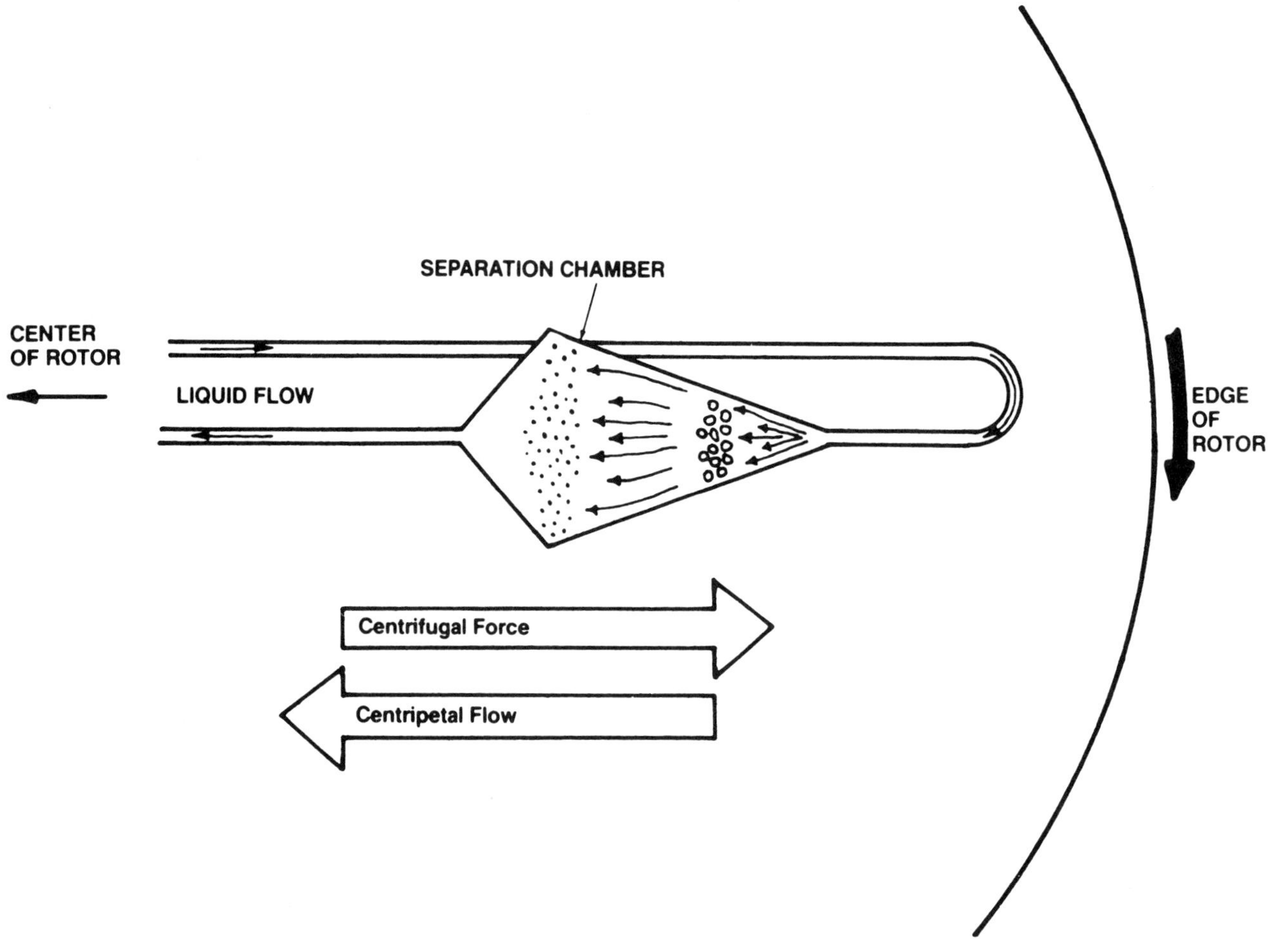

Fig. 12.6. Separation chamber of elutriator rotor.

PBSA (sterile)
0.25% trypsin (sterile)
Syringe or gradient harvester (sterile)
24-well plates or microtitration plates (sterile)
Refractometer or density meter
Hemocytometer or cell counter

Protocol

1. Prepare gradient: (a) Prepare two media, one regular culture medium and one with 20% Percoll; (b) adjust the density of the Percoll solution to 1.10 g/ml and its osmotic strength to 290 mOsm/kg; (c) mix the two media in varying proportions to give the desired density range (e.g., 1.020–1.100 g/ml) in 10 or 20 steps; and (d) layer one step over another, building up a stepwise density gradient in a 25-ml centrifuge tube (see Fig. 12.8). Gradients may be used immediately or left overnight.

   Alternatively, place medium containing Percoll of density 1.085 g/ml in a tube and centrifuge at 20,000 *g* for 1 hr. This generates a sigmoid gradient (Fig. 12.9), the shape of which is determined by the starting concentration of Percoll, the duration and centrifugal force of the centrifugation, the shape of the tube, and the type of rotor.

   A continuous linear gradient may be produced by mixing, for example, 1.020 g/ml with 1.08 g/ml Percoll in a gradient-forming device (Fig. 12.8d) (Fisons, Pharmacia, Buchler).
2. Trypsinize cells and resuspend in medium plus serum. Check that they are singly suspended.
3. Layer up to 2×10^7 cells in 2 ml medium on top of the gradient.
4. The tube may be allowed to stand on the bench for 4 hr or centrifuged for 20 min at between 100 and 1,000 *g*.
5. Collect fractions using a syringe or a gradient harvester (Fig. 12.10) (Fisons). Fractions of 1 ml may be collected into a 24-well plate or 0.1 ml into microtitration plates. Samples should be

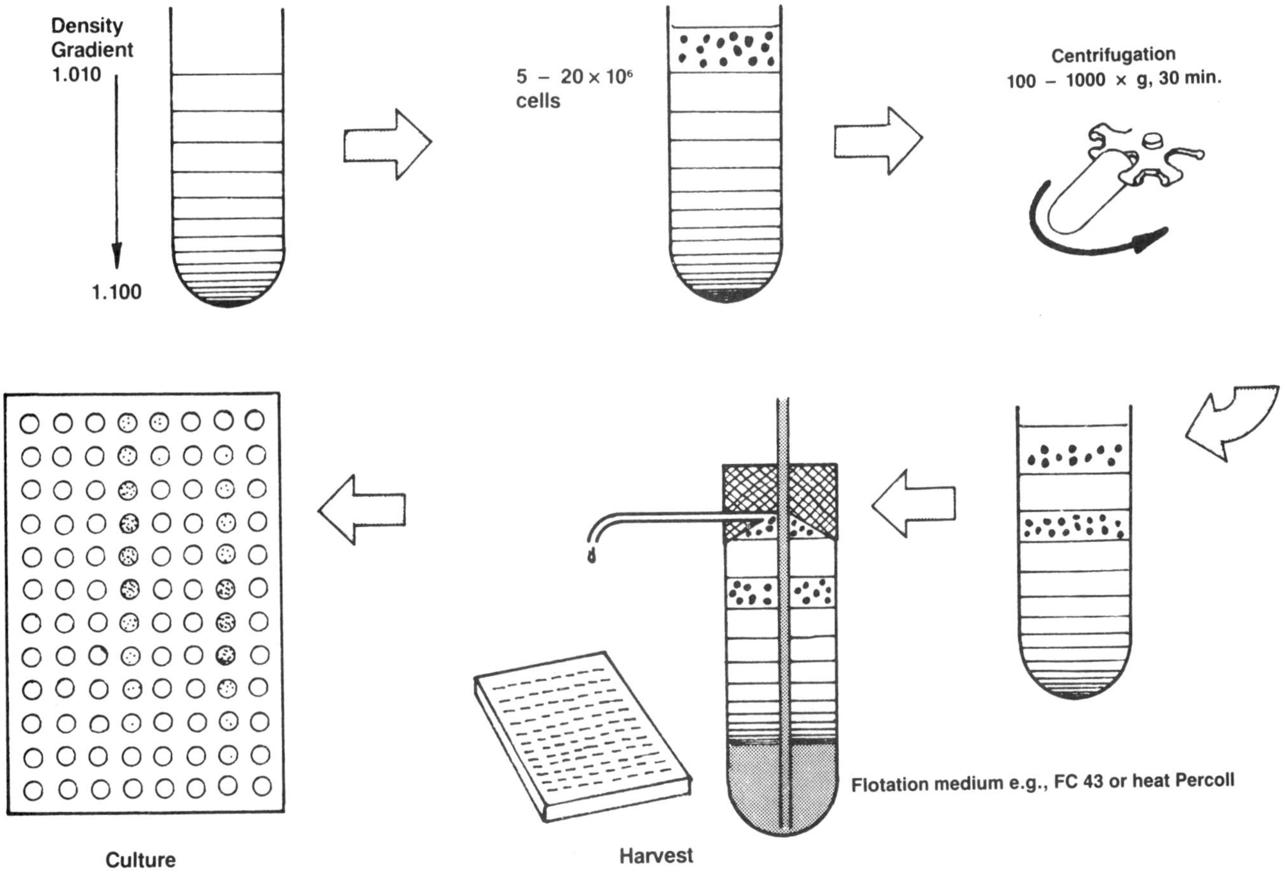

Fig. 12.7. Cell separation by isopyknic centrifugation.

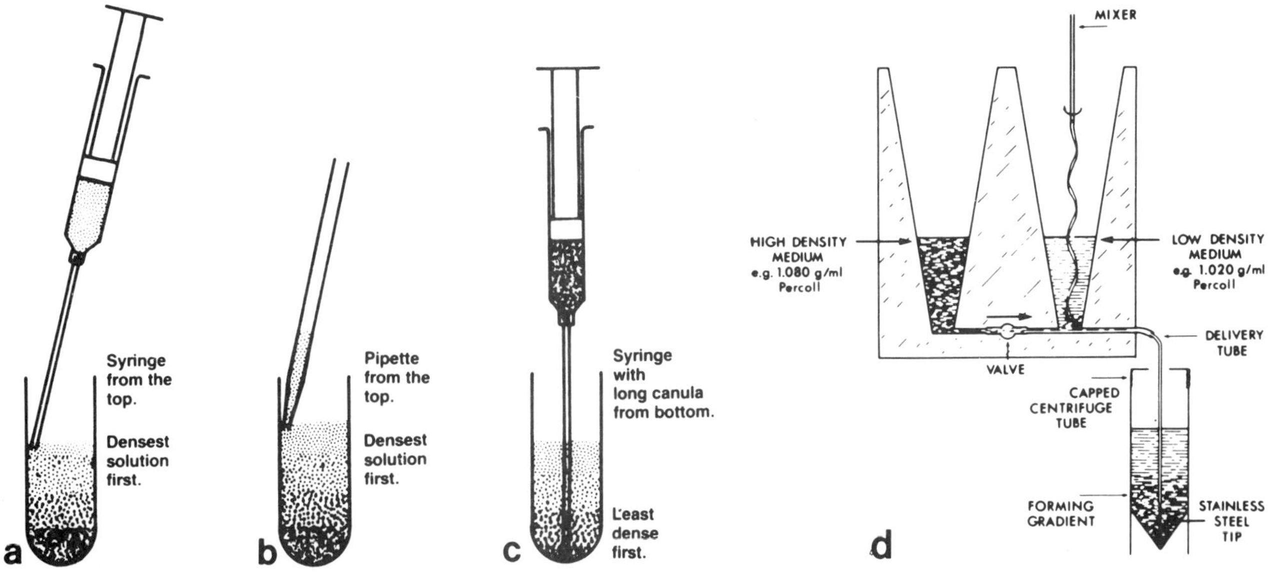

Fig. 12.8. Layering density gradients (a) by syringe from the top, (b) by pipette from the top, (c) by syringe from the bottom, and (d) by gradient mixing device (Buchler).

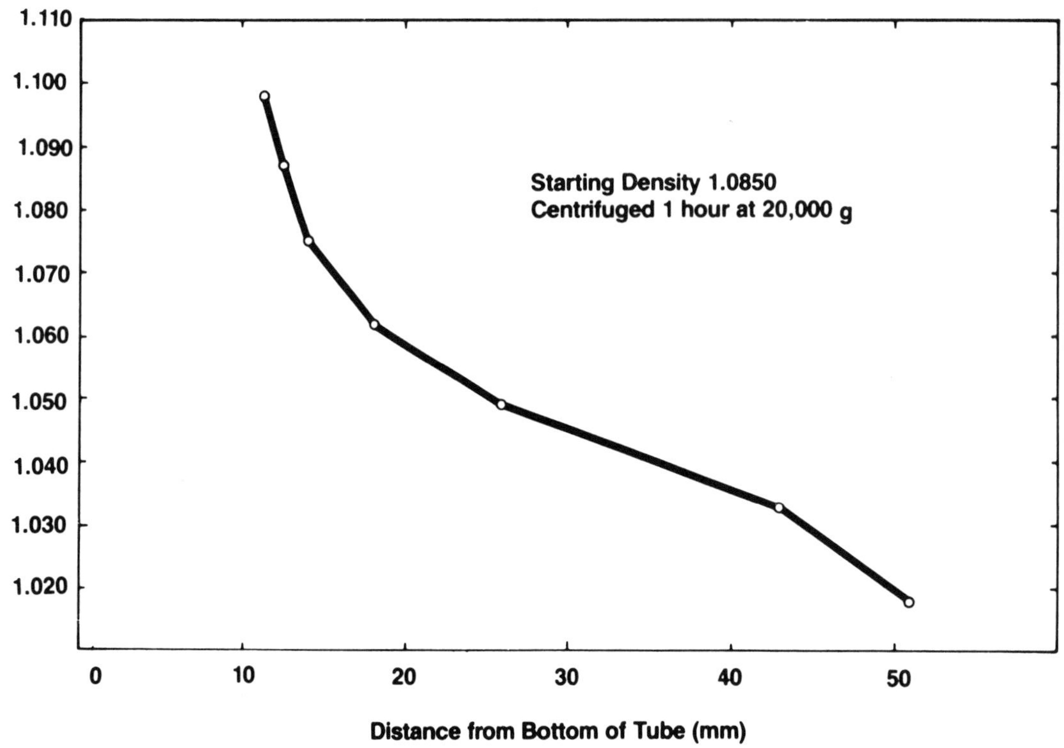

Fig. 12.9. *Gradient generated by spinning Percoll at 20,000 g for 1 hr.*

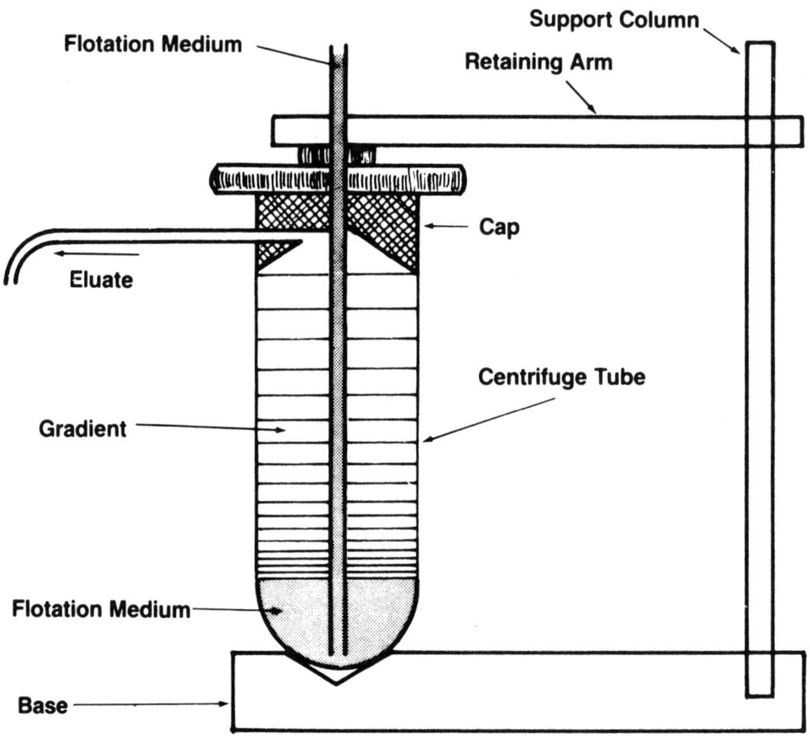

Fig. 12.10. *Gradient harvester (Fisons/M.S.E.). Fluorochemical FC43 is pumped down the inlet tube to the bottom of the gradient and displaces the gradient and cells upward and out through the delivery tube. (After an original design by Dr. G.D. Birnie.)*

taken at intervals for cell counting and determination of the density (ρ) of the gradient medium. Density may be measured on a refractometer (Hilger) or density meter (Paar).

6. Add equal volume of medium to each well and mix (to ensure cells settle to bottom of well). Change the medium to remove the Percoll after 24–48 hr incubation.

Variations

Position of cells. Cells may be incorporated into the gradient during formation by centrifugation. Only one spin is required, although spinning the cells at such a high *g* force may damage them.

Other media. Ficoll is one of the most popular media, as it, like Percoll, can be autoclaved. It is a little more viscous at high densities and may cause agglutination of some cells. Metrizamide (Nygaard), a nonionic derivative of metrizoate, which is a radio-opaque-iodinated substance used in radiography (Isopaque, Hypaque, Renografin) and in lymphocyte purification (e.g., Lymphoprep) (see Chapter 23), is less viscous at high densities [Rickwood and Birnie, 1975] but may be incorporated into some cells (Fig. 12.11), as is Isopaque [Splinter et al., 1978]. Where such media are used, cells should always be layered on top of the gradient and not mixed in during formation.

Marker beads. Pharmacia manufactures colored marker beads of standard densities that may be used to determine the density of regions of the gradient.

Isopyknic sedimentation is quicker than velocity sedimentation at unit gravity and gives a higher yield of cells for a given gradient volume. It is ideal where clear differences in density exist between cells. Cell density may be affected by the gradient medium (e.g., metrizamide, Fig. 12.11; by the position of the cells in the growth cycle, and by serum, Fig. 12.12). This type of separation can be done on any centrifuge, as high *g* forces are not required, and can even be performed at 1 *g*.

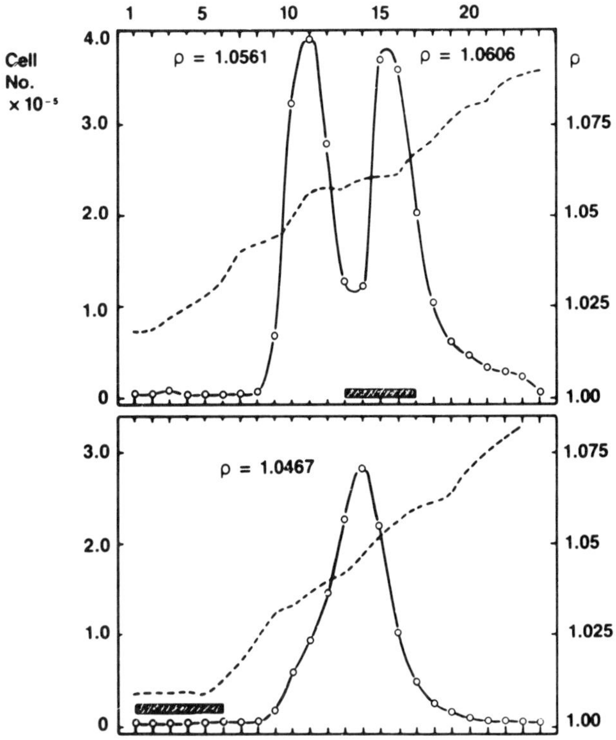

Fig. 12.11. Incorporation of metrizamide (Nygaard) into cells during isopyknic centrifugation. MRC-5 cells (human diploid embryonic lung fibroblasts) were layered in metrizamide-containing medium of the appropriate density in the center of the gradient or in medium alone at the top of the gradient. After centrifugation, the cells were eluted and counted, and samples were taken from each fraction to determine the density. [Reproduced from Freshney, 1976a, with permission of the publisher.] Slashed bars, position of cells at start.

FLUORESCENCE-ACTIVATED CELL SORTING

This technique [Herzenberg et al., 1976; Watson and Erba, 1992] operates by projecting a single cell stream through a laser beam in such a way that the light scattered from the cells is detected by a photomultiplier and recorded (Figs. 12.13, 12.14). If the cells are pretreated with a fluorescent stain (e.g., propidium iodide or Chromomycin A3 for DNA) or fluorescent antibody, the fluorescence emission excited by the laser is detected by a second photomultiplier tube. This information is processed and displayed as a two- (Fig. 12.15) or three-dimensional graph on an oscilloscope. If specific coordinates are then set to delineate sections of the display, the cell sorter will divert cells with the properties that would place them within these coordinates (e.g., high or low light scatter, high or low fluorescence) into a receiver tube placed below the cell stream. The cell stream is deflected by applying a charge to it as it passes between two oppositely charged plates. The charge is applied briefly and at a set time after the cell has cut the laser beam such that only one cell is deflected into the receiver. A low cell concentration in the cell stream is required such that the gap between cells is sufficient to prevent two cells being deflected together [Watson and Erba, 1992].

All cells having similar properties will be collected into the same tube. A second set of coordinates may be set and a second group of cells collected simultaneously into a second tube by changing the polarity of the cell stream and deflecting the cells in the oppo-

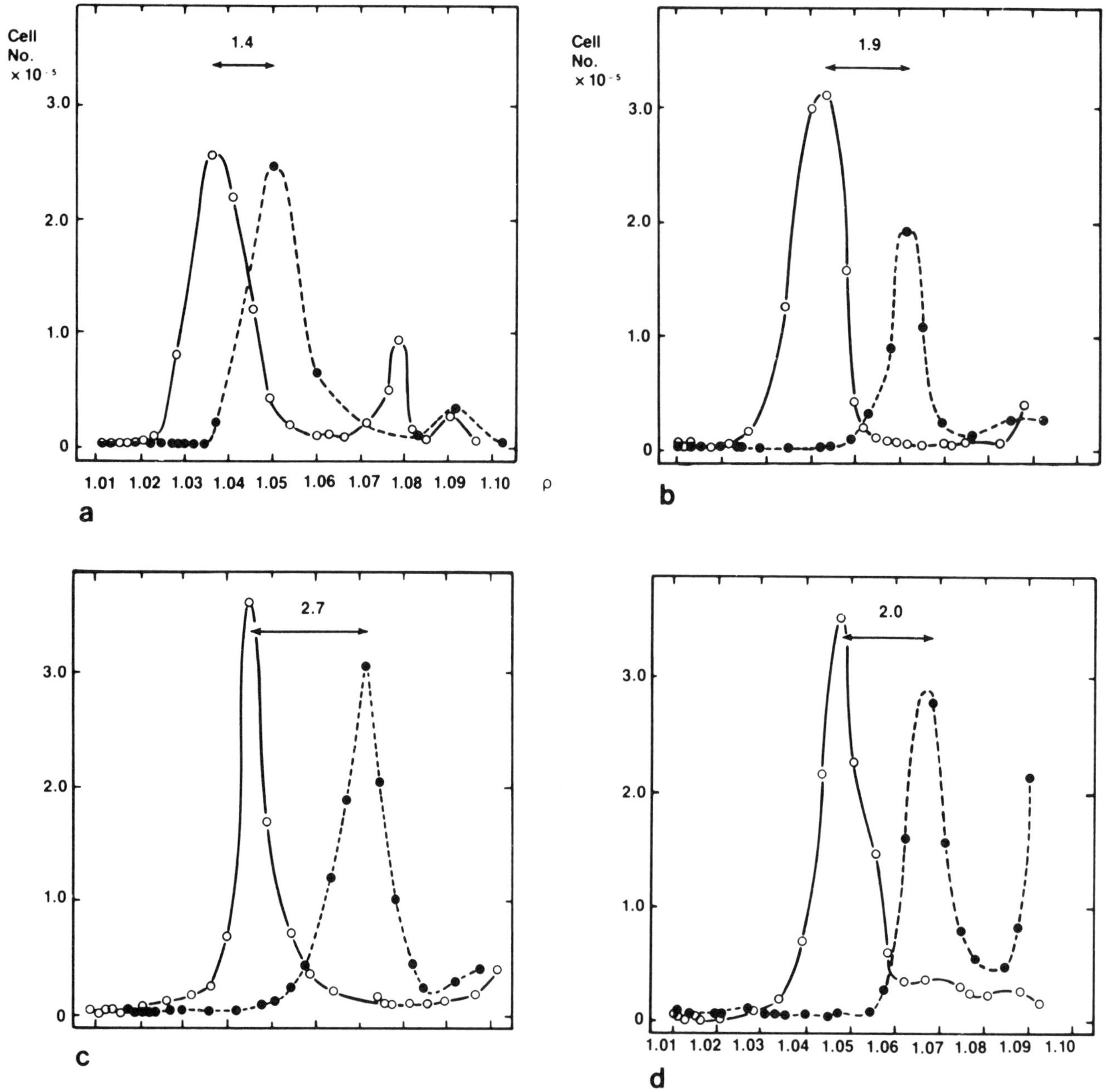

Fig. 12.12. Sedimentation profiles of HeLa and MRC-5 cells, centrifuged to equilibrium in gradients of metrizamide in culture medium. a,b. Cells taken from log phase of growth. c,d. Cells taken from plateau phase. Gradients in (a) and (c) contained no serum; those in (b) and (d), 10% fetal bovine serum. Numbers over the arrows are the differences in density between the peaks, multiplied by 100. Solid circles, HeLa; open circles, MRC-5. [From Freshney, 1976a, with permission of the publisher.]

site direction. All remaining cells will be collected in a central reservoir.

This method may be used to separate cells with any differences that may be detected by light scatter (e.g., cell size) or fluorescence (e.g., DNA, RNA, protein, enzyme activity, specific antigens) and has been used for a wide range of cell types. It has probably been used most extensively for hemopoietic cells [Battye and Shortman, 1991] where disaggregation into the obligatory single cell suspension is relatively simple, but has also been used for solid tissues, e.g., lung [Aitken et al., 1991]. It is an extremely powerful tool but limited by cell yield (about 10^7 cells is a reasonable maximum) and the very high cost of the instrument (approx-

Fig. 12.13. *Fluorescence-Activated Cell Sorter (FACS). This is the Becton Dickinson version of the flow cytophotometer. a. Flow chamber panel on left and computer/readout on right. b. Closeup of flow chamber compartment (see also Fig. 12.14).*

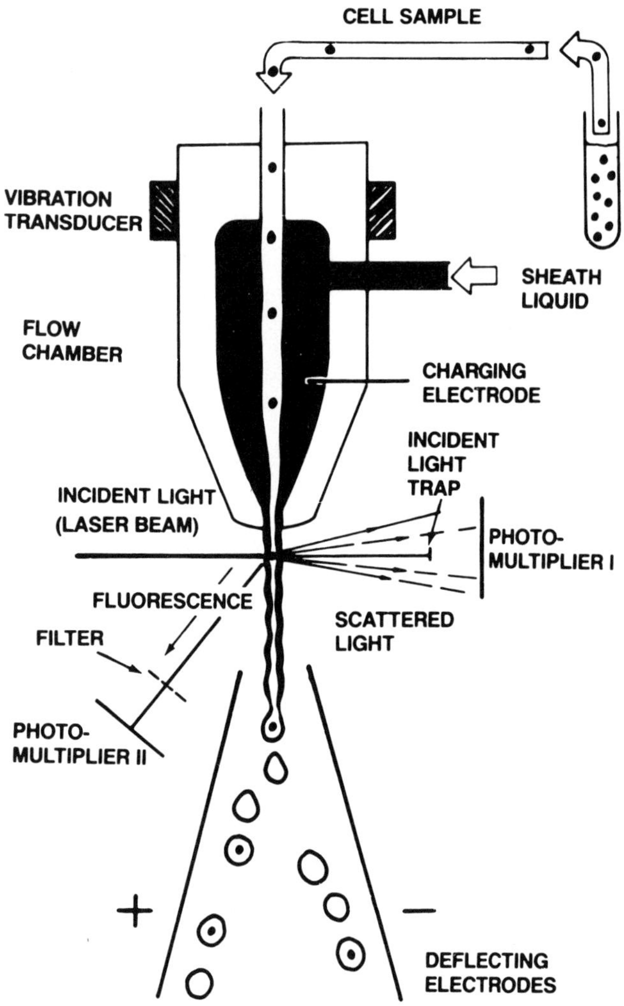

Fig. 12.14. *Principle of operation of flow cytophotometer (see text). [Reproduced from Freshney et al., 1982a, with permission of the publisher.]*

imately $200,000). It also requires a full-time, skilled operator.

New generation machines (e.g., FacsMate, Becton Dickinson) are simpler to operate but are designed principally as analytical tools and are usually not equipped to collect cells.

OTHER TECHNIQUES

The many other techniques that have been used successfully to separate cells are too numerous to describe in detail. They are summarized below and listed in Table 12.1.

Electrophoresis either in a Ficoll gradient [Platsoucas et al., 1979] or by curtain electrophoresis; the second technique is probably more effective and has been used to separate kidney tubular epithelium [Kreisberg et al., 1977].

Affinity chromatography on antibody [Varon and Manthorpe, 1980; Au and Varon, 1979] or plant lectins [Pereira and Kabat, 1979] bound to nylon fiber [Edelman, 1973] or Sephadex (Pharmacia). These techniques appear to be useful for fresh blood cells but less so for cultured cells.

Counter current distribution [Walter, 1975, 1977; Sharpe, 1988] has been used to purify murine ascites tumor cells, with reasonable viability.

Panning has been used successfully with a number of different cell types [Silvestri et al., 1991; Murphy et al., 1992]. A cell type-specific antibody is conjugated to the bottom of a petri dish and, when the mixed cell population is added to the dish, the cells to which the antibody is directed will attach rapidly to the bottom of the dish. The remainder can then be removed. This can be used positively to select a specific subset of cells, which can be released subsequently by mechanical detachment or light trypsinization, or negatively, to remove unwanted cells. The following protocol has been provided by Tessa Holyoake, CRC Beatson Laboratories, Garscube Estate, Bearsden, Glasgow, G61 1BD, Scotland.

Immune "Panning" Technique for Separation of Human Hemopoietic Progenitor Cells

Human hemopoietic progenitor cells can be separated by various techniques that include fluorescence-activated cell sorting (FACS) [Battye and Shortman, 1991], magnetic activated cell sorting (MACS) [Berloncello et al., 1991], the use of monoclonal antibody coated immunomagnetic microspheres [Gaudernack et al., 1986], and immune panning. Panning is the simplest of these methods and has been applied over the last decade to separate different cellular types on the basis of their immune phenotype. The vast majority of panning methods are derived from the work of Wysocki and Sata [1978], who developed panning for separation of lymphocyte populations. The CD34 antigen is present on immature hemopoietic precursor cells and all hemopoietic colony-forming cells (CFU-C) in bone marrow and blood, including lineage restricted (CFU-GM, BFU-E) and pluripotent progenitors (CFU-GEMM, CFU-Mix, CFU-Blast). Isolation of $CD34^+$ cells enables analysis of the direct effects of cytokines in the absence of accessory cells. In addition, the successful analysis of the molecular events that control hemopoietic progenitor cell development depends on purity and uniformity of target cells. Panning can be adapted for positive selection or negative depletion of target cells.

This protocol describes the use of AIS MicroCellector T-25 cell culture flasks (Applied Immune Sciences, CA) for the selection of $CD34^+$ cells from bone marrow mononuclear cells (BMMNC) or peripheral blood

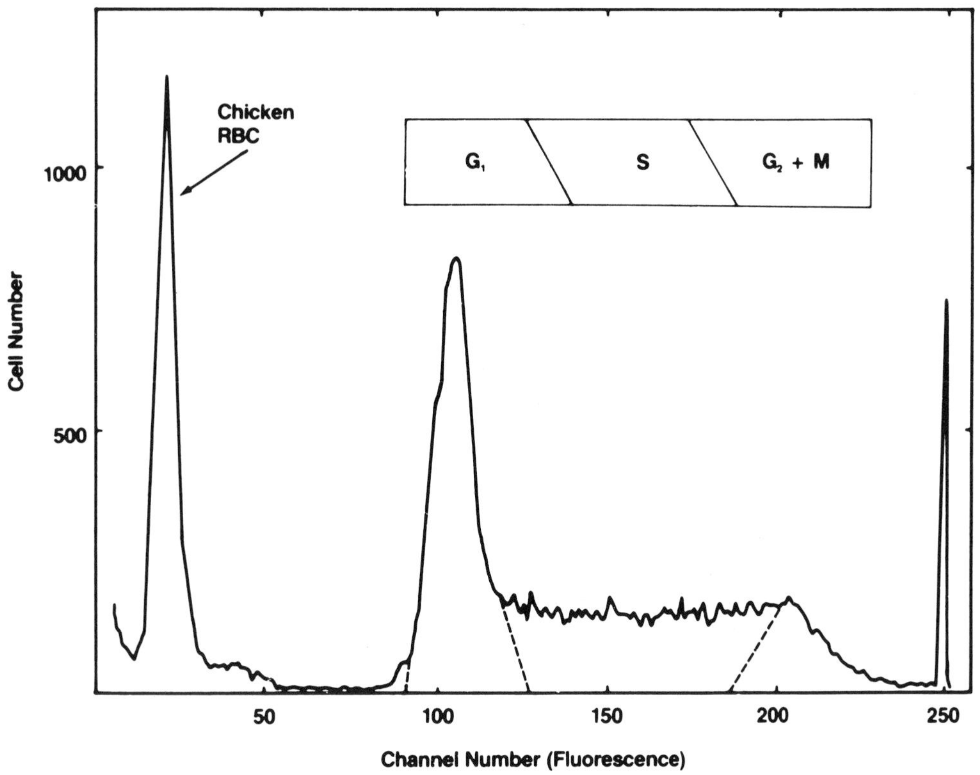

Fig. 12.15. Printout from FACS II. Friend (murine erythroleukemia) cells were fixed in methanol as a single cell suspension and stained with Chromomycin A_3. The cell suspension was then mixed with chicken erythrocytes, fixed, and stained in the same way, and run through the FACS II. The printout plots cell number on the vertical axis and channel number (fluorescence) on the horizontal axis. Since fluorescence is directly proportional to the amount of DNA per cell, the trace gives a distribution analysis of the cell population by DNA content. The lowest DNA content is found in the chicken erythrocytes, included as a standard. The major peak around channel 100 represents those cells as the G_1 phase of the cell cycle. Cells around channel 200 are, therefore, G_2 and metaphase cells (double the amount of DNA per cell) and cells with intermediate values are in S (DNA synthetic) phase (see also Chapters 13 and 23). Cells accumulated in channel 250 are those where DNA value is off scale or cells that have formed aggregates. (Courtesy of Dr. B.D. Young.)

MNC (PBMNC). From 0.5% to 3% of BMMNC are CD34 positive and this figure may be greater in cord blood. Soybean agglutinin (SBA) flasks are used as a pre-enrichment step. SBA binds to galactose and other sugar residues on cells and is very effective in removing certain T-, B-, myeloid, fibroblast, and stromal cells. From 70% to 90% of BMMNC bind to SBA flasks, leaving a nonadherent fraction that is 5–15% CD34 positive. From 70% to 100% of human progenitor cells are recovered in the SBA^- fraction.

Materials

Sterile:

AIS MicroCELLector T-25 cell culture flasks, both SBA and CD34 (Applied Immuno Sciences)
15-ml centrifuge tubes
PE (EDTA is used in all buffers to prevent platelet aggregation)
Fetal bovine serum (FBS)
Ficoll-Hypaque (or equivalent)
Filter, 0.2 μm
Human gamma globulins: Cohn fraction II, III (Sigma) or Gamimune N (Miles, Inc./Cutter Biological)
PE (see Reagent Appendix), 150 ml per experiment
PE containing 0.5% human gamma globulins (PEGG), 10 ml:

1. Dilute human gamma globulin to 0.5% in PBS.
2. Heat solution to 56°C for 30 min.
3. Cool to 2–10°C.
4. Add EDTA to bring final concentration to 1 m*M*.

5. Filter sterilize.
Make up fresh each time.

0.5% human gamma globulin functions by preventing nonspecific adherence of Fc receptor positive cells.

PE containing 10% FBS, 15 ml per experiment

Note. (i) All manipulations must be performed under a laminar flow biosafety hood.

(ii) Flasks must be used within 2 hr of priming.

(iii) Avoid scratching the binding surface of the flasks.

(iv) The binding surface must remain moist during priming.

Protocol

Priming procedure

1. Add 10 ml PE to the SBA flask and swirl for 30 s to moisten all interior surfaces.
2. Incubate the flask, binding surface down, for 1 hr at room temperature (RT).
3. After 1 hr, shake the flask and aspirate PE.
4. Add a further 10 ml PE to the flask, shake for 30 s, then aspirate PE. Repeat this washing procedure twice more and leave the last PE wash in the flask until cells are ready to be added.
5. Just prior to initiating the cell separation procedure, aspirate PE.

Preparation of BMMNC or PBMNC

1. Isolate the MNC fraction from bone marrow or peripheral blood using a Ficoll-Hypaque step gradient (see Chapter 23).
2. Determine cell count and assess viability (must be >90%).
3. Resuspend cells in 4 ml PE containing 0.5% human gamma globulins.
4. Incubate cell suspension for 15 min at room temperature.

Selection procedure for SBA$^-$ fraction

1. Pipette the cell suspension up and down to create a homogeneous preparation and introduce slowly into the SBA flask. For high-efficiency removal of target cells, load no more than 2×10^7 total cells per flask.
2. Incubate for 1 hr at room temperature.
3. At the end of 1 hr, gently rock the flask from side to side for 10 s to resuspend the nonadherent cells.
4. Remove the nonadherent cells and retain.
5. Rinse the flask twice using 4 ml PE and add to the nonadherent cell suspension.

Selection procedure for CD34$^+$ fraction

1. Centrifuge the nonadherent SBA$^-$ cell fraction for 10 min at 1,200 g
2. Resuspend in 4 ml PEGG.
3. After priming the CD34 flask, add the 4 ml nonadherent cell suspension.
4. Incubate for 1 hr at RT and remove nonadherent CD34$^-$ fraction exactly as for SBA flask.
5. Add 4 ml PE containing 10% FBS to the flask with the adherent cells.
6. Firmly hit the narrow side of the flask with your hand two to three times to release the adherent cells.
7. Remove the released cell suspension and rinse the flask twice using PE plus 10% FBS to recover additional cells. Ten-percent FBS improves recovery of detached CD34$^+$ cells.

SBA$^-$ CD34$^+$ cells should be >90% viable and enriched 10–50-fold in CFU-C activity. These cells, when analyzed using flow cytometry, show low to medium forward scatter and low 90° right scatter properties. The expected recovery for clonogenic cells should be around 30%.

Variations. (1) Goat anti-mouse IgG (Fab specific, Sigma) at 250 μg/ml is added to 100-mm tissue culture petri dishes (5 ml/dish) and incubated overnight at 4°C. Dishes are washed before use. After a pre-enrichment step, add My10 antibody (CD34) (Becton Dickinson) to the nonadherent cells and incubate for 30 min at 4°C. This cell suspension is then added to the anti-mouse Ig-coated petri dishes and incubated for 60 min at 4°C. This technique allows positive selection of adhering CD34$^+$ cells by indirect immune adherence [Silvestri et al., 1991].

(2) Anti-immunoglobulin (Ig) plates are prepared by incubating culture plates with affinity-purified goat anti-mouse Ig (Zymed) overnight at 4°C. MNC are incubated for 30 min with various monoclonal antibodies for negative depletion. This cell suspension is then added to the Ig plates and incubated for 1 hr. The procedure is performed twice to ensure removal of all antibody-labeled cells. This procedure allows negative depletion rather than positive selection of target cells and can be adapted for murine progenitor cells [Imagawa et al., 1989].

Magnetic Sorting

Magnetic sorting can be achieved by conjugating specific antibody to ferritin beads (Dynabeads, Dynal), mixing the cell suspension with the beads, and then running the suspension past a magnet that draws the cells which have attached to the beads onto the side of the separating chamber. The cells and beads are released when the current is switched off, and the cells may be separated from the beads by trypsinization or vigorous pipetting. Several cell types have been separated by this method, including bone-marrow purging of leukemic cells [Trickett et al., 1990] and kidney tubular epithelium [Pizzonia et al., 1991].

As so many techniques exist, it is difficult for the novice to know where to start. It is best to start with a simple technique such as velocity sedimentation at unit gravity or density gradient centrifugation, and, if necessary, the two may be used in tandem, in a two-stage fractionation. However, if a specific cell surface phenotype can be predicted, then panning on coated dishes or separation via ferritin beads may be the simplest and most effective. If there are still problems of resolution or yield, then it may be necessary to employ high-technology methods such as centrifugal elutriation, or fluorescence-activated cell sorting (FACS). Of all of these, FACS will probably give the purest cell population, based on quite stringent criteria.

Where a high-purity cell suspension is required and selective culture is not an option, then it will probably be necessary to employ a two- or even multistep fractionation, analogous to the purification of proteins. In many such procedures, density gradient separation is used as a first step, with panning or ferritin bead as a second, and the final purification on a flow cytometer, such as has been used in the isolation of hemopoietic stem cells [Broxmeier, 1993].

CHAPTER 13

Characterization

INTRODUCTION

Some of the methods in general use for cell line characterization are listed in Table 13.19 (see also Hay et al., [1992]). There are four main requirements for cell line characterization:

(1) Correlation with the tissue of origin: (a) identification of the lineage to which the cell belongs; (b) position of the cells within that lineage, i.e., the stem, precursor, or differentiated status; (c) whether transformed or not;

(2) Monitoring for genetic instability and phenotypic variation (see Chapter 17);

(3) Checking for cross-contamination (see Chapter 16) and confirmation of species of origin;

(4) Identification of specific cell lines within a group from the same origin, selected cell strains, or hybrid cell lines, all of which require demonstration of features unique to that cell line or cell strain.

Species Identification

Chromosomal analysis (see below) is one of the best methods for distinguishing between species. Isoenzyme electrophoresis is also a good diagnostic test and is quicker than chromosomal analysis, but requires the appropriate apparatus and reagents (see below). In practice a combination of the two is often used and will give unambiguous results [Hay, 1992]. Recently, techniques have been introduced for "chromosome painting," i.e., using combinations of specific molecular probes to hybridize to individual chromosomes (see below). These probes will identify individual chromosome pairs and are species-specific. The availability of probes is limited to a few species at present, and most are either mouse or human, but it is a good method for distinguishing between human and mouse chromosomes in potential cross-contaminations and interspecific hybrids.

Certain chromosome banding patterns can also be used to distinguish human and mouse chromosomes (see below).

Lineage or Tissue Markers

Cell surface antigens. These markers are particularly useful in sorting hemopoietic cells [Visser and De Vries, 1990] and have also been effective in discriminating epithelium from stroma with antibodies such as anti-EMA [Heyderman et al., 1979] and anti-HMFG 1 and 2 [Burchell and Taylor-Papadimitriou, 1989] and neuroectodermally derived cells (anti-2AB5) [Dickson et al., 1983] from other germ layer-derived cells.

TABLE 13.1. Major Methods for Characterization of Cell Lines and Cell Strains

Criterion	Method	Reference
Karyotype	Chromosome spread with banding	Rothfels and Siminovitch [1958], Rooney and Czepulkowski [1986]
Isoenzyme analysis	Agar or starch gel electrophoresis	Hay [1992]
Cell surface antigens	Immunohistochemistry	Hay [1992], Burchell et al. [1983, 1987]
Cytoskeleton	Immunocytochemistry with antibodies to specific cytokeratins; two-dimensional electrophoresis	Lane [1982], Moll et al. [1982]
DNA fingerprint	Restriction enzyme digest; PAGE; satellite DNA probes	Hay [1992], Jeffreys et al. [1985]

Intermediate filament proteins. These are among the most widely used lineage or tissue markers [Lane, 1982; Ramaekers et al., 1982]. Glial fibrillary acidic protein (GFAP) for astrocytes [Bignami et al., 1980] and desmin [Bochaton-Piallat et al., 1992; Brouty-Boyé et al., 1992] for muscle are the most specific, while cytokeratin will mark epithelial cells and mesothelium. Vimentin, though usually restricted to mesodermally derived cells *in vivo*, can appear in other cell types *in vitro*.

Differentiated products. Hemoglobin for erythroid cells, myosin or tropomyosin for muscle, melanin for melanocytes, and serum albumin for hepatocytes are among the best examples of specific cell type markers, but, like all differentiation markers, depend on the complete expression of the differentiated phenotype.

Enzymes. Three parameters are available in enzymic characterization: (1) the constitutive level (i.e., in the absence of inducers or repressors); (2) the response to inducers and repressors; and (3) isozymic differences (see below). Creatine kinase BB isozyme is characteristic of neuronal and neuroendocrine cells, as is neuron-specific enolase; lactic dehydrogenase is present in most tissues but in different isozymic forms, and a high level of tyrosine aminotransferase, inducible by dexamethasone, is generally regarded as specific to hepatocytes.

Special functions. Transport of inorganic ions and water is characteristic of some epithelia such that, grown as monolayers, they will produce "domes," hemicysts in the monolayer caused by transport of water from the medium to the underside of the monolayer [Rabito et al., 1980](Fig. 13.1g,h). This is found in many types of absorptive and secretory epithelia [Abaza et al., 1974; Lever, 1986]. Other specific functions that can be expressed *in vitro* include muscle contraction and depolarization of nerve cell membrane.

Regulation. Although differentiation is usually regarded as an irreversible process, the level of expression of many differentiated products is under the regulatory control of environmental influences such as hormones, matrix, and adjacent cells (see Chapter 14). Hence the measurement of specific lineage markers may require preincubation of the cells in, for example, a hormone such as hydrocortisone, specific growth factors, or growth of the cells on extracellular matrix of the correct type. Maximum expression of both tyrosine aminotransferase in liver cells and glutamine synthetase in glia require prior induction with dexamethasone. Glutamine synthetase is also repressed by glutamine, so glutamate should be substituted in the medium 48 hr before assay [DeMars, 1957].

Lineage fidelity. Although many of the markers described above have been claimed as lineage markers, they are more properly regarded as tissue or cell type markers, as they are often more characteristic of the function of the cell than its embryologic origin. Cytokeratins occur in mesothelium and kidney epithelium, although both of these derive from the mesoderm. Neuron-specific enolase and creatine kinase BB are expressed in neuroendocrine cells of the lung, although these are now recognized to derive from the endoderm and not from neuroectoderm as one might expect of neuroendocrine-type cells.

Unique Markers

Unique markers include specific chromosomal aberrations, enzymic deficiencies, isoenzymes, and drug resis-

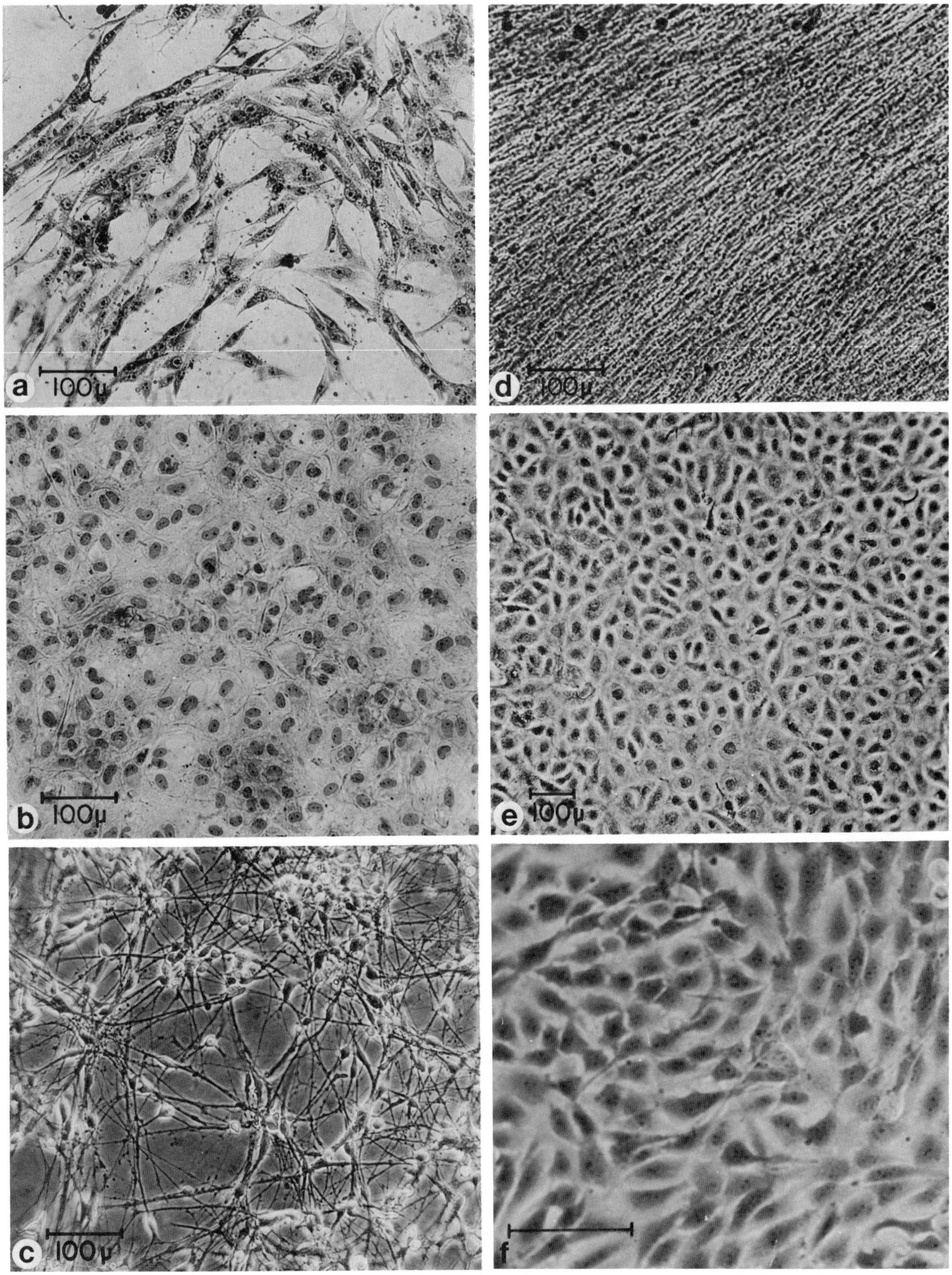

Fig. 13.1. *Examples of variations in cell morphology. a. BHK-21 (baby hamster kidney fibroblasts), clone 13, in log growth. The culture is not confluent, and the cells are well spread and randomly oriented (although some orientation is beginning to appear). b. Cells of an epithelial-like morphology from fetal human intestine (FHI). c. Astrocytes from human astrocytoma. This pattern is quite characteristic but is lost as the cells are passaged, and a morphology not unlike (b),*

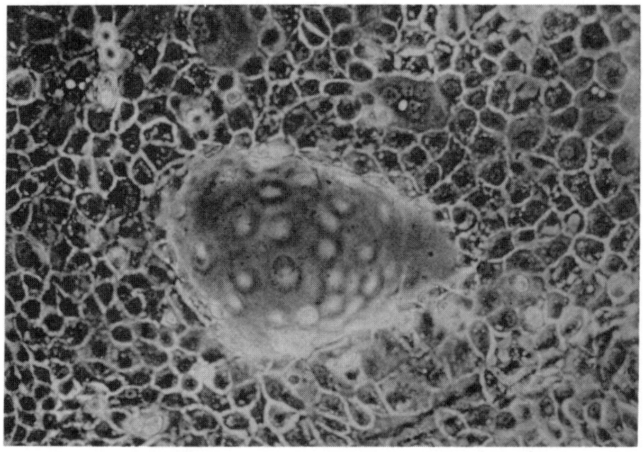

g

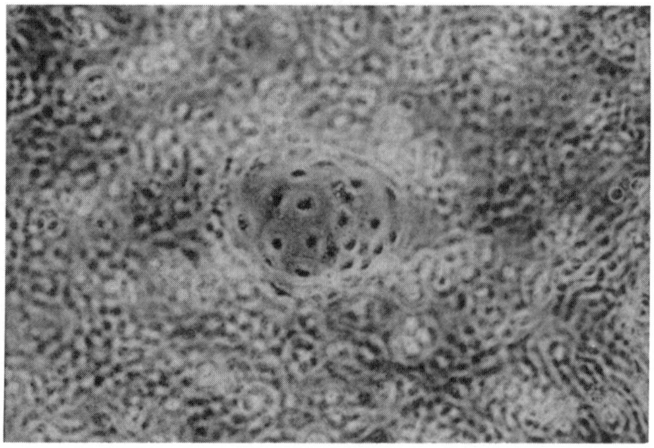

h

Fig. 13.1. (Continued)
(e), or (f) develops. d. Plateau-phase BHK-21 C13 cells. The cells are smaller, more highly condensed, and have assumed a parallel orientation with each other. e. Bovine aortic endothelium. Similar regular appearance to (b), though cells are more closely packed. f. Again a similar pavement-like appearance as found in (b) and (c) but now produced by 3T3 cells, mouse fibroblasts. With experience, these cell types may be distinguished, but their similarity underlines the need for criteria for identification other than morphology. g. "Dome" formed in an epithelial monolayer by downward transport of ions and water, lower focus (on monolayer). h. Upper focus (top of "dome") from (g). (a,b,d) Giemsa stained. (c,e–h) Phase contrast. (e) by courtesy of Dr. P. Del Vecchio.

tance, but the most specific of these is DNA fingerprinting (see below, Stacey protocol).

Transformation

Transformation is dealt with in Chapter 15.

MORPHOLOGY

Observation of morphology is the simplest and most direct technique used to identify cells. It has, however, certain shortcomings that should be recognized. Most of these are related to the plasticity of cellular morphology in response to different culture conditions; e.g., epithelial cells growing in the center of a confluent sheet are usually regular, polygonal, and with a clearly defined edge, while the same cells growing at the edge of a patch may be more irregular, distended, and, if transformed, may break away from the patch and become fibroblastoid in shape. Subconfluent fibroblasts from hamster kidney or human lung or skin assume multipolar or bipolar shapes and are well spread on the culture surface, but at confluence they are bipolar and less well spread. They also form characteristic parallel arrays and whorls that are visible to the naked eye. Mouse 3T3 cells and human glial cells grow like multipolar fibroblasts at low cell density but become epithelial-like at confluence (Fig. 13.1). Alterations in the substrate [Gospodarowicz et al., 1978b; Freshney, 1980], and the constitution of the medium [Coon and Cahn, 1966] can also affect cellular morphology. Hence, comparative observations should always be made at the same stage of growth and cell density in the same medium, and growing on the same substrate (see Fig. 13.1).

The terms "fibroblastic" and "epithelial" are used rather loosely in tissue culture and often describe the appearance rather than the origin of the cells. Thus a bipolar or multipolar migratory cell, the length of which is usually more than twice its width, would be called "fibroblastic," while a monolayer cell that is polygonal, with more regular dimensions, and that grows in a discrete patch along with other cells, is usually regarded as "epithelial." However, where the identity of the cells has not been confirmed, the terms "fibroblast-like" or "fibroblastoid" and "epithelial-like" or "epithelioid" should be used.

Frequent brief observations of living cultures, preferably with phase-contrast optics, are more valuable than infrequent stained preparations studied at length. They will give a more general impression of the cell's morphology and its plasticity and will also reveal differences in granularity and vacuolation that bear on the health of the culture. Unhealthy cells often become granular and then display vacuolation around the nucleus (see Fig. 10.1).

It is useful to keep a set of photographs for each cell line as a record in case a morphological change is suspected. This record can be supplemented with photographs of stained preparations and an autoradiograph from a DNA fingerprint.

Staining

A polychromatic blood stain, such as Giemsa, provides a convenient method of preparing a stained culture. The recommended procedure is as follows.

Outline

Fix the culture in methanol and stain directly with Giemsa. Wash and examine wet.

Materials

BSS
Undiluted Giemsa stain
Methanol
Deionized water

Protocol

1. Remove medium and discard.
2. Rinse monolayer with BSS and discard rinse.
3. Add BSS:methanol, 1:1, 5 ml per 25 cm^2. Discard 50% methanol/BSS mixture and replace with fresh methanol. Leave for 10 min.
4. Rinse monolayer and discard 50% methanol/BSS mixture and replace with fresh methanol. Leave for 10 min.
5. Discard methanol and replace with fresh anhydrous methanol, rinse monolayer, and discard methanol.
6. At this point, the flask may be dried and stored or stained directly. It is important that staining should be done directly from fresh anhydrous methanol even with a dry flask. If the methanol is poured off and the flask is left for some time, water will be absorbed by the residual methanol and will inhibit subsequent staining. Even "dry" monolayers can absorb moisture from the air.
7. Add neat Giemsa stain, 2 ml per 25 cm^2, making sure the entire monolayer is covered and remains covered.
8. After 2 min, dilute stain with 8 ml water and agitate gently for a further 2 min.
9. Displace stain with water so the scum that forms will be floated off and not left behind to coat the cells. Wash vigorously in running tap water until any pink cloudy background stain (precipitate) is removed but stain is not leached out of cells.
10. Pour off water, rinse in deionized water, and examine on microscope while monolayer is still wet. Store dry and rewet to examine.

Note. Giemsa staining is a simple procedure giving a good high-contrast polychromatic stain, but precipitated stain may give a spotted appearance to the cells. This occurs (1) due to oxidation at the surface of the stain forming a scum, and (2) throughout the solution, particularly on the surface of the slide when water is added. Washing off stain by replacement, rather than pouring off or removing slides, is designed to prevent slides from coming in contact with scum. Vigorous washing at the end is designed to remove precipitate left on the slide.

Culture Vessels for Cytology: Monolayer Cultures

(1) Regular 25-cm^2 flasks or 50-mm petri dishes.

(2) Coverslips (glass or Thermanox [Lux]) in multiwell dishes (see Fig. 7.4), petri dishes, or Leighton tubes (Bellco, Costar; see Figs. 7.5, 7.6, 13.2a).

(3) Microscope slides in 90-mm petri dishes or with attached multiwell chambers (Lab-Tek, Bellco; Fig. 13.2b).

(4) Petriperm dishes (Heraeus), cellulose acetate or polycarbonate filters, Melinex, Thermanox, and Teflon-coated coverslips have all been used for EM cytology studies. Some pretreatment of filters or Teflon may be required (e.g., gelatin, collagen, fibronectin, Matrigel, or serum coating; see Chapter 7).

(5) The Gabridge chamber/dish (Bionique, Northumbria Biologicals) can be used with a variety of different plastic membranes or standard glass or plastic coverslips, allowing culture directly on a thin, optically clear surface for high-resolution microscopy (Fig. 13.2).

Suspension Culture

The following are four ways of preparing cytological specimens from suspension cultures.

Smear (as used in the preparation of blood films).

Materials

Concentrated cell suspension
Serum
Microscope slides
Methanol

Protocol

1. Place a drop of cell suspension 10^6 cells/ml or more, in 50–100% serum, on one end of a slide. Dip the end of a second slide into the drop and move it up the first slide, distributing a thin film of cells on the slide (Fig. 13.3).
2. Dry off quickly and fix in methanol.

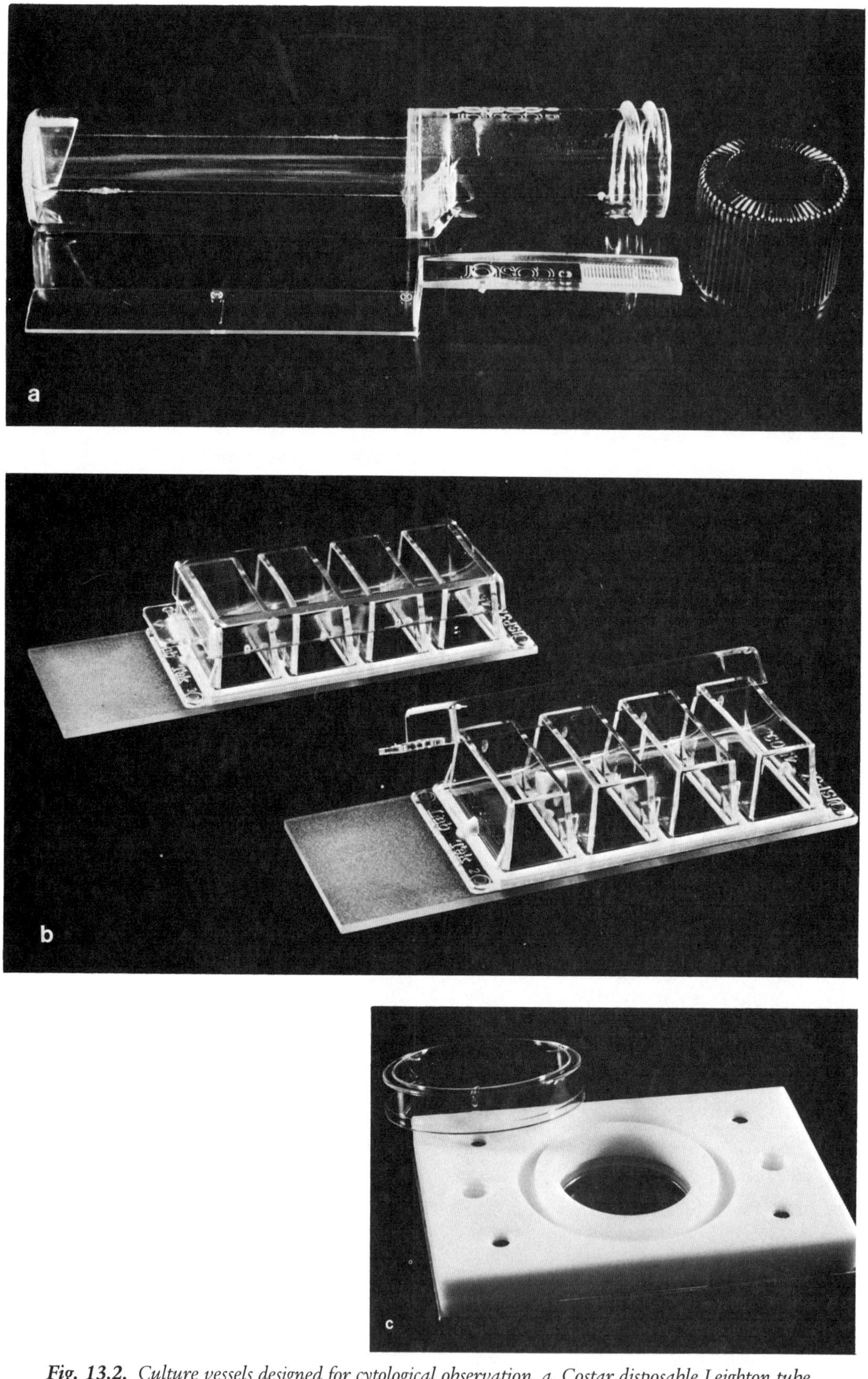

Fig. 13.2. *Culture vessels designed for cytological observation. a. Costar disposable Leighton tube with coverslip with handle for easy retrieval. b. Lab-Tek plastic chambers on regular microscope slide. One, two, four, eight, and sixteen chambers per slide are available (Nunc). c. Reusable chamber to take a regular glass coverslip (Bionique) (Courtesy of Dr. M. Gabridge).*

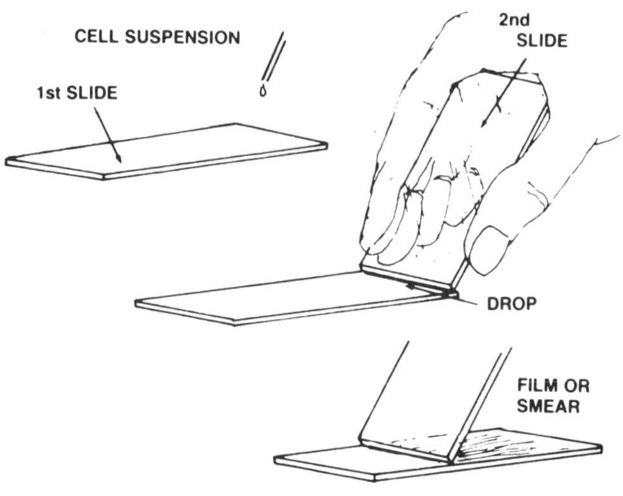

Fig. 13.3. Preparing a cell smear. Top: A drop of cell suspension in serum or serum-containing medium is placed on a slide. Center and bottom: A second slide is used to spread the drop.

◇ Gloves should be worn if human or other primate cells are being handled.

Centrifugation. The Cytospin (Shandon) (Fig. 13.4) is a centrifuge with sample compartments, specially designed to spin cells down onto a microscope slide, and located within an enclosed rotor.

Materials

Same as for smear preparation.

Protocol

1. Place approximately 100,000 cells in 250 μl of medium in at least two sample blocks.
2. Switch on and spin the cells down onto the slide at 100 *g* for 5 min.
3. Dry off slide quickly and fix in MeOH.

   Note. Some centrifuges (e.g., Damon-IEC) have centrifuge buckets designed for preparing cytological preparations (Fig. 13.5). Fixation in this case is performed *in situ*. The design of the Cytocentrifuge chamber appears to confer advantages on cell recovery.

   ◇ Rotor must be sealed during spinning if human or other primate cells are being used.

Drop technique. Same as for chromosome preparation (see below) but omit Colcemid and hypotonic treatment.

Filtration. This technique is used in exfoliative cytology (see manufacturer's instructions for further details: Gelman, Millipore, Nuclepore).

Materials

Filters (e.g., 25-mm Nuclepore, 0.5-μm porosity)
Filter holder stand (Gelman, Nuclepore, Millipore)
Cell suspension (~10^6 cells in 5–10 ml medium with 20% serum)
20 ml BSS
50 ml methanol
Vacuum pump or tap siphon
Giemsa stain
Mountant (DePeX or Permount)

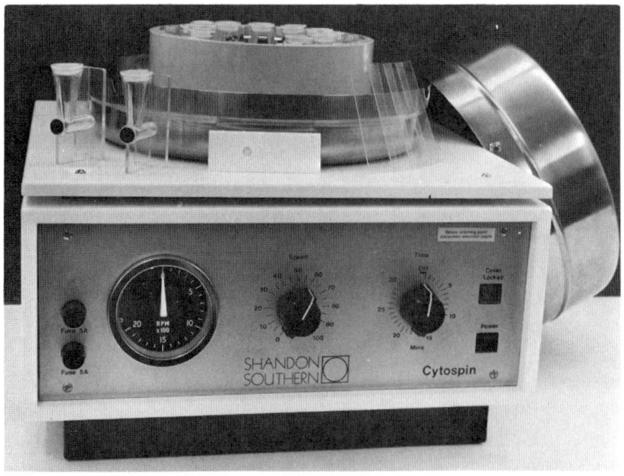

Fig. 13.4. Shandon Cytospin. Centrifuge for making slide preparations from cell monolayers. (Courtesy of Shandon Scientific.)

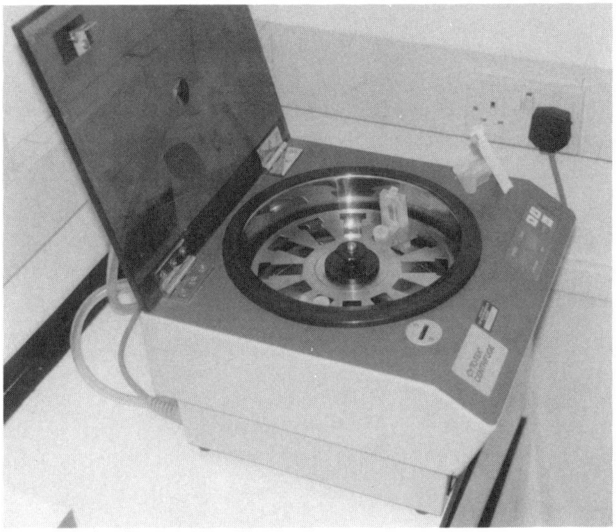

Fig. 13.5. Cytotek centrifuge (Miles Scientific). Similar principle to Shandon Cytospin but with polypropylene slide chambers that prevent contact between the slide, wick, and cell suspension until centrifugation commences.

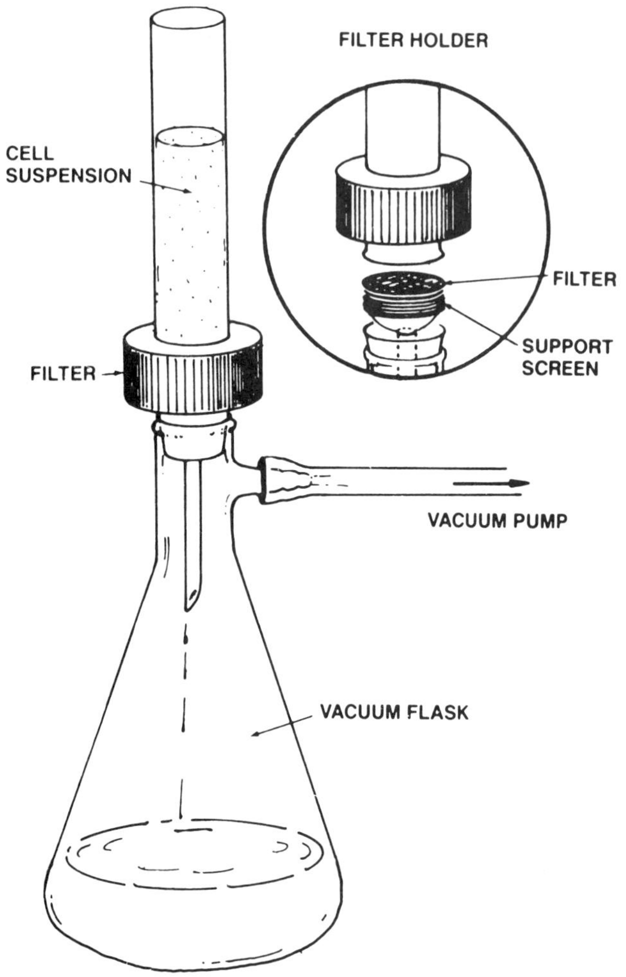

Fig. 13.6. Filter assembly for cytological preparation (see text).

Protocol

1. Set up filter assembly (Fig. 13.6) with 25-mm diameter, 0.5-μm porosity polycarbonate filter.
2. Draw cell suspension onto filter using a vacuum pump. Do not let all the medium run through.
3. Add 10 ml BSS gently when cell suspension is down to 2 ml.
4. Repeat when BSS is down to 2 ml.
5. Add 10 ml methanol to BSS and repeat until pure methanol is being drawn through filter.
6. Switch off vacuum before all the methanol runs through.
7. Lift out filter and air dry.
8. Stain filter in Giemsa and dry.
9. Mount on a slide in DePeX or Permount by pressing the coverslip down to flatten the filter.

Photography

There are two major frame sizes you may wish to consider: 35 mm, best for routine color transparencies and high-volume black and white, and 3½ × 4½ in (12 × 9 cm) Polaroid with positive/negative film (type 665) for low-volume black and white. With Polaroid, you obtain an instant result and know that you have the record without having to develop and print a film. The unit cost is high, but there is a considerable saving in time. However, Polaroid films generally have relatively low contrast and the color saturation is usually not as good as conventional film. Polaroid film also has a much shorter shelf-life.

Specific instructions are supplied with microscope cameras, but a few general guidelines may be useful.

1. Choose and load the film and set the film speed on the exposure meter before bringing out the culture. Make sure the culture is free of debris, e.g., change the medium on a primary culture before photography.
2. Choose the appropriate field quickly, avoiding imperfections or marks on the flask (always label on the side of the flask or dish and not on the top or bottom). Rinse medium over the upper inside surface if condensation has formed and let it drain down before attempting to photograph.
3. Focus the eyepieces to your own eyes, then focus the microscope. Focus the camera eyepiece, if separate, and, finally, critically refocus the microscope.
4. Turn up the light, check the focus and exposure, and expose; then turn down the light immediately to avoid overheating the culture.

   Note. An infrared filter may be incorporated to minimize overheating.
5. If Polaroid, check exposure on finished print and repeat if necessary at a different setting. If 35 mm, bracket the exposure by rephotographing at half and double the exposure.
6. Return the culture to the incubator.
7. Label Polaroid prints immediately. If 35 mm, keep a record against the frame number; otherwise the prints or slides will be difficult to identify.
8. Photograph a stage micrometer slide, at the same settings, to provide a scale bar when processed with your photographs.
9. File photographs in a readily accessible way, e.g., albums or filing cabinet sheets with transparent pockets.
10. Reproduction of photographs for publication is best done on high-contrast, glossy, double-weight paper. Reproduction for theses and reports, where several copies are required, can be achieved quite satisfactorily by laser photocopying. If 2–4 prints are blocked, the unit cost compares very favorably with photographic printing, the time taken is much less, the quality is good, and binding can be achieved without the problems of mounted photographs.

CHROMOSOME CONTENT

Chromosome content is one of the most characteristic and well-defined criteria for identifying cell lines and relating them to the species and sex from which they were derived. See the Committee on Standardized Genetic Nomenclature for Mice [1972] for mouse karyotype; Committee for a Standardized Karyotype of *Rattus norvegicus* [1973] for rat, and Paris Conference [1971 (1975)], or An International System for Human Cytogenetic Nomenclature [1978] for human. There is also an *Atlas of Mammalian Chromosomes* [Hsu and Benirschke, 1967], although it predates chromosome banding (see below). Chromosome analysis can also distinguish between normal and malignant cells, as the chromosome number is more stable in normal cells (except in mice, where the chromosome complement of normal cells can change quite rapidly after explantation into culture).

Chromosome Preparations [Rothfels and Siminovitch, 1958; Rooney and Czepulkowski, 1986]

Outline

Cells arrested in metaphase and swollen in hypotonic medium are fixed and dropped on a slide, stained, and examined (Fig. 13.7).

Materials

Cultures of cells in log phase
10^{-5} M colcemid in BSS
PBSA
0.25% crude trypsin
Centrifuge tubes
Low speed centrifuge
Hypotonic solution:
0.04 M KCl
0.025 M sodium citrate

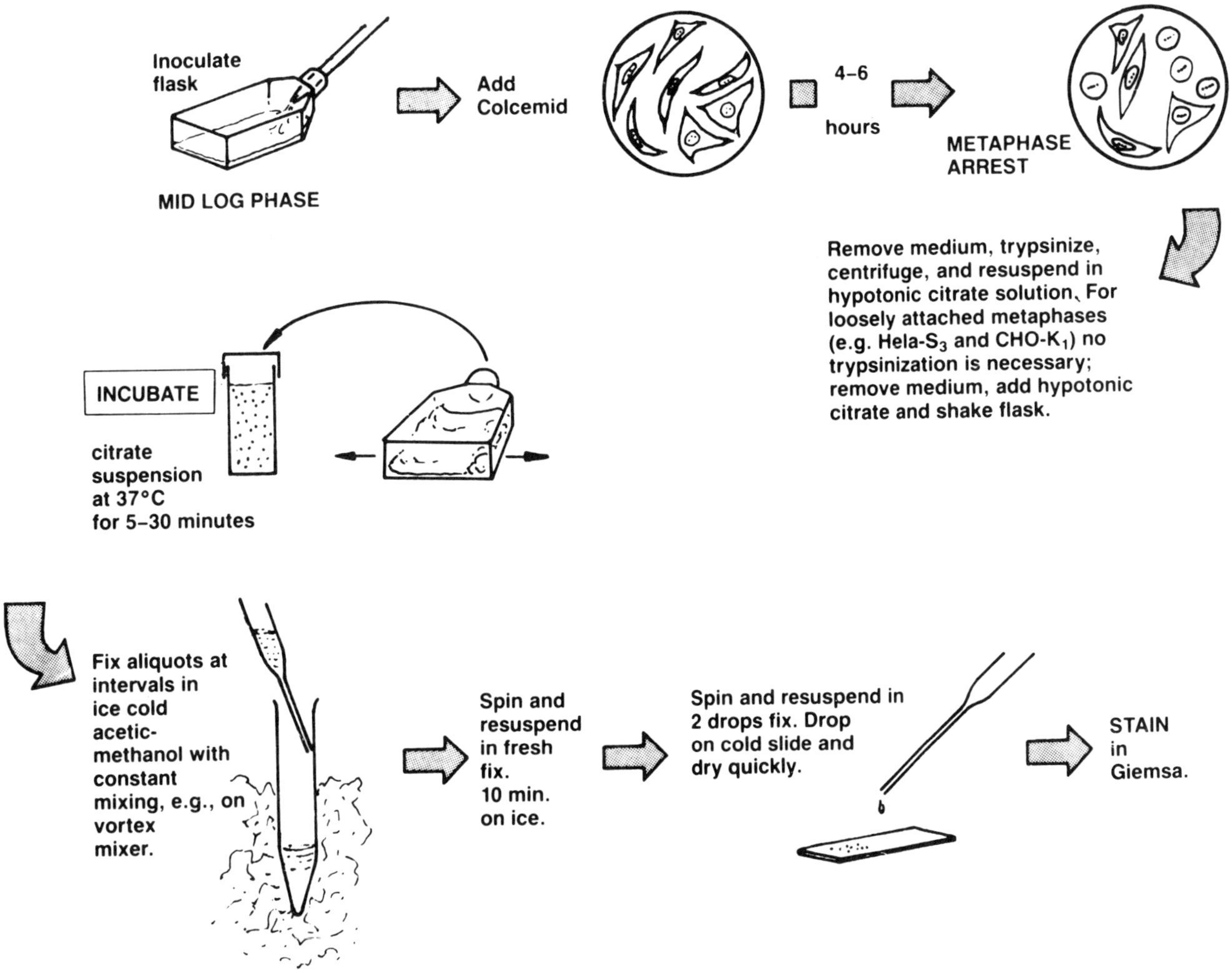

Fig. 13.7. Preparation of chromosome spreads.

Fixative:
1 part glacial acetic acid plus 3 parts anhydrous methanol or ethanol, made up fresh and kept on ice
Vortex mixer
Pasteur pipettes
Ice
Slides
Giemsa stain
Slide dishes
00 coverslips
DePeX or Permount

Protocol

1. Set up 75-cm^2 flask culture at between 2×10^4 and 5×10^4 cells/ml (4×10^3 and 1×10^4 cells/cm^2) in 20 ml.
2. Approximately 3–5 d later, when cells are in the log phase of growth, add colcemid (10^{-7} M, final) to the medium already in the flask.
3. After 4–6 hr, remove medium gently, add 5 ml 0.25% trypsin, and incubate 10 min.
4. Centrifuge cells in trypsin and discard supernatant trypsin.
5. Resuspend the cells in 5 ml of hypotonic solution and leave for 20 min at 36.5°C.
6. Add equal volume of freshly prepared ice-cold acetic methanol, mixing constantly, and then centrifuge at 100 *g* for 2 min.
7. Discard the supernatant mixture, "buzz" the pellet on a vortex mixer, and slowly add fresh acetic methanol with constant mixing.
8. Leave 10 min on ice.
9. Centrifuge for 2 min at 100 *g*.
10. Discard supernatant acetic methanol and resuspend pellet with "buzzing" in 0.2 ml acetic methanol, to give a finely dispersed cell suspension.
11. Draw one drop into the tip of a Pasteur pipette and drop on a cold slide. Let the drop run down the slide as it spreads.
12. Dry off rapidly over a beaker of boiling water and examine on microscope. If cells are evenly spread and not touching, prepare more slides at same cell concentration. If piled up and overlapping, dilute two- to fourfold and make a further drop preparation. If satisfactory, prepare more slides. If not, dilute further and repeat.
13. Stain with Giemsa: (a) Immerse slides in neat stain for 2 min; (b) place dish in sink and add approximately 10 V water, allowing surplus stain to overflow from top of slide dish; (c) leave for a further 2 min; (d) displace remaining stain with running water and finish by running slides individually under tap to remove precipitated stain (pink, cloudy appearance on slide); and (e) check staining under microscope. If satisfactory, dry slide thoroughly and mount 00 coverslip in DePeX or Permount.

Chromosome Banding

This group of techniques [see Rooney and Czepulkowski, 1986] was devised to enable individual chromosome pairs to be identified where there is little morphological difference between them [Wang and Fedoroff, 1972, 1973]. For Giemsa banding, the chromosomal proteins are partially digested by crude trypsin, producing a banded appearance on subsequent staining. Trypsinization is not required for quinacrine banding. The banding pattern is characteristic for each chromosome pair (Fig. 13.8).

A protocol for chromosome banding has been contributed by Marie Ferguson-Smith and is presented in Chapter 23 under "Culture of Amniocytes."

Chromosome Analysis

(1) Count chromosome number per spread for between 50 and 100 spreads (need not be banded). Closed-circuit television or a camera lucida attachment may help. You should attempt to count all the mitoses that you see and classify them (a) by chromosome number or (b), if counting is impossible, as "near diploid uncountable" or "polyploid uncountable." Plot the results as a histogram (see Fig. 12.2).

(2) Prepare karyotype. Photograph about 10 or 20 good spreads of banded chromosomes and print on 20 × 25 cm high-contrast paper. Cut out the chromosomes, sort into sequence, and stick down on paper (see Figs. 13.9, 13.10).

Variations.

Metaphase block. (1) Vinblastine, 10^{-6} M, may be used instead of colcemid. (2) Duration of the metaphase block may be increased to give more metaphases for chromosome counting, but chromosome condensation will increase, making banding very difficult.

Collection of mitosis by "shake-off" technique. Some cells, e.g., CHO and HeLa, detach readily when in metaphase. This allows trypsinization for collection of metaphases to be eliminated. (1) Add colcemid. (2) Remove carefully and replace with hypotonic citrate/KCl. (3) Shake the flask to dislodge cells in metaphase either before or after incubation in hypotonic medium. (4) Fix as before.

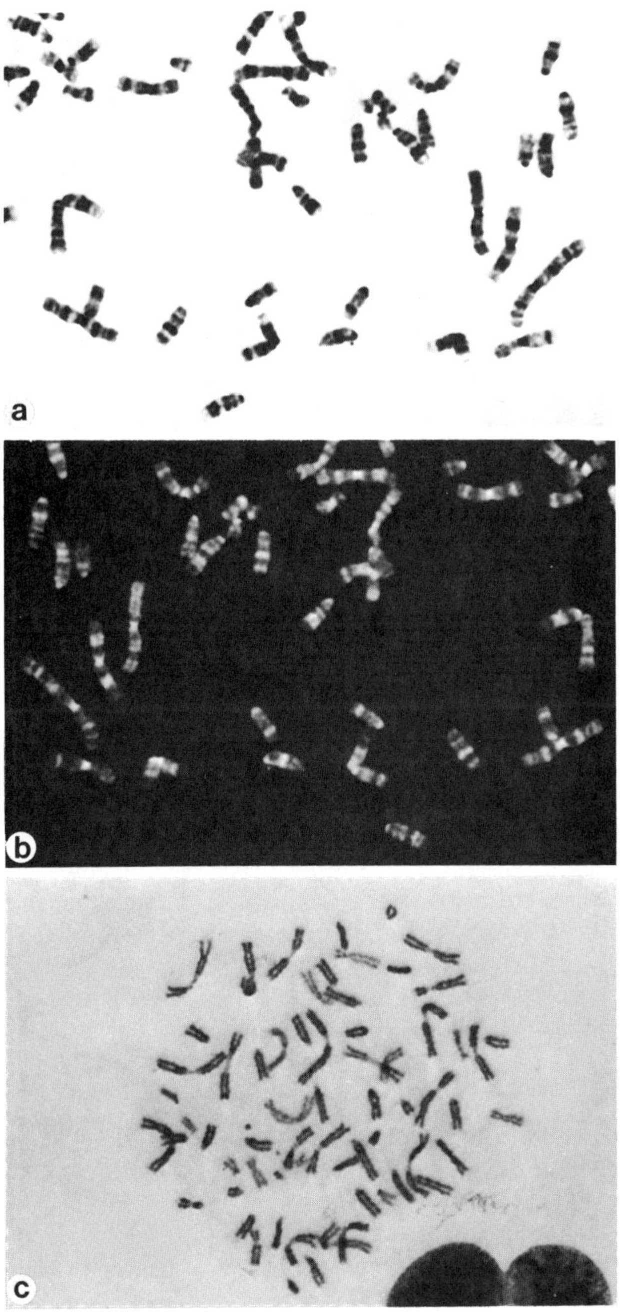

Fig. 13.8. *Chromosome staining. a. Human chromosomes banded by standard trypsin-Giemsa technique. b. Same preparation as (a) stained with Hoechst 33258. c. Human/mouse hybrid stained with Giemsa at pH 11. Human chromosomes are less intensely stained than mouse. Several human/mouse chromosomal translocations can be seen. (Courtesy of Dr. R.L. Church.)*

Hypotonic treatment. Substitute 0.075 M KCl alone or HBSS diluted to 50% with distilled water. Duration of hypotonic treatment may be varied from 5 min to 30 min to reduce lysis or increase spreading.

Spreading. There are perhaps more variations at this stage than any other, all designed to improve the degree and flatness of the spread. They include: (1) dropping cells onto slide from a greater height. Clamp the pipette and mark the position for the slide using a trial run with fixative alone. (2) Flame drying. Dry slide after dropping cells by heating over a flame or actually burn off the fixative by igniting the drop on the slide as it spreads (this may make banding more difficult later). (3) Ultracold slide. Chill slide on solid CO_2 before dropping on cells. (4) Refrigerate fixed cell suspension overnight before dropping. (5) Drop cells on a chilled slide (e.g., steep in cold alcohol and dry off), then place over a beaker of boiling water. (6) Tilt slide or blow drop across slide as it spreads.

Banding. (1) Giemsa-banding: use trypsin + EDTA rather than trypsin alone. (2) Q-banding [Caspersson et al., 1968]: stain in 5% (w/v) quinacrine dihydrochloride in 45% acetic acid, rinse, and mount in deionized water at pH 4.5 [Lin and Uchida, 1973; Uchida and Lin, 1974]. (3) C-banding: this technique emphasizes the centromeric regions. The fixed preparations are pretreated for 15 min with 0.2 *N* HCl, 2 min with 0.07 *N* NaOH, and then treated overnight with SSC (either 0.03 M sodium citrate, 0.3 M NaCl, or 0.09 M sodium citrate, 0.9 M NaCl) before staining with Giemsa stain [Arrighi and Hsu, 1974] (see also Chapter 23).

Techniques have been developed for discriminating between human and mouse chromosomes, principally to aid the karyotypic analysis of human and mouse hybrids. These include fluorescent staining with Hoechst 33258, which causes mouse centromeres to fluoresce more brightly than human [Hilwig and Gropp, 1972; Lin et al., 1974] alkaline staining with Giemsa ("Giemsa-11") [Bobrow et al., 1972; Friend et al., 1976] and *in situ* hybridization (see below and Chapter 23).

Chromosome counting and karyotyping will allow species identification of the cells and, when banding is used, will distinguish cell line variation and marker chromosomes. Banding and karyotyping is time-consuming and chromosome counting with a quick check on gross chromosome morphology may be sufficient to confirm or exclude a suspected cross-contamination.

Chromosome Painting

With the advent of fluorescently labeled probes that will bind to specific regions, and even specific genes, on the chromosomes, it has become possible to locate genes, identify translocations, and determine the species of origin of chromosomes. This technique employs *in situ* hybridization technology, which can also be used for extrachromosomal and cytoplasmic localization of specific nucleic acid sequences, such as might be used to localize specific messenger RNA spe-

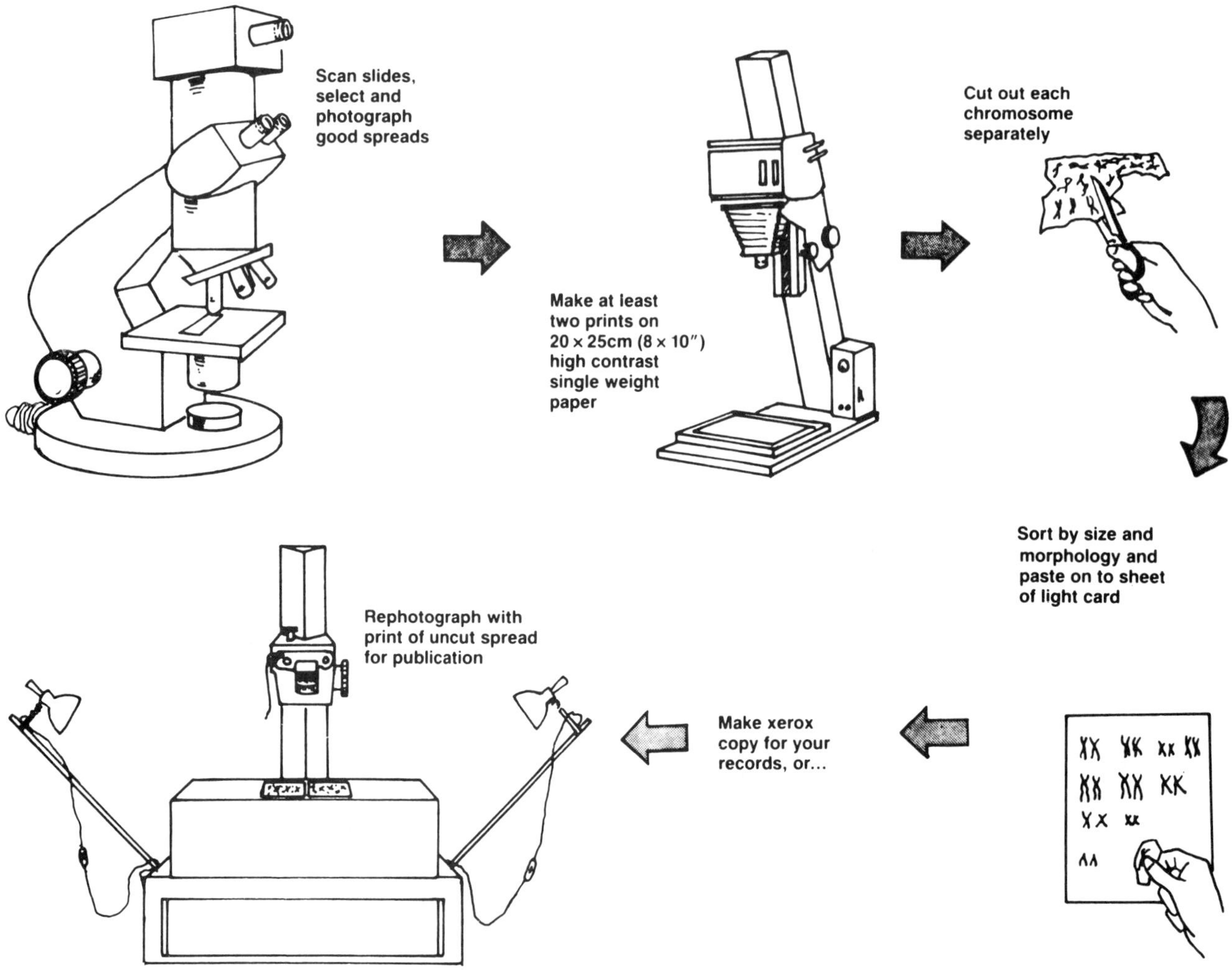

Fig. 13.9. Steps in the preparation of a karotype.

cies. See Chapter 23 for protocols for *in situ* hybridization and chromosome painting.

DNA CONTENT

The amount of DNA per cell is relatively stable in normal cell lines such as human fibroblasts and glia, and chick and hamster fibroblasts, but varies in cell lines from the mouse and from many neoplasms. DNA can be measured by microdensitometry of Feulgen-stained cells [Pearse, 1968] or by ethidium bromide fluorescence and microfluorimetry [Watson and Erba, 1992]. The advent of flow cytometry (fluorescence-activated cell sorting; see Chapter 12) has made the assay of DNA per cell much more quantitative and reproducible (see Fig. 12.15 and Chapter 18), although the generation of the necessary single cell suspension will, of course, destroy the topography of the specimen. Analysis of DNA content is particularly useful in the characterization of transformed cells that are often aneuploid and heteroploid (see Chapter 15).

DNA Hybridization

Using specific molecular probes, it is possible to hybridize to specific regions within the DNA (Southern blotting) and detect the hybrids by radioisotopic, fluorescent, or luminescent labels [Sambrook et al., 1989; Frederick et al., 1993]. Extracted DNA, whole or cut with restriction endonucleases, electrophoresed, blotted onto nitrocellulose, and hybridized to a particular probe or set of probes, will provide information about species-specific regions or amplified regions of the DNA that are characteristic to that cell line. Thus, strain-specific gene amplifications, such as amplification of the dihydrofolate reductase (DHFR) gene, may be detected in cell lines selected for resistance to methotrexate [Biedler et al., 1972], amplification of the MDR gene in vinblastine-resistant cells [Schoenlein et al., 1992], overexpression of a specific oncogene, or on-

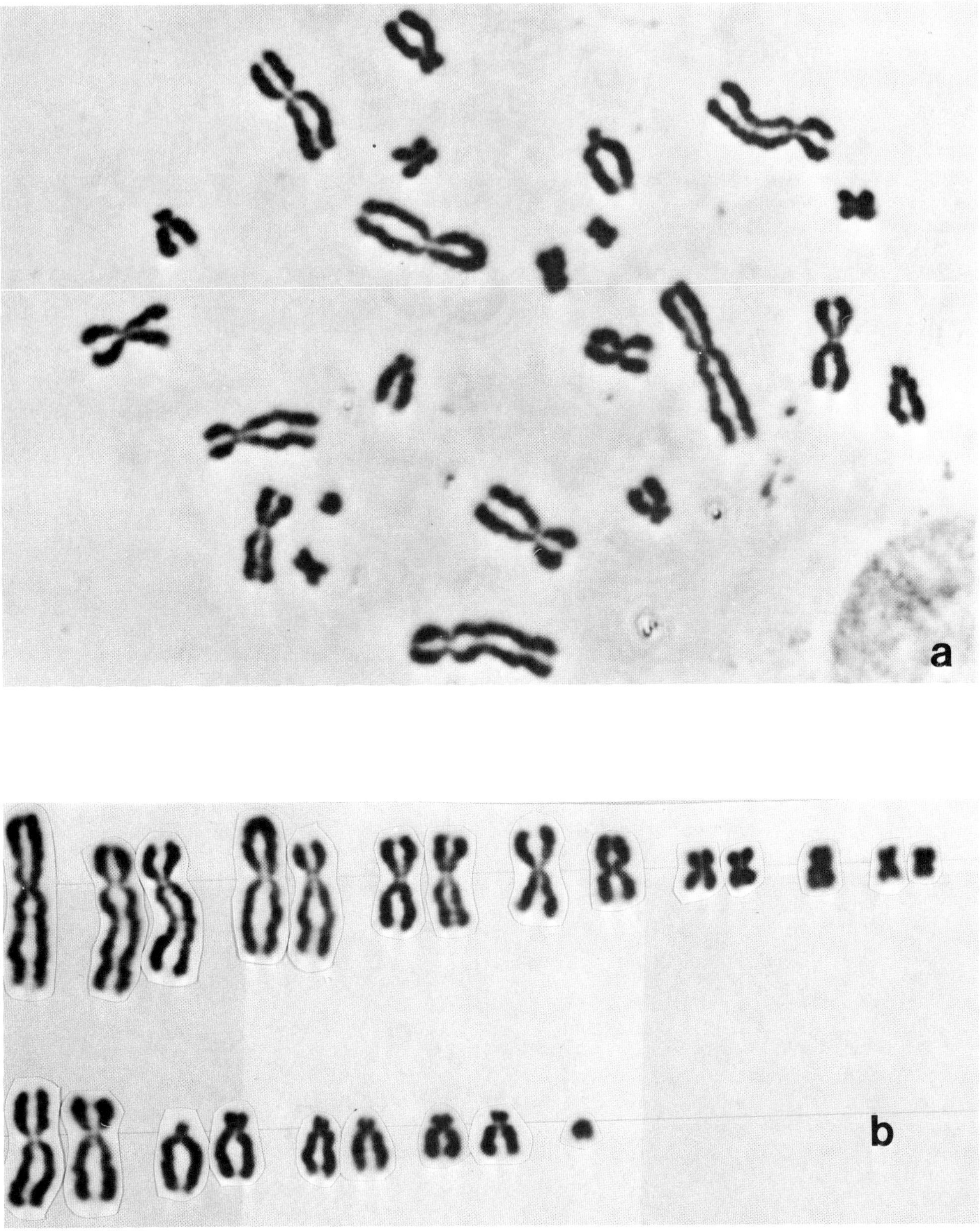

Fig. 13.10. *Example of karyotype. Chinese hamster cells recloned from the Y-5 strain of Yerganian and Leonard [1961] (Acetic-orcein stained).*

cogenes in transformed cell lines [Bishop, 1991; Hames and Glover, 1991; Weinberg, 1991], or deletion, or loss of heterozygosity in suppressor genes [Witkowski, 1990; Marshall, 1991].

It is also possible to label a cell strain for future identification by transfecting in a reporter gene, such as β-galactosidase, and detect it by Southern blotting or, more likely in this particular case, by chromogenic assay for the gene product.

Techniques associated with gene technology have expanded so much over recent years, it would be inappropriate to include them here, and the reader is recommended to such texts as Sambrook et al. [1989] and Frederick et al. [1993].

DNA Fingerprinting

DNA contains regions known as satellite DNA that are apparently not transcribed. The functions of the regions are not understood and they may be purely structural or may provide a reservoir of potentially coded regions for further genetic evolution. Regardless of function, these regions are not highly conserved, since they are not transcribed, and give rise to regions of hypervariability. When the DNA is cut with specific endonucleases, it may be probed with cDNAs that hybridize to these hypervariable regions. These probes were originated by Jeffreys et al. [1985], but more are now becoming available (see below and Stacey et al. [1992]). Electrophoresis reveals variations in fragment length that are specific to the individual from which the DNA was derived. When analyzed by polyacrylamide electrophoresis, each individual's DNA gives a specific hybridization pattern as revealed by autoradiography with radioactive probes. These have come to be known as DNA fingerprints and are cell line specific, except where more than one cell line has been derived from one individual.

These DNA fingerprints appear to be quite stable in culture, and cell lines from the same origin, but maintained separately in different laboratories for many years, still retain the same or very similar DNA fingerprints. This has become a very powerful tool to determine the origin of a cell line, if the original cell line or DNA from it or from the individual has been retained. Furthermore, if a cross-contamination is suspected, this can be confirmed or denied by fingerprinting the cells and all potential contaminants.

The following protocol for DNA fingerprinting has been provided by Glyn Stacey of the European Collection of Animal Cell Cultures at CAMR, PHLS, Porton, Salisbury, England.

Multilocus DNA Fingerprinting of Cell Lines

Principle

Since the first report of multilocus DNA fingerprinting by Jeffreys et al. [1985], many probes have been developed for analysis of polymorphic loci called variable number tandem repeats or VNTRs. In multilocus DNA fingerprinting the probe is used under conditions that allow cross-hybridization with a variety of different repetitive DNA families. Thus the final result represents polymorphic information from many parts of the genome and is therefore inherently sensitive to change in a cell line's genomic DNA complement. Use of repetitive sequences in the M13 phage genome as a probe for identity testing has been demonstrated in animals, plants, and bacteria and it therefore represents a very wide-ranging authentication technique that has the added advantage that pure probe DNA is available commercially at low cost.

Preparation of Southern blots of cell line DNA

Outline

DNA is extracted from cells digested with restriction enzyme, immobilized on nylon membrane by the Southern blot technique [Southern, 1980], and hybridized with labeled DNA from the M13 phage.

Materials

Nonsterile:

PBS: phosphate-buffered saline (pH 7.2)

Hinfl enzyme and buffer (e.g., GIBCO BRL, Life Technologies)

Agarose gels: 0.7% agarose (Type 1A, Sigma) in 1× TBE, 1 $mg.1^{-1}$ ethidium bromide

◇ Ethidium bromide is a potential carcinogen.

5x TBE (stock solution): 54 g Trisbase (Sigma), 27.5 g boric acid (Sigma), and 20 ml 0.5 M Na_2 EDTA (Sigma)

6× loading buffer: 0.25% (w/v) bromophenol blue, 0.25% (w/v) xylene cyanol, and 40% (w/v) sucrose in distilled water

Acid wash: 500 ml 0.1 M HCl

Alkaline wash: 500 ml 0.2 M sodium hydroxide 0.6 M NaCl

Neutralizing wash: 500 ml 0.5 M Tris pH 7.6/1.5 M NaCl

2× SSC: 1 in 10 dilution of stock 20× SSC

20× SSC (stock solution): 175.3 g NaCl (Analar, BDH) and 88.2 g sodium citrate (Analar, BDH) dissolved in distilled water to a final volume of 11 at final pH 7.2

Protocol

1. Wash approximately 10^7 cells twice in PBS and recover as a pellet.
2. Extract the high molecular weight genomic DNA and check its integrity by minigel electro-

phoresis of a small aliquot as described in step 6 (below). A method for DNA extraction that has been used routinely for cell lines and avoids the use of phenol is described in Stacey et al. [1992].

3. Quantify the DNA by UV spectroscopy using a 1 in 50 dilution of DNA [Sambrook et al., 1989]. Quantification may also be achieved by running dilutions of DNA samples in a minigel in parallel with DNA standards (e.g., Hoeffer).
4. Mix 5 μg DNA with 40 u of HinfI enzyme, 3 μl 10× enzyme buffer and sterile distilled water to a total volume of 30 μl.
5. Incubate at 37°C overnight. Remix and microfuge the digest after 1 hr.
6. Test that the digestion is complete by agarose minigel electrophoresis of 1 μl digest in 9 μl electrophoresis buffer (1× TBE) and 2 μl 6× loading buffer. HinfI digests should show a low molecular weight smear of DNA fragments (<5 kb) with no residual genomic DNA band (20–30 kb).
7. To the remainder of the digest, add 6 μl 6× loading mix and load it into an analytical agarose gel (0.7% in 1× TBE, 1 mg/l ethidium bromide at least 20 cm long) in parallel with a HindIII digest marker and a digest of 5 μg DNA from a standard cell line (e.g., HeLa). Electrophorese at approximately 3 Vcm^{-1} until the 2.3-kb HindIII fragment has run the length of the gel (i.e., after 17–20 hr). Photograph the gel against a white ruler over a UV transilluminator as a record of migration distance.
8. Treat the separated DNA fragments in the gel with acid wash (15 min), alkaline wash (30 min), and finally a neutralizing wash (30 min). The agarose gel can then be blotted onto a nylon membrane (Hybond N, Amersham) most simply by capillary action using 20× SSC transfer buffer as described by Sambrook et al. [1989]. The membrane is then rinsed in 2× SSC, dried, and wrapped in Saran Wrap (Dow).
9. Fix DNA fragments to the membrane by exposing the DNA to ultraviolet light (302 nm) for 5 min using a UV-transilluminator (TM20, UVP Inc).
10. Fixed membranes should be stored in dry polythene bags prior to hybridization.

Preparation of labeled M13 phage DNA

Outline

M13mp9 DNA is labeled by random primer extension [Feinberg and Vogelstein, 1982] and purified on a Sephadex column.

Materials

Nonsterile:

Template: M13mp9 ssDNA (0.25 ng/μl Boehringer Mannheim) diluted 1:4 in distilled water
1 M HEPES (pH 6.6)
DTM reagent: dATP, dCTP, dTTP at 100 μm each in 250 m*M* Tris-HCl, 25 m*M* $MgCl_2$, 50 m*M* 2-mercaptoethanol (pH 8.0).
◊ 2-mercaptoethanol is toxic.
OL reagent: 90 O.D. units of oligodeoxyribonucleotides in 1 m*M* Tris, 1 m*M* EDTA (pH 7.5)
Labeling buffer: LS buffer (100:100:28 of 1 M HEPES:DTM:OL)
Bovine serum albumin (20 mg/ml molecular biology grade, GIBCO BRL Life Technologies)
α-^{32}PdGTP (3Ci/mMol Amersham International plc).
◊ ^{32}P is a source of high-energy β particles, to which exposure should be minimized. Always use gloves and 1-cm Perspex screens.
Klenow enzyme (1 μg/μl, sequencing grade, GIBCO BRL Life Technologies)
Sephadex column: (e.g., NICK column, Pharmacia)
Column buffer: 10 m*M* Tris-HCl pH 7.5, 1 m*M* EDTA

Protocol

1. Mix 2 μl diluted M13 DNA with 11 μl distilled water in a sterile microtube.
2. Place the tube in a boiling water bath for 2 min.
3. Transfer tube directly to ice for 5 min.
4. Add 11.4 μl LS buffer and 0.5 μl BSA.
5. Behind a Perspex screen add 10 μl ^{32}PdGTP and 2 μl Klenow enzyme.
6. Mix and briefly centrifuge.
7. Incubate at room temperature overnight behind a Perspex screen.
8. Carefully pipette the contents of the tube onto a Sephadex column previously equilibrated with 10 ml column buffer.
9. Apply 2× 400 μl of column, collecting the eluate from the second 400 μl in a microtube as the purified probe.
   ◊ Radioactivity in the column will remain high due to retention of unincorporated nucleotide.
10. Boil the purified probe for 2 min and chill on ice before mixing with hybridization solution.

Hybridization

Outline

Labeled M13mp9 DNA is hybridized to Southern-blotted cell DNA and, after stringency washing, autoradiography visualizes the pattern of fragments binding the M13 sequences [Westneat et al., 1988].

Materials

Nonsterile

Prehybridization/Hybridization solutions: 0.263 M disodium hydrogen phosphate 7% sodium dodecyl sulfate, 1 m*M*

EDTA, 1% BSA fraction V (Boehringer Mannheim GmbH)

Stringency wash: 2× SSC

20× SSC (stock solution): 175.3 g sodium chloride (Analar BDH) and 88.2 g sodium citrate (Analar BDH) dissolved in distilled water to a final volume of 1 l at final pH 7.2.

Protocol

1. Add membranes to 200 ml (for up to three membranes) prewarmed prehybridization solution at 55°C in a polythene sandwich box.
2. Shake at 55°C for 4 hr.
3. Transfer membranes to 150 ml prewarmed hybridization solution in a second sandwich box at 55°C with the boiled and chilled probe added.
4. Shake at 55°C overnight (minimum of 16 hr).
5. Transfer membranes (up to two) to 1 l stringency wash and shake at room temperature for 15 min.
6. Repeat step 6 (see above).
7. Replace previous wash with 1 l prewarmed (55°C) stringency wash plus 0.1% SDS and shake at 55°C for 15 min.
8. Rinse membranes in 1× SSC at room temperature.
9. Allow membranes to air dry on filter paper until just moist, then wrap in Saran Wrap and measure surface counts using a Geiger-Muller monitor prior to autoradiography at −80°C with two films (e.g., Fuji RX) in an autoradiography cassette. Using standard x-ray-developing chemicals (e.g., 1 + 4 dilutions of CD15 developer [5 min] and CD40 fixer [5 min] from Genetic Research Instrumentation Ltd.). Develop the first film after 24 hr to estimate the final exposure time required. Figure 13.9 shows typical fingerprints produced by the method described.

   ◇ In the UK, procedures involving radioactive materials should be performed according to the Protection of Persons Against Ionising Radiation Arising from Any Work Activity [1985]. Consult your local rules and safety officer before beginning such work.

Discussion. This simple method of fingerprinting using M13 phage DNA provides useful fingerprint patterns for cell lines from a wide range of animals, including insects. Some cultures may show fingerprints with high molecular weight bands that are poorly resolved and it may be helpful in such cases to interpret only those bands in a molecular weight range of about 3–15 kb. The method may be refined by using the 0.78-kb Cla1-Bsm1 M13 DNA fragment, which contains the most useful probe sequences. Nonradioactive probe-labeling methods can also be applied [Medeiros et al., 1988].

Other useful multilocus fingerprint probes include the probes 33.15 and 33.6 [Jeffreys et al., 1985] and oligonucleotide probes such as $(GTG)_5$ [Alie et al., 1986], both of which are available commercially from ICI Cellmark Diagnostics and Fresnius, respectively. Nevertheless, M13 phage DNA is the most readily available and the cheapest multilocus fingerprinting probe.

A variety of single locus probes and polymerase chain reaction methods for VNTR analysis are also proving very useful for the identification of cell cultures. However, multilocus methods have the potential to detect changes in genetic structure at many loci throughout the genome [Thacker et al., 1988] and are generally more versatile for the analysis of multiple species.

The fingerprinting method described in the above protocols will provide a cheap and straightforward route to useful quality control for cell lines from a wide range of species. Such testing enables simultaneous confirmation of identity and exclusion of cross-contamination. These are essential criteria to ensure consistent experimental data and to avoid wasted time and money due to mislabeled or cross-contaminated cultures.

RNA AND PROTEIN

Histochemical reactions and flow cytophotometry also enable measurement of RNA and protein per cell [Watson and Erba, 1992]. These are prone to considerable fluctuations, but in some cases, the ratio of RNA:DNA or protein:DNA may be found characteristic of the cell type if measured under standard culture and assay conditions. However, the greatest specificity in RNA analysis is found in hybridization techniques such as Northern blotting [Sambrook et al., 1989; Frederick et al., 1993] using radioactive, fluorescent, or luminescent probes. This can be carried out at the cellular level, as in FISH, or by *in situ* hybridization with radioactive probes (see Chapter 23).

Qualitative analysis of total cell protein (see Chapter

18) will reveal differences between cells when whole cells, or cell membrane extracts, are run on two-dimensional gels [O'Farrell, 1975]. This produces a characteristic "fingerprint" similar to polypeptide maps of protein hydrolysates, but contains so much information that interpretation can be difficult. It is possible to scan these gels by computerized densitometry (Molecular Dynamics), which will attribute quantitative values to each spot and can normalize each fingerprint to set standards and interpret positional differences in spots. Labeling the cells with [^{32}P], [^{35}S] methionine, or a combination of ^{14}C-labeled amino acids, followed by autoradiography, may make analysis easier, but it is not a technique suitable for routine use unless the technology for preparing two-dimensional gels is currently in use in the laboratory.

ENZYME ACTIVITY

Specialized functions *in vivo* are often expressed in the activity of specific enzymes, e.g., urea cycle enzymes in liver, alkaline phosphatase in endothelium. Unfortunately, many enzyme activities are lost *in vitro* for the reasons discussed in Chapter 2 and are no longer available as markers of tissue specificity. Liver parenchyma loses arginase activity within a few days and cell lines from endothelium lack high-alkaline phosphatase activity. However, some cell lines do express specific enzymes such as tyrosine aminotransferase in the rat hepatoma HTC cell lines [Granner et al., 1968]. When looking for specific marker enzymes, the constitutive (uninduced) level and the induced level should be measured and compared with a number of control cell lines. Glutamyl synthetase activity, for instance, characteristic of astroglia in brain, is increased several fold when the cells are cultured in the presence of glutamate instead of glutamine [DeMars, 1957]. Induction of enzyme activity will require specialized conditions for each enzyme and these may be obtained from the literature. Common inducers are glucocorticoid hormones such as dexamethasone, polypeptide hormones such as insulin and glucagon, or alteration in substrate or product concentrations in the medium, as in the example above with glutamyl synthetase.

Isoenzymes

Enzyme activities can also be compared qualitatively between cell strains, due to polymorphisms among species and, sometimes, among races or individuals within a species. These so-called isoenzymes or isozymes may be separated chromatographically or electrophoretically and the distribution patterns (zymograms) found to be characteristic of species or tissue. Paul and Fottrell [1961] demonstrated differences in esterase zymograms between normal and malignant mouse cells and human cells (Fig. 13.11), and O'Brien et al. [1977] have described a number of very useful isozymic markers for human cell lines after electrophoresis of cell extracts [see also Macy, 1978, and protocol, below].

Electrophoresis media include agarose, cellulose acetate, starch, and polyacrylamide. In each case, a crude enzyme extract is applied to one point in the gel and a potential difference applied across the gel. The different isozymes migrate at different rates and can be detected later by staining with chromogenic substrates. Stained gels may be read directly by eye and photographed, or scanned with a densitometer.

The following protocol for isoenzyme analysis has been contributed by Marvin L. Macy, American Type Culture Collection, 12301 Park Lawn Drive, Rockville, MD.

Principle

Determination of the electrophoretic mobility of only three enzymes, purine nucleoside phosphorylase (NP; E.C.2.4.2.1) glucose-6-phosphate dehydrogenase (G6PD; E.C.1.1.1.49), and lactate dehydrogenase (LDH; E.C.1.1.1.27), is adequate to provide an effective means for identification of species of origin of cultured cells in the majority of instances [Montes de Oca et al., 1969]. Cell lines representing 37 taxonomic groups are easily identifiable from each other by comparison of their NP, G6PD, and LDH mobility differences as determined by vertical starch gel electrophoresis.

Outline

Harvest cells, wash in PBS, resuspend at 5×10^7 cells/ml and prepare crude extract. Recover supernatant, aliquot, and store at $-70°C$. Prepare vertical starch gel apparatus, add cell extract, and electrophorese. Slice gel horizontally, apply enzyme stains, incubate, wash, and examine.

Materials

Nonsterile:

Starch, gel electrophoresis apparatus (Haake Buchler)
Electrostarch (Electrostarch)
DC power supply, 160 V output (Pharmacia, Biorad)
Gel-slicing device (Haake Buchler)
Eppendorf tubes, 1.5 ml
Microcentrifuge, Eppendorf (Brinkmann)
Stirring hot plate (Cole Palmer, Baird & Tatlock)
Staining boxes (Haake Buchler)

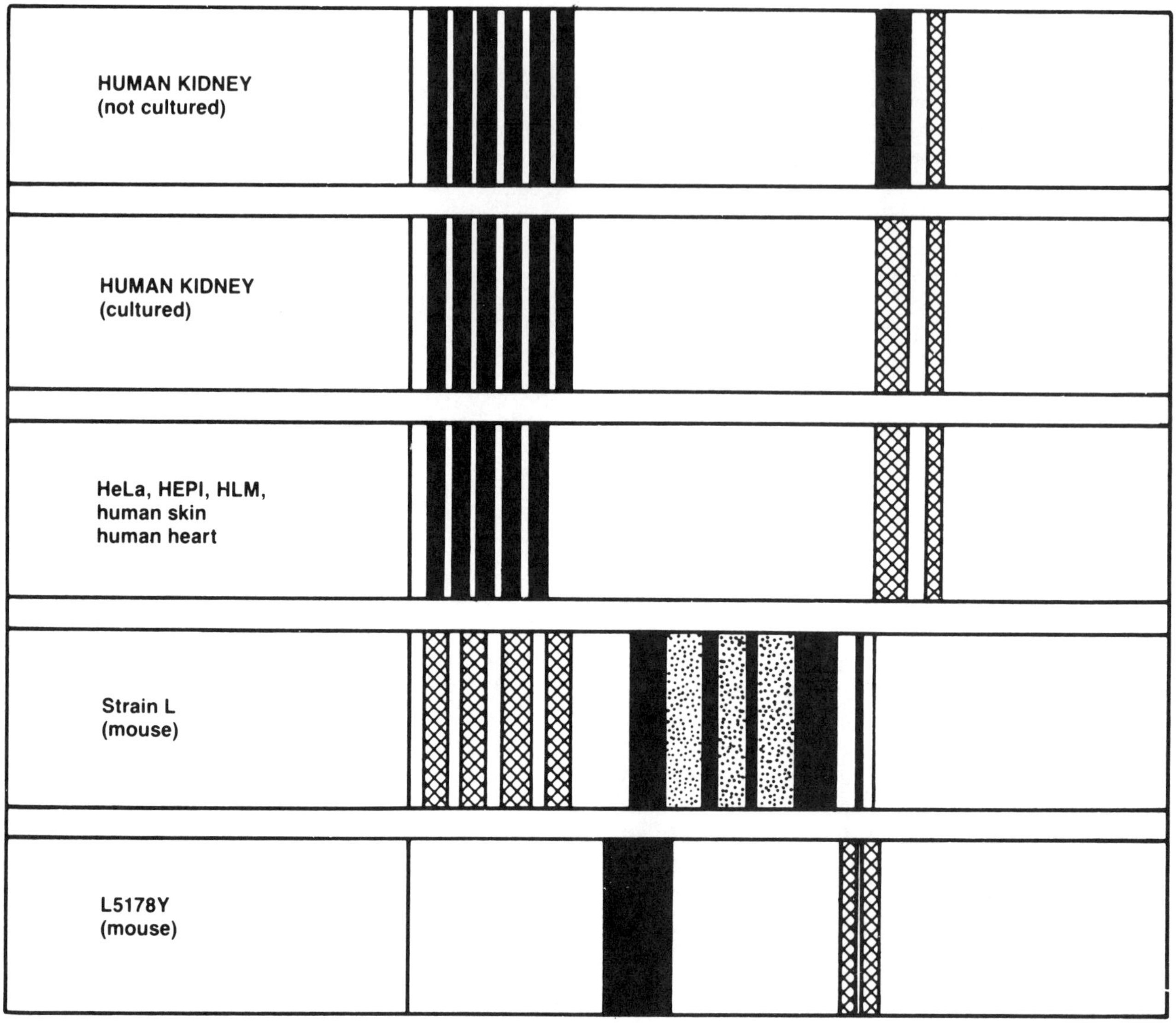

Fig. 13.11. *Isozyme patterns revealed by electrophoresis. Starch gel zymogram for esterase. [From Paul and Fottrell, 1961].*

Syringe, 50 μl (Hamilton)
Gloves (Zetex, Johnson & Johnson)
Petrolatum (Petroleum jelly, Vaseline)
TEB chamber buffer: 0.18 M Tris, 4 m*M* EDTA, disodium salt, 0.1 M boric acid, pH 8.6, to 1 l.
For LDH and G6PDH, add 4 ml 5 m*M* NADP, mix, and place at 4°C.

Protocol

1. Harvest cells, resuspend in PBS, and count viable cells.
2. Centrifuge at 300 *g* for 5–10 min to pellet cells and decant supernate. Cell pellet may be stored at −70°C at this point.
3. Resuspend at 5×10^7 cells/ml in a 1:15 mixture of Triton X-100 and 0.9% NaCl solution (pH 7.1) containing 6.6×10^{-4} M EDTA. Aspirate the solution using a small-bore pipette for several minutes until the cell membranes clump together.
4. Transfer the homogenate to Eppendorf tubes (1.5 ml) and spin at top speed (8,733 *g*) for 2 min in a microfuge. Recover the supernate, dispense in desired volume aliquots (50 μl), and store at −70°C.
5. In a 1,000-ml filter flask, add 60 g electrostarch in 500 ml 0.1 × TEB buffer and heat over burner with shaking (◇ wear gloves) until mixture thickens and begins to bubble (90°C). Apply a vacuum for about 1 min to completely degas,

seen by lack of solution rise and removal of gas bubbles. Remove vacuum and allow gel to cool to 60°C with continuous swirling. For LDH and G6PD, add 1 ml 5 m*M* NADP and mix.

6. Pour into gel mold (260 × 160 × 6 mm) and allow to cool for 1–3 min. Place glass plate over top and seal edges with molten starch.
7. Leave at room temperature for 30–60 min, then place at 4°C. Gel is ready to use in 4 hr and is usable for 24 hr.
8. Thaw cell extracts and spin in microfuge for 1 min. Melt about 100 ml of petrolatum. Gently remove slot former and apply 40-μl sample in each slot. Cover sample chamber with melted petrolatum and allow to harden (5 min).
9. Remove end pieces of gel mold and place the mold into the bottom electrode chamber and secure. Add 600 ml of cold chamber buffer to cathode top chamber and 400 ml to anode, bottom chamber. Connect the output terminal electrodes from the power supply to the upper (black) and lower (red) electrode chamber leads. Run the gel at 160 V for 16–18 hr at 4°C.
10. After electrophoresis, remove gel from mold, cut off and discard 5 cm from both ends, slice the gel horizontally into three equal sections using the gel-cutting device, separate, and place cut side up in staining boxes.
11. Each stain solution is mixed with 25 ml of boiled 2% noble agar cooled to 45–50°C.

    For NP, combine 20 ml H_2O, 5 ml 0.1 M $NaH_2PO_4H_2O$, 50 mg inosine, 0.1 ml xanthine oxidase (100 mg/ml), 1 ml MTT (5 mg/ml), and 2 ml PMS (1 mg/ml).

    For G6PD, combine 5 ml H_2O, 5 ml 0.5 M Tris pH 7.5, 5 ml 0.025 M glucose-6-phosphate, 5 ml 0.1 M $MgCl_2$, 5 ml 0.005 M NADP, 1 ml MTT, and 2 ml PMS.

    For LDH, combine 10 ml H_2O, 5 ml 0.5 M Tris pH 7.5, 5 ml 1.0 M sodium lactate, 5 ml NAD (10 mg/ml), 1 ml MTT, and 2 ml PMS.
12. Cover the entire gel surface with the individual noble agar–stain mixtures and incubate at 37°C in the dark. Most enzyme bands appear in 1–3 hr. After development, carefully wash off the agar–stain mixture with cold tap water.

Analysis

Record results obtained both in terms of number and intensity of enzyme bands with reference to the distance of band migration from the point of origin for each sample.

Variations

Cell extracts can be prepared by ultrasonication, freezing and thawing rapidly three times, or treatment with octyl alcohol [Macy, 1978].

Isoenzyme analysis of polymorphic gene enzyme systems in cell culture of human origin can also be performed for intraspecies identification [Povey et al., 1976; O'Brien et al., 1977]. The resolution of these allelic isoenzymes yields an allozyme genetic signature and provides a phenotype frequency product as a means of identifying individual human cell lines [O'Brien et al., 1980; Wright et al., 1981].

Polymorphic isoenzymes have also been reported for cell lines of murine origin as a means of identifying inbred strains of mice [Nichols and Ruddle, 1973; Krog, 1976].

Isoenzyme analysis has proved to be a great value in determining the chromosomal constitution of somatic cell hybrids (see Chapter 23) [Meera Khan, 1971; Nichols and Ruddle, 1973; Van Someren et al., 1974; Harris and Hopkinson, 1976].

Isoenzyme analysis for species identification can also be performed using an agarose gel system [Halton et al., 1983]. A complete system (AuthentiKit System) is available from Innovative Chemistry, Inc., Marshfield, MA. The principles of the agarose system are similar to those described above. The kit uses smaller precast gels for separation, includes premeasured lyophilized incubation mixtures (step 11 above), and assays four additional enzyme substrates. In addition to the three enzymes mentioned above, the system includes aspartate aminotransferase (E.C.2.6.1.1), malate dehydrogenase (E.C.1.1.1.37), mannose phosphate isomerase (E.C.5.3.1.8), and peptidase B (E.C.3.4.11.4). This system allows easy screening of up to six cell lines per gel for seven genetic markers in less than 3 hr [Hay, 1992]. The kit is not yet suitable for intraspecies evaluations.

ANTIGENIC MARKERS

This area has expanded more than any other in recent years due largely to the availability of affinity-purified and monoclonal antibodies, many of which are now supplied by commercial companies. Regardless of the source of the antibody, it is essential to be certain of the specificity of the antibody by using appropriate control material. This is true for monoclonal antibodies and polyclonal antisera alike; a monoclonal antibody is highly specific for a particular epitope but the generality or specificity of the expression of the epitope must still be demonstrated.

Immunological staining may be direct, i.e., the specific antibody is itself conjugated to a fluorochrome such as fluorescein or rhodamine or to horseradish peroxidase or alkaline phosphatase, and used to stain the specimen directly, followed by direct observation on a fluorescence microscope or further development with a chromogenic peroxidase or phosphatase substrate and observation on a regular microscope. Alternatively, an indirect method may be used where the primary (specific) antibody is used in its native form to bind to the antigen in the specimen and this is followed by treatment with a second antibody, raised against the immunoglobulin of the first antibody. The second antibody may be conjugated to a fluorochrome [Coons and Kaplan, 1950; Kawamura, 1969] or peroxidase or alkaline phosphatase for subsequent visualization [Avrameas, 1970; Taylor, 1978].

Various methods have been employed to enhance the sensitivity of detection of these methods, particularly the peroxidase-linked methods. The commonest of these is the peroxidase-antiperoxidase or PAP technique, where a further tier is added by reacting with a peroxidase complex containing antibody from the same species as the primary antibody [Sternberger, 1970]. This is bound to the free valency of the second antibody. Even greater sensitivity has been obtained by using a biotin-conjugated second antibody with a streptavidin complex carrying peroxidase or alkaline phosphatase (Amersham) or gold-conjugated second antibody with subsequent silver intensification (Janssen).

Indirect Immunofluorescence

Outline

Fix cells and treat sequentially with first and second antibodies. Examine by UV light.

Materials

Culture grown on glass coverslip, chambered slide (GIBCO-Life Techologies), or polystyrene petri dish
Freshly prepared fixative: 5% acetic acid in ethanol–place at -20°C
Primary antibody diluted 1:100–1:1,000 in culture medium with 10% FBS
Second antibody raised against the species of the first, e.g., if first antibody was raised in rabbit then the second should be from a different species, e.g., goat anti-rabbit immunoglobin; the second antibody should be conjugated to fluorescein or rhodamine
HBSS or PBS without phenol red
Mountant: 50% glycerol in PBS or HBSS without phenol red and containing a fluorescence-quenching inhibitor (Vecta)

Protocol

1. Wash coverslip with cells in HBSS or PBS and place in suitable dish, e.g., 24-well plate for 13-mm coverslip.
2. Place at −20°C for 10 min and add cold fixative for 20 min.
3. Remove fixative, wash in HBSS or PBS, and add 1 ml normal swine serum and leave at room temperature for 20 min.
4. Rinse in PBS or HBSS, drain on tissue, and place inverted on a 50-μl drop of diluted primary antibody. Place at 37°C for 30 min, room temperature for 1–3 hr or overnight at 4°C. For the last, antibody may be diluted 1:1,000.
5. Rinse in PBS or HBSS and transfer to second antibody diluted 1:20 for 20 min at 37°C.
6. Rinse in PBS or HBSS and mount in 50% glycerol in PBS with fluorescence bleaching retardant (Vecta).
7. Examine on fluorescence microscope.

Variations. For cell surface or particularly fixation-sensitive antigens, treat with antibodies first, then postfix as above. Where a glass substrate is used, cold acetone may be substituted for acid ethanol.

Indirect peroxidase. Substitute peroxidase-conjugated antibody at stage 5 and then transfer to peroxidase substrate (diaminobenzidine-stained preparations can be dehydrated and mounted in DePeX but ethyl carbazole must be mounted in glycerol as above).

PAP. Use unconjugated second antibody at stage 5 and then transfer to diluted PAP complex (1:100) (most immunobiological suppliers, e.g., Dako, Vecta) in PBS or HBSS for 20 min. Rinse and add peroxidase substrate as above, incubate, wash, and mount.

Specific cell surface antigens are usually stained in living cells (at 4°C in the presence of sodium azide to inhibit pinocytosis) while intracellular antigens are stained in fixed cells, sometimes requiring light trypsinization to permit access of the antibody to the antigen.

HLA and blood-group antigens can be demonstrated on many human cell lines and serve as useful characterization tools, especially where the donor patient profile is known [Espmark and Ahlqvist-Roth, 1978; Stoner et al., 1980; Pollack et al., 1981].

DIFFERENTIATION

Many of the characteristics described under antigenic markers or enzyme activities may also be regarded as markers of differentiation, and as such they can help to correlate cell lines with their tissue of origin. Other examples of differentiation, and as such highly specific markers of cell line identity, are given in Table 2.2, and the appropriate assays for these properties may be derived from the references cited (see also Chapter 20).

While much of the interest in characterization is related to the study of specialized functions and their relationship to the cells' behavior *in vivo*, these techniques are also important in confirming the identity of a cell line and excluding the possibility of cross-contamination (see Chapter 16).

CHAPTER 14

Induction of Differentiation

As discussed in Chapter 2, when cells are cultured and propagated as a cell line, the resultant cell phenotype is often different from the characteristics predominating in the tissue from which it was derived. This is due to several factors, many of them as yet undefined, that regulate the geometry, growth, and function *in vivo* but that are absent from the tissue culture environment. Before considering these systematically it is first necessary to state what is meant by *differentiation* in this context. The term is used here to define the process leading to the expression of phenotypic properties characteristic of the functionally mature cell *in vivo*. It does not imply that the process is complete or that it is irreversible. There will be processes such as the cessation of DNA synthesis in the erythroblast nucleus that are not normally reversible, but, for simplicity, no attempt is made here to distinguish these from reversible processes such as the induction of albumin synthesis in hepatocytes, which is lost under certain culture conditions but may be reinduced.

Differentiation is used here to describe the combination of constitutive and adaptive properties found in the mature cell. *Commitment*, on the other hand, will imply an irreversible progression from a stem cell to a particular defined lineage endowing the cell with the potential to express a limited repertoire of properties either constitutively or when induced to do so.

Terminal differentiation implies that a cell has progressed down a particular lineage to a point at which the mature phenotype is fully expressed and beyond which the cell cannot progress. In principle this need not exclude cells that can revert to a less differentiated phenotype and resume proliferation, such as a fibrocyte, but in practice the term tends to be reserved for cells like neurons, skeletal muscle, or keratinized squames where differentiation is irreversible.

Dedifferentiation has been used to describe the loss of the differentiated properties of a tissue when it becomes malignant or when it is grown in culture. As these are complex processes with several contributory factors, including cell death, selective overgrowth, adaptive responses, and loss of certain phenotypic properties, the term should be used with great caution, or not at all. When used correctly, dedifferentiation means the loss by a cell of the specific phenotypic properties associated with the mature cell. When it occurs, it is, most probably, either an *adaptive* process, implying that the differentiated phenotype may be regained given the right inducers (see also Chapter 2), or *selective*, implying that a precursor cell has been selected because of its greater proliferative potential. This latter does not preclude the possibility that the precursor cell may be induced to mature to the fully differentiated cell given the correct environmental conditions. Where the wrong lineage has been selected, however, no amount of induction can bring back the required phenotype, and this deficit has often been erroneously attributed to dedifferentiation in the past.

STAGES OF COMMITMENT AND DIFFERENTIATION

There are two main pathways to differentiation in the adult organism (Fig. 14.1). Typically, a small population of totipotent or pluripotent undifferentiated stem cells gives rise to committed precursor cells that will progress toward terminal differentiation, losing their capacity to divide as they reach the terminal stages. This gives rise to a fully mature, differentiated cell that normally will not divide. Alternatively, cells such as fibrocytes may respond to a local reduction in cell density and/or the presence of one or more growth factors by losing some of their differentiated properties, e.g., collagen synthesis, and re-entering the cell cycle. When the tissue has regained the appropriate cell density by division, cell proliferation stops and differentiation is reinduced.

The first process is used where continual renewal is required, such as in the hemopoietic system, the skin, and the gastric mucosa, while the second is used where regeneration is not continual and requires quick mounting in response to trauma, such as in wound repair or liver regeneration. In the first, amplification is possible by having several cell divisions at each precursor stage but takes longer than the rapid recruitment of a higher proportion of the total cell population into a limited number of divisions available in the second.

PROLIFERATION AND DIFFERENTIATION

As differentiation progresses, cell division is reduced and eventually lost. In most cell systems, cell proliferation is incompatible with the expression of differentiated properties. Tumor cells can sometimes break this restriction, and in melanoma, for example, melanin continues to be synthesized while the cells are proliferating. Even in these cases, however, synthesis of the differentiated product increases when division stops.

There are severe implications for this relationship in culture, where the major objective for many years has been the propagation of cell lines and the production of large numbers of cells for biochemical or molecular analysis. Where the incompatibility of differentiation and proliferation is maintained, it is not surprising to find that the majority of cell lines do not express fully differentiated properties.

This fact was noted many years ago by the exponents of organ culture (see Chapter 22), who set out to retain three-dimensional, high cell density tissue architecture and to prevent dissociation and selective overgrowth of undifferentiated cells. However, although of considerable value in elucidating cellular interactions regulating differentiation, organ culture has always suffered from the inability to propagate large numbers of identical cultures, particularly if large numbers of cells are required, and the heterogeneity of the sample, assumed to be essential for the maintenance of the tissue phenotype, has in itself made the ultimate biochemical analysis of pure cell populations and of their responses extremely difficult.

Hence, in recent years, there have been many attempts to reinduce the differentiated phenotype in pure populations of cells by re-creating the correct environment and, by doing so, defining individual influences exerted on the induction and maintenance of differentiation. This usually implies cessation of cell division and creation of an interactive high-density cell population as in histotypic or organotypic culture. This is discussed in greater detail below.

COMMITMENT AND LINEAGE

Progression from a stem cell to a particular pathway of differentiation usually implies a rapid increase in commitment with advancing stages of progression (see Figs. 2.3 and 14.2). A hemopoietic stem cell after commitment to lymphocytic differentiation will not change lineage at a later stage and adopt myeloid or erythrocytic characteristics. Similarly, a primitive neuroectodermal stem cell, once committed to become a neuron, will not change to a glial cell. This is not to say that if the inductive environment is altered early enough (before commitment) that a cell can alter its destiny or even adopt a mixed phenotype under artifactual or pathological circumstances. Commitment may therefore be regarded as the point between the stem cell and a particular precursor stage, where a cell or its progeny can no longer transfer to a separate lineage.

Many claims have been made in the past for cells transferring from one lineage to another. Perhaps the best substantiated of these is the regeneration of the amphibian lens by recruitment of cells from the iris [Clayton et al., 1980; Cioni et al., 1986]. Since the iris can be fully differentiated and still regenerate lens, this has been proposed as proper transdifferentiation. It is, however, one of the few examples, and most other

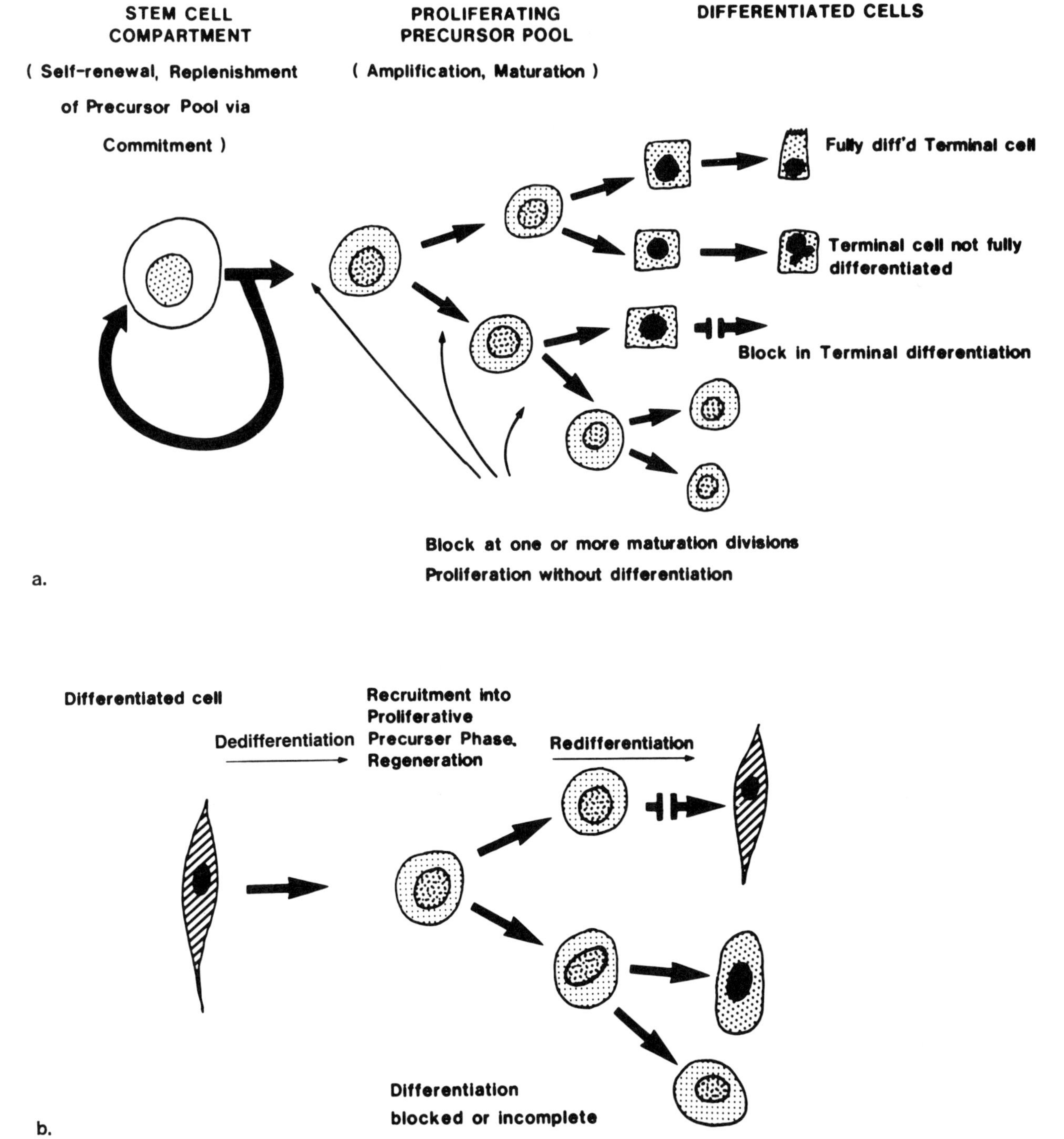

Fig. 14.1. Alternative pathways for cell differentiation. a. Generation from stem cell. b. Generation by recruitment from differentiated cells. The blocks refer to points at which expression of differentiation may be inhibited in vitro, *but could apply equally to blocks found in neoplasia. (Reproduced from Freshney [1985] by permission of the publisher.)*

claims have been from tumor cell systems where the origin of the tumor population may not be clear. Small cell carcinoma of the lung has been found to alter to squamous or large cell carcinoma following recurrence after relapse from chemotherapy. Whether this implies that one cell type, the Kulchitsky cell [de Leij et al., 1985], presumed to give rise to small cell lung carcinoma, changed its commitment or whether the tumor originally derived from a multipotent stem cell and on recurrence progresses down a different route is still not

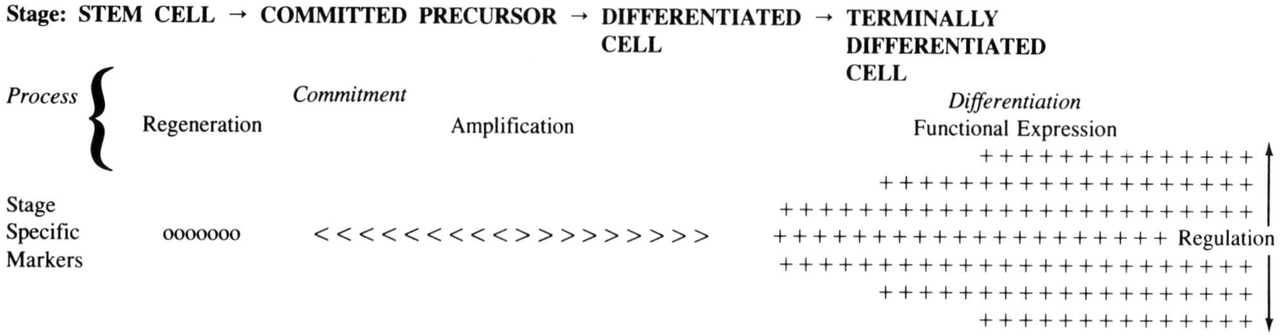

Fig. 14.2. Stages in cell differentiation. Stem cells are capable of regeneration without commitment or commitment division, leading to a precursor cell that is still capable of cell proliferation and hence amplification of the lineage. Successive divisions lead to further differentiation, in the presence of the appropriate inducer environment, until, finally, a stage may be reached where no further division is possible. This is terminal differentiation. Phenotypic expression may be regulated quantitatively at terminal differentiation from the constitutive to the induced state by the action of hormones, metabolites, etc.

clear [Gazdar et al., 1983; Goodwin et al., 1983; Terasaki et al., 1984]. Similarly, the K562 cell line was isolated from a myeloid leukemia, but subsequently was shown capable of erythroid differentiation [Andersson et al., 1979b]. Rather than a committed myeloid precursor converting to erythroid, it seems more likely that the tumor arose in the common stem cell known to give rise to both erythroid and myeloid lineages. For some reason, as yet unknown, continued culture favored erythroid differentiation rather than the myeloid features seen in the original tumor and early culture. In some cases, again in cultures derived from tumors, a mixed phenotype may be generated. The C_6 glioma of rat expresses both astrocytic and oligodendrocytic features and these may be demonstrated simultaneously in the same cells.

In general, however, these cases are unusual and restricted to tumor cultures. Most cultures from normal tissues, although they may differentiate in different directions, once committed will not alter to a different lineage. This raises the question of the actual status of cell lines derived from normal tissues. This has been dealt with in Chapter 2 and the conclusion reached that most cultures are derived from (1) stem cells, or progenitor cells, which may differentiate in one or more different directions, e.g., lung mucosa, which can become squamous or mucin-secreting depending on the stimuli; (2) committed precursor cells, which will stay true to lineage; or (3) differentiated cells such as fibrocytes, which may dedifferentiate and proliferate, but still retain lineage fidelity. Some mouse embryo cultures, loosely called fibroblasts (e.g., the various cell lines designated 3T3), probably more correctly belong to (1) as they can be induced to become adipocytes, muscle cells, and endothelium as well as fibrocytes.

Many cell lines may have different degrees of commitment depending on the "stemness" or precursor status of the cells from which they were derived; but unless the correct environmental conditions are reestablished and proliferation is discouraged they will remain at the same position in the lineage. Conversely, some cell populations, e.g., bronchial epithelium, will spontaneously mature in regular serum-containing media and require defined conditions to remain proliferative.

There are now some well-described examples where progenitor cells, e.g., the O2A common progenitor of the oligodendrocyte and type 2 astrocyte in the brain, which will remain as a proliferating progenitor cell in a mixture of PDGF and bFGF, will differentiate into an oligodendrocyte in the absence of growth factors or serum, or into a type 2 astrocyte in fetal bovine serum or a combination of ciliary neurotropic factor (CNTF) and bFGF [Raff, 1990; Richardson et al., 1990] (see Barnett protocol, Chapter 20). Similarly, cardiac muscle cells remain undifferentiated and proliferative in serum and bFGF, but differentiate in the absence of serum [Goldman and Wurzel, 1992], and primitive germ cell cultures will differentiate spontaneously unless kept in the undifferentiated proliferative phase by bFGF, SCF (stem cell factor, Steel factor, *kit* ligand), and LIF (lymphocyte inhibitory factor) [Matsui et al., 1992].

Hence, with the advent of more defined media, it is gradually becoming possible to define the correct inducer environment that will maintain cells in a stem-like, or precursor status, or that will induce the cells to differentiate.

MARKERS OF DIFFERENTIATION

Before studies of differentiation, its properties, and the regulation of its expression can be made, marker properties must be defined that will allow differentiation to be recognized. Markers expressed early and retained throughout subsequent maturation stages are generally regarded as lineage markers, e.g., intermediate filament proteins such as the cytokeratins (epithelium) [Moll et al., 1982] or glial fibrillary acidic protein (astrocytes) [Eng and Bigbee, 1979; Bignami et al., 1980]. Markers of the mature phenotype representing terminal differentiation are more usually specific cell products or enzymes involved in the synthesis of these products, e.g., hemoglobin in an erythrocyte, serum albumin in a hepatocyte, transglutaminase [Schmidt et al., 1985] or involucrin [Parkinson and Yeudall, 1992] in a differentiating squame, or glycerol phosphate dehydrogenase in an oligodendrocyte [Breen and De Vellis, 1974] (see Table 2.2). These properties are often expressed well after commitment and are more likely to be reversible and under adaptive control by hormones, etc.

Differentiation should be regarded as the expression of one or preferably more than one of these marker properties. While lineage markers are helpful in confirming cell identity, the expression of the functional properties of the mature cells is the best criterion for terminal differentiation.

INDUCTION OF DIFFERENTIATION

There are four main parameters governing the control of differentiation, summarized in Figure 14.3.

Soluble Inducers (Table 14.1)

Soluble inducers include established endocrine hormones such as hydrocortisone, glucagon, and thyroxine (or triiodotyrosine), paracrine factors released by one cell and influencing adjacent cells, which as yet are poorly characterized (e.g., TGFβ from platelets, prostaglandins, NGF, glia maturation factor) [Lim and Mitsunobu, 1975], alveolar maturation factor [Post et al., 1984], and interferons [e.g., Pfeffer and Eisenkraft, 1991], vitamins such as vitamin D and retinoic acid, and inorganic ions, particularly Ca^{2+}, where high Ca^{2+} promotes keratinocyte differentiation, for example (see Chapter 20).

Nonphysiological inducers (Table 14.2). Friend observed that mouse erythroleukemia cells treated with DMSO (to induce the production of Friend leukemia virus) turned red due to the production of hemoglobin [Rossi and Friend, 1967]. Subsequently it was demonstrated that many other cells, e.g., neuroblastoma, myeloma, and mammary carcinoma, also responded to DMSO by differentiating. Many other compounds have now been added to this list of nonphysiological inducers–hexamethylene bisacetamide, N-methyl acetamide, sodium butyrate, benzodiazepines, whose action may be related to that of DMSO, and a range of cytotoxic drugs such as hydroxyurea, cytosine arabinoside, and mitomycin C (see Table 14.2).

The action of these compounds is as yet unclear but may be mediated by changes in membrane fluidity (particularly the polar solvents like DMSO and the anesthetics and tranquilizers), by their influence as lipid intercalators on enzymes of signal transduction such as protein kinase C (PKC) or phospholipase D (PLD), or by alterations in DNA methylation or histone acetylation. Induction of differentiation by polar solvents such as DMSO may be phenotypically normal but the induction by cytotoxic drugs may also induce gene expression unrelated to differentiation [McLean et al., 1986].

Tumor promoters such as phorbol meristate acetate (PMA) have been shown to induce squamous differen-

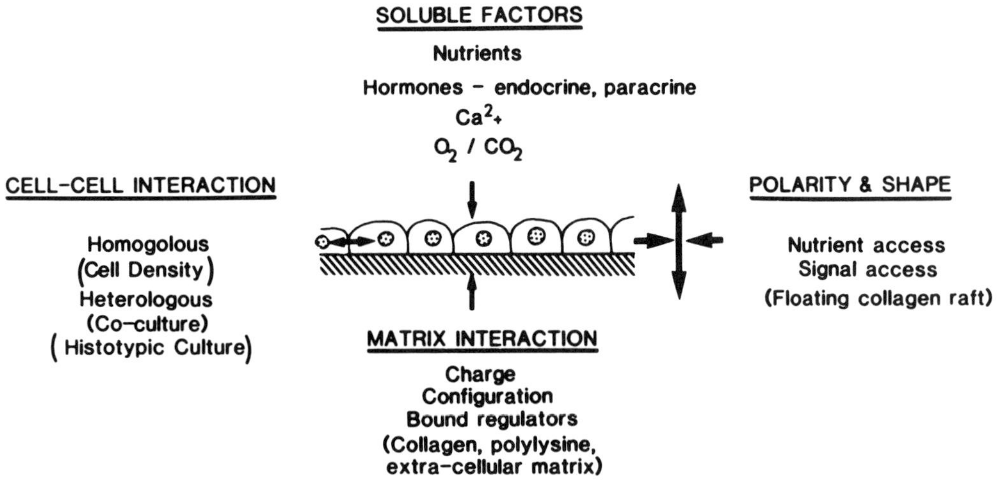

Fig. 14.3. Parameters controlling expression of differentiation in vitro. *(Reproduced from Freshney [1985] by permission of the publisher.)*

TABLE 14.1. Soluble Inducers of Differentiation–Physiological

Inducer	Cell type	Reference
Steroid and related:		
Hydrocortisone	Glia glioma	[McLean et al., 1986]
	Hepatocytes	[Granner et al., 1968]
	Mammary epithelium	[Stockdale and Topper, 1966]
	Myeloid leukemia	[Sachs, 1978]
Retinoic acid	Bronchial epithelium	[Wu and Wu, 1986]
	Tracheal epithelium	[Klann and Marchok, 1982]
	Melanoma	[Lotan and Lotan, 1980; Meyskens and Fuller, 1980]
	Myeloid leukemia	[Breitman et al., 1980]
Peptides:		
Melanotropin	Melanocytes	[Fuller and Meyskens, 1981]
Thyrotropin	Thyroid	[Chambard et al., 1983]
Erythropoietin	Erythroblasts	[Goldwasser, 1975]
Prolactin	Mammary epithelium	[Stockdale and Topper, 1966; Rudland et al., 1982]
Insulin	Mammary epithelium	[Stockdale and Topper, 1966; Rudland et al., 1982]
Nerve growth factor	Neurons	[Levi-Montalcini, 1979]
Glia maturation factor	Glial cells	[Lim and Mitsunobu, 1975]
Fibrocyte-pneumocyte factor	Type II pneumocytes	[Post et al., 1984]
Interferon-α	Type II pneumocytes	(Freshney and McCormick, in preparation)
	HL60, myeloid leukemia	[Kohlhepp et al., 1987]
Interferon-γ	Neuroblastoma	[Wuarin et al., 1991]
CNTF	Type 2 astrocytes	[Raff, 1990]
KGF	Keratinocytes	[Aaronson et al., 1991]
	Prostatic epithelium	[Cunha and Young, 1992; Yan et al., 1993]
HGF	Kidney (MDCK)	[Bhargava et al., 1992; Li et al., 1992]
	Hepatocytes	[Montesano et al., 1991]
TGFβ	Bronchial epithelium	[Masui et al., 1986b]
	Melanocytes	[Fuller and Meyskens, 1981]
Prostaglandins		
Vitamins:		
Vitamin E	Neuroblastoma	[Prasad et al., 1980]
Vitamin D_3	Myeloma	[Murao et al., 1983]
Minerals:		
Ca^{2+}	Keratinocytes	[Boyce and Ham, 1983]

tiation in bronchial mucosa although not in bronchial carcinoma [Willey et al., 1984a,b]. Although these are not normal regulators *in vivo*, they bind to specific receptors and activate signal transduction, such as the activation of PKC [Dotto et al., 1985].

Cell Interaction

Homologous cell interaction occurs optimally at high cell density. It may involve gap junctional communication [Finbow and Pitts, 1981], where metabolites, second messengers such as cyclic AMP, diacylglycerol (DAG), Ca^{2+}, or electrical charge may be communicated between cells. This interaction probably harmonizes the expression of differentiation within a population of similar cells rather than initiating its expression.

The presence of homotypic cell–cell adhesion molecules, such as the CAMs or cadherins, which are Ca^{2+} dependent, provides another mechanism by which contacting cells may interact. The adhesion molecules

TABLE 14.2. Soluble Inducers of Differentiation–Nonphysiological

	Cell type	Reference
Planar-polar compounds:		
DMSO	Murine erythroleukemia	[Rossi and Friend, 1967]
	Myeloma	[Tarella et al., 1982]
	Neuroblastoma	[Kimhi et al., 1976]
	Mammary epithelium	[Rudland et al., 1982]
Sodium butyrate	Erythroleukemia	[Andersson et al., 1979b]
	Colon cancer	[Augeron and Laboisse, 1984; Chung et al., 1985]
N-methyl acetamide	Glioma	[McLean et al., 1986]
N-methyl formamide, dimethyl formamide	Colon cancer	[Dexter et al., 1979]
Hexamethylenebis-acetamide	Erythroleukemia	[Osborne et al., 1982]
Benzodiazepines	Erythroleukemia	[Clarke and Ryan, 1980]
Cytotoxic drugs:		
Cytosine arabinoside	Myeloid leukemia	[Takeda et al., 1982]
Mitomycin C; anthra-cyclines	Melanoma	[Raz, 1982]
Signal transduction modifiers:		
Cyclic AMP	Oat cell cancer	[Tsuji et al., 1976]
TPA (phorbol meris-tate acetate, PMA)	Bronchial epithelium	[Willey et al., 1984a,b]
	Myeloma	[Rovera et al., 1979]
	Neuroblastoma	[Spinelli et al., 1982]

promote interaction primarily between like cells via identical, reciprocally acting, extracellular domains, and appear to have signal transduction potential via phosphorylation of the intracellar domains [Doherty et al., 1991].

Heterologous cell interaction, e.g., between mesodermally and endodermally or ectodermally derived cells, is responsible for initiating and promoting differentiation. During and immediately following gastrulation in the embryo, and later during organogenesis, mutual interaction between cells originating in different germ layers promotes differentiation [Grobstein, 1953a,b; Cooper, 1965]. For example, when endodermal cells form a diverticulum from the gut and proliferate within adjacent mesoderm, the mesoderm induces the formation of alveoli and bronchiolar ducts and is itself induced to become elastic tissue [Taderera, 1967; Wessells, 1977].

The extent to which this process is continued in the adult is not clear, but evidence from epidermal maturation suggests that the underlying dermis is required for the formation of keratinized squames with fully cross-linked keratin [Bohnert et al., 1986]. Dissimilar cells do not form gap junctions readily, so the exchange of information may occur at the cell surface by an effector/receptor type of interaction, by modification of the intercellular matrix (see below), or by the transmission of paracrine factors.

Hepatocyte growth factor (HGF), one of the family of heparin-binding growth factors (HBGF) [Nakamura et al., 1987, 1989] (Table 7.8), has been shown to be released from fibroblasts, such as the MRC-5, and induce tubule formation in the MDCK continuous cell line from dog kidney [Li,Y., et al., 1992], while KGF (another HBGF) is produced by dermal fibroblasts and influences epidermal differentiation [Aaronson et al., 1991; Gumbiner, 1992]. Both of these factors are made only by fibroblasts and are prime candidates for paracrine differentiation factors. The role of factors such as FGF, Int-2, KGF, TGF-β, and activin in morphogenesis [Jessell and Melton, 1992] now seems clearly established, and the branching of epithelial ducts has been shown to be under the control of activin, HGF, and epimorphin [Gumbiner, 1992; Hirai et al., 1992].

The release of paracrine factors may be under the control of systemic hormones in some cases. It has been demonstrated that type II alveolar cells in the lung produce surfactant in response to dexamethasone *in vivo. In vitro* experiments have shown that this is dependent on the steroid binding to receptors in the stroma, which then releases a peptide to activate the alveolar cells [Post et al., 1984]. Similarly, the response of epithelial cells in the mouse prostate to androgens is mediated by stroma, as receptor-deficient epithelium (testicular feminization mutant, *tfm*) will respond in co-culture with normal stroma, while normal epithelium will not respond in co-culture with *tfm* stroma [Lasnitzki and Mizuno, 1979; Cunha, 1984]. The differentiation of the intestinal enterocyte, which is stimulated by hydrocortisone, also requires underlying stromal fibroblasts [Kédinger et al., 1987] and, in this case, modification of the extracellular matrix between the two cell types is implicated [Simon-Assman et al., 1986]. The FGF-like peptide KGF (keratinocyte growth factor) has been shown to be at least one component of the interaction in the prostate [Yan et al., 1993].

Cell–Matrix Interactions

Surrounding the surface of most cells is a complex mixture of glycoproteins and proteoglycans that is almost certainly highly specific for each tissue and even parts of a tissue. Reid [1990] has shown that the construction of artificial matrices from different constituents can regulate gene expression. Addition of liver-derived matrix material will induce expression of the albumin gene in hepatocytes. Furthermore, collagen has been found to be essential for the functional expression of many epithelial cells [Yang et al., 1979; Flynn et al., 1982; Burwen & Pitelka, 1980] and for endothelium to mature into capillaries [Folkman and Haudenschild, 1980]. The RGD motif (arginine-glycine-valine) in matrix molecules appears to be the receptor interactive moiety in many cases [Yamada, 1991]. Small polypeptides containing this sequence effectively block matrix-induced differentiation, implying that the intact matrix molecule is required [Pignatelli and Bodmer, 1988].

Attempts to mimic matrix effects by use of synthetic macromolecules have been partially successful using poly-D-lysine to promote neurite extension in neuronal cultures (see Chapter 20), but it would seem that there is still a great deal to learn about the specificity of matrix interactions. It is unlikely that charge alone is sufficient to mimic the more complex signals demonstrated in many different types of matrix interaction, but charge alterations probably allow cell attachment and spreading, and under these conditions the cells may be capable of producing their own matrix.

It has been shown that endothelium [Kinsella et al., 1992] and many epithelial cell types will differentiate more effectively on Matrigel [Kibbey et al., 1992], a matrix material produced by the Engelberth Holm Swarm (EHS) sarcoma, made up predominantly of laminin but also containing collagen and proteoglycans. This is a useful technique but has the problem of introducing another biological variable to the system. Defined matrices are required but, as yet, these have to be made in the laboratory and are not commercially available. While fibronectin, laminin, collagen, and a number of other matrix constituents are available commercially, the specificity probably lies largely in the proteoglycan moiety, within which there is the potential for wide variability, particularly in the number, type, and distribution of the sulfated glycosaminoglycans, such as heparan sulfate [Poole, 1986].

The extracellular matrix may also play a role in modulation of growth factor activity. It has been suggested that the matrix proteoglycans, particularly heparan sulfate proteoglycans (HSPGs), may bind certain growth factors, such as GM-CSF [Damon et al., 1989; Luikart et al., 1990], and make them more available to adjacent cells. Transmembrane HSPGs may also act as low-affinity receptors for growth factors and transport them to the high-affinity receptors [Klagsbrun and Baird, 1991].

Polarity and Cell Shape

Studies with hepatocytes [Sattler et al., 1978] showed that full maturation required the growth of the cells on collagen gel and the subsequent release of the gel from the bottom of the dish using a spatula or bent Pasteur pipette. This allowed shrinkage of the gel and an alteration in cell shape from flattened to cuboidal or even columnar. Accompanying or following shape change, and also possibly due to access to medium through the gel, the cells developed polarity, visible by electron microscopy; when the nucleus became asymmetrically distributed, nearer the bottom of the cell, an active Golgi complex formed and secretion toward the now-apical surface was observed.

A similar establishment of polarity has been demonstrated in thyroid epithelium by Mauchamp [Chambard et al., 1983], using a filter well assembly. In this case the lower (basal) surface generated receptors for thyroid-stimulating hormone (TSH) and secreted triiodotyrosine and the upper (apical) surface released thyroglobulin. More recent studies by Guguen-Guillouzo and Guillouzo [1986] (see Chapter 20) with hepatocytes and by Jetten and Smets [1985] with bronchial epithelium have suggested that floating collagen may not be essential, but this point is not yet fully resolved.

DIFFERENTIATION AND MALIGNANCY

It is frequently observed that, with increasing progression of cancer, histology of the tumor indicates poorer differentiation, and from a prognostic standpoint patients with poorly differentiated tumors will generally have a lower survival rate than those with differentiated tumors. It has also been stated that cancer is principally a failure of cells to differentiate normally. It is therefore surprising to find that many tumors grown in tissue culture can be induced to differentiate (Table 14.2). Indeed, much of the fundamental data on cellular differentiation has been derived from the Friend murine leukemia, mouse and human myeloma, hepatoma, and neuroblastoma. Nevertheless, there appears to be an inverse relationship between the expression of differentiated and malignancy-associated properties, even to the extent that the induction of differentiation has often been proposed as a mode of therapy [Dexter et al., 1979; Spremulli and Dexter, 1984; Freshney, 1985].

Apparently tumor cells may often retain the ability to respond to inducers of differentiation, though not always those active on normal cells. Friend erythroleukemia responds to DMSO but not to erythropoietin, while normal bone marrow [Gross and Goldwasser, 1971] and fetal mouse liver [Cole and Paul, 1966] respond to erythropoietin, but as yet there are no reports of induction by DMSO.

If tumor cells, or at least a proportion of them, respond to differentiation induction, it may be thought of as strange that they do not do so *in vivo*. The reasons for this are complex, but partly attributable to the overexpression of the "growth mode phenotype" due to overexpression of oncogenes and reduced or abnormal expression of tumor suppressor genes. This cannot be the only reason, however, as these genes are still expressed *in vitro*. Part of the reason may be the failure of normal cell-cell and cell-matrix interaction. The matrix is modified by the inability of the tumor cell to produce the correct constituents, and the increased production of extracellular proteases and glycosidases, which further modifies the matrix. The stromal cells may also be modified, due to the action of transforming growth factors, such as TGFα. Recent experiments in the author's laboratory have shown that lung fibroblasts exposed to tumor cells are no longer able to support steroid-induced differentiation of the lung epithelium (Freshney, McCormick, and Muir, unpublished observations).

Whatever the clinical outcome, tumor cells remain useful models for the study of differentiation and the production of specialized products.

PRACTICAL ASPECTS

It is clear that, given the correct environmental conditions, and assuming that the appropriate cells are present, partial or even complete differentiation is achievable in cell culture. The conditions required for individual cell types are not all elaborated and would be difficult to review here, but some indications are given in Chapter 20. As a general approach to promoting differentiation, as distinct from cell proliferation and propagation, the following may be suggested:

(1) Select the correct cell type by use of appropriate isolation conditions and medium (see Chapters 12 and 20).

(2) Grow to high cell density ($>10^5/cm^2$) on the appropriate matrix. This may be collagen of a type appropriate to the site of origin of the cells with or without fibronectin or laminin, or more complex tissue-derived [Reid and Rojkind, 1979], cell-derived [Gospodarowicz et al., 1980], Matrigel [Hartley and Yablonka-Reuveni, 1990; Schwarz et al., 1990; Ghosh et al., 1991; Kibbey et al., 1992; Kinsella et al., 1992], or synthetic (e.g., poly-D-lysine for neurons [Yavin and Yavin, 1980]) matrix.

(3) Change to a differentiating medium rather than propagation medium, e.g., for epidermis increase Ca^{2+} to around 3 m*M*, for bronchial mucosa increase the serum concentration (see Chapter 20). For other cell types, this may require definition of the growth factors appropriate to maintaining cell proliferation and those responsible for inducing differentiation (e.g., see glial culture, Chapter 20).

(4) Add differentiation-inducing agents such as glucocorticoids, retinoids, vitamin D_3, DMSO, hexamethylene bisacetamide (HMBA), prostaglandins, or peptide-differentiating factors such as bFGF, EGF, KGF, HGF, TGFβ, interferons, NGF, or melanocyte-stimulating hormone (MSH) as appropriate for the type of cell (see Tables 14.1, 14.2).

(5) Add interacting cell type during growth phase ([2] above), or induction phase ([3] and [4] above), or both. Selection of the correct cell type is not always obvious, but lung fibroblasts for lung epithelial maturation [Post et al., 1984; Speirs et al., 1991], glial cells for neuronal maturation [Lindsay, 1979], and bone marrow adipocytes for hemopoietic cells (see Chapter 20) are some of the better characterized examples.

(6) Floating the culture on detached collagen rafts [Sattler et al., 1978] or in a filter well [Chambard et al., 1983] may be advantageous, particularly for certain epithelia.

Preparation of Collagen Gel

The following was adapted from a protocol by Ted Ebendal, Department of Zoology, University of Uppsala, Sweden, based in turn on the method of Elsdale and Bard [1972].

Principle

Collagen is soluble in liquid solution at low pH and salt concentration. Increasing the pH to 7.4 and increasing the ionic strength causes the collagen to gel.

Outline

Dissolve rat-tail tendons in 0.5 M acetic acid and dialyze against one-tenth-strength culture medium, pH 4.0. Dilute dialyzed collagen with culture medium to form gel.

Materials

10× BME (Eagle's Basal Medium)
7.5% solution of $NaHCO_3$
L-glutamine (200 m*M*)
Fetal calf serum
Distilled water
0.142 M NaOH
0.5 M acetic acid

Protocol A. To prepare collagen

1. Cut off rat tail and soak in 70% EtOH for 1 min.
2. Fracture progressively from the tip and pull out the tendons. Handle with sterile gloves and keep tendons sterile.
3. Dissolve tendons from 20 tails in 200 ml of 0.5 M sterile acetic acid and filter through sterile muslin gauze.
4. Dialyze against 4 l sterile 1:10 Eagle's BME for 24 hr.
5. Repeat stage 4, adjusting the pH of the medium to 4.0 beforehand.
6. Centrifuge to clarify, 17,000 *g* for 24 hr or 50,000 *g* for 2 hr.

Protocol B. To prepare gel

1. Dilution medium–mix the following components in a glass tube on an ice bath to prepare approximately 5 ml of gel:
   455 μl 10× BME
   112 μl 7.5% $NaHCO_3$
   50 μl L-glutamine
   55 μl fetal bovine serum (if 1% serum in the gel is desired; for 10% serum take 555 μl)
   383 μl distilled water (minus the volume of water in the NaOH solution). Immediately before use add 50–150 μl of a 0.142 M NaOH solution, the exact amount needed to raise the pH to 7.4, as indicated by the change in color of the medium. This must be tested in advance for each batch of collagen solution that will be used.
2. Transfer 0.8 ml of the collagen solution to a 10-ml glass tube on ice (one tube for each dish).
3. To prepare the gel, take 0.21 ml of the dilution medium (for a gel intended to contain 10% serum, take 0.31 ml) and mix thoroughly with 0.80 ml collagen solution (avoid blowing air bubbles, which might be trapped in the gel) using a wide-bore pipette.
4. Transfer this final mixture to the culture dish and allow the gel to set.

Variations

Collagen gel may be derivatized by carboimide to increase its adherence to the plastic substrate (see below). Collagen is available commercially (ICN Flow; Collaborative Research/Becton Dickinson; Collagen Corporation).

The following protocol for coating surfaces with cross-linked collagen has been contributed by Jeffrey D. Macklis, Department of Neurology and Program in Neuroscience, Harvard Medical School, and Mental Retardation Research Center, Children's Hospital, Boston, MA 02115.

Principle

A new type of collagen surface for culture of nervous system cells was described by Macklis et al. [1985], which allowed extended culture survival, improved microscopy, and dry storage of coated culture dishes. Collagen was derivatized to plastic culture dishes by a cross-linking reagent, 1-cyclohexyl-3-(2-morpholinoethyl)-carbidiimide-metho-p-toluenesulfonate (carbodiimide); comparison to conventional ammonia-polymerized or adsorbed surfaces showed superior culture viability and improved optical characteristics. Simple covalent bonding of collagen fibrils to active groups on tissue culture plastic is described below. A large supply of coated dishes can be prepared in a single 5-hr session and stored for later use.

Outline

Prepare stock collagen solution, then dilute into an aqueous solution of carbodiimide. Coat dishes, incubate, wash, air dry, sterilize under UV, and use or store dry.

Materials

Collagen solution in dilute acetic acid at protein concentration of approximately 500 μg/ml, which can be purchased commercially or prepared by extraction from rat tails by the method of Bornstein and Murray [1958]
Carbodiimide (Aldrich Chemical Co.)
Tissue culture plates (Falcon, Becton Dickinson)
Double-distilled water, sterilized by autoclaving

Protocol

1. Place approximately 2 μg of carbodiimide in each of several 15-ml sterile medium tubes. Seal and store at 4°C until used.
2. Add 14 ml of sterile, double-distilled water at room temperature to each tube containing carbodiimide to be used (each prepared tube will coat 15 35-mm culture dishes.

3. Vortex each tube for approximately 10 s and set aside.
4. Add 1 ml of stock collagen solution to one tube containing carbodiimide solution (approximately 130 μg/ml) and rapidly vortex until uniform.
5. Rapidly transfer collagen-carbodiimide solution to dishes, generously covering the bottoms (approximately 1 ml in a 35-mm dish). The rapidity of transfer minimizes derivitization to the solution tube and maximizes early contact with the dish.
6. Incubate dishes at 25°C for 3 hr.
7. Wash three times with sterile, double-distilled water.
8. Air dry at room temperature for 1 hr.
9. Sterilize under ultraviolet irradiation for 1 hr.
10. Use dishes immediately or store dry for later use.

All of these factors may not be required and the sequence they are presented in is meant to imply some degree of priority. Scheduling may also be important; e.g., matrix generally turns over slowly so prolonged exposure may be important, while some hormones may be effective in relatively short exposures. Furthermore, the response to hormones may depend on the presence of the appropriate extracellular matrix, cell density, or heterologous cell interaction, so these components may be required to be stabilized prior to drug exposure.

Not all types of differentiation may be reproducible in culture, but it seems likely that given the isolation and survival of the correct cell types and the elaboration of the correct inducers, many of which may be tissue-derived peptides analogous to growth factors, production of functionally mature cells of many more types may be feasible in the near future.

CHAPTER 15

The Transformed Phenotype

WHAT IS TRANSFORMATION?

In microbiology, where the term was first employed in this context, *transformation* implies a change in phenotype dependent on the uptake of new genetic material. Although this is now possible in mammalian cells (see Chapter 23), it has been called "transfection" to distinguish it from transformation, which, in tissue culture, implies spontaneous or induced permanent phenotypic change that does not necessarily involve the uptake of new genetic material. Although transformation can arise from infection with a transforming virus such as polyoma and from incorporation of new genomic DNA, it can also arise spontaneously or following exposure to a chemical carcinogen. The primary alteration is usually considered to be genetic and irreversible, although recent studies have shown that some of the phenotypic properties of transformed cells can be restored to normal by chemical or hormonal inducers (see Chapter 14). Transformation is associated with three major classes of phenotypic change, one or all of which may be expressed in one cell strain:

Immortalization is the acquisition of an infinite life-span, presumed to be due to the deletion or mutation of one or more senescence genes [Pereira-Smith and Smith, 1988] or overexpression or mutation of one or more oncogenes that override the action of the senescence gene(s). Somatic hybridization experiments between finite and immortal cell lines usually generate finite life-span hybrids, suggesting that the senescence genes are dominant.

Immortalization, *per se*, does not imply either failure in growth control or the development of malignancy.

Aberrant growth control may arise simultaneously with immortalization, but can also occur as a distinct subsequent event. Cell lines such as the 3T3 group are immortal but still maintain contact inhibition of cell motility and density limitation of cell proliferation (see below). Immortalized cell lines can subsequently lose growth control, presumably by an additional genetic event(s), grow to higher saturation densities, clone in agar, and grow on confluent monolayers of homologous cells (see below). These lines exhibit lower serum or growth factor dependence and usually form clones with a higher efficiency, and are assumed to have acquired some degree of autonomous growth control by overexpression of oncogenes or by deletion of suppressor genes. Growth control becomes autocrine, i.e., the cells secrete mitogens for which they possess receptors, or the cells express receptors or stages in signal transduction that are permanently active and unregulated.

Although immortalization does not necessarily imply a loss of growth control, many cells progress readily from immortalization to aberrant growth, due perhaps to genetic instability intrinsic to the immortalized genotype.

The third development is *malignancy*, which is not adequately described *in vitro*, as it is essentially an *in vivo* phenomenon. Malignancy implies that the cells have developed the capacity to generate invasive tumors if implanted *in vivo* into an isologous host, or transplanted as a xenograft into an immune-deprived animal. While the development of malignancy can be recognized as a discrete phenotypic event, it often accompanies the development of aberrant growth control, suggesting that some of the lesions responsible for aberrant growth control also cause malignancy. Obvious candidates would be a deficit in cell–cell interaction depriving the cell of control of proliferation (density limitation of cell proliferation) and of motility control (contact inhibition).

Transformation is therefore a complex series of events, much like carcinogenesis *in vivo*, leading to a loss of growth control and positional control and immortalization. It is not clear as yet whether immortalization is a necessary prerequisite for aberrant growth control and malignancy, and, in fact, cell lines cultured from tumors frequently undergo further transformation and give rise to continuous cell lines, implying that the original isolate was malignant (from a malignant tumor and exhibiting aberrant growth control) but still required an additional step to become immortalized. The evidence at present, therefore, suggests that transformation can be separated into at least two distinct events, immortalization and acquisition of autonomous growth control, not neccessarily in that order, that one predisposes toward the other, and that, of the two, acquisition of autonomous growth control is most likely to be associated with malignancy.

This has significant implications both for the study of cancer and the utilization of continuous cell lines in biotechnology, where quality-assurance measures would prefer that immortalization were not equated with malignant transformation. That does seem to be the case, although immortalized cell lines are undoubtedly closer to a malignant phenotype than finite cell lines.

There are several properties associated with transformation *in vitro* (Table 15.1), the most important one being immortalization. Most normal cells have a finite life-span of 20–100 generations (see Chapter 2) but some cells, notably those from rodents and from most tumors, can produce continuous cell lines with an infinite life-span. The rodent cells are karyotypically normal at isolation and appear to go through a crisis after about 12 generations, in which most of the cells die out, but a few survive with an enhanced growth rate and give rise to a continuous cell line.

TABLE 15.1. Properties of Transformed Cells[a]

Growth characteristics	Immortal
	Anchorage independent: clone in agar, may grow in stirred suspension
	Loss of contact inhibition
	Growth on confluent monolayers of homologous cells "focus" formation (see Fig. 15.1)
	Reduced density limitation of growth: high saturation density, high growth fraction at saturation density (see Chapter 18, Fig. 18.8)
	Low serum requirement
	Growth factor independent
	High plating efficiency
	Shorter population doubling time
Genetic properties	High spontaneous mutation rate
	Aneuploid
	Heteroploid
	Overexpressed oncogenes
	Deleted suppressor genes
Structural alterations	Modified actin cytoskeleton
	Loss of cell surface-associated fibronectin
	Increased lectin agglutination
	Modified extracellular matrix
	Altered cell adhesion molecules
	Disruption in polarity
Neoplastic properties	Tumorigenic
	Angiogenic
	Enhanced protease secretion, e.g., plasminogen activator
	Invasive

[a]It is not implied in this table that all of these properties are expressed in transformed cell lines, but there is a higher probability of them occurring than in normal, finite cell lines. Furthermore, the degree of expression of individual properties will vary greatly among different lines and depend on the degree or type of transformation (see above).

If continuous cell lines from mouse embryos (e.g., the various 3T3 cell lines) are maintained at a low cell density and not allowed to remain at confluence for any length of time, they remain sensitive to contact inhibition and density limitation of growth [Todaro and Green, 1963] (see below). If, however, they are allowed to remain at confluence for extended periods, foci of cells appear with reduced contact inhibition, begin to pile up, and will ultimately overgrow (Fig. 15.1b).

The fact that these cells are not apparent at low densities, or when confluence is first reached, suggests that they arise *de novo*, by a further transformation event. They appear to have a growth advantage and

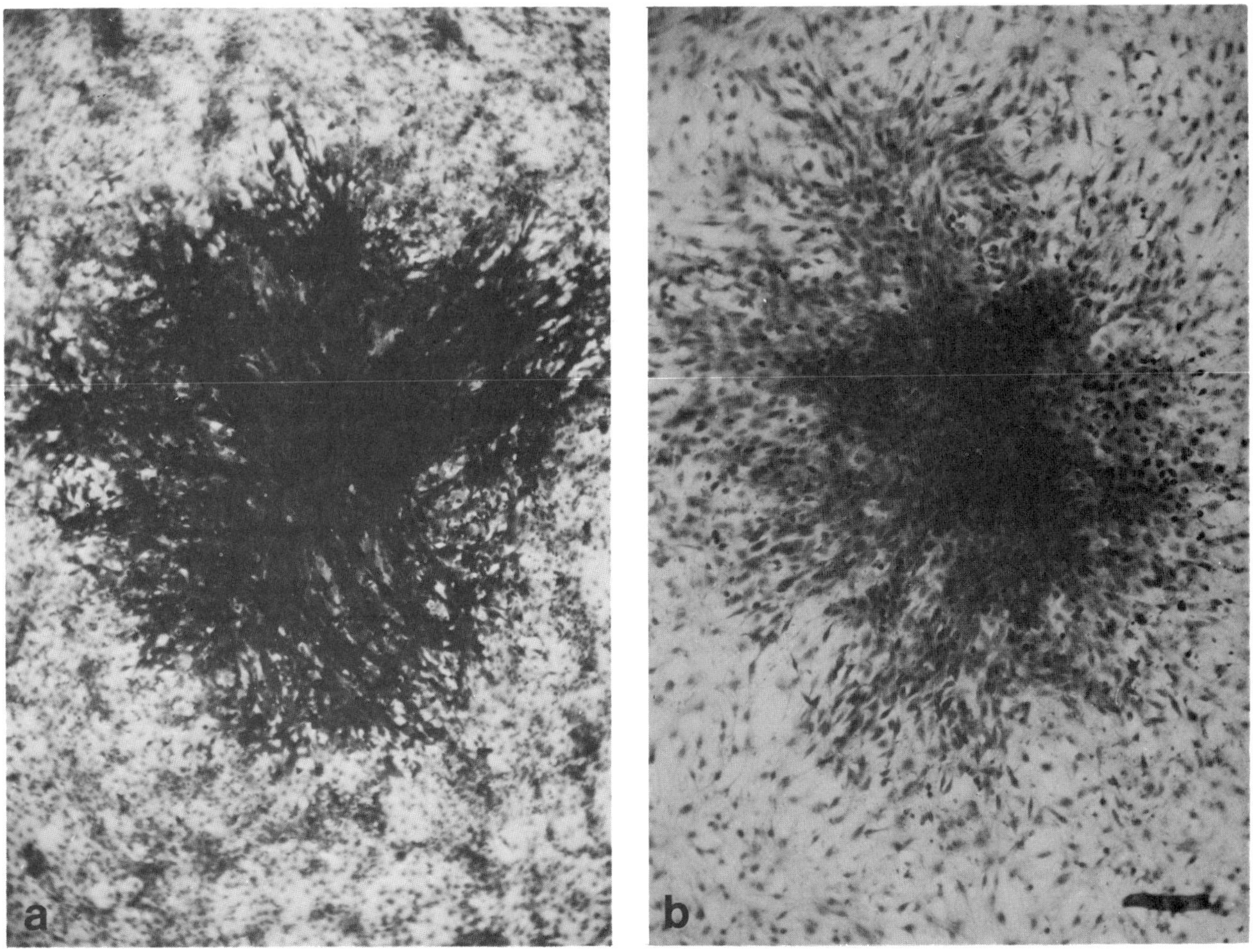

Fig. 15.1. Transformation foci in a monolayer of normal, contact-inhibited NIH3T3 mouse fibroblasts. a. NIH3T3 mouse fibroblasts transformed by transfection with bovine papilloma virus DNA cloned in bacterial plasmid pAT-153, coprecipitated with $Ca_3(PO_4)_2$. b. Spontaneous transformant arising when cells maintained at a high density. (Courtesy of D. Spandidos, photographs by M. Freshney.)

subsequent subcultures will rapidly be overgrown by the randomly growing cell. This cell type is often found to be *tumorigenic.*

Studies on cell lines from normal human urothelium also indicate that *in vitro* transformation is a progressive, multistep event [Christensen et al., 1984]. Normal urothelial cell cultures are made up of predominantly diploid cells that are nonangiogenic, noninvasive, and nontumorigenic. Over a period of up to 2 years these lines become first angiogenic, then aneuploid, and ultimately tumorigenic and invasive.

Hence transformation is apparently a multistep process often culminating in the production of neoplastic cells [Quintanilla et al., 1986]. It is therefore strange to find that cell lines from malignant tumors, presumably already "transformed," can undergo further "transformation" with increased growth rate, reduced anchorage dependence, more pronounced aneuploidy, and immortalization. Such a transformation has been observed in a squamous lung carcinoma cell line in the author's laboratory, and the resultant continuous cell line actually lost tumorigenicity in nude mice.

This suggests that a series of steps, not necessarily coordinated or interdependent and not necessarily individually tumorigenic, is required for malignant transformation. The same set of properties need not be expressed in every tumor and progression may imply expression of new properties or deletion of old ones that may induce metastasis or even spontaneous remission. The process may be likened to a game of cards such as gin rummy, where the cards need not be obtained in any specific sequence and several different sets will allow the player to "go out." There are, therefore, several steps in transformation, including increased immortalization and tumorigenicity (see Table 15.1); the sequence may be determined by environmental selective pressure. *In vitro*, where there is little restriction on growth imposed, the events need not necessarily follow in the same sequence.

Consequently, transformation is not as easily defined in cell culture as it is in microbiology and should be used with caution and preferably qualified. Three distinct steps can be observed *in vitro*, (1) transformation into a continuous cell line, normally called *immortalization*, (2) loss of growth control, and (3) *malignant* or *neoplastic* transformation, implying the ability to grow as an invasive tumor *in vivo*.

Since malignancy as such cannot be demonstrated *in vitro*, we are obliged to use a number of properties associated with cells from malignant tumors grown *in vitro*. In the present discussion, I describe markers that might be used in cell identification and do not imply a causal relationship between these properties and the expression of malignancy *in vivo*, but clearly many of the properties discussed have an obvious functional relationship to malignancy.

Two approaches have been used to explore malignancy-associated properties: (1) Cells have been cultured from malignant tumors and characterized. (2) Transformation *in vitro* with a virus, or a chemical carcinogen, or transfection with oncogenes, has produced cells that were tumorigenic and that could be compared with the untransformed cells. The second system provides transformed clones of the same lineage, which can be shown to be malignant, and they can be compared with untransformed clones, which are not. Unfortunately, many of the characteristics of cells transformed *in vitro* have not been found in cells derived from spontaneous tumors. Ideally, tumor cells and equivalent normal cells should be isolated and characterized. Unfortunately, there have been relatively few instances where this has been possible, and even then, although the cells may belong to the same lineage, their position in that lineage is not always clear (see Fig. 2.3) and comparison not strictly justified.

At best, there are a number of generally accepted properties that can be recognized in many tumor cells *in vitro*. Many occur in normal cells, confirming that malignancy is not the expression of abnormal characteristics *de novo* but rather the inappropriate and uncontrolled expression of normal properties, and none is common to all neoplastic cell lines.

ANCHORAGE INDEPENDENCE

Many of the properties associated with neoplastic transformation *in vitro* are the result of cell surface modifications [Hynes, 1974; Nicolson, 1976; Bruynell et al., 1990], e.g., changes in the binding of plant lectins [Willingham and Pastan, 1975; Reddy et al., 1979; Laferte and Loh, 1992] and in cell surface glycoproteins [Hynes, 1976; Van Beek et al., 1978; Warren et al., 1978; Lloyd et al., 1979], which may be correlated with the development of invasion and metastasis *in vivo*. Fibronectin (large extracellular transformation-sensitive [LETS] protein) is lost from the surface of transformed fibroblasts [Hynes, 1973; Vaheri et al., 1976]. This may contribute to a decrease in cell–cell and cell–substrate adhesion [Yamada, 1991; Reeves, 1992] and to a decreased requirement for attachment and spreading for the cells to proliferate.

Transformed cells may lack specific CAMs, e.g., L-CAM, which, when transfected back into the cell, regenerate the normal noninvasive phenotype [Mege et al., 1989], and as such they may be recognized as tumor suppressor genes. Others may be overexpressed, such as N-CAM in small cell lung cancer [Patel et al., 1989], where the extracellular domain is subject to alternative splicing [Rygaard et al., 1992]. The expression of and degree of phosphorylation of integrins may also change [Watt, 1991], potentially altering cytoskeletal interactions, regulation of gene transcription (see Chapter 14), the substrate adhesion of the cells, and the relationship between cell spreading and cell proliferation.

In addition, loss of cell–cell recognition, a product of reduced adhesion, leads to a disorganized growth pattern and loss of density limitation of growth (see below). This results in the ability of cells to grow detached from the substrate, either in stirred suspension culture or suspended in semisolid media such as agar or Methocel. There is an obvious analogy with detachment from the tissue in which a tumor arises and the formation of metastases in foreign sites, but the validity of this analogy not clear.

Suspension Cloning

Macpherson and Montagnier [1964] were able to demonstrate that polyoma-transformed BHK21 cells could be grown preferentially in soft agar, while untransformed cells cloned very poorly. Subsequently, it has been shown that colony formation in suspension is frequently enhanced following viral transformation. The position regarding spontaneous tumors is less clear, however, in spite of the fact that Freedman and Shin [1974; Kahn and Shin, 1979] demonstrated a close correlation between tumorigenicity and suspension cloning in Methocel. Although Hamburger and Salmon [1977] have shown that many human tumors contain a small percentage of cells (<1.0%) that are clonogenic in agar, we [Freshney and Hart, 1982] and others [Laug et al., 1980] have shown that a number of normal cells will also clone in suspension (see Fig. 11.11) with equivalent efficiency. Since normal fibroblasts are among these cells that will clone in suspension, the value of this technique for assaying for the presence of tumor cells in short-term cultures from human tumors is in some doubt. It remains a valuable technique for assaying neoplastic transformation *in vitro* by tumor viruses and was used extensively by Styles [1977] to assay for carcinogenesis.

The technique of cloning in suspension is described in Chapter 11. Variations with particular relevance to the assay of neoplastic cells are in the choice of suspending medium. It has been suggested [Neugut and Weinstein, 1979] that agar may only allow the most highly transformed cells to clone, while agarose (lacking sulfated polysaccharides) is less selective. Montagnier [1968] was able to show that untransformed BHK21 cells, which would grow in agarose but not in agar, could be prevented from growing in agarose by the addition of dextran sulfate.

Contact Inhibition and Density Limitation of Growth

The loss of contact inhibition (see Chapter 10) may be detected morphologically by the formation of a disoriented monolayer of cells or rounded cells in foci within the regular pattern of normal surrounding cells. This is illustrated in Figure 15.1, where 3T3 cells transformed by bovine papilloma virus DNA are compared with spontaneous transformants. Cultures of human glioma show a disorganized growth pattern and exhibit reduced density limitation of growth by growing to a higher saturation density than normal glial cell lines [Freshney et al., 1980a,b]. As variations in cell size will influence the saturation density, the increase in the labeling index with [^{3}H]-thymidine at saturation density (see Chapter 18) is a better measurement of reduced density limitation of growth. Human glioma, labeled for 24 hr at saturation density with [^{3}H]-thymidine, gave a labeling index of 8%, while normal glial cells gave 2% [Guner et al., 1977].

Outline

The culture is grown to saturation density in nonlimiting medium conditions and the percentage of cells labeling with [^{3}H]-thymidine determined autoradiographically.

Materials

Culture of cells ready for subculture
PBSA
0.25% trypsin
24-well plates containing 13-mm coverslips
Growth medium
9-cm petri dishes (bacteriological grade, one per coverslip)
Maintenance medium (no serum or growth factors) containing 37 KBq/ml (1.0 μCi/ml) [^{3}H]-thymidine, 74 GBq/mmol (2Ci/mmol)

Protocol

1. Trypsinize cells and seed 10^5 cells/ml into 24-well plate, 1 ml/well, each well containing a 13-mm-diameter coverslip.
2. Incubate in humidified CO_2 incubator for 1–3 d.
3. Transfer the coverslips to 9-cm bacteriological grade petri dishes containing 20 ml medium and return to CO_2 incubator.
4. Continue culturing, changing medium every 2 d after cells become confluent on the coverslips. Trypsinize and count cells from two coverslips every 3–4 d. As cells become denser on the coverslip, it may be necessary to add 200–500 units/ml crude collagenase to the trypsin to achieve complete dissociation of the cells for counting.
5. When cell growth ceases, i.e., two sequential counts show no significant increase, add 2.0 ml 37 KBq/ml (1.0 μCi/ml) [^{3}H]-thymidine, 74 GBq/mmol (2 Ci/mmol), and incubate for a further 24 hr.

   ◊ *Note*. Handle [^{3}H]-thymidine with care. Although a low-energy β-emitter, it localizes to DNA and can induce radiolytic damage. Wear gloves, do not handle in horizontal laminar flow but in a biohazard or cytotoxic drug handling cabinet (see Chapter 6), and discard waste liquids and solids by the appropriate route specified in the local rules governing the handling of radioisotopes.
6. Transfer coverslips back to a 24-well plate and trypsinize the cells for autoradiography (see Chapter 23). They may be fixed in suspension and dropped on a slide as for chromosome preparations (without the hypotonic treatment), centrifuged onto a slide using a cytocentrifuge (e.g., Shandon Cytospin), or trapped on Millipore or Nuclepore filters by vacuum filtration (see Chapter 13).

   Note. It is necessary to trypsinize high-density cultures for autoradiography because of their thickness and the weak penetration of β-emission from ^{3}H (mean path length in water is approximately 1 μm). Labeled cells in the underlying layers would not be detected by the radiosensitive emulsion due to the absorption of the β-particles by the overlying cells. If the cells remain as a monolayer at saturation density, this step may be omitted, and autoradiographs prepared by mounting the coverslips, cells uppermost, on a microscope slide.

Analysis. Count the number of labeled cells as a percentage of the total. Scan the autoradiographs under the microscope and count the total number of cells and the proportion labeled in representative parts of the slide. A suggested scanning pattern for a circular array of cells (such as would be produced by the drop technique or the Cytospin) is given in Figure 18.6.

Variations. Cells in DNA synthesis may also be labeled with bromodeoxyuridine (BUdR) and subsequently detected by antibody to BUdR-labeled DNA (Dako). Human cycling cells can also be labeled with the Ki67 monoclonal antibody (Dako), which binds to DNA polymerase. While there is generally good agreement between [^{3}H]-thymidine and BUdR labeling, Ki67 will label more cells, as the antigen is present throughout cycle and not restricted to S-phase.

Growth of cells at high density in nonlimiting medium can also be achieved by growing the cells in a filter well (Falcon, Costar, Millipore), choosing a filter diameter substantially below that of the dish (e.g., Costar 8-mm filter in 24-well plate).

Growth on Confluent Monolayers

Aaronson et al. [1970] showed that transformed mouse 3T3 cells and human fibrosarcoma cells were able to form colonies on confluent monolayers of normal cells, while normal fibroblasts were not. In the author's laboratory, cells cultured from both normal and malignant breast tissue formed colonies on contact-inhibited monolayers of human fetal intestinal cells (FHI) and on normal mouse embryo fibroblasts (STO) [Freshney et al., 1982b] (Fig. 11.8), as do normal epidermal cells (Fig. 15.2). Rheinwald and Green [1975] had previously shown that normal epidermal cells grew on 3T3 monolayers. In some cases, therefore, where an appropriate control cell can be tested, this technique may discriminate between malignant and normal cells, but this may require that the normal cells are of the same lineage as the malignant cells. We have found that human glioma will grow on confluent glial feeder layers, but cells derived from normal glia will not (see Fig. 11.9), while some normal glial cell lines will grow on other feeder layers, e.g., fetal lung fibroblasts and FHI.

Outline

Prepare cells as for cloning and seed onto confluent monolayers of contact-inhibited cells (see Fig. 11.3).

Materials

14 × 25-cm^2 flasks
Growth medium
PBSA
0.25% trypsin
Cultures of tumor cells
Tubes for dilutions
Hemocytometer or cell counter

Protocol

1. Seed 14 25-cm^2 flasks with feeder cell (3T3 or other cell line that will become contact inhibited after reaching confluence).
2. Feed cultures until all available substrate is covered with cells and mitosis is greatly reduced. If a cell line is used that does not arrest at confluence, treat cells during exponential growth with mitomycin C or irradiate (see Chapter 11) and reseed sufficient cells to produce confluent monolayers directly with no spaces between cells.
3. Trypsinize putative tumor cells and suitable control cells (e.g., equivalent normal cells) and dilute to 10^4, 10^3, and 10^2 cells/ml in 20 ml and replace medium in each pair of feeder-layer flasks with 5 ml of cell suspension, and with medium alone in two flasks.
4. Seed two flasks at each concentration with each cell type without feeder layers.
5. Incubate for 2–3 wk with medium changes every 2 d.
6. Wash, fix, and stain (see Chapter 13).

Analysis. Count foci of tumor cells (usually morphologically distinguishable) and express as percentage of number of cells seeded. Compare with flasks with no feeder layers.

Variations. It is possible to have proliferation of cells seeded on confluent monolayers without the formation of discrete colonies. In our experience, human

Fig. 15.2. Epidermal cells from a skin biopsy of a benign nevus growing on a confluent monolayer of fetal human intestinal epithelial cells (FHS74Int) [Owens et al., 1974]. The large, dark-staining, circular colonies resemble keratinocytes. The smaller colonies may have been melanocytes (fibroblasts do not normally form colonies readily in this system), but no further characterization was done on them. The lower flask was treated with 1.0 mM dibutyryl cyclic AMP, inhibiting the small colony type but not the large (putative keratinocytes). Cyclic AMP stimulates growth of keratinocytes (see Chapter 20) and may have done so here (top flask, four colonies; bottom flask, eight colonies).

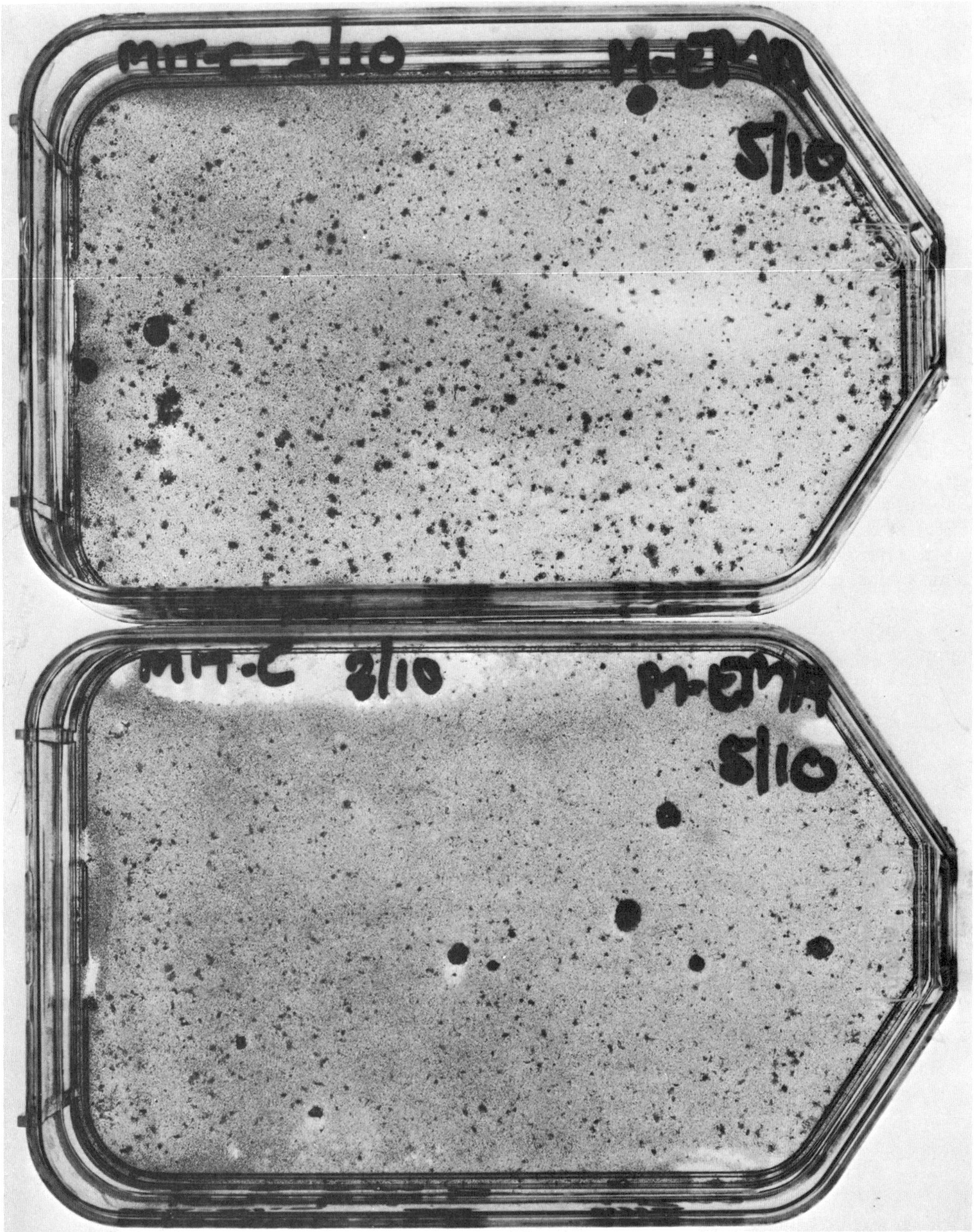
MIT-C 2/10
M-EMA
5/10
MIT-C 2/10
M-EMA
5/10

fibrosarcoma and glioma seeded on confluent monolayers of fetal human intestine do not form colonies, but still proliferate, infiltrating the monolayer as they do so. In order to quantify this type of growth, DNA synthesis is first inhibited in the monolayer (see Chapter 11) by treatment with mitomycin C or by irradiation. Growth of cells seeded onto this layer can then be monitored by measuring the incorporation of [^{3}H]-thymidine by scintillation counting or autoradiography (see Chapters 10 and 24).

GENETIC ABNORMALITIES

Whether mutation is the prime cause of neoplastic transformation or not, chromosomal abnormalities and variations in DNA content per cell are found frequently in cells derived from malignant tumors.

Chromosomal Aberrations

Both changes in ploidy and increases in the frequency of individual chromosomal aberrations can be found [Biedler, 1976; Croce, 1991]. Figure 15.3 demonstrates variations in chromosome number found in human glioma and melanoma in culture. Chromosome analysis is described in Chapter 13 [see also Sandberg, 1982]. The frequency of chromosomal rearrangement can be determined by the sister chromatid exchange assay [Venitt, 1984] (see also Chapter 9).

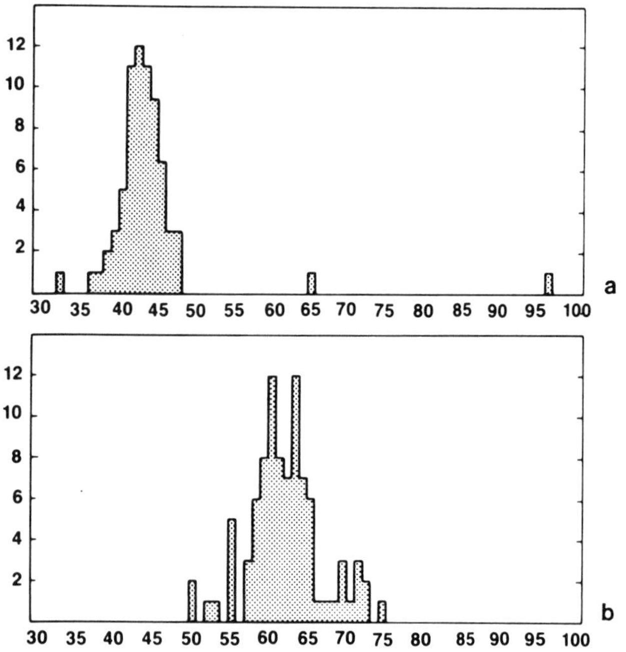

Fig. 15.3. Chromosome counts for cells cultured from human anaplastic astrocytoma (a) and human metastatic melanoma (b). The normal human diploid chromosome number is 46.

DNA content. Microdensitometry of Feulgen-stained preparations [Wright and Dendy, 1976; Stolwijk et al., 1987] and flow cytometry [Traganos et al., 1977] show that the DNA content of tumor cells may vary from the normal. DNA analysis does not substitute for chromosome analysis, however, as cells with an apparently normal DNA content can yet have an aneuploid karyotype. Deletions and polysomy may cancel out, or translocations may occur without net loss of DNA.

Some specific aberrations are associated with particular types of malignancy [Croce, 1991]. The first of these to be documented was the Philadelphia chromosome in chronic myeloid leukemia (trisomy 13). Subsequently translocations of the long arms of chromosomes 8 and 14 were found in Burkitt's lymphoma [Lebeau and Rowley, 1984]. Several other leukemias also express similar and different translocations [Mark, 1971]. Meningiomas often have consistent aberrations and small cell lung cancer frequently has a 3p2 deletion [Wurster-Hill et al., 1984]. These aberrations constitute tumor-specific markers that can be extremely valuable in cell line characterization and confirmation of neoplasia.

CELL PRODUCTS AND SERUM DEPENDENCE

Transformed cells have a lower serum dependence than their normal counterparts [Temin, 1966; Eagle et al., 1970]. Lindgren et al. [1975] showed that a short exposure to serum would trigger glioma cells into cycle, while normal glial cells required serum to be present throughout G_1.

A possible explanation for the low serum dependence of tumor cells lies in the demonstration of Todaro and DeLarco [1978], and others, that tumor cells may secrete their own growth factors. While the production of these polypeptides is not assayed as easily as some of the foregoing properties, the increasing availability of specific antibodies against them could bring this approach within the reach of any laboratory with the appropriate immunological expertise.

These factors have been collectively described as autocrine growth factors. Implicit in this definition is (1) the cell produces the factor; (2) the cell has receptors for it; and (3) the cell responds to it by entering mitosis. Some of these factors may have an apparent transforming activity on normal cells (e.g., TGFα) binding to the EGF receptor and inducing mitosis [Richmond et al., 1985], although, unlike true transformation (see above), this is probably reversible. They also cause non-transformed cells to adopt a transformed phenotype and grow in suspension [Todaro and DeLarco, 1978].

Tumor cells can also produce many hemopoietic

growth factors such as the interleukins 1, 2, and 3, along with colony-stimulating factor (CSF) [Fontana et al., 1984; Metcalf, 1985a]. It has been proposed [Cuttitta et al., 1985] that some factors such as bombesin and vasoactive intestinal peptide (VIP), hitherto believed to be ectopic hormones produced by lung carcinomas, may in fact be autocrine growth factors. Production of such factors implies expression of the malignancy-associated phenotype and can also help in lineage classification.

Autonomous growth control is also achieved in transformed cells by expression of modified receptors, such as the *erb*-B2 oncogene product, modified G protein, such as mutant *ras*, or overexpression of genes regulating stages in signal transduction (e.g., *src* kinase) or transcriptional control (e.g., *myc, fos,* or *jun*) [Bishop, 1991]. In many cases the gene product is permanently active and is unable to be regulated. Overexpression of the genes can be detected by antibody staining or immunoblotting for the protein product or Northern blotting for mRNA. In some cases, the oncogene product, e.g., *erb*-B2 or activated Ha-*ras*, can be distinguished from the normal product (EGF receptor, normal *ras*) *qualitatively* as well as quantitatively, by specific antibodies.

Tumor Angiogenesis Factor (TAF)

Tumor cells also release a factor (or factors) capable of inducing blood vessel proliferation [Folkman, 1992]. Fragments of tumor, pellets of cultured cells, or cell extracts, implanted on the surface of the chorioallantoic membrane (CAM) of the hen's egg, promote an increase in vascularization that is apparent to the naked eye 6–8 days later (Fig. 15.4). Since this assay is not readily quantified, stimulation of cell migration [Gullino, 1985] or proliferation [Freshney, 1985] in monolayer cultures of vascular endothelium may provide the basis for a more quantitative assay.

Plasminogen Activator

Other cell products that can be recognized are proteolytic enzymes [Mahdavi and Hynes, 1979], long since associated with theories of invasive growth [Liotta, 1987]. Since proteolytic activity may be associated with the cell surface of many normal cells and is absent on some tumor cells, an equivalent normal cell must be used as a control when using this criterion. Plasminogen activator (PA) is higher in some cultures from human glioma than in cultures from normal brain [Hince and Roscoe, 1980] (Fig. 15.5), and others have shown previously that PA is associated with many different tumors [Rifkin et al., 1974; Nagy et al., 1977]. PA may be measured by clarification of a fibrin clot or release of free soluble ^{125}I from [^{125}I] fibrin [Unkless et al., 1974; Strickland and Beers, 1976]. A simple chromogenic assay has also been developed by Whur et al. [1980].

It has been proposed that, for some carcinomas, soluble urokinase-like PA (uPA) is elevated more than tissue-type PA (tPA) [Markus et al., 1980; Duffy et al., 1990], so it is informative to couple the above chromogenic assay with immunoblotting to determine the proportion of each type of PA.

INVASIVENESS

An advantage of the CAM assay using tumor cells to induce angiogenesis is that the subsequent histology may reveal whether the tumor cells have penetrated the underlying basement membrane. Easty and Easty [1974] showed that invasion of the CAM could be demonstrated in organ culture, and others [Hart and Fidler, 1978] have attempted, with some limited success, to construct a chamber capable of quantitating the penetration of tumor cells across the CAM. Mareel et al. [l979] developed an *in vitro* model for invasion, using chick embryo heart fragments co-cultured with reaggregated clusters of tumor cells. Invasion appears to be correlated with the malignant origin of the cells, is progressive, and causes destruction of the host tissue. The application of this technique to human tumor cells shows a good correlation between malignancy and invasiveness in the assay [de Ridder and Calliauw, 1990].

The embryonic chick heart technique has been used extensively but is difficult to quantify and requires skilled histological interpretation. This has led others [Repesh, 1989; Schlechte et al., 1990] to develop filter-well techniques based on penetration of filters coated with Matrigel or some other extracellular matrix constituent. Penetration into the gel, or through to the distal side of the filter, is rated as invasiveness, and rated histologically by number of cells and distance moved, or by prelabeling the cells with ^{125}I and counting the radioactivity on the distal side of the filter or bottom of the dish. It is more readily quantitated but lacks the presence of normal host cells in the barrier, normally associated with invasion *in vivo*. Penetration of Matrigel is likely to be a measure of matrix degradation and reflect the production of proteases or glycosidases by the cells. It is not clear how cells that do not make their own degradative enzymes but rely on the production of proteases induced in the stroma will perform in these assays.

TUMORIGENESIS

The only definition of malignancy that is generally accepted is the demonstration of the formation of invasive or metastasizing tumors *in vivo*. Transplantable

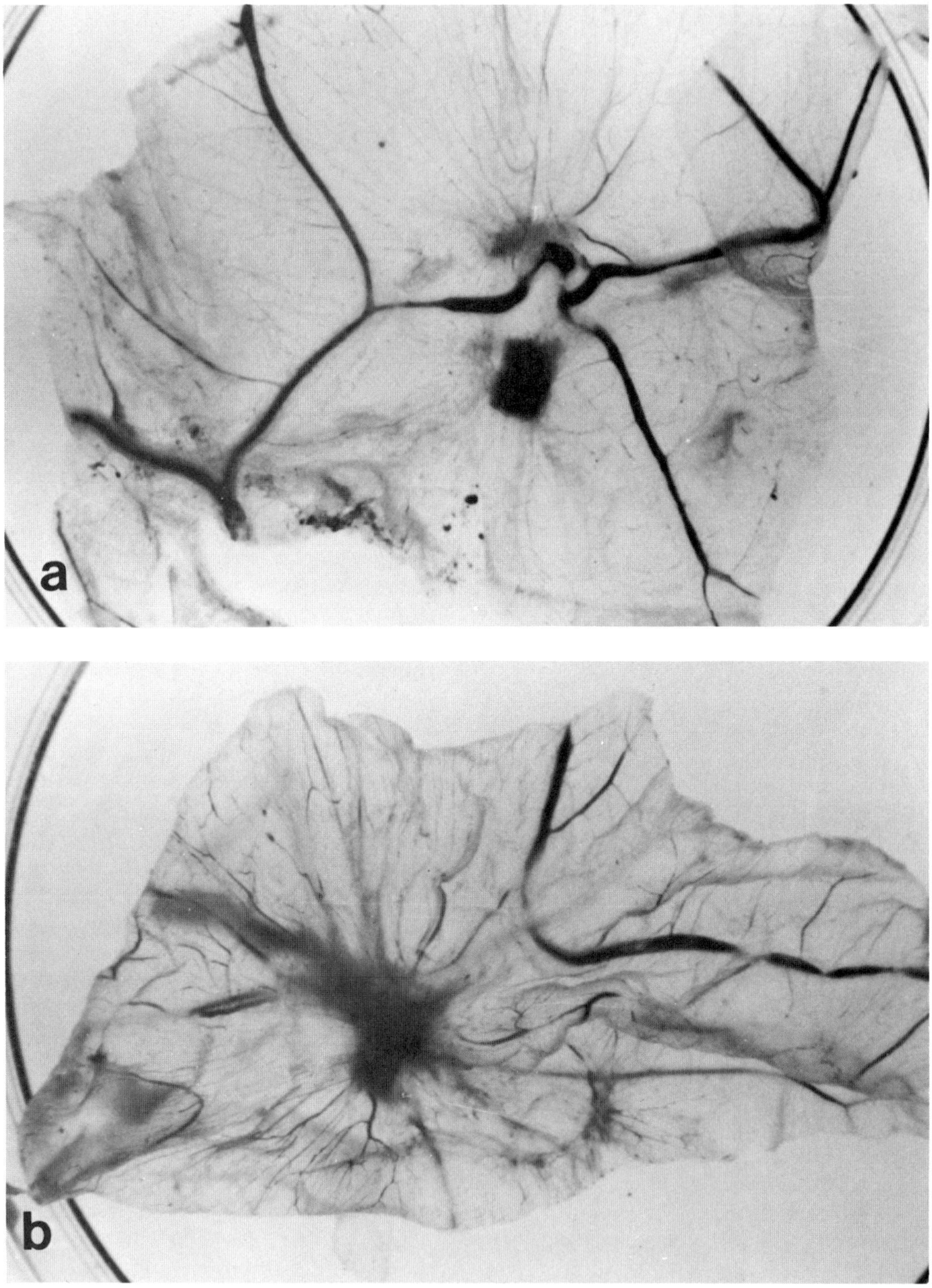

Fig. 15.4. Induction of angiogenesis in chick chorioallantoic membrane by tumor cell extract. A crude extract of Walker 256 carcinoma cells absorbed into sterile filter paper was placed on the chorioallantoic membrane at 10 d incubation, and the membrane was removed 2 wk later. a. Control. b. Walker 256 extract. (Courtesy of Margaret Frame.)

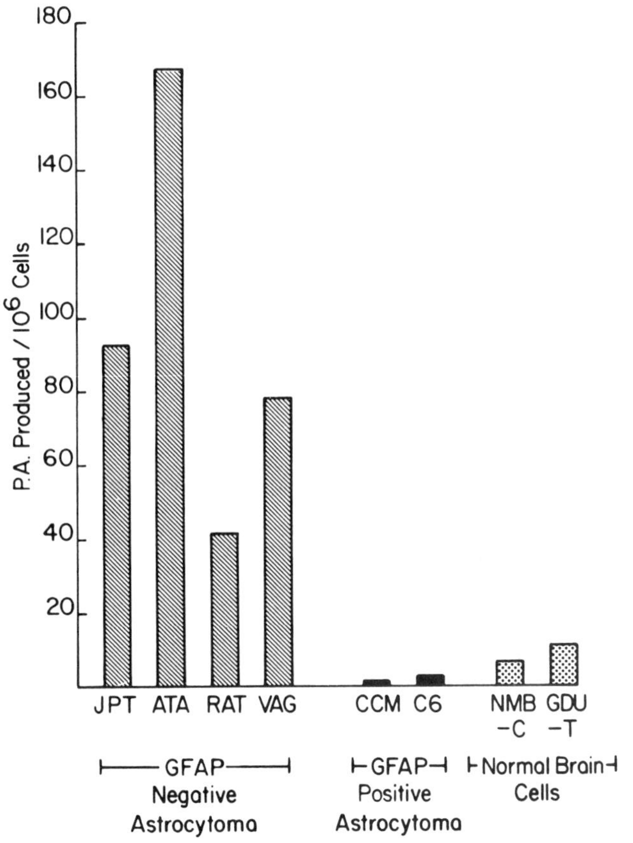

Fig. 15.5. Plasminogen activator (PA) produced by tumor cells in vitro *(arbitrary units). The four gliomas, JPT, ATA, RAT, and VAG, were all higher than cells cultured from normal brain (NMB-C, GDU-T). It was also found that the only cells to produce the differentiated glial marker glial fibrillary acidic protein, CCM and C6, had the lowest PA of all.*

tumor cells (10^6 or less) injected into isogeneic hosts will produce invasive tumors in a high proportion of cases, while 10^6 normal cells of similar origin will not. To study the tumorigenicity of human tumors, a number of models have been developed using immune-suppressed or immune-deficient host animals. The genetically athymic "nude" mouse [Giovanella et al., 1974] and thymectomized irradiated mice [Bradley et al., 1978; Selby et al., 1980] have both been used extensively as hosts for xenografts. The take rate varies, however, and many clearly defined tumor cell lines and tumor biopsies have failed to produce tumors as xenografts, and frequently those that do then fail to metastasize, although they may be invasive locally. Take rates can be improved by sublethal irradiation of the host nude mouse (30–60 Gy), by using asplenic athymic (*scid*) mice, or by implanting cells in Matrigel [Pretlow et al., 1991]. In spite of the frequency of false negatives, it remains a good indicator of malignancy.

CHAPTER 16

Contamination

Maintenance of asepsis is still one of the most difficult challenges to the newcomer to tissue culture. Part of this is due to awkwardness during early training, and experience will eventually cure the problem. However, there are certain situations where even the most experienced worker will suffer contamination. Potential routes of contamination are listed in Table 16.1, along with some suggestions on how to avoid them. As a general principle, given that the reagents used are sterile and equipment is in proper working order, contamination depends on the interaction of operator technique with environmental conditions. Where the skill and level of care of the operator is high, and the atmosphere is clean, dust-free, and still, contaminations as a result of manipulation will be rare. If the environment deteriorates (e.g., as a result of building work or a seasonal increase in humidity), or the operator's technique declines (omission of one or more apparently unnecessary steps), the probability of infection increases. If both happen simultaneously, or sequentially, then the results can be catastrophic.

If this is considered in graphic form (see Fig. 5.1), the maintenance of good technique may be represented by the top graph line with occasional lapses shown as a downward peak, and the quality of the environment may be depicted by the bottom graph, with occasional, sporadic increases in risk such as a contaminated plate opened accidentally in the area, or dust generated from equipment maintenance. Provided that the two graphs are kept well apart (good technique, high-quality environment) then the coincidence of a lapse in technique and an environmental breakdown will be rare. If, however, there is a decline in technique or in the environment, the frequency will increase, and if both deteriorate, contamination will be regular and widespread.

TYPES OF MICROBIAL CONTAMINATION

Bacteria, yeasts, fungi, molds, and mycoplasmas all appear as contaminants in tissue culture, and where protozoology is carried on in the same laboratory, some protozoa can infect cell lines. Usually the species or type of infection is not important unless it becomes a frequent occurrence. It is only necessary to note the type (bacterial rods or cocci, yeast, etc.), how detected, the location where the culture was last handled, and the operator's name. If a particular type of infection recurs frequently, it may be beneficial to have it identified to help to find the origin. [For more detailed screening procedures for microbial contamination, see

TABLE 16.1. Routes of Contamination

Source	Route or cause	Prevention[a]
Manipulations, pipetting, dispensing, etc.	Nonsterile surfaces and equipment. Spillage on necks and outside of bottles and on work surface. Touching or holding pipettes too low down, touching necks of bottles, screw caps. Splash-back from waste beaker. Sedimentary dust or particles of skin settling on culture or bottle. Hands or apparatus held over open dish or bottle.	Clear work area of items not in immediate use. Swab regularly with 70% alcohol. Do not pour if it can be avoided. Dispense or transfer by pipette, autodispenser, or transfer device (see Chapter 4). If you pour: (1) do so in one smooth movement, (2) discard the bottle that you pour from, and (3) wipe up any spillage with sterile swab moistened with 70% alcohol. Discard into beaker with funnel or, preferably, by drawing off into reservoir with vacuum pump (Figs. 4.5, 5.4). Do not work over (vertical laminar flow and open bench) or behind and over (horizontal laminar flow) an open bottle or dish.
Solutions	Nonsterile reagents and media. Dirty storage conditions. Inadequate sterilization procedures. Poor commercial supplier.	Filter or autoclave before use. Monitor performance of autoclave with recording thermometer or sterility indicator (see Trade Index). Check integrity of filters after use (bubble-point). Test all solutions after sterilization.
Glassware and screw caps	Dust and spores from storage. Ineffective sterilization, e.g., sealed bottles preventing ingress of steam.	Dry-heat sterilize or autoclave before use. Do not store unsealed for more than 24 hr. Check oven and autoclave regularly and monitor each load
Tools, instruments, pipettes	Contact with nonsterile surface or material. Invasion by insects, mites, or dust. Ineffective sterilization.	Sterilize by dry heat before use: monitor performance of oven. Resterilize instruments (70% alcohol, burn and cool off) during use. Do not grasp part of instrument or pipette which will later pass into culture vessel. Do not store for more than 24 hr, unless sealed with tape.
Culture flasks, media bottles in use	Dust and spores from incubator or refrigerator. Dirty storage or incubation conditions. Media under cap spreading to outside of bottle.	Use screw caps in preference to stoppers. Wipe flasks and bottles with 70% alcohol before using. Flame necks and caps (without opening) before placing in laminar flow hood. Cover cap and neck with aluminum foil during storage or incubation.
Room air	Drafts, eddies, turbulence, dust, aerosols.	Clean filtered air. Reduce traffic and extraneous activity. Wipe floor and work surfaces regularly.
Work surface	Dust, spillage.	Swab before, after, and during work with 70% alcohol. Mop up spillage immediately.
Operating–hair, hands, breath, clothing	Dust from skin, hair, or clothing dropped or blown into culture. Aerosols from talking, coughing sneezing, etc.	Wash hands thoroughly. Keep talking to a minimum and face away from work if you do. Avoid working with a cold or throat infection or wear a mask. Tie back long hair, or wear a cap. Do not wear the same lab coat as in general lab area or animal house. Change to gown or apron.

(*continued*)

TABLE 16.1. *(Continued)*

Source	Route or cause	Prevention[a]
Hoods	Perforated filter. Filter change needed. Spillages, particularly in crevices or below work surface.	Check filters regularly for holes and leaks. Check pressure drop across filter. Clear around and below work surface regularly.
Tissue samples	Infected at source or during dissection.	Do not bring animals into tissue culture lab. Incorporate antibiotics in dissection fluid (see Chapter 9). Dip large tissue samples in 70% alcohol for 30 s.
Incoming cell lines	Contaminated at source or during transit.	Handle alone, after all other sterile work is finished, swab down bench or hood carefully after use, and do not use until next morning. Check for contamination by growing for 2 wk without antibiotics. (Keep duplicate in antibiotics at first subculture.) Check for contamination visually, by phase-contrast microscopy and Hoechst stain for mycoplasma. Using indicator cell allows screening before first subculture.
Mites, insects, and other infestation in wooden furniture, benches, incubators, and on mice, etc., taken from animal house.	Entry of mites, etc., into sterile packages.	Seal all sterile packs. Avoid wooden furniture if possible, use plastic laminate, one-piece, or stainless steel benchtops. If wooden furniture is used, seal with polyurethane varnish or wax polish and wash regularly with disinfectant. Keep animals out of tissue culture lab.
Anhydric incubators	Growth of molds and bacteria on spillages.	Wipe up any spillage with 70% alcohol on a swab. Clean out regularly.
CO_2, humidified incubators	Growth of molds and bacteria in humid atmosphere on walls and shelves. Spores, etc., carried on forced-air circulation.	Clean out weekly with detergent and 70% alcohol. Enclose open dishes in plastic boxes with tight-fitting lids (but do not seal). Swab with 70% alcohol before opening. Fungicide or bacteriocide in humidifying water (but check first for toxicity).

[a]A one-to-one relationship between prevention and cause is not intended throughout this table. Preventative measures are interactive and may relate to more than one cause.

Fogh, 1973; McGarrity, 1982; Cour et al., 1979; Hay et al., 1979; Hay, 1992].

Characteristic features of microbial contamination are as follows: (1) Sudden change in pH, usually a decrease with most bacterial infections, very little change with yeast until contamination is heavy, and sometimes an increase in pH with fungal contamination. (2) Cloudiness in medium, sometimes with a slight film or scum on the surface or spots on the growth surface that dissipate when the flask is moved. (3) When examined on a low-power microscope (~ ×100), spaces between cells will appear granular and may shimmer with bacterial contamination (Fig. 16.1c). Yeasts appear as separate round or ovoid particles that may bud off smaller particles (Fig. 16.1a). Fungi produce thin filamentous mycelia (Fig. 16.1b) and sometimes denser clumps of spores. With toxic infection, some deterioration of the cells will be apparent. (4) On high-power microscopy (~ ×400), it may be possible to resolve individual bacteria and distinguish between "rods" and cocci. At this magnification, the shimmering seen in some infections will be seen to be due to mobility of the bacteria. Some bacteria form clumps or associate with the cultured cells. (5) If a slide preparation is made, the morphology of bacteria can be resolved more clearly at ×1,000, but this is not usually necessary. Microbial infection may be confused with precipitates of media constituents, particularly protein, or with cell debris, but can be distinguished by their regular particulate morphology. Precipitates may be crystalline or globular and irregular and are not usually as uniform in size. If in doubt, plate out a sample of medium on nutrient agar (see Chapter 8). (6) Mycoplasmal infections (Fig. 16.1d–f) cannot be detected

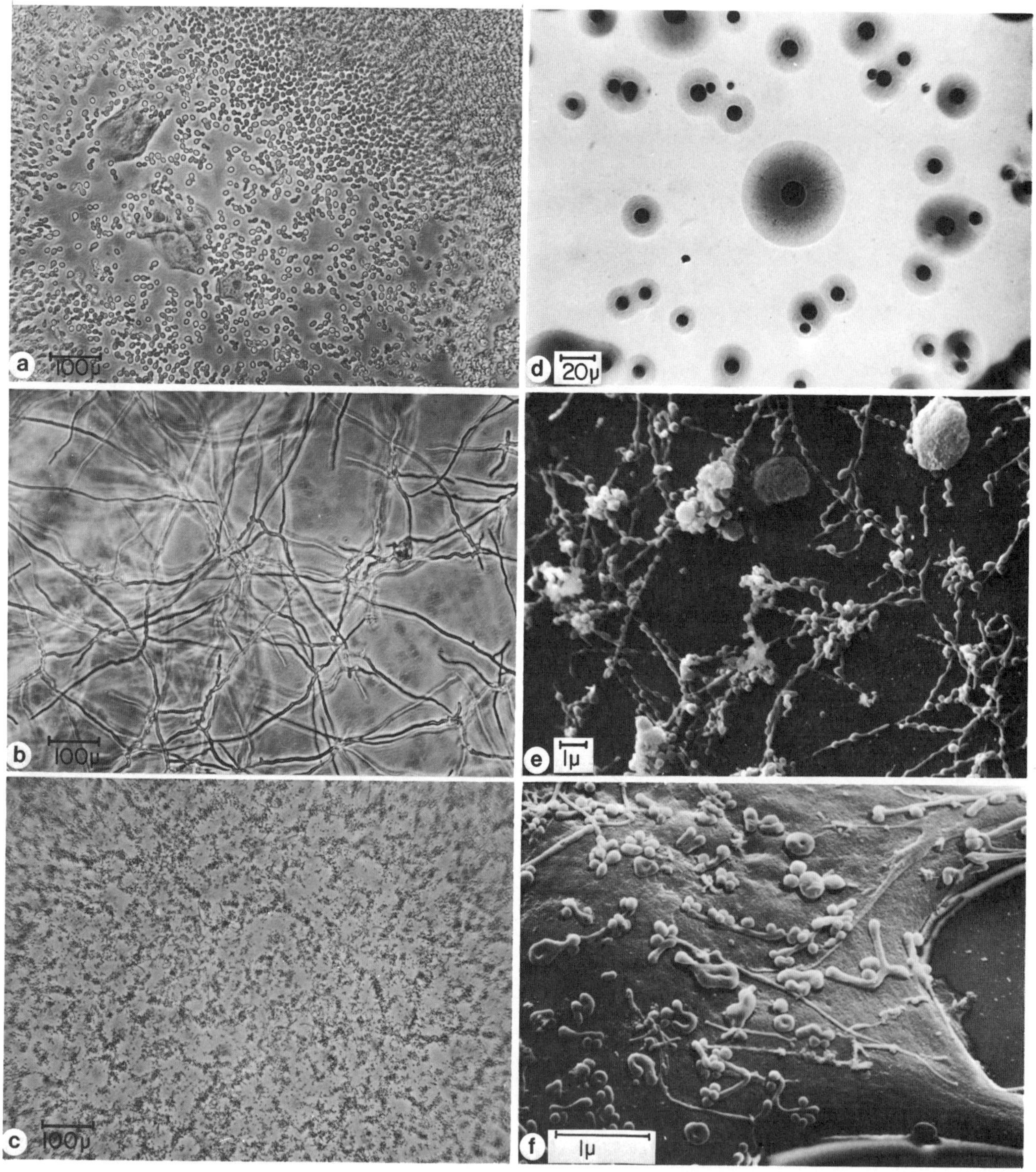

Fig. 16.1. Examples of micro-organisms found as contaminants of cell cultures (magnification approximately ×100). a. Yeast. b. Mold. c. Bacteria. d. Mycoplasma colonies growing on special nutrient agar (not as seen in cell culture [see Fig. 16.2]). e,f. Scanning electron micrograph of mycoplasma growing on the surface of cultured cells. (d–f, courtesy of Dr. M. Gabridge.)

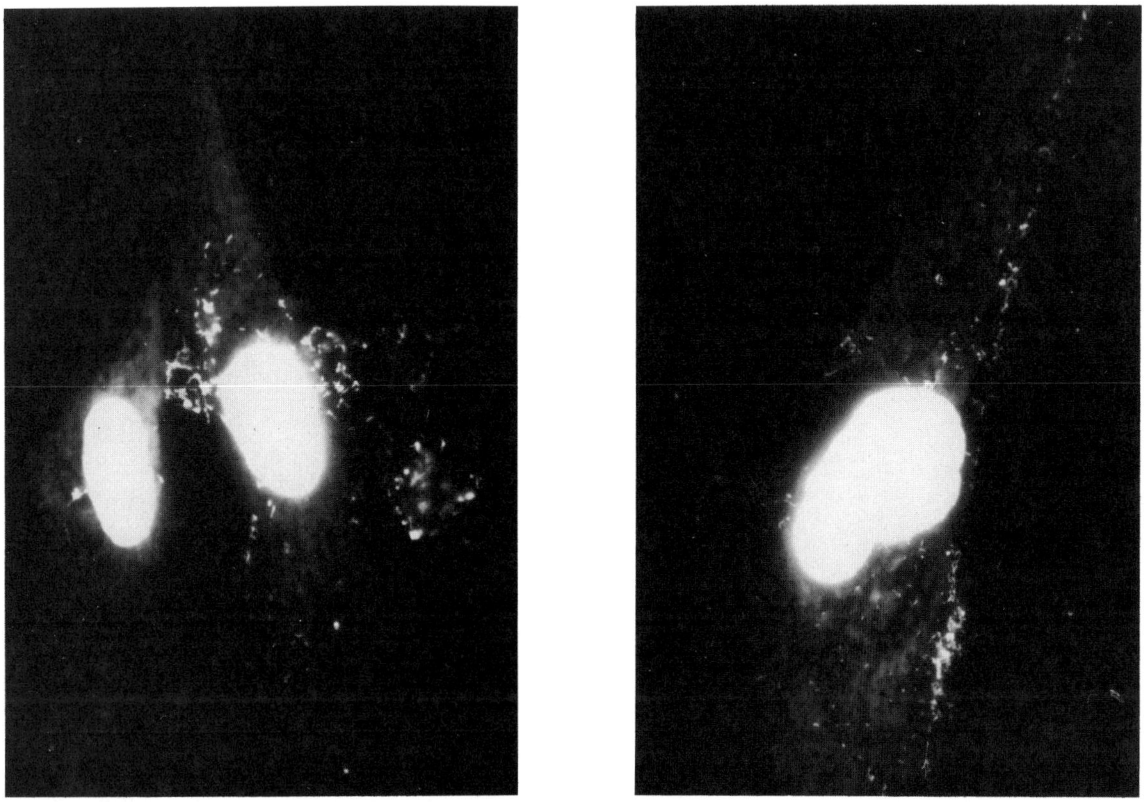

Fig. 16.2. Human normal diploid lung fibroblasts, infected with mycoplasma, stained with Hoechst 33258. The nuclei fluoresce brightly from cellular DNA and extranuclear fluorescence due to mycoplasma is also apparent. Light cytoplasmic staining is present in these preparations, but its diffuse nature makes it easily distinguishable from the bright particulate or filamentous staining of the mycoplasma. Usually, there is no cytoplasmic background. The lack of nuclear detail is due to the overexposure of the photographs necessary to reveal the mycoplasma.

by naked eye other than by signs of deterioration in the culture. The culture must be tested specially by fluorescent staining, orcein or Giemsa staining, autoradiography, or microbiological assay (see below). Fluorescent staining of DNA by Hoechst 33258 [Chen, 1977] is the easiest and most reliable method (see below) and reveals mycoplasmal infections as fine particulate or filamentous staining over the cytoplasm at ×500 magnification (Fig. 16.2). The nuclei of the cultured cells are also brightly stained by this method, as are any other microbial contaminations.

It is important to appreciate that mycoplasmas do not always reveal their presence with macroscopic alterations of the cells or media. Many mycoplasma contaminants, particularly in continuous cell lines, grow slowly and do not destroy host cells. However, they can alter the metabolism of the culture in subtle ways. As mycoplasmas take up thymidine from the medium, infected cultures show abnormal labeling with [^{3}H]-thymidine. Immunological studies can also be totally frustrated by mycoplasmal infections, as attempts to produce antibodies against the cell surface may raise antimycoplasma antibodies. Mycoplasmas can alter cell behavior and metabolism in many other ways [Barile, 1977; McGarrity, 1982], so there is an absolute requirement for routine, periodic assays for possible covert contamination of all cell cultures, particularly continuous cell lines.

Monitoring Cultures for Mycoplasmas

Superficial signs of chronic mycoplasmal infection include reduced rate of cell proliferation, reduced saturation density [Stanbridge and Doersen, 1978], and agglutination during growth in suspension. Acute infection causes total deterioration with perhaps a few resistant colonies. "Resistant" colonies and resulting cell lines are not necessarily free of contamination and may carry a chronic infection.

Fluorescent Technique for Detecting Mycoplasmas

Principle

The cultures are stained with Hoechst 33258, a fluorescent dye, which binds specifically to DNA [Chen,

1977]. Since mycoplasmas contain DNA, they can be detected readily by their characteristic particulate or filamentous pattern of fluorescence on the cell surface and, if the contamination is heavy, in surrounding areas.

Outline

Fix and stain subconfluent cultures or smears and look for fluorescence other than in the nucleus.

Monolayer Cultures

Materials

Hoechst 33258 stain, 50 ng/ml in BSS without phenol red (BSS-PR) or PBSA
PBSA
Deionized water
Fixative: freshly prepared acetic methanol (1:3, cold)
Mountant: 50% glycerine in 0.044 M citrate, 0.11 M phosphate buffer, pH 5.5

Protocol

1. Seed culture at regular passage density (2×10^4–10^5 cells/ml, 4×10^3–2.5×10^4 cells/cm^2) and incubate at 36.5°C until they reach 20–50% confluence. Allowing cultures to reach confluence will inhibit staining and impair subsequent visualization of mycoplasma. Cultures may be grown on glass coverslips in a multiwell plate, in chamber slides (Nunc), or in 35-mm petri dishes without coverslips (see Chapter 13).
2. Remove medium and discard.
3. Rinse monolayer with BSS-PR or PBSA and discard rinse.
4. Add fresh BSS-PR or PBSA diluted 50:50 with fixative, rinse monolayer, and discard rinse.
5. Add pure fixative, rinse, and discard.
6. Add more fixative, ~0.5 ml/cm^2, and leave for 10 min.
7. Remove fixative and discard.
8. Dry monolayer completely if to be stored. (Samples may be accumulated at this stage and stained later.)
9. If proceeding directly, wash off fixative with deionized water and discard wash.
10. Add Hoechst 33258 in BSS-PR or PBSA and leave 10 min at room temperature.
11. Remove stain and discard.
12. Rinse monolayer with water and discard rinse.
13. Mount a coverslip in a drop of mountant and blot off surplus from edges of coverslip.
14. Examine by epifluorescence with 330/380-nm excitation filter, and LP 440-nm barrier filter.

Suspension Cultures and Infected Media

Suspension cultures are a little more difficult to handle, as the cells do not spread as well in cytological preparations and give less flat cytoplasm for critical examination, and, as preparations often involve centrifuging the cells onto a slide, they are often contaminated with particulate debris, some of which can contain DNA and stain positively with Hoechst 33258. To overcome this problem, to reveal low-level contaminations in resistant cell lines, and to screen potentially infected media, many laboratories now use an indicator cell line such as 3T6 or Vero cells.

The indicator cell is seeded in test medium from the cells to be screened, neat or diluted 50:50 with fresh medium, or mixed 50:50 with test cell suspension, and grown for 72 hr. If the test medium contains antibiotics it should be removed after 24 hr and culture of the monolayer continued for a further 72 hr in fresh antibiotic-free medium.

The monolayer is then fixed and stained as above. Control monolayer cells unexposed to test media should be stained and examined in parallel.

This method has proved useful in screening holding medium or primary culture medium where the culture cannot be sacrificed, as well as suspension cultures. It also has the advantage that a constant indicator cell is used, so the operator becomes more familiar with its normal appearance. Furthermore, if a cell line, such as 3T6, is chosen for its known capacity to support mycoplasmas, then low-level or cryptic contaminations may become more apparent.

Analysis. Check for extranuclear fluorescence. Mycoplasmas give pinpoints or filaments of fluorescence over the cytoplasm and sometimes in intercellular spaces. The pinpoints are close to the limits of resolution with a 50× objective (0.1–1.0 μm) and are usually regular in size. Not all of the cells will necessarily be infected, so most of the preparation should be scanned before declaring the culture uninfected.

Fluorescence outside the nucleus can be observed in uninfected cultures where there is evidence of cell damage, e.g., primary cultures or cells recently recovered from frozen storage. Usually the fluorescent particles in this case are irregular in size and shape and disappear following subculture. Use of an indicator cell line helps to eliminate this problem (see above).

Sometimes a light, uniform staining of the cytoplasm is observed, probably due to RNA. This tends to fade on storage, and examination the next day (after storage dry and in the dark) usually gives clearer results. This artifact never has the sharp punctate or filamentous appearance of mycoplasma, and can be distinguished fairly readily after some experience in observation.

If there is any doubt regarding the interpretation of this test, it should be repeated, preferably with an indicator cell line, allowing about 1 wk for any low-level infection to increase. During this time, quarantine the suspect cell line.

Alternative Methods

Several other methods have been reported for the detection of mycoplasmal infections such as the detection of mycoplasma-specific enzymes like arginine deiminase or nucleoside phosphorylase [see Schneider and Stanbridge, 1975; Levine and Becker, 1977], or toxicity with 6-methylpurine decoscyribase (Mycotect BRL), but the DNA-fluorescence method is simpler and, though not specific for mycoplasmas, will detect any DNA-containing infection, which is, after all, the prime objective. Several other methods have been reported, however, that are of general importance. The first depends on microbiological culture of the organism, although it is best not attempted unless you have the necessary expertise, as these organisms are quite fastidious. The cultured cells are seeded into mycoplasma broth [Taylor-Robinson, 1978], grown for 6 days, and plated out onto special nutrient agar [Hay, 1992]. Colonies form in about 8 days and can be recognized by their size (~200 μm diameter) and their characteristic "fried-egg" morphology-dense center with lighter periphery (Fig. 16.1d). Commercial kits for microbiological detection (Mycotrim) are available from Hana Media.

If the microbiological culture method is used for mycoplasma detection, it is necessary to grow a known strain of mycoplasma at the same time as a positive control. It is this aspect that discourages most laboratories, which do not wish to introduce live mycoplasma deliberately unless there is proper accommodation available where this work can be carried on totally separate from the rest of tissue culture work.

While the use of selective culture conditions and examination of colony morphology enables the species of mycoplasma to be identified, the microbiological culture method is much slower and more difficult to perform than the fluorescence technique. Commercial screening for mycoplasma is available (e.g., Flow Laboratories, Microbiological Associates), using microbiological culture. Specific monoclonal antibodies are now available that allow the characterization of mycoplasma infections.

Other methods use staining with aceto-orcein or Giemsa stain (see under "Staining," Chapter 13). In both cases, particulate cytoplasmic staining is regarded as indicative of mycoplasmal infection, but both are more difficult to interpret than the fluorescent method and can give false positives due to nonspecific precipitation of stain.

Molecular probes specific to mycoplasma DNA can be used in Southern blot analysis to detect infections by conventional molecular hybridization techniques. A kit is available, using similar probes and a simple single-step separation of hybrid DNA for subsequent counting by scintillation counting (Gen Probe, Laboratory Impex Ltd.).

One other method that has been used quite successfully is autoradiography with [^{3}H]-thymidine [Nardone et al., 1965]. The culture is incubated overnight with 4 KBq/ml (~0.1 μCi/ml) high specific activity [^{3}H]-thymidine and an autoradiograph prepared (see Chapter 23). Grains over the cytoplasm are indicative of infection (see Fig. 23.12) and this can be accompanied by a lack of nuclear labeling due to the thymidine being trapped at the cell surface by the mycoplasma.

Eradication of Mycoplasma Infection

If mycoplasma is detected in a culture, the first and overriding rule is that it should be discarded for autoclaving or incineration. In exceptional cases, e.g., if the infected line is irreplaceable, an attempt may be made to decontaminate it. This should be done only by an experienced operator and the work must be carried out totally separate from the rest of tissue culture, i.e., in a separate, dedicated hood, preferably in a separate room.

Several agents have activity against mycoplasma, including kanamycin, gentamycin, tylosin [Friend et al., 1966], polyanethol sulfonate [Mardh, 1975], and 5-bromouracil in combination with Hoechst 33258 and UV light [Marcus et al., 1980]. Co-culture with macrophages [Schimmelpfeng et al., 1968], animal passage [VanDiggelen et al., 1977], or cytotoxic antibodies [Pollock and Kenney, 1963] can also be effective in some cases. However, the most successful agents have been Mycoplasma Removal Agent (MRA) (ICN-Flow) [Gignac et al., 1992], ciprofloxacin [Mowles, 1988], and BM-Cycline (Boehringer-Mannheim).

Suspected cultures should be rinsed five times with PBSA, trypsinized, and reseeded at the usual concentration for subculture, in medium containing antibiotic at the manufacturer's recommended concentration. The medium containing the antibiotic is changed every 2 days until the next subculture. The cells should be subcultured a total of three times into antibiotic-containing medium, and then subcultured a further three times without antibiotic before testing again.

It must be stressed that this operation should not be undertaken unless it is absolutely essential, and then must be in experienced hands and in isolation. It is far safer to discard infected cultures.

DETECTION OF MICROBIAL CONTAMINATION

Potential sources of contamination are listed in Table 16.1, along with the precautions that should be taken to avoid them. Even in the best laboratories, however, contaminations do arise, so the following procedure is recommended.

1. Check for contamination at each handling by eye and on microscope. Every month check for mycoplasmas.
2. If a contamination is suspected but not obvious and cannot be confirmed *in situ*, clear the hood or bench of all but your suspected culture and one can of Pasteur pipettes. Because of the potential risk to other cultures, this is best done after all your other culture work is finished. Remove a sample from the culture and place on a microscope slide (Kovaslides are convenient for this as they do not require a coverslip). Check on microscope, preferably by phase contrast. If contamination is confirmed, discard pipettes, swab hood or bench with 70% alcohol containing a phenolic disinfectant, and do not use until the next day.
3. Note the nature of contamination, etc., on record sheet.
4. If new contamination (not a repeat and not widespread), discard (a) culture, (b) medium bottle used to feed it, and (c) any other bottle, e.g., trypsin, that has been used in conjunction with this culture. Discard into disinfectant, preferably in a fume hood, and outside the tissue culture area.
5. If new and widespread (i.e., in at least two different cultures), discard all media and stock solutions, trypsin, etc.
6. If similar contamination is repeated, check stock solutions for contaminations by (a) incubation alone or in nutrient broth (see Chapter 8) or (b) plating out on nutrient agar (Oxoid, Difco) (see Chapter 8). (c) If (a) and (b) fail and contamination is still suspected, incubate 100 ml, filter through 0.2-μm filter, and plate out filter on nutrient agar.
7. If contamination is widespread, multispecific, and repeated, check sterilization procedures, e.g., temperature of ovens and autoclaves, particularly in the center of load, times of sterilization, packaging, storage (e.g., unsealed glassware [see Chapter 8] should be resterilized every 24 hr), and integrity of aseptic room and laminar flow hood filters.
8. Do not attempt to decontaminate cultures unless they are irreplaceable. If necessary, decontaminate by (a) washing five times in BSS containing a higher than normal concentration of antibiotics (see DBSS in Reagent Appendix) and by (b) adding antibiotics (as in DBSS) to medium for three subcultures. If possible, the infection should be tested for sensitivity to a range of individual antibiotics. (c) Remove antibiotics and culture without for a further three subcultures. (d) Recycle (b) and (c) twice. (e) Culture for 2 months without antibiotics to check that contamination has been eliminated. Check by phase-contrast microscopy and Hoechst staining (see above).

The general rule should be, however, that contaminated cultures are discarded, and that decontamination is not attempted unless it is absolutely vital to retain the cell strain. Complete decontamination, especially with mycoplasmas, is difficult to achieve and attempts to do so may produce hardier, antibiotic-resistant strains of the contaminant.

A major source of contamination stems from the use of humid incubators. There is no requirement for a high humidity unless open vessels are being used, and sealed flasks are better propagated in an anhydric incubator or hot room. If there is a need to gas flasks with CO_2, this is better done from a cylinder, and the flasks sealed and placed in a normal incubator. Using permeable caps (see Chapter 4) minimizes the risk but increases the unit cost and still exposes the flask to a higher risk atmosphere. If flasks are maintained in a CO_2 incubator with slack or permeable caps, it is preferable to keep the incubator dry and use a different incubator for open plates. The CO_2-monitoring system will need to be recalibrated if the incubator is used dry.

There are, however, many situations where a humid incubator must be used. To reduce the contamination risk, an incubator should be selected with an interior all of which is readily accessible and can be cleaned easily. Cleaning should be carried out regularly using 10% Rocal or an equivalent nontoxic, antifungal cleaner. The frequency will depend; monthly may suffice for a clean area with filtered room air, but this will need to be increased for a greenfield site, where the spore count is higher, or during building work or renovation. During use, any spillage must be mopped up immediately and contaminated cultures removed as soon as they are detected.

Copper-lined incubators have reduced fungal growth but are usually about 20–30% more expensive. Placing copper foil in the humidifier tray has a similar effect, but only in the tray, and will not protect the walls. There are a number of fungal retardants that have been used, including copper sulfate, riboflavin, sodium dodecyl sulfate (SDS), and Rocal. Rocal is a proprietary fungicidal cleaner and is used at 2%. Comparison of colony formation in incubators with and without Rocal show no toxic effect. Many of these retardants are detergents, so it is important not to have a CO_2 or air line bubbling through liquid containing them or it will foam.

Any fungicide will only protect the tray; there is no substitute for regular cleaning!

Persistent Contamination

Many laboratories have suffered from periods of contamination that seem to be refractory to all the remedies suggested in Table 16.1. There is no easy answer to this, other than to follow the recommendations above in a logical and analytical fashion, paying particular attention to changes in technique, new staff, new suppliers, new equipment, or insufficient maintenance of laminar flow hoods or other equipment. Typically, an increase in contamination rate stems from an increased spore count in the atmosphere, poorly maintained incubators, a contaminated cold room or refrigerator, or a minor intermittent fault in a sterilizing oven or autoclave.

Constant use of antibiotics is also a potential source of chronic contamination. Many organisms are inhibited by antibiotics but not killed. They will, therefore, persist in the culture, undetected for most of the time, but periodically surfacing when conditions change, or due to intrinsic host–parasite-type population fluctuations. It is essential that your cultures are maintained for at least part of the time in antibiotic-free conditions, otherwise cryptic contaminations will persist, their origins will be difficult to determine, and their elimination will be impossible.

A slight change in practices, introduction of new staff, or an increase in activity as more people use the facility can all contribute to an increase in the rate of contamination. Procedures must remain stringent, even if the reason is not always obvious to the operator, and alterations in routine should not be made casually. If strict practices are maintained, contamination may not be eliminated entirely, but it will be detected early.

CROSS-CONTAMINATION

During the history of tissue culture, a number of cell strains have evolved with very short doubling times and high plating efficiencies. Although these properties make such cell lines valuable experimental material, they also make them potentially hazardous for cross-infecting other cell lines. The extensive cross-contamination of many cell lines with HeLa cells [Gartler, 1967; Nelson-Rees and Flandermeyer, 1977] is now well known, but many operators are still unaware of the seriousness of the risk. To avoid cross-contamination:

(1) Obtain cell lines from a reputable cell bank where appropriate characterization has been performed (see Chapter 17 and Reagent Appendix) or perform the necessary characterization yourself as soon as possible (see Chapter 13).

(2) Do not have bottles or flasks of more than one cell line open simultaneously.

(3) Handle rapidly growing lines, such as HeLa, on their own and after other cultures.

(4) Never use the same pipette for different cell lines.

(5) Never use the same bottle of medium, trypsin, etc., for different cell lines.

(6) Whenever possible, do not put a pipette back into a bottle of medium, trypsin, etc., after it has been in a culture flask containing cells. Add medium and any other reagents to the flask first and then add the cells last.

(7) Do not use unplugged pipettes, even micropipettes, for routine maintenance.

(8) Check the characteristics of the culture regularly and suspect any sudden change in morphology, growth rate, etc. Confirmation of cross-contamination may be obtained by DNA fingerprinting (see this chapter, above) [Stacey et al., 1992], karyotype [Nelson-Rees and Flandermeyer, 1977] or isoenzyme analysis [O'Brien et al., 1977, 1980] (see Chapter 13).

CONCLUSIONS

Check living cultures regularly for contaminations using normal and phase-contrast microscopy, and for mycoplasmas by fluorescent staining of fixed preparations.

Do not maintain all cultures routinely in antibiotics. Grow at least one set of cultures of each cell line in the absence of antibiotics for a minimum of 2 wk at a time to allow cryptic contaminations to become overt.

Do not attempt to decontaminate a culture unless it is irreplaceable and then do so under strict quarantine.

Quarantine all new lines that come into your laboratory until you are sure that they are uncontaminated.

Do not share media or other solutions among cell lines or among operators. Check cell line characteristics (see Chapter 13) periodically to guard against cross-contamination.

It cannot be overemphasized that cross-contaminations can and do occur. It is essential that the above precautions be taken and regular checks of cell strain characteristics be made.

New cell lines should be characterized, preferably by DNA fingerprinting, as soon after isolation as possible.

CHAPTER 17

Instability, Variation, and Preservation

It is a fundamental property of living organisms that they diversify to provide sufficient variation for selection to be a useful mechanism in the adaptation of the species to its environment. It is, therefore, not surprising that cells in culture behave in a similar fashion. Spontaneous variation and selection occur as with any micro-organism.

During the evolution of a cell line from a primary culture and during subsequent maintenance as a cell line or purified cell strain, there is evidence of both phenotypic and genotypic instability. This arises as a result of variations in culture conditions, selective overgrowth of constituents of the cell population, and genomic variations.

Since the constitution of a culture may vary from time to time, it is important (1) to standardize the culture conditions; (2) to select a period in the life history of the cell line when variation is at a minimum; (3) to select a pure cloned and characterized cell strain if possible; and (4) to preserve a seed stock to recall into culture at intervals, to maintain consistency.

ENVIRONMENT

Environment has been discussed in Chapter 7. Once the appropriate conditions have been adopted, they should be adhered to, as alterations in medium, serum, substrates, etc., will alter phenotypic expression. Test batches of serum, select one that has the required properties, and reserve enough to be sufficient for 6 months to 1 year. Repeat the process before changing to a new batch (see also Chapter 7 and below).

SELECTIVE OVERGROWTH, TRANSFORMATION, AND SENESCENCE

Following isolation of a primary culture, the predominant phenotype may change, as cells with a higher proliferative capacity will tend to overgrow the more slowly dividing and nondividing cells (see Chapters 2, 9, 11, 20, and 21). As an example, consider a culture taken from carcinoma of the colon. Following explantation, these cultures can be predominantly epithelial; but after the first subculture, the epithelial cells steadily lose ground, and the fibroblasts take over. Preserving the epithelial component during this phase is a major problem and requires various selective culture and separation techniques (see Chapters 11, 12 and 20). Occasionally, however, a transformed line may appear without using selective conditions. This may be a minority component of the original culture that has undergone transformation and now has increased growth capacity (shorter doubling time, infinite survival), and will ultimately outgrow the fibroblasts, which grow more slowly and have a finite life-span in culture.

To counter cell line variation, finite cell lines should be used between certain generation limits; e.g., human diploid fibroblasts gain a fair degree of uniformity by the 10th generation and remain fairly stable up to about the 30th. Beyond 40 generations, senescence can be anticipated, although some lines may survive longer.

The culture should be grown through the first five or six generations until sufficient cells have been produced to freeze down a seed stock (see below).

GENETIC INSTABILITY

Evidence of genetic rearrangement can be seen in chromosome counts (see Fig. 2.2) and karyotype analysis. While the mouse karyotype is made up exclusively of small telocentric chromosomes, several metacentrics are apparent in many continuous murine cell lines (Robertsonian fusion). Furthermore, while virtually every cell in the animal has the normal diploid set, this is more variable in culture. In extreme cases, e.g., continuous cell strains such as HeLa-S3, less than half of the cells will have exactly the same karyotype, i.e., they are *heteroploid*.

Most continuous cell strains, even after cloning, contain a range of genotypes that are constantly changing. As transformation often involves chromosomal rearrangement, it is possible that it can only occur in cells with the capacity for chromosomal alterations. Alternatively, transformation may cause genetic instability to arise in a previously stable genotype. Hence, transformed continuous lines retain a capacity for genetic variation that is not apparent *in vivo* or in many finite cell lines. There are two main causes of genetic variation: (1) the spontaneous mutation rate appears to be higher *in vitro*, associated, perhaps, with the high rate of cell proliferation, and (2) mutant cells are not eliminated unless their growth capacity is impaired. It is not surprising that phenotypic variation will arise as a result of this genetic variation. Minimal deviation rat hepatoma cells, grown in culture, express tyrosine aminotransferase activity constitutively and may be induced further by dexamethasone [Granner et al., 1968], but, as can be seen from Figure 17.1, subclones of a cloned strain of H4-II-E-C3 [Pitot et al., 1964] differed both in the constitutive level of the enzyme and in its capacity to be induced by dexamethasone.

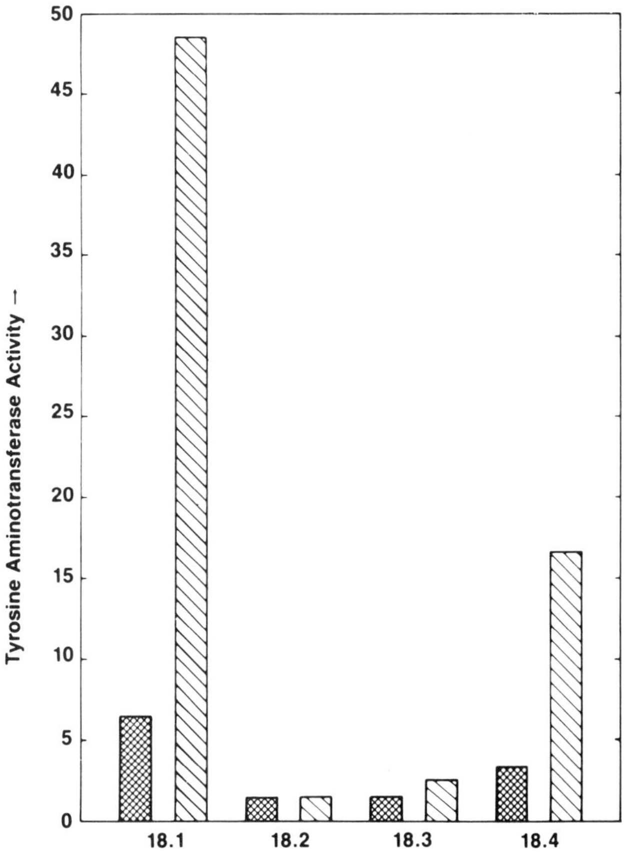

Fig. 17.1. Variation in tyrosine aminotransferase activity between four subclones of clone 18 of a rat minimal deviation hepatoma cell strain. H-4-II-E-C3 cells were cloned; clone 18 was isolated, grown up, and recloned, and the second-generation clones were assayed for tyrosine aminotransferase activity with and without pretreatment of the culture with dexamethasone. Cross-hatching, basal level; hatching, induced level. Data provided by J. Somerville.

PRESERVATION

In order to minimize genetic drift in continuous cell lines, to avoid senescence or transformation in finite cell lines, and to guard against accidental loss by contamination or failure of incubators, reagents, etc., it is now common practice to freeze aliquots of cells to be thawed out at intervals as required.

Selection of Cell Line

A cell line is selected with the required properties. If it is a finite cell line, it is grown to between the fifth and tenth population doubling to create sufficient bulk of cells for freezing. Continuous cell lines should be cloned (see Chapter 11) and an appropriate clone selected and grown up to sufficient bulk to freeze. Prior to freezing, the cells must be characterized (see Chapter 13) and checked for contamination (see Chapter 16), particularly cross-contamination.

Continuous cell lines have the advantage that they will survive indefinitely, grow more rapidly, and can be cloned more easily; but they may be less stable genetically. Finite cell lines are usually diploid or close to it, stable between certain passage levels, but are harder to clone, grow more slowly, and will eventually die out or transform (Table 10.3).

Standardization of Medium and Serum

The type of medium used will influence the selection of different cell types and regulate their phenotypic expression (see Chapters 2, 7, 14, and 20). Consequently, once a medium has been selected, standardize on that medium and, preferably, on one supplier if it is being purchased ready-made.

Variation in serum. Considerable variation [see Olmsted, 1967; Honn et al., 1975] may be anticipated between batches of serum, resulting from differing methods of preparation and sterilization, different ages and storage conditions, and variations in animal stocks from which the serum was derived, including strain differences, pasture, climate, and so on. It is important to select a batch, use it for as long as possible, and replace it, eventually, with one as similar as possible (see Chapter 7).

The best method of eliminating serum variation is to convert to a serum-free medium (see Chapter 7), although, unfortunately, serum-free formulations are not yet available for all cell types, and the conversion may be costly and time-consuming. Serum substitutes (see Chapter 7) may offer greater consistency, are generally cheaper than serum or growth factor supplementation, but do not offer the control over the physiological environment afforded by serum-free medium.

Cell Freezing

When a cell line has been produced, or a cloned cell strain selected, with the desired characteristics and free of contamination, then a *seed stock* should be stored frozen.

For most cell lines, particularly those in constant and general use, the seed stock should be protected and not made available for general issue. A *using stock* should be frozen, and ampules from this issued to individuals as required. Individuals requiring stocks over a prolonged period should then freeze down their own *personal user stocks*, which may be discarded when the work is finished or the person leaves. When the using stock becomes depleted, it may be replenished from the seed stock. When the seed stock becomes depleted, it should be replenished before any other ampules are issued, and with the minimum increase in generation number from the first freezing.

Storage in liquid nitrogen (Fig. 17.2; see also Chapter 4) is currently the most satisfactory method of preserving cultured cells. The cell suspension, preferably at a high concentration, should be frozen slowly, at 1°C per min [Leibo and Mazur, 1971; Harris and Griffiths, 1977] in the presence of a preservative such as glycerol or dimethyl sulfoxide [Lovelock and Bishop, 1959]. The frozen cells are transferred rapidly to liquid nitrogen when they reach −50°C or below and are stored immersed in the liquid nitrogen or in the gas phase above the liquid. When required, the cells are thawed rapidly and reseeded at a relatively high concentration to optimize recovery. If liquid nitrogen storage is not available, cells may be stored in a conventional freezer. The temperature should be as low as possible, but significant deterioration may yet occur even at −70°C. Little deterioration is found at −196°C [Green et al., 1967].

Outline

The culture is grown to late log phase and a high cell density suspension is prepared and frozen slowly with a preservative (Fig. 17.3). When required, aliquots are thawed rapidly and reseeded at high cell density.

Materials

For freezing:
Sterile:
Cultures to be frozen
If monolayer: PBSA and 0.25% crude trypsin
Growth medium
Dimethyl sulfoxide or glycerol
Glass or plastic ampules
Nonsterile:
Hemocytometer or cell counter
If glass ampules are used, ampule sealer with gas/O_2 burner
Canes or racks for storage
Cotton wool, polystyrene box, or plastic foam insulation tube
Forceps
Protective gloves
For thawing:
Sterile:
Culture flask
Centrifuge tube (if centrifugation required)
Growth medium
Pipettes, 1 ml, 10 ml
Syringe and 19-g needle (if glass ampule)
Nonsterile:
Protective gloves and face mask
Bucket of water at 37°C with lid
Forceps
70% alcohol
Swab
1% naphthalene black (amido black)

Protocol

1. Check culture for (a) healthy growth, (b) freedom from contamination, and (c) presence of

specific characteristics required of the line for subsequent use (viral propagation, differentiation, antigenic constitution, etc.).

2. Grow up to late log phase and, if monolayer, trypsinize and count—if suspension, count and centrifuge.
3. Resuspend at approximately 5×10^6–2×10^7 cells/ml in culture medium containing serum and a preservative such as dimethyl sulfoxide (DMSO) or glycerol at a final concentration of 5–10%. The preservative must be pure and free from contamination. DMSO should be colorless. Glycerol should be not more than 1 yr old, as it may become toxic after prolonged storage.

   It is *not* advisable to place ampules on ice in an attempt to minimize deterioration of the cells. A delay of up to 30 min at room temperature is not harmful when using DMSO and is beneficial when using glycerol.

   ◇ DMSO is a powerful solvent. It will leach impurities out of rubber and some plastics and should, therefore, be kept in its original stock bottle or in glass tubes with glass stoppers. It can also penetrate many synthetic and natural membranes, including *skin* and rubber gloves [Horita and Weber, 1964]. Consequently, any potentially harmful substances in regular use (e.g., carcinogens) may well be carried into the circulation through the skin and even through rubber gloves. DMSO should always be handled with caution because of its known hazardous potential, particularly in the presence of any toxic substances.
4. Dispense cell suspensions into 1- or 2-ml prelabeled glass or plastic ampules and seal (Figs. 17.4, 17.5).

   ◇ Glass ampules are still widely used but must be perfectly and quickly sealed. If sealing takes too long, the cells will heat up and die, and the air in the ampule will expand and blow a hole in the top (Fig. 17.5d). If the ampule is not perfectly sealed, it may inspire liquid nitrogen during freezing and storage in the liquid phase of the nitrogen freezer and will subsequently explode (violently) on thawing, or it may become infected.

   It is possible to check for leakage by placing ampules in a dish of 1% methylene blue in 70%

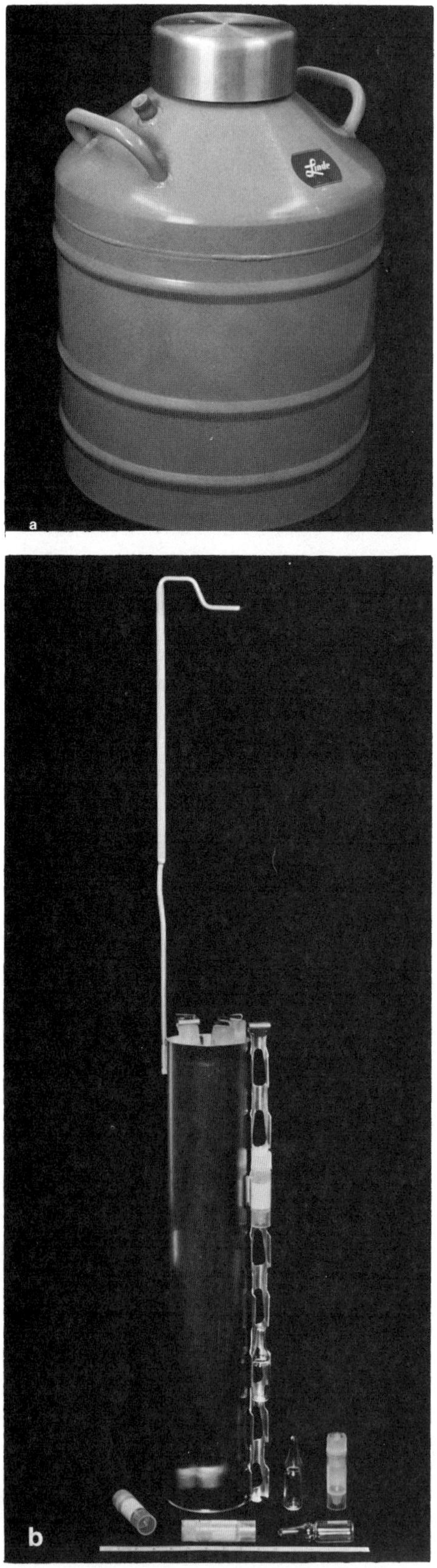

Fig. 17.2. *Liquid nitrogen freezers. a. Narrow-necked freezer with storage on canes in canisters (b). c. Narrow-necked freezer with storage in drawers (d); these freezers are tending to replace the older wide-necked drawer and rack storage systems (see also Fig. 17.8).*

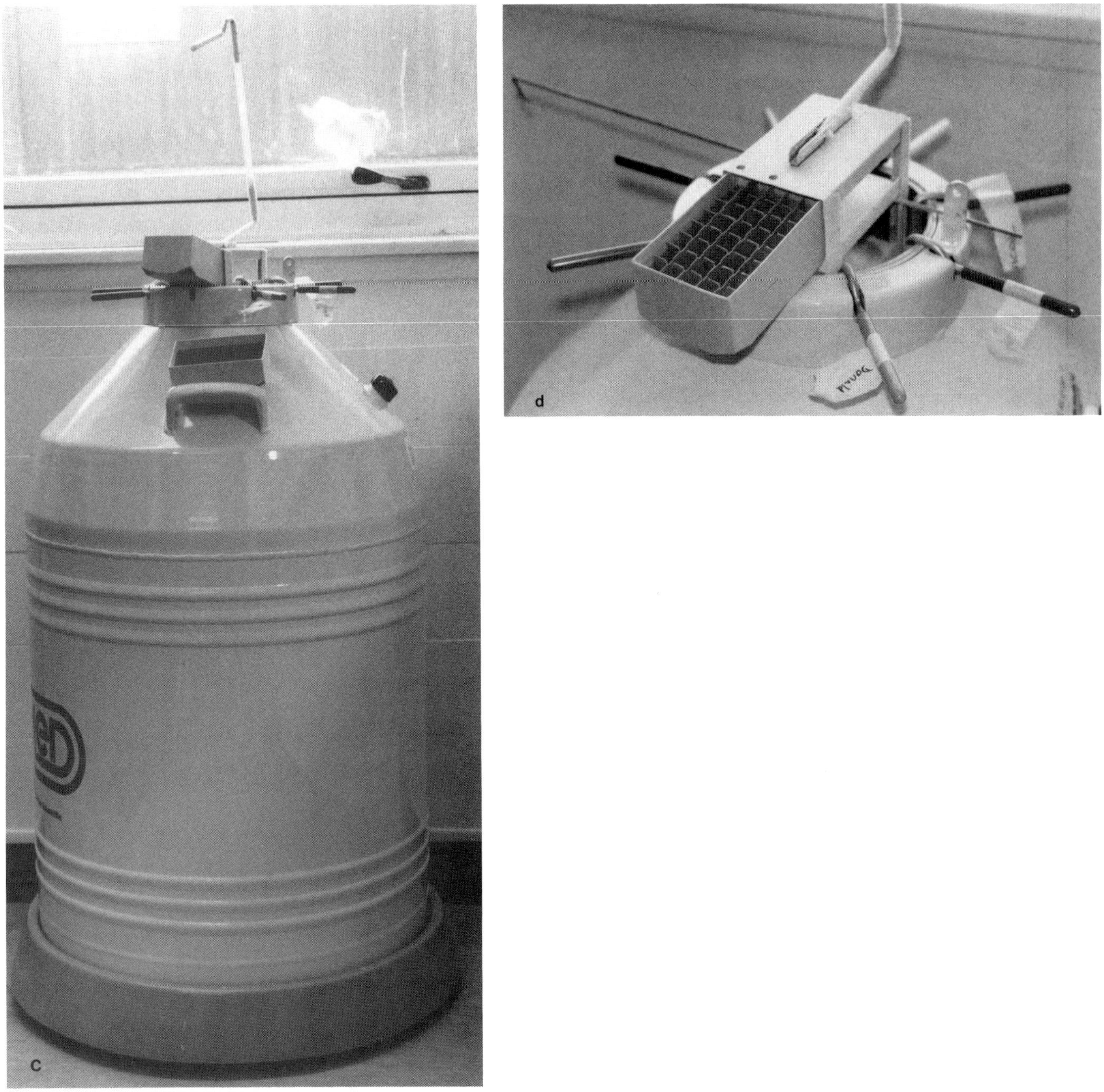

Fig. 17.2. (Continued)

alcohol at 4°C for 10 min before freezing. If the ampules are not properly sealed, the methylene blue will be drawn in and the ampule should be discarded. The ampules may need to be relabeled after this procedure. This step is inconvenient and time-consuming and may only be necessary when sealing glass ampules for the first few times. When experience is obtained, a well-sealed ampule may be recognized by the appearance of the tip (Fig. 17.5f).

Plastic ampules (e.g., Nunclon) are unbreakable but must be sealed with the correct torsion on the screw cap or they may leak. They are of a larger diameter and taller than equivalent glass ampules. Check that they will fit in the canes or racks used for storage. (Special canes for plastic ampules are available.)

Plastic ampules may be more convenient for the average experimental and teaching laboratory but repositories and cell banks generally

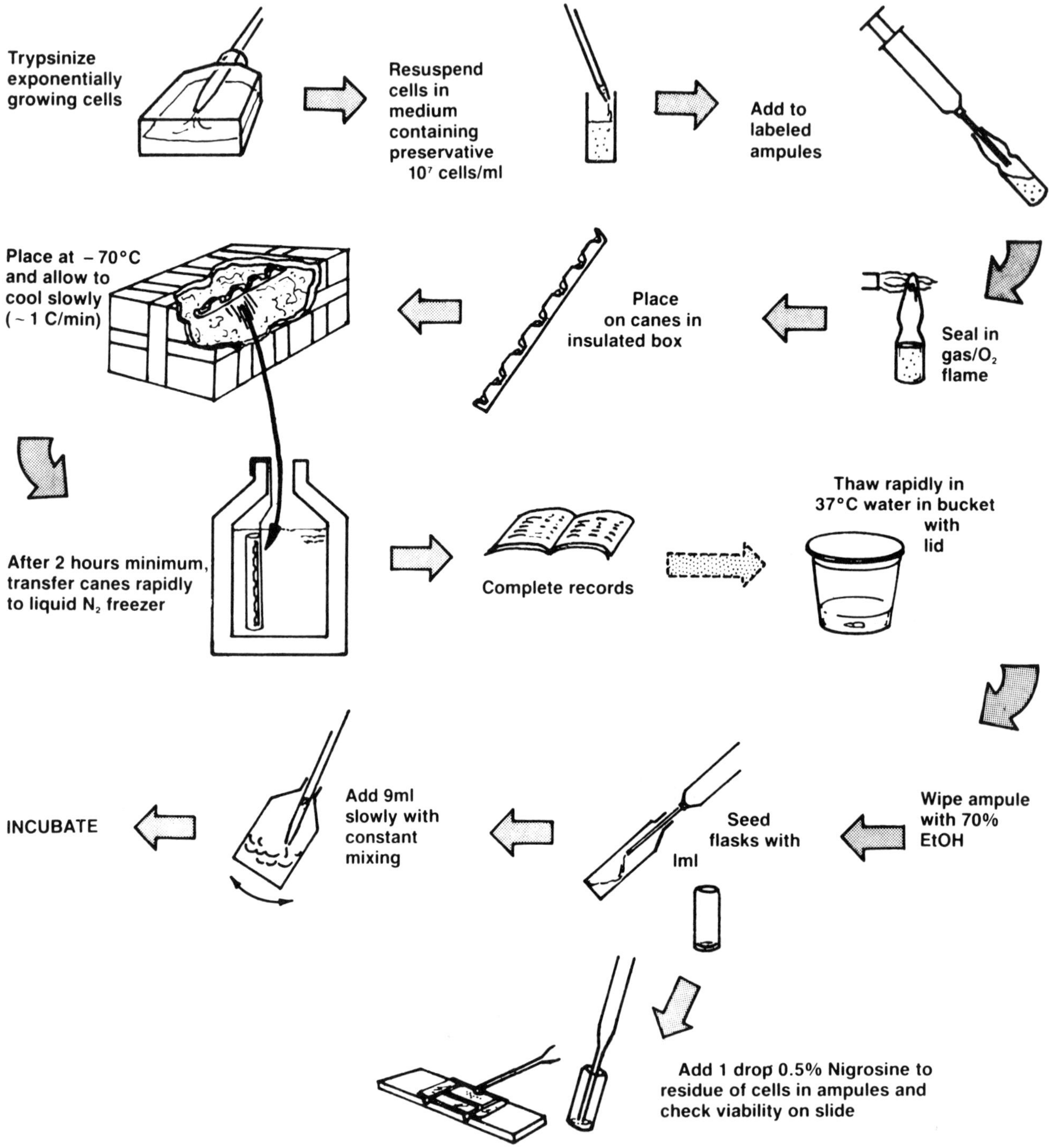
Trypsinize exponentially growing cells
Resuspend cells in medium containing preservative 10⁷ cells/ml
Add to labeled ampules
Seal in gas/O₂ flame
Place on canes in insulated box
Place at −70°C and allow to cool slowly (~1 C/min)
After 2 hours minimum, transfer canes rapidly to liquid N₂ freezer
Complete records
Thaw rapidly in 37°C water in bucket with lid
Wipe ampule with 70% EtOH
Seed flasks with 1ml
Add 9ml slowly with constant mixing
INCUBATE
Add 1 drop 0.5% Nigrosine to residue of cells in ampules and check viability on slide

Fig. 17.3. Preparation of cells for freezing and subsequent recovery after storage.

Fig. 17.4. *Semiautomatic ampule sealer (Kahlenburg-Globe). Ampules are placed between the rollers and the flat plate by the operator's left hand. As the jig carrying the burner is moved forward, the ampule rotates in the flame and is finally ejected against the plate at the bottom of the picture.*

prefer glass ampules, as the long-term storage properties of glass are well characterized and, correctly performed, sealing is absolute.

5. Place ampules on canes for canister storage or leave loose for drawer storage. Lay on cotton wool in a polystyrene foam box with a wall thickness of ~15 mm. This, plus the cotton wool, should provide sufficient insulation such that the ampules will cool at 1°C/min when the box is placed at −70°C or −90°C in a regular deep freeze or insulated container with solid CO_2. Tubular foam pipe insulation can also be used for ampules on canes (Fig. 17.7c).

   Most cultured cells survive best at this cooling rate; but if recovery is low, try a faster rate (i.e., less insulation) [Leibo and Mazur, 1971].

   Programmable coolers are available to control the cooling rate (Fig. 17.6), usually by sensing the temperature of the ampule and running in liquid N_2 at the correct rate, to achieve a preprogrammed cooling rate. They are, however, very expensive and have few advantages unless you wish to vary the cooling rate [e.g., Foreman and Pegg, 1979].

   Taylor Wharton market a special neck plug that carries ampules and places them in the neck of the freezer at the desired level to obtain a set cooling rate (Fig. 17.7a,b). There is also a container supplied by Nalgene, which, when filled with isopropanol and placed at −70°C, will freeze 1.0-ml ampules at 1°C/min.

6. When the ampules have reached −70°C (1.5–2 hr after placing at −70°C if starting from 20°C ambient), transfer to liquid N_2 freezer. This must be done quickly, as the cells will deteriorate rapidly if the temperature rises above −50°C.

   ◇ Protective gloves and face mask should be used when handling liquid nitrogen.

7. When ampules are safely located in freezer, make sure that the appropriate entries are made in the freezer index (Tables 17.1, 17.2). Records should contain (a) a freezer index showing what is in each part of the freezer and (b) a cell strain index, describing the cell line, its designation, what its special characteristics are, and where it is located. This may be done on a conventional card index, but a computerized data file will give superior data storage and retrieval. Material stored on discs or tape must have back-up copies on disc, tape, or hard-copy printout.

   The ampules should also carry a label with the cell strain designation and preferably the date

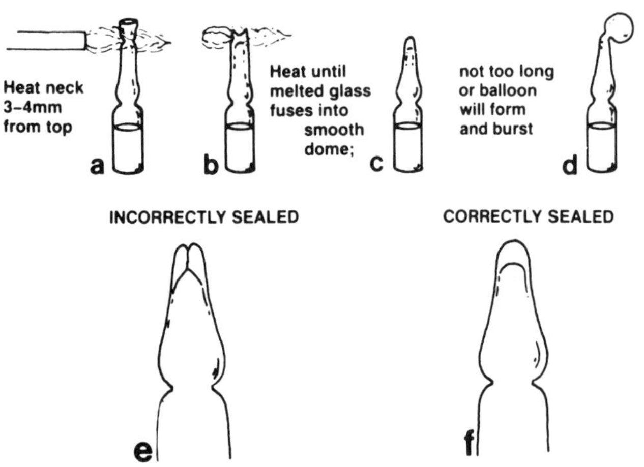

Fig. 17.5. Appearance of ampules during and after sealing. a. Tip melting. b. Molten glass folds inward. c. Sides coalesce to form dome. d. Overheating can cause air in the ampule to blow out the glass, forming a balloon, or bursting it completely. e. Incorrectly sealed ampule; fine capillary hole left. f. Correctly sealed ampule; glass fused evenly inside and outside.

and user's initials, although this is not always feasible in the available space. Remember, cell cultures stored in liquid nitrogen may well outlive you! They can easily outlive your stay in a particular laboratory. The record therefore should be readily interpreted by others and sufficiently comprehensive so that the cells may be of use to others.

There are four main types of liquid nitrogen storage systems (Fig. 17.8), based on whether the storage vessel is wide-necked or narrow-necked, and whether storage

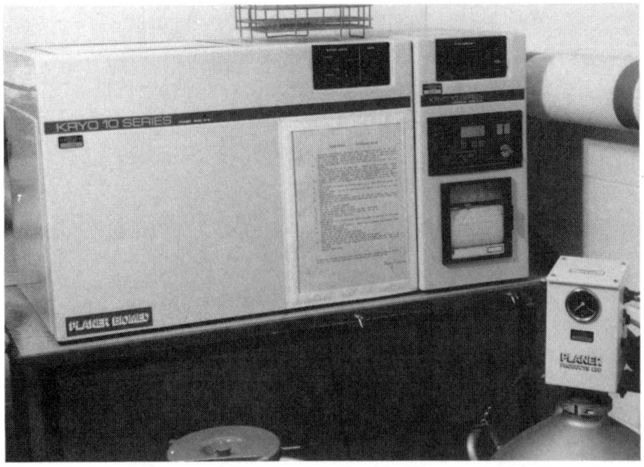

Fig. 17.6. Programmable coolers capable of regulating the cooling rate before and after freezing. Ampules are placed in an insulated box and the cooling rate regulated by injecting liquid N_2 into the box at a rate determined by a sensor in the box and a preset program in the console unit (Planer Biomed).

is in the vapor or liquid phase. Wide-necked freezers are chosen for ease of access and maximum capacity, narrow for economy (slow evaporation rate). Storage in the vapor phase eliminates the explosion risk with sealed ampules, while storage under the liquid means the container can be filled and the liquid nitrogen will therefore last longer. Storage capacities and static holding times (time to boil dry without being opened) are given in manufacturer's literature.

It is possible to effect a compromise by selecting a relatively narrow-necked freezer but still use a tray system for storage. Freezers are available with inventory control based on square-array storage trays, mounted on racks that are accessed by the same system as the cane and canister of conventional narrow-necked freezers. These freezers do not have the storage capacity of the cane and canister, but have equivalent holding times and a honeycomb storage array that many people prefer.

◇ Biohazardous material *must* be stored in the gas phase and teaching and demonstrating are best done with gas-phase storage. Above all, if liquid-phase storage is used, the user must be made aware of the explosion hazard of both glass and plastic, and wear a face shield or goggles.

Thawing

Protocol

1. When you wish to recover cells from the freezer, check the index, label your culture flask, then retrieve the ampule from the freezer, check that it is the correct one, and place it in 10 cm of water at 37°C in a 5-l bucket with a lid.

   ◇ A face shield or protective goggles and gloves must be worn. If the ampule has been stored in the gas phase, a lid is not necessary; but glass or plastic ampules stored under liquid nitrogen may inspire the liquid and, on thawing, will explode violently. A plastic bucket with a lid is therefore essential in this case to contain any explosion.
2. When thawed, swab the ampule with 70% EtOH and open. (Prescored glass ampules are available; these are much easier to open.) Transfer the contents to a culture flask. The dregs in the ampule may be stained with naphthalene black to determine viability (see Chapter 19).

   Add medium slowly to the cell suspension: 10 ml over about 2 min added dropwise at the start and then a little faster, gradually diluting the cells and preservative. This is particularly important with DMSO, where sudden dilution can cause severe osmotic damage and reduce survival by half.

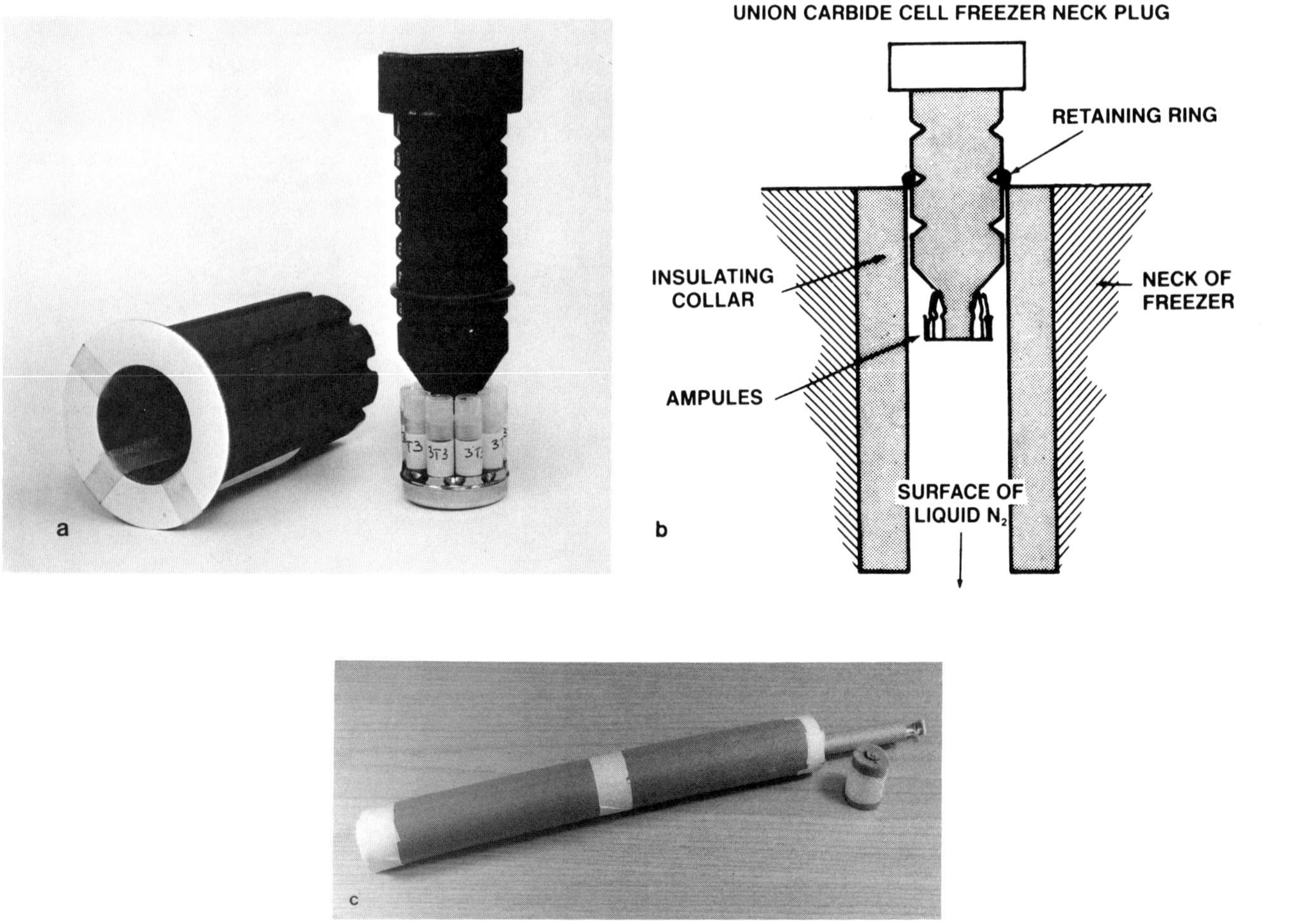

Fig. 17.7. Modified neck plug for narrow-necked freezers, allowing controlled cooling at different rates (Taylor Wharton). a. Inner unit separated from outer unit, showing ampules with 3T3 cells in the "basket." b. Section of freezer neck with modified neck plug in place. The "O" ring is used to set the height of the ampules within the neck of the freezer. The lower the height, the faster the cooling. c. Insulated freezing container made from domestic pipe lagging.

Variations. In some cases, e.g., L5178Y mouse lymphoma with DMSO, the preservative may be too toxic for the cell to survive on thawing (M. Freshney, personal communication). In these cases, dilute the cells slowly, as above, and centrifuge for 2 min at 100 *g*, discard the supernatant medium with preservative, and resuspend the cells in fresh medium for culture.

Whole flasks may be frozen by growing the cells to late log phase, adding 10% DMSO to the medium, and placing the flask in an expanded polystyrene container of 15-mm wall thickness [Ohno et al., 1991]. The insulated container is placed in a −70°C to −90°C freezer and will freeze at approximately 1°C/min. Survival is good for several months as long as the flask in its container is not removed from the freezer. Twenty-four-well plates may also be frozen in the same manner [Ure, 1992] and can be used to store large numbers of clones during evaluation procedures.

The number of cells frozen should be sufficient to allow for 1:10 or 1:20 dilution on thawing to dilute out the preservative but still keep the cell concentration higher than at normal passage; e.g., for cells subcultured normally at 10^5/ml, 10^7 should be frozen in 1 ml, and the whole seeded, after thawing into 20 ml, giving 5×10^5/ml (5x the normal seeding density) and diluting the preservative from 10% to 0.5%, at which concentration it is less likely to be toxic. Residual preservative may be diluted out as soon as the cells start to grow (for suspension cultures) or the medium changed as soon as the cells have attached (for monolayers).

The dye exclusion viability and the approximate take (e.g., proportion attached versus those still floating af-

TABLE 17.1. Record Cards for Cells in Frozen Storage

Position:

Freezer no.________ Cannister/section no.__________ Tube/drawer no.__________

Cell strain/line___________ Freeze date__________ Frozen by___________

No. ampules frozen________ No. cells/ampule___________ in___________ml

Growth medium___________ Serum__________ Conc.__________ Freeze medium___________

Method of cooling_________________ Cooling rate____________

Thaw date	No. of amp.	No. left	Seeding: Conc.	Seeding: Vol.	Seeding: Medium	Take*: Dye exclusion	Take*: % Attached by 24 h	Take*: Cloning efficiency	Notes

*Only one parameter need be used

TABLE 17.2. Record Cards for Cells in Frozen Storage—Cell Line/Cell Strain Card

Cell clone designation | Freezing date | Location

Origin:
Species________Normal/Neoplastic Adult/fetal/NB
Tissue________Site_____ Gen/Pass No.___
Author________Lab/ref__________
Morphology________ Mode of growth________
Special characteristics e.g., karyotype, isoenzymes,
differentiation markers, virus sensitivity, antigens, products:

____

____

____

(Further details on back of card)
Mycoplasma:
Date of test___ Method_______ Result_____
Normal maintenance:
Subculture
Frequency_______ Min. cell conc_______ Agent_______
Medium change
Frequency___ Whole/part vol_____Medium_____Serum____%_
Gas phase_______ Buffer_________ pH_____ Matrix?_________
Any other special conditions___________________________

____

Freezing instructions:
Rate_____ Preservative______%___
Thawing instructions:
Thaw rapidly (37°C), fresh water bath, dilute to 5/10/20/50 ml.
Centrifuge to remove preservative: Yes_____ No_____
Notes or special requirements:

____

Biohazard precautions:

____

Name of person completing card__________Date_____

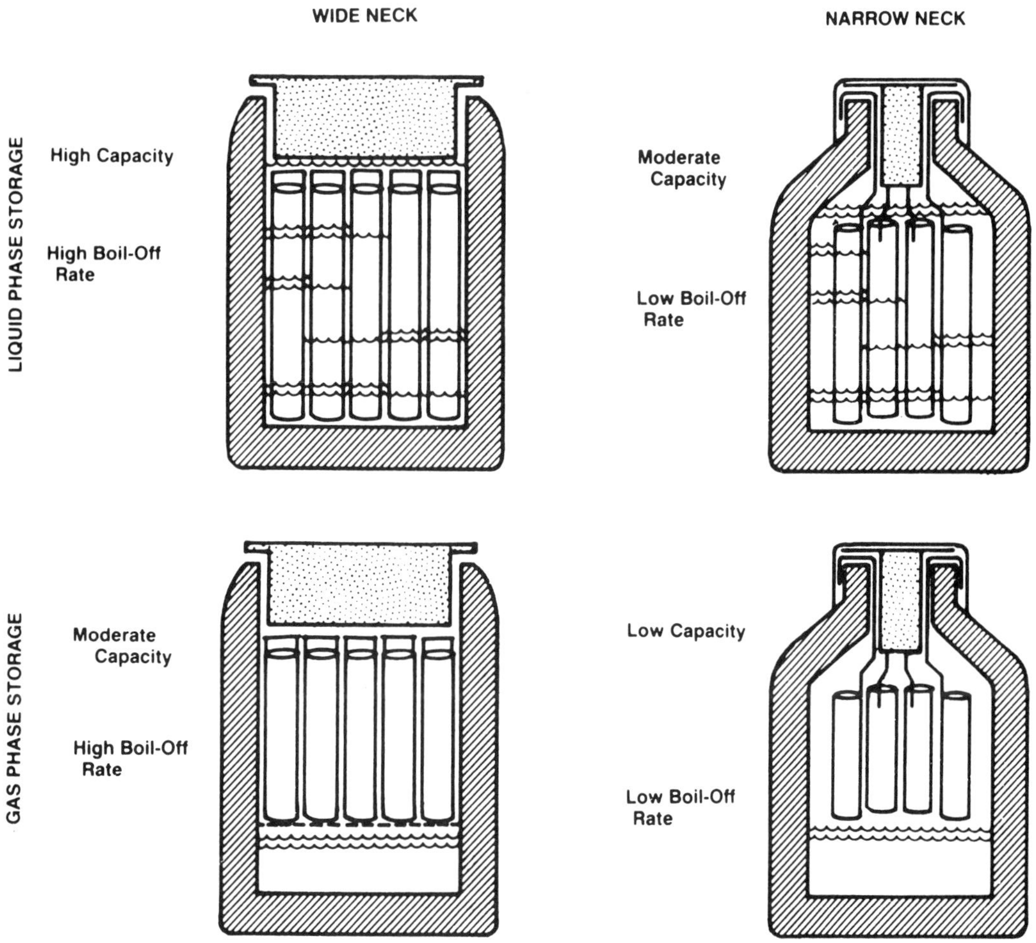

Fig. 17.8. *Variations in freezer design and usage.*

ter 24 hr) should be recorded on the appropriate record card or file to assist in future thawing. One ampule should be thawed from each batch as it is frozen, to check that the operation was successful.

During a prolonged series of experiments, lasting more than 6 months, stock cultures should be replaced from the freezer at regular intervals, say every 3 months, thereby preserving the properties of the cell line.

Although variability cannot be eradicated entirely from cultured cell lines, it may be minimized by adopting the procedures described above. In summary: (1) Select finite lines around the fifth generation, or clone, isolate, and grow up a continuous cell strain; (2) confirm characteristics of the strain; (3) check for microbial and cross-contamination; (4) freeze down a large number of ampules; (5) thaw one ampule to test survival; and (6) replace stock regularly from freezer.

CELL BANKS

To assist in the distribution of standardized, characterized cell lines and to provide secure repositories for valuable material, a number of cell banks have been set up in different parts of the world (Table 17.3). Since many cell lines may come under patent restrictions, particularly hybridomas and other genetically modified cell lines, it has also been necessary to provide patent repositories with limited access.

As a general rule it is preferable to obtain your initial seed stock from a reputable cell bank, where the necessary characterization and quality control will have been done. Furthermore, it is a good plan to submit valuable cultures to a cell bank for preservation in addition to maintaining your own frozen stock, as this will protect you against loss of your own lines and allow distribution to others. If you feel that your cells should not be distributed, then they can be banked with that restriction placed on them.

In addition to the provision of frozen cell stocks, some cell banks also make their catalogue information available on line or on disc. There are also databanks that can be accessed on line and that maintain information on cell lines held by subscribers in their own laboratories (Table 17.4). This provides a vast increase

TABLE 17.3. Cell Banks[a]

United States & Canada	American Type Culture Collection (ATCC), 12301 Parklawn Drive, Rockville, MD 20852
	National Institute of General Medical Sciences (NIGMS), Human Genetic Mutant Cell and National Institute on Aging Cell Culture Repositories, Coriell Institute for Medical Research, Copewood Street, Camden, NJ 08103
	Repository for Human Cell Strains, and Cell Repository for Neuromuscular Diseases, Montreál Children's Hospital Research Institute, McGill University, 2300 rue Tupper Street, Montreal, Quebec H3H 1P3, Canada
Europe	European Collection for Animal Cell Cultures (ECACC), PHLS/CAMR, Porton, Salisbury, England
	European Collection for Biomedical Research, Dept. of Cell Biology & Genetics, Erasmus University, P.O. Box 1783, Rotterdam, Netherlands
	Collection Nationale des Cultures des Microorganismes, Institute Pasteur, 25 rue du Dr Roux, F-75724 Paris Cédex 15, France
	Human Genetic Cell Repository, Hospices Civils de Lyon, Hôpital Debrousse, 29 rue Soeur Bourier, F-69322 Lyon Cédex 05, France
	Tumorbank, Institut für Experimentelle Pathologie (dkfz), Deutsche Krebsforschungszentrum, In Neuenheimer Feld 280, Postfach 101949, D-6900 Heidelberg 1, Germany
	Centro Substrati Cellulari, Instituto Zooprofilattico Sperimentale della Lombardia e dell'Emilia, Via A.Biandri 7, I-25100 Brescia, Italy
	National Bank for Industrial Microorganisms and Cell Cultures (NBIMCC), Blvd. Lenin 125 BL 2, V Floor, Sofia, Bulgaria
	National Collection of Agricultural & Industrial Microorganisms (NCAIM), Dept. of Microbiology, University of Horticulture, Somloi ut 14-16, H-1118 Budapest, Hungary
Japan	Japanese Cancer Research Resources Bank (JCRB), National Institute of Hygienic Sciences, Kami-Yoga, Setagaya-Ku, Tokyo
	General Cell Bank, Institute of Physical and Chemical Research of RIPEN, Saitama
	Institute for Fermentation, Osaka, 17-85, Juson-honmachi 2-chome, Yodawa-ku, Osaka 532, Japan
Australia	Commonwealth Serum Laboratories, 45 Poplar Road, Parkville, Victoria 3052

[a]Further information available from ATCC or ECACC, addresses as in table. Most of this information has been obtained from Doyle et al. [1990].

in material potentially available, but it must be remembered that cell lines obtained from other laboratories will vary significantly in the amount of characterization that they have had and in the quality of their maintenance.

Transporting Cells

Cultures may be transferred from one laboratory to another as frozen ampules, shipped in solid carbon dioxide, in a thick-walled polystyrene foam container. When cells are shipped in solid CO_2, the carrier must be informed. Usually cells will remain frozen for up to 3 days if properly packed, but if they thaw slowly, their viability will decline rapidly.

Alternatively, cells may be shipped as a growing culture. The flask should be filled to the top with medium, taped securely around the neck with a stretch-type waterproof adhesive tape, and packed within a rigid container, packaged, in turn, in a polythene bag and then in plastic foam or bubble wrap. Place a "fragile" label on the box and instructions, in large letters, not to freeze it!

With the universal deterioration in postal services, it is better to ship cells via a carrier, who should be

TABLE 17.4. Databanks

ATCC Catalogue	Address as in Table 17.3 EMail: BT TYMNET 42:CDT0004
ECACC Catalogue	Address as in Table 17.3 EMail: TELECOM GOLD 75:DBI0222
CERDIC	2° CAI-Solarex, Avenue des Maurettes, F-06270 Villeneuve-Loubet, France EMail: TELECOM GOLD 75:DBI0098
CODATA	CODATA Secretariat, 51 Boulevard de Montmorency, F-75061 Paris TELECOM GOLD 75:DBI0010
Culture de Cellules Eucaryotes. Repertoire des Utilisateurs.	M. Adolphe, D. Gourdji, A. Tixier-Vidal, R. Robineaux, Inserm Publications, 101 Rue de Tolbiac, 75654 Paris, Cedex 13
Fundação Tropical de Pesquisas e Tecnologia "André Tosello"	Rue Latino Coelho, 1301, 13.000 Campinas, SP, Brazil EMail: BT TYMNET 42:CDT0094
Hybridoma Data Bank (HDB)	Information from ATCC, ECACC, RIKEN, and CERDIC
Collection of Microorganisms	Institute of Physiology and Biochemistry of Microorganisms, Academy of Sciences, Pushchino-na-oke, 142292 Moscow Region, Russia
Japan Federation for Culture Collections (JFCC)	NODAI Research Institute Culture Collection, Tokyo University of Agriculture, 1-1-1 Sakuragaoka, Setagaya-ku, Tokyo 156, Japan
LSRIS	RIKEN, 2-1 Hirosawa, Wako, Saitama 351-01, Japan EMail: BT TYMNET 42:CDT0007
Information Centre for European Culture Collections	Mascheroder Weg 1b, D-3300 Braunschweig, Germany EMail: TELECOM GOLD 75:DBI0274
MSDN	Secretariat, Institut of Biotechnology, Cambridge University, 307 Huntingdon Road, Cambridge CB3 0JX, UK EMail: TELECOM GOLD 75:DBI0005; JANET: msdn@phx.cam.ac.uk
CLDB (Cell Line Data Base)	Interlab Project Users Service, Servizio Biotecnologie, Istituto Nazionale per la Ricerca sul Cancro, viale Benedetto XV, 10, I-16132 Genova, Italy Fax: 39 10 352888 EMail: ipus@istge.ist.unige.it

Most of this information has been obtained from Doyle et al. [1990].

briefed as to the contents of the package and the urgency of delivery. International mailings will be required to negotiate customs controls and here it is good to have an agent nominated who is familiar with this type of importation and can speed the package through customs. Cells shipped to the United States are required to be quarantined and tested by the Bureau of Agriculture, so it is better to arrange this through one of the cell banks, ECACC or ATCC.

Warn the recipient before shipping, enclose details of the cell line and its maintenance, and fax these details to the recipient before the cells are shipped.

CHAPTER 18

Quantitation and Experimental Design

SELECTION OF CELL LINE

The selection of a cell line, or lines, is a key decision at the initiation of a program of experiments. This will be governed mainly by the specific properties that are required for your work, but there are other general considerations to take into account:

(1) Is there a continuous cell line that expresses the right functions? A continuous cell line will generally be easier to maintain, will grow faster, will clone more easily, and will produce a higher cell yield per flask (see Chapter 10, Table 10.3).

(2) Is it important whether the line is malignantly transformed or not? If it is, it might be possible to obtain a permanent line that is still not malignant.

(3) Is species important? Nonhuman lines have fewer biohazard restrictions and have the advantage that the original tissue may be more accessible.

(4) What growth characteristics do you require?

Population doubling time
Saturation density (yield per flask)
Cloning efficiency
Growth fraction
Ability to grow in suspension

(5) If you have to use a finite cell line, are there sufficient stocks available, or will you have to generate your own line(s)?

(6) Can the line be made to express the right characteristics?

(7) How well characterized is the line, if it exists already, or, if not, can you do the necessary characterization?

(8) If you are using a mutant, transfected, transformed, or abnormal cell line, is there a normal equivalent available should it be required?

(9) How stable is the line? Has it been cloned? Can you clone it if it has not, and how long would this take to generate sufficient frozen and usable stocks?

EXPERIMENTAL DESIGN

Experiments in tissue culture should be designed following the same principles as in any other discipline. Appropriate controls must be performed for each variable examined, and observations should be made in triplicate, or a higher number of replicates if intrinsic variability is expected to be high. In general, however, relatively few replicates are required in cell line work.

Replicate cultures may be performed in multiwell

plates if all are to be harvested simultaneously. If not, separate dishes, plates, tubes, or flasks will be required.

It is important to monitor the consistency of media and sera used. The same medium should be used throughout a particular series, preferably from the same supplier, and the same batch number, although this may not be possible if the experiments extend over a long period. Serum should be from the same, pretested, batch, and before changing, a new batch must be tested and reserved. Similarly, culture vessels should be from the same supplier, although there is now little variation among the main producers of culture vessels, and the ratio of medium to surface area should be kept constant.

Incubation temperatures should be carefully monitored, and the CO_2 tension checked regularly.

GROWTH PHASE

In the design of cell culture experiments it is important to be aware of the growth state of the culture, as well as the qualitative characteristics of the cell strain or cell line. Cultures will vary significantly in many of their properties between exponential growth and stationary phase. It is, therefore, important to take account both of the status of the culture at the initiation of an experiment and at the time of sampling, and the effects of the duration of an experiment on the transition from one state to another. Adding a drug in the middle of the exponential phase and assaying later may give different results, depending on whether the culture is still in exponential growth when harvested or whether it has entered plateau. Cells that have entered plateau have a greatly reduced growth fraction, a different morphology, may be more differentiated, and may become polarized. They generally tend to secrete more extracellular matrix and may be more difficult to disaggregate.

Quantitation of growth is also important in routine maintenance as an important element in monitoring the consistency of the culture and knowing the best time to subculture, the optimum dilution, and the estimated plating efficiency at different cell densities. Testing medium, serum, new culture vessels or substrates, etc., all require quantitative assessment.

The first part of this chapter considers the various methods commonly used to quantify cells in culture, such as cell counting, biochemical determinations, and cytometry. The second part deals with the kinetics of cell growth *in vitro* as determined by studying the growth cycle, the cell cycle, plating efficiency, and the growth fraction.

CELL COUNTING

While estimates can be made of the stage of growth of a culture from its appearance under the microscope, standardization of culture conditions and proper quantitative experiments are difficult unless the cells are counted before and after, and preferably during, each experiment.

Hemocytometer

The concentration of a cell suspension may be determined by placing the cells in an optically flat chamber under a microscope (Fig. 18.1). The cell number within a defined area of known depth is counted and the cell concentration derived from the count.

Materials

Hemocytometer (Improved Neubauer)
PBSA
0.25% crude trypsin
Medium
Tally counter
Pasteur pipette
Microscope

Protocol

1. Prepare the slide: (a) Clean the surface of the slide with 70% alcohol, taking care not to scratch the semi-silvered surface. (b) Clean the coverslip and, wetting the edges very slightly, press it down over the grooves and semi-silvered counting area (see Fig. 18.1). The appearance of interference patterns ("Newton's rings"–rainbow colors between coverslip and slide like those formed by oil on water) indicates that the coverslip is properly attached, thereby determining the depth of the counting chamber.
2. Trypsinize monolayer or collect sample from suspension culture. Approximately 10^6 cells/ml minimum are required for this method, so the suspension may need to be concentrated by centrifuging (100 g for 2 min) and resuspending in a measured smaller volume.
3. Mix the sample thoroughly and collect about 20 μl into the tip of a Pasteur pipette or micropipette. Do not let the fluid rise in a Pasteur pipette or cells will be lost in the upper part of the stem.
4. Transfer the cell suspension immediately to the edge of the hemocytometer chamber and let the suspension run out of the pipette and be drawn under the coverslip by capillarity. Do not

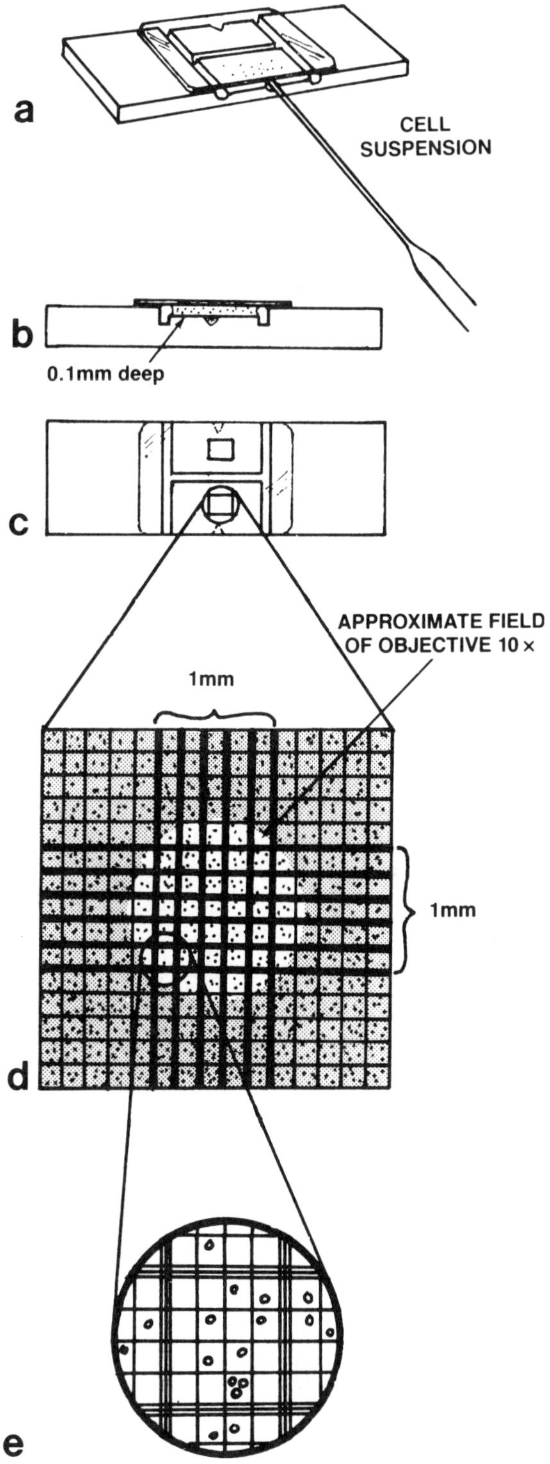

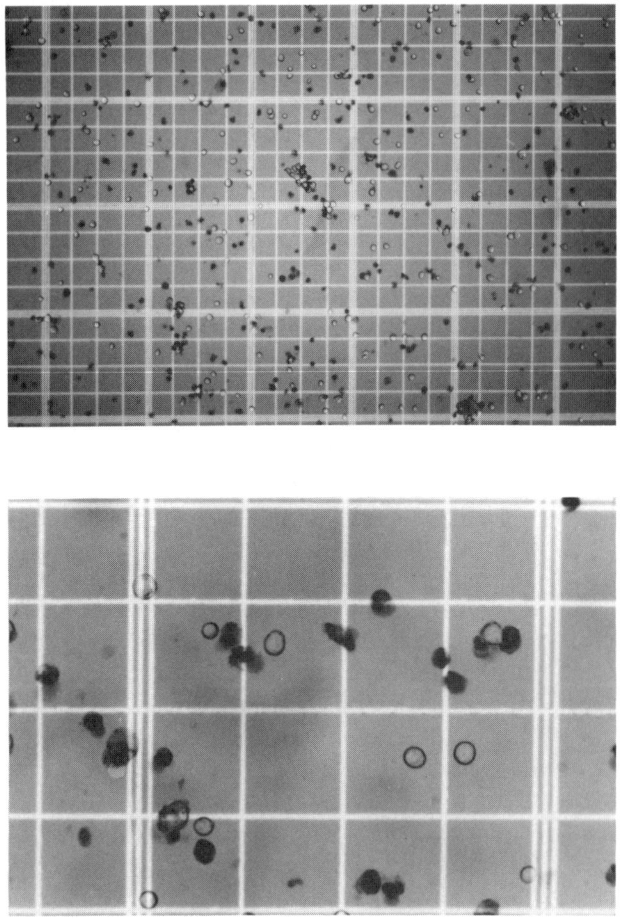

Fig. 18.1. *Hemocytometer slide (Improved Neubauer). a. Adding cell suspension to the assembled slide. b. Longitudinal section of slide showing position of cell sample in 0.1-mm-deep chamber. c. Top view of slide. d. Magnified view of total area of grid. Light central area is that area which would be covered by the average 10× objective (depending on field of view of eyepiece). This covers approximately the central 1 mm² of the grid. e. Magnified view of one of the 25 smaller squares, bounded by triple parallel lines making up the 1-mm² central area. This is subdivided by single grid lines into the 16 smallest squares to aid counting. f. Low-power (10× objective) microphotograph showing 20 of the 25 smaller squares of a slide loaded with cells pretreated with naphthalene black (amido black) (see text). Viable cells are unstained and clear with a refractile ring around them; nonviable cells are dark and have no refractile ring. g. High-power (40× objective) microphotograph of one of the smaller squares, bounded by three parallel lines and containing 16 of the smallest squares. The distance between each set of triple lines is 200 μm.*

overfill or underfill the chamber or its dimensions may change due to alterations in surface tension; the fluid should run to the edges of the grooves only. Reload the pipette and fill the second chamber if there is one.

5. Blot off any surplus fluid (without drawing from under the coverslip) and transfer the slide to the microscope stage.
6. Select 10× objective and focus on grid lines in chamber (see Fig. 18.1). If focusing is difficult

because of poor contrast, close down the field iris or make the lighting slightly oblique by tilting the mirror or offsetting the condenser.

7. Move the slide so that the field you see is the central area of the grid and is the largest area that you can see bounded by three parallel lines. This area is 1 mm^2. Using a standard 10× objective, it will almost fill the field or the corners will be slightly outside the field, depending on the field of view (see Fig. 18.1).
8. Count the cells lying within this 1-mm^2 area, using the subdivisions (also bounded by three parallel lines) and single grid lines as an aid to counting. Count cells that lie on the top and left-hand lines of each square but not those on the bottom or right-hand lines to avoid counting the same cell twice. For routine subculture, attempt to count between 100 and 300 cells per mm^2; the more cells that are counted, the more accurate the count becomes. For more precise quantitative experiments, 500–1,000 cells should be counted.
9. If there are very few cells (<100/mm^2), count one or more additional squares (each 1 mm^2) surrounding the central square.
10. If there are too many cells (>1,000/mm^2), count only five small squares (each bounded by three parallel lines) across the diagonal of the larger (1-mm) square.
11. If the slide has two chambers, move to the second chamber and do a second count. If not, rinse the slide and repeat the count with a fresh sample.

Analysis. Calculate the average of the two counts, and derive the concentration of your sample as follows:

$$c = n/v$$

where c = cell concentration (cells/ml), n = number of cells counted, and v = volume counted (ml). For the Improved Neubauer slide, the depth of the chamber is 0.1 mm, and assuming only the central 1 mm^2 is used, v is 0.1 mm^3 or 10^{-4} ml. The formula becomes:

$$c = n \times 10^4.$$

Hemocytometer counting is cheap and gives you the opportunity to see what you are counting. If the cells are mixed previously with an equal volume of a viability stain (see below and Fig. 18.1f,g), a viability determination may be performed at the same time. The procedure is, however, rather slow and prone to error both in the method of sampling and the size of samples and requires a minimum of 10^6 cells/ml.

Most of the errors occur by incorrect sampling and transfer of cells to the chamber. Make sure the cell suspension is properly mixed before you take a sample, and do not allow the cells time to settle or adhere in the tip of the pipette before transferring to the chamber. Ensure also that you have a single cell suspension, as aggregates make counting inaccurate. Larger aggregates may enter the chamber more slowly or not at all. If aggregation cannot be eliminated during preparation of the cell suspension (see Table 10.4), lyse the cells in 0.1 M citric acid containing 0.1% crystal violet at 37°C for 1 hr and count the nuclei [Sanford et al., 1951].

Electronic Particle Counting

Although a number of different automatic methods have been developed for the counting of cells in suspension, the system devised originally by Coulter Electronics is the one most widely used. Briefly, cells drawn through a fine orifice change the current flow through the orifice, producing a series of pulses that are sorted and counted.

Coulter counter (Fig. 18.2). There are two main components of the system (Fig. 18.3): (1) an orifice tube connected to a pump and a mercury manometer by a two-way valve (Fig. 18.3a); and (2) an amplifier, pulse height analyzer, and scaler connected to two electrodes–one in the orifice tube and one in the sample beaker–the current to them controlled by switch points on the mercury manometer.

When the two-way valve is turned vertically, the mercury manometer and orifice tube are connected to the pump (Fig. 18.3a). While liquid is drawn through the orifice, generating a signal on the cathode-ray oscilloscope and enabling visual analysis of the cell suspension, the mercury is drawn up the manometer to a preset level determined by the negative pressure generated by the pump. When the valve is restored to the horizontal position (Fig. 18.3b), the pump is disconnected, and the mercury manometer is connected directly to the orifice tube. As the mercury returns to equilibrium, fluid carrying the cell suspension is drawn through the orifice.

As the mercury travels along the tube, it passes two switch points; the first starts the count cycle, the second stops it. The mercury displacement between the two switches is 0.5 ml, hence 0.5 ml of cell suspension is drawn in through the orifice during the count cycle.

As each cell passes through the orifice, it changes the resistance to the current flowing through the orifice by an amount proportional to the volume of the cell. This generates a pulse ($amps^{-1}$) that is amplified and counted. Since the size of the pulse is proportional to the volume of the cell or particle passing through the ori-

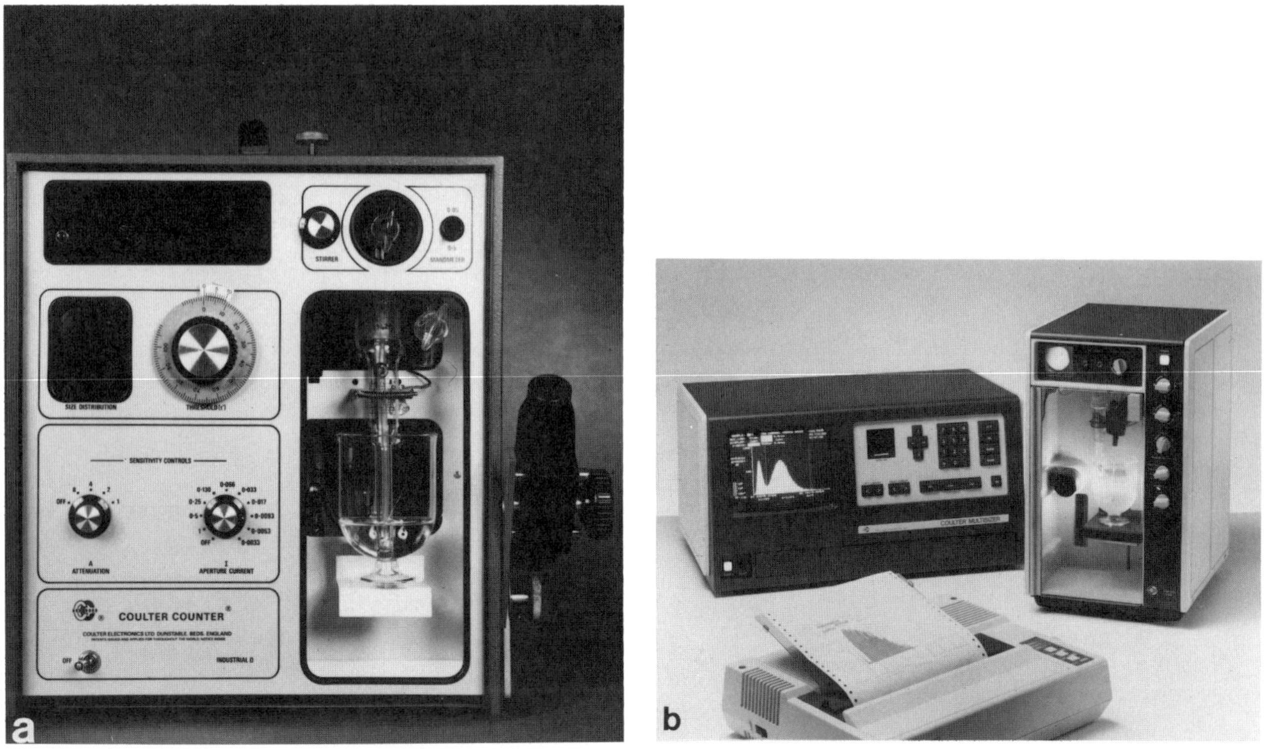

Fig. 18.2. Coulter electronic cell counter and cell sizer. a. D Industrial, suitable for routine laboratory use for cell counting. b. Multisizer. Can be used for cell counting, but is primarily intended for cell sizing.

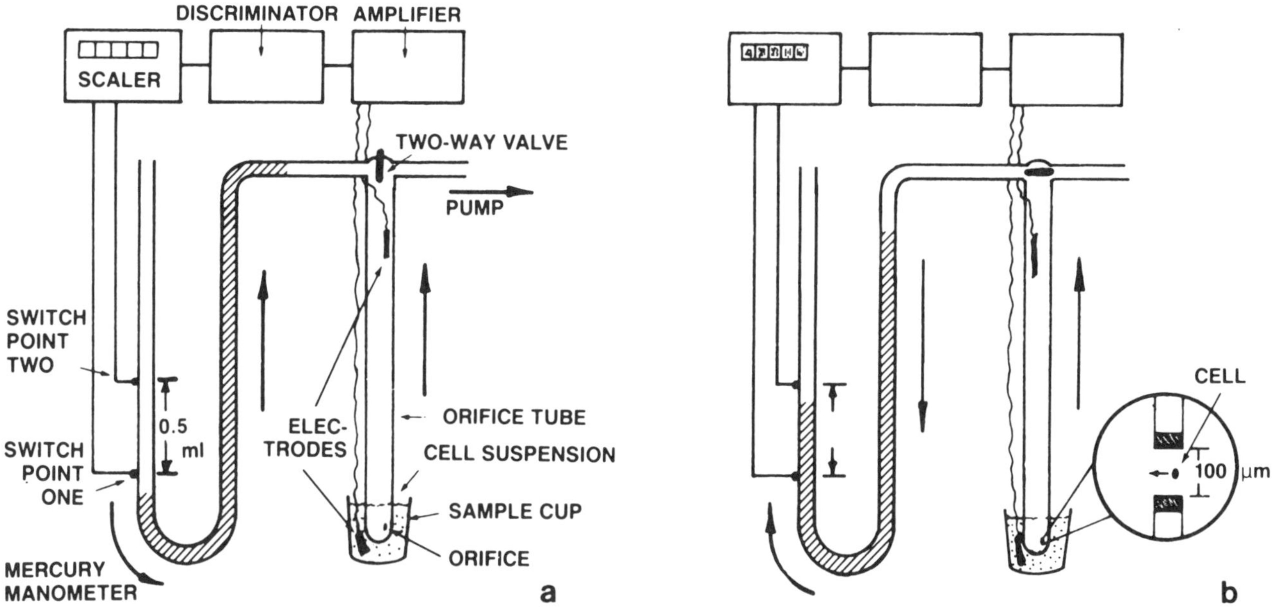

Fig. 18.3. Principle of operation of electronic particle counter (based on Coulter counter). a. Mercury manometer connected to pump, drawing mercury up to starting position. b. Mercury returning to equilibrium, drawing sample though orifice, and activating count cycle. Inset: magnified view of orifice in section.

fice, a series of signals of varying pulse height are generated. A threshold control is set on the front panel to eliminate electronic noise and fine particulate debris but to retain pulses derived from cells (see below). This setting controls a pulse height analyzer circuit between the amplifier and the scaler that only allows pulses to pass to the scaler above the preset threshold.

Operation of Coulter Cell Counter (Model DI)

Outline

A sample of cells is diluted in electrolyte (physiological saline or PBS), placed under the orifice tube, and counted by drawing 0.5 ml of the sample through the counter.

Materials

Culture
PBSA
0.25% crude trypsin
Medium
Counting cup
Counting fluid (see Reagent Appendix)

Protocol

1. Trypsinize cell monolayer or collect sample from suspension culture. The cells must be well mixed and singly suspended.
2. Dilute the sample of cell suspension 1:50 in 20 ml counting fluid in a 25-ml beaker or disposable sample cup. An automatic dispenser will speed up this dilution and improve reproducibility.

   Note. Dispensing counting fluid rapidly can generate air bubbles that will be counted as they pass through the orifice. Consequently, the counting fluid should stand for a few moments before counting. If the fluid is dispensed first and cells added second, this problem is minimized.
3. Mix well and place under tip of orifice tube, ensuring that the orifice is covered and that the external electrode lies submerged in the counting fluid in the sample beaker.
4. Set the two-way valve vertically until red light comes on and then extinguishes (mercury has passed both switch points).
5. Clear the display.
6. Restore two-way valve to the horizontal position and allow count to proceed (red light comes on).
7. When the red light extinguishes, the count is finished. Note the count and replace the sample with fresh counting fluid.

Analysis. The counter takes 0.5 ml of the 1:50 dilution, so multiply the final count on the readout by 100 to give the concentration per milliliter of the original cell suspension.

Problems. (1) Count stops, will not start, or counts slowly (i.e., takes longer than 25 s):

Orifice clogged. Free with the tip of finger or fine brush.

(2) Count lower than expected, orifice blocks frequently. Cell suspension aggregated:

Disperse cells by pipetting original sample vigorously, redilute, and proceed.

(3) High electrical activity on oscilloscope screen but will not count:

Electrode out of beaker or disconnected.

(4) Red light comes on but will not go out, or will not come on at all:

(a) Blocked orifice as in (1); (b) insufficient negative pressure. Pump has failed or leakage in tubing. Check pump and connections.

(5) Gurgling sound from pump:

Counting fluid from waste reservoir has been drawn into pump:

Switch off, disconnect, and dry out pump. Relubricate, reconnect, and start again. The level in the waste reservoir should be checked regularly so that it may be emptied before the pump becomes contaminated.

(6) High background (a) line or radio interference, usually from electrical equipment (motors, fluorescent lights, incubators):

Check and eliminate by fitting suppressors to equipment. Line filters are available but are not always effective. Check grounding (earthing) of counter, particularly the case.

(b) Particulate matter in counting fluid:

Filter through disposable Millex or equivalent filter.

Calibration. To set the threshold in the correct position, perform a series of counts on the cell suspension, moving the threshold up in increments of 5 or 10 units from zero to 100. Plotting the counts against the threshold settings should produce a curve with a plateau (Fig. 18.4a). If the plateau is too short, decrease the attenuation and repeat. If the curve is too drawn out and does not reach a plateau, increase the attenuation.

The shape of the plateau can also be controlled by the aperture current. However, increasing the aperture current will increase the heat generated in the sample, so this should be done only if changing the attenuation will not generate a plateau. Given the size of mammalian cells, it is unlikely that a plateau will not be reached by varying the attenuation within intermedi-

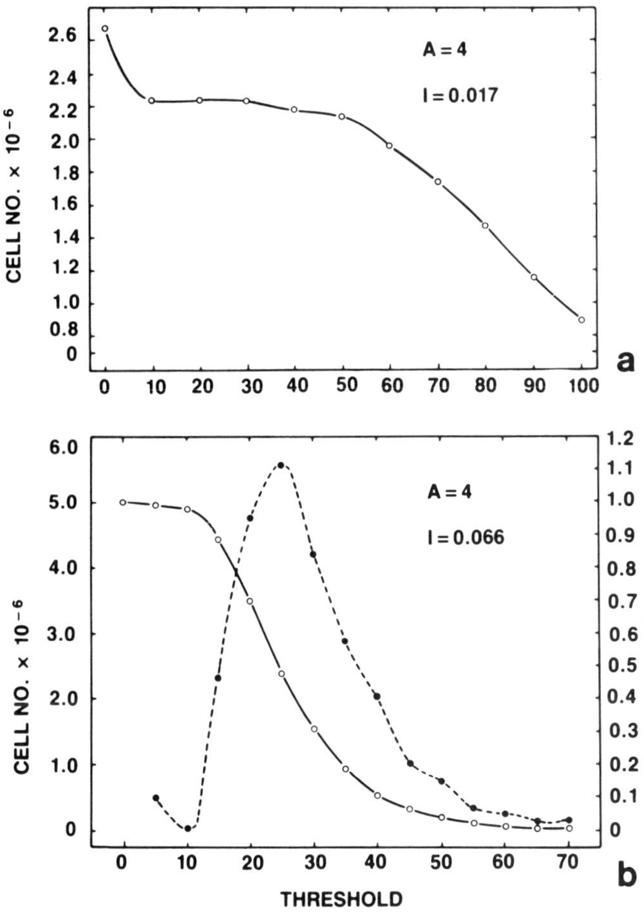

Fig. 18.4. *Calibration of cell counter for a specific cell type. a. A sample of LS cells was counted at a range of threshold values. The correct threshold setting is that equivalent to the center of the plateau, i.e., 20. b. Repeat counts (higher cell concentration) with aperture current reset to 0.066. Plateau is now between zero and ten (unsuitable for routine counting) and cell count falls to near zero by threshold setting 70. By differentiating the first curve (solid line) a cell size analysis is obtained (dotted line). (Counters with automatic simultaneous pulse height analysis will plot this curve automatically; see Fig. 18.2b.) A = attenuation. I = aperture current.*

ate settings on the aperture current control, but if in doubt contact the Coulter technical service.

The generation of a plateau depends on the existence of a homogeneous population of cells, as found in most continuous cell lines. Primary cultures, on the other hand, can be very heterogeneous, and it may not be possible to recognize a significant plateau. What is observed is a series of small overlapping plateaus, each related to one population of cells, and that could be resolved either by cell-sizing apparatus, or by differentiating the "plateau" curve (subtracting one count from the previous count and plotting the difference against the threshold setting). In substance, what this means is that the population is too heterogeneous to be displayed as a single plateau (or single peak on the differentiated count). To make a satisfactory count from such a population, it is necessary to count the cells by hemocytometer, making the decision whether to count all cells or just nucleated cells (see above), or the size threshold must be determined, using latex beads or pollen grains, below which no viable, reproductive, cells can be anticipated, and only cells above this threshold are counted.

Cell Sizing

If the amplification is set such that the curve of cell counts versus threshold reaches a value at 100, which is <5% of the count at 5, then the bulk of the cell population has been analyzed and differentiating the curve (subtracting each count from the previous count at a lower threshold setting) will give a plot of relative size distribution (Fig. 18.4b). If standard latex particles are treated in the same way, a standard curve is obtained and the absolute cell size may be derived from the formula

$$v = KAIT$$

where v = cell volume, A = attenuation setting (inverse of amplification), I = current setting, T = threshold setting, and K = constant. Derive K from standard latex particles and substitute in above equation to derive v for your cell sample.

Counters are available with automatic cell-sizing capabilities (Fig. 18.2b). They operate on the same principle but instead of controlling the threshold manually, the pulses from the amplifier are fed into a pulse height analyzer and sorter, which gives an instant readout on an oscilloscope and will print the size distribution histogram on an X-Y recorder. A counter with this facility will cost about three or four times as much as the simple version described above but makes cell sizing faster, easier, and more accurate.

Electronic cell counting is rapid and has a low inherent error due to the high number of cells counted. It is prone to misinterpretation, however, as cell aggregates, dead cells, and particles of debris of the correct size will all be counted indiscriminately. The cell suspension should be examined carefully before dilution and counting.

Electronic particle counters are expensive, but if used correctly, they are very convenient and give greater speed and accuracy to cell counting. There are now several such instruments available, but the Coulter D (Industrial) remains one of the cheapest.

Stained Monolayers

There are occasions when cells cannot be harvested for counting or are too few to count in suspension. This situation is encountered with some multiwell plates or Terasaki plates. In these cases, the cells may be fixed and stained *in situ* and counted by eye on a microscope. Since this procedure is tedious and subject to high operator error, isotopic labeling or estimation of total DNA or protein (see below) is preferable, though these measurements may not correlate directly with cell number, e.g., if the ploidy of the cell varies. A rough estimate of cell number per well can also be obtained by staining the cells with crystal violet and measuring absorption on a densitometer. This method has also been used to calculate the number of cells per colony in clonal growth assays [McKeehan et al., 1977]. Staining cells with Coomassie blue, sulforhodamine B [Boyd, 1989; Skehan et al., 1989], or MTT [Plumb et al., 1989] also gives an estimation of cell number, given that linearity has been demonstrated previously in a standard plot of absorption against cell number (see Chapter 19). MTT staining has the advantage that it will only stain viable cells.

CELL WEIGHT

Wet weight is seldom used unless very large cell numbers are involved, as the amount of adherent intracellular liquid gives a high error. As a rough guide, however, there are about 2.5×10^8 HeLa cells (14–16 μm diameter) per gram wet weight, about $8-10 \times 10^8$ cells/g for murine leukemias (e.g., L5178Y murine lymphoma or Friend murine erythroleukemia) (11–12 μm diameter) and about 1.8×10^8 for human diploid fibroblasts (16–18 μm diameter) (see Table 18.1).

Dry weight is, similarly, seldom used, as salt derived from the medium contributes to the weight of unfixed cells and fixed cells will lose some of their low molecular weight intracellular constituents and lipids. However, an estimate of dry weight can be derived by interferometry [Brown and Dunn, 1989].

DNA CONTENT

In practice, apart from cell number, DNA and protein are the two most useful measurements for quantifying the amount of cellular material. DNA may be assayed by several fluorescence methods, including the reactions with DAPI [Brunk et al., 1979] or Hoechst 33258 [Labarca and Paigen, 1980]. DNA can also be measured by its absorbance at 260 nm, where 50 μg/ml has an O.D. of 1.0. Because of interference from other cellular constituents, the direct absorbance method is only useful for purified DNA. The following is a relatively simple and straightforward assay for DNA.

Hoechst 33258 Method [Labarca and Paigen, 1980]

Principle

The fluorescence emission of Hoechst 33258 at 458 nm is increased by interaction of the dye with DNA at pH 7.4 and in high salt to dissociate the chromatin protein. This method gives a sensitivity of 10 ng/ml, but requires intact double-stranded DNA.

Outline

Cells or tissue are homogenized in buffer and sonicated. Aliquots are mixed with H33258 and the fluorescence measured.

Materials

Buffer: 0.05 M $NaPO_4$, 2.0 M NaCl pH 7.4 containing 2×10^{-3}M EDTA
H33258: 1 μg/ml in buffer for DNA above 100 ng/ml and 0.1 μg ml for 10–100 ng/ml

Protocol

1. Homogenize cells in buffer, 10^5/ml for 1 min, using a Potter homogenizer.
2. Sonicate for 30 s.
3. Dilute 1:10 in Hoechst + buffer.
4. Read fluorescence at 356 nm excitation and 492 nm emission using calf thymus DNA as a standard.

DAPI Method

The following protocol was contributed by Edith Schwartz (based on the method of Brunk et al., 1979), Departments of Orthopedic Surgery and Physiology,

TABLE 18.1. Relationship Between Cell Size, Volume, and Mass

Cell type	Diameter (μm)	Volume (μm³)	Cells/g × 10⁻⁶ Calculated	Cells/g × 10⁻⁶ Measured
Murine leukemia, e.g., L5178Y or Friend	11–12	800	1,250	1,000
HeLa	14–16	1,200	800	250
Human diploid fibroblasts	16–18	2,500	400	180

Tufts University School of Medicine, 136 Harrison Avenue, Boston, MA 02111.

Principle

DAPI (4′,6-diamidino-2-phenylindole) bound to DNA fluoresces when excited by long-wavelength UV.

Outline

Lysed cells are mixed with DAPI in the dark and fluorescence measured on a fluorimeter.

Materials

DNA buffer solution: 0.01 M NaCl + 0.005 M HEPES buffer, pH 7.0

DNA standard: prepare a stock solution of DNA by dissolving at room temperature 80 mg of calf thymus DNA into a final volume of 10 ml of distilled water. Divide this solution into 1-ml aliquots into Pyrex test tubes and store them at −20°.

DAPI stock (4 ,6-diamidino-2-phenylin-dole): dissolve 300 mg DAPI in 1 ml DNA buffer solution. This solution is 100×.

Working solution of DAPI: to prepare the working solution of DAPI, add 100 µl of DAPI stock solution to 900 µl of DNA buffer solution and mix well. This yields 1 ml of DAPI (10×). Add 0.5 ml of DAPI (10×) to 4.5 ml of DNA buffer solution to give 5 ml of DAPI working solution.

Assay dilution buffer: 0.1 M sodium acetate, pH 6.2

Protocol

1. Prepare a set of blanks and standards (in triplicate) containing from 0 to 0.8 µg DNA.
2. Prepare test samples by suspending 5×10^4–5×10^5 cells in 1 ml of 0.1 M sodium acetate buffer, pH 6.2. Sonicate for a total of 30 s.
3. Add 150 µl sonicated cell suspension in triplicate to 850 µl of DNA buffer solution.
4. Vortex standards and tests and blanks (no DNA).
5. *In the dark*, add 50 µl of DAPI working solution to each sample and standard. Vortex each tube immediately after adding DAPI solution. Cover tubes with foil.
6. Let the tubes stand in the dark for 30 min.
7. Measure the fluorescence of standards and tests against mean of blanks at 372 nm excitation and 454 nm emission wavelengths.
8. Plot fluorescence (%) versus g DNA in standards. Use these values to determine DNA content of test samples.

PROTEIN

Protein content is widely used for estimating total cellular material and can be used in growth experiments or as a denominator in expression of specific activity of enzymes, receptor content, or intracellular metabolite concentrations. Proteins in solubilized cells can be estimated directly by measuring absorbance at 280 nm, and in this case, as the bulk of the cell's dry weight is protein, interference from nucleic acids and other constituents is minimal. Absorbance at 280 nm can detect down to 100 µg or about 2×10^5 cells.

Colorimetric assays are more sensitive, and among these the Lowry method [Oyama and Eagle, 1956], the Bradford reaction with Coomassie blue [Bradford, 1976], a sensitive test available from Pierce (see Suppliers List), and sulforhodamine B [Skehan et al., 1989] are the most widely used.

Solubilization of Sample

Both methods rely on a final colorimetric step and must be carried out on clear solutions. Cell monolayers and cell pellets may be dissolved in 0.5–1.0 *N* NaOH by heating to 100°C for 30 min or leaving overnight at room temperature. Alternatively, with 0.3 *N* NaOH and 1% sodium lauryl sulfate, solution is complete after 30 min at room temperature.

Lowry Method

The cell sample is precipitated with ice-cold 10% trichloracetic acid to remove the amino acid pools and dissolved in alkali, and protein is estimated colorimetrically after sequential addition of alkaline copper sulfate and Folin Ciocalteau reagent [after Lowry et al., 1951].

Principle

In a two-step reaction, Folin's reagent reacts with the aromatic amino acids in protein, after treatment with alkaline copper, to give a blue color.

Outline

Solubilize sample in 0.1 *N* NaOH, add alkaline copper solution and then diluted Folin's solution, and read in a spectrophotometer.

Materials

2% Sodium carbonate in 0.1 *N* NaOH (A)

0.5% copper sulfate in 1% sodium citrate solution (B)

1 ml B:50 ml A–make up with constant mixing and use on same day (C)

1 ml Folin Ciocalteau:1.3 ml water (D)

BSA standard solution 50 µg/ml in 0.1 *N* NaOH

1 *N* NaOH

Protocol

1. Dissolve cell pellet in 1.0 *N* NaOH and dilute to 0.1 *N*.
2. Add 1.0 ml of reagent C to 200 μl of protein sample, mix, and leave 10 min.
3. Add 100 μl of reagent D, with constant mixing, and leave 40 min.
4. Read on a spectrophotometer at 700 nm against a reagent blank and with a BSA standard, 50 μg/ml.

Protein measurements by the Lowry method are dependent on the presence of tyrosine and phenylalanine residues and will give underestimates if the frequency of these amino acids is low, as in nuclear histone proteins. For this reason, the Bradford method, which is not dependent on specific amino acids, has become more popular.

Bradford Method

Samples may be solubilized as before, although removal of amino acid pools in 10% trichloracetic acid is not necessary. The Bradford assay is more sensitive than Lowry's and only 50–100,000 cells are required. Color is generated in one step with a short incubation and read within 30 min.

Principle

Coomassie blue undergoes a spectral change on binding to protein in acidic solution. This gives a more sensitive assay than Lowry, independent of aromatic amino acid frequency in the protein.

Outline

Protein is dissolved and mixed with color reagent and the O.D. read after 10 min.

Materials

0.1% sodium dodecyl sulfate (SDS) in water or 0.3 *N* NaOH
0.01% in Coomassie Brilliant Blue G-250 in 4.7% EtOH and 85% (w/v) phosphoric acid: dissolve 100 mg Coomassie blue in 50 ml 95% EtOH, add 100 ml 85% phosphoric acid, and dilute to 1 l.

Protocol

1. Solubilize protein (1–20 μg) or cells (around 10^6) in 0.1% sodium dodecyl sulfate in water or 0.3 *N* NaOH.
2. Add 1.0 ml Coomassie blue to 100 μl protein solution, mix, and let stand for 10 min.
3. Read on spectrophotometer at 595 nm against a reagent blank and with a BSA standard curve (1–50 μg/ml).

Variations. Reagents available in kit from BioRad.

RATES OF SYNTHESIS

Protein Synthesis

Colorimetric assays measure the total amount of protein present at any one time. Sequential observations over a period of time may be used to measure net protein accumulation or loss (protein synthesized–protein degraded), while the rate of protein synthesis may be determined by incubating with radioisotopically labeled amino acid, such as ^{3}H-leucine or ^{35}S-methionine, and measuring (e.g., by scintillation counting) the amount of radioactivity incorporated per 10^6 cells or per milligram protein over a set period.

◇ Radioisotopes must be handled with care and according to local regulations governing permitted amounts, authorized work areas and disposal, and so forth.

Materials

Sterile:
Cell culture, 10^4–10^6 cells in, e.g., 24-well plate
^{3}H-leucine, 2 MBq/ml (~50 μCi/ml) in culture medium without serum (specific activity unimportant as this will be determined by the leucine concentration in the medium)
Nonsterile:
1% sodium dodecyl sulfate in 0.3 *N* NaOH
Scintillation vials
Eppendorf tubes
Scintillation fluid with 10% water tolerance, e.g., Instagel or Ecoscint

Protocol

1. Incubate culture to appropriate cell density.
2. Remove from incubator and add prewarmed solution of radioisotope in medium or BSS, 1:10, e.g., 100 μl per 1 ml/well.
3. Return to incubator as rapidly as possible.
4. Incubate for 4–24 hr.

   Note. Different proteins turn over at different rates. This protocol is not aimed at any specific subset but at total protein in rapidly proliferating cells. When assaying protein synthesis in a cell line for the first time, check that the rate of synthesis is linear over the incubation time chosen. A lag may be encountered if the amino acid pool is slow to saturate.

5. Remove from incubator and withdraw medium carefully from wells into radioactive liquid waste.
6. Wash cells gently with cold HBSS or PBS.
   Note. Some monolayer cultures may detach during washing, particularly some loosely adherent continuous cell lines such as HeLa-S_3. In this case add methanol, after removing isotope, to fix monolayer. Leave for 10 min, remove carefully, and dry monolayer (see also Chapter 19).
7. Place plates on ice. Add 10% trichloracetic acid, at 4°C, to remove unincorporated precursor. Leave 10 min.
8. Repeat step 7 twice, 5 min each.
9. Wash with MeOH and dry.
10. Add 0.5 ml 0.3 *N* NaOH, 1% SDS, and leave 30 min at room temperature.
11. Mix contents well and transfer to scintillation vials.
12. Add 5 ml scintillant and count.
    Note. Biodegradable scintillants, e.g., Ecoscint, should be used in preference to toluene- or xylene-based scintillants, as they are less toxic to handle and can be poured down the sink with excess water provided the levels of radioactivity fall within the legal limits.
    For suspension cultures spin (1,000 *g*, 10 min) at step 5 to remove medium and at step 6 and after steps 7 and 8; omit step 9.

DNA Synthesis

Measurements of DNA synthesis are often taken as representative of cell proliferation (see also Chapter 19). Incorporation of [^{3}H]-thymidine (^{3}H-TdR) or [^{3}H]-deoxycytidine are the usual precursors used. Exposure may be for short periods (0.5–1 hr) for rate estimations or for 24 hr or more to measure accumulated DNA synthesis where the basal rate is low, e.g., in high-density culture. ^{3}H-TdR should not be used for incubations longer than 24 hr or at high specific activities, as radiolysis of DNA will occur due to the short path length of β-emission ($\leq$1 μm) from decaying tritium releasing energy within the nucleus and causing DNA strand breaks. If prolonged incubations or high specific activities are required, ^{14}C-TdR or ^{32}P should be used.

Materials

Sterile:

Cells at suitable stage

^{3}H-TdR, 0.4 MBq/ml (~10 μCi/ml), 100 μl for each ml culture medium

Note. Some media, e.g., Ham's F10 and F12, contain thymidine, which will ultimately determine the specific activity of added ^{3}H-TdR. Allowance will have to be made for this when judging the amount of isotope to add. While 40 KBq/ml (~1.0 Ci/ml) may be sufficient for most media, 0.2 MBq/ml (~5 μCi/ml) should be used with F10 or F12.

Nonsterile:

10% trichloracetic acid (on ice), 6 ml for each 1 ml culture

HBSS, ice cold, 2 ml per ml culture

2 *N* perchloric acid or 0.3 *N* NaOH, 1% sodium duodecyl sulphate (SDS), 0.5 ml per 1 ml culture

MeOH, 1 ml per 1 ml culture

Scintillation vials

Scintillant (10× vol of perchloric acid or NaOH/SDS)

Outline

Label cells with ^{3}H-TdR, extract DNA, and count radioactivity on scintillation counter.

Protocol

1. Grow culture to desired density (usually mid-log phase for maximum DNA synthesis or plateau for density-limited DNA synthesis [see Chapter 15]).
2. Add ^{3}H-TdR, 40 KBq/ml (~1.0 μCi/ml), 2 MBq/mmol ($\leq$50 Ci/mol) in HBSS.
3. Incubate for 1–24 hr as required.
4. Remove radioactive medium carefully and discard into liquid radioactive waste.
5. Wash carefully with 2 ml HBSS and add 2 ml ice-cold 10% trichloracetic acid for 10 min. Fix in MeOH first if cells are loosely adherent (see "Protein Synthesis," above).
6. Repeat trichloracetic acid washes twice, 5 min each.
7. Add 0.5 ml 2 *N* perchloric acid, place on hot plate at 60°C for 30 min, and allow to cool.
8. Collect perchloric acid, transfer to scintillant, and count.
9. The residue may be dissolved in alkali for protein determination as above. If protein determinations are not required, the whole monolayer can be dissolved in 1% SDS in 0.3 *N* NaOH and counted. In this case, and perhaps anyway, replicate cultures should be set up to provide cell counts to allow calculation of DNA synthesis by cell number, if DNA synthesis per milligram protein is not suitable.

For suspension cultures spin (100 *g*, 10 min) at steps 4, 5, and 6, mix on vortex mixer to disperse pellet before each wash, 10% trichloracetic acid and 2 *N* perchloric acid, at step 7. Spin after step 7 (1,000 *g*, 10 min) to separate precipitate (for protein estimation if required) and supernatant (for scintillation counting).

◇ ^{3}H-thymidine represents a particular hazard because of the induced radiolytic damage mentioned above. Take care to avoid accidental ingestion, injection, or inhalation of aerosols. Work in a biohazard cabinet or on the open bench but not in horizontal laminar flow where aerosols will be blown directly at you.

Incubation with isotopic precursor can provide several different types of data, depending on the incubation conditions and subsequent processing. Incubation followed by a short wash in ice-cold BSS will give a measure of uptake and, if carried out over a few minutes' duration, will give a fair measure of unidirectional flux. In uptake experiments, incorporation into acid-insoluble precursors such as protein or DNA is assumed to be minimal, due to the short incubation time, and only the acid-soluble pools are counted by extraction into cold 10% trichloracetic acid. In longer incubations, 2–24 hr, it is assumed that the precursor pools become saturated. Equilibrium levels may be measured by cold trichloracetic acid extraction and incorporation into polymers measured by extraction with hot 2 *N* perchloric acid (DNA), cold dilute alkali (RNA), or hot 1.0 *N* NaOH (protein).

PREPARATION OF SAMPLES FOR ENZYME ASSAY AND IMMUNOASSAY

As the amount of cellular material available from cultures is often too small for efficient homogenization, other methods of lysis are required to release soluble products and enzymes for assay. It is convenient either to set up cultures of the necessary cell number in sample tubes (see below) or to trypsinize a bulk culture and aliquot cells into assay tubes. In either case the cells should be washed in HBSS, to remove serum, and lysis buffer added. The lysis buffer is chosen to suit the assay, but if unimportant, 0.15 M NaCl or PBS may be used. If the product to be measured is membrane bound, add 1% detergent (Na deoxycholate, Nonidet P40) to the lysis buffer. If the cells are pelleted, resuspend in the buffer by vortex mixing. Freeze and thaw the preparation three times by placing in EtOH containing solid CO_2 ($\leq$-90°C) for 1 min and then in 37°C water for 2 min (longer for samples greater than 1 ml). Finally spin at 10,000 *g* for 1 min (e.g., in Eppendorf centrifuge) and collect supernatant for assay.

Alternatively, whole extract may be assayed for enzyme activity and the insoluble material removed by centrifugation later if necessary.

REPLICATE SAMPLING

As cultured cells can be prepared in a uniform suspension in most cases, the provision of large numbers of replicates for statistical analysis is often unnecessary. Usually three replicates are sufficient, and for many simple observations, e.g., cell counts, duplicates may be sufficient.

Many types of culture vessel are available for replicate monolayer cultures (see Chapter 4) and the choice is determined (1) by the number of cells required in each sample and (2) by the frequency or type of sampling. For example, if incubation time is not a variable, replicate sampling is most readily performed in multiwell plates such as microtitration plates or 24-well plates. If, however, samples are collected over a period of time, e.g., daily for 5 days, then constant removal of a plate for daily processing may impair growth in the rest of the wells. In this case, replicates are best prepared in individual tubes or 4-well plates. Plain glass or tissue culture treated plastic test tubes may be used, though Leighton tubes are superior, as they provide a flat growth surface. Alternatively, if optical quality is not critical, flat-bottomed glass specimen tubes and even glass scintillation vials may be good containers. If glass vials or tubes are used, they must be washed as tissue culture glassware (Chapter 8) and not returned to tissue culture after use with scintillant.

Sealing large numbers of vials or tubes can become tedious, so many people seal tubes with vinyl tape rather than screw caps. Such tapes can also be color coded to identify different treatments.

Handling suspension cultures is generally easier, as the shape of the container and its surface charge are less important. Multiple sampling can also be performed on one culture. This is done conveniently by sealing the bottle with a silicone rubber membrane closure (Pierce) and sampling via a syringe and needle (remembering to replace the volume removed with an equal volume of air).

Data Analysis

Analysis of data from cultured cells is not necessarily different from the way the data would be handled from any other system. However, as indicated above, the production of large amounts of data in cell culture experiments is relatively easy, particularly when dealing with microtitration plates, where several hundred data points can be generated without a great deal of effort. Handling these data will depend on their form, but the way that the data are to be processed will influence the choice of parameter. While cell counting is the accepted method for generating data to construct a growth curve, this does not lend itself to expansion to, say, determining 50 growth curves to measure the response of one or more cell lines to growth factors and combinations. However, if a colorimetric endpoint is chosen, then absorbance (e.g., MTT assay; see next chapter) or fluorescence emission (e.g., sulforhodamine [Boyd, 1989]) can be used and the plate analyzed on an enzyme-linked immunosorbent assay (ELISA) plate reader.

A radiometric endpoint, e.g., [^{3}H]-thymidine, used in conjunction with microtitration plates can now be determined by simultaneous measurement of the whole plate in a specially designed scintillation counter (Canberra Packard, Pharmacia Wallac). Similarly, large numbers of sample tubes can be read automatically by γ- or β-counting, using robotic systems.

In these and many other systems, the ability to generate large amounts of data has been made possible by creation of multiple replicate analysis systems that can lend themselves to many applications. As with ELISA analysis, the rate-limiting step is no longer the generation of the data but in its analysis. It is important, therefore, when choosing a parameter of measurement to suit the culture system that some thought be given to the amount of data generated and how that will be handled. The easiest approach is to direct the data into a computer, either via a network or to a dedicated PC. A number of companies now market programs that will display and analyze data from microtitration plate assays (not just ELISA). These include titration curves, enzyme kinetics, and binding assays. With the necessary skill, and using a spreadsheet for importation of the data, you may be able to set this up for yourself; alternatively, consultant advice is often available from the suppliers of plate readers (Molecular Devices, Alpha Laboratories, Lab Systems, Biorad).

GROWTH CYCLE

As described in Chapter 10, following subculture, cells will progress through a characteristic growth pattern of lag phase, exponential or "log" phase, and stationary or "plateau" phase (see Fig. 10.3). The log and plateau phases give vital information about the cell line, the population doubling time during log growth, and the maximum cell density achieved in plateau (saturation density). Measurement of the population doubling time is used to quantify the response of the cells to different inhibitory or stimulatory culture conditions such as variations in nutrient concentration, hormonal effects, or toxic drugs. It is also a good monitor of the culture during serial passage and enables the calculation of cell yields and the dilution factor required at subculture.

It must be emphasized that the population doubling time is an average figure and describes the net result of a wide range of division rates, including zero, within the culture.

Single time points are unsatisfactory for monitoring growth, without knowing the shape of the growth curve. A reduced cell count after, say, 5 days could be caused by a reduced growth rate of some or all the cells, a longer lag period implying adaptation or cell loss (difficult to distinguish), or a reduction in saturation density. This is not to say that growth curves are of no value. They can be useful for a rapid screen, and once the response being monitored is fully characterized and the type of response predictable, e.g., an increased doubling time, then single time point observations may be sufficient. Growth curves are particularly useful for the determination of saturation density, although growth at saturation density should be assessed by the labeling index with ^{3}H-thymidine (see below and Chapter 15).

The preferred method for analyzing growth and survival at lower cell densities is by clonal growth analysis (see Chapter 19). This technique will reveal differences in growth rate within a population and will distinguish between alterations in growth rate (colony size) and survival (colony number). It should be remembered, however, that cells may grow differently as isolated colonies at low cell densities. Fewer cells will survive even under ideal conditions, and all interaction is lost until the colony starts to form. Heterogeneity in clonal growth rates reflects differences in growth capacity between lineages within a population, but these need not necessarily be expressed in an interacting monolayer at higher densities where cell communication is possible.

The population doubling time (PDT) derived from a growth curve should not be confused with the cell cycle or generation time. The PDT is an average figure for the population and subject to the reservations stated above. The cell cycle time or generation time is measured from one point in the cell cycle until the same point is reached again (see below) and refers only to the growing cells in the population, while the PDT is influenced by nongrowing and dying cells. PDTs vary from 12 to 15 hr in rapidly growing mouse leukemias like the L1210, to 24–36 hr in many adherent continuous cell lines, and up to 60 or 72 hr in slow-growing finite cell lines.

A growth cycle is performed each time the culture is passaged and can be analyzed in more detail, as described below.

Outline

Set up a series of cultures at three different cell concentrations and count the cells at daily intervals until they "plateau."

Materials

Sterile:
Cell culture
24-well plates
100 ml growth medium
0.25% crude trypsin (for monolayer cultures only)

Nonsterile:
Plastic box to hold plates
CO_2 incubator or CO_2 supply to gas box

Protocol

1. Trypsinize cells as for regular subculture (see above).
2. Dilute cell suspension to 10^5 cells/ml, 3×10^4 cells/ml, and 10^4 cells/ml, in 25 ml medium for each concentration.
3. Seed three 24-well plates, one at each cell concentration, with 1 ml per well. Add cell suspension slowly from the center of the well so that it does not swirl around the well. Similarly, do not shake the plate to mix the cells, as the circular movement of medium will concentrate the cells in the middle of the well.
4. Place in a humid CO_2 incubator or sealed box gassed with CO_2.
5. After 24 hr, remove plates from incubator and count the cells in three wells of each plate: (a) remove medium completely from wells to be counted; (b) add 1 ml trypsin/EDTA to each well; (c) incubate with trypsin/EDTA; and (d) after 15 min, disperse cells in trypsin/EDTA and transfer 0.4 ml to counting fluid and count on cell counter.

   Note. Hemocytometer counting may be used but may be difficult at lower cell concentrations. Reduce trypsin volume to 0.1 ml and disperse cells carefully without frothing, using a micropipette, and transfer to hemocytometer.
6. Return plate to incubator as soon as cell samples in trypsin are removed. The plate must be out of the incubator for the minimum length of time to avoid disruption of normal growth.
7. Repeat sampling at 48 and 72 hr as in steps 5 and 6.
8. Change medium at 72 hr or sooner if indicated by pH drop (see above).
9. Continue sampling at daily intervals for rapidly growing cells (doubling time 12–14 hr) but reduce frequency of sampling to every 2 days for slowly growing cells (doubling time >24 hr) until plateau is reached.
10. Keep changing medium every 1, 2, or 3 days as indicated by pH.

Analysis. (1) Calculate cell number per well, per ml of culture medium (same figure), and per cm^2 of available growth surface in well. (Stain one or two wells [see Chapter 13] at each density to determine whether distribution of cells in wells is uniform and whether they grow up the sides of the well.)

(2) Plot cell density (per cm^2) and cell concentration (per ml), on a log scale, against time on a linear scale (Fig. 10.3).

(3) Determine the lag time, population doubling time, and plateau density (see below and Fig. 10.3).

(4) Establish which is the appropriate starting density for routine passage. Repeat growth curve at intermediate cell concentrations if necessary.

Variations. (1) Different culture vessels may be used, e.g., 25-cm^2 flasks, although more cells and medium will be required, or flat-bottomed glass sample tubes. Individual tubes have the advantage that the rest are not disturbed when samples are removed for counting.

(2) Frequency of medium changing may be altered.

(3) Different media or supplements may be tested.

Suspension Cultures

1. Add cell suspension in growth medium to wells at a range of concentrations as for monolayer.
2. Sample 0.4 ml at intervals as per trypsin samples. Alternatively, seed two 75-cm^2 flasks with 20 ml for each cell concentration and sample 0.4 ml from each flask daily or as required. Mix well before sampling and keep flasks out of incubator for the minimum length of time. Do not feed cultures during growth curve.

The growth cycle (Fig. 10.3) is conventionally divided into three phases:

The Lag Phase

This is the time following subculture and reseeding during which there is little evidence of an increase in cell number. It is a period of adaptation during which the cell replaces elements of the glycocalyx lost during trypsinization, attaches to the substrate, and spreads out. During spreading, the cytoskeleton reappears and its reappearance is probably an integral part of the spreading process. Enzymes, such as DNA polymerase, increase, followed by the synthesis of new DNA and structural proteins. Some specialized cell products may disappear and not reappear until cessation of cell proliferation at high cell density.

The Log Phase

This is the period of exponential increase in cell number following the lag period and terminating one or two doublings after confluence is reached. The length of the log phase depends on the seeding density, the growth rate of the cells, and the density at which cell proliferation is inhibited by density. In the log phase the growth fraction is high (usually 90–100%) and the culture is in its most reproducible form. It is the optimal time for sampling since the population is at its most uniform and viability is high. The cells are, however, randomly distributed in the cell cycle and, for some purposes, may need to be synchronized (see Chapter 23).

The Plateau Phase

Toward the end of the log phase, the culture becomes confluent—i.e., all the available growth surface is occupied and all the cells are in contact with surrounding cells. Following confluence, the growth rate of the culture is reduced, and in some cases, cell proliferation ceases almost completely after one or two further population doublings. At this stage, the culture enters the plateau (or stationary) phase, and the growth fraction falls to between 0 and 10%. The cells may become less motile; some fibroblasts become oriented with respect to one another, forming a typical parallel array of cells. "Ruffling" of the plasma membrane is reduced, and the cell both occupies less surface area of substrate and presents less of its own surface to the medium. There may be a relative increase in the synthesis of specialized versus structural proteins and the constitution and charge of the cell surface may be changed.

The phenomenon of cessation of motility, membrane ruffling, and growth was described originally by Abercrombie and Heaysman [1954] and designated *contact inhibition*. It has since been realized that the reduction of the growth of normal cells after confluence is reached is not due solely to contact but may also involve reduced cell spreading [Stoker et al., 1968; Folkman and Moscona, 1978], depletion of nutrients, and, particularly, growth factors [Dulbecco and Elkington, 1973; Stoker, 1973; Westermark and Wasteson, 1975] in the medium [Holley et al., 1978]. The term *density limitation* of growth has, therefore, been used to remove the implication that cell–cell contact is the major limiting factor [Stoker and Rubin, 1967], and "contact inhibition" is best reserved for those events resulting directly from cell contact, i.e., reduced cell motility and membrane activity, resulting in the formation of a strict monolayer and orientation of the cells with respect to each other.

Cultures of normal simple epithelial and endothelial cells will stop growing after reaching confluence and remain as a monolayer. Most cultures, however, with regular replenishment of medium will continue to proliferate (although at a reduced rate) well beyond confluence, resulting in multilayers of cells. Human embryonic lung or adult skin fibroblasts, which express contact inhibition of movement, will continue to proliferate, laying down layers of collagen between the cell layers until multilayers of six or more cells can be reached under optimal conditions [Kruse et al., 1970]. They still retain an ordered parallel array, however. The terms "plateau" and "stationary" are not strictly accurate, therefore, and should be used with caution.

Cultures that have transformed spontaneously or have been transformed by virus or chemical carcinogens will usually reach a higher cell density in the plateau phase than their normal counterparts [Westermark, 1974] (Fig. 18.5). This is accompanied by a higher growth fraction and loss of density limitation. Plateau in these cultures is equilibrium between cell proliferation and cell loss. These cultures are often *anchorage independent* for growth—i.e., they can easily be made to grow in suspension (see "Density Limitation of Growth" in Chapter 15; also see Chapter 2).

The construction of a growth curve from cell counts performed at intervals after subculture enables the measurement of a number of parameters that should be found to be characteristic of the cell line under a given set of culture conditions. The first of these is the duration of the *lag period* or "lag time" obtained by extrapolating a line drawn through the points on the

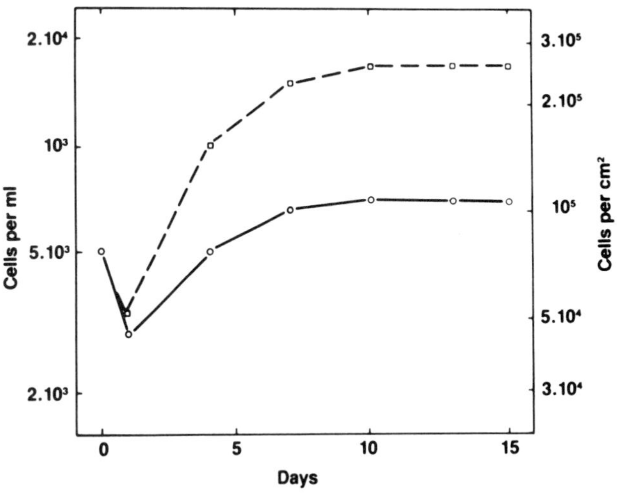

Fig. 18.5. Difference in plateaus (saturation densities) attained by cultures from normal brain (circles, solid line) and a glioma (squares, broken line). Cells were seeded onto 13-mm coverslips and 48 hr later the coverslips were transferred to 9-cm petri dishes with 20 ml growth medium, to minimize exhaustion of the medium.

exponential phase until it intersects the seeding or inoculum concentration (see Fig. 10.3), and reading off the elapsed time since seeding equivalent to that intercept. The second is the *doubling time*, i.e., the time taken for the culture to increase twofold in the middle of the exponential, or "log," phase of growth. This should not be confused with the *generation time* or *cell cycle time* (see below), which are determined by measuring the transit of a population of cells through the cell cycle until they return to the same point in the cell cycle.

The last of the commonly derived measurements from the growth cycle is the *plateau level* or *saturation density*. This is the cell concentration in the plateau phase and is dependent on cell type and frequency of medium replenishment. It is difficult to measure accurately, as a steady state is not achieved as easily as in the log phase. Ideally the culture should be perfused; but a reasonable compromise may be achieved by growing the cells on a restricted area, say a small-diameter coverslip (15 mm) in a large-diameter petri dish (90 mm) with 20 ml of medium replaced daily (see Chapter 15). Under these conditions, medium limitation of growth is minimal, and cell density exerts the major effect. Counting the cells under these conditions gives a more accurate and reproducible measurement. "Plateau" does not imply complete cessation of cell proliferation but represents a steady state where cell division is balanced by cell loss.

With normal cells a "steady state" may be achievable by not replenishing the growth factors in the medium. In this case cells are seeded and grown and plateau reached without changing the medium. Clearly, the conditions used to attain "plateau" must be carefully defined.

The maximum cell concentration in suspension cultures, which are not limited by available substrate, is usually limited by available nutrients. By fortifying the medium with a higher concentration of amino acids, Pirt and others [Birch and Pirt, 1971; Blaker et al., 1971] were able to obtain a maximum cell concentration of 5×10^6 cells/ml for LS cells, far in excess of what can be achieved with attached cells.

PLATING EFFICIENCY

When cells are plated out as a single cell suspension at low cell densities (2–50 cells/cm²), they will grow as discrete colonies (see Chapter 11). When these are counted the results are expressed as the plating efficiency:

$$\text{No. of cells formed/No. of colonies seeded} \times 100 = \text{Plating efficiency.}$$

If it can be confirmed that each colony grew from a single cell, this term becomes the *cloning efficiency*. Strictly according to the definition, plating efficiency measurements are derived from counting colonies over a certain size (usually 16–50 cells) growing from a low inoculum of cells, and the term should not be used for the recovery of adherent cells after seeding at higher cell densities (e.g., 2×10^4 cells/cm²). This is more properly called the *seeding efficiency*:

$$\text{No. of cells recovered/No. of cells seeded} \times 100 = \text{Seeding efficiency.}$$

It should be measured at a time when the maximum number of cells has attached but before mitosis starts. This provides a crude measurement of recovery in, for example, routine cell freezing or primary culture.

Clonal Growth Assay by Dilution Cloning

The protocol for dilution cloning [Puck and Marcus, 1955] is given in Chapter 11. When colonies have formed, remove medium, rinse carefully in BSS, and fix and stain colonies in crystal violet (see Chapter 13).

Analysis. (1) Count colonies and calculate plating efficiency. Magnifying viewers (e.g., Fig. 18.6) help to make counting easier.

(2) The size distribution of the colonies may also be determined (e.g., to assay the growth-promoting ability of a test medium or serum; see Chapter 7) by counting the number of cells per colony or by densitometry. Fix and stain the colonies with crystal violet, and measure absorption on a densitometer [McKeehan et al., 1977].

Automatic Colony Counting

If the colonies are uniform in shape and quite discrete, they may be counted on an automatic *colony counter* (e.g., New Brunswick, Artek, Micromeasurements Ltd.), which scans the plate with a conventional TV camera and analyzes the image to give an instantaneous readout of colony number. A size discriminator gives size analysis based on colony diameter (not always proportional to cell number, as cells may pile up in the center of the colony).

Though expensive, these instruments can save a great deal of time and make colony counting more objective. They will not cope well with colonies that overlap or have irregular outlines.

LABELING INDEX

If a culture is labeled with [³H]-thymidine, cells that are synthesizing DNA will incorporate the isotope.

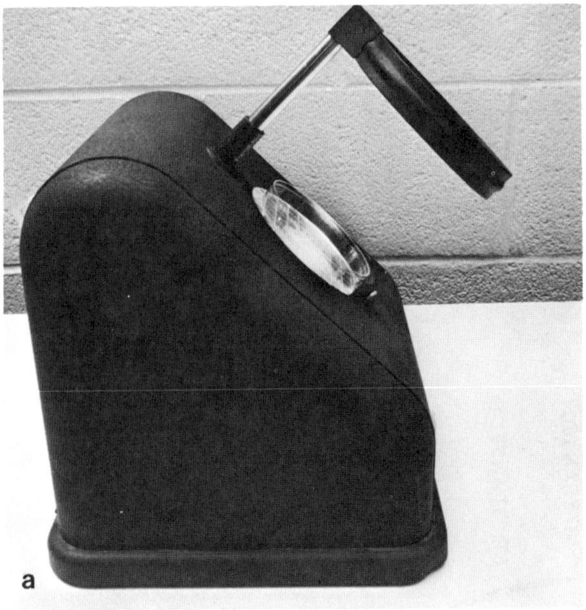

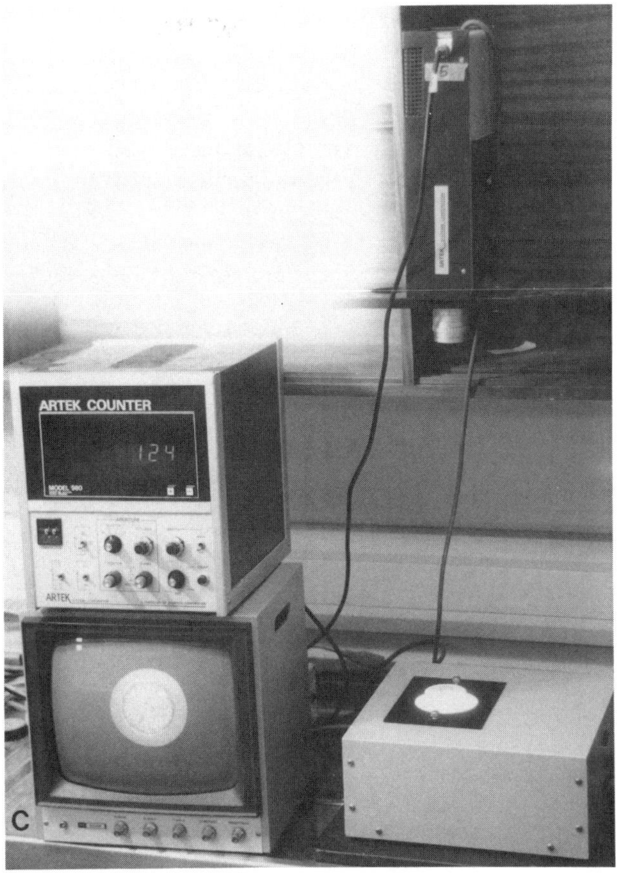

Fig. 18.6. *a. Simple magnifying colony counter. Versions are available with an electronically activated marking pen that records the count automatically. b. Bellco projection viewer. Magnifies the plate and allows more discrimination in scoring colonies. c. Automated colony counter (Artek). A stained petri dish of clones is placed on the light box (lower right), scanned by the camera (top right), displayed on the monitor (bottom left), which indicates the area being scanned, and flags the colonies being counted. Center left is the analyzer with threshold, sensitivity, and geometry settings, displaying the colony count.*

The percentage of labeled cells, determined by autoradiography [e.g., Westermark, 1974; Macieira-Coelho, 1973] (see Chapter 23 and Fig. 23.12a), is known as the labeling index (L.I.). Measurement after a 30-min exposure to [^{3}H]-thymidine shows a large difference between exponentially growing cells (L.I. = 10–20%) and plateau cells (L.I. $\leq$1%). Since the L.I. is very low in plateau, exposure times may have to be increased to 24 hr. With this length of label, normal cells can be shown to have a lower labeling index with thymidine than neoplastic cells (see Chapter 15).

Outline

Grow cells to appropriate density, label with [^{3}H]-thymidine for 30 min, wash, fix, remove unincorporated precursor, and prepare autoradiographs.

Materials

Sterile:

Culture of cells

Multiwell plate(s) containing 13-mm Thermanox coverslips

PBSA

0.25% crude trypsin
Growth medium
[^{3}H]-thymidine, 2.0 MBq/ml (~50 μCi/ml), 75 GBq/mmol (~2 Ci/mmol)
Nonsterile:
Hemocytometer or cell counter
HBSS
Acetic methanol (1 part glacial acetic acid to 3 parts methanol), icecold, freshly prepared
Microscope slides
DePeX
10% trichloracetic acid (TCA), ice cold
Deionized water
Methanol

Protocol

1. Set up cultures at 2 × 10^4/ml–5 × 10^4/ml in 24-well plates containing coverslips.

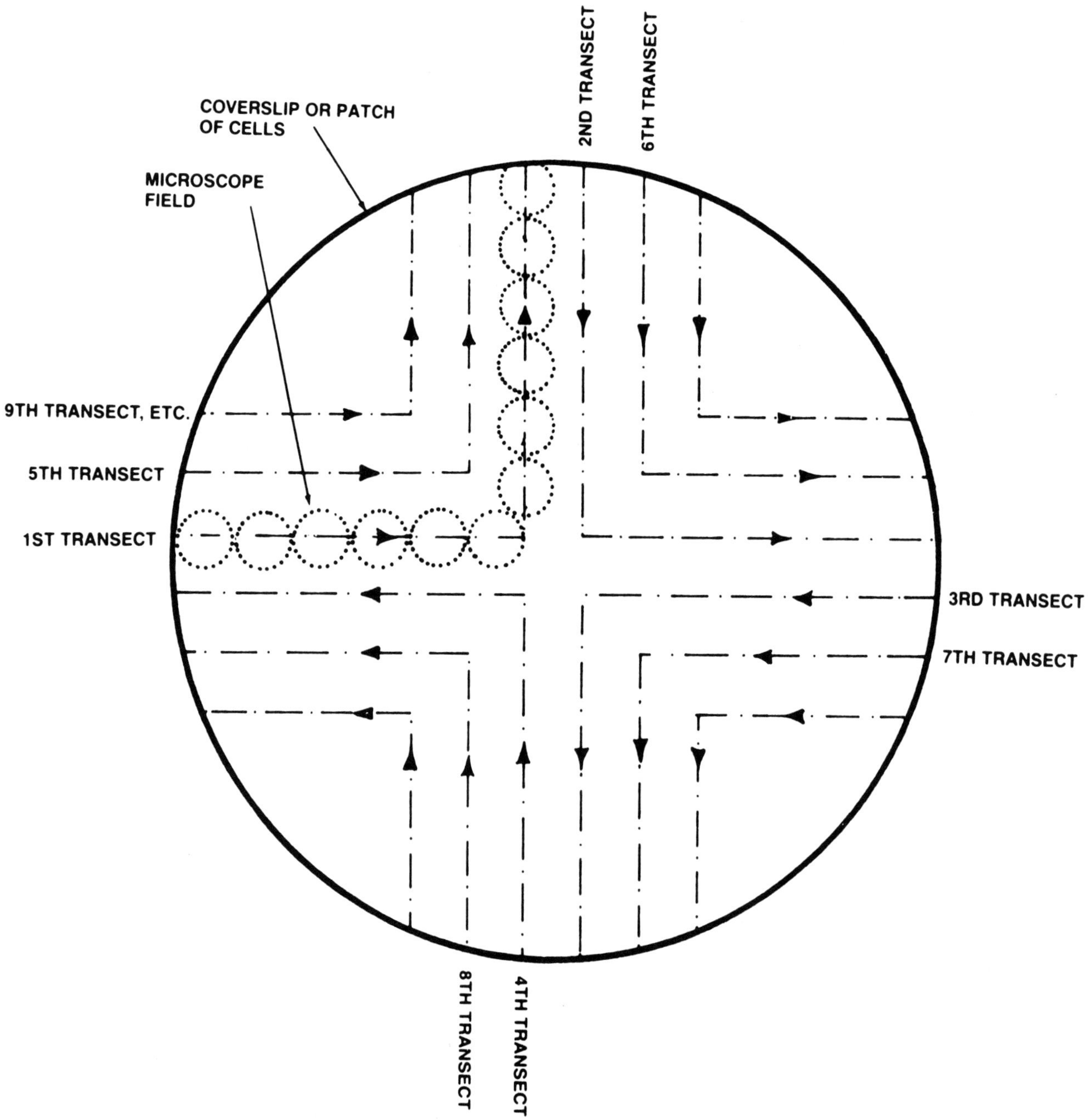

Fig. 18.7. *Scanning pattern for analysis of cytological preparations. Each dotted circle represents one microscope field and the whole, greater, circle, the extent of the specimen (e.g., a coverslip, culture dish or well, or spot of cells on a slide). Guide lines can be drawn with a nylon-tipped pen.*

2. Allow cells to attach, start to proliferate (48–72 hr), and grow to desired cell density.
3. Add [^{3}H]-thymidine to medium, 100 KBq/ml (~5 μCi/ml), and incubate for 30 min.
   Note. Some media, e.g., Ham's F10 and F12, contain thymidine. In these cases, the concentration of radioactive thymidine must be increased to give the same specific activity in the medium. In prolonged exposure to high specific activity, [^{3}H]-thymidine causes radiolysis of the DNA. This can be reduced by using [^{14}C]-thymidine or [^{3}H]-thymidine of a lower specific activity.
4. Remove labeled medium and discard (◇ care, radioactive!). Wash coverslip three times with HBSS. Lift the coverslip off the bottom (but not right out of the well) at each wash to remove isotope from underneath.
5. Add 1:1 HBSS: acetic methanol, 1 ml per well and remove immediately.
6. Add 1 ml acetic methanol at 4°C and leave 10 min.
7. Remove coverslips and dry at fan.
8. Mount coverslip cells uppermost on a microscope slide.
9. When mountant is dry (overnight), place slides in 10% trichloracetic acid at 4°C in a staining dish and leave 10 min. Replace trichloroacetic acid twice during this extraction. This removes unincorporated precursors.
10. Rinse in deionized water, then in methanol, and dry.
11. Prepare autoradiograph (see Chapter 23).

Analysis. Count percentage labeled cells. To cover a representative area, follow the scanning pattern illustrated in Figure 18.7.

Growth Fraction

If cells are labeled with [^{3}H]-thymidine for varying lengths of time up to 48 hr, the plot of labeling index against time increases rapidly over the first few hours and then flattens out to a very low gradient, almost a plateau (Fig. 18.8). The level of this plateau, read against the vertical axis, is the growth fraction of the culture, i.e., the proportion of cells in cycle at the time of labeling.

Outline

Label culture continuously for 48 hr, sampling at intervals for autoradiography.

Protocol

As for labeling index except that at step 3, incubation should be carried out for 15 min, 30 min, 1, 2, 4, 8, 24, and 48 hr.

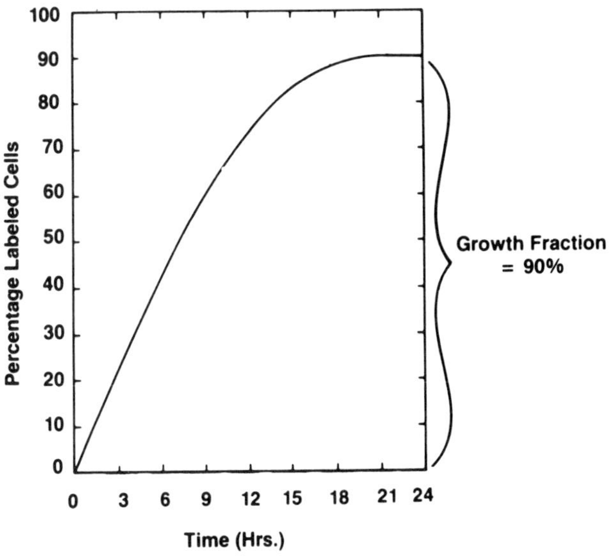

Fig. 18.8. Determination of the growth fraction. Cells are labeled continuously with [^{3}H]-thymidine and the percentage labeled cells determined at intervals by autoradiography (see text).

Analysis. Count labeled cells as a percentage of the total using scanning pattern from previous protocol. Plot labeling index against time (Fig. 18.8).

Note. Autoradiographs with ^{3}H can only be prepared where the cells remain as a monolayer. If they form a multilayer, they must be trypsinized after labeling and slides prepared by the drop technique or by cytocentrifugation (see Chapter 13), as the energy of β-emission from ^{3}H is too low to penetrate an overlying layer of cells.

A labeling index can also be determined by labeling cells with BUdR, which becomes incorporated into DNA. This can be detected subsequently by immunostaining with an anti-BUdR antibody (Dako). Results from this method are generally in agreement with [^{3}H]-thymidine [Khan et al., 1991].

Mitotic Index

The mitotic index is the fraction or percentage of cells in mitosis, determined by counting mitoses in stained cultures as a proportion of the whole population. The scanning pattern should be as in Figure 18.7.

Division Index

A number of antibodies, such as Ki67 (Dako), are able to stain cells in the division cycle. These antibodies are raised against proteins expressed during the cell cycle but not in resting cells. Some are directed against DNA polymerase, but the epitopes for others are not known.

Cells are stained by immunofluorescence or immunoperoxidase (see Chapter 13), and the proportion of stained cells determined. This will give a higher index than either mitotic counting or DNA labeling, as the whole cycle is responsive to stain. It gives a particularly useful indication of the growth fraction.

CELL CYCLE TIME (GENERATION TIME)

To determine the length of the cell cycle and its constituent phases, cells are labeled with [^{3}H]-thymidine for 30 min, the label is removed, and the appearance of the label in mitotic cells is determined autoradiographically at 30-min intervals up to 48 hr. The plot of the percentage of labeled mitoses against time takes the form of Figure 18.9, for which the cell cycle time and the length of its constituent phases may be derived.

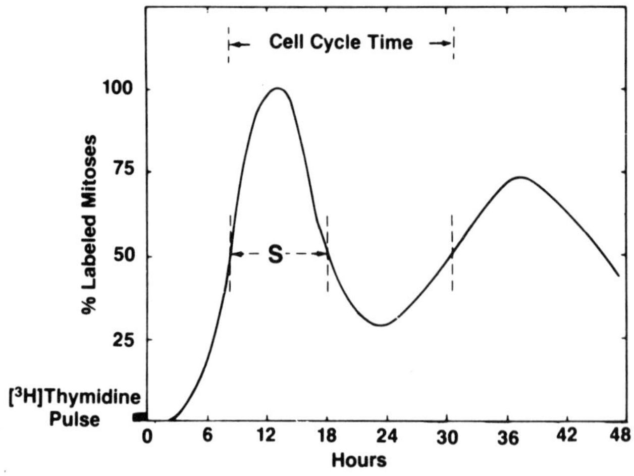

Fig. 18.9. Determination of cell cycle time. The percentage of labeled mitosis relative to the total number of mitoses is plotted against time after removal of the thymidine. The time between the midpoint of the first ascending curve and the second rising curve is the cell cycle time. [After Van't Hof, 1973; Macieira-Coelho, 1973; for further discussion, see Quastler, 1963; Van't Hof, 1968.]

CYTOMETRY

Histochemical and other cytological techniques (see also Chapter 13) can measure the amounts of enzyme, DNA, RNA, protein, or other cellular constituents *in situ*. It is difficult, however, to interpret the results of such techniques in a quantitative fashion. The introduction of flow cytometry [Herzenberg et al., 1976; Shapiro, 1988; Watson, 1991; Watson and Erba, 1992], while losing the relationship between cytochemistry and morphology, has added a new dimension to the measurement of cellular constituents and activities [Kurtz and Wells, 1979]. The potential amount of information that may be obtained about the constituent cells of a population is so vast that the problem becomes one of intellectual interpretation rather than data collection.

For a further description of flow cytometry, see Chapter 12.

CHAPTER 19

Measurement of Viability and Cytotoxicity

INTRODUCTION

Once a cell is explanted from its normal *in vivo* environment, the question of viability, particularly in the course of experimental manipulations, becomes fundamental. Previous chapters have dealt with the status of the cultured material relative to the tissue of origin and how to quantify changes in growth and phenotypic expression. None of this data is acceptable unless the great majority of the cells are shown to be viable. Furthermore, many experiments carried out *in vitro* are for the sole purpose of determining the potential cytotoxicity of compounds under study, either because they are being used as a pharmaceutical or cosmetic and must be shown to be nontoxic, or because they are designed as anticancer agents, when cytotoxicity may be crucial to their action.

Current legislation demands that new drugs, cosmetics, food additives, etc., go through extensive cytotoxicity testing before they are released. This usually involves a large number of animal experiments, which are very costly and raise considerable public concern. There is therefore much pressure, both humane and economic, to perform at least part of cytotoxicity testing *in vitro*. The introduction of specialized cell lines, as well as the continued use of long-established cultures, may make this a reasonable proposition.

Cytotoxicity is a complex event *in vivo*, where its expression may be manifest in a wide spectrum of effects, from simple cell death, as in the toxic effects of anticancer drugs on both the cells of the tumor and normal cells of the bone marrow, skin, or gut, to complex metabolic aberrations such as neuro- or nephrotoxicity, where no cell death may occur, only functional change. Definitions of cytotoxicity will tend to vary depending on the nature of the study, whether cells are killed or simply have their metabolism altered. While an anticancer agent may be required to kill cells (though it need not), the proof of the absence of toxicity of another pharmaceutical may require more subtle analysis of minor metabolic changes or an alteration in cell–cell signaling such as might give rise to an inflammatory or allergic response.

All of these assays oversimplify the events that they measure and are employed because they are cheap, easily quantified, and reproducible. However, it has become increasingly apparent that they are inadequate for modern drug development, where greater emphasis on molecular target specificity and precise metabolic regulation is required. Gross tests of cytotoxicity are still required but there is a growing need for them to be supplemented with more subtle tests of metabolic perturbation. Perhaps the most obvious of these is the induction of an inflammatory or allergic response, which need not imply cytotoxicity of the allergen or inflammatory agent, but which is still one of the hardest endpoints to demonstrate *in vitro*.

It is not possible here to define all the requirements of a cytotoxicity assay, so I will concentrate on those aspects that influence growth or survival. Growth is generally taken to be regenerative potential, measured by clonal growth, net change in population size, as in a growth curve (see Chapters 10 and 18), or a change in cell mass (total protein or DNA) or gross metabolic activity such as respiration or DNA, RNA, or protein synthesis. Survival can be an instantaneous measurement, usually plasma-membrane integrity, or can use the same growth parameters as above, not to measure growth *per se* but to say that capacity for growth implies survival.

LIMITATIONS OF *IN VITRO* METHODS

It is important that any measurement can be interpreted in terms of the *in vivo* response, or at least with the understanding that clear differences exist between *in vitro* and *in vivo* measurements.

Measurement of toxicity *in vitro* is a purely cellular event as presently carried out. It would be very difficult to re-create the complex pharmacokinetics of drug exposure, for example, *in vitro*, and there will usually be significant differences in drug exposure time and concentration, rate of change of concentration, metabolism, tissue penetration, clearance, and excretion. Although it may be possible to simulate these parameters, e.g., using multicellular tumor spheroids for drug penetration, most studies concentrate on a direct cellular response. They thereby gain their simplicity and reproducibility.

Many nontoxic substances become toxic after metabolism by the liver; in addition, many substances that are toxic *in vitro* may be detoxified by liver enzymes. For testing *in vitro* to be accepted as an alternative to animal testing, it must be demonstrated that potential toxins reach the cells *in vitro* in the same form as they would *in vivo*. This may require additional processing by purified liver microsomal enzyme preparations [McGregor et al., 1988] or co-culture with activated hepatocytes [Guillouzo, 1989; Frazier, 1992].

The nature of the response must also be considered carefully. A toxic response *in vitro* may be measured by changes in cell survival or metabolism (see below), while the major problem *in vivo* may be a tissue response, e.g., an inflammatory reaction or fibrosis. For *in vitro* testing to be more effective, construction of models of these responses will be required, perhaps utilizing cultures reassembled from several different cell types and maintained in the appropriate hormonal milieu.

It should not be assumed that complex tissue and even systemic reactions cannot be simulated *in vitro*. Assays for the inflammatory response, teratogenic disorders, or neurological dysfunctions may be feasible *in vitro*, given a proper understanding of cell–cell interaction and the interplay of endocrine hormones with local paracrine and autocrine factors.

NATURE OF THE ASSAY

The choice of assay will depend on the agent under study, the nature of the response, and the particular target cell. Assays can be divided into two major classes: (1) an immediate or short-term response such as an alteration in membrane permeability or a perturbation of a particular metabolic pathway, and (2) long-term survival, either absolute, usually measured by the retention of self-renewal capacity, or survival in altered state, e.g., expressing genetic mutation(s) or malignant transformation.

Short-Term Assays—Viability

Assays of this type are used to measure the proportion of viable cells following a potentially traumatic procedure such as primary disaggregation (Chapter 9), cell separation (Chapter 12), or freezing and thawing (Chapter 17).

Dye exclusion viability tends to overestimate viability, e.g., 90% of cells thawed from liquid nitrogen may exclude trypan blue but only 60% prove to be capable of attachment 24 hr later.

Most viability tests rely on a breakdown in membrane integrity determined by the uptake of a dye to which the cell is normally impermeable (e.g., trypan blue, erythrosin, or naphthalene black) or the release of a dye or isotope normally taken up and retained by viable cells (e.g., diacetyl fluorescein or ^{51}chromium).

Dye exclusion. Viable cells are impermeable to naphthalene black, trypan blue, and a number of other dyes [Kaltenbach et al., 1958].

Outline

A cell suspension is mixed with stain and examined by low-power microscopy.

Materials

Cells
PBSA

0.25% trypsin
Growth medium
Hemocytometer
Viability stain (e.g., 0.4% trypan blue or 1% naphthalene black in PBSA or HBSS)
Pasteur pipettes
Microscope
Tally counter

Protocol

1. Prepare cell suspension at a high concentration ($\sim 10^6$ cells/ml) by trypsinization or centrifugation and resuspension.
2. Take a clean hemocytometer slide and fix the coverslip in place (see Chapter 18).
3. Mix one drop of cell suspension with one drop (trypan blue) or four drops (naphthalene black) of stain, transfer to the edge of the coverslip, and allow to run into the counting chamber.
4. Leave 1–2 min (do not leave longer or viable cells will deteriorate and take up stain).
5. Place on microscope under a 10× objective.
6. Count the number of stained cells and the total number of cells.
7. Wash hemocytometer and return to box.

Analysis. Calculate the percentage of unstained cells. This is the percentage viability by this method. If the volumes of cell suspension and stain are measured accurately at step 3, this method of viability determination can be incorporated into the hemocytometer cell-counting protocol.

Dye uptake. Viable cells take up diacetyl fluorescein and hydrolyse it to fluorescein, to which the cell membrane of live cells is impermeable [Rotman and Papermaster, 1966]. Live cells fluoresce green; dead cells do not. Nonviable cells may be stained with ethidium bromide or propidium iodide and will fluoresce red. Viability is expressed as the percentage of cells fluorescing green. This method may be applied to flow cytometry (see Chapters 12 and 18).

Outline

Stain cell suspension in a mixture of propidium iodide and diacetyl fluorescein and examine by fluorescence microscopy or flow cytometry.

Materials

Single cell suspension
Fluorescein diacetate 10 μg/ml in HBSS
Fluorescence microscope
Propidium iodide 500 μg/ml
Filters:
fluorescein: excitation 450/590 nm, emission LP 515 nm
propidium iodide: excitation 488 nm, emission 615 nm

Protocol

1. Prepare cell suspension as for dye exclusion above but in medium without phenol red.
2. Add fluorescent dye mixture 1:10 to give final concentration of 1 μg/ml diacetyl fluorescein and 50 μg/ml propidium iodide.
3. Incubate at 36.5°C for 10 min.
4. Place a drop of cells on a microscope slide, add a coverslip, and examine by fluorescence.

Analysis. Cells fluorescing green are viable, those fluorescing red nonviable. Express viability as percentage fluorescing green as a proportion of the total.

Chromium release. Reduced $^{51}Cr^{3+}$ is taken up by viable cells and oxidized to $^{51}Cr^{2+}$, to which the membrane of viable cells is impermeable [Holden et al., 1973; Zawydiwski and Duncan, 1978]. Dead cells release the $^{51}Cr^{2+}$ into the medium. A reduction in viability is detected by γ-counting aliquots of medium from cultures labeled previously with $Na_2{}^{51}CrO_4$ for released ^{51}Cr. The test works well for a few hours, but over longer periods spontaneous release of ^{51}Cr may be a problem.

Analysis. Express counts released as a percentage of total (medium + cells) and plot against time.

This method allows comparison of different toxic stimuli but does not give an absolute figure for percentage viable cells. It is often used to measure cytotoxic T-lymphocyte activity.

Metabolic tests. Alterations in glycolysis and respiration [Dickson and Suzangar, 1976], enzyme activity [DiPaolo, 1965], and incorporation of labeled precursors [Freshney et al., 1975] have all been used to measure response to potentially toxic stimuli. Although these are often interpreted as viability or survival assays, they are not and should be interpreted solely as metabolic responses, specific to the parameter measured. Application to cell survival is limited, but comparison with clonogenic survival is possible if culture is continued for two to three population doublings after removal of drug (see below). Protocols for measuring precursor uptake and total protein or DNA are given in Chapter 18.

Long-Term Tests—Survival

While short-term tests are convenient and usually quick and easy to perform, they only reveal cells that are dead (i.e., permeable) at the time of the assay. Frequently, cells subjected to toxic influences, e.g., antineoplastic drugs, will only show an effect several hours or even days later. The nature of the tests required to measure viability in these cases is necessarily different, since by the time the measurement is made, the dead cells may have disappeared. Long-term tests are often used to demonstrate the metabolic or proliferative capacity of cells after, rather than during, exposure to a toxic influence. The objective is to measure survival rather than short-term toxicity, which may be reversible.

The ability of cells to survive a toxic insult has been the basis of most cytotoxicity assays. In this context, survival implies the retention of regenerative capacity and is usually measured by plating efficiency, as would be the case with bacteria or other micro-organisms. Unfortunately, many animal cells have poor plating efficiencies, particularly normal cells, freshly isolated, so a number of alternatives have been devised for assaying cells at higher densities, e.g., in microtitration plates. None of these tests measures survival directly. Instead the net increase in cell number (growth curve), the increase in total protein or DNA, or the residual ability to synthesize protein or DNA is determined. "Survival" in these cases is defined as the retention of metabolic or proliferative ability by the cell population as a whole; such assays cannot discriminate between a reduction in metabolic or proliferative activity per cell and a reduced number of cells.

Plating efficiency, as described in Chapter 18, is the best measure of survival and proliferative capacity, provided that the cells plate with a high enough efficiency that the colonies can be considered representative. Though not ideal, anything over 10% is usually acceptable.

Since the colony number may fall at high toxic concentrations, it is usual to compensate by seeding more cells so that approximately the same number of colonies form at each concentration. This removes the risk of cell concentration influencing survival and improves statistical reliability. In addition, cells should be plated on a preformed feeder layer, the density of which ($5 \times 10^3/cm^2$) greatly exceeds that of the cloning cells, where the plating efficiencies of controls is <100%. A typical survival curve is prepared as follows:

Outline

Treat cells with experimental agent at a range of concentrations for 24 hr. Trypsinize, seed at low cell density, and incubate for 1–3 wk. Stain and count colonies.

Materials

Sterile:
25-cm^2 flasks
6- or 9-cm petri dishes labeled on rim side of base
PBSA
0.25% trypsin
Growth medium
Agent to be tested at 5x maximum concentration to be used, dissolved in growth medium–check pH and osmolality, adjust if necessary

Nonsterile:
Hemocytometer or cell counter
HBSS
Methanol
Crystal violet, 1% (BDH)

Protocol

1. Prepare a series of cultures in 25-cm^2 flasks, three for each of six agent concentrations, and three controls. Seed cells at 5×10^4/ml in 4-ml growth medium and incubate for 48 hr, by which time the cultures will have progressed into log phase (see above).
2. Prepare 50 ml of agent to be tested in regular growth medium. Check pH and osmolality and adjust if necessary. Dilute agent 1:5 by adding 1 ml to first flasks, mix, transfer 1 ml to second flasks, mix, and so on, completing serial dilution. Remove 1 ml from lowest concentration and discard.
3. Return to incubator.
4. If the agent is slow-acting or partially reversible, repeat steps 2 and 3 twice; i.e., expose cultures to the agent for 3 d, replacing the agent daily by changing the medium. With fast-acting agents, 1 hr exposure will be sufficient.
5. Remove medium from each group of three flasks in turn, trypsinize cells, and seed at required density for clonal growth (see Chapter 11), diluting all cultures by the same amount as the control. If toxicity is expected, increase the seeding density at higher agent concentrations to keep the number of colonies forming in the same range. In addition, plate cells onto a feeder layer (see Chapter 11) of the same cells irradiated or treated with mitomycin C (see Chapter 11), if plating efficiency of controls is substantially less than 100%.
6. Incubate until colonies form: fix, in absolute methanol, stain for 10 min in 1% crystal violet (stain may be reused).
7. Wash in tap water, drain, and dry inverted.
8. Count colonies >50 cells (>5 generations).

Analysis. (1) Plot relative plating efficiency (plating efficiency as a percentage of control) against drug concentration (Fig. 19.1) (*survival curve*).

(2) Determine ID_{50} or ID_{90}: the concentration promoting 50% or 90% inhibition of colony formation.

(3) Complex survival curves may be compared by calculating the area under the curve.

Variations.

Concentration of agent. A wide concentration range, in log increments, e.g., 10^{-6} M, 10^{-5} M, 10^{-4} M, 10^{-3} M, 0, should be used for the first attempt and a narrower range (log or linear), based on the indications of the first, for subsequent attempts.

Invariate agent concentrations. Some conditions tested cannot easily be varied, e.g., testing the quality of medium, water, or an insoluble plastic. In these cases, the serum concentrations should be varied, as serum may have a masking effect on minor toxic effects.

Duration of exposure to agent. Some agents act rapidly; others, more slowly. Exposure to ionizing radiation, for example, need only be a matter of minutes, sufficient to achieve the required dose, while testing some antimetabolic drugs may take several days for a measurable effect.

Duration of exposure (T) and drug concentration (C) are related, although $C \times T$ is not always a constant. Prolonging exposure can increase sensitivity beyond that predicted by $C \times T$ due to cell cycle effects and cumulative damage.

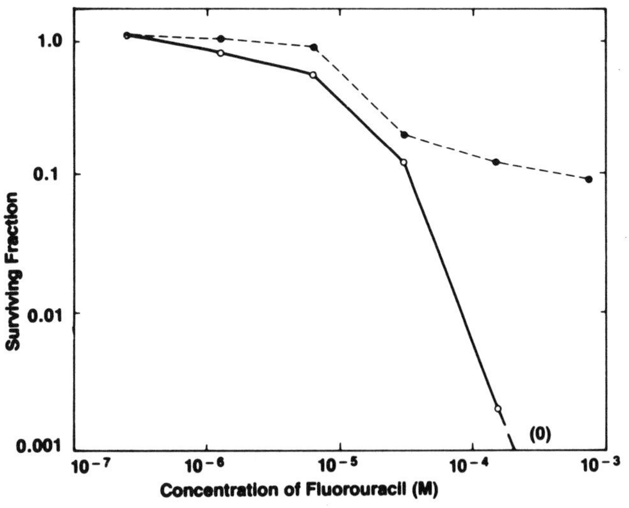

Fig. 19.1. Survival curve from plating efficiency. Human glioma cells were plated out in the presence (dotted line) and absence (solid line) of a feeder layer after treatment with various concentrations of 5-fluorouracil. A 10% resistant fraction is apparent at 10^{-4} M drug only in the presence of a feeder layer. In the absence of the feeder layer, the small number of colonies constituting the resistant fraction was unable to survive alone.

Time of exposure to agent. Where the agent is soluble and expected to be toxic, the above procedure should be followed; but where the quality of the agent is unknown, stimulation is expected, or only a minor effect is expected (e.g., 20% inhibition rather than 100-fold or more), the agent may be incorporated during clonal growth rather than at preincubation.

Cell density. The density of the cells during exposure to an agent can alter its response; e.g., HeLa cells are less sensitive to the alkylating agent mustine at high cell densities [Freshney et al., 1975]. In this kind of experiment, the cell density should be varied in the preincubation phase, during exposure to drug.

Colony size. Some agents are cytostatic but not cytotoxic and during continuous exposure may reduce colony size without reducing colony number. In this case the size of the colonies should be determined by densitometry [McKeehan et al., 1977], automatic colony counting or image analysis, or visually counting the number of cells per colony (see also Chapter 18).

For *colony counting*, the threshold number of cells per colony (e.g., 32 as above) is purely arbitrary, and assumes that most of the colonies are greatly in excess of this. Colonies should be grown until quite large (>10^4 cells), when growth of larger colonies will tend to slow down and smaller, but still viable, colonies will tend to catch up.

For *colony sizing*, harvest earlier, before growth rate in larger colonies has slowed down, and score all colonies.

Solvents. Some agents to be tested above have low solubilities in aqueous media, and it may be necessary to use an organic solvent. Ethanol, propylene glycol, and dimethyl sulfoxide have been used for this purpose but may themselves be toxic. Use the minimum concentration of solvent to obtain solution. The agent may be made up at a high concentration in, for example, 100% ethanol, then diluted gradually with BSS, and finally diluted into medium. The final concentration of solvent should be <0.5%, and a solvent control must be included.

Take care when using organic solvents with plastics or rubber. It is better to use glass with undiluted solvents and only to use plastic where the solvent concentration is <10%.

While plating efficiency is one of the best methods for testing survival, it should only be used where the cloning efficiency is high enough for colonies that form to be representative of the whole cell population.

Ideally this means that controls should plate at 100% efficiency. In practice, this is seldom possible and control plating efficiencies of 20% or less are often accepted.

Plating efficiency tests are also time-consuming to set up and analyze, particularly where a large number of samples is involved, and the duration of each experiment may be anything from 2 to 4 wk.

Microtitration Assays

The introduction of multiwell plates revolutionized the approach to replicate sampling in tissue culture. They are economical to use, lend themselves to automated handling, and can be of good optical quality. The most popular is the 96-well microtitration plate, each well having 28–32-mm^2 growth area and capacity for 0.1 or 0.2 ml medium and up to 10^5 cells.

They may be used for cloning, for antibody, virus, and drug titration, for cytotoxicity assays of potential toxins, and for numerous other applications. The following example illustrates the use of microtitration plates in the assay of anticancer drugs but would be applicable with minor modification to any cytotoxicity assay.

Outline

Microtitration plate cultures are exposed to a range of drug concentrations during the log phase of growth and viability determined, several days after drug removal, by measuring incorporation of [^{35}S]-methionine (Fig. 19.2).

Materials

Sterile:

Culture of cells
PBSA
0.25% trypsin
Growth medium
96-well microtitration plates
Test solution
Sealing film (ICN Flow, Conway) (Inner surface sterile by manufacture though not deliberately sterilized)
Growth medium containing 4.0 KBq/ml (~0.1 μCi/ml) [^{35}S]-methionine
HBSS

Nonsterile:

Hemocytometer or cell counter
Methanol
10% trichloroacetic acid (TCA)
Scintillation fluid
x-ray film (Kodak Royal)
Dark box
Silica gel
Black light-tight bag
−70°C freezer
Photographic developer and fixer

Protocol

1. Trypsinize cells (see Chapter 10). Seed microtitration plates at 10^3 cells/well, 0.1 ml/well (~3,000 cells/cm^2). Set up one duplicate plate for cell counting.
2. Place plate in CO_2 incubator with loose-fitting lid for 30 min to equilibrate with CO_2 (see Chapter 7).
3. Prepare drug dilutions in 1.0-ml aliquots. Dilute in complete medium and prepare similar aliquots of medium with no drug. The number of dilution steps and the volume required in each aliquot will depend on the way the plate is subdivided (Fig. 19.3). One milliliter is sufficient for duplicates, changed daily for 3 d, 0.1 ml per well.
4. Remove from incubator. Return to aseptic area and quickly seal plate with self-adhesive film. Return plates to incubator for 48–72 hr.
5. Count the cells on one plate daily, throughout the assay:

   (a) Remove plate from incubator and swab with 70% alcohol.

   (b) Cut film round eight wells and peel off film.

   (c) Remove medium completely from these wells and add 0.1 ml of trypsin.

   (d) Incubate for 15 min at 36.5°C and then disperse the cells in the trypsin.

   (e) Pool the cell suspensions from four wells, dilute to 20 ml, and count on cell counter.

   If a cell counter is not available, do not trypsinize; rinse the cells in BSS and fix in methanol. At the end of the experiment the cells may be stained with crystal violet and counted by eye on a microscope, or an estimate of cell number made by staining with coomassie blue and reading the plate on a plate reader.

   Sulforhodamine is a fluorescent dye that stains protein and can also be used to estimate the amount of protein (i.e., cells) per well on a plate reader with fluorescence detection [Boyd, 1989].

   Note. The cells must remain in exponential growth throughout (see Chapter 10). If the cell growth curve shows that the cells are moving into stationary phase, proceed directly to step 16.
6. Change medium in remaining wells and return plate to incubator, equilibrate with CO_2 for 30 min, and reseal.

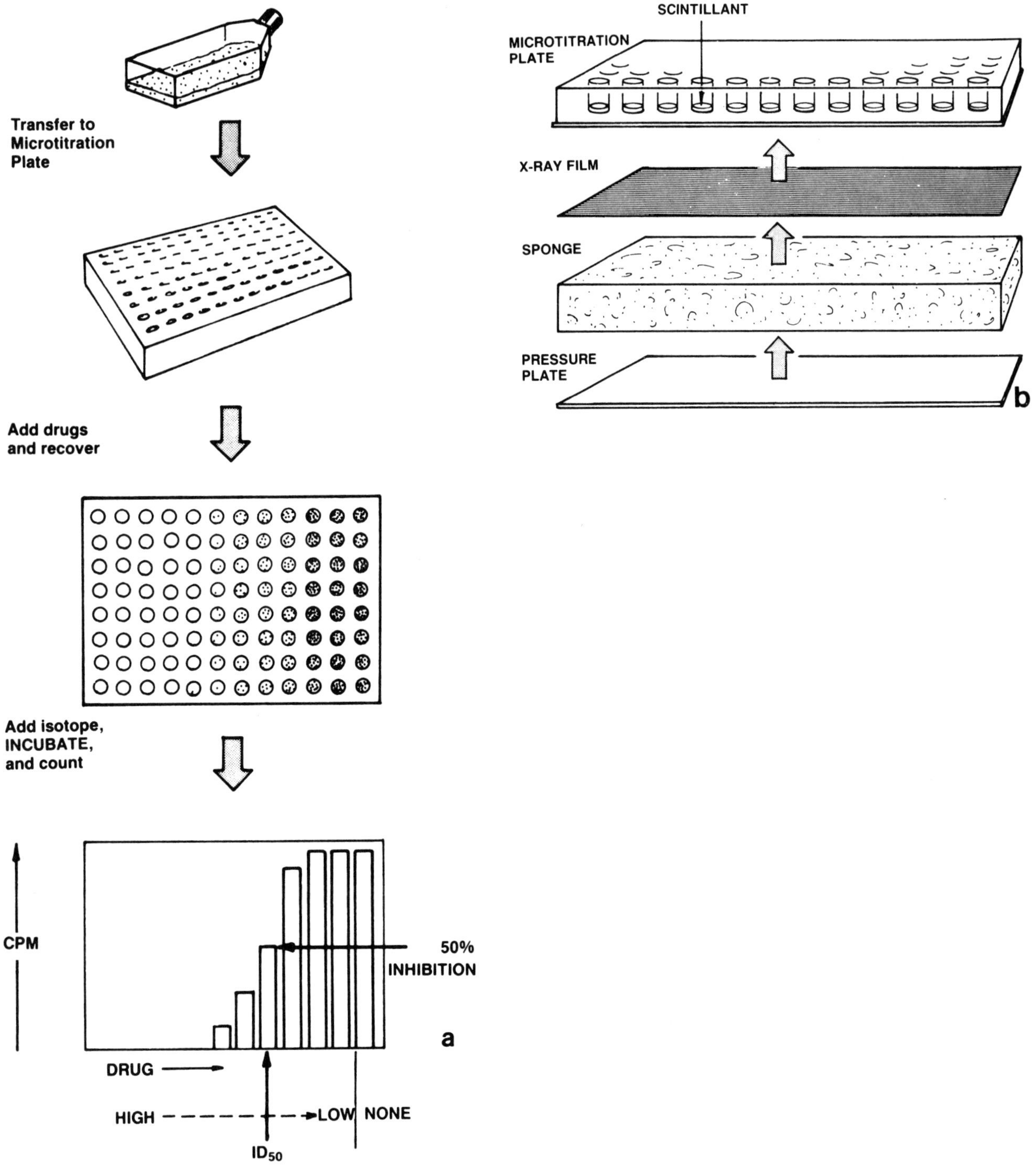

Fig. 19.2. *Microtitration assay for cytotoxicity. a. Stages of assay. Quantitation in this figure is represented by CPM but could equally be A_{570} after exposure to MTT (see text), or any one of several other parameters measuring total cells or cell protein per well. b. Measurement of incorporated isotope by autofluorography. This is a simple and inexpensive technique that may be quantified on a densitometer, but radioisotope detectors are available that will count emission from radioactive isotopes directly (Canberra-Packard, LKB-Wallac).*

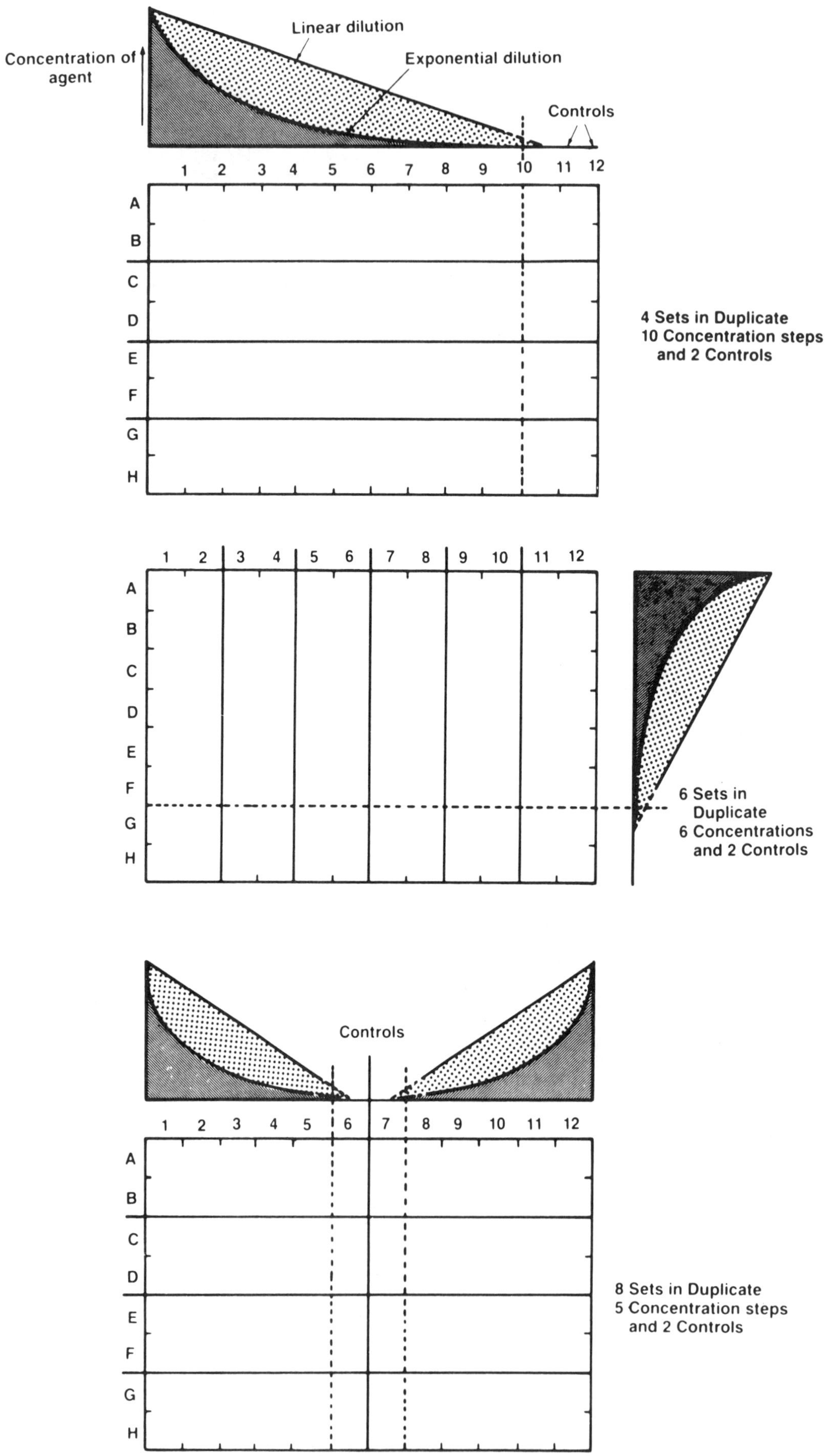
Linear dilution
Concentration of agent
Exponential dilution
Controls
1 2 3 4 5 6 7 8 9 10 11 12
A B C D E F G H
4 Sets in Duplicate
10 Concentration steps
and 2 Controls
1 2 3 4 5 6 7 8 9 10 11 12
A B C D E F G H
6 Sets in Duplicate
6 Concentrations
and 2 Controls
Controls
1 2 3 4 5 6 7 8 9 10 11 12
A B C D E F G H
8 Sets in Duplicate
5 Concentration steps
and 2 Controls

7. Add drugs 48–72 hr after seeding plate and replace daily if exposure is to exceed 24 hr.
   (a) Remove drug plate from incubator, swab with 70% alcohol, and gently peel off film.
   (b) Remove medium, two rows at a time, and add drug dilutions, the second row duplicating the first. There are a number of different ways of dividing up the plate to give different numbers of dilutions and replicates (see Fig. 19.3).
   (c) Equilibrate with CO_2 in incubator for 30 min as in steps 2 and 3 above, reseal with film, and incubate for 24 hr.
   (d) Repeat steps 8, 9, and 10 twice more to give three complete days' exposure to the drug. (Do cell counts and change the medium on accompanying plate each time the drugs are renewed [see above].)
8. After drug exposure, remove drug medium and wash wells by gently adding and removing 0.1 ml medium three times ("dumping" the medium by inverting the plate can detach cells and increases the chance of contamination). Finally, leave 0.1 ml of fresh medium in each well.
9. Count samples from accompanying plate and feed remaining wells.
10. Incubate for a further 5 d, changing the medium on the second or third day.
11. Change medium and count samples on cell count plate at the time the medium is changed on the drug plate and at the end.
12. Remove medium (drug plate only) and add 0.1 ml medium containing 4 KBq/ml (~0.1 μCi/ml) [^{35}S]-methionine. (The specific activity is unimportant, as this is controlled by the methionine in the culture medium.)
13. Incubate for 3 hr.
14. Remove isotope and wash plate by submerging it in BSS, rubbing the wells with a gloved finger or comb to promote entry of BSS into wells. Rapid removal of medium by pouring off or flicking the plate may dislodge cells.
15. Repeat BSS wash twice; do not pour off previous wash from wells.
16. Immerse plate in 100% methanol and rub as in step 8.
17. Repeat twice in fresh methanol and leave for 10 min in final bath of methanol.
18. Pour off methanol and dry plate at fan.
19. Add ice-cold 10% trichloracetic acid to the plate from a wash bottle. Fill the wells and stand on ice for 5 min. Remove trichloracetic acid and repeat twice more with fresh trichloracetic acid.
20. Wash in methanol and dry.
21. Add 50 μl scintillation fluid (e.g., Ecoscint) and dry down in a flat film onto the cells by centrifuging the plate for 1 hr at 20°C.
22. Bind dry plate with x-ray film (see Fig. 19.2b) (under dark-red safelight) and seal in dark box with desiccant such as silica gel.
23. After 2–14 d, open and remove film, under safelight conditions, develop for 10 min in D19, wash in tap water, fix in photographic fixer for 5 min, wash, and dry.

Analysis. If the titration point is obvious, the plate may be read by eye. If not, scan plate on a densitometer, and determine ID_{50} (Fig. 19.4).

Variations.

Duration. As for plating efficiency (see above), some agents may act more quickly, and the exposure period and recovery may be shortened.

Sampling. When trying the assay at first, it may be desirable to sample ([^{35}S]-methionine labeling and autofluorograph) on each day of drug exposure and recovery (Fig. 19.5). If a stable ID_{50} is reached earlier, then the assay may be shortened.

Endpoint. [^{35}S]-methionine labeling and autofluorography were chosen for speed and ease of analysis, and because active protein synthesis implies that the cells are still alive. Other alternatives are possible, however, including direct staining and cell counting *in situ* or by densitometry, fluorimetric DNA assay [Kissane and Robbins, 1958; Brunk et al., 1979], measurement of dehydrogenase activity [DiPaolo, 1965], or labeling with [^{3}H]-thymidine (DNA synthesis), [^{3}H]-uridine (RNA synthesis), or other isotopes and analysis by scintillation counting or autofluorography. One percent sodium salicylate may be substituted for conventional scintillant.

Two types of scintillation counting are available, capable of counting the contents of microtitration plate wells. The cellular contents may be aspirated onto filters by trypsinization and suction transfer onto glass fiber filters and the filters dried and counted in scintillant (Canberra-Packard), or scintillant may be added

Fig. 19.3. *Variations in layout of microtitration plate for assay of a dose-response curve. The graphs represent agent concentrations in linear (stippled) or exponential (shaded) dilutions.*

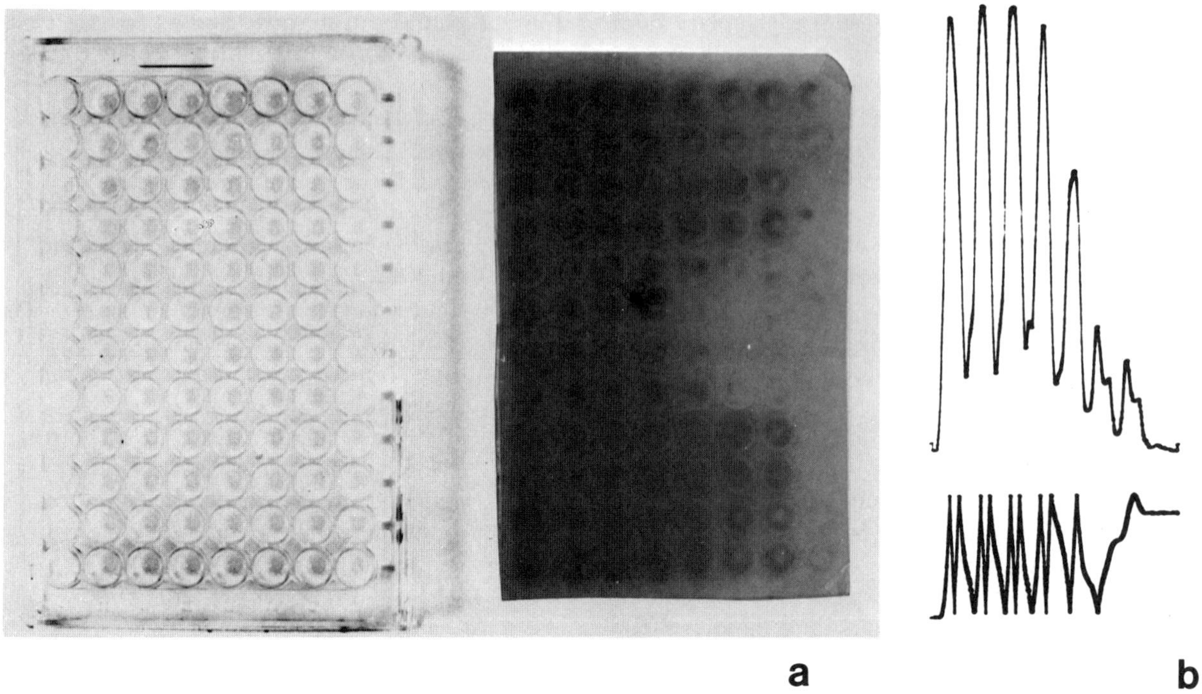

Fig. 19.4. Autofluorograph from isotopically labeled cultures in a microtitration plate (a) and densitometer scan (b, upper trace) from one row.

to the plate directly and the whole plate counted on a specially adapted counter (Wallac, Canberra-Packard).

In recent years microtitration assays of cytotoxicity have been dominated by the use of MTT reduction in the determination of viable cells at the end of the assay. The following protocol has been provided by Jane Plumb of the CRC Department of Medical Oncology, University of Glasgow, Scotland.

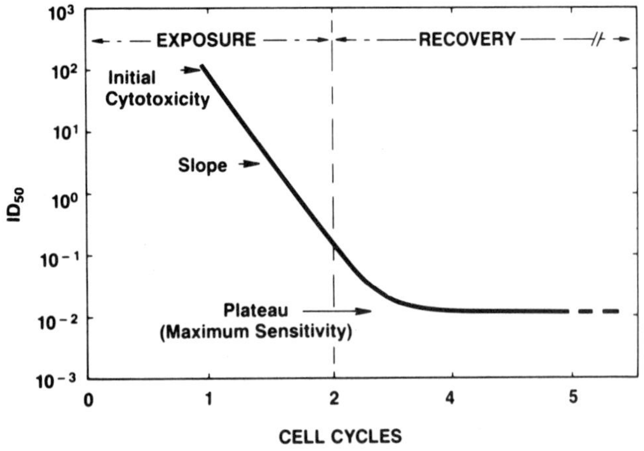

Fig. 19.5. Time course of the fall in ID_{50}. Idealized curve for an agent with a progressive increase in cytotoxicity with time but eventually reaching a maximum effect after three cell cycles. Not all cytotoxic drugs will conform to this pattern.

MTT-Based Cytotoxicity Assay

The endpoint of a microtitration assay is usually an estimate of cell number. While this can be direct by cell counts or by indirect methods such as isotope incorporation, MTT reduction as a cell viability measurement [Mosmann, 1983] is now widely chosen as the optimal endpoint [Cole, 1986; Alley et al., 1988]. MTT is a yellow water-soluble tetrazolium dye that is reduced by live, but not dead, cells to a purple formazan product that is insoluble in aqueous solutions. However, a number of factors can influence the reduction of MTT [Vistica et al., 1991]. The following assay has been shown to give the same results as a standard clonogenic assay [Plumb et al., 1989].

Principle

Cells in exponential phase of growth are exposed to a cytotoxic drug. The duration of exposure is usually determined as the time required for maximal damage to occur but is also influenced by the stability of the drug. After removal of the drug, the cells are allowed to proliferate for two to three doubling times in order to distinguish between cells that remain viable and are capable of proliferation and those that remain viable but cannot proliferate. Surviving cell numbers are then determined indirectly by MTT dye reduction. The amount of MTT-formazan produced can be determined spectrophotometrically once solubilized in a suitable solvent.

Outline

Monolayer cultures in microtitration plates are incubated in a range of drug concentrations. The drug is removed and plates are fed daily for two to three cell population doublings, when they are fed again and MTT is added to each well. Plates are incubated in the dark for 4 hr, the medium and MTT removed, and the water-insoluble MTT–formazan crystals dissolved in DMSO. A buffer is added to adjust the final pH and absorbance is recorded in an ELISA plate reader.

Materials

Sterile:

Microtiter plates (Linbro, ICN Flow)
Growth medium
Tip box, autoclavable (ICN-Flow)
Pipette tips (ICN-Flow)
Petri dishes, 5 cm and 9 cm (Sterilin)
Universal containers, 30 ml and 100 ml (Sterilin)
Trypsin (0.25% + EDTA, 1 m*M*, in PBS)
3-(4,5-dimethylthiazol-2-yl)-2,5-diphenyltetrazolium bromide (MTT, Sigma), 50 mg/ml, filter sterilized
Sorensen's glycine buffer (0.1 M glycine, 0.1 M NaCl adjusted to pH 10.5 with 1 M NaOH)

Nonsterile:

Plastic lunch box
Multichannel pipette (Costar)
Dimethyl sulfoxide (DMSO)
DMSO dispenser (optional, Well-fill, Denley)
ELISA plate reader (Bio-Rad or ICN-Flow)
Plate carrier for centrifuge (for cells growing in suspension)

Protocol

(a) Plating out cells

1. Trypsinize a subconfluent monolayer culture and collect cells in growth medium containing serum.
2. Centrifuge suspension (200 *g*, 5 min) to pellet cells, resuspend in growth medium, and count cells.
3. Dilute cells to $2.5-50 \times 10^3$ cells/ml, depending on the growth rate of the cell line, and allowing 20 ml per microtiter plate.
4. Transfer cell suspension to a 9-cm petri dish, and, with a multichannel pipette, add 200 μl into each well of the central 10 columns of a flat-bottomed 96-well plate (80 wells per plate), starting with column 2 and ending with column 11, giving $0.5-10 \times 10^3$ cells/well.
5. Add 200 μl of growth medium to the eight wells in columns 1 and 12. Column 1 will be used to blank the plate reader; column 12 helps to maintain the humidity for column 11 and minimizes the "edge effect."
6. Put plates in a plastic lunch box and incubate in a humidified atmosphere at 37°C for 1–3 d such that cells are in the exponential phase of growth at the time of drug addition.

   For nonadherent cells, prepare a suspension in fresh growth medium. Dilute the cells to $5-100 \times 10^3$ cells/ml and plate out only 100 μl of the suspension into round-bottomed 96-well plates. Drug is added immediately to these plates.

(b) Drug addition

1. Prepare a serial fivefold dilution of the cytotoxic drug in growth medium to give eight concentrations. This should be chosen such that the highest concentration kills most of the cells and the lowest kills none of the cells. Once the toxicity is known, a smaller range of concentrations can be used. Normally, three plates are used for each drug to give triplicate determinations within one experiment.
2. For adherent cells, remove the medium from the wells in columns 2 to 11. This can be achieved with a hypodermic needle attached to a suction line.
3. Feed the cells in the eight wells in columns 2 and 11 with 200 μl of fresh growth medium; these cells are used as the controls.
4. Add the cytotoxic drug to the cells in columns 3 to 10. Only four wells are needed for each drug concentration such that rows A–D can be used for one drug and rows E–H for a second drug.
5. Transfer the drug solutions to 5-cm petri dishes and add 200 μl to each group of four wells with a four-tip micropipette.
6. Return plates to the lunch box and incubate for a defined exposure period.

   For nonadherent cells, the drug dilution is prepared at twice the desired final concentration and 100 μl is added to the 100 μl of cells already in the wells.

(c) Growth period

1. Remove the medium, at the end of the drug exposure period, from all wells containing cells and the cells fed with 200 μl of fresh medium. Centrifuge plates containing nonadherent cells (200 *g*, 5 min) to pellet cells before removal of the medium using a fine-gauge needle to prevent disturbance of the cell pellet.
2. Feed plates daily for 2–3 cell population doubling times.

(d) Estimation of surviving cell numbers

1. Feed plate with 200 μl of fresh medium at the end of the growth period and add 50 μl of MTT to all wells in columns 1 to 11.
2. Wrap plates in aluminum foil and incubate for 4 hr in a humidified atmosphere at 37°C. This is a minimum incubation time and plates can be left for up to 8 hr.
3. Remove the medium and MTT from the wells and dissolve the remaining MTT–formazan crystals, adding 200 μl of DMSO to all wells in columns 1 to 11.
4. Add glycine buffer (25 μl per well) to all wells containing DMSO.
5. Record absorbance at 570 nm immediately, since the product is unstable. The wells in column 1, which contained medium, MTT, but no cells, are used to blank the plate reader.

Analysis. Plot graph of absorbance (Y-axis) against drug concentration (X-axis). The mean absorbance reading from the wells in columns 2 and 11 is used as the control absorbance and the ID_{50} concentration is determined as the drug concentration required to reduce the absorbance to half that of the control (cf. Fig. 19.2). ID_{10} or ID_{90} values can be determined in a similar manner. The absorbance values in columns 2 and 11 should be the same. Occasionally they are not, and this is taken to indicate uneven plating of cells across the plate.

Variations. A similar assay can also be used to determine cellular radiosensitivity [Carmichael et al., 1987b]. MTT can be used to determine cell numbers after a variety of treatments other than cytotoxic drug exposure such as growth factor stimulation. However, in each case it is essential to ensure that the treatment itself does not affect the ability of the cell to reduce the dye.

Microtitration offers a method whereby large numbers of samples may be handled simultaneously but with relatively few cells per sample. The whole population is exposed to the agent and viability determined metabolically, by staining or by counting cells from a few hours to several days later.

In practice it may not matter which criterion is used for viability or survival at the end of the assay; it is the design of the assay (drug exposure, recovery, cell density, growth rate, etc.) that is most important.

A variety of automated handling techniques are available (autodispensers, diluters, cell harvesters, and densitometers [see Fig. 6.2]), reducing the time required per sample. The volume of medium required per sample is less than one-fiftieth of that required for cloning, though the cell number is approximately the same. Microtitration, however, is unable to distinguish between differential responses between cells within a population and the degree of response in each cell–e.g., a 50% inhibition could mean that 50% of the cells respond or that each cell is inhibited by 50%.

A comparison of ID_{50}s derived by microtitration and plating efficiency assays showed a strong correlation between the two methods (Fig. 19.6) for the assay of antineoplastic drugs.

A significant feature of microtitration assays, particularly with a colorimetric or radiometric endpoint, is the generation of large amounts of data, often in a format readily analyzed by computer. As was discussed in the previous chapter, there are now a number of different software programs (Packard, Wallac, Molecular Devices, Biorad, Lab Systems) that can be used to generate tabular and graphical output from your data. It is important, however, that a system be used whereby the operator can scan the raw data as well as the data-reduced endpoint, e.g., an ID_{50}, a V_{max}, or a binding constant. Computer analysis will save a great deal of time but it will make different assumptions or corrections to deal with aberrant data points, which will not be apparent unless the raw data is available for scrutiny.

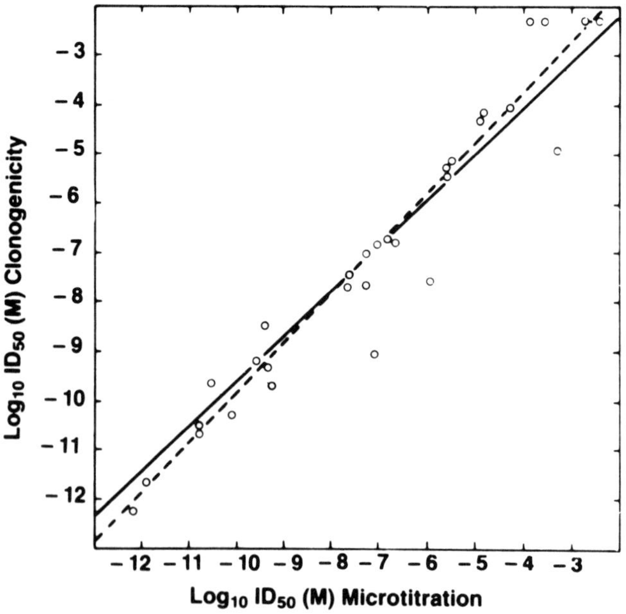

Fig. 19.6. Correlations between microtitration and cloning in the measurement of the ID_{50} (solid line) of a group of five cell lines from human glioma and six drugs (vincristine, bleomycin, VM-26 epidophyllotoxin, 5-fluorouracil, methyl CCNU, mithramycin). Most of the outlying points were derived from one cell line that later proved to be a mixture of cell types. The dotted line is the regression with the data points from the heterogeneous cell line omitted. [From Freshney et al., 1982a.]

Metabolic Tests

The distinction between a metabolic test and a survival test is that a survival test examines the number of cells or amount of cellular activity remaining after prolonged exposure to an agent or short exposure and prolonged recovery in the absence of the agent. What is measured is the ability of the cell to survive, and in most cases, to continue to proliferate. In a metabolic test, on the other hand, the direct effect of the agent on one or more metabolic pathways is being measured. If, in the above protocols for microtitration assay, the labeling with ^{35}S-methionine or assay of MTT reduction is carried out during or immediately after the first 24 hr of drug exposure, this becomes a metabolic test. There are many such tests, ranging from simple observation of the inhibition of pH depression to more specific tests depending on precursor incorporation or measurement of enzyme activity. They may be short-term (30 min or so) or long-term (several days).

In the context of cytotoxicity and viability testing, it should be kept in mind that effects on metabolism measured in the presence of an agent must be interpreted only as such. To establish an irreversible effect on cell survival, culture must be continued in the absence of the agent.

Drug Interaction

The investigation of cytotoxicity often involves the study of the interaction of different drugs, readily investigated by microtitration systems where several different ratios of interacting drugs can be examined simultaneously. Analysis of drug interaction can be performed using an *isobologram* to interpret the data [Steel, 1979; Berenbaum, 1985]. A rectilinear plot implies an additive response, while a curvilinear plot implies synergy if the curve dips below the predicted line and antagonism if above.

ANTICANCER DRUG SCREENING

The measurement of viability by clonal growth analysis and microtitration has been described in detail above and in Chapter 18. One of the major applications of such tests is in the development of new anticancer agents where comparison of survival curves in clonogenicity assays with L1210, P388, Hep-2, and, more recently, with cell lines derived from human bronchial carcinoma and other human tumors can give comparative figures for relative cytotoxicity. Drug testing *in vitro* does not allow for the modification of drugs by liver metabolism en route to the target tissue, so some workers have included liver microsomal enzyme preparations in the culture medium to activate drugs such as cyclophosphamide [Sladek, 1973], or cocultured with hepatocytes [Donato et al., 1990].

Drug screening for the identification of new anticancer drugs can be a tedious and often inefficient method of discovering new active compounds. The trend is now more toward monitoring effects on specific molecular targets. However, there have been attempts to improve screening by adopting rapid, easily automated assays like those based on the determination of viable cell number by staining with MTT [Mosmann, 1983; Carmichael et al., 1987a; Plumb et al., 1989]. To cut down on manipulations still further, the MTT incubation step has been omitted and the endpoint determined by total protein using sulforhodamine B [Boyd, 1989]. While this is quicker and easier to do, it should be remembered that nonviable, and certainly nonreplicative, cells will still stain, so the assay should be used as a time course, measuring rate of change rather than absolute amount, or at least confirmed where activity is detected, using a more reliable indicator such as clonogenicity or MTT reduction.

Predictive Testing

The possibility has often been considered that measurement of the chemosensitivity of cells derived from a patient's tumor might be used in designing a chemotherapeutic regime for the patient [Freshney, 1978]. This has never been exhaustively tested, although small-scale trials have been encouraging [Hamburger and Salmon, 1977; Bateman et al., 1979; Hill, 1983; Thomas et al., 1985; Von Hoff, 1986]. What is required now is the development of reliable and reproducible culture techniques for the common tumors such as breast, lung, and colon, such that cultures of pure tumor cells capable of cell proliferation over several cell cycles may be prepared routinely. Assays might then be performed in a high proportion of cases, within 2 weeks of receipt of the biopsy. So far this test has not been possible, but developments with defined media (see Chapter 7) may have brought this closer [Carney et al., 1981].

The major problem is, however, logistic. The number of patients whose tumors will grow *in vitro* sufficiently to be tested, who can be expected to respond, and whose response can be followed up is extremely small. Hence it has proved difficult to use any *in vitro* test as a predictor of response or even to prove its reliability. Correlation of insensitivity *in vitro* with nonresponders is high, but few clinicians would withhold chemotherapy because of an *in vitro* test, particularly when the agent in question would probably not be used alone.

Culture Systems

Most of the procedures described above depend on the exposure of cells in exponential growth in convention-

TABLE 19.1. Types of Culture Systems for Cytotoxicity Assay

System	Exposure	Recovery period	Endpoint	Use
Monolayer clones	Continuous	None	Colony no. and size	Chronic exposure. Low-grade toxin, e.g., tap-water impurities or pharmaceuticals presumed to be nontoxic
Monolayer clones	1–24^h pulse	None or 3 cell cycles (for delayed effects, repair, or recovery)	Colony no. and size	Standard survival assay from acute shock
Exponential monolayer, e.g., in microtitration plates or flasks	1–24^h pulse, repeated if necessary	Approx. 3 cell cycles (cells must remain in exponential growth in controls)	^{3}H-leucine incorporation; ^{3}H-thymidine incorporation (but not with base analogues); MTT reduction (absorb. change); Stain, sulfarhodamine, neutral red, Coomassie blue	Large-scale screening of potential toxins, anticancer drugs, drug combinations, doses
Spheroids	Pulsed or continuous	None	Spheroid volume (growth delay); trypsinize and clone	Drug penetration, solid tumor modeling, homologous cell interaction
Filter wells or mixed capillary bed perfusion	Pulsed or continuous (perfused)	None	Metabolic (e.g., isotopic precursor incorporation; clonogenicity)	Solid tissue simulation, complex tissue responses, heterologous cell interactions

al monolayer culture. There are, however, a number of variations that can be made to the culture system, depending on whether maximum sensitivity, chronic exposure, or *in vivo* simulation is required. These types of culture systems must be considered along with the duration of exposure (see above) and stability of the agent, as this will influence sensitivity and *in vivo* simulation (Table 19.1). At one extreme, cloned cultures exposed continuously to a stable toxin will demonstrate higher sensitivity than a high-density culture pulsed with a brief exposure to an unstable toxin. The first may be appropriate, for example, for monitoring trace toxins in tap water where the keynote is chronic exposure and the effects minor, while the second may approximate more closely to intravenous bolus injection of a pharmaceutical, such as an anticancer drug.

The types of culture systems that have been used for cytotoxicity assay are almost as numerous as the workers in the field, and as a result a major problem has emerged in comparing data between laboratories. A number of suggestions are made in Table 19.1 based on practical examples.

Most tests to date have relied on cell survival, cell growth, or reductions in DNA, RNA, or protein synthesis as indicators of toxicity. However, there are many other ways that toxicity may be expressed *in vivo* and not recognized *in vitro*. One of the major *in vivo* indicators is inflammation; as yet there is no satisfactory *in vitro* model for the inflammatory response. However, as more is learned about primary cellular products that may elicit inflammatory or other more general responses, it may become possible to assay such products directly (e.g., by immunoassay) or indirectly by an appropriate bioassay with, for example, bone marrow stem cells (see below). Until such time as this situation can be achieved, a substantial amount of toxicity testing will continue *in vivo*.

MUTAGENICITY

The determination of mutagenicity is a task more amenable to culture, as it represents a cellular response. Metabolism by enzymes of the liver and the gastrointestinal tract may still be required, however ["S9 mix"; see Venitt et al., 1984; McGregor et al., 1988].

Assaying for mutagenicity can be performed by a variety of standard genetic techniques [Venitt and Parry, 1984], e.g., an increase in the frequency of occurrence of such characterized mutants as the absence of thymidine kinase (TK^-), which makes cells resistant to bromodeoxyuridine (BUdR), or hypoxanthine guanine phosphoribosyl transferase deficiency ($HPRT^-$), making the cell resistant to 8-azaguanine [Littlefield, 1964b]. As TK^- or $HPRT^-$ cells will not clone in hypoxanthine, aminopterin, and thymidine (HAT) medium but revertant mutants will, the reversion rate of TK or HPRT deficiencies assayed by cloning in HAT medium can also be used to assay for frame-shift mutations. Double strand breaks in DNA can lead to exchange of chromatid fragments between homologous chromatids. These exchanges, sister chromatid exchanges or SCEs, can be detected by incorporating BUdR into the chromatids and subsequently staining with Giemsa and Hoechst 33258 [Dean and Danford, 1984; Aghamohammadi and Savage, 1989; Boyes et al., 1990].

The following protocol for assaying mutagenicity has been contributed by Jane Cole and David Beare of the MRC Cell Mutation Unit, University of Sussex, Falmer, Brighton BN1 9RR, UK.

Mutagenesis, Selection of Mutants, and Estimation of Mutant Frequency

A variety of systems are available for selecting mutants in cultured animal cells [reviewed in Cole et al., 1990] (see Chapter 11), most based on loss of functional gene product (enzyme) in the mutant cell. The spectrum of mutations detected depends on the particular system of choice; for example, those based on the nonessential enzymes of the nucleotide salvage pathways, e.g., hypoxanthine guanine phosphoribosyltransferase (HPRT) or TK, detect a wide variety of mutational events, from point mutations to complete gene deletion. In contrast, ouabain, which kills cells by binding to the essential cell membrane Na^+K^+-dependent ATPase, detects only specific point mutations at the ouabain binding site. These differences in target size and range of mutational events detected are reflected in the spontaneous mutant frequencies seen in a given system, which vary from $\sim 1 \times 10^{-6}$ for rare events like ouabain resistance to 10^{-4} for TK mutants, where many events from point mutations to loss of a whole chromosome may give rise to the mutant phenotype [Cole et al., 1991].

Principle of Detecting Mutants and Determining Mutant Frequency

Cells are plated at a high density in *selective medium* containing an inhibitor in which only mutants are able to survive and form a colony, and at low cell density in *nonselective medium* to determine plating efficiency in the absence of the inhibitor. The plates are incubated for sufficient time for colonies to form. Colonies on both sets of plates are scored, and the mutant frequency (MF) at the locus of interest calculated from the formula:

MF per clonable cell = CE in selective medium/CE in nonselective medium.

Before a detailed protocol for a "typical" mutation assay is described, a number of points applicable to *all* systems should be noted:

(1) For mutant frequency estimation, and particularly for subsequent analysis of mutant clones, it is *essential* that a single cell suspension should be obtained before plating, and that adequate precautions are taken to ensure the single-cell origin or mutant colonies. Unless this is done, recloning of mutants should be undertaken before further analysis.

(2) The maximum cell density in selective medium depends on the system of choice. Too high a cell density may result in either (a) incomplete kill of nonmutants or (b) loss of mutants due to metabolic cooperation (in the case of attached cells, e.g., V79 Chinese hamster cells or human fibroblasts), or overcrowding (in the case of suspension cultures such as mouse lymphoma or human lymphoblastoid cell lines).

(3) For an accurate estimate of mutant frequency, sufficient cells should be plated in the selective medium to obtain *at least* 10 (preferably 100) spontaneous mutant colonies. As cell density per milliliter is limited by practical constraints (2 above) it may be necessary to increase the number of plates to fulfill this requirement.

(4) After mutagen treatment, it is essential that sufficient time be allowed for the cellular expression of the newly reduced mutant phenotype (the *expression time*). Expression time depends on the system of choice and may be as short as 24 hr (e.g., ouabain resistance) or as long as 1 to 2 wk or more (e.g., *hprt* mutations).

For detailed discussions of these and many other important aspects of quantitative mutagenesis, the following references are strongly recommended: Arlett et al. [1989]; Robinson et al. [1989]; Cole et al. [1990].

Two protocols for mutagenesis are given, first for the most widely used gene mutation, which involves selecting TK mutants in a $TK^{+/-}$ mouse lymphoma cell line using the selective agent trifluorothymidine (TFT); and second for the selection of HPRT mutants in freshly isolated human lymphocytes using selection in 6-thioguanine (6TG).

Protocol for the L5178Y $TK^{+/-}$ Mouse Lymphoma Assay (MLA)

Materials

L5178Y $TK^{+/-}$ (3.7.2c) mouse lymphoma cells (obtainable from D. Clive, Burroughs Wellcome

Co., Research Triangle Park, NC 27709, USA, or J. Cole, MRC Cell Mutation Unit, University of Sussex, Falmer, Brighton, BN1 9RR, UK)

RPMI 1640 medium, HEPES buffered, Dutch Modification (Life Technologies or Imperial Laboratories)

Heat-inactivated (56°C for 30 min) horse serum (ICN Flow), pretested for supporting good suspension culture growth and high plating efficiency of mutant and nonmutant clones. Stored frozen in 100-ml aliquots at −20°C.

L-glutamine, 200 m*M* (GIBCO), stored at −20°C

Penicillin/streptomycin, 10,000 u/μg per ml (Life Technologies), stored frozen at −20°C

Stock solution of sodium pyruvate: 2 g dissolved in 100 ml of water, filter sterilized, stored at 4°C

Stock TFT solution: 400 μg/ml, dissolved under subdued lighting in RPMI 1640 medium, filter sterilized, stored at −20°C in 5-ml aliquots in foil-wrapped containers

Rat liver S9 fraction. Standard protocols are available for S9 preparation. Alternatively, S9, prepared under specified conditions at 30 mg/ml protein, may be obtained from the Robens Institute, University of Surrey, Guildford, Surrey, GU2 5XH, UK. Store in liquid nitrogen.

Co-factor mix: Dissolve 336.5 mg NADP and 760.5 mg glucose-6-phosphate (disodium salt) in 100 ml ice-cold RPMI 1640, filter sterilize, and store in 10-ml aliquots at −20°C.

Glass or plastic tissue culture flasks

96-well flat-bottom microtitration trays (Nunc) (nontissue culture grade are adequate)

8-channel 200-μl multipipetter (Dynatech Laboratories, autoclavable)

Suspension culture medium (R10): RPMI 1640 medium supplemented with 10% heat-inactivated horse serum, 100 u/ml penicillin, 100 μg/ml streptomycin, 200 μg/ml sodium pyruvate and 2 m*M* L-glutamine

THMG: R10 supplemented with *T*hymidine (9 μg/ml), *H*ypoxanthine (15 μg/ml), *M*ethotrexate (0.3 μg/ml), and *G*lycine (22.5 μg/ml); store at −20°C.

Cloning medium (R20): 20% horse serum, 2 m*M* L-glutamine, 100 u/ml penicillin, 100/μg streptomycin, and 200 μg/ml sodium pyruvate.

Protocol (a): Suspension Culture Growth

Incubate cells in a 5% CO_2/air humidified incubator and maintain in exponential growth by daily dilution in R10 to 10^5 cells/ml in 75-cm² (maximum volume 40 ml) or 175-cm² (maximum volume 100 ml) tissue culture vessels. A population doubling time ~11 hr should be achieved in these conditions.

Variations

(1) Fischer's medium may be used in place of RPMI 1640 medium.

(2) Cells in suspension culture may be constantly mixed by growth in gassed culture flasks on roller or shaker apparatus, when Pluronic F68, 1 g/l (BASF, Wyandotte), should be added to the medium to prevent damage to the cells. Under these conditions a doubling timed of 8–10 hr may be achieved.

Protocol (b): THMG Treatment

Five to seven days before mutagen treatment, L5178Y $tk^{+/-}$ 3.7.2c should be purged of pre-existing $tk^{-/+}$ mutants by 24 hr culture in medium containing THMG.

Protocol

1. Centrifuge and resuspend 10^7 cells in THMG-supplemented medium and incubate at 10^5 cells/ml for 24 hr.
2. Count the cell density, centrifuge, discard the supplemented medium, and incubate the cells at 10^5 cells/ml in R10 medium supplemented with THG only for 24 hr.
3. Count, dilute in R10 to 10^5 cells/ml, and incubate for a further 24 hr, when there should be a total of ~3–4 × 10^8 cells.
4. Cryopreserve the THMG-treated cells by standard procedures (4–5 × 10^6/vial) (see Chapter 17).
5. Three to four days before a mutation experiment, thaw 1–3 vials as appropriate of THMG-treated cells and culture in R10 to provide sufficient cells.

Protocol (c): Mutagen Treatment

Standard conditions should be used for mutagen treatment. Thus, exponentially growing cells at 5 × 10^5/ml in reduced serum (5%) should be used. Treatment should be in the presence (for mutagens requiring exogenous metabolic activation) and absence (for direct-acting mutagens) of S9. For accurate estimation of spontaneous and induced MF and subsequent statistical analysis of the data obtained, it is essential that sufficient cells (at least 10^7) should survive treatment and be subcultured during the expression time. It is also highly recommended that duplicate control and mutagenized cultures should be treated and maintained throughout the experiment.

(1) Preparation of cells before treatment

On the day of the experiment, count the cells in suspension culture, centrifuge, and resuspend sufficient cells in ½ conditioned, ½ fresh R10 at 10^6 cells/ml. Dispense 10-ml aliquots in plastic centrifuge tubes and incubate for ~30 min while steps (b) and (c) are undertaken.

(2) Preparation of S9 and co-factor mix
Thaw sufficient S9 and co-factor mix immediately before use and hold on ice.

(3) Preparation of test compound
Test compounds should be dissolved in RPMI 1640 medium or suitable solvent (e.g., DMSO) at 100× final concentration immediately prior to use.

◇ Known or suspected mutagens should be handled at all times throughout the treatment period and subsequent disposal under strictly controlled safety conditions.

(4) Treatment

1. To each centrifuge tube from (a) above, add the appropriate amount of RPMI 1640 medium (without serum) so that the final serum content will be 5% and cell density 5×10^5/ml (see Table 19.2).
2. Fix 1 part ice-cold S9:9 parts ice-cold co-factor mix.
3. For exposure in the absence of S9, add 0.2 ml solvent or test compound at 100× concentration.
4. For exposure in the presence of S9, add 2 ml S9/co-factor mix and 0.2 ml solvent or test compound (final S9 concentration, 1%).
5. Tightly cap and seal centrifuge tubes and incubate at 37°C for 4 hr, shaking at regular intervals (e.g., every 15 min).
6. At the end of the treatment period, centrifuge cultures, remove supernatant, resuspend pellet in R10, and wash twice; finally, resuspend cells in 20 ml R10, count using a hemocytometer, and adjust the cell density to 2×10^5 cells/ml in R10. Do not discard any cells. Remove 1 ml to estimate cloning efficiency at the end of treatment (see step 4, below) and incubate remaining cells.

Variations

1. Co-factor mix consisting of NADP (2 mg/ml) and sodium isocitrate (11.25 mg/ml) may be used.
2. The recommended S9 concentration, 1%, may not be optimal for all mutagens. It may be varied from 0.5 to 10%, but the proportion of S9:co-factors should be kept constant.
3. Four hours' treatment time in the presence of S9 is generally considered optimal. For direct-acting mutagens in the absence of exogenous metabolic activation, treatment times of 30 min–24 hr may be considered, depending on the half-life of the compound. Longer times are particularly suitable for relatively insoluble compounds.

(e) Estimating the toxic effect of the mutagen immediately after treatment

1. Serially dilute 1 ml removed from each culture (above) to provide 20 ml at 10 cells/ml in R20. Dispense at 200 µl per well using a multichannel pipetter into one 96-well microtiter tray (at two cells per well) for estimation of CE immediately after treatment.
2. Incubate for 10 days and score *negative* wells in which no colony has grown.

(f) Subculture during the expression period
An expression time of 2–3 d is sufficient for the selection of newly induced TK mutants. During this time, the cells in the control and treated cultures should be resuspended every 24 hr by pipetting, and diluted in R10 to provide a *minimum* of 50 ml at 2×10^5cells/ml.

Variations
During the expression time, the cells may be grown in static suspension culture in 250-ml vessels, or in shaken cultures in medium supplemented with Pluronic F68.

(g) Selection of mutants at the end of the expression period

1. Vigorously resuspend the cells in each control and treated culture by pipetting to produce a

TABLE 19.2. Dilution of S9 and Test Compound in Medium

Culture	Milliliters cell suspension at 10^6/ml in R10	Milliliters RPMI	Milliliters S9/Co-factor mix	Solvent or test compound at 100× conc.
No S9	10	9.8	–	0.2
1% S9	10	7.8	2	0.2

single cell suspension, and count the cells using a hemocytometer to estimate the cell concentration. When large volumes of cells are used, it may be difficult to obtain a single cell suspension by pipetting. This can be overcome by centrifuging a suitable volume of the cell suspension and resuspending the pellet in the same medium.

2. Dilute the cells in R20 to provide 81 ml at 1×10^4 cells/ml.
3. Remove 1 ml and serially dilute to provide 20 ml at 10 cells/ml in R20. Dispense at 200 μl/well using a multichannel pipetter onto one 96-well microtiter tray (= 2 cells per well) for estimation of CE in nonselective medium.
4. Thaw stock TFT solution just before use, and add 0.8 ml to 80 ml cells at 1×10^4 per ml in R20 (final TFT concentration 4 μg/ml). Dispense at 200 μl per well into four 96-well plates (= 2×10^3 cells/well) to estimate CE in the TFT-containing selective medium in which only tk mutants will grow.
5. Incubate the plates at 37°C in a humidified 5% CO_2-in-air incubator for 10 d.
6. Using a low-powered dissection microscope, count the *negative* wells (i.e., the wells in which no colony has grown).

Variation

Cells may be cloned in 9-cm petri dishes in semisolid agar. In this case, colonies are scored using an automatic colony counter (e.g., Artek). Colony growth, particularly the small mutant colonies discussed in step 8 below, is dependent on agar source and batch and should be carefully presented. Detailed methods are available in Turner et al. [1984].

(h) Calculation of cloning, efficiency, and mutant frequency

Calculate the CE in nonselective (one plate, 96 wells) and selective (four plates, 384 wells) medium, using the zero term of the Poisson distribution [P(O)] method:

P(O) = number of empty wells/total wells plated

CE = ln P(O)/ number of cells per well

Mutant frequency per viable cell − CE in nonselective medium/CE in selective medium

Example: Non-selective plate: total negative wells = 14; cells per well = 2

$$P(O) = 14/96 \quad CE = -\ln(0.1458)/2 = 96.3\%$$

TFT selective plates: total negative wells = 63; cells per well = 2,000

$$P(O) = 321/384 \quad CE = -\ln(0.8359)/2{,}000 = 8.95 \times 10^{-5}$$

Mutant frequency to TFT − resistance per clonable cell = $8.95 \times 10^{-5}/0.963 = 9.30 \times 10^{-5}$

(i) Scoring "small" and "large" tk mutants

Two types of TFT-resistant mutants are selected in the MLA: (1) "Large" colonies that grow at the normal rate, and have been found to have point mutations or deletions within the tk gene; (2) "small" colonies that have variable, reduced growth rates and characteristically have chromosome mutations, including *visible* aberrations involving chromosome 11b, which carries the tk^+ allele [Hozier et al., 1992]. The two colony types are easily distinguished and may be scored separately, providing added information on the mode of action and nature of the mutagen. For the agar cloning method, colony size is undertaken using the automated colony counter [Moore et al., 1985]. When mutations are cloned in microtiter trays, if small and large colonies are to be scored separately it may be necessary to reduce the cell density per well to ensure clonality at high induced mutant frequencies, and thus avoid masking the presence of small colonies.

Variations

A similar assay to the MLA, using TK6 human lymphoblastoid cells detecting large and small colony tk mutants, has been described [Liber and Thilly, 1982]. Protocols based on the same principles, for use with attached cells such as V79 [Jenssen, 1984] and CHO [Li et al., 1987] Chinese hamster cells or human fibroblasts [Cole and Arlett, 1984] are also available using standard tissue culture techniques. All of these cell lines may be used to select mutants at a variety of genetic loci, but expression time and cell density must be taken into consideration in each case. Mutagenesis may also be undertaken with freshly isolated (G_0) human lymphocytes, or with mitogen-stimulated dividing T-lymphocytes. Here the techniques involved are more complex, and below we give a protocol for obtaining consistently high cloning efficiency of human T-lymphocytes and for the selection of mutants at the *hprt* locus.

Principle of the Method

Mononuclear cells are separated from whole blood, and the lymphocytes stimulated to divide by the mitogen phytohemagglutinin (PHA). The cells are cloned in the presence of irradiated lymphoblastoid cells and the T-cell growth factor interleukin-2 (IL-2). Cloning efficiencies of 50–60% of the cells plated (which include B-cells and some monocytes as well as T-cells) have consistently been obtained in our laboratory from several hundred blood samples from a variety of donors; thus a high percentage of T-lymphocytes are forming colonies *in vitro* under the conditions we describe.

Materials

Overnight Culture Medium

Dutch-modified, HEPES-buffered RPMI 1640 medium (Life Technologies) supplemented with 10% heat-inactivated (56°C for 30 min) human AB serum (National Blood Transfusion Centre) pretested for supporting good colony growth and used as pooled batches from ~6 donors; 2 m*M* L-glutamine (Sigma); 0.2 mg/ml sodium pyruvate (filter sterilized) (Sigma); 100 u/ml penicillin (GIBCO); and 100 μg/ml streptomycin (Life Technologies)

2× Cloning Medium

Dutch modified, HEPES-buffered RPMI 1640 medium supplemented with 10% heat-inactivated pooled human AB serum, 40% HL-1 (Ventrex), 1% PHA (Wellcome, HA15), 400 u/ml recombinant interleukin-2 (rIL-2) (Cetus), 4 m*M* L-glutamine, 0.4 mg/ml sodium pyruvate, 200 u/ml penicillin, 200 μg/ml streptomycin. (Protocol variation: HL-1 may be omitted, but has been found to improve the consistency of the assay.)

2× Cloning Medium + Irradiated Feeder Cells

Lethally irradiate (40 Gy) EBV-transformed Lesch/Nyhan (*hprt*$^-$) lymphoblastoid B-cells (GM1899A or RJK853). Pellet the irradiated feeder cells (400 g for 10 min) and resuspend at a density of 1×10^6 cells/9.0 ml 2× cloning medium.
Variations: (1) TK6 lymphoblastoid cells may also be used but do not support such good colony growth in our experience. (2) Mitomycin C may be used in place of lethally irradiation to inactivate feeder cells. The cells should be cultured overnight in medium containing 2 μg/ml mitomycin C (Sigma), followed by centrifugation and three washes before use.

Dilution Medium

Dutch-modified, HEPES-buffered RPMI 1640 medium supplemented with 10% heat-inactivated pooled human AB serum

Culture Vessels

80-cm^2 tissue culture flasks (Nunc), and 96-well flat-bottomed (tissue culture grade) microtitration plates (Nunc)

10^{-4} M 6-thioguanine (6-TG)

Prepare a stock of 334 μg/ml 6-TG (Sigma) in 0.5% sodium carbonate solution, filter sterilize. Dilute this 334 μg/ml stock 1:20 in RPMI 1640 to give 1×10^{-4} M. Final concentration for mutant selection is 5×10^{-6} M.
Variation: It is essential that the 6-TG should effectively kill all nonmutants. Final selective concentrations of 1 or 2×10^{-5} M 6-TG are sometimes recommended. In our experience, similar MFs are obtained.

Protocol for a Standard Eight-Plate Assay

1. Set up overnight cultures. Culture ~20×10^6 peripheral blood mononuclear cells (MNCs), separated from whole blood by standard procedures (see Chapter 23), for 16–24 hr in overnight culture medium at a density of 1×10^6 cells/ml in 80-cm^2 tissue culture flasks. Lay the flasks flat on the incubator shelf so that the macrophage/monocyte cells are removed by adhesion to the flask, leaving mainly lymphocytes (both B- and T-cells) in suspension.
2. Cell count and first dilution. Count the cells in suspension using a hemocytometer. Dilute to 2×10^5 cells/ml with dilution medium. For a standard eight-plate assay, 81 ml at 2×10^5 cells/ml is the minimum requirement for each culture.
3. Nonselective plates. Remove 1 ml at 2×10^5 cells/ml and serially dilute to 30 cells/ml.

$$\underset{10^5/\text{ml}}{(1\text{ ml} + 1\text{ ml})} \rightarrow \underset{10^4/\text{ml}}{(0.5\text{ ml} + 4.5\text{ ml})}$$
$$\rightarrow \underset{10^3/\text{ml}}{(0.5\text{ ml} + 4.5\text{ ml})} \rightarrow \underset{10^2/\text{ml}}{(1\text{ ml} + 9\text{ ml})}$$
$$\rightarrow \underset{30/\text{ml}}{(3.75\text{ ml} + 8.75\text{ ml})}$$

   Make up a nonselective plating medium as follows:

Diluted MNCs (at 30 cells/ml)	10.00 ml
RPMI 1640 medium	1.00 ml
2× cloning medium + irradiated feeder cells	9.00 ml

   Plate out on a 96-well plate at 200 μl/well (= 3 cells/well).
4. Prepare selective plating medium for 6-TG selection plates as follows:

MNCs (at 2×10^5/ml)	80.00 ml
10^{-4}M 6-TG solution	8.00 ml
2 × cloning medium + irradiated feeder cells	72.00 ml

   Plate out on eight 96-well plates at 200 μl/well (2×10^4 cells/well).

5. Incubation and scoring of plates. Tape the 96-well plates together (to reduce evaporation) and incubate in a 37°C humidified CO_2 incubator on a ~5° sloping shelf. After 13 d, turn the plates around (180°). Score the plates for colonies after a further 2–4 d incubation using an inverted microscope.
6. Calculation of mutant frequencies (MF).

$$P(0) = \text{Number of negative wells/Total number of wells}$$
$$CE = -\ln\,[P(0)\ \text{nonselective plates}]/(3\ \text{cells/well})$$
$$MF = -\ln\,[P(0)\ \text{TG plates}]/(2 \times 10^4\ \text{cells/well})$$
$$\text{MF/Clonable cell} = \text{MF/CE}$$

Variations

1. FBS may be used in place of human AB serum, but batch testing is required. The presence of HL-1 medium is essential if FBS is used [O'Neill et al., 1987].
2. If FBS is used, the cells may be cultured for 36–40 hr in "overnight culture medium" containing PHA at 0.5% before cloning. No cell division takes place during this period. The cells are cloned as described above in medium containing FBS in place of human serum and 0.125% PHA.
3. The protocol described above provides an estimate of the *hprt* mutant frequency in *circulating* T-cells. G_0 lymphocytes, or mass cultures of mitogen-stimulated T-lymphocytes, may be mutagen treated and cloned at appropriate expression times essentially as described for the MLA. However, it must be noted that expression time for newly induced mutants in lymphocytes is complicated by variable growth rates in these cells.

CARCINOGENICITY

The potential for *in vitro* testing for carcinogenesis is considerable [Berky & Sherrod, 1977], for this is one area where *in vivo* testing is far from adequate. The models are poor and the tests often take weeks or even months to perform. However, the development of a satisfactory test is hampered (1) by the lack of a universally acceptable criterion for malignant transformation *in vitro*, and (2) by the inherent stability of human cells used as targets.

The most generally accepted tests so far assume that carcinogenesis, in most cases, is related to mutagenesis (see above). This is the basis of the Ames tests [Ames, 1980], where bacteria are used as targets and activation can be carried out using liver microsomal enzyme preparations. This test has a high predictive value but nevertheless dissimilarities in uptake, susceptibility, and the type of cellular response have led to the introduction of alternative tests using mammalian and human cells as targets.

Some of these tests are mutagenesis assays also, using suspensions of L5178Y lymphoma cells as targets and the induction of mutations or reversion, or cytological evidence of sister chromatid exchange, as evidence of mutagenesis. Others [Styles, 1977] have used transformation as an endpoint, assaying clonogenicity in suspension (anchorage independence; see Chapter 15) as a criterion for transformation. Critics of these systems say that both use cells that are already transformed as targets; even the BHK21-C13 cell used by some workers is a continuous cell line and may not be regarded as completely normal. Furthermore, the bulk of the common cancers arise in epithelial tissues, which have so far been difficult to grow routinely, and not in connective tissue cells.

It would appear that at present the Ames test may be an appropriate first-line test for screening potential carcinogens. The next step to improve on this is not obvious and may need the demonstration of a reliable and reproducible culture system with epithelial cells before carcinogenicity screening can be transferred routinely to mammalian cells.

Guillouzo et al. [1981] have proposed that such a system should incorporate normal hepatocytes in coculture for their metabolic activity.

The final hurdle is the general acceptance of an appropriate criterion for malignant transformation (see Chapter 15). No one criterion may be sufficient and two or three may be required. Demonstration of increased oncogene expression, amplification, or the presence of increased or altered oncogene products may provide reliable criteria, in some cases functionally related to the carcinogen. However, as the proposals gain in complexity the resistance to relinquishing the Ames test increases and *in vitro* carcinogenesis assays remain the experimental tool of those studying the mechanism of carcinogenesis.

INFLAMMATION

There is an increasing need for tissue culture testing to reveal the inflammatory responses that are likely to be induced by pharmaceuticals and cosmetics with topical application, or xenobiotics that may be inhaled or ingested. This is an area that is only at the early stages of development but that bears great promise for the future. It is a sensitive topic in more ways than one.

Animal-rights groups are naturally incensed at the needless use of large numbers of animals to test new cosmetics that have little benefit but commercial advantage to the manufacturer, particularly when the testing of substances such as shampoos involves the Draze test, where the compound is added to the rabbit's eye. More important, clinically, is the apparent increase in allergenic responses produced by pharmaceuticals and xenobiotics. These are little understood and poorly controlled, largely due to the absence of a simple reproducible *in vitro* test.

Since the advent of filter-well technology, several models for skin and cornea have appeared [Braa and Triglia, 1991; Triglia et al., 1991] (see also Kahn, Chapter 20), utilizing the facility for co-culture of different cell types that the filter-well system provides. In these systems, the interaction of an allergen or irritant with a primary target, e.g., epidermis, is presumed to initiate a paracrine response, which triggers the release of a cytokine from a second, stromal, component, e.g., dermis. This cytokine can then be measured by ELISA technology to monitor the degree of the response. Although still in the early stages of development, kits for the measurement of irritant responses are available ($Skin^2$, Organogenesis Corp.). A protocol for corneal culture suitable for modeling irritant responses in the eye has been provided by Carolyn Cahn (see Chapter 20).

It would seem that this may be a major area of development with the real prospect that allergen screening from patients' own skin may become possible, and that, ultimately, analysis of GI tract will reveal responsible allergens for irritable bowel syndrome. In each case, and in many others, there is the possibility of specific mechanistic studies into the processes of abnormal cell interaction that typify many allergic and degenerative diseases.

CHAPTER 20

Culture of Specific Cell Types

It will be apparent from the discussion in previous chapters that the expression of specialized functions in culture is controlled by the nutritional constitution of the medium, the presence of hormones and other inducer or repressor substances, and the interaction of the cells with the substrate and other cells. Reviews of specialized culture techniques for specific cell types can be found in Barnes et al. [1984a–d], Jakoby and Pastan [1979], Kruse and Patterson [1973], Sato [1979, 1981], Pollard and Walker [1990], Freshney [1992], and Freshney et al. [1993] (see also Table 2.2).

The development of techniques for cell line immortalization (see also Chapter 15) has meant that it has been possible to generate continuous cell lines from a number of finite cell lines from untransformed tissue [Chang et al., 1982; Klein et al., 1990; Steele et al., 1992; Wyllie et al., 1992]. In many cases the differentiated properties are lost, but by using a switchable promoter (e.g., temperature sensitivity) it may prove possible to recover the differentiated phenotype. The development of the transgenic mouse carrying the large T gene of SV40 has opened up a wide range of possibilities [Yanai et al., 1991], as cells cultured from these animals are already immortalized but still retain some differentiated functions.

It is encouraging to find that it is becoming possible to purchase cultures of specialized cells, with the appropriate selective media. Epidermal keratinocytes, melanocytes, endothelial cells, dermal fibroblasts, and mammary epithelium are marketed by Clonetics and Promocell. The cost is naturally very high, but it is to be hoped that as demand increases, the cost will come down. Skin cultures are also available for cytotoxicity and inflammation research from Organogenesis and Advanced Tissue Sciences. These are prepared in filter wells by combining keratinocytes with dermal fibroblasts and collagen supported by a nylon net in a so-called "skin equivalent." Other tissues have also been prepared in a similar way for toxicity studies, in particular the cornea (see below, protocol by Kahn).

My main purpose here is to present an outline of some of the techniques that are available for culturing different tissue or cell types and to exemplify the diversity of cell types that can be cultured. It is useful to classify these anatomically, although there is considerable overlap in the techniques used.

A number of specialized procedures have now been devised and some representative examples have been contributed by experts in each area. It is assumed in the following protocols that the basic prerequisites of the cell biology laboratory, as specified in Chapter 4, will be available. Consequently items such as inverted microscopes, bench centrifuges, and water baths will not be reiterated in each Materials section. It is also assumed that all reagents and materials are sterile.

EPITHELIAL CELLS

Epithelial cells are often responsible for the recognized functions of an organ, e.g., controlled absorption in kidney and gut, secretion in liver and pancreas, and gas exchange in lung. They are also of interest as models of differentiation and stem cell kinetics (e.g., epidermal keratinocytes) and are among the principal tissues where the common cancers arise. Consequently, culture of various epithelial cells has been a focus of attention for many years. The major problem in the culture of pure epithelium has been the overgrowth of the culture by stromal cells such as connective tissue fibroblasts and vascular endothelium. Most of the variations in technique are aimed at preventing this, by nutritional manipulation of the medium or alterations in the culture substrate (Table 20.1) to promote the growth of the undifferentiated epidermis and, preferably, the stem cells. Subsequent modifications may then be employed to enhance epithelial differentiation, though perhaps at the expense of proliferation.

Factors contained in serum, many of them derived from platelets, have a strong mitogenic effect on fibroblasts, and tend to inhibit epithelial proliferation by inducing terminal differentiation. Consequently, one of the most significant developments in the isolation and propagation of specialized cell cultures has been the development of selective, serum-free media, supplemented with specific growth factors as appropriate.

Isolation of epithelial cells from donor tissue is best performed with collagenase (see Chapter 9), as this disperses the stroma but leaves the epithelial cells in small clusters, which favors their subsequent survival.

Epidermis

Rheinwald and Green [1975] showed that murine and human epidermal keratinocytes could be cultured selectively on feeder layers of irradiated 3T3 cells and could mature to form differentiated squames [Green, 1977]. Basal cell carcinoma can also be cultured by this method [Rheinwald and Beckett, 1981]. Alteration of the constituents of the medium enabled Peehl and Ham [1980] and Tsao et al. [1982] to culture keratinocytes from human foreskin selectively without feeder layers or serum, and others have shown that reductions in pH [Eisinger et al., 1979], Ca^{2+} [Peehl and Ham, 1980], and temperature [Miller et al., 1980] may all contribute to improved selective growth of epidermal keratinocytes. Addition of hydrocortisone (1.4×10^{-7} M) [Peehl and Ham, 1980], 10^{-10} M cholera toxin, or 10^{-6} M isoprenaline (isoproterenol) and epidermal growth factor (10 ng/ml) [Rheinwald and

TABLE 20.1. Inhibition of Fibroblastic Overgrowth

Method	Agent	Tissue	Reference
Selective detachment	Trypsin	Fetal intestine, cardiac muscle, epidermis	Polinger [1970], Owens et al. [1974], Milo et al. [1980]
	Collagenase	Breast carcinoma	Lasfargues [1973]
Selective attachment and substrate modification	Polyacrylamide	Various tumors	Jones and Haskill [1973, 1976]
	Teflon	Transformed cells	Parenjpe et al. [1975]
	Collagen (pigskin)	Epidermis	Freeman et al. [1976]
Confluent feeder layers	Mouse 3T3 cells	Epidermis	Rheinwald and Green [1975]
	Fetal human intestine	Breast epithelium, normal and malignant	Stampfer et al. [1980]
		Colon carcinoma	Freshney et al. [1982b]
Selective media	D-valine	Kidney	Gilbert and Migeon [1975, 1977]
	Cis-OH-proline	Cell lines	Kao and Prockop [1977]
	Ethylmercurithiosalicylate	Neonatal pancreas	Braaten et al. [1974]
	Phenobarbitone	Liver	Fry and Bridges [1979]
	MCDB 153	Epidermis	Boyce and Ham [1983]
	Antimesodermal antibody	Squamous carcinomas	Edwards et al. [1980]
		Colonic adenoma	Paraskeva et al. [1985]
	MCDB 170	Breast	Hammond et al. [1984]
	Low Ca^{2+}	Epidermal melanocytes	Naeyaert et al. [1991]

Green, 1977] to the medium has made continued serial subculture possible over many cell generations [Green et al., 1979]. When mouse sublingual epidermal cultures were grown on collagen rafts (see also under "liver") at the gas–liquid interface, complete histological maturation was possible [Lillie et al., 1980].

The following protocol for the culture of epidermal cells has been contributed by Norbert E. Fusenig, Division of Carcinogenesis and Differentiation, German Cancer Research Center, Im Neuenheimer Feld 280, 6900 Heidelberg, Germany.

Principle

Protease digestion of skin samples leads to a separation between epidermis and dermis with the split level being partly beneath and partly above the basal cell layer depending on tissue sample and incubation temperature. Single cells are obtained by mechanical dispersion.

Keratinocytes of animal and human skin have been grown in primary culture and for limited numbers of subcultures using a variety of substrata, culture media, and additives including feeder layers [for review, see Holbrock and Hennings, 1983; Karasek, 1983; Fusenig, 1986]. Cultures tend to stratify, forming differentiating multilayers when grown at normal (1.4 m*M*) Ca^{2+} concentrations, but stay essentially as "undifferentiated" monolayers in media with low (below 0.1 m*M*) Ca^{2+} concentrations. Maintenance at a high cell density or at low density in the presence of confluent feeder cells (e.g., 3T3) (see Chapter 11) helps to reduce fibroblast contamination.

Outline

Separate dermis from epidermis by trypsin digestion, collect keratinocytes by pipetting epidermis, and propagate in serum-free low Ca^{2+} medium or on growth-arrested fibroblast/3T3 feeder cells in normal Ca^{2+} with serum.

Materials

Sterile:

Tissue culture media:

(1) Eagle's MEM with a 4x concentration of vitamins and amino acids (essential and nonessential) (4xMEM) and 10–15% heat-inactivated fetal bovine serum (FBS) [Fusenig and Worst, 1975]

(2) The same 4xMEM medium with Ca^{2+}-free HBSS, Chelex- treated FCS [Hennings et al., 1980], and the Ca^{2+} concentration adjusted by added $CaCl_2$ or normal FBS to a final concentration of 0.08 and 0.10 m*M* for mouse and human cells, respectively

(3) FAD medium [Wu et al., 1982] consisting of a mixture of 1 part Ham's F12 and 3 parts of DMEM, enriched with adenine (1.8×10^{-4} M), cholera toxin (10^{-10} M), EGF (10 ng/ml), hydrocortisone (0.4 μg/ml), and 5% FCS. All media contain antibiotics (penicillin 100 u/ml and Streptomycin 50 μg/ml)

Trypsin (1:250), 0.3%, 0.2%

HBSS

DBSS (HBSS + antibiotics; see Reagent Appendix)

PBSA

100-mm bacteriological grade petri dishes

Scalpels, curved forceps

Protocol

1. *Specimens:* Neonatal as well as juvenile foreskin samples from surgery are commonly used. Larger samples can be obtained from postmortem (up to 48 hr) abdominal skin. To prevent infection of cultures, biopsies should be rinsed 5–10 times in DBSS (see Reagent Appendix) or larger samples can be incubated for 30 min in betadine solution (10% iodine) followed by two rinses in PBS for 10 min each.
2. To provide better and consistently good access of the enzyme to the epithelial–mesenchymal border zone, split-thickness skin is prepared by a Castroviejo dermatome (Storz Intrument) set to 0.1–0.2 mm. The skin specimens are dissected into pieces of equal size (approximately 1 × 2 cm) with scalpels. Alternatively, subcutaneous tissue and part of the dermis can be eliminated mechanically. For this purpose the specimens are placed epidermis-side down into dry bacterial plastic petri dishes and irrigated with a few drops of PBS. The subcutaneous and lower dermal tissue is cut off as much as possible with curved scissors.
3. The tissue samples are rinsed again (2–5×) in sterile Ca^{2+}- and Mg^{2+}-free PBS. Split-thickness skin samples are floated on 0.3% trypsin, 0.025% EDTA in PBSA (pH 7.4) for 30 min at 37°C. Separation of full-thickness skin, with subcutaneous tissue and part of the dermal tissue removed mechanically, is better when the skin is floated on ice-cold 0.2% crude trypsin at 4°C for 15–48 hr (depending on the thickness of the skin). The pH of the trypsin has to be monitored by added indicator dye (e.g., phenol red) to avoid pH shift leading to altered enzyme activity and loss in cell viability.
4. When the first detachment of epidermis is visible at the cut edges of skin samples, the pieces are placed (dermis-side down) in 100-mm plastic petri dishes irrigated with 5 ml complete

culture medium including serum. With two fine curved forceps the epidermis is gently peeled off and pooled in a 50-ml centrifuge tube containing 20 ml complete culture medium. Viable keratinocytes are detached from the epidermal parts by vigorous pipetting and sieving through nylon gauze (100-μm mesh).

5. The remaining dermal part is gently scraped with curved forceps on its epidermal (upper surface) to remove loosely attached basal cells. The isolated cells from dermal and epidermal parts are combined, passed through a nylon gauze (100-μm mesh), washed twice in culture medium by centrifugation at 100 *g* for 10 min, and counted for total and viable (trypan blue excluding) cells (see Chapter 19). Scraping of the dermal surface yields higher cell numbers, but will also give rise to more contamination by mesenchymal cells. Moreover, when trypsinization is performed at 37°C, scraping of the upper surface of the dermis does not substantially increase cell yield, since splitting occurs mostly at the basal lamina.
6. Cells are plated at 37°C in medium (1) or (3) at $1-5 \times 10^5$ cells/cm^2 and left undisturbed for 1–3 d, depending on species and isolation procedure, for attachment and spreading. In order to obtain higher proliferative activities and yield higher cell numbers with lower seeding densities, seed $1-2 \times 10^4$ cells/cm^2 on dishes containing growth-arrested mesenchymal feeder cells. Instead of murine 3T3 cells, as originally described by Rheinwald and Green [1975], normal human skin fibroblasts, or capillary endothelial cells, can be used, after irradiation with 70 Gy or treatment with mitomycin C [Limat et al., 1989].
7. When cells have attached, cultures are extensively rinsed with medium to eliminate nonattached dead and differentiated cells and cultivation is either continued in medium (1) or (3) for long-term growth of stratified cultures or changed to medium (2) and continued at low Ca^{2+} concentration.
8. The shift to low Ca^{2+} culture medium (2) is done by rinsing 1–3-day-old primary cultures twice with low-Ca^{2+} medium and further incubating cultures in this medium. Remaining cell aggregates or differentiated cells sticking to the monolayer detach in this medium within a few days so that subsequent cultures are essentially monolayer.
9. For subcultivation, cultures in medium (1) and (3) are first incubated in 0.05–0.1% EDTA to initiate cell detachment for 10–30 min until cells start to round up, visible by the enlargement of intercellular spaces. Final detachment is achieved by incubation in 0.1% trypsin and 0.05% EDTA and pipetting; the latter procedure alone is sufficient for cultures in medium (2).
10. Human and mouse cells can be grown in all three media for several months. While mouse cells can only be subcultured once or twice, human cells can be passaged for four and seven times in media (2) and (3), respectively [see Boukamp et al., 1988].
11. Cultured cells have to be characterized for their epidermal (epithelial) nature to exclude contamination by mesenchymal cells. This is best achieved using cytokeratin-specific antibodies for epithelium and antivimentin for mesenchyme-derived cells. Further identification criteria for other epidermis-derived cells in keratinocyte cultures are available by histochemical and immunocytochemical methods [for review, see Fusenig, 1986].

Variations

Keratinocytes can also be grown at clonal density cocultured with x-irradiated 3T3 cells [Rheinwald and Green, 1975], at reduced Ca^{2+} concentrations with fibroblast conditioned medium [Yuspa et al., 1981], and in defined serum-free medium [Tsao et al., 1982].

In order to provide more *in vivo*-like growth conditions, "organotypic" culture systems have been developed by seeding the cells on collagen gels or pieces of dermis and lifting these supports to the air–medium interface [for review, see Prunieras et al., 1983; Fusenig, 1986]. Under these improved growth conditions, keratinocytes express many aspects of growth and differentiation of the epidermis *in vivo* that are less absent or pronounced in submerged cultures on plastic.

Growth under mesenchymal influence can be provided by transplantation of cell suspensions or intact cultures *in vivo* or by recombining cultures with mesenchyme *in vitro* [for review, see Fusenig, 1986, 1992]. This leads to an almost complete expression of growth and differentiation characteristics of normal epidermis [Bohnert et al., 1986].

Alternative methods have been described for cell isolation in large quantities and further purification with Ficoll density gradients [Fusenig and Worst, 1975] or with Percoll gradients [Brysk et al., 1981], methods that are particularly useful for newborn rodent epidermis. For descriptions of other procedures, see Yuspa et al. [1980] and Fusenig [1986].

Combined cultures of epidermis and stroma, mounted on mesh filters, are now commercially available (Skin[2]), and have been suggested as models for irritancy and inflammation. Similarly, cultures have been prepared from cornea with a similar objective. The following protocol for culture of corneal epithelium has been provided by Carolyn Kahn, Gillette Medical Evaluation Laboratories, 401 Professional Drive, Gaithersburg, MD 20879.

Corneal Epithelial Cells: Culture in Serum-Free Medium of Normal Human Corneal Cells and Cell Lines With Extended Life-Span

Principle

The corneal epithelium directly contacts the external environment and is the first tissue compromised in ocular injury. Animal models are most frequently used to model the human ocular surface. The following system is being developed for *in vitro* toxicological investigations, which can complement *in vivo* studies.

The corneal epithelium *in vivo* is oxygenated by diffusion from the tear film; in addition it undergoes frequent mitosis. These two characteristics, coupled with the availability of donor cornea tissue, combine to make the corneal epithelium a good starting material for the generation of primary cultures.

One may take advantage of extant sources for tissue acquisition; a sophisticated eye-banking system exists to provide donor corneas for corneal transplants. Eye-bank tissue, which is downgraded after 3 days of storage, is then made available to researchers. This is due to the limited *in vitro* viability of the endothelial cell layer. While the endothelium is labile, the epithelium retains its generative capacity for extended periods of time in storage at 4°C, making expired tissue suitable as a source of viable epithelium.

Outline

Corneal tissue is placed on collagen, allowed to adhere, and the outgrowth is expanded and propagated on fibronectin-collagen-coated surfaces. Early passage cultures are transfected with SV40 early-region genes or infected with intact Ad12-SV40 hybrid virus. Large T-antigen apparently confers an extended life-span.

Materials

Biocoat six-well plate precoated with rat-tail collagen, type I (Collaborative Research, CA)
Keratinocyte serum-free medium (KGM; Clonetics, Irvine, CA) containing 0.15 m*M* calcium, human epidermal growth factor (0.1 ng/ml), insulin (5 μg/ml), hydrocortisone (0.5 μg/ml), and bovine pituitary extract (30 μg/ml)
Dulbecco's phosphate-buffered saline (GIBCO, Grand Island, NY) (PBSA)
Fetal bovine serum (GIBCO, Grand Island, NY)
Fibronectin/Collagen (FNC) (Bethesda Research Faculty and Facility, Ijamsville, MD). This solution consists of fibronectin, 10 μg/ml, collagen, 35 μg/ml, with bovine serum albumin (BSA), 100 μg/ml, added as a stabilizer.
0.05% trypsin-EDTA (Life Technologies)
Eagle's Minimal Essential Medium (MEM/Life Technologies)
Vero cells (ATTC CCL 81)
Lipofectin (Life Technologies)
DMSO (Sigma)

Protocol: (a) Primary cultures

1. Place donor corneas epithelial-side up on a sterile surface and cut into 12 triangular-shaped wedges, using a single cut of the scalpel and avoiding any sawing motion. Careful handling of the cornea in this manner decreases damage to the collagen matrix of the stroma and minimizes liberation of fibroblasts.
2. Turn each corneal segment epithelial-side down and place four segments in each well of a six-well tray (precoated with rat-tail collagen, type I, Biocoat).
3. Press each segment down gently with forceps to ensure good contact between the tissue and the tissue culture surface and allow tissue to dry for 20 min.
4. Place one drop of keratinocyte serum-free medium carefully upon each segment and incubate overnight at 37°C in 5% CO_2. Although the donor corneas received from the eye bank are stored in antibiotic-containing medium (either McCarey-Kauffman or Dexsol), all manipulations are performed under antibiotic-free conditions.
5. The following day, add 1 ml of medium to each well.

   During the initial culture period, emigration of cells is observed only from the limbal region of the cornea. No cells are observed to migrate away from the central cornea or the sclera. Fibroblast outgrowth is minimized by utilizing a serum-free medium low in calcium (0.15 m*M*) and minimizing disruption of the collagen matrix.
6. Remove the tissue segment with forceps 5 d after the explantation of the donor cornea slice, and add 3 ml of medium.

   After donor tissue is removed with sterile forceps, adherent cells continue to proliferate and, within 2 wk from the time of establishment of the culture, confluent monolayers form, displaying the typical cobblestone morphology associated with epithelia. The yield is approximately 6 $\times 10^6$ cells/cornea.

(b) Propagation

1. Following the initial outgrowth period, feed cultures twice per week.
2. At 70–80% confluence, rinse cells in Dulbecco's phosphate-buffered saline (PBSA) and release with trypsin/EDTA (0.05% trypsin, 0.53 m*M* EDTA) for 4 min at 37°C.
3. Stop the reaction with 10% FBS in PBSA.
4. Wash cells (centrifugation followed by resuspension in KGM), count, and plate at 1 $\times 10^4$ cells/cm^2 onto tissue culture surfaces coated with FNC.
5. Incubate at 37°C in 95% air, 5% CO_2.
6. Exchange culture medium with fresh medium 1 d after trypsinization and reseeding.

   Immediately after passage, cells appear more spindle shaped, are refractile, and are highly migratory. Within 7 d, control cultures become 70–80% confluent, continue to display a cobblestone morphology, and if allowed to become postconfluent, the cultures retain the ability to stratify in discrete areas.

   Although corneal epithelial cultures can be expanded until P5 (approximately 9–10 population doublings), most of the proliferation occurs between passages one and three. Approximate yields are 1.25 $\times 10^6$ cells/cornea. Senescence always ensues by P5.

Development of Continuous Cell Lines

In order to provide a larger supply of cells for experimental purposes, it is possible to develop lines of human corneal epithelium (HCE) with extended lifespan.

1. Grow early-passage cultures of corneal epithelium to 60% confluence in four 25-cm^2 flasks.
2. Remove medium from flasks and add 5 ml of medium containing Ad12-SV40 hybrid virus, at a multiplicity of infection of approximately 100, to three of the flasks.

   ◊ Follow local guidelines for potentially biohazardous material (see Chapter 6).
3. Incubate cells with virus overnight at 37°C and change the medium the following day, and twice weekly thereafter. The control flask receives medium only.
4. Passage cultures as they approach confluence, 2–5 days after inoculation.

   The control flask begins to undergo senescence by passage 5, while successfully infected cultures develop foci of growth between weeks 4 and 6. Successfully infected cultures continue to propagate past P5.

To assess whether lines are shedding virus, culture samples of supernatant with Green Monkey kidney (Vero) cells using the method of Rhim et al. [1981].

1. Seed Vero cells in six-well plates at 20% confluence in Earle's Minimal Essential Medium (EMEM) containing 5% FBS.
2. Following 24 hr incubation, remove medium and add 0.5 ml "spent" filtered medium from each culture, along with 2.5 ml of fresh medium, to a test well. Each plate contains one control well that received "spent" medium from a primary culture.
3. Feed Vero cultures with fresh medium twice weekly for 21 d, at which time cytopathic effects (CPE) are scored. CPE is defined as any culture that appears different from the control.

Early-passage cultures can also be transfected using the method of Felgner et al. [1987]. Twenty-five-cm^2 flasks at 60% confluence are exposed to plasmid RBV-T, a SV40 ori-construct containing the SV40 early-region genes and the Rous sarcoma virus long terminal repeat, as the transfecting agent.

1. Mix Lipofectin (30 μg/50 μl distilled water) 1:1 with plasmid DNA (10 μg/50 μl distilled water) in polystyrene tubes.
2. After gentle agitation, allow the mixture to stand at room temperature for 15 min.
3. Add 2 ml of medium to the flask and 0.1 ml of Lipofectin mixture dropwise to each 25-cm^2 flask of epithelial cells.
4. Incubate at 37°C overnight and replace the medium. Feed cultures twice weekly thereafter. Control cultures receive Lipofectin only.

There are no morphological differences observed between the virally infected and transfected cultures. Both are large T antigen immunoreactive. The confluent monolayers formed by the primary cultures and the HCE lines appear similar upon phase contrast photomicroscopy.

Long-Term Storage of Cells

Cells can be stored frozen in liquid nitrogen.

1. Disperse healthy cultures with trypsin and pellet gently (60 *g* $\times$ 5 min).
2. Resuspend in medium containing 10% DMSO and 10% FBS.
3. Aliquot at 1–2 $\times 10^6$ cells/ml.
4. Freeze at a rate of 1°C/min using a controlled-

temperature freezing apparatus (see Chapter 17).

5. Thaw cells in 37°C water (see Chapter 17) until all but a small bit of ice remains. Add warm FBS to the vial and dilute the contents into a 15-ml tube filled with warm serum-free medium.
6. Wash the cells gently by centrifugation and plate at 1×10^4 cell/cm^2.

Phenotypic development *in vitro*. Both primary cultures and HCE lines retain phenotypic characteristics of corneal epithelium *in situ*. They continue to synthesize collagenase and express EGF receptors and corneal-specific cytokeratins, although the level of expression under the current culture conditions is less than that observed *in situ*. When cultured upon collagen membranes at air–liquid interfaces, the morphology and barrier function of corneal epithelium is fairly well preserved. Stratified membranes develop that are able to inhibit the diffusion of Na-fluorescein. Air–liquid interface cultures survive for 2 wk *in vitro* and have been used to investigate injury and repair mechanisms. Studies are underway to reconstitute cells into more complete three-dimensional tissue models by supplying corneal fibroblasts in a collagen gel.

Human corneal epithelial cells propagated *in vitro* may provide a suitable model to explore basic cell biological mechanisms as well as toxicological phenomena. Although the primary cultures are adequate for *in vitro* studies, HCE lines with extended life-span provide a reliable source of material that can be shared among laboratories.

Breast

Milk [Buehring, 1972; Ceriani et al., 1979] and reduction mammoplasty are suitable sources of normal ductal epithelium from breast, the first giving purer cultures of epithelial cells. Growth on confluent feeder layers of fetal human intestine [Stampfer et al., 1980; Freshney et al., 1982b] represses stromal contamination with both normal and malignant tissue (see Fig. 11.8), and optimization of the medium [Stampfer et al., 1980; Smith et al., 1981; Hammond et al., 1984] enables serial passage and cloning of the epithelial cells. Cultivation in collagen gel allows three-dimensional structures to form; these correlate well with the histology of the original donor tissue [Yang et al., 1979, 1980, 1981].

As with epidermis, cholera toxin [Taylor-Papadimitriou et al., 1980] and EGF [Osborne et al., 1980] stimulate the growth of epithelioid cells from normal breast *in vitro*.

The hormonal picture is more complex. Many epithelial cells survive better with insulin added to the culture (1–10 IU/ml) in addition to hydrocortisone ($\sim 10^{-8}$ M). The differentiation of acinar breast epithelium in organ culture requires hydrocortisone, insulin, and prolaelin [Stockdale and Topper, 1966], and requirements for estrogen, progesterone, and growth hormone have been demonstrated in cell culture [Klevjer-Anderson and Buehring, 1980].

The following protocol for the culture of cells from human milk was contributed by Joyce Taylor-Papadimitriou, Imperial Cancer Research Fund, Lincoln's Inn Fields, London, UK.

Principle

Early-lactation and postweaning milk, which give the highest cell yield, contain clumps of epithelium that can proliferate in culture [Buehring, 1972; Taylor-Papadimitriou et al., 1980]. Primary cultures, grown in hormone-supplemented human serum containing medium, give cell lines of limited life-span but clonogenic [Stoker et al., 1982]. These are eventually overtaken by nonepithelial "late milk" cells [McKay and Taylor-Papadimitriou, 1981].

Outline

Cells centrifuged from early-lactation milk are grown in the presence of endogenous macrophages in an enriched medium. These may be subcultured with a mixed protease chelating solution.

Materials

Sterile:

Nunc plastic dishes, 5 cm

Universal containers or 20–50-ml centrifuge tubes

Growth medium RPMI 1640 (GIBCO)

Fetal calf serum (FCS, Flow Labs)

Human serum (HuS; outdated pooled serum from blood banks; Australian antigen negative)

Stock solutions:

Insulin (Sigma), 1 mg/ml in 6 m*M* HCl

Hydrocortisone, 0.5 mg/ml in physiological saline

Cholera toxin (Schwartz-Mann), 50 μg/ml in physiological saline

Note. Serum and stock solutions of insulin, hydrocortisone, cholera toxin, pancreatin, and trypsin should be kept at −20°C

Trypsinization solution (TEGPED):

10 ml 0.5% EGTA (Sigma, ethylene glycol-bis-(β-aminoethylether) N'N' tetraacetic acid) in PBSA

4 ml 0.02% EDTA (Sigma, diaminoethane tetraacetic acid) in PBSA

4 ml 0.2% trypsin (Difco) in Hanks' balanced salt solution (HBSS)

2 ml 1.0% pancreatin (Difco) in HBSS

Growth medium:

RPMI 1640 containing 15% FCS, 10% HuS, cholera

toxin 50 ng/ml, hydrocortisone 0.5 μg/ml, insulin 1 μg/ml

Protocol

Milk collection: Milk (2–7 d postpartum) can best be collected on hospital wards. The breast is swabbed with sterile H_2O and the milk manually expressed into a sterile container. Five to 20 ml are usually obtained per patient. The milks are pooled and diluted 1:1 with RPMI 1640 medium to facilitate centrifugation.

Primary cultures:

1. Spin diluted milk at 600–1,000 g for 20 min. Carefully remove supernatant, leaving some liquid so as not to disturb the pellet.
2. Wash the pelleted cells two to four times with RPMI containing 5% FCS until supernatant is not turbid.
3. Resuspend the packed cell volume in growth medium and plate 50 μl packed cells in 5-cm dishes (Nunc) in 6 ml growth medium. Incubate at 37°C in 5% CO_2.
4. Change medium after 3–5 d and thereafter twice weekly. Colonies appear around 6–8 d and expand to push off the milk macrophages, which initially act as feeders.

Subculture of milk cells: Incubate in TEGPED (1.5 ml per 5-cm plate) at 37°C for 5–15 min, depending on the age of the culture, to produce a single-cell suspension that can be diluted one-third in fresh medium for replating (after washing free of enzyme mixture).

Variations

It is convenient to use the macrophages that are already present in the milk as feeders, and these are gradually lost as the epithelial colonies expand. However, macrophages can be removed by absorption to glass, and in that case other feeders must be added. Irradiated or mitomycin-treated 3T6 cells (see Chapter 11) show the best growth-promoting activity [Taylor-Papadimitriou et al., 1977b]. Analogues of cyclic AMP can be used to replace the cholera toxin [Taylor-Papadimitriou et al., 1980], although this is not possible with macrophage feeders, which are killed by the analogues.

Uses and application of milk epithelial cell culture. Milk cultures provide cells from the fully functioning gland and allow definition of phenotypes by immunological markers [Chang and Taylor-Papadimitriou, 1983]. They have been successfully transformed by SV40 virus [Chang et al., 1982] and provide an important source of normal cells for comparison with breast cancer cell lines and for transfection with oncogenes.

Cervix

A modification of the epidermal culture technique can be used for the propagation of cervical epithelium [Stanley and Parkinson, 1979].

The following protocol for the culture of epithelial cells from cervical biopsy samples has been contributed by Margaret Stanley, Department of Pathology, University of Cambridge, UK.

Principle

Cervical keratinocytes can be grown in serial culture at clonal density using a modification [Stanley and Parkinson, 1979] of the method described for epidermal keratinocytes [Rheinwald and Green, 1975].

Outline

Single cell suspensions, from enzymatically disaggregated epithelium from punch or wedge cervical biopsies, are inoculated into flasks or plates together with lethally inactivated Swiss 3T3 cells and grown in serum and growth factor supplemented medium. When the keratinocyte colonies that arise contain 1,000 cells or more, the cultures can be trypsinized and passaged again using fibroblast feeder support.

Materials

Sterile:

Transport medium: Dulbecco's modification of Eagle's medium (DMEM) supplemented with 10% fetal calf serum, 100μg/ml gentamicin sulfate and 10 μg/ml amphotericin

Trypsin solution for disaggregation of epithelial cells: 0.25% trypsin (v/v), 0.01% EDTA as the disodium salt pH 7.4 in PBS

Keratinocyte culture medium (KCM): DMEM supplemented with 10% fetal bovine serum, 0.5 μg/ml hydrocortisone, 10^{-10}M cholera toxin

Fetal bovine serum: not all batches support growth adequately. Serum samples should be tested and a large batch bought of suitable quality.

Epidermal growth factor (EGF): Sigma 100 mg dissolved in 1 ml sterile distilled water. Aliquot in 100 μl lots and store at -20°C. Working stocks 100 μl of EGF at 100 μg/ml in 10 ml medium/serum. Store at 4°C. Use at a final concentration of 10 ng/ml.

Cholera toxin (CT): Sigma. Add 1.18 ml sterile distilled water to 1 mg CT to give a 10^{-5} M solution. Store at +4°C.

Working stocks 100 μl 10^{-5} M CT in 10 ml

medium with serum. Filter sterilize and keep at +4°C. Use at a final concentration of 10^{-10} M.
Hydrocortisone: Sigma. Dissolve 1 mg hydrocortisone in 1 ml 50% ethanol/water (V/V). Aliquot into 100-μl lots and keep at −20°C. Use at a final concentration of 0.5 μg/ml.
Tissue culture petri dishes, 60 mm and 90 mm diameter
Forceps, rat toothed
Curved iris scissors
Disposable scalpels, No. 22 blade
Pipettes
Centrifuge tubes
Nonsterile:
Hemocytometer
Radiation source, e.g., x-rays or ^{60}Co

Swiss 3T3 Fibroblasts

A large master stock of cells should be prepared and frozen down in individual ampules of 10^6 cells. Cells should not be used for more than 20 passages. Grow 3T3s in DMEM/10% calf serum in 175-cm^2 tissue culture flasks. Inoculate cells at 1.5 × 10^4 cells/cm^2. Medium change after 2 d. Passage every 4–5 d. To avoid low-level contamination, maintain one master flask of cells on antibiotic-free medium; these cells are then used at each passage to inoculate the flasks required for that week's feeder cells.

Feeder layers are inactivated by irradiating with 60 Gy (6,000 rad) either from an x-ray or ^{60}Co source. Irradiated cells (XR-3T3) may be kept at 4°C for 3–4 d.

In the absence of an irradiation source, inactivate feeder cells with mitomycin C. Expose 3T3 cells growing in monolayer to mitomycin C, 400 μg/ml for 1 hr at 37°C. Trypsinize off the treated cells, wash the cell pellet 2× with fresh medium/serum, resuspend at a suitable cell concentration in complete medium, and use.

Protocol

1. Remove the cervical biopsy from transport medium and wash 2–3x with 5 ml sterile PBS containing gentamicin sulfate 50 μg/ml and amphotericin 5 μg/ml.
2. Place the biopsy, epithelial surface down, on a sterile culture dish. Using a disposable scalpel fitted with a No. 22 blade, cut and scrape away as much of the muscle and stroma as possible, leaving a thin opaque epithelial strip.
3. Mince the epithelial strip *finely* with curved iris scissors.
4. Add 10 ml trypsin/EDTA (prewarmed to 37°C) to the epithelial mince and transfer to a sterile glass universal containing a plastic coated magnetic flea.
5. Add a further 5–10 ml of trypsin/EDTA.
6. Place the universal on a magnetic stirrer in an incubator or hot room at 37°C and stir slowly for 30–40 min.
7. Allow the suspension to stand at room temperature for 2–3 min.
8. Remove the supernatant containing single cells and filter through a stainless steel mesh into a 50-ml centrifuge tube (see Chapter 9). Add 10 ml complete medium to this.
9. Add a further 15 ml warm trypsin EDTA to the fragments in the universal and repeat the above procedure twice. Combine the trypsin supernatants, spin in a bench centrifuge at 1,000 rpm (80 *g*) for 5 min.
10. Remove the supernatant; add 10 ml of complete medium to the pellet, resuspend vigorously to give a single cell suspension, and count the cells in a hemocytometer. Assess cell viability with trypan blue exclusion.
11. Dilute the cervical cell suspension with KCM and plate out at 2 × 10^4 cells/cm^2 together with 10^5 cells/cm^2 lethally inactivated 3T3 cells (i.e., 10^5 cervical cells: 5 × 10^5 XR-3T3/60-mm dish).
12. Incubate the cultures at 37°C in 5% CO_2.
13. Seventy-two hours after the initial plating, change the medium; replace with complete medium supplemented with EGF at 10 ng/ml. Check the cultures microscopically to ensure that the feeder layer is adequate. Add further feeder cells if necessary.
14. Change the medium twice weekly; EGF should be present in the medium except when the cells are initially plated. Keratinocyte colonies become visible by days 8–12 and by days 14–16 should be visible to the naked eye. Cultures should be passaged at this time.

Passage

1. Remove the medium from the cell layer and remove the feeders by rinsing rapidly with 0.01% EDTA. Wash twice with PBS.
2. Add to each culture dish enough prewarmed trypsin/EDTA to cover the cell sheet. Leave at 37°C until the keratinocytes have detached: This should be checked microscopically. Do not leave the cells in trypsin for more than 20 min.
3. Remove the cell suspension from the plate and transfer to a sterile centrifuge tube.
4. Rinse the growth surface with complete medium and add to the suspension. Mix with a 10-ml pipette.

5. Spin at 1,000 rpm for 5 min.
6. Remove the supernatant, add 10 ml complete medium, and resuspend vigorously with a 10-ml pipette to achieve a single cell suspension.
7. Count the cells in a hemocytometer.
8. Cells may then be replated on inactivated 3T3 cells and grown as described above or frozen down for recovery later.

Cervical keratinocytes grown as described can be used for a range of investigations, including papillomavirus carcinogenesis, differentiation studies, and response to mutagens and carcinogens.

Gastrointestinal Tract

Culture of normal gut lining epithelium has not been extensively reported, although there are numerous reports in the literature of continuous lines from human colon carcinoma [Tom et al., 1976; Bergerat et al., 1979; Kim et al., 1979; Noguchi et al., 1979; Van der Bosch et al., 1981]. Colorectal carcinoma cells plated on confluent feeder layers of FHI (see below and Chapter 11) form colonies that apparently disappear but reappear 8–10 weeks later as nodules in the monolayer (Fig. 20.1). These nodules will increase and can be subcultured with or without feeder layers and, in several cases, have given rise to continuous cell lines, some of which produce carcinoembryonic antigen (CEA) and sialomucin, markers of neoplasia in human colon [Freshney et al., 1982b].

Owens et al. [1974] were able to culture cells from fetal human intestine as a finite cell line (FHS 74 Int). Similar results have been obtained in the author's laboratory, where more vigorous growth was observed in the epithelial cell component of these cultures and fibroblastic cells were eventually diluted out, giving rise to a finite cell line, FHI. Unfortunately these lines are often difficult to maintain and this line has subsequently been lost.

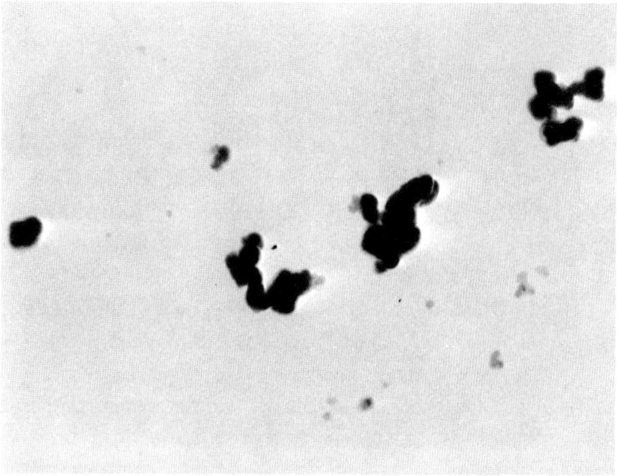

Fig. 20.1. *Colonies of colonic epithelium shed into medium after several weeks of cultivating colon carcinoma cells on a confluent feeder layer (FHI). (Courtesy of Dr. J.M. Russell.)*

The following protocol has been condensed from Paraskeva and Williams [1992], which should be consulted for greater detail and alternative methodology and applications.

Preparation of tissue specimens for cell culture is usually started within 1–2 hr of removal from the patient. If this is impossible, fine cutting of the tissue into small pieces (1–2 mm) with scalpels and storage overnight at 4°C in washing medium can also prove successful.

In general, adenomas are usually digested enzymatically and cancers can be dealt with simply by cutting with surgical blades. In the situation in which a well-differentiated colorectal cancer does not readily release tumor cells when the specimen is out, it can be digested enzymatically.

Materials

Sterile:

Growth medium: DMEM containing 2 m*M* glutamine and supplemented with 20% fetal bovine serum (batch selection is essential), hydrocortisone sodium succinate (1 μg/ml), insulin (0.2 u/ml), 2 m*M* glutamine, penicillin (100 u/ml), and streptomycin (100 μg/ml).

Washing medium: growth medium with 5% FBS, 200 u/ml penicillin and 200 μg/ml streptomycin, and 50 μg/ml gentamycin. Gentamycin is kept in the primary culture for at least the first week and then removed.

Digestion solution: DMEM and antibiotics, as described for the washing solution, with collagenase (1.5 mg/ml, Worthington type IV), hyaluronidase (0.25 mg/ml; Sigma type I), and between 2.5–5% FBS. Although we use Worthington collagenase, "Sigma" culture grades can also be tried.

Dispase for subculture: Prepare dispase (a neutral protease, Boehringer, Grade 1) at 2 u/ml in DMEM containing 10% FBS, glutamine, penicillin, and streptomycin. Sterile filter the solution and store at −20°C. If a precipitate forms on thawing, the solution should be centrifuged and the active supernatant should be removed and used.

Setting Up Primary Culture: (a) Enzyme Digestion Protocol

1. Wash tumor specimens four times in washing medium and mince in a small volume of the same medium, just enough to cover the tissue (do not allow the tissue to dry out). Mince the tissues with crossed surgical blades or sharp scis-

sors to tissue fragments of approximately 1 mm^3.

2. After cutting, wash the tissue again four times (the number of washings can be varied with experience depending on whether there are contamination problems) by bench centrifugation (300 g for 3 min) and resuspension.
3. After washing, put the tumor specimen into the digestion solution (see below) and rotate at 37°C, usually overnight (approximately 12–16 hr). The time in digestion is not critical, because it is a mild process, but it is important not to let the digestion medium become acid during the procedure. Such a result would indicate that it has been left too long or too much tissue was put in the volume of digestion mixture. Very approximately 1 cm^3 of tumor tissue is put into 20–40 ml of digestion solution.

(b) Nonenzymic Tissue Preparation
Adenomas almost invariably need digestion with enzymes. However, with carcinomas it is often found that during the cutting of the tumor with blades into 1-mm pieces, small clumps of tumor cells are released from the tumor tissue into the washing medium. In this case the following procedure can be carried out.

Protocol

1. Remove the washing medium containing released clumps of cells and separate the cells into large and small clumps by allowing them to settle by gravity for a few minutes in a centrifuge tube and remove the supernatant phase.
2. Put the remaining tissue pieces either directly into culture or rotate gently for 30–60 min in washing medium to release more small clumps, which can be removed and plated put into culture separately from the remaining larger pieces of tissue.
3. Wash all samples three times before putting into culture.

Standard primary culture conditions. Growth medium in 25-cm^2 flasks coated with collagen type IV in the presence of Swiss 3T3 feeder cells (approximately 1×10^4 cells/cm^2) at 37°C in a 5% CO_2/95% air incubator. Change culture medium twice weekly.

Inoculate epithelial tubules and clumps of cells derived from tissue specimens into flasks in 4 ml of medium per 25 cm^2. They start to attach to the substratum, and epithelial cells migrate out within 1–2 d. Most of the tubules and small clumps of epithelium attach within 7 d, but the larger organoids can take up to 6 wk to attach, although they will remain viable during this period.

Attachment of epithelium during primary culturing and subculturing is more reproducible and efficient when cells are inoculated onto collagen-coated flasks (see Chapters 7 and 14), and significantly better growth is obtained with 3T3 feeders than without.

When the epithelial colonies expand in size to several hundred cells per colony, they become less dependent on 3T3 feeders and addition of further feeders is not necessary.

Subculture and propagation. Most colorectal adenoma primary cultures and adenoma-derived cell lines cannot, at present, be passaged by routine trypsin/EDTA procedures [Paraskeva et al., 1984, 1989]. Trypsinization to single cells of the cultured adenoma cells with 0.1% in 0.1% EDTA will result in extremely poor or even zero growth. Instead dispase is used.

1. Add dispase to cell monolayer, just enough to cover the cells (~2.5 ml/25-cm^2 flask) and leave 40–60 min for primary cultures and 20–40 min for cell lines.
2. Once the epithelial layers begin to detach (they do so as sheets rather than single cells), pipette to help detachment and disaggregation into smaller clumps.
3. Wash and replate under standard culture conditions. It may take several days for clumps to attach, so replace medium carefully at feeding.

One of the most reproducible and successful methods for the short-term primary culture of *normal adult colonic epithelium* is the following, described previously by Buset et al. [1987].

Protocol

1. Biopsies of about 1–3 mm^3 are taken with biopsy forceps to sample only the mucosal layer and not the muscle layer.
2. After fine mincing with two scalpels, digest the tissue in a serum-free medium containing bovine serum albumin, neuraminidase, hyaluronidase, and collagenase for 10 min at 37°C and a further 1 hr at room temperature [Buset et al., 1987].
3. Wash the digested tissue and seed in serum-free NCTC 168 medium (K.C. Biologicals, Lenexa, KS) supplemented with ethanolamine, phosphoethanolamine, hydrocortisone, ascorbic acid, transferrin, insulin, epidermal growth factor, pentagastrin, and deoxycholic acid [Buset et al., 1987].

The best substrate for cell attachment is a mixture of ungelled collagen I and bovine serum albumin. However, under these conditions adult normal colonic epithelial cells still have a very short life of no more than 4 d.

Alternative methods for culture of tumor cells. Brattain and co-workers [1983] used C3H/10T½ cells as feeder layers for improved establishment of human colon cancers in tissue culture and for the enhancement of growth of continuous colon cell lines. McCoy's 5A medium supplemented with 10% heat-inactivated FBS was used for these studies.

For the successful primary culture and isolation of adenoma and carcinoma cell lines, Wilson et al. [1987] used a type I collagen gel with basic medium of MEM with Earle's salts supplemented with 0.1 m*M* nonessential amino acids, 2% FBS, insulin, transferrin, sodium selenite, hydrocortisone, and triiodothyronine. Optimal seeding efficiency of the adenomatous cells was obtained by using 1 part medium to 2 parts medium conditioned for 48 hr in either primary explant culture or subcultures of the cell line being passaged. Subculture was carried out by multiple 3-min exchanges (usually 4) of 2 ml of HBSS containing 0.04% EDTA at 37°C. Initial split ratios were not greater than 1:2.

Kirkland and Bailey [1986] obtained clumps of tumor cells by gently teasing pieces of carcinomas. No attempt was made to further investigate the released cell clumps, which were cultured in DMEM, 10% fetal bovine serum, and antibiotics. Primary cultures were then subcultured only when areas of tumor cell growth became very confluent, and for the first few passages, the entire contents of the 25-cm² flask were transferred to a fresh culture flask. Cells were subcultured with trypsin (Worthington, Biochemicals; 3x crystallized and dialyzed) 0.05% (w/v) in EDTA, but no attempt was made to produce a single cell suspension; instead clumps of tumor cells were transferred to fresh culture vessels at subculture. Fibroblasts were continually removed either mechanically by scraping cultures or by differential trypsinization with 0.025% trypsin (w/v) in EDTA. By these techniques, a high success rate for the isolation of new carcinoma cell lines was achieved [Kirkland and Bailey, 1986].

Serum-free conditions for the culture of some human colorectal cancer cell lines have been described [Fantini et al., 1987; Murakami and Masui, 1980], but these are not generally suitable for newly isolated carcinoma cultures, which require serum.

Fibroblast contamination. Several techniques can be employed to deal with fibroblast contamination and one or a combination of the following techniques can be used.

1. Physically remove a well-isolated fibroblast colony by scraping with a sterile blunt instrument (e.g., plunger of a sterile syringe). Care has to be taken to wash the culture up to six times to remove any fibroblasts that have detached to prevent them reseeding and reattaching to the flask.
2. Differential trypsinization can be attempted with the carcinomas [Kirkland and Bailey, 1986].
3. Dispase preferentially (but not exclusively) removes the epithelium during passaging and leaves behind most of the fibroblastic cells attached to the culture vessel [Paraskeva et al., 1984]. During subculture, cells that have been removed with dispase can be preincubated in plastic Petri dishes for 2–6 hr to allow the preferential attachment of any fibroblasts that may have been removed with the epithelium. Clumps of epithelial cells still floating can be transferred to new flasks under standard culture conditions. This technique takes advantage of the fact that fibroblasts in general attach much more quickly to plastic than clumps of epithelial cells, so that a partial purification step is possible.
4. The use of a conjugate between anti-Thy-1 monoclonal antibody and the toxin ricin [Paraskeva et al., 1985]. Thy-1 antigen is present on colorectal fibroblasts but not colorectal epithelial cells; therefore the conjugate kills contaminating fibroblasts but shows no signs of toxicity toward the epithelium, whether adenoma- or carcinoma-derived. This conjugate is unfortunately not yet available commercially.
5. Reducing the concentration of serum to about 2.5–5% in situations in which there are heavy concentrations of fibroblastic cells. It is worth remembering that the normal fibroblasts do have a finite growth *in vitro* and that using the combination of techniques above will eventually push the cells through so many divisions that they will senesce.

Liver

Although cultures from adult liver do not express all the properties of liver parenchyma, there is little doubt that the correct lineage of cells may be cultured. So far, attempts at generating proliferating cell lines have not been particularly successful, but functional hepatocytes can be cultured under the correct conditions [Guguen-Guillouzo, 1992].

The development of the correct conditions for perfusing liver *in situ* with collagenase and hyaluronidase [Berry and Friend, 1969] (see Chapter 9) provided a technique whereby a good viable suspension of liver parenchymal cells could be plated out with high purity and form a viable monolayer, and reports such as those of Malan-Shibley and Iype [1981] suggest that some epithelial cultures may be propagated.

Some of the most useful continuous liver cell lines were derived from Reuber H35 [Pitot et al., 1964] and Morris [Granner et al., 1968] minimal deviation hepatomas of the rat. Induction of tyrosine aminotransferase in these cell lines with dexamethasone proved to be a valuable model for studying the regulation of enzyme adaptation in mammalian cells [Granner et al., 1968; Reel and Kenney, 1968].

Pitot and others have demonstrated that, as with epidermis, greater functional expression can be obtained by culturing liver parenchymal cells on free-floating collagen sheets [Michalopoulos and Pitot, 1975; Sirica et al., 1980]. The cells are seeded in medium onto a preformed collagen gel on the base of a petri dish. After the cells have attached, the gel is released from the base of the dish with a bent Pasteur pipette or spatula, allowing it to float freely in the medium. This permits access of nutrients to the cell from above and below. It is possible that diffusion of nutrients and metabolites via the collagen layer is analogous to the situation *in vivo* where epithelial cells usually lie on basement membrane, and may be important in establishing a necessary polarity in the cells.

The following protocol for the culture of isolated adult hepatocytes has been contributed by Christiane Guguen-Guillouzo, Hopital de Pontchaillou, INSERM U49, Rue Henri le Guilloux, 35033 Rennes, France.

Principle

Proteolytic enzymes such as collagenase, which are relatively noncytotoxic, when perfused into the liver through the vessels and capillaries at an adequate flow rate, will disrupt intercellular junctions, and will digest the connective framework within 15 min, if the liver is previously cleared of blood and depleted of Ca^{2+} by washing with calcium-free buffer. Hepatocytes are selected from the cell suspension by two or three differential centrifugations.

Outline

Introduce a cannula in the portal vein or a portal branch, wash the liver with a calcium-free buffer (15 min), perfuse with the enzymatic solution (15 min), collect and wash the cells, and count the viable hepatocytes.

Materials

Sterile:

Tygon tube (ID 3.0 mm; OD 5.0 mm)
Disposable scalp vein infusion needles, 20G (Dubernard Hospital Laboratory, Bordeaux, France)
Sewing thread for cannulation
Graduated bottles and petri dishes
Surgical instruments (sharp, straight, and curved scissors and clips)
2 × 1-ml disposable syringes
Heparin (Roche)
Nembutal (Abbot 5%)
Calcium-free HEPES buffer pH 7.65: 160.8 m*M* NaCl; 3.15 m*M* KCl: 0.7 m*M* $Na_2HPO_4.12H_2O$, 33 m*M* HEPES, sterilization by 0.22-μm Millipore filters and storage at 4°C (2 months)
Collagenase (Sigma grade I; Boehringer 103578)
Collagenase solution: 0.025% collagenase; 0.075% $CaCl_22H_2O$ in calcium-free HEPES buffer pH 7.65; preparation and sterilization by filtration just before use
L-15 Leibovitz medium

Nonsterile:

Chronometer
Peristaltic pump (10 to 200 rpm)
Water bath

Protocol for Isolation of Rat Hepatocytes [Guguen-Guillouzo and Guillouzo, 1986]

1. Warm the washing HEPES buffer and collagenase solution in a water bath (usually approximately 38–39°C to achieve 37°C in the liver). Oxygenation is not necessary.
2. Set the pump flow rate at 30 ml/min.
3. Anesthetize the rat (180–200 g) by intraperitoneal injection of nembutal (100 μl/100 g) and inject heparin into the femoral vein (1,000 IU).
4. Open the abdomen, place a loosely tied ligature around the portal vein approximately 5 mm from the liver, insert the cannula up to the liver, and ligate.
5. Rapidly incise the subhepatic vessels to avoid excess pressure and start the perfusion with 500 ml calcium-free HEPES buffer, at a flow rate of 30 ml/min; verify that the liver whitens within a few seconds.
6. Perfuse 300 ml of the collagenase solution at a flow rate of 15 ml/min for 20 min. The liver becomes swollen.
7. Remove the liver and wash it with HEPES buffer; after disrupting the Glisson capsule, disperse the cells in 100 ml L-15 Leibovitz medium.
8. Filter through two-layer gauze or 60–80-μm nylon mesh, allow the viable cells to sediment for 20 min (usually at room temperature), and discard supernatant (60 ml) containing debris and dead cells.

9. Wash three times by slow centrifugations (50 g for 40 s) to remove collagenase, damaged cells, and nonparenchymal cells.
10. Collect the hepatocytes in Ham's F12 or Williams' E medium enriched with 0.2% bovine albumin (grade V, Sigma) and 10 μg/ml bovine insulin (80–100 ml).

Analysis
Determine cell yield and viability by the well-preserved-refringent shape or the trypan blue exclusion test (0.2% w/v) (usually 400 to 600 × 10^6 viable cells with a viability of more than 95%).

Isolation of hepatocytes from other species. The basic two-step perfusion procedure [Seglen, 1975] can be used for obtaining hepatocytes from various rodents, including mouse, rabbit, guinea pig, or woodchuck, by adapting the volume and the flow rate of the perfused solutions to the size of the liver. The technique has been adapted for the human liver [Guguen-Guillouzo et al., 1982] by perfusing a portion of the whole liver (usually at 1.5 l HEPES buffer and 1 l collagenase solution at 70 and 30 ml/min, respectively) or biopsies (15–30 ml/min, depending on the size). A complete isolation into a single cell suspension can be obtained by an additional collagenase incubation at 37°C under gentle stirring for 10 to 20 min (especially for human liver) [Guguen-Guillouzo and Guillouzo, 1986]. Fish hepatocytes can be obtained by cannulating the intestinal vein and incising the heart to avoid excess pressure. Perfusion is performed at room temperature at a flow rate of 12 ml/min.

Applications. Isolated parenchymal cells can be maintained in suspension for 4–6 hr and used for short-term experiments. They survive for a few days when seeded in the nutrient medium supplemented with 10^{-6} M dexamethasone on plastic culture dishes (7 × 10^5 viable cells/ml). A survival of several weeks is obtained by seeding the cells onto a biomatrix [Rojkind et al., 1980]. However, they rapidly lose their specific functions. A high stability (2 months) can be obtained by co-culturing hepatocytes with rat liver epithelial cells presumed to derive from primitive biliary cells [Guguen-Guillouzo et al., 1983]. Hepatocytes are also more stable when seeded on Matrigel [Bissell et al., 1987] and are capable of undergoing from one to three rounds of cell division when cultured in a medium supplemented with EGF and pyruvate [McGowan, 1986] or with nicotinamide and EGF [Mitaka et al., 1991].

Pancreas

There has not been the same effort expended on culture of pancreas as in liver. Pahlman et al. [1979] and Lieber et al. [1975] described neoplastic cell lines from exocrine pancreas; and Wallace and Hegre [1979] produced epithelial monolayers from fetal rat pancreas by a primary explant technique. These cultures remained free of fibroblasts for several days, and contained many endocrine cells (see also under "Endocrine Cells," this chapter).

The following protocol for the culture of pancreatic acinar cells has been contributed by Robert J. Hay and Maria das Gracas Miranda, American Type Culture Collection, 12301 Parklawn Drive, Rockville, MD 20852.

Principle
Fractionated populations of pancreatic acinar epithelia are inoculated onto collagen-coated culture dishes, and two-dimensional aggregate colonies are allowed to develop for subsequent study [Hay, 1979; Ruoff and Hay, 1979].

Outline
Dissociate guinea pig pancreatic tissue, filter the mixed cell suspension through cheesecloth or nylon sieves, and layer the suspension over a BSA solution. After three sequential centrifugation/fractionation steps, collect the cell pellet, inoculate onto collagen-coated culture vessels, incubate, and observe.

Materials
Sterile:
Conical flask, 25 ml (Bellco) siliconized
Pipettes, 5 or 10 ml wide-bore (Bellco)
Büchner funnel
Dialysis tubing (spectro-por)
250-ml sidearm flask
50-ml polypropylene centrifuge tubes
Cheesecloth
F12K tissue culture medium with 20% bovine calf serum (F12K-CS20)
Collagenase (GIBCO)
Trypsin 1:250 (Difco)
Dextrose (Difco, No. 0155-174)
HBSS
HBSS without Ca^{2+} and Mg^{2+} (HBSS-DVC)
Bovine serum albumin, fraction V, (Sigma)
Collagen-coated culture dishes (see Chapter 7)
Collagenase solution: Dissolve collagenase in 1X HBSS-DVC (pH 7.2) to give 1,800 U/ml. Adjust to pH 7.2 and dialyze using a 12-KDa exclusion membrane for 4 hr at 4°C, against 1× HBSS-DVC containing 0.2% glucose. The HBSS is discarded and this step is repeated for 16 to 18 hr with fresh HBSS-DVC. Filter, sterilize, and store in 10–20-ml aliquots at −70°C or below.
Trypsin solution: Trypsin 1:250, 0.25% in citrate

buffer; 3 g/l trisodium citrate; 6 g/l NaCl, 5 g/l dextrose; 0.02 g/l phenol red; pH 7.6.

Nonsterile:

Agitating water bath (Backer model M5B#11-22-1)

Centrifuge

Protocol

1. Make up the dissociation fluid proper first before use by mixing 1 part of collagenase with 2 parts of trypsin solution.
2. Aseptically remove the entire pancreas (0.5–1.0 g) and place in F12K-CS20. Trim away mesenteric membranes and other extraneous matter, and mince into 1–3-mm^3 fragments.
3. Transfer to a siliconized, 25-ml conical flask in 5 ml of prewarmed dissociation fluid. Agitate at about 120 rpm for 15 min at 37°C in a shaker bath. Repeat this dissociation step two or three times with fresh fluid until most of the tissue has been dispersed.
4. After each dissociation, allow large fragments to settle and transfer the supernate to a 50-ml polypropylene centrifuge tube with approximately 12 ml of cold F12K-CS20 to neutralize the dissociation fluid. Spin down at 600 rpm for 5 min and resuspend pellet in 5 to 10 ml of cold F12K-CS20. Keep cell suspension in ice.
5. Pool the cell suspensions and pass through several layers of sterile cheesecloth in a Büchner funnel. Apply light suction by inserting the funnel stem into a vacuum flask during filtration. Take an aliquot for cell quantitation.

   Generally, 1 to 2 $\times 10^5$ cells/mg of tissue are obtained at this step with 90% to 95% viability.
6. Layer 5 $\times 10^7$ to 5 $\times 10^8$ cells from the resulting fluid to the surface of 2 to 4 columns consisting of 35 ml of 4% cold BSA in HBSS-DVC (pH 7.2) in polypropylene 50-ml centrifuge tubes. Centrifuge at 600 rpm for 5 min. This step is critical to achieve good separation of acinar cells from islet, ductal, and stromal cells of the pancreas. Discard the supernate and repeat this fractionation step twice more, pooling and resuspending the pellets in cold F12K-CS20 after each step.
7. Collect the cell pellets in 5 to 10 ml cold F12K-CS20. Take an aliquot for counting. Yields of 2–5 $\times 10^4$ cells/mg of tissue are obtained, with 80–95% of the total being acinar cells. Inoculation densities can be 3 $\times 10^5/cm^2$. Cells adhere as aggregate colonies within 72 hr.

Variations

After cells have adhered, F12K-C20 can be replaced by serum-free MEM with insulin (10 ng/ml), transferrin (5 ng/ml), selenium (9 ng/ml), and EGF (21 ng/ml). This medium supports pancreatic cell growth for at least 15 d without altering cell morphology.

This method has been applied for studies with guinea pig and human (transplant donor) tissues. Addition of human lung irradiated feeder-fibroblasts produces a marked stimulation (up to 500%) in ^{3}H-thymidine incorporation and prolongs survival at least by a factor of two.

Kidney

The kidney is a structurally complex organ in which the system of nephrons and collecting ducts is made up of numerous functionally and phenotypically distinct segments. This segmental heterogeneity is compounded by cellular diversity that has yet to be fully characterized. Some tubular segments possess several morphologically distinct cell types. In addition, evidence points to rapid adaptive changes in cell ultrastructure that may correlate with changes in cell function [Stanton et al., 1981]. This structural and cellular heterogeneity presents a challenge to the cell culturist interested in isolating pure or highly enriched cell populations. The difficulty of the problem is further compounded for studies of the human kidney, where form and access to the specimen may make some manipulations such as vascular perfusion difficult or impossible.

Several approaches have been used successfully to culture the cells of specific tubular segments. Density gradient methods are now commonly used to isolate enriched populations of enzyme-digested tubule segments, and are particularly effective in establishing proximal tubule cell cultures from experimental animals [Taub et al., 1989]. Specific nephron or collecting duct segments can also be isolated by microdissection, then explanted to the culture substrate. This method, developed using experimental animals [Horster, 1979] has been applied to culture of human kidney [Wilson et al., 1985] and the cyst wall epithelium of polycystic kidneys [Wilson et al., 1986]. Immunodissection [Smith and Garcia-Perez, 1985] and immunomagnetic separation [Pizzonia et al., 1991] methods have also been developed to isolate specific nephron cell types on the basis of expression of cell type-specific ectoantigens. These elegant methods hold considerable promise for the study of specific kidney cell types in health and disease, but as yet have been applied almost exclusively to studies on experimental animals. The limited (and unscheduled) availability and inconsistent form (e.g., excised pieces, damaged vasculature, lengthy postnephrectomy period) of human donor kidneys make progressive enzymatic dissociation a more practical means to isolate human kidney cells for culture.

A number of methods for primary culture of human kidney have been reported [Detrisac et al., 1984; States et al., 1986; McAteer et al., 1991]. The following protocol was contributed by James McAteer, Stephen Kempson and Andrew Evan, Indiana University School of Medicine, Indianapolis, IN 46202-5120.

Principle

Primary culture of tissue fragments excised from the outer cortex of human kidney provides a means to isolate cells that express many of the functional characteristics of the proximal tubule [Kempson et al., 1989]. Progressive enzymatic dissociation and crude filtration yields single cells and small aggregates of cells that when seeded at high density give rise to a heterogeneous epithelial-enriched population. Large numbers of cells can be harvested, making it practical to establish multiple replicate primary cultures or to propagate cells for frozen storage. Experience with the method shows that the functional characteristics of such primary and subcultured cells are reproducible for kidneys from different donors, and that with proper handling of specimens good cultures can be derived from kidneys following even a lengthy post-nephrectomy period.

Outline

Tissue fragments are excised from the outer cortex, minced, washed, and incubated (with agitation) in collagenase-trypsin solution. The tissue is periodically pipette-triturated and the dissociated cells collected and pooled. The cell suspension is filtered through size-limiting screen to remove undigested fragments. The cells are then washed free of enzyme, suspended in serum-supplemented medium, and seeded to culture-grade plastic.

Materials

Sterile:

Growth medium DMEM/F12 (JRH Biosciences)
Fetal bovine serum (Sterile Systems)
BSS
0.1% collagenase (Worthington Type IV), 0.1% trypsin (1:250, Sigma) solution in 0.15 M NaCl
0.05 mg/ml DNAase in saline
Trypsin-EDTA solution: 0.1% trypsin (1:250), 0.03% EDTA (culture grade, Sigma)
Scalpels, tissue forceps
Nitex screen (160 μm) (Tetko)
Culture dishes, flasks (Corning)
Tubes, sterile 50 ml
Pipettes, 10 ml

Nonsterile:

Orbital shaker (Bellco)
Centrifuge, Centra 4-B (IEC)

Protocol: (a) Tissue Dissociation

1. Cut 5–10-mm-thick coronal slices of kidney and wash fragments in chilled basal medium.
2. Excise fragments from outer cortex and use crossed blades to mince the tissue into 1–2-mm^3 pieces.
3. Transfer approximately 5 ml of tissue fragments to a 50-ml tube containing 20-ml warm collagenase-trypsin solution. Secure the tube to the platform of an orbital shaker within a 37°C incubator.
4. Incubate with gentle agitation for 1 hr.
5. Discard and replenish the enzyme solution.
6. Subsequently, collect the supernatant at 20-min intervals following addition of 20 ml basal medium containing 0.05 mg/ml DNase and gentle trituration through a 10-ml pipette (10 cycles).
7. Dilute collected supernatant with an equal volume of complete culture medium and hold on ice.
8. Repeat (five or more times) this cell collection procedure until fragments are spent.
9. Pool the harvested supernatants and aliquot to 50-ml tubes.
10. Sediment by centrifugation (1,200 rpm, 15 min) and resuspend each pellet with 45 ml complete medium containing DNAase.
11. Filter the suspension through Nitex cloth (160 μm).

(b) Cell Plating, Subculture, Cryopreservation

The isolation protocol yields cell aggregates as well as single cells. As such, cell counting is unreliable. To propagate cells for subculture or cryopreservation, use 15 ml to seed 1 75-cm^2 flask. Subculture NHK-C cells by routine methods. Dissociation with trypsin-EDTA produces a monodisperse suspension suitable for hemocytometer counting. Seed culture-grade dishes at approximately 1×10^5 cells per cm^2. Cryopreservation is performed by routine methods using complete medium containing 10% dimethyl sulfoxide.

Comments and safety precautions. A prominent variable in this method is the quality of the kidney specimen itself. Procedures for tissue procurement are impossible to standardize. Donor tissue commonly includes segments of kidney collected at surgical nephrectomy (e.g., for renal cell carcinoma) and intact kidney that is judged unsuitable (e.g., anomalous vasculature) for transplantation. These specimens are collected under different conditions (e.g., with or without perfusion, varying periods of warm ischemia, different duration of the postnephrectomy period). In addition, differences in the age and health of the donor ensure that no two specimens are equivalent. Although not all kidneys yield satisfactory cultures, this protocol has been effective for a wide variety of specimens, including fresh surgical specimens and intact, perfused kidney held on wet ice for nearly 100 hr.

◇ Work involving any tissue or fluid specimen of human origin requires precautions to avoid unprotected contact with blood-borne pathogens.

Applications. NHK-C cells exhibit predominantly the functional characteristics of the renal proximal tubule, including a parathyroid hormone (PTH)-inhibitable Na^+-dependent inorganic phosphate transport system [Kempson et al., 1989]. They also show phlorizin-sensitive Na^+-dependent hexose transport, exhibit a proximal tubule-like pattern of hormonal stimulation of cAMP (PTH responsive, vasopressin insensitive), and express several proximal tubule brush border enzymes (maltase, leucine aminopeptidase, gamma-glutamyl transpeptidase). NHK-C cells have also been used to demonstrate the specificity of phosphonoformic acid, an inhibitor of Na^+-Pi co-transport [Yusufi et al., 1986], and serve as a model of oxidant injury in the renal tubule [Andreoli and McAteer, 1990].

Bronchial and Tracheal Epithelium

There are a number of reports of primary culture of alveolar, bronchial, and tracheal epithelium [e.g., Steele et al., 1978; Fraser and Venter, 1980; Stoner et al., 1980, 81], including the use of floating collagen [Geppert et al., 1980] and pigskin [Yoshida et al., 1980], and Steele et al. [1978] were able to produce nontumorigenic continuous cell lines by treating tracheal epithelium with a phorbol ester. More recently Lechner and LaVeck [1985] have developed a low-serum medium for clonal growth of normal lung (see Chapter 20) and Carney et al. [1981] have developed a serum-free medium supplemented with hydrocortisone, insulin, transferrin, estrogen, and selenium (HITES medium) for small cell carcinoma of lung and with modifications for large cell and adenocarcinoma (see Chapters 7 and 21). These media are selective and do not support stromal cells.

The following protocol for the isolation and culture of normal human bronchial epithelial cells from autopsy tissue was contributed by Moira A. LaVeck and John F. Lechner, National Institutes of Health, Bethesda, MD. John Lechner's current address is Inhalation Toxicology Research Institute, P.O. Box 5890, Albuquerque, NM 87185, USA.

Principle

In the presence of serum, normal human bronchial epithelial (NHBE) cells cease to divide and, furthermore, terminally differentiate. The serum-free medium that has been optimized for NHBE cell growth does not support lung fibroblast cell replication, thus permitting the establishment of pure NHBE cell cultures [Lechner and LaVeck, 1985].

Outline

This procedure involves first explanting fragments of large airway tissue in a serum-free medium (LHC-9) for initiating and subsequently propagating fibroblast-free outgrowths of NHBE cells; four subculturings and 30 population doublings is routine.

Materials

Sterile:

Culture medium (sterile, filtered):
L-15 (GIBCO)
LHC-9 [Lechner and LaVeck, 1985] (see Tables 7.5 and 7.6)
HB [Lechner and LaVeck, 1985] (see Reagent Appendix)
Bronchial tissue from autopsy of noncancerous donors
Plastic tissue culture dishes (60 and 100 mm)
Mixture of human fibronectin 10 μg/ml (Collaborative Research)/collagen (Vitrogen 100, 30 μg/ml; Collagen Corp.)/crystallized bovine serum albumin, BSA, 10 μg/ml (Miles Biochemical) in LHC basal medium (Biofluids Inc.) (FN/V.BSA) [Lechner and LaVeck, 1985]
Scalpels No. 1621 (Becton Dickinson)
Surgical scissors
Half-curved microdissecting forceps
Pipettes (10 and 25 ml)
0.02% Trypsin (Cooper Biomedical), 0.02% EGTA (Sigma), 1% polyvinylpyrrolidine (U.S. Biochemical Corp.) solution (sterile, filtration) [Lechner and LaVeck, 1985]
High O_2 gas mixture (50% O_2, 45% N_2, 5% CO_2)

Nonsterile:

Gloves (human tissue can be contaminated with biologically hazardous agents)
Humidified CO_2 incubator at 36.5°C
Controlled atmosphere chamber (Bellco No. 7741)
Rocker platform (Bellco No. 7740)
Phase contrast inverted microscope

Protocol

1. Coat culture dish with 1 ml of FN/V/BSA mixture per 60-mm dish and incubate in a humidified CO_2 incubator at 36.5°C for at least 2 hr

(not to exceed 48 hr). Vacuum aspirate the mixture and fill the dish with 5 ml culture medium.
2. Aseptically dissected lung tissue from noncancerous donors, autopsied within the previous 12 hr, is placed into ice-cold L-15 medium for transport to the laboratory, where the bronchus is further dissected from the peripheral lung tissues.
3. Before culturing, scratch a square centimeter area at one edge of the surface of the culture dishes using a scalpel blade.
4. Open the airways (submerged in L-15 medium) with surgical scissors and cut (slice, not saw) with a scalpel into two pieces, 20 × 30 mm.
5. Using a scooping motion to prevent damage to the epithelium, pick up the moist fragments and place epithelium-side up onto the scratched area of the 60-mm dish. Remove medium and incubate at room temperature for 3 to 5 min to allow time for fragments to adhere to the scratched areas of the dishes.
6. Add 3 ml of HB medium to each dish and place in a controlled-atmosphere chamber. Flush chamber with a high O_2 gas mixture and place on a rocker platform. Rock chamber at 10 cycles per minute, causing the medium to intermittently flow over the epithelial surface. Incubate rocking tissue fragments at 36.5°C, changing the medium and atmosphere after 1 d and at 2-d intervals for 6–8 d. This step improves subsequent explant cultures by reversing ischemic damage to the epithelium that occurred from time of death of the donor until the tissue was placed in ice-cold L-15 medium.
7. Before explanting, scratch seven areas of the surface of each 100-mm culture dish with a scalpel. Coat the culture dish surfaces with the FN/V/BSA mixture and aspirate as before.
8. Cut the moist ischemia-reversed fragments into 7 × 7-mm pieces and explant epithelium-side up on the scratched areas. Incubate at room temperature without medium for 3 to 5 min as before.
9. Add 10 ml of LHC-9 medium to each dish and incubate explants at 36.5°C in a humidified 5% air/CO_2 incubator. Replace spent medium with fresh every 3 to 4 d.
10. After 8 to 11 d of incubation, when epithelial cell outgrowths radiate from the tissue explants more than 0.5 cm, transfer the explants to new scratched and FN/V/BSA coated culture dishes to produce new outgrowths of epithelial cells. This step can be repeated up to seven times with high yields of NHBE cells.
11. Incubate the postexplant outgrowth cultures in LHC-9 medium for an additional 2 to 4 d before trypsinizing (with trypsin/EGTA/PVP solution) for subculture or for experimental use.

Prostate

Propagation of prostatic cell lines was not reported before the introduction of serum-free selective media (see below). There have been reports of primary culture systems [Weber et al., 1974; Franks, 1980], but these were not successfully subcultured.

The following protocol for the primary culture of rat prostate epithelial cells has been contributed by W.L. McKeehan, W. Alton Jones Cell Science Center, Lake Placid, NY.

Principle

Isolated normal rat prostate epithelial cells are a valuable model to study the cell biology of maintenance, growth, and function of normal prostate under investigator-controlled and -defined conditions [McKeehan et al., 1984; Chaproniere and MeKeehan, 1986]. Normal cells serve as the control cell type out of which prostate adenocarcinoma cells arise. Conditions similar to those described here have also been useful for primary culture of epithelial cells for transplantable rat tumors. Key features of this method are specific support of epithelial cells by improved nutrient medium and hormone-like growth factors, while concurrently inhibiting fibroblast outgrowth due to deficient or inhibitory properties of the medium.

Outline

Remove and prepare prostates for cell culture (30 min). Incubate with collagenase (1 hr) and collect single cells and small aggregates of cells (1 hr). Inoculate and culture cells to monolayer (7 d).

Materials

Sterile
60-mm petri dishes (glass or plastic)
Type I collagenase (Sigma)
Media salt solution (MSS) [McKeehan et al., 1984]
Penicillin
Kanamycin
Scissors
Syringe and 14-g cannula
25-ml Erlenmeyer flasks
1-mm wire mesh screen made to fit a 50-ml conical centrifuge tube
Fetal calf or horse serum
50-ml plastic conical centrifuge tubes
Nylon screen filters of 253, 150, 100, and 41 μm to fit 50-ml conical tubes

Growth medium WJAC404 [McKeehan et al., 1984; Chaproniere and McKeehan, 1986]
25-well plastic tissue culture dishes
Cholera toxin
Dexamethasone
Epidermal growth factor (EGF)
Ovine or rat prolactin
Insulin
Partially purified [McKeehan et al., 1984; Chaproniere and MeKeehan, 1986] or purified [Crabb et al., 1986] prostatropin (prostate epithelial cell growth factor)
Nonsterile:
Shaking water bath at 37°C
Centrifuge
Hemocytometer or Coulter counter
10–12-week-old male rats

Protocol

1. Aseptically remove desired lobes of the prostate and place in a sterile 60-mm petri dish.
2. Trim fat from lobes and weigh.
3. Add 2 ml collagenase at 675 U/ml in MSS containing 100 U/ml penicillin and 100 μg/ml kanamycin.
4. Mince with scissors to approximately 1-mm pieces, small enough to fit through a 14-g cannula.
5. Using a syringe and 14-g cannula, transfer minced tissue fragments to a sterile 25-ml Erlenmeyer flask. Add collagenase to 1 ml per 0.1 g original wet tissue weight.
6. Incubate for 1 hr at 37°C on a shaking water bath.
7. Aspirate the digested suspension three times through a 14-g cannula and then pass the suspension through a coarse (1-mm) wire mesh screen fitted to 50-ml plastic conical tubes to remove debris and undigested material. Rinse through with an equal volume of MSS containing 5% whole fetal calf serum or horse serum.
8. Collect cells by centrifugation at 100 *g* for 5 min at 4°C.
9. Resuspend pellet in 5 ml of MSS plus 5% serum and pass the suspension successively through nylon screen filters of mesh sizes 253, 150, 100, and 41 μm. Wash each screen with 5 ml of MSS plus 5% serum.
10. Collect cells by centrifugation and resuspend in 5 ml of nutrient medium WAJC404.
11. Count cells and adjust concentration to 4×10^6 cells/ml.
12. Inoculate 50 μl containing 2×10^5 cells into each well of a 24-well plate (area = 2 cm^2) containing 1 ml medium WAJC 404 and 10 ng/ml cholera toxin, 1 μM dexamethasone, 10 ng/ml EGF, 1 μg/ml prolactin, 5 μg/ml insulin, 10 ng/ml prostatropin, 100 U/ml penicillin, and 100 μg/ml streptomycin. Partially purified sources of prostatropin can be substituted as described in McKeehan et al. [1984] and Chaproniere and McKeehan [1986].
13. Incubate in a humidified atmosphere of 95% air and 5% CO_2 at 37°C. Change medium at days 3 and 5. Cells should be near confluent by day 7.

Variations

This procedure can be applied to culture of human prostate epithelial cells with modifications described in Chaproniere and McKeehan [1986].

The purified growth factor, prostatropin, is identical to heparin-binding fibroblast growth factor type one, previously called acidic FGF or HBGF-1 [Burgess and Maciag, 1989]. The consensus nomenclature is now FGF-1. Molecular characterization of new members of the FGF ligand family, and the FGF receptor family in prostate cells, has revealed that prostate epithelial cells express a specific splice variant (FGF-R2IIIb) of one of the four FGF receptor genes. The specific receptor recognizes FGF-1 and stromal cell-derived FGF-7 (also called keratinocyte growth factor) but not FGF-2 (previously called bFGF or HBGF-2), described in Yan et al. [1993]. Prostate stromal cells express only the FGF-R2 gene, which recognizes FGF-1 and FGF-2 but not FGF-7. Since stromal cells respond to both FGF-1 and FGF-2, prostatropin/FGF-1 can be substituted for FGF-7 in the above procedure to provide an additional selection for epithelial cells. FGF-2 will not substitute for FGF-1 or FGF-7 in the above procedure. If purified FGF-1 is used in the protocol, its activity can be potentiated by addition of 10–50 μg per ml of heparin.

MESENCHYMAL CELLS

I include here those cells which are derived from the embryonic mesoderm, but exclude the hemopoietic system, which is discussed below. This group includes the structural and vascular cells.

Connective Tissue

Connective tissue cells are generally regarded as the weeds of the tissue culturist's garden. They survive most mechanical and enzymatic explantation techniques and may be cultured in many of the simplest media, such as Eagle's basal medium.

Although cells loosely called fibroblasts have been

isolated from many different tissues and assumed to be connective tissue cells, the precise identity of cells in this class remains somewhat obscure. Fibroblast lines, e.g., 3T3 from mouse, produce types I and III collagen and release it into the medium [Goldberg, 1977]. While collagen production is not restricted to fibroblasts, synthesis of type I in relatively large amounts is characteristic of connective tissue. However, 3T3 cells can also be induced to differentiate into adipose cells [Kuriharcuch and Green, 1978]. It is possible that cells may transfer from one lineage to another under certain conditions, but such transdifferentiation has rarely been confirmed. It is more likely that mouse embryo fibroblastic cell lines are primitive mesodermal cells (Franks, personal communication) that may be induced to differentiate in more than one direction.

Human, hamster, and chick fibroblasts are morphologically distinct from mouse fibroblasts, as they assume a spindle-shaped morphology at confluence, producing characteristic parallel assays of cells distinct from the pavement-like appearance of mouse fibroblasts. The spindle-shaped cell may represent a more highly committed precursor and may be more correctly termed a fibroblast. NIH3T3 cells may become spindle shaped if allowed to remain at high cell density.

It has also been suggested that fibroblastic cell lines may be derived from vascular pericytes, connective tissue-like cells in the blood vessels, but in the absence of the appropriate markers, this is difficult to confirm.

It is clearly possible to cultivate cell lines, loosely termed fibroblastic, from embryonic and adult tissues, but these should not be regarded as identical or classed as fibroblasts without confirmation with the appropriate markers. Collagen, type I, is one such marker. Thy I antigen has also been used [Raff et al., 1979], although this may also appear on some hemopoietic cells.

Cultures of fetal or adult fibroblasts can be prepared by any of the techniques described in Chapter 9.

Adipose Tissue

Although it may be difficult to prepare cultures from mature rat cells, differentiation may be induced in cultures of mesenchymal cells (mouse 3T3) by maintenance of the cells at a high density for several days [Kuriharcuch and Green, 1978]. An adipogenic factor in serum appears to be responsible for the induction.

Muscle

Myoblasts from the three main categories of muscle may be grown in culture. Skeletal and cardiac myoblasts may be prepared from chick embryo as described in Chapter 9 [see also Konigsberg, 1979]. Yaffe [1968] and others have described the stages of differentiation in these cultures.

Cardiac myoblast cells will also progress through differentiation *in vitro* and can be seen to contract rhythmically a few days after explantation from the embryo, although they tend to lose this capacity with continued subculture. Polinger [1970] used the differential rate of attachment to reduce fibroblastic contamination of primary cultures.

Smooth muscle cells may be cultured from blood vessels following disaggregation in trypsin or collagenase [Ross, 1971; Burke and Ross, 1977]. Gospodarowicz et al. [1976] described cloned cell lines from bovine aorta derived from scraped or collagenase-treated tissue. Yasin et al. [1981] obtained cell lines from adult skeletal muscle and showed that they retained specialized markers.

Muscle cells may be identified by a number of antigenic markers, including myosin and tropomyosin. Creatine phosphokinase activity increases as muscle cells differentiate [Richler and Yaffe, 1970; Yaffe, 1971]. The most obvious property of all is spontaneous contraction, which is observed in both skeletal and cardiac muscle.

In common with other cells with excitable membranes (and some hemopoietic cells), muscle cells may be stained selectively with the fluorescent dye merocyanine 540 [Easton et al., 1978].

The following protocol for the culture of skeletal muscle from rat has been contributed by Jane Plumb, CRC Department of Medical Oncology of the University of Glasgow, Alexander Stone Building, Garscube Estate, Bearsden, Glasgow, G61 1BD, Scotland.

Principle

Skeletal muscle cultures can be established from either embryonic or newborn animal tissues. The muscle is dissected free of skin and bone and disaggregated by enzymatic digestion. Myoblasts are plated out at a high density to minimize fibroblast growth and fusion of the myotubes occurs within 5 to 6 days.

Outline

Remove muscle tissue from animals, mince, and incubate with collagenase for 24 hr. Disperse, remove collagenase, add medium, and seed cultures.

Materials

Sterile:

PBSA + antibiotics (see Chapter 9)

Serum-free medium (Ham's F10: DMEM, 50:50)

Growth medium: as above plus 10% fetal bovine serum

Collagenase (2,000 u/ml) (Worthington, CLS; Sigma, 1A)
3–4-day-old rats
Pipettes
Scissors, forceps, scalpels
Universal containers or 50-ml centrifuge tubes
Petri dishes
25-cm^2 culture flasks
Primaria petri dishes, 50 mm (Beckton Dickinson)
Nonsterile:
Plastic box
Centrifuge

Protocol

1. Kill rats by cervical dislocation and wash thoroughly with 70% alcohol. Transfer to a sterile paper towel.
2. Use sterile instruments to remove skin. Start with a transverse cut around the body in the abdominal region. Then peel skin back over the fore- and hindlimbs.
3. Cut off fore- and hindfeet and remove fore- and hindlimbs at shoulder and hip joints, respectively. Place limbs in PBSA + antibiotics in a sterile container and keep on ice. Repeat procedure for up to four rats at a time.
4. Transfer container to a laminar flow hood. Remove PBSA and place limbs in a fresh sterile container. Wash four times with 20 ml PBSA and antibiotics.
5. Place the limbs in a sterile petri dish and add 5 ml of serum-free medium. Place 5 ml of serum-free medium in a second petri dish.
6. Remove fat from limbs and dissect muscle tissue away from the bone. Transfer tissue to the second petri dish.
7. Mince muscle tissue with crossed scalpels into pieces about 1 mm^3 and transfer medium and tissue to a sterile container.
8. Allow tissue to settle for about 5 min and then remove medium. Add 10 ml of serum-free medium and pipette medium plus tissue into a 25-cm^2 culture flask. Add 1.0 ml collagenase (2,000 u/ml) and incubate for 24 hr at 37°C.
9. Pipette vigorously to disperse tissue and transfer suspension to a sterile universal container or centrifuge tube. Centrifuge for 5 min at 500 *g*.
10. Remove supernatant and resuspend pellet in culture medium containing fetal bovine serum (10%) and antibiotics (not streptomycin since this may inhibit myoblast fusion) [Moss et al., 1984]. Seed cultures at a high density in Primaria petri dishes. Tissue from four rats is sufficient for 30 50-mm dishes.
11. Put dishes in a plastic box and place in a CO_2 incubator at 36.5°C. Feed daily.

Characterization of Cultures
Fusion of the myoblasts occurs after 4 to 5 d and can be observed with the aid of a phase-contrast microscope. Provided that the cells are plated at a high density, fibroblast contamination is less than 20% of the total cell population. Fusion can be quantified by monitoring the production of the myosin heavy chain [Moss and Strohman, 1976].

Variations
Cultures of skeletal muscle from chick embryos can be established in a similar manner [Konigsberg, 1979; Walker et al., 1979]. Fibroblast contamination can be reduced further by preplating the culture for 1 hr in a standard tissue culture flask prior to culture in Primaria petri dishes. However, this procedure reduces the overall yield of muscle cells.

The following protocol for the culture of human adult skeletal muscle cells was contributed by G. Barlovatz-Meimon, S. Bonavaud, and Ph. Thibert, Faculté de Science, Université Paris XII, Val de Marne, Avenue Général de Gaulle, F-94010 Creteil, France.

Principle
It is possible to culture myogenic cells from adult skeletal muscle of several species under conditions in which they continue to express at least some of their differentiated traits. These cells (called satellite cells) mimic, partially, the first steps of skeletal muscle differentiation. They proliferate and migrate randomly on the substratum, then align and finally undergo a fusion process to form multinucleated myotubes [Yaffe, 1971; Campion, 1984; Hartley and Yablonka-Reuveni, 1990]. Although three to four passages can be performed by means of trypsinization, subculture is no longer possible once differentiation (i.e., fusion) has taken place. For the same reason, proliferation is very difficult to estimate once some cells have started to fuse.

The procedure described here is an enzymatic method of digestion of a muscle biopsy. Primary cultures from human healthy muscle biopsies are highly enriched in myogenic cells, as at least 85% positivity to desmin by immunostaining was observed at day 10 after seeding.

Outline
Primary cultures can be grown easily in Ham's F12 medium supplemented with 20% fetal calf serum (FCS). Without modifying the culture conditions,

these cells proliferate and differentiate by fusing to form multinucleated myotubes, confirming the myogenicity of the cultivated cells.

Materials

Sterile:

Transport medium: Ham's F12 (Seromed 08101) without $NaHCO_3$, with 20 m*M* HEPES (Serva 25245), 75 ml
Growth medium: Ham's F12 with 20% fetal bovine serum (GIBCO 011-06290), 200 ml
Pronase solution: 0.15% pronase (Sigma P-6911) and 0.03% EDTA (Merck 8418) with 20 IU/ml penicillin, and 20 μg/ml streptomycin (0.4% of Eurobio PES 3000), in Ham's F12 or PBS, 100 ml
PBSA, 100 ml
100-μm nylon mesh
Scalpels, long, fine scissors
9-cm petri dishes for dissection
Flasks, 25 cm^2, 4
Centrifuge tubes, 20 ml or universal containers, 5

Nonsterile:

Hemocytometer

Protocol: (a) Dissociation

1. Trim off nonmuscle tissue from the biopsy with a scalpel and rinse in PBSA.
2. Cut the biopsy into fragments parallel to the fibers and wash in PBSA prior to weighing the biopsy.
3. The fragments are placed parallel to each other in the lid of a petri dish, cut into thinner cylinders and then finally into 1-mm^3 pieces without crushing the tissue. The final cutting can be done in a tube with long scissors, again avoiding crushing.
4. Rinse with PBSA and let the pieces settle; discard supernatant.
5. Digest for 1 hr at 37°C in pronase solution (use 15 ml for a biopsy of 1 to 3 mg). Shake tube gently at 10–15-min intervals.
6. Triturate with a pipette at the end of incubation. Medium should become increasingly opaque as more and more cells are released.
7. Let the fragments settle to the bottom by gravity, forming pellet P1 and supernatant S1.
8. Filter S1 through 100-μm nylon mesh, into a 20-ml centrifuge tube, and shake or pipette gently to resuspend.
9. Centrifuge 8–10 min at 350 *g*. Discard supernatant by aspiration.
10. Resuspend pellet very, very gently by means of a rubber bulb pipette in precisely 10 ml of growth medium, and count on a hemocytometer.
11. Dilute the suspension in growth medium to seed culture flasks with about 15,000 cells/ml. 100,000 to 200,000 cells/g are obtained from healthy donor biopsies.
12. Add 15 ml digestion medium to P1, leave 30 min in water bath at 37°C, and shake periodically.
13. Pipette the suspension to disaggregate and then filter through nylon mesh, and rinse filter with 20 ml growth medium.
14. Centrifuge the suspension for 8–10 min at 350 *g*, count, and seed as before.
15. Transfer the flasks to a 37°C humidified 5% CO_2 air incubator.

◊ The rest of biopsy, tubes, pipettes, plates, etc., should be treated with hypochlorite.

(b) Maintenance of Cultures

Medium is changed very gently 24 hr after seeding, and then every 3–4 d. The evolution of these cultures is mainly to differentiate. The timing of the three phases for human muscle cells is about 4–6 d for the proliferation peak; then the cells align at about day 8; and then, around day 10 to 12, the increase in cell fusion and myotubes formation is observed. Nevertheless one must keep in mind that some cells may differentiate earlier and that some of them will still proliferate when the majority of the culture is undergoing differentiation.

Trypsinizing primary cultures is performed as follows:

1. Add a film of EDTA gently to cells and remove immediately.
2. Add trypsin solution (0.25%) as a "film" on cells.
3. When cells detach, add 5 to 10 ml of complete growth medium, pass very gently in and out of pipette, and then spin down for 10 min.
4. Count an aliquot and seed at the chosen concentration.

(c) Proliferation and Differentiation Indexes

The growth curves of human myogenic cell cultures obtained in Ham's F12 medium supplemented with 20% FCS show the three traditional phases: the lag phase, the exponential phase, and the plateau, which corresponds to the onset of fusion. The last, evaluated in terms of number of nuclei incorporated into myotubes or in terms of fusion index (percent of nuclei incorporated into myotubes/total number of nuclei), commences usually around day 8 after plating and rises dramatically around day 10. According to the sample, this chronology can gain or lose one day.

Hence differentiation, expressed as the number of nuclei per myotube/cm^2, may be observed morphologically. But the differentiation process can also be monitored by the use of biochemical markers such as the sarcomeric proteins, enzymes involved in differentiation such as creatine phosphokinase and its time-dependent muscle-specific isoform shift, or the appearance of α-actin [Buckingham, 1992].

A family of genes, the best known being MyoD, was shown to activate muscle-specific gene expression in myogenic progenitors [for review, see Buckingham, 1992]. The process of myogenic cell differentiation involves the activation of a variety of other genes with concurrent changes in cell surface adhesive properties [Dodson et al., 1990], or the recently shown requirement of cell surface plasminogen activator urokinase and its receptor [Quax et al., 1992].

Variations and Applications

The explant–re-explant method has been developed by the group of Askanas [Askanas et al., 1990; Pegolo et al., 1990]. It has proved reliable for a long time, especially for studies on muscle defects.

When the aim is the study of the mechanism of differentiation, one has the choice between the explant and the enzymatic methods followed by clonal or nonclonal culture or by three-dimensional cultivation. If the aim of the work is to obtain a large number of cells for biochemical or molecular studies, then the enzymatic method seems to be preferable.

Cartilage

Coon and Cahn [1966] described a technique (see also Chapter 11) for the cultivation of cartilage-synthesizing cells from chick embryo somites. Cahn and Lasher [1967] later used this system for analysis of the involvement of DNA synthesis as a prerequisite for cartilage differentiation. Chondrocytes respond to stimulation of growth by both EGF and FGF [Gospodarowicz and Mescher, 1977] but ultimately lose their differentiated function [Benya et al., 1978].

The following protocol for the culture of human chondrocyte cultures has been contributed by Edith Schwartz, Departments of Orthopedic Surgery and Physiology, Tufts University School of Medicine, 136 Harrison Avenue, Boston, MA 02111.

Principle

Chondrocytes are embedded in a dense matrix of proteoglycans, which must be digested by sequential enzymatic treatment to release the relatively small cellular compartment of the tissue.

Outline

Finely chopped fragments of cartilage are treated twice with hyaluronidase, trypsin, and collagenase to remove the matrix and make them available to more prolonged collagenase digestion. The cells that are then released are propagated by conventional monolayer techniques in slightly alkaline Ham's F12 medium with an elevated Mg^{2+} concentration.

Materials

Sterile:

Gey's balanced salt solution (GBSS) pH 7.0; if contamination becomes a problem, add antibiotics (100 U/ml penicillin, 100 μg/ml streptomycin or 50 μg/ml gentamycin, with or without 100 μg/ml mycostatin) to digestion mixture

Ham's F12 with 12% fetal bovine serum, 2.3 m*M* Mg^{2+}, 100 U/ml penicillin, 100 μg/ml streptomycin SO_4, pH 7.6 serum-free medium: as above, without serum

0.5% testicular hyaluronidase in GBSS with 100 U/ml penicillin and 100 μg/ml streptomycin

0.2% collagenase in GBSS

0.2% trypsin (Sigma type IX) (for tissue digestion) in GBSS

0.25% trypsin (Sigma type XI) in Ca^{+-}- and Mg^{+-}-free Tyrode's (to be used for cell passage only)

Protocol

1. Transfer cartilage to a 50-cm sterile glass petri dish.
2. Cover cartilage with GBSS.
3. With the use of two scalpels (one No. 20 blade and one No. 11 blade), cut cartilage into 1–2-mm^3 segments.
4. Transfer cut segments to a second glass petri dish. Cover with GBSS.
5. When all segments have been cut and combined, remove GBSS and cover tissue with 4 ml of hyaluronidase for 5 min at room temperature.
6. Remove the hyaluronidase solution and replace with 8 ml fresh hyaluronidase solution for an additional 10 min at room temperature.
7. Remove the hyaluronidase solution at the end of this time and wash the tissue fragments two times with 5 ml of GBSS.
8. At the end of this procedure, use GBSS to transfer the cartilage pieces to a 25-ml Codex screw-top tube.
9. Remove the GBSS and wash the tissue with 2 ml of trypsin solution. Discard wash.
10. Wash the tissue again with an additional 2 ml of trypsin solution and discard wash.
11. Incubate the tissue with 4 ml of trypsin at 37°C for 30 min with stirring.
12. Remove and discard the trypsin solution after 30 min and wash the tissue fragments twice with 5 ml GBSS at room temperature.

13. Wash the tissue fragments for 5 min with 2 ml of 0.2% collagenase dissolved in GBSS. Discard collagenase wash.
14. Incubate the tissue with 4 ml of collagenase for 30 min at 37°C.
15. Remove and discard the supernatant.
16. Add an additional 4 ml of collagenase to the tissue sample and incubate at 37°C for 90 min.
17. Remove the supernatant and centrifuge at 600 *g* for 8 min to pellet the cells.
18. Resuspend the cells in 8 ml Ham's F12 medium with 20% FCS and centrifuge at 180 *g* for 1 min to sediment undigested matrix particles.
19. Remove the supernatant to another tube and centrifuge it at 600 *g* for 10 min to sediment the cells.
20. Resuspend the cells in 8 ml of F12 medium with 12% FBS and inoculate a 75-cm^2 flask.
21. Add an additional 4 ml of F12 medium with 12% FBS to bring the final volume to 12 ml.
22. Repeat steps 16–21 two or three more times as warranted by the amount of tissue in the sample.

Bone

Although bone is mechanically difficult to handle, thin slices treated with EDTA and digested in collagenase [Bard et al., 1972] give rise to cultures of osteoblasts that have some functional characteristics of the tissue. Antiserum against collagen has been used to prevent fibroblastic overgrowth without inhibiting the osteoblasts [Duksin et al., 1975]. Propagated lines have been obtained from osteosarcoma [Smith et al., 1976; Weichselbaum et al., 1976] but not from normal osteoblasts.

The following protocol for the culture of bone cells has also been contributed by Edith Schwartz.

Principle

Bone culture suffers from the inherent problem that the hard nature of the tissue makes manipulation difficult. However, conventional primary explant culture or digestion in collagenase and trypsin releases cells that may be passaged in the usual way.

Materials

Ham's F12 medium with 12% fetal bovine serum, 2.3 m*M* Mg^{2+}, 100 U/ml penicillin, and 100 μg/ml streptomycin SO_4

Trypsin solution to be used for cell passage: dissolve 125 mg trypsin (Sigma type XI) in 50 ml of Ca^{2+}- and Mg^{2+}-free Tyrode's solution. Adjust to pH 7.0 and filter sterilize. Place aliquots of 5 and 10 ml in Pyrex tubes and store at −20°C.

Digestion solution for the isolation of osteoblasts:

Solution A: 8.0 g NaCl, 0.2 g KCl, 0.05 g NaH_2PO_4 H_2O in 100 ml distilled water

Solution B (Collagenase–trypsin solution): dissolve 137 mg Collagenase (type I, Worthington Biochemicals) and 50 mg trypsin (Sigma, type III) in 10 ml solution A. Adjust to pH 7.2 and then bring to 100 ml with distilled water. Filter sterilize and distribute into 10-ml aliquots that are stored at −20°C.

Scalpels

Forceps

9-cm petri dishes

25- or 75-cm^2 flasks (Corning, Falcon, Nunc)

Explant Cultures

Outline

As described in Chapter 9, small fragments of tissue are allowed to adhere to the culture flask by incubation in a minimal amount of medium. The adherent explants are then flooded and the outgrowth monitored.

Protocol

1. Obtain bone specimens from the operating room.
2. Rinse the tissue several times at room temperature with sterile saline.
3. If the bone cannot be used immediately, cover the bone specimen with sterile Ham's F12 medium containing 12% fetal calf serum and penicillin and streptomycin sulfate. The bone may be stored overnight at 4°C.
4. The next morning, rinse the tissue with Tyrode's solution containing penicillin (100 U/ml) and streptomycin sulfate (100 μg/ml).
5. Place the bone in a petri dish and with the use of scalpel and forceps, remove trabeculae and place in a second petri dish.
6. Add 10 ml Tyrode's solution over the excised trabeculae. Rinse the trabeculae several times with Tyrode's solution until blood and fat cells are removed.
7. To initiate explant cultures, prepare a 25-cm^2 flask by preincubating with 2 ml of complete medium for 20 min to equilibrate the medium with the gas phase. Adjust the pH as necessary with CO_2, 4.5% $NaHCO_3$, or HCl.
8. Cut the trabeculae into fragments of 1–3 mm^3.
9. Remove the preincubation medium from the flask and add 2.5 ml of fresh Ham's F12 medium to the flask. Transfer between 25 and 40 fragments of trabeculae to the flask.
10. With the flask in an upright position, slide the

explant pieces with the aid of an inoculating loop along the base of the flask and distribute the explant pieces evenly.

11. Permit the flask to remain upright for 15 min at 37°C.
12. Slowly restore the flask into a normal horizontal position. The explant pieces will stick to the bottom of the flask.
13. Leave the flask in the horizontal position at 37°C for 5 to 7 d. After this period, check for outgrowth and replace the medium of the flask with fresh medium.

    To avoid detachment of the explant pieces, lift the flask slowly into the vertical position before carrying from the incubator to the hood or to the microscope.
14. To maintain cultures, change medium two times per week.
15. When confluency is reached, the explant pieces are removed, the cell layer is trypsinized, and the cells are isolated by centrifugation and seeded into flasks or wells.

Monolayer Cultures From Disaggregated Cells
Outline
Trabecular bone is dissected down to 2–5 mm^3 and digested in collagenase and trypsin. Suspended cells are seeded into flasks in F12 medium.

Protocol

1. Wash trabecular bone specimens repeatedly with Tyrode's solution to remove the fat and blood cells. The trabeculae are excised with scalpel and forceps under sterile conditions.
2. After collecting as much bone as possible, wash the remaining blood and fat cells away by rinsing three times with Tyrode's solution.
3. Wash the cut trabeculae with F12 medium containing fetal calf serum.
4. Place the bone pieces in a small sterile bottle with a magnetic stirrer and add 4 ml of digestion solution (this should cover the bone specimens).
5. Stir the solution containing bone fragments at room temperature for 45 min.
6. Remove the suspension of released cells and discard, since these cells are most likely to contain fibroblasts.
7. Add a second aliquot of 4 ml of digestion solution to the bone fragments and stir the mixture at room temperature for 30 min.
8. Collect the digestion solution from bone fragments and centrifuge for 2 min at 580 *g* at room temperature.
9. After removing the supernatant, suspend the cells in 4 ml of Ham's F12 medium with 20% fetal calf serum and count the cells.
10. Centrifuge at 580 *g* for 10 min and resuspend the cells in 4 ml of complete medium. This will become the inoculum.
11. Preincubate 75-cm^2 flasks for 20 min with 8 ml of complete Fl2 medium to equilibrate with the gas phase.
12. Remove the preincubation medium and add 2 ml of complete F12 medium.
13. Add 4 ml of medium containing the cell suspension. The inoculum should contain 6–10,000 cells per cm^2 of surface area.
14. Finally, add an additional 6 ml of Ham's F12 medium to give a total volume of 12 ml.
15. In the interim, add an additional 4 ml of digestion solution to the remaining bone pieces and repeat the digestion for 30 min. The released cells are harvested and, if necessary, the digestion step is repeated several more times. With large amounts of bone, the digestion period can be increased to 1–3 hr. Cell counts are performed after each digestion period and the released cells are used to inoculate a different flask.

Passage of Cells in Culture

1. Remove the explant pieces.
2. Remove the medium and rinse the cell layer with PBS, 0.2 ml/cm^2.
3. Add trypsin to the flask, 0.1 ml/cm^2, and incubate at 37°C until the cells have detached and separated from one another. This is monitored under the microscope. In general, a 10-min incubation is sufficient.
4. Transfer the released cells to a centrifuge tube with an equal volume of Ham's F12 medium with 20% fetal calf serum (FCS).
5. Centrifuge at 600 *g* for 5 min.
6. Discard the supernatant and resuspend the cells in complete medium by gentle repeated pipetting.
7. Set one aliquot aside for the determination of cell concentration and another for DNA determination (see Chapter 18).
8. Inoculate the remaining cells into culture flasks or wells that have previously been equilibrated with medium. The cells should reattach within 24 hr.

Endothelium

Endothelium has been successfully cultured by collagenase perfusion of bovine aorta (Fig. 13.1e) [Gospodarowicz et al., 1976, 1977; Schwartz, 1978] and human umbilical vein [Gimbrone et al., 1974], trypsinization of white matter from rat cerebral cortex [Phillips et al., 1979], and microdissection of adrenal cortex [Folkman et al., 1979].

Endothelium can be characterized by the presence of factor VIII antigen [Booyse et al., 1975], type IV collagen [Howard et al., 1976], Weibel-Palade bodies [Weibel and Palade, 1964], and, sometimes, the formation of tight junctions, although the last feature is not always demonstrated readily in culture.

Endothelial cultures are good models for contact inhibition and density limitation of growth, as cell proliferation is strongly inhibited after confluence is reached [Haudenschild et al., 1976].

Much interest has been generated in endothelial cell culture because of the potential involvement of endothelial cells in vascular disease, blood vessel repairs, and angiogenesis in cancer. Folkman and Haudenschild [1980] described the development of three-dimensional structures resembling capillary blood vessels derived from pure endothelial lines *in vitro*. Growth factors, including angiogenesis factor derived from Walker 256 cells *in vitro*, play an important part in maintaining proliferation and survival, so that secondary structures can be formed.

The following protocol for the culture of large vessel endothelial cells has been contributed by Bruce Zetter, Department of Surgery, Harvard Medical School, Boston, MA.

Principle

Endothelial cells comprise a single cell layer at the inner surface of all blood vessels. The vessels most commonly used to obtain cultured endothelial cells are the bovine aorta [Booyse et al., 1975], bovine adrenal capillaries [Folkman et al., 1979], rat brain capillaries [Bowman et al., 1981], human umbilical veins [Jaffe et al., 1973], and human dermal [Davison et al., 1983] and adipose [Kern et al., 1983] capillaries. Although all endothelia share some properties, significant differences exist between the endothelial cells of large and small blood vessels [Zetter, 1981].

Outline

Endothelial cells released by collagenase incubated within the blood vessel are cultured on a gelatin-coated substrate in medium supplemented with mitogens (human and bovine capillary) or without mitogens (bovine large vessel).

Materials

Aseptically isolated blood vessels, preferably in 10-cm sections, approximately 5-mm diameter. If asepsis cannot be guaranteed, clamp both ends and dip in 70% alcohol for 30 s.

Collagenase: 0.25% crude collagenase, ~200 U/mg, in PBSA containing 0.5% bovine serum albumin

Dulbecco's modified Eagle's medium (see Chapter 7) supplemented with 10% calf serum and antibiotics (see Chapter 9)

10-cm tissue culture grade petri dishes or 75-cm^2 flasks coated with 1.5% gelatin in PBSA: incubate overnight in gelatin solution, remove gelatin, add medium with serum, and incubate until cells are ready

Two clamps or hemostats, 25 mm

Sharp scissors, 50 mm

Protocol: Isolation of Endothelial Cells

1. Ligature one end of a 10-cm section of blood vessel 2–10-mm diameter to a 5-ml plastic syringe.
2. Wash vessel with 20 ml PBSA to remove blood.
3. Run in collagenase solution until it appears at bottom end, clamp lower end with hemostat, and incubate at room temperature for 10 min.
4. Cut vessel above clamp with sharp scissors and collect the collagenase in a 10-cm Petri dish.
5. Rinse lumen of vessel with 10 ml PBSA and add to collagenase.
6. Centrifuge pooled digest and wash at 100 *g* for 5 min.
7. Wash twice by resuspending in medium and centrifuging.
8. Resuspend final pellet in growth medium and seed into gelatin-coated dishes or flasks, approximately one 10-cm section of blood vessel, 5-mm diameter, per 75-cm flask or 10-cm diameter dish.
9. Subculture by conventional trypsinization (see Chapter 10).

Identification

This is proved by production of factor VIII [Hoyer et al., 1973], angiotensin-converting enzyme [Del Vecchio and Smith, 1981], uptake of acetylated low-density lipoprotein [Voyta et al., 1984], presence of Weibel-Palade bodies [Weibel and Palade, 1964], and expression of endothelial specific cell surface antigens [Parks et al., 1985].

Variations

Human endothelial cells are cultured in medium 199 with 10–20% human serum, 25 μg/ml hypothalamus-derived endothelial mitogen (BTI, Stoughton, MA) and 90 μg/ml heparin [Thornton et al., 1983]. Capillary endothelial cells also require direct addition of an endothelial mitogen or conditioned medium from a tumor culture [Folkman et al., 1979].

NEUROECTODERMAL CELLS

Neurons

Nerve cells appear to be more fastidious in their choice of substrate than most other cells [Nelson and Lieberman, 1981]. They will not survive well on untreated glass or plastic but will demonstrate neurite outgrowth in collagen [Ebendal and Jacobson, 1977; Ebendal, 1979] and poly-D-lysine [Yavin and Yavin, 1980]. Neurite outgrowth is encouraged by a polypeptide nerve growth factor (NGF) [Levi-Montalcini, 1964, 1979] and a factor secreted by glial cells [Barde et al., 1978; Lindsay, 1979] immunologically distinct from NGF.

Cell proliferation has not been found in cultures of neurons even with cells from embryonic stages where mitosis was apparent *in vivo*. Much of the work on nerve cell differentiation has, therefore, been performed on neuroblastoma cell lines [Augusti-Tocco and Sato, 1969; Liebermann and Sachs, 1978; Littauer et al., 1979] or on glial–neuronal hybrids [Minna et al., 1972; Minna and Gilman, 1973]. This remains an intriguing area with many unsolved problems.

The following protocol for the monolayer culture of cerebellar neurons has been contributed by Bernt Engelsen and Rolf Bjerkvig, Department of Cell Biology and Anatomy, University of Bergen, Årstadveien 19, N-5009 Bergen, Norway.

Principle

Cerebellar granule cells in culture provide a well-characterized neuronal cell population suited for morphological and biochemical studies [Messer, 1977; Drejer et al., 1983; Kingsbury et al., 1985]. The cells are obtained from the cerebella of 7- or 8-day-old rats, and initial growth inhibition of non-neuronal cells is obtained by a short addition of cytosine arabinoside to the cultures.

Outline

The cerebella from four to eight neonates are cut into small cubes and trypsinized for 15 min at 37°C in Hanks' balanced salt solution. The cell suspension is seeded in poly-L-lysine coated culture wells/flasks.

Materials

Sterile:

Dulbecco's modification of Eagle's medium containing:
- 10% heat-inactivated fetal calf serum
- 30 m*M* glucose
- L-glutamine 293.2 mg/l
- 24.5 m*M* KCl
- 100 mU/l insulin (Sigma I-1882)
- 7 μM p-aminobenzoic acid (Sigma A-3659)
- 100 μg/ml gentamycin

Hanks' balanced salt solution with 3 g/l BSA (HBSS)
Poly-L-lysine, mol. wt. >300,000 (Sigma P-1524)
Cytosine arabinoside (Cytostar, Upjohn; powder)
Trypsin (type II) (0.025% in HBSS)
Silicone (Aquasil, Pierce 42799)
35-mm tissue culture petri dishes
Scalpels, scissors, and forceps
Pasteur pipettes and 10-ml pipettes
12- and 50-ml sterile test tubes

Nonsterile:

Water bath

Siliconization of Pasteur Pipettes

Dilute the Aquasil solution in distilled deionized water to a 0.1–1% concentration. Dip the pipettes into the solution or flush on the inside. Air dry for 24 hr, or for several minutes at 100°C, and sterilize by dry heat.

Poly-L-lysine Treatment of Culture Dishes

Dissolve the poly-L-lysine in distilled water (10 mg/l) and sterilize by filtration. Add 1 ml of poly-L-lysine solution to each of the 35-mm petri dishes. Remove the poly-L-lysine solution after 10–15 min and add 1–15 ml of culture medium. Place the culture dishes in the incubator (minimum 2 hr) until seeding the cells.

Protocol

1. Dissect out the cerebella aseptically and place them in HBSS. Mince the tissue with scalpels into small cubes approximately 0.5 mm^3.
2. Transfer to test tubes (12 ml) and wash three times in HBSS. Allow the tissue to settle to the bottom of the tubes between each washing.
3. Add 10 ml of 0.025% trypsin (in HBSS) to the tissue and incubate in a water bath for 15 min at 37°C.
4. Transfer the trypsinized tissue to a 50-ml test tube and add 20 ml of growth medium to stop trypsin action. Shear tissue by trituration through a siliconized Pasteur pipette, until a single cell suspension is obtained.
5. Let the cell suspension stay in the test tube for 3–5 min, allowing small clumps of tissue to settle to the bottom of the tube. Remove these clumps with a Pasteur pipette.
6. Centrifuge the single cell suspension at 200 g for 5 min and aspirate off the supernatant.
7. Resuspend the pellet in growth medium and seed the cells at a concentration of $2.5–3.0 \times 10^6$ cells/dish.
8. After 2–4 d (best results usually after 2 d), incubate the cultures with 5–10 μM cytosine ar-

abinoside for 24 hr. Then change to ordinary culture medium.

Analysis

Neurons can be identified by immunological characterization using neuron-specific enolase antibodies or by using tetanus toxin as a neuronal marker. Astrocyte contamination can be quantified by using glial fibrillary acidic protein as a marker.

Variations

A single cell suspension can be obtained by mechanical sieving through nylon meshes of decreasing diameter, or by sequential trypsinization (i.e., 3 × 5-min trypsin treatment). Instead of HBSS, Puck's solution, Krebs, or other buffers with glucose can be used.

The following protocol for aggregating cultures of brain cells has been contributed by Rolf Bjerkvig, Department of Cell Biology and Anatomy, University of Bergen, Årstadveien 19, N-5009 Bergen, Norway.

Principle

Aggregating cultures of fetal brain cells have been extensively used to study neural cell differentiation [Seeds, 1971; Trapp et al., 1981; Bjerkvig et al., 1986a]. The aggregating cells follow the same developmental sequence as observed *in vivo*, leading to an organoid structure consisting of mature neurons, astrocytes, and oligodendrocytes. A prominent neuropil is also formed. In tumor biology, the aggregates can be used to study brain tumor cell invasion *in vitro* [Bjerkvig et al., 1986b].

Outline

Brains from fetal rats at day 17–18 of gestation are removed and prepared as a single cell suspension. Brain aggregates are formed by overlay cultures in agar-coated multiwells. The cells in the aggregates will form a mature organoid brain structure during a 20-day culture period.

Materials

Sterile:

Dulbecco's modification of Eagle's medium containing 10% heat-inactivated newborn calf serum, four times the prescribed concentration of nonessential amino acids, L-glutamine 293.2 mg/l, penicillin 100 U/ml, streptomycin 100 μg/ml
Phosphate-buffered saline (PBS) with Ca^{2+} and Mg^{2+}
Trypsin type II (0.025% in PBSA)
Agar (Difco)
Multiwell tissue culture dishes (24 wells; Nunc)
10-cm petri dishes
12-ml test tubes
Pasteur pipettes and 5-ml pipettes
Scalpels, scissors, and surgical tweezers
Two 100-ml Erlenmeyer flasks

Nonsterile:

Water bath

Medium-Agar Coating of Microwells

Prepare a 3% stock solution (3 g agar in 100 ml PBSA) in an Erlenmeyer flask. Heat the flask in boiling water until the agar is dissolved. Place an empty Erlenmeyer flask in boiling water and add 10 ml of hot agar solution. Then slowly add warm complete growth medium to the flask until a medium-agar concentration of 0.75% is reached. Add 0.5 ml of warm medium-agar solution to each well in the multiwell dish. Wait until the agar has cooled. The multiwell dishes can be stored in a refrigerator for 1 wk.

Protocol

1. Dissect out aseptically the whole brains from a litter of fetal rats at day 17–18 gestation, and place the tissue in a 10-cm petri dish containing PBSA.
2. Mince the tissue with scalpels into small cubes, ~0.5 cm^3.
3. Transfer the tissue to a test tube and wash three times in PBSA. Allow the tissue to settle to the bottom of the tube between each washing.
4. Add 5 ml of trypsin solution and incubate in a water bath for 5 min at 37°C.
5. Shear tissue by trituration through a Pasteur pipette approximately 20 times.
6. Allow to settle for 3 min and transfer the clump-free milky cell suspension to a test tube containing 5 ml of growth medium.
7. Add 5 ml of new trypsin to the undissociated tissue and repeat the trypsinzation and dissociation procedure twice more.
8. Spin the cell suspension at 200 *g* for 5 min.
9. Aspirate supernatant, resuspend, and pool cells in 10 ml of growth medium.
10. Count the cells and add 3 × 10^6 cells to each agar-coated well. The volume of the overlay suspension should be 1 ml.
11. Place the multidish in a CO_2 incubator for 48 hr.
12. Remove the aggregates to a sterile 10-cm petri dish and add 10 ml growth medium.
13. Transfer larger aggregates individually to new agar-coated multiwells by using a Pasteur pipette.
14. Change medium every third day by carefully removing and adding new overlay medium.

    During 20 days in culture, the aggregates will become spherical and develop into an organoid structure.

Analysis

The next step is fixation and embedding in paraffin or epon for histological or electron microscopic evaluation. Oligodendrocytes, astrocytes, and neurons are identified by transmission electron microscopy or by immunohistochemical localization of myelin basic protein, glial fibrillary acidic protein, and neuron-specific enolase, respectively.

Variations

A single cell suspension can be obtained by mechanical sieving through steel or nylon meshes [Trapp et al., 1981]. Reaggregation cultures can also be obtained using a gyratory shaker. A speed (about 70 rpm) is selected such that the cells are brought into vortex, thereby greatly increasing the number of collisions between cells. This movement also prevents cell attachment to the culture flasks.

Glia

Greater success has been obtained in culturing glial cells from avian, rodent, and human brain. Embryonic and adult brain give cultures by trypsinization [Pontén and Macintyre, 1968], collagenase digestion (see Chapter 9), and primary explant [Bornstein and Murray, 1958], which closely resemble glia. Astrocytic markers can be demonstrated for several subcultures, although there is only one report that cell lines from human adult normal brain lines express the most specific marker, glial fibrillary acidic protein (GFAP) [Gilden et al., 1976]. It is our experience that while some glial properties remain (high-affinity γ-aminobutyric acid and glutamate uptake, glutamine synthetase activity), GFAP is lost [Frame et al., 1980]. Oligodendrocytes do not readily survive subculture, but Schwann cells from optic nerve have been subcultured using cholera toxin as a mitogen [Raff et al., 1978; Brockes et al., 1979]. Cultures of human glioma can also be prepared by mechanical disaggregation, trypsinization, or collagenase digestion [Pontén, 1975; Freshney, 1980] (see Fig. 13.1c). The right temporal lobe from human males appears to be marginally better than other regions of the brain [Westermark et al., 1973], but most give a good chance of success. The glia/glioma system provides a good model for comparing normal and neoplastic cells under the same conditions.

The following protocol for the culture of central nervous system (CNS) glial cells has been contributed by Susan Barnett, Departments of Neurology and Medical Oncology, University of Glasgow, CRC Beatson Laboratories, Garscube Estate, Bearsden, G61 1BD, Glasgow, Scotland.

There are two main types of CNS glial cells, the astrocyte and the oligodendrocyte. The astrocyte is believed to give metabolic support to the neurons present in the brain and the oligodendrocyte produces the myelin sheath that insulates the axon of the neuron. Cultures made from the rat optic nerve produce mixed populations of glial cells, comprised of two types of astrocytes, oligodendrocytes and progenitor cells, which, in the appropriate culture conditions, can differentiate into oligodendrocytes or type-2 astrocytes. This progenitor cell has therefore been termed the oligodendrocyte-type-2 astrocyte (O-2A) progenitor cell [Raff et al., 1983]. The second type of astrocyte is termed the type-1 astrocyte and does not originate from the O-2A progenitor cell but from another, as yet undefined, precursor cell.

O-2A lineage cells. Study of the growth and differentiation of glial cells from rat optic nerve has defined the antigenic phenotype of the progenitor and its progeny and many of the factors that regulate differentiation [Raff, 1990; Richardson et al., 1990]. Many of the identifying antibodies are available commercially. Two growth factors, PDGF and bFGF, promote the growth and self-renewal of the O-2A progenitor cell but inhibit its differentiation [Bögler et al., 1990]. This allows long-term propagation of O-2A progenitor cells, and an increase in cell number prior to changing the culture conditions for selective differentiation into oligodendrocytes or type-2 astrocytes.

Type-1 astrocyte. The cortical astrocyte in newborn rat brain is one of the simplest glial cells to culture. The preparation can be treated with Ara C (cytosine arabinoside) to produce a homogeneous population of astrocytes. The resultant monolayers are useful for the collection of astrocyte-conditioned medium (ACM), which contains a cocktail of mitogenic growth factors for other glial cells. One is PDGF, a known mitogen for O-2A progenitor cells that synchronizes the cell's inherent division and differentiation with that seen normally *in vivo* [Raff et al., 1988]. ACM also contains another as yet undefined growth factor that is mitogenic to olfactory bulb glial cells [Barnett, 1993].

Olfactory bulb glial cells. The olfactory bulb is the only mammalian CNS tissue that can support neuronal regeneration throughout life. It is thought that these properties might be associated with an as yet unpurified, novel glial cell that resides in the outer layer of the olfactory bulb called the olfactory nerve ensheathing cell (ONEC), which shares properties with both CNS and PNS glial cells. These glial cells have become of particular importance, as they may be crucial in supporting the outgrowth of neurons.

Materials

Sterile:

DMEM medium (GIBCO) containing 5% or 10% fetal calf serum (DMEM5FB/10FB), 4.5 g/l glucose, and supplemented with 25 μg/ml gentamycin (Life Technologies), 0.0286% BSA Pathocyte (ICN), 0.5 μg/ml bovine pancreatic insulin (Sigma), 100 μg/ml human transferrin, 0.2 μM progesterone, 0.10 μM putrescine, 0.45 μM -L-thyroxine, 0.224 μM selenium, and 0.49 μM 3,3′,5-triiodo-L-thyronine (all from Sigma) [DMEM-BS, Michler-Stucke & Bottenstein, 1982]

Leibowitz medium (L-15) + 25 μg/ml gentamycin (ICN-Flow)

HBSS, Ca^{2+} and Mg^{2+} free (GIBCO)

0.02% Na-EDTA in Ca^{2+}- and Mg^{2+}-free HBSS (EDTA)

Collagenase, stock of 13 mg/ml in L-15 (~155 u/mg: Worthington Biochemical Corporation)

0.25% trypsin (Sigma)

Soybean trypsin inhibitor (0.52 mg/ml), DNAse (0.04,mg/ml), and bovine serum albumin (3 mg/ml) in DMEM (SD)

Cytosine arabinoside (Ara-C: Sigma) stock of 1 m*M* in DMEM

Poly-L-lysine (PLL; Sigma, <100,000 MW), stock 4 mg/ml, dilute 1:300 in sterile double-distilled water; use at a final concentration of 13.3 mg/ml

Monoclonal antibodies: O4 (Dr. I. Sommer, Department of Neurology, Institute of Neurological Sciences, Southern General Hospital, Glasgow G51 4BF, Scotland); Anti-galactocerebroside [anti-GalC, Ranscht et al., 1982]

Second antibodies: anti-mouse IgM-fluorescein and anti-mouse IgG3-phycoerythrein (Southern Biotechnology Assoc.)

Curved and straight forceps (size 7), curved dissecting scissors, scalpel

1-ml plastic syringe and 21G, 23G, and 25G needles

15-ml polypropylene conical, and No. 2058 polystyrene round-bottomed 12 × 75 mm tubes (for FACS sorting), both with caps; 75-cm² and 25-cm² flasks; 24-well plates; Petri dish, 35 mm

Nonsterile:

Dissecting board and pins

70% ethanol in a spray bottle

Sprague-Dawley rats, 2–7 day old

For characterization: Antinerve growth factor receptor (NGFr) (Boehringer Mannheim)

Anti-GFAP antibody, Dakopatts

Anti-GalC [Ranscht et al., 1982]

Anti-O4 (contact Ilse Sommer, Dept. of Neurology, Institute of Neurosciences, Southern General Hospital, Glasgow, Scotland)

A2B5 [Eisenbarth et al., 1979]

Access to a flow cytometer, e.g., FACS (Becton Dickinson), will be necessary to purify olfactory bulb glial cells.

Preparation of Plasticware

Pipette enough PLL to coat the appropriate dish/flask and incubate 0.5 hr or overnight. Remove all the PLL and leave the dish/flask to air dry in the hood before use. Dry PLL-coated dishes can be stored for at least a week.

Dissections

Dissections (Fig. 20.2) are carried out in perinatal rat pups (newborn–7 d), as this yields the greatest number of cells. The same protocol could apply for older animals but tissue dissection may be difficult, as the skull is harder to dissect and more myelin is present in the tissue. For astrocytes, use newborn–1-d-old rats; for olfactory bulb glial and optic nerve preparations, 7-d-old rats [Barnett et al., 1993].

Protocol

1. Perinatal rats are killed by decapitation. For older animals (>2 wk old) cervical dislocation should be carried out first, followed by decapitation.
2. Pin the head dorsal-side up onto a dissecting board and spray with 70% ethanol.
3. Dip all dissecting instruments in 70% ethanol prior to use and shake dry. Remove the skin from the head using sharp curved scissors and make a circular cut to remove the top of the skull, revealing the brain and the two olfactory bulbs at the nose tip (Fig. 20.1).
4. Using curved forceps, gently remove the tissue required for dissociation.

For a Cortex Preparation

1. Remove the cerebellum and olfactory bulbs and then remove the cortex, flip it over, and place in a petri dish containing 1 ml of L-15 + gentamycin. Use 2.5 animals for each 75-cm² flask.
2. Remove the central tissue, followed by the edge tissue and hippocampus, until a flat, white, butterfly-shaped piece of tissue remains. Flip this over and peel off the meninges.
3. Chop up the cortex using a sterile scalpel and place in a universal container with collagenase. For 15 animals, place the cortices in 2 ml of L-15 and 1 ml of collagenase and incubate at 37°C for 30 min.
4. Add 1 ml of 0.25% trypsin and incubate for a further 20 min.
5. Centrifuge the cells at 800 *g* for 5 min and then remove supernatant.

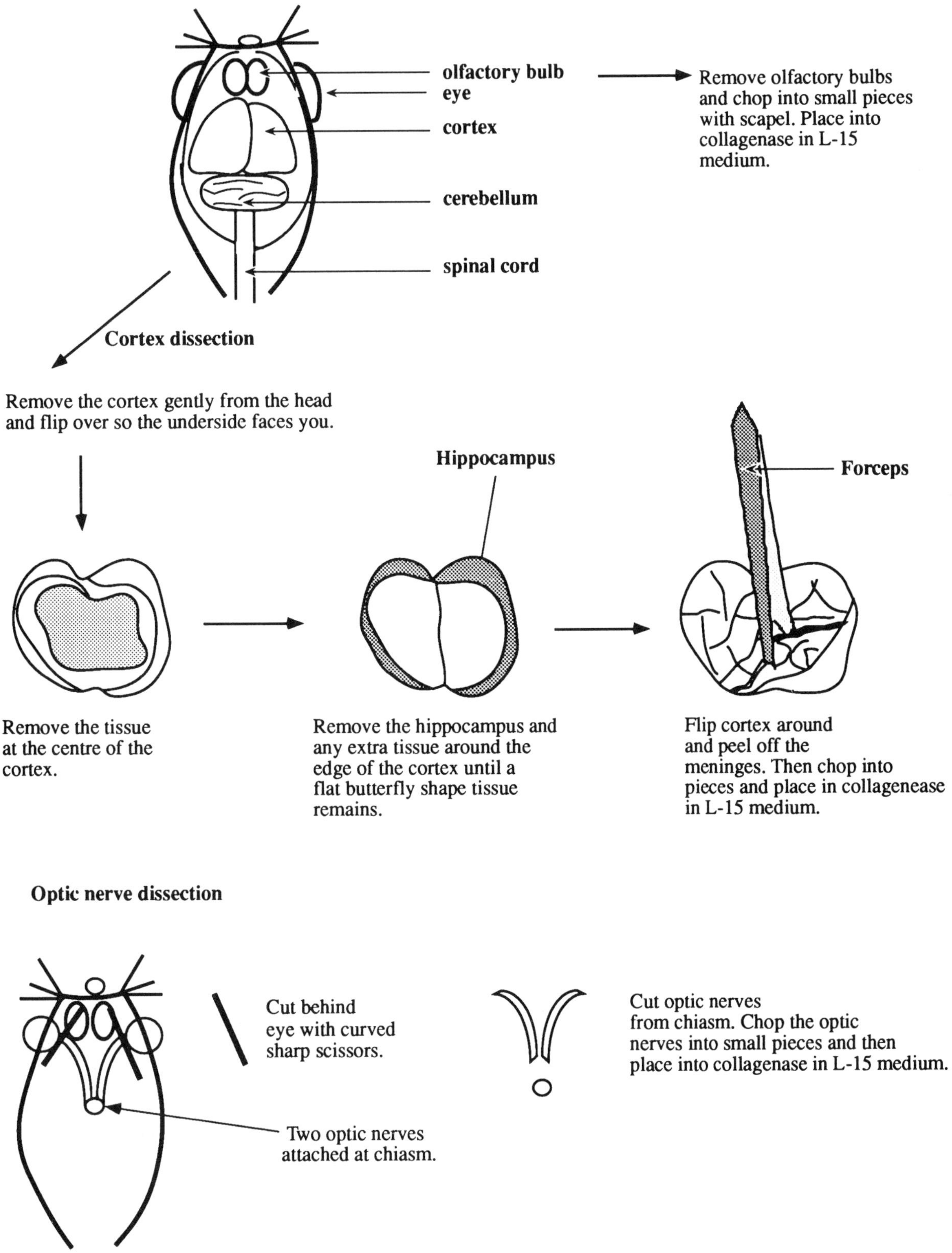
olfactory bulb
eye
cortex
cerebellum
spinal cord
Remove olfactory bulbs and chop into small pieces with scapel. Place into collagenase in L-15 medium.
Cortex dissection
Remove the cortex gently from the head and flip over so the underside faces you.
Hippocampus
Forceps
Remove the tissue at the centre of the cortex.
Remove the hippocampus and any extra tissue around the edge of the cortex until a flat butterfly shape tissue remains.
Flip cortex around and peel off the meninges. Then chop into pieces and place in collagenease in L-15 medium.
Optic nerve dissection
Cut behind eye with curved sharp scissors.
Cut optic nerves from chiasm. Chop the optic nerves into small pieces and then place into collagenase in L-15 medium.
Two optic nerves attached at chiasm.

Fig. 20.2. Schematic diagram of CNS tissue dissection. (Courtesy of Susan Barnett.)

6. Add 1 ml EDTA and 1 ml of 0.25% trypsin and incubate for a further 20 min.
7. Add 1 ml of SD, then centrifuge the cells at 1,000 g for 5 min and remove supernatant. Add 5 ml DMEM-10%FCS and triturate 5x through a 5-ml pipette. If clumps remain, triturate the cells through a 21G needle once.
8. Count the cells and plate at $1.5 \times 10^7 - 2 \times 10^7$ cells/75-cm^2 flask. The total volume of the plating medium should be 10 ml of DMEM10FB.
9. Feed cultures the next day to remove debris. Keep feeding twice a week until confluent (approx. 2 wk).

   The astrocytes are the flat fibroblast-like cells, and contaminating oligodendrocytes and O-2A progenitors will grow on the top. To remove these top cells:

   (a) Shake the cells on a rocker platform overnight.

   (b) The next day, feed with DMEM10FB containing 2×10^{-5}M Ara C.

   (c) Incubate overnight to kill the actively dividing cells.

   (d) Feed again the next day with DMEM10FB. The monolayers should consist of pure type-1 astrocytes.

For Cultures of O-2A Progenitors, Oligodendrocytes, and Type-2 Astrocytes

1. Dissect out the optic nerve (Fig. 20.2). For each 25-cm^2 flask, use 20 animals. A pair of optic nerves will yield $\sim 2 \times 10^4$ cells, and for each coverslip plate 7,000 cells in 25 μl in the center of the coverslip.
2. Reveal brain as described above. Using curved forceps, lever out the whole brain gently and you will see underneath two fine white optic nerves.
3. Cut around the eyeball with curved scissors and then gently pull out the two optic nerves by picking up the chiasm.
4. Place in a petri dish containing 5 ml L-15.
5. Under a dissecting microscope, gently tease or cut the chiasm away from the two nerves. Clear away any meninges, place the optic nerves in a drop of L-15, and chop into small pieces with a scalpel blade.
6. Follow dissociation procedure as in steps 3–7 above, but use smaller volumes. For 10–15 animals, initial collagenase step can be carried out for 1 hr at 37°C, in a 1:1 solution of enzyme and L-15, to give a total volume of 500 μl.
7. After 1 hr, add 500 μl trypsin to the collagenase mix, incubate for 25 min at 37°C, centrifuge at 800 g, and then add a 1:1 volume of EDTA and trypsin to give a total volume of 500 μl, and reincubate for 25 min at 37°C.
8. After enzymatic treatment, add 500 μl of SD and then centrifuge at 1,000 g for 5 min.
9. Remove supernatant and add 500 μl of Ca^{2+}/Mg^{2+}-free HBSS. Triturate through a 23G, 25G, and lastly 29G needle twice to dissociate cells. Take care and treat gently, creating very few bubbles, as the cells are very fragile. If clumps remain, allow to settle and take off top 200 μl to a fresh bijoux.
10. Triturate the remaining clumps through the 29G needle and repeat this procedure twice, adding more HBSS as necessary. This will protect the already dissociated cells from too much damage by repeated passage through a small needle.
11. Count cells and plate onto coverslips (7000) or into a flask (approx. $1.5 \times 10^5 - 2 \times 10^5$ cells).
12. Feed in DMEM-BS media containing 10 ng/ml of both bFGF and PDGF for proliferation of O-2A progenitors [Bögler et al., 1990] or plain DMEM-BS media for oligodendrocytes or DMEM10FB for type-2 astrocytes [Raff et al., 1983]. To enhance viability, plate cells in ACM overnight [Noble and Murray, 1984].

For Olfactory Bulb Glia Cells

1. Dissect the olfactory bulb from the head as illustrated in Fig. 20.2.
2. Chop up the bulbs and place in 500 μl collagenase and 1 ml of L-15 at 37°C for 30–45 min. One of the antigens that defines the olfactory bulb glial cells is trypsin sensitive (nerve growth factor receptor, NGFr) so the cells must be dissociated in collagenase alone.
3. To purify the olfactory glial cells from astrocytes and neurons, the cells are labeled with two cell surface antibodies (O4 and anti-GalC) followed by fluorescein and phycoerythrein conjugated second antibodies (IgM and IgG3, respectively) and the O4+GalC- cells sorted on a flow cytometer (FACS, Becton Dickinson) [see Barnett, 1993, for details of method]. Approximately 30 coverslips can be made from 20 animals.
4. Cells are fed with ACM thrice a week.

Characterization of all these glial cells has been described in several papers and a range of antibodies are available (Table 20.2). The main thing to consider when culturing glial cells is that they are fragile cells that need gentle handling and generally like to be closely associated with other glial cells. When placing these cells on coverslips, keep the volume small so that the cells stay quite dense. Glial cell immunofluorescence will produce some beautiful photographs.

TABLE 20.2. Characterization of Glial Cells by Immunostaining

	GFAP	Gal-C	O4	A2B5	NGF_r
O2-A	−	−	−	+	−
Type-1 astrocyte	+	−	−	−	−
Type-2 astrocyte	+	−	−	+	−
Oligodendrocyte	−	+	+[a]	+[a]	−
ONEC	+/−[c]	−	+[b]	−	+

[a]Mature oligodendrocytes lose A2B5, become $GalC^+$ and $O4^+$ only.
[b]ONECs lose O4 in culture.
[c]GFAP may be fibrous or amorphous. ONECs gain GFAP expression in culture.

A number of gliomas have been cultured from rodents, among which the C_6 deserves special mention [Benda et al., 1968]. This cell line expresses the astrocytic marker, glial fibrillary acidic protein, in up to 98% of cells [Freshney et al., 1980a] but still carries the enzymes glycerol phosphate dehydrogenase and 2'3' cyclic nucleotide phosphorylase [Breen and De Vellis, 1974], both of which are oligodendrocytic markers. This appears to be an interesting example of a precursor cell tumor that can mature along two distinct phenotype routes simultaneously.

Linser and Moscona [1980] separated the Müller cells of the neural retina from pigmented retina and neurons and demonstrated that full functional development could not be achieved unless the Müller cells (astroglia) were recombined with neurons from the retina. Neurons from other regions of the brain were ineffective.

Endocrine Cells

The problems of culturing endocrine cells [O'Hare et al., 1978] are similar to the culture of any other specialized cell but accentuated because the relative number of secretory cells may be quite small. Sato and colleagues [Buonassisi et al., 1962; Sato and Yasumura, 1966] cultured functional adrenal and pituitary cells from rat tumors by mechanical disaggregation of the tumor [Zaroff et al., 1961] and regular monolayer culture. The functional integrity of the cells was retained by intermittent passage of the cells as tumors in rats [Buonassisi et al., 1962; Tashjian et al., 1968]. These lines are now fully adapted to culture and can be maintained without animal passage [Tashjian, 1979], and in some cases in fully defined media [Hayashi and Sato, 1976].

Fibroblasts have been reduced in cultures of pancreatic islet cells by treatment with ethylmercurithiosalicylate [Braaten et al., 1974] and have also been purified by density gradient centrifugation [Prince et al., 1978] and by centrifugal elutriation [Bretzel et al., 1990]. These cells apparently produce insulin but not as propagated cell lines.

Pituitary cells, which continue to produce pituitary hormones for several subcultures, have been isolated from the mouse [De Vitry et al., 1974], but in our experience, normal human pituitary cells do not survive well and even pituitary adenoma cells gradually lose the capacity for hormone synthesis.

Melanocytes

Pigment cells were cultured successfully by Coon [Coon and Cahn, 1966] from chick pigmented retina and propagated over many generations. As with the chick embryo cartilage cells, a fraction derived from embryo extract was required for the functional differentiation of these cells.

Until recently, other normal pigment cells have proven difficult to culture although cultures were obtained from human uveal melanocytes [Meyskens et al., 1980]. Pigment cells from skin do not survive readily without the appropriate growth factors (see below), although cultures can be obtained from melanomas with a reasonable degree of success [Creasey et al., 1979; Mather and Sato, 1979a,b]. Primary melanomas are often contaminated with fibroblasts, but since they can be cloned on confluent feeder layers of normal cells (see Chapter 11; Fig. 15.2) [Creasey et al., 1979; Freshney et al., 1982b] purification may be possible. In general, however, greater success is obtained with secondary growth from lymph nodes, or from distant metastatic recurrences. Sato [1979] has described conditions for serum-free culture of cell lines from human and murine melanoma.

The following protocol for the culture of human melanocytes has been contributed by Barbara A. Gilchrest, Department of Dermatology, Boston University School of Medicine, 80 East Concord St., Boston, MA 02118-2394.

Principle

The greater substrate dependency of cultured keratinocytes has been utilized to obtain preferential melanocyte attachment and growth in a hormone-supplemented medium containing bFGF [Naeyaert et al., 1991]. The system has been modified to obviate the problem of keratinocyte contamination while supporting good melanocyte proliferation and pigment production in the absence of tumor promoters or chemotherapeutic agents with minimal serum supplementation [Gilchrest et al., 1984; Wilkins et al., 1985]. With conventional serum supplementation (5–20%), melanocyte growth is far better than reported in other systems and fibroblast overgrowth is controlled by reducing the Ca^{2+} concentration to 0.03 m*M* [Naeyert et al., 1991].

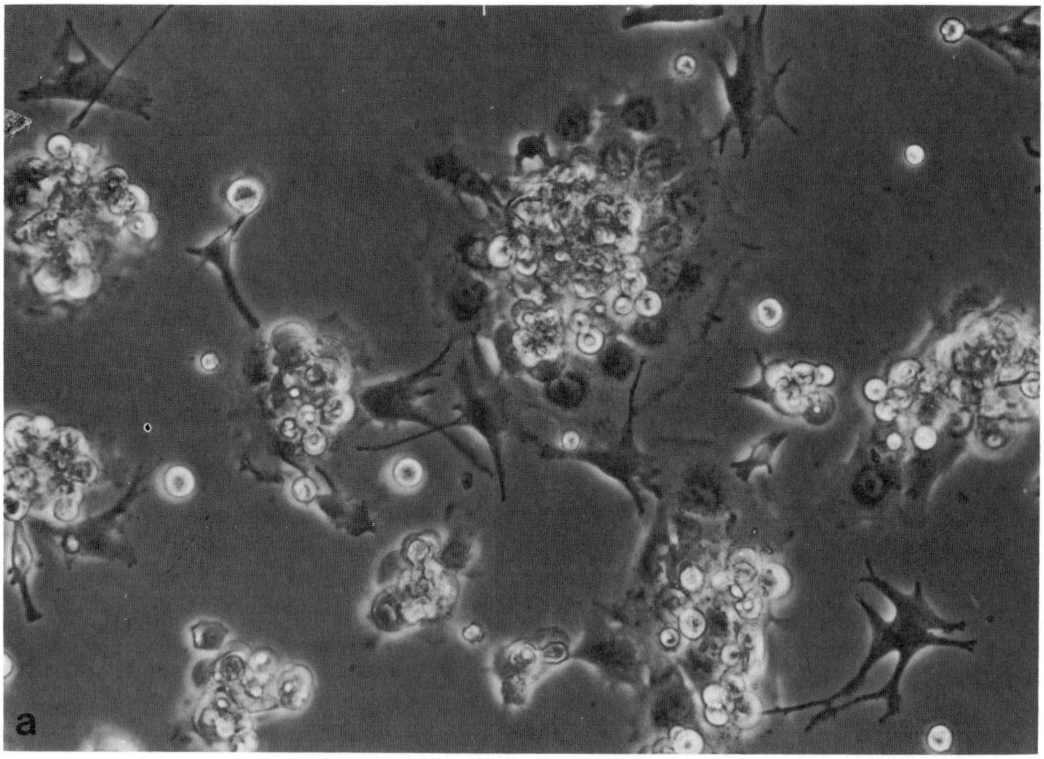

Fig. 20.3. Melanocyte cultures. a. Culture of newborn foreskin-derived melanocytes 48 hr after inoculation. Note multiple keratinocyte colonies with central stratification and tightly apposed epithelial cells at the periphery. Melanocytes are relatively small dark dendritic cells, most of them in contact with the keratinocyte colonies via dendritic projections (phase contrast, ×640). b. Secondary cultures of newborn and (c) adult epidermal melanocytes. Newborn cells tend to be polygonal under these culture conditions, while adult cells are more dendritic. Note larger cell size compared to melanocytes immediately after establishment of the culture (a). (Phase contrast, ×640.)

Outline

Epidermis, stripped from small fragments of skin following cold trypsinization, is dissociated in EDTA and cultured in serum-free medium supplemented with bFGF.

Materials

Sterile:

PBSA
Fetal bovine serum
Hormone-supplemented medium:
  Medium 199 (GIBCO 400-1200), 93.1 ml
  EGF (Collaborative Research), 0.1 ml, 10 μg/ml stock concentration, 10 ng/ml final concentration
  Transferrin (Sigma), 0.1 ml, 10 mg/ml stock, 10 μg/ml final
  Insulin (Sigma), 0.1 ml, 10 mg/ml stock, 10 g/ml final
  Triiodothyronine (Collaborative Research), 0.1 ml, 10^{-6}M stock, 10^{-9}M final
  Hydrocortisone (Calbiochem), 0.5 ml, 200× stock, 1.4×10^{-6} M final
  Cholera toxin (Calbiochem), 1.0 ml, 10^{-8} M stock, 10^{-9} M final
  Basic FGF (Amgen), 5.0 ml, 200 ng/ml stock, 10 ng/ml final
0.25% trypsin/0.1% EDTA (GIBCO 610-5050)
Tissue culture dishes (Falcon, Becton Dickinson), 100-, 60-, and 35-mm diameter
Pipettes
15- and 50-ml centrifuge tubes

Nonsterile:

Humidified incubator (37°C, 8% CO_2)
Water bath
Coulter counter or hemocytometer

Protocol: Establishing the Primary Culture

Day 1.

1. Rinse skin specimen in 70% ethanol, then twice in PBSA.
2. Transfer tissue to a sterile 100-ml dish, epidermal-side down.
3. With dissecting scissors, excise subcutaneous fat and deep dermis.

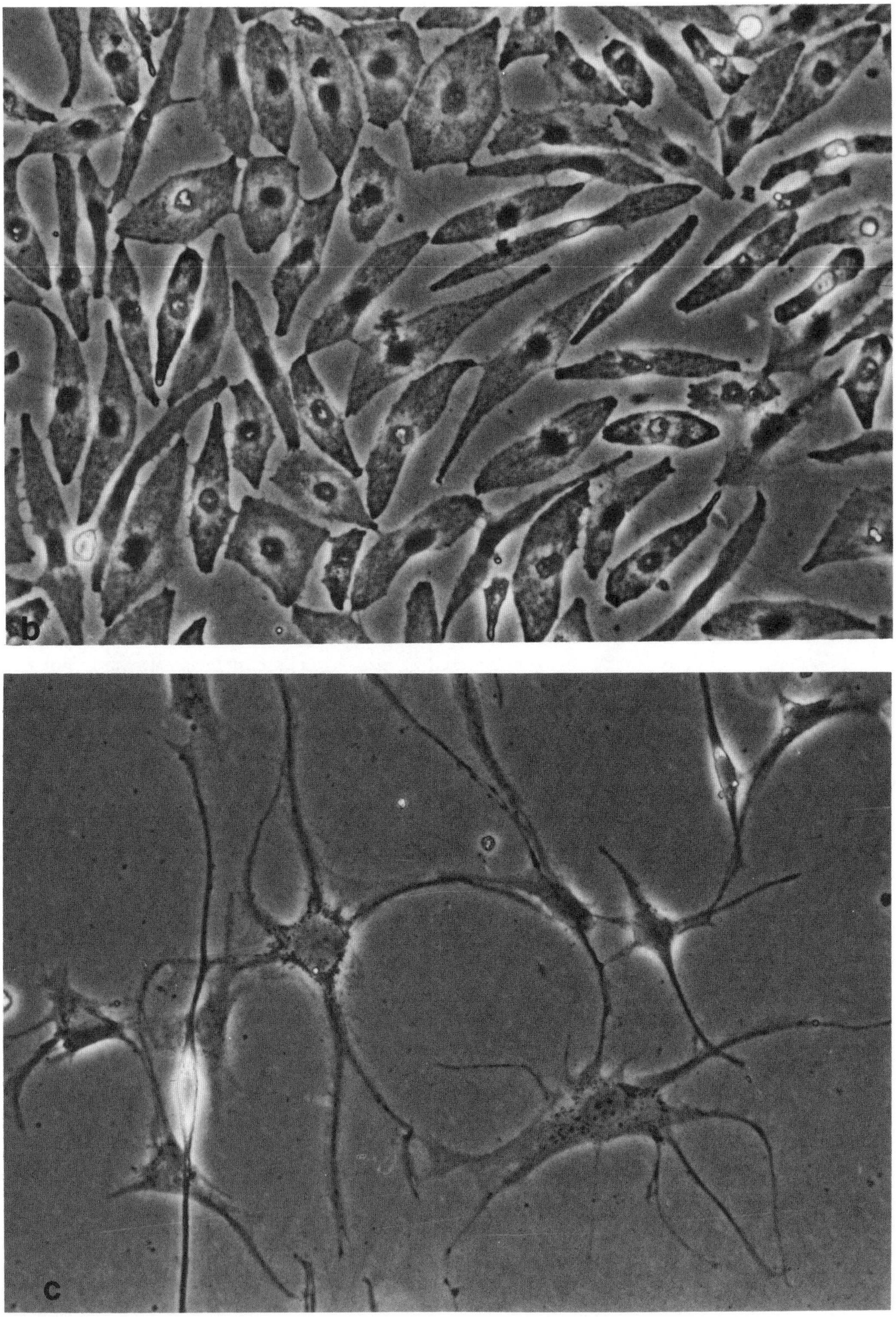

Fig. 20.3. (*Continued*)

4. Cut remaining tissue into 5 × 5 mm pieces with a scalpel, rolling blade over tissue (do not use a sawing action).
5. Transfer tissue fragments to cold 0.25% trypsin in a 15-ml centrifuge tube.
6. Incubate tube at 4°C for 18–24 hr.

Day 2.

1. Gently tap centrifuge tube to dislodge fragments settled in the bottom, then rapidly pour tube contents into a 60-mm dish.
2. Using forceps, transfer tissue fragments individually to a dry 100-mm dish, epidermal-side down. Gently roll each tissue fragment against the dish. The epidermal sheet should adhere and allow clean separation of the dermis with forceps. Discard dermal fragments.
3. Transfer all epidermal sheets to a sterile 15-ml centrifuge tube containing 5 ml 0.02% EDTA, taking care to place each epidermal sheet in the EDTA solution, not on the plastic wall of the centrifuge tube.
4. Gently vortex tube to disintegrate epidermal sheets into a single cell suspension.
5. Centrifuge cells for 5 min at 350 g, then aspirate supernatant.
6. Resuspend pellet in serum-free melanocyte medium.
7. Determine cell count using a hemocytometer and inoculate 10^6 cells (approximately $2-4 \times 10^4$ melanocytes) per 35-mm dish in 2 ml of hormone-supplemented medium containing 2% serum to facilitate attachment.
8. Refeed fresh serum-free medium twice weekly.

Days 3–30.
At 24 hr, cultures will contain primary keratinocytes with scattered melanocytes (Fig. 20.3a). Keratinocyte proliferation should cease within several days and colonies should begin to detach during the second week. By the end of the third week, only melanocytes should remain (Fig. 20.3b,c). In most cases, cultures attain near confluence and are ready to passage within 1 month.

Subcultivation

1. Gently rinse culture dish twice with 0.02% EDTA. Add 1 ml 0.25% trypsin/0.1% EDTA and incubate at 37°C. Examine dish under phase microscopy every 5 min to detect cell detachment.
2. When most cells have detached, inactivate trypsin with soybean trypsin inhibitor or 1 ml medium containing 10% serum. (Melanocytes maintained under serum-free conditions greatly benefit from a "serum kick" at the time of cultivation.)
3. Pipette dish contents to ensure complete melanocyte detachment. Aspirate and centrifuge for 5 min at 350 g. Aspirate supernatant, resuspend cells in melanocyte medium containing 2% serum, and replate at $2-4 \times 10^4$ cells per 35-mm dish. Serum is required for good melanocyte attachment (>75%) to plastic dishes, but can be omitted if dishes are coated with fibronectin or type I/III collagen [Gilchrest et al., 1985].
4. Refeed twice with serum-free or serum-containing melanocyte medium. Serum-containing medium will give a greater melanocyte yield but encourages the growth of fibroblasts. Reducing the calcium concentration to 0.03 m*M* has no effect on proliferation of melanocytes but inhibits fibroblast overgrowth [Naeyaert et al., 1991].

Confirmation of Melanocytic Identity
Melanocyte cultures may be contaminated initially with keratinocytes and at any time by dermal fibroblasts. Both forms of contamination are rare in cultures established and maintained by an experienced technician/investigator, but are common problems for the novice. Melanocytic identity of the cultured cells can be confirmed with moderate certainty by frequent examination of cultures under phase microscopy, presuming familiarity with the respective cell morphologies (Fig. 20.3b,c). More definitive identification is provided by electron microscopic examination, dopa staining, or immunofluorescent staining with S100 antibody.

HEMOPOIETIC CELLS

There have been three major milestones in the area of hemopoietic cells. Bradley and Metcalf [1966], Pluznik and Sachs [1965], and McCulloch and co-workers [Wu et al., 1968] developed techniques for cloning normal hemopoietic precursor cells in agar or Methocel (see also Chapter 11), in the presence of colony-stimulating factor(s) [Burgess and Metcalf, 1980]. The colonies matured during growth and could not be subcloned, implying that the colony-forming unit (CFU) was a precursor cell that was not regenerated in culture. This cell, the CFU-C ("colony-forming unit-culture") is distinct from the CFU-S (spleen colony-forming unit), which is a pluripotent stem cell present in colonies forming in the spleens of sublethally irradiated mice after bone marrow reconstitution [Wu et al., 1968].

Hence, suspension colonies, which contain cells of only one lineage, survive only as primary cultures that lose repopulation efficiency and cannot be subcultured. Granulocytic colonies are the most common; but under the appropriate conditions (see below), lymphoid [Choi and Bloom, 1970] and erythroid [Stephenson et al., 1971] colonies can be produced.

Golde and Cline [1973] obtained survival in a liquid culture system of normal and neoplastic leukocytes at high cell densities but with abundant medium by placing the cells in a small diffusion chamber immersed in medium. Dexter et al. also demonstrated, in a liquid culture system, that lymphoid, granulocytic, and erythroid stem cells could be propagated from bone marrow if a bone marrow culture was first prepared and allowed to form a monolayer and to act as a feeder layer for a later, second bone marrow primary culture [Dexter et al., 1977, 1979].

The third major development that occurred over several years between the earlier suspension cloning and Dexter's liquid culture system was the development of a number of functional cell lines from hemopoietic cells. Human lymphoblastic cell lines in both B- and T-cell lineage were developed by Moore et al. [1967] and subsequently Epstein-Barr virus has been found to be implicated in the ability of these cell lines to become permanent. A number of myeloid cell lines have also been developed from murine leukemias [Horibata and Harris, 1970] and, like some of the human lymphoblastoid lines [Collins et al., 1977], have been shown to make globulin chains, and in some cases, complete α- and γ-globulins (see Chapter 23, "Production of Monoclonal Antibodies"). Some of these lines can be grown in serum-free medium [Iscove and Melchers, 1978]. T-cell lines require T-cell growth factors (e.g., Interleukin IL-2) [Burgess and Metcalf, 1980; Gillis and Watson, 1981], and B-cell growth factors have also been described [Howard et al., 1981; Sredni et al., 1981; see also Freshney et al., 1993].

Originally human lymphoblastoid cell lines were derived by culturing peripheral lymphocytes from blood at very high cell densities ($\sim 10^6$/ml), usually in deep culture (>10 mm) [Moore et al., 1967]. A monolayer culture appeared in the cell pellet at the bottom of the culture tube and eventually cells were shed into suspension and started to proliferate. This could be detected by the pH drop and the cells were then subcultured. The cell concentration was kept high initially, but eventually these cells adapted to regular culture conditions and could be passaged at 10^5 cells/ml or less. More recently the development of cell lines has become easier by the use of irradiated spleen cells, antigenic stimulation (for T-cell lines), and T- and B-cell growth factors [Paul et al., 1981; Schnook et al., 1981].

Erythroid cell lines have also been cultured from the mouse. Rossi and Friend [1967] demonstrated that a mouse RNA virus (the "Friend virus") could cause splenomegaly and erythroblastosis in infected mice. Cell cultures taken from minced spleens of these animals could, in some cases, give rise to continuous cell lines of erythroleukemia cells. All of these cell lines are transformed by what is now recognized as a complex of defective and helper virus derived from Moloney sarcoma virus [Ostertag and Pragnell, 1978, 1981]. Some cell lines can produce virus that is infective *in vivo* but not *in vitro*, and the cells can also be passaged as solid tumors or ascites tumors in DBA2 or BALB-C mice.

Treatment of cultures of Friend cells with a number of agents, including DMSO, sodium butyrate, isobutyric acid, and hexamethyl-bis-acetamide, promotes erythroid differentiation [Friend et al., 1971; Leder and Leder, 1975]. Untreated cells resemble undifferentiated proerythroblasts, while treated cells show nuclear condensation, reduction in cell size, and an accumulation of hemoglobin to the extent that centrifuged cell pellets are red in color. Evidence for differentiation can also be demonstrated by staining for hemoglobin with benzidine, isolating globin-specific messenger RNA, and fluorescent antibody detection of spectrin, a specific cell-surface constituent of erythrocytes, on the surface of stimulated cells [Eisen et al., 1977].

Andersson et al. [1979c] have shown that the human leukemic cell line K562 can also be induced to differentiate with sodium butyrate and hemin, though not with DMSO.

Macrophages may be isolated from many tissues by collecting the cells that attach during enzymatic disaggregation. The yield is rather low, however, and a number of techniques have been developed to obtain larger numbers of macrophages. Mineral oil or thioglycollate broth [Adams, 1979] may be injected into the peritoneum of a mouse, and 3 days later the peritoneal washings contain a high proportion of macrophages.

If necessary, macrophages may be purified by their ability to attach to the culture substrate in the presence of proteases, as above. They can only be subcultured with difficulty because of their insensitivity to trypsin. Methods have been developed using hydrophobic plastics, e.g., Petriperm dishes (Heraeus).

There are some reports of propagated lines of macrophages, mostly from murine neoplasia [Defendi, 1976]. Normal mature macrophages do not proliferate, although it may be possible to culture replicating precursor cells by the method of Dexter (see below).

Long-Term Bone Marrow Cultures

The following protocol for the long-term culture of bone marrow has been contributed by E. Spooncer and T.M. Dexter, Paterson Laboratories, Christie Hospital, Manchester, England.

Principle

By culturing whole bone marrow the relationship between the stroma and stem cells is maintained and, in the presence of the appropriate hemopoietic cell and stromal cell interactions, proliferation of stem cells and specific progenitor cells can be maintained over several weeks [Dexter et al., 1984; Spooncer et al., 1992]. Progenitor cells from fresh marrow or long-term cultures may be assayed by clonogenic growth in soft agar [Metcalf, 1985a; Heyworth and Spooncer, 1992] or in mice [Till and McCulloch, 1961].

Outline

Marrow is aspirated into growth medium and maintained as an adherent cell multilayer for at least 12 wk and up to 30 wk. Stem cells, maturing, and mature myeloid cells are released from the adherent layer into the growth medium. Granulocyte/macrophage progenitor cells can be assayed in soft gels.

Materials

(All reagents must be pretested to check their ability to support the growth of the cultures.)

Sterile:

Fischer's medium (GIBCO) supplemented with 50 U/ml penicillin and 50 μg/ml streptomycin and containing 16 m*M* (1.32 g/l) $NaHCO_3$

Growth medium: as above, 100-ml aliquots supplemented with 10^{-6} M hydrocortisone, sodium succinate, and 20% horse serum (hydrocortisone sodium succinate made up as 10^{-3} M stock in Fischer's medium and stored at −20°C)

1-ml sterile syringes with 21G needles

Gauze, swabs, scissors, forceps

25-cm² tissue culture flasks

Five mice: (C57Bl/6 × DBA/2)F_1 bone marrow performs well in long-term culture, but marrow from some strains, e.g., CBA, does not [Greenberger, 1980]

Protocol: Long-Term Bone Marrow Cultures

1. Kill donor mice by cervical dislocation.
2. Wet the fur with 70% alcohol, remove both femurs. Collect 10 femurs in a petri dish on ice containing Fischer's medium. One femur contains 1.5–2.0 × 10^7 nucleated cells.
3. In a laminar flow cabinet, clean off any remaining muscle tissue using gauze swabs. Hold the femur with forceps and cut off the *knee* end. The 21G needle should fit snugly into the bone cavity. Cut off the other end of the femur as close to the end as possible. Insert the tip of the bone into a 100-ml bottle of growth medium and aspirate/depress the syringe plunger several times until all the bone marrow is flushed out of the femur. Repeat with the other nine bones.
4. Disperse to a suspension by pipetting the large marrow cores through a 10-ml pipette. There is no need to disaggregate small cell clumps.
5. Dispense 10-ml aliquots of the cell suspension into 25-cm² tissue culture flasks, swirling the suspension often to ensure even distribution of the cells in the 10 cultures.
6. Gas the flasks with 5% CO_2 in air and tighten the caps.
7. Incubate the cultures horizontally at 33°C.
8. Feed the cultures weekly:
   (a) Agitate the flasks *gently* to suspend the loosely adherent cells.
   (b) Remove 5 ml of growth medium, including the suspension cells, taking care not to touch the layer of adherent cells with the pipette.
   (c) Add 5 ml of fresh growth medium to each flask; to avoid damage do not dispense the medium directly onto the adherent layer.
   (d) Gas the cultures and replace in the incubator.

Analysis

Cells harvested during feeding can be investigated by a range of methods, including morphology, CFC assays (see below), and the *in vivo* CFU-S assay for stem cells [Till and McCulloch, 1961].

Variations

Mouse erythroid [Dexter et al., 1981], B-lymphoid [Whitlock et al., 1984], and human long-term cultures [Gartner and Kaplan, 1980; Coutinho et al., 1992] have been grown.

The following protocol for GM-CFC assay has also been contributed by E. Spooncer and T.M. Dexter.

Principle

Hemopoietic progenitor cells may be cloned in suspension in semisolid media, in the presence of the appropriate growth factor(s) (Table 20.3) [Metcalf, 1985a; Heyworth and Spooncer, 1992]. Pure or mixed colonies will be obtained depending on the potency of the stem cells isolated.

Outline

Fresh bone marrow or supernatant cells from long-term cultures are diluted in a growth factor–enriched plating medium, mixed with melted agar, and plated out. Colonies form in suspension in the gelled agar.

TABLE 20.3. Growth Factors in Myelopoiesis[a]

Growth factor[b]	Synonyms	Species	Target cells								
			CFU-S	CFC-MIX	GM-CFC	BFU-E	EOS-CFC	Meg-CFC	Mast	CFU-E	Mature myeloid
IL-3	Multi-CSF	Mouse	+	+	+	+	+	+	+	±	+
	HCGF BPA	Human Gibbon	(Insufficient data)								
GM-CSF	MGI-GM Pluripoietin α	Mouse Human	–	+	+	+	+	+	–	–	+
G-CSF	MGI-G	Mouse	–	–	+(G)	–	–	–	–	–	+
	Pluripoietin	Human	NA	(+)	+(G)	(+)	(+)	(+)	–	–	+
M-CSF	MGI-M CSF-1	Mouse Human	–	–	+(M)	–	–	–	–	–	+
EOS-DF	IL4	Mouse	–	–	–	–	+	–	–	–	+
Erythro-poietin		Human Monkey Mouse	NA	–	–	–	–	–	–	+	–

[a]These growth factors have all been molecularly cloned from at least one species. Other "factors" have been described that have not yet been cloned. –, no reported effect; +, direct stimulation; (+), indirect stimulation; G, granulocyte; M, macrophage; NA, not applicable.
[b]General references: Burgess and Nicola [1983], Metcalf [1985b, 1986], Whetton and Dexter [1986]; IL-3, Hapel et al. [1985], Kindler et al. [1986], Lord et al. [1986], Metcalf et al. [1986], Yang et al. [1986]; GM-CSF, Gabrilove et al. [1986], Gough et al. [1984], Nicola et al. [1983], Sachs [1982]; G-CSF, Nicola et al. [1985], Sachs [1982], Souza et al. [1986]; M-CSF, Das and Stanley [1982], Kawasaki et al. [1985], Sachs [1982], Stanley and Guilbert [1981]; EOS-DF, O'Garra et al. [1986]; erythropoietin, Jacobs et al. [1985].

Materials

Sterile:

Iscove's medium (GIBCO) 340 mOsm/kg
Horse serum
Mice
Syringes, 21-gauge needles, scissors, swabs, and forceps
Granulocyte/macrophage colony-stimulating factor (GM-CSF) or interleukin 3 (IL-3). GM-CSF is prepared by conditioning medium for 2 d with mouse lung tissue (lung-CM) and IL-3 by conditioning medium with WEHI-3B cells (WEHI-CM), or other IL-3 producing cells, for 3–6 d. (Both these growth factors have been molecularly cloned and recombinant material is commercially available.)
3.3% Noble agar (Difco) in double-distilled water, sterilized by boiling 10-ml tubes in which to mix plating medium
Tissue culture–grade plastic petri dishes

Nonsterile:

White cell diluting fluid (WCFD): 3% glacial acetic acid (nonsterile) colored with gentian violet

GM-CFC Assay: Protocol

1. Flush the marrow cells into Fischer's medium as described in steps 1–3 above.
2. Count the nucleated cells in a hemocytometer, using WCFD, and adjust to the appropriate concentration. Fresh bone marrow cells should be plated at about 5×10^4/ ml to produce 50–150 colonies/plate.
3. A convenient volume of plating mixture to prepare is 3.3 ml. Mix the following:
   Cells (10× final concentration), 10% v/v
   Horse serum, 20% v/v
   WEHI-CM, 10–15% v/v, or lung-CM, 10–15% v/v
   Iscove's medium to make 90% of final volume
   Warm the plating mixture to above 20°C.
4. Melt the agar in a boiling water bath and add 10% v/v per plating mixture.
5. Rapidly mix the plating mixture thoroughly to ensure the agar is evenly distributed, and dispense 1-ml aliquots into triplicate 35-mm petri dishes using an automatic pipettor, e.g., Gilson P1000. Swirl the plates gently so the agar mixture covers the base of the plate.
6. Place the agar cultures in the refrigerator for about 5 min to set the agar.
7. Place the cultures at 37°C in a humidified atmosphere of 5% CO_2 in air.
8. After 7 d, score the colonies using a stereomicroscope (magnification about 25x). A colony is classed as a group of more than 50 cells. It may have a very compact form, a diffuse form, or it may have a compact center with a diffuse halo. These colony types are likely to be granulocytic, monocytic, and mixed granulocyte/monocyte,

respectively. However, in order to classify colony types correctly, the colonies must be picked out, disaggregated, cytocentrifuged onto slides (see Chapter 13), and stained.

Variations

CFC assays exist for mouse multipotent-CFC, erythroid-CFC, megakaryocyte-CFC, B-cell-CFC, eosinophil-CFC, and "fibroblast'-CFC (a component of the hemopoietic environment) [Dexter et al., 1984; Golde, 1984; Metcalf, 1985a; Friedenstein et al., 1992; Heyworth and Spooncer, 1992]. A similar range of human CFC can also be grown [Testa, 1985; Coutinho et al., 1992].

GONADS

Culture of germ cells has on the whole been disappointing. Ovarian granulosa cells can be maintained and are apparently functional in primary culture [Orly et al., 1980], but specific functions are lost on subculture. A cell line started from Chinese hamster ovary (CHO-KI) [Kao and Puck, 1968] has been in culture for many years but its identity is still not confirmed. Although epithelioid at some stages of growth, it undergoes a fibroblastic-like modification when cultured in dibutyryl cyclic AMP [Ilsie and Puck, 1971].

Cellular fractions from testis have been separated by velocity sedimentation at unit gravity, but prolonged culture of these has not been reported. The TM4 is an epithelial line from mouse testis although its differentiated features have not been reported, and Sertoli cells have also been cultured from testis [Mather, 1979].

MINIMAL DEVIATION TUMORS

Several cell lines have been derived from the Reuber and Morris hepatomas of the rat [Pitot et al., 1964; Granner et al., 1968] (see above) adrenal cortex and pituitary [Sato and Yasumura, 1966] (as described above) and provide a valuable, if rare, source of continuous cell lines with differentiated properties.

TERATOMAS

When cells from an embryo are implanted into the adult, e.g., under the kidney capsule, these can give rise to tumors known as teratomas. Teratomas also arise spontaneously when groups of embryonic cells or single cells are carried over into the adult, often at an inappropriate site.

Artificially derived teratomas have been used extensively to study differentiation [Martin and Evans, 1974; Martin, 1975, 1978], as they may develop into a variety of different cell types (muscle, bone, nerve, etc.). Growth of teratoma cells on feeder layers of, for example, SCI mouse fibroblasts, will proliferate but not differentiate, whereas when grown on gelatin without feeder layer, or in nonadherent plastic dishes, nodules form that eventually differentiate.

C H A P T E R 21

Culture of Tumor Tissue

Culture of cells from tumors, particularly spontaneous human tumors, presents similar problems to the culture of specialized cells from normal tissues. The tumor cells must be separated from normal connective tissue cells, preferably by provision of a selective medium that will sustain tumor cell growth but not that of normal cells. While the development of such media for normal cells has advanced considerably (see Chapters 7 and 20), progress in tumor culture has been limited by variation both among and within samples of tumor tissue, even from the same tumor type. It is often surprising to find that tumors that grow *in vivo*, largely as the result of their apparent autonomy from normal regulatory controls, fail to grow *in vitro*.

There are many possible reasons for the failure of some tumor cultures to survive. The nutritional balance may be wrong, or attempts to remove stroma may actually deprive the tumor cells of a matrix, or of nutritional or informational stimuli necessary for survival. Alternatively, dilution of tumor cells to provide sufficient nutrient per cell may dilute out autocrine growth factors produced by the cells. Strictly speaking, truly autocrine factors should be independent of dilution if they are secreted onto the surface of the cell and are active on the same cell, but it is possible that some so-called autocrine growth factors are in fact often paracrine, i.e., they act on adjacent cells, not only on the cell releasing them. Hence a closely interacting population is required. Interaction with certain types of stromal cell may substitute if the stroma are able to make the requisite growth factors either spontaneously or in response to the tumor cells.

It may be incorrect to assume that the growth factor dependence of a tumor cell is similar to the normal cells of the tissue from which it was derived. Tumor cells may produce endogenous autocrine growth factors, such as TGF-α, and the provision of exogenous growth factors, such as EGF, may compete for the same receptor. Furthermore, the response of a tumor cell to a growth factor, or hormone, will depend on what other growth factors are present, some of which may be tumor cell derived, and on the status of the cell. A normal cell, capable of expressing growth and senescence suppressor genes, may respond differently from a cell where one or more of these genes is inactive or mutated and antagonistic, growth-promoting, oncogenes are overexpressed.

So there are many possible reasons why a tumor cell population may respond differently to the nutritional and mitogenic environment optimized for normal cells of the same lineage. More information is required on differences in the nutritional requirements of tumor cells, but, given the heterogeneity of tumors, the task is a daunting one. As the potential therapeutic benefit to be derived from a knowledge of the nutritional requirements of tumor cells is not likely to be great, greater emphasis is being placed on the response of tumor cells to growth factors and the differences in signal transduction, where the potential therapeutic benefit may be greater.

Optimization of nutritional conditions has been restricted to certain specific types of tumor for which well-characterized continuous cell lines are available, and the result is the generation of selective media for cells such as HeLa, the small cell lung cancer cell lines, such as NCI-H69, and basal cell carcinoma of skin. This means that some continuous cell lines may be maintained serum free, but, unfortunately, the selective conditions developed may be specific to continuous cell lines from that tumor, or even to that continuous cell line.

Finally, it is probable that most cells in a tumor have a limited life-span due to genetic aberration, terminal differentiation, or natural senescence. Dilution into culture may reduce the numbers and their interaction with other cells, such that survival is impossible. Cells from multicellular animals, unlike prokaryotes, do not survive readily in isolation. Even a tumor is still a multicellular organ and may require continuing cell interaction for survival. The lethality of the tumor to the host lies in its uncontrolled infiltration and uncontrolled colonial growth, but the origin of the bulk of the cell population may reside in a relatively small population of transformed stem cells, so small that its dilution subsequent to explantation deprives it of some prerequisites for survival.

In sum, the problem is either to create the correct, defined nutritional and hormonal environment or, failing that, to provide a sustaining environment as yet undefined but nevertheless able to permit the survival of the appropriate or representative population. There has been a continuing trend to use serum and feeder layers in order to get the cells to grow, and only a few tumors have responded to culture serum free. As many transformed cells are not inhibited by TGF-β, there has not been the same need to eliminate serum, other than to repress fibroblastic growth, which remains a major problem. The trend will probably continue to adapt medium designed for equivalent normal cells.

SAMPLING

Isolation of cell cultures from normal tissue presents a sampling problem, more in evidence in subsequent culture, in producing cultures representative of a particular cell type. Tumor cell culture, in addition to selecting the appropriate cell type, be it gastric epithelium or neuroblast, must also separate this cell type from its normal equivalent and prevent the overgrowth of connective tissue or vascular cells, which tend to predominate in conventional culture systems. Furthermore, while any section of gastric epithelium may be regarded as representative of that particular zone of the gastric mucosa, tumor tissue, dependent as it is on genetic variation and natural selection for its development, is usually heterogeneous and composed of a series of often diverse subclones displaying considerable phenotypic diversity. Ensuring representativity in cultures derived from this heterogeneous population is difficult and can never be guaranteed unless sampling is totally representative and survival is 100%. Since this is practically impossible, the average tumor culture is a compromise. Assuming representative subpopulations have been retained, and are able to interact, the corporate identity may be similar to the original tumor.

The problem of selectivity is accentuated when sampling is carried out from secondary metastases, which often grow better, but may not be typical either of the primary or all other secondaries.

In view of these almost overwhelming problems facing tumor culture, it is almost surprising that the field has produced any valid data whatsoever. In fact it has, and this may result from (1) the aforementioned autonomy of tumor populations, which may have allowed their proliferation under conditions where normal cells would not multiply; (2) the increased size of the proliferative pool in tumors, which is larger than that of most normal tissues; (3) the ability of tumor cells to give rise to tumors as xenografts in immune-deprived mice; and (4) the propensity of malignantly transformed cells to give rise to continuous immortalized cell lines more frequently than normal cells. This last feature, more than any other, has allowed extensive research to be carried out on tumor cell populations, even on supposedly "normal" differentiation processes, in spite of the uncertainty of their relationship to the tumor from which they were derived.

The uncertainty of the status of continuous cell lines remains, but nevertheless they have provided a valuable source of human cell lines for molecular and virological research. The question of whether they represent advanced stages of progression of the tumor whose development has been accelerated in culture, or a cryptic stem cell population, or a purely *in vitro* artifact is still to be resolved. They are certainly distinct from most early-passage tumor cultures and should be regarded as a valuable resource, albeit genetically and phenotypically distinct from early-passage cell lines, but predominantly of the genotype of the parental cell from which they were derived. Their immortality is more likely due to deletion or suppression of genes inducing senescence [Pereira-Smith and Smith, 1988] than to overexpression of genes conferring malignancy *per se*, but this point has yet to be fully resolved.

DISAGGREGATION

Some tumors such as human ovarian carcinoma, some gliomas, and many transplantable rodent tumors are readily disaggregated by purely mechanical means such as pipetting or sieving (see Chapter 9), which may also help to minimize stromal contamination, as stromal cells are often more tightly locked in fibrous connective tissue. With many of the common human carcinomas, however, the tumor incorporates large amounts of fibrous stroma, making mechanical disaggregation difficult. In these cases, enzymatic digestion has proved preferable. Trypsin has often been used, although its effectiveness against fibrous connective tissue is limited and its toxicity to some epithelial tumor cells may be high.

In the author's laboratory, crude collagenase has been used successfully with several different types of tumor, although disaggregation also releases many stromal cells, requiring selective culture techniques for their elimination. Collagenase exposure may be carried out over several hours or even days in complete growth medium (see Chapter 9).

Necrosis is also a problem of tumors not usually encountered with normal tissue. Usually attachment of viable cells allows necrotic material to be removed on subsequent feeding, but if the amount is large and not easily removed at dissection it may be advisable to use a Ficoll/Hypaque separation (see Chapter 23) to remove necrotic cells.

PRIMARY CULTURE

Some cells, e.g., macrophages, attach to the substrate during collagenase digestion but may be removed by transferring to a fresh flask when the collagenase is removed. The adherent cells may be retained if required. The reseeded cells will contain many stromal cells (fibroblasts and endothelium principally), some of which may be removed by a second transfer to a fresh vessel in 2–4 hr, as the tumor cells, particularly clusters of malignant epithelium, are often less adhesive and take longer to attach. This is generally only partially successful, however, and it will usually require selective culture conditions for their complete removal. These selective conditions will need to replace any of the culture requirements provided by the contaminating stroma.

Physical separation techniques have also been used to remove stromal contaminants [e.g., Sykes et al., 1970; Green et al., 1980] (see also Chapter 12), but in general these are only suitable if the cells are to be used immediately, as stromal overgrowth usually follows in the absence of selective conditions.

Cloning is a method that suggests itself, but there are several limitations. Tumor cells in primary culture often have poor plating efficiencies. Furthermore, because of the heterogeneity of tumor cell populations, several clones must be isolated to be at all representative. By the time a clone has grown to sufficient numbers to be of potential analytical value, it may have changed considerably and even become heterogeneous itself due to genetic instability. Cloned isolates from a tumor should be studied collectively and even in coculture for a meaningful interpretation.

There has also been some difficulty in propagating cell lines from primary clones, particularly by the suspension method. It may be that although they are "clonogenic," few of them are really stem cells, or, if they are, they mature spontaneously due to the suspension mode of growth and lose their regenerative capacity. Nevertheless, cloned lines from tumors would be valuable material for studying tumor clonal diversity and interaction, and they represent a key area of study for future investigation.

CHARACTERIZATION

Isolation of cells from tumors may give rise to several different types of cell line. Besides the neoplastic cells, connective tissue fibroblasts, vascular endothelium and smooth muscle cells, infiltrating lymphocytes, granulocytes and macrophages, and elements of the normal tissue in which the neoplasia arose can all survive explantation. The hemopoietic components seldom form cell lines, although this has been demonstrated with small cell carcinoma of the lung, where the confusion is serious since this carcinoma also tends to produce suspension cultures. Macrophages and granulocytes are so strongly adherent and nonproliferative that they are generally lost at subculture. Smooth muscle does not propagate readily without the appropriate conditions, so the major potential contaminants of tumor cultures are fibroblasts, endothelium, and the normal equivalents of the neoplastic cells.

Of these contaminants, the major problem lies with the fibroblasts, which grow readily in culture in any case, but may also respond to tumor-derived factors and grow even faster. Similarly, endothelial cells, particularly in the absence of fibroblasts, may respond to tumor-derived factors and proliferate readily. The role of normal equivalent cells is harder to define, as their similarity to the neoplastic cells has made the appropriate experiments difficult to analyze. Characterization must therefore define lineage. Endothelium is factor VIII positive, contact inhibited, and sensitive to density limitation of growth. Fibroblasts have a characteristic spindle-shaped morphology, are density limited for growth (though less so than endothelium), have a finite life-span of 50 generations or so, make type I collagen, and are rigidly diploid.

The cells from the normal tissue, phenotypically equivalent to the tumor cells, are harder to identify and eliminate. In general they will be diploid, though some normal epithelia become aneuploid with time in culture (in excess of 1 year), and tumor cells may be close to diploid. They are usually anchorage dependent and will have a finite life-span (although again there are cases of normal epithelial cell lines becoming continuous) [Christensen et al., 1984; Boukamp et al., 1988]. Normal cells will have the same lineage markers and general morphological characteristics, some behavioral and some biochemical; genetic or molecular parameters will have to be employed to distinguish them and separate them from their neoplastic derivatives.

From a behavioral aspect, the ability of neoplastic cells to grow on a preformed monolayer of the same cell type is a good criterion for cell recognition, a potential model for separation, and a potential feeder layer to sustain the tumor cells. Glioma, for example, will grow readily (better than on plastic in some cases) on a preformed monolayer of normal glial cells [Macdonald et al., 1985] (see Fig. 11.9), but their normal counterparts will not, and the same may be true for hepatoma cells and skin carcinomas.

Normal cells tend to have a low growth fraction (labeling index with [^{3}H]-thymidine exposure; see Chapters 15 and 18) at saturation density, while neoplastic cells continue to grow faster postconfluence. Maintenance of cultures at high density can sometimes provide conditions for overgrowth of the neoplastic cells.

Many of these properties have been reviewed in Chapter 15; the major objective here is to provide criteria that might favor selection as well as identification. One of the more useful among these is the capacity for growth on confluent homologous or even heterologous feeder layers in media supplemented by a lower serum concentration. One of the most successful methods of culturing carcinoma cells has been to seed these onto a preformed monolayer of normal cells, usually mouse embryo fibroblasts (see Chapter 11), although fetal human intestinal epithelium has been very successful for breast epithelium [Lan et al., 1981; Freshney et al., 1982b]. Subsequent identification of cells derived from the colonies that form is probably best achieved by chromosome analysis, cloning in suspension in, for example, agar, measuring their angiogenic capacity, urokinase-like plasminogenic activator activity, and transforming ability for normal rodent fibroblasts [Todaro and DeLarco, 1978] (see Chapter 15).

A major problem in characterization of many tumor lines is that the criteria employed are often negative, e.g., lack of differentiation markers; hence the appearance of reliable malignancy markers becomes doubly important.

DEVELOPMENT OF CELL LINES

Propagation of primary cultures into cell lines is often difficult. Primary cultures of carcinoma cells do not always take readily to trypsin passage and many of the cells in the primary culture may not be capable of propagation due to genetic or phenotypic aberrations, terminal differentiation, or nutritional insufficiency. Nevertheless some tumor cultures can be subcultured and this often opens up major possibilities. It implies that the neoplastic cell has not been overgrown, probably has a faster growth rate than contaminating normal cells, and, if necessary, is available for cloning or other selective culture methods (see Chapters 11 and 12).

One of the major advantages of subculture is amplification. Replicate cultures can be prepared for characterization and assay of specific parameters such as genomic alterations, changes in gene expression, chemosensitivity, or invasiveness. Disadvantages include evolution away from the phenotype of the tumor due to inherent genetic instability and selective adaptation to the culture environment.

Continuous Cell Lines

One major criterion for the neoplastic origin of a culture is its capacity to form a continuous cell line. The constituent cells of this line are normally aneuploid, heteroploid, insensitive to density limitation of growth, less anchorage dependent, and often tumorigenic. The relationship to the primary culture and the parent tumor is still difficult to assess, as such cells are not always typical of the tumor population (and need not even be tumorigenic). They may be either (1) an artifact generated by adaptation to culture, made possible by the unstable genotypic characteristics of tumor cells, or (2) a specific subset or stem cell population of the tumor. Currently, the second seems more likely, as their origin is suggestive of colonies forming from a minor subset, with cell culture merely providing the appropriate conditions for their expansion.

There is no indication, so far, that tumor cells are necessarily immortalized, but merely that their intrinsic instability increases the probability that they will give rise to continuous cell lines. Whether immortality is causally related to neoplasia, or a consequence of its intrinsic genetic instability, is still far from clear.

Nevertheless, the capacity to form continuous cell lines is a useful criterion for a malignant origin, and some authors maintain that their characteristics (tumorigenicity, type of tumor, chemosensitivity, etc.) remain constant [Tveit and Pihl, 1981; Minna et al., 1983]. In any event they provide useful experimental material, although the time required for their evolution makes immediate clinical application difficult.

GENERAL METHOD

There are too many methods for the culture of tumors for them to be reported here in detail, so one frequently successful method is presented with reference to other specific techniques.

Principle

Since trypsin is often toxic and yet many tumors require prolonged enzymatic digestion, collagenase is employed in complete growth medium for prolonged periods. The disaggregation is gentle, nonmechanical, and can be carried out in full nutritional support. Prolonged incubation facilitates disaggregation of the tumor, but may restrict survival of normal anchorage-dependent cells. Collagenase is active in the presence of serum that may be a necessary additive to the culture medium, but is less effective in disaggregating epithelium than stroma and hence allows for potential separation by sedimentation of partially disaggregated epithelial cell clusters from the more fully disaggregated stroma.

Outline

Tumor tissue is chopped finely with crossed scalpels and incubated in collagenase in complete growth medium for 1–5 d; collagenase is removed by centrifugation and fractions cultured after separation by sedimentation at unit gravity. For full details of this technique, see Chapter 9 under "Collagenase."

SELECTIVE CULTURE

Four main approaches have been adopted to select tumor cells in primary culture: selective media (Table 7.7), selective substrates, confluent feeder layers (see also Chapter 11), and suspension cloning (see Chapters 11 and 15).

Selective Media

There are only a few media that have been developed as selective agents for tumor cells, due to the problems of variability and heterogeneity described above. HITES medium [Carney et al., 1981; Table 7.6] is one exception and may owe its success to the production of peptide growth factors by small cell lung cancer, for which the medium was developed. A proportion, but not all, of small cell lung cancer biopsies will grow in pure HITES; others will survive with a low serum supplement (2.5%). HITES medium is modified RPMI 1640 with *h*ydrocortisone, *i*nsulin, *t*ransferrin, *e*stradiol, and *s*elenium. Of these, selenium, insulin, and transferrin are probably the most important, and are to be found in many serum-free formulations (see Chapter 7).

The same group [Brower et al., 1986] has also produced a selective medium for adenocarcinoma, reputedly suitable for lung, colon, and, potentially, many others. It is also based on RPMI 1640 supplemented with selenium, insulin, and transferrin but also incorporates hydrocortisone, EGF, triiodotyrosine, BSA, and sodium pyruvate (Table 21.1).

A simplified version of this medium, RPMI 1640 plus selenium, insulin, and transferrin supplemented with 2.5% fetal bovine serum, may be suitable for a number of tumors with minimal stromal overgrowth. A selective, serum-free medium has also been reported for urinary bladder epithelium [Messing et al., 1982].

Other types of selective media depend on metabolic inhibition of fibroblastic growth and are not specifically optimized for any particular tumor type (see Chapter 11). They have not been found to be generally effective, with the exception of the use of a monoclonal antibody against fibroblasts by Edwards et al. [1980] and Paraskeva et al. [1985]. This has proved useful in establishing cultures from laryngeal and colon cancer.

TABLE 21.1. Serum-Free Medium for Non-Small Cell Lung Cancer Cells

Basal medium	RPMI 1640
Supplements:	
Insulin	20 μg/ml
Transferrin	10 μg/ml
Hydrocortisone	5.0 E-8
Sodium selenite	2.5 E-8
EGF	10 ng/ml
BSA (optional)	5.0 mg/ml
Na pyruvate (in addition to basal medium)	5.0 E-4
Glutamine	2.0 E-3
Triiodothyronine	1.0 E-10

From Brower et al. [1986].
Precoating of the substrate with collagen and fibronectin (see Chapter 7) is recommended with this medium.

Selective Substrates

As transformed cells are less anchorage dependent for growth (see Chapter 15), substrates with reduced adhesive properties have been employed for the selective culture of tumor cells (see also Chapter 11). Jones and Haskill [1973, 1976] found repression of fibroblastic overgrowth on polyacrylamide, as did Parenjpe et al. [1975] using Teflon.

Culture in HITES medium often gives rise to aggregate cells growing in suspension that, in itself, helps to separate tumor from stroma. Dr. Morag McCallum of the Victoria Hospital, Glasgow, observed a similar situation with cultures of breast carcinoma, separated from the biopsy by vigorous manual shaking (personal communication). These tend to grow in suspension as aggregates when cultured in supplemented MCDB 170 (see Table 7.6). The proportion of stromal cells released is less than by enzymatic digestion and those that are released tend to adhere to the culture vessel, allowing subsequent separation from the floating tumor aggregates. The only disadvantage of this technique is that it may select for more anaplastic tumors, and against very fibrous tumors.

Selective attachment can also be promoted using the panning technique, where the substrate is coated with a selective antibody, either for the cell to be isolated or for the cell to be eliminated (see Chapter 12).

Confluent Feeder Layers

This technique (see Figs. 11.3, 11.8), perhaps more than any other, has been applied successfully to many types of tumor. Smith and others [Lan et al., 1981] used confluent feeder layers of fetal human intestine, FHS74Int, to grow epithelial cells from mammary carcinoma, using media conditioned by other cell lines, although later reports from this and Ham's laboratory suggest that selective culture in MCDB 170 is a more suitable approach [Hammond et al., 1984]. In the author's laboratory this technique has been used successfully with breast, colon, and lung carcinoma, employing human fetal intestinal epithelium or mouse 3T3 or STO embryonic fibroblasts. Basal cell carcinoma has also been grown on a similar feeder layer system as for epidermis [Rheinwald and Beckett, 1981].

This technique relies on the prevention of fibroblastic overgrowth by a preformed monolayer of other contact-inhibited cells. It is not selective against normal epithelium, as normal epidermis and normal breast epithelium both form colonies on confluent feeder layers. Results from glioma [MacDonald et al., 1985], however, suggest that selection against equivalent normal cells may be possible on a homologous feeder layer. Glioma grown on normal glial feeder layers should lose any normal glial contaminants. By the same argument, breast carcinoma seeded on confluent cultures of normal breast epithelium, e.g., milk cells (see Chapter 20), could become free of any contaminating normal epithelium.

Confluent feeder layers can be highly selective and may even remove some elements of the neoplastic population. Colon carcinoma seeded on normal human fetal intestinal epithelium feeder layers virtually disappeared during early primary culture but if maintained for 2–3 months gave rise to colonies of apparently transformed cells (CEA producing, aneuploid, and heteroploid). Hence only a very small proportion of the initial inoculum gave rise to the eventual culture. Although this raises questions of representativity, it is interesting to speculate that these may be neoplastic stem cells.

Outline (Fig. 21.1)

Feeder cells are treated in midexponential phase with mitomycin C and reseeded to give a confluent monolayer. Tumor cells, dissociated from the biopsy by collagenase digestion, or from a primary culture with trypsin, are seeded onto the confluent layer. Colonies may form in 3 weeks to 3 months from epithelial tumors. Fibrosarcoma and gliomas do not always form colonies but may infiltrate the feeder layer and gradually overgrow.

Materials (All sterile)

Feeder cells (e.g., 3T3, STO, 10T½ or FHS 74 Int)
Mitomycin C (Sigma) 10 μg/ml
Growth medium
Collagenase, 2,000 U/ml, CLS grade (Worthington) or equivalent
Trypsin, 0.25% in PBSA
Tumor biopsy or primary culture
Forceps, scalpels, petri dishes for dissection as for primary culture

Protocol

1. Grow up feeder cells to 80% confluence in six 75-cm^2 flasks.
2. Add mitomycin C to give 5 μg/10^6 cells.
   It is advisable to do a dose-response curve with mitomycin C when using feeder cells for the first time to confirm that this dose allows the feeder layer to survive for 2–3 wk but does not permit further replication in the feeder layer after about two doublings at most. Treat the cells in 25-cm^2 flasks as above, and trypsinize and reseed entire contents into 75-cm^2 in 20 ml fresh medium, grow for 3 wk, feeding two times per week. Stain and check for surviving colonies.
3. Incubate overnight (18 hr) in mitomycin C, remove, wash monolayer, and grow for a further 24–48 hr.

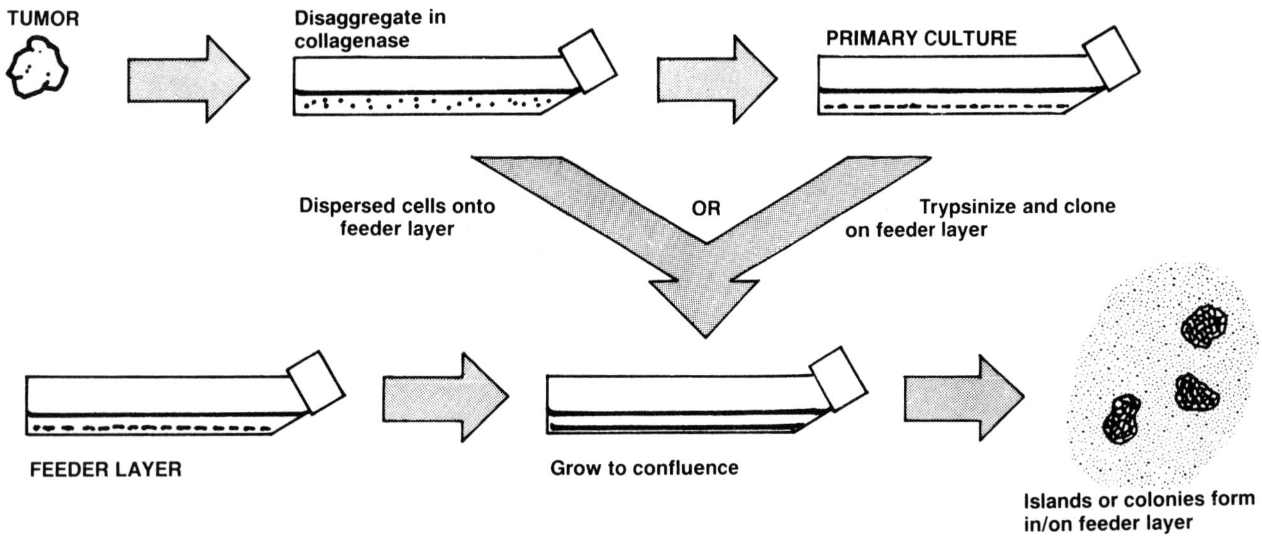

Fig. 21.1. *Selective growth on confluent feeder layers. Colonies of epithelial cells will form on confluent fetal intestinal epithelium (FHS 74 Int), normal human glia, or irradiated 3T3 or STO cells. Selection is against stromal components but not against normal epithelium.*

4. Trypsinize and reseed in 25-cm^2 flasks at 5×10^5 cells/ml ($10^5/cm^2$) and leave 24 hr.
5. If using biopsy material, biopsy should be dissected and placed in collagenase at the same time as step 2.
6. Seed cell suspension from biopsy, approximately 20–100 mg/flask, into two of the 25-cm^2 flasks in 6 ml total. Remove 1 ml and add to 4 ml medium in second pair of flasks. The third pair of flasks is kept as a control to guard against feeder cells surviving the mitomycin C treatment.

If using a primary culture, trypsinize or dissociate in collagenase, 200 U/ml final, and seed onto feeder layer at 10^5/ml in two flasks and 10^4/ml in two flasks.

If glioma or fibrosarcoma, colonies may not appear and surviving tumor will only be confirmed by subculturing without feeder layer (by which time contaminating normal cells should have been eliminated).

It is essential to confirm the species of origin of any cell line derived by this method to guard against accidental contamination from resistant cells in the feeder layer. This can be done by chromosomal analysis and lactate dehydrogenase isoenzyme electrophoresis (see Chapter 13).

Suspension Cloning

Transformation of cells *in vitro* leads to an increase in their clonogenicity in agar [Macpherson and Montagnier, 1964]; tumorigenicity has also been shown to correlate with cloning in Methocel [Freedman and Shin, 1974]. Since cells may be cloned in suspension directly from disaggregated tumors [Hamburger and Salmon, 1977], or at least colonies may grow (they may not be clones) in preference to normal stromal cells, this would seem to be a potentially selective technique. However, the colony-forming efficiency is very low (often <0.1%) and it is not easy to propagate colonies isolated from agar. On the whole this has not been a successful method for generating cell lines and the method has had greater exploitation in the assay of primary cultures from tumor biopsies (see Chapter 19).

Suspension cloning is described in detail in Chapter 11 and the Hamburger and Salmon modification in Chapter 19. These methods are also reviewed in Dendy and Hill [1983]. However, as this has not been a major method used in establishing cell lines it is not discussed further here.

Histotypic Culture

Organ culture and histotypic culture in general is discussed in Chapter 22. Apart from organ culture itself, which is not fundamentally different for tumor tissue, other methods with particular application to tumor culture are spheroid culture and filter-well culture.

Spheroids

The technique for generating spheroids is described in Chapter 23 and their use in drug assay in Chapter 19. Their particular relevance to this chapter stems from the inability of normal stroma to form spheroids or even to become incorporated in tumor-derived spheroids. Hence cultures from tumors allowed to form spheroids on nonadhesive substrates like agarose will tend to overgrow their stromal component.

Some cultures from breast and small cell carcinoma

of lung often generate spheroids or other nonspheroidal cellular aggregates that float off into suspension (see above). These may be collected from supernatant medium, leaving the stroma behind. These spheroids or organoids do not always appear soon after culture and can sometimes take weeks or even months to form, suggesting a derivation from a minority cell population in the tumor.

Spheroid generation does not arise in all tumor cultures, but has also been described in neuroblastoma, melanoma, and glioma. Its potential has not been fully explored, however, as the bulk of attention has been given to either forming attached monolayers or suspension colonies in agar for assay purposes. The selectivity described above for PTFE and polyacrylamide might be extended to simpler substrates such as nontissue culture-grade polystyrene to favor the selection of anchorage-independent cells.

Xenografts

When cultures are derived from human tumors, scarcity of material and the infrequency of rebiopsy means that it is difficult to make several attempts with the same tumor, using different selective techniques. Some tumors can be made to grow in immune-deprived animals [Rofstad, 1991], which makes much greater amounts of tumor available. It has sometimes been found that cultures can be initiated more easily from xenografts than from the parent biopsy, but whether this is due to the availability of more tissue, progression of the tumor, or modification by the heterologous host (e.g., by murine retroviruses) is not clear.

Two main types of host are in current use, the genetically athymic nude mouse, which is T-cell deficient [Giovanella et al., 1974] or neonatally thymectomized animals subsequently irradiated and treated with cytosine arabinoside [Selby et al., 1980; Fergusson et al., 1986]. The first are expensive to buy and difficult to rear but maintain the tumor for longer. Thymectomized animals are more trouble to prepare but cheaper and easier to provide in large numbers. They will, however, regain immune competence and ultimately reject the tumor after a few months.

Take rates can be enhanced by using mice that are asplenic as well as athymic, genetically (*scid* mice) or by splenectomy, or by sublethal irradiation of nude mice. Implantation with fibroblasts or Matrigel has also been reported to improve tumor take (Stanley, personal communication).

If access to a nude mouse colony is available, or facilities exist for neonatal thymectomy and irradiation, this is a step to be considered. Although only a small proportion of tumors may take, the resulting tumor will probably be easier to culture and repeated attempts may be made.

However, particular care must be taken, as with isolation from mouse feeder layers, to ensure that the cell line ultimately surviving is human and not mouse by proper characterization with isoenzyme and chromosome analysis.

Freezing

It is often difficult to take advantage of a large biopsy and utilize all the valuable material that it provides. It is possible in these cases to preserve the tissue by freezing.

Outline

Chop tumor, expose to DMSO, and freeze.

Materials

Sterile:

Biopsy

Instruments (scalpels, forceps, dishes, etc., as for primary culture)

Dissection BSS

Growth medium with antibiotics (see above)

Dimethyl sulfoxide (Self-sterilizing if placed in a sterile container)

Protocol

1. Chop tumor (after removing necrotic, fatty, and fibrous tissue) into about 3-4-mm pieces and wash in DBSS as for primary culture.
2. Place four or five pieces in each ampule.
3. Add 1 ml growth medium + 10% DMSO and leave for 30 min at room temperature.
4. Freeze at 1°C/min (see Chapter 17) and transfer to liquid N_2.
5. To thaw, place in 37°C water (with appropriate precautions; see Chapter 17).
6. Swab thoroughly in alcohol, open ampule, and remove half of medium.
7. Replace medium slowly with fresh medium without DMSO.
8. Gradually replace all of medium with DMSO-free medium, transfer to petri dish, and proceed as for regular primary culture, but allowing twice as much material per flask.

CHAPTER 22

Three-Dimensional Culture Systems

When tissue culture was first developed, it was based on the explantation of whole fragments of tissue with a view to studying them as tissues in isolation. However, it was observed that cells often grew out from these tissues in a sheet to form a monolayer on the supporting glass substrate. This divergence in growth properties—growth as a migratory and potentially proliferative monolayer versus residence within the original explant—set the pattern for future divergence in approach to the culture of animal cells. One school of thought believed in the retention of histological structure and the possibility of organotypic function, while the other looked toward the biology of individual cells. The former required retention of histotypic structure, cell interaction, and histological characteristics and enabled the study of developmental problems such as embryonic induction, *in vitro* modeling of malignant invasion, and hormonal control of morphogenesis and differentiation. The latter gave rise to propagated cell lines that laid down the basis for most of modern cellular and molecular biology and its insight into the regulation of gene expression.

Now, although the potential for propagated cell lines is far from exhausted, many people are reverting to the notion that nutritional completeness and hormonal supplementation are inadequate in themselves to recreate full structural and functional competence in a given cell population. The vital missing factor is cell interaction and the signaling capacity that it entails. A hormone may activate a specific pathway in cell A, and it or a different hormone activate a different commitment in cell B. Both, alone, may lead to individual modifications in phenotypic expression, but if allowed to interact, a cascade of interactions may occur dependent on the cells being in association. Alveolar cells of the lung will only synthesize and release surfactant in response to hormonal stimulation of adjacent fibroblasts [Post et al., 1984]; prostate epithelium response to stromal signals is in turn activated by androgen binding to the stroma [Cunha, 1984]. Epithelium differentiates in response to matrix constituents often determined jointly by the epithelium on one side and connective tissue stroma on the other, as may be the case with the interaction between epidermis and dermis *in vitro* [Bohnert et al., 1986]. Hence the whole integrated tissue may easily, and understandably, respond differently to simple ubiquitous signals, not because of the specificity of the signal or the receptor capacity but because of the quality of the microenvironment encoded in the juxtaposition of one cell type with a specific correspondent. As in human society, the response of one individual to an exogenous stimulus will be dictated as much by the spatial and temporal relationship with other individuals as by their endogenous make-up. Likewise a primitive neural crest cell may become a neuron, an endocrine cell, or a teratoma, dependent on its ultimate location, its interaction with adjacent cells, and its response, mediated by neighboring cells, to hormonal stimuli.

In essence this preamble establishes that while some cell functions, such as cell proliferation, glycolysis, respiration, and gene transcription, proceed in isolation, their regulation as related to a functioning multicellular organism must depend ultimately on the interaction among cells of the appropriate lineage, the appropriate stage in that lineage, and in the interaction between cells of different lineages occupying the came microenvironment. This suggests that if you want to study cell biology, cell lines are fine, but if you want to learn something of the integrated function, or dysfunction, of whole organs, a histotypic or organotypic model will be required.

There are three major ways to approach this goal. One is to accept the cellular distribution within the tissue, explant it, and maintain it as an organ culture. The second is to purify and propagate individual cell lineages, study them alone under conditions of homologous cell interaction, recombine them, and study their mutual interactions. This has given rise to three major types of technique: *Organ culture*, where whole organs, or representative parts, are maintained as small fragments in culture and retain their intrinsic distribution, numerical and spatial, of participating cells; *histotypic culture*, where propagated cell lines are grown to high density in a three-dimensional matrix alone; or *organotypic culture*, in which cells of different lineages are recombined in experimentally determined ratios and spatial relationships to re-create a component of the organ under study.

Organ culture seeks to retain the original structural relationship of different and similar cells and hence their interactive function, in order to study the effect of exogenous stimuli on further development [Lasnitzki, 1992]. This may even be achieved by separating the constituents and recombining them, as in the now classical experiments of Grobstein and Auerbach and others in organogenesis [Auerbach and Grobstein, 1958; Cooper, 1965; Wessells, 1977; Jessell and Melton, 1992]. Organotypic culture represents the synthetic approach, whereby a three-dimensional, high-density culture is regenerated from isolated (and preferably purified and characterized) lineages of cells that are then recombined, their interaction studied, and, in particular, their response to exogenous stimuli characterized.

"Exogenous stimuli" may be regulatory hormones, nutritional conditions, or xenobiotics. In each case the response, and the justification of this approach, will be different from the responses of a pure cell type in isolation, grown at a low cell density.

Several types of system have been described to study isolated, whole, undisaggregated tissue, recombinations of tissues, or purified cell lineages. As these may provide models for quite distinct types of investigation, each is described separately.

ORGAN CULTURE

Gas and Nutrient Exchange

A major deficiency in tissue architecture in organ culture is the absence of a vascular system, limiting the size (by diffusion) and potentially also the polarity of the cells within the organ culture. When cells are cultured as a solid mass of tissue, gaseous diffusion and the exchange of nutrients and metabolites becomes limiting. The dimensions of individual cells cultured in suspension or as a monolayer are such that diffusion is rapid but aggregates of cells beyond about 250 μm diameter (5,000 cells) start to become limited by diffusion and at or above 1.0-mm ($\sim 2.5 \times 10^5$ cells) central necrosis is often apparent. To alleviate this problem, organ cultures are usually placed at the interface between the liquid and gaseous phases to facilitate gas exchange while retaining access to nutrients. Most systems achieve this by positioning the explant on a raft or gel exposed to the air, but explants anchored to a solid substrate can also be aerated by rocking the culture, exposing it alternately to liquid medium or gas phase [Nicosia et al., 1983; Lechner & LaVeck, 1985] (see also LaVeck and Lechner protocol, Chapter 20), or by using a roller bottle or tube (see Chapter 23) to the same end.

Anchorage to a solid substrate can lead to the development of an outgrowth from the explant and resultant alterations in geometry, although this can be minimized using a nonwettable surface. One of the advantages of culture at the gas–liquid interface is that the explant retains a spherical geometry if the liquid is maintained at the correct level: too deep and gas exchange is impaired; too shallow and surface tension will tend to flatten the explant and promote outgrowth.

Increased permeation of oxygen can also be achieved by using increased O_2 concentrations up to pure oxygen or by using hyperbaric oxygen. Certain tissues, e.g., thyroid [De Ridder and Mareel, 1978] and prostate, trachea, and skin [Lasnitzki, 1992], particularly from newborn or adult, may benefit from elevated O_2 tension but often this is at the risk of O_2-induced toxicity. This may have to be determined for each tissue type under study.

Increasing the O_2 tension will not facilitate CO_2 release or nutrient/metabolite exchange, so the benefits of increased oxygen may be overridden by other limiting factors.

Structural Integrity

Structural integrity, above other considerations, was and is the main reason for adopting organ culture as an *in vitro* technique in preference to cell culture. While cell culture utilizes cells dissociated by mechanical or enzymic techniques, or spontaneous migration, organ culture deliberately maintains the cellular associations found in the tissue. Initially this was selected to facilitate histological characterization, but ultimately it was discovered that certain elements of phenotypic expression were only found if cells were maintained in close association.

It is now recognized that associated cells do exchange signals via junctional communications (gap junctions), via paracrine hormones, and via surface information exchange. This is most striking during organogenesis, but is probably also required for the maintenance of fully mature tissues.

Hence maintenance of the structural integrity of the original tissue may preserve the correct homologous and heterologous cellular interactions present in the original tissue, and maintain the correct chemical configuration of the extracellular matrix.

Growth and Differentiation

It has been stated previously (Chapter 14) that there appears to be a relationship between growth and differentiation such that differentiated cells no longer proliferate. It is also possible that cessation of growth may in itself contribute to the induction of differentiation, if only by providing a permissive phenotypic state receptive to exogenous inducers of differentiation.

Because of the rules of density limitation of growth, and the physical restrictions imposed by their geometry, most organ cultures do not grow, or if they do, proliferation is limited to the outer cell layers. Hence the status of the culture is permissive to differentiation and, given the appropriate cellular interactions and soluble inducers (see Chapter 14), should provide an ideal environment for differentiation to occur.

Limitations of Organ Culture

In view of the advantages described above, it is perhaps surprising that organ culture is not more popular. The reasons relate largely to the development of biochemical and molecular criteria for *in vitro* behavior, particularly the monitoring of differentiation. These criteria have been adopted because they are more readily quantified and generally more objective than histological criteria, although they lack the resolution of histological techniques where local response in a minority of cells can be detected.

Biochemical monitoring requires reproducibility between samples that is less easily achieved in organ culture than in propagated cell lines. This is due to the sampling variation in preparing an organ culture, to minor differences in handling and geometry, and to variations in cell type heterogeneity between cultures.

Organ cultures are also more difficult to prepare than replicate cultures from a passaged cell line and do not have the advantage of a characterized reference stock to which they may be related. Preparation is labor-intensive and as a result the yield of usable tissue is often too low to be of value in biochemical or molecular assays. Furthermore, as the reacting cell population may be a minor component, it is difficult to analyze the biochemical nature of the response and attribute it to the correct cell type other than by autoradiographic, histochemical, or immunocytochemical techniques that tend to be more qualitative than quantitative.

Organ cultures cannot be propagated and hence each experiment requires recourse to the original donor tissue.

Organ culture is essentially a technique for studying the behavior of integrated tissues rather than isolated cells. It is precisely in this area that the future understanding of the control of gene expression (and ultimately of cell behavior) in multicellular organisms may lie, but the limitations imposed by the organ culture system are such that recombinant systems between purified cell types may contribute more at this particular stage. However, there is no doubt that organ culture has contributed a great deal to our understanding of developmental biology and tissue interactions and that it will continue to do so in the absence of adequate synthetic systems.

Types of Organ Culture

As the technique has been dictated largely by the requirement to place the tissue at a location allowing optimal gas and nutrient exchange, most techniques place the tissue at the gas–liquid interface. This has been achieved by placing fragments of tissue on semisolid gel substrates of agar [Wolff and Haffen, 1952] or clotted plasma [Fell and Robison, 1929] or on a raft of microporous filter, lens paper, or rayon supported on a stainless steel grid or adherent to a strip of Perspex or Plexiglas [see Fig. 22.1 and Lasnitzki, 1992].

Two techniques have successfully departed from this method, the use of stirred cultures of small tissue fragments [Mareel et al., 1979; Bjerkvig et al., 1986a,b] (see Chapter 21) and the use of rocking or rolled cultures where the tissue is anchored to a substrate and subjected alternately to liquid culture medium and the gas

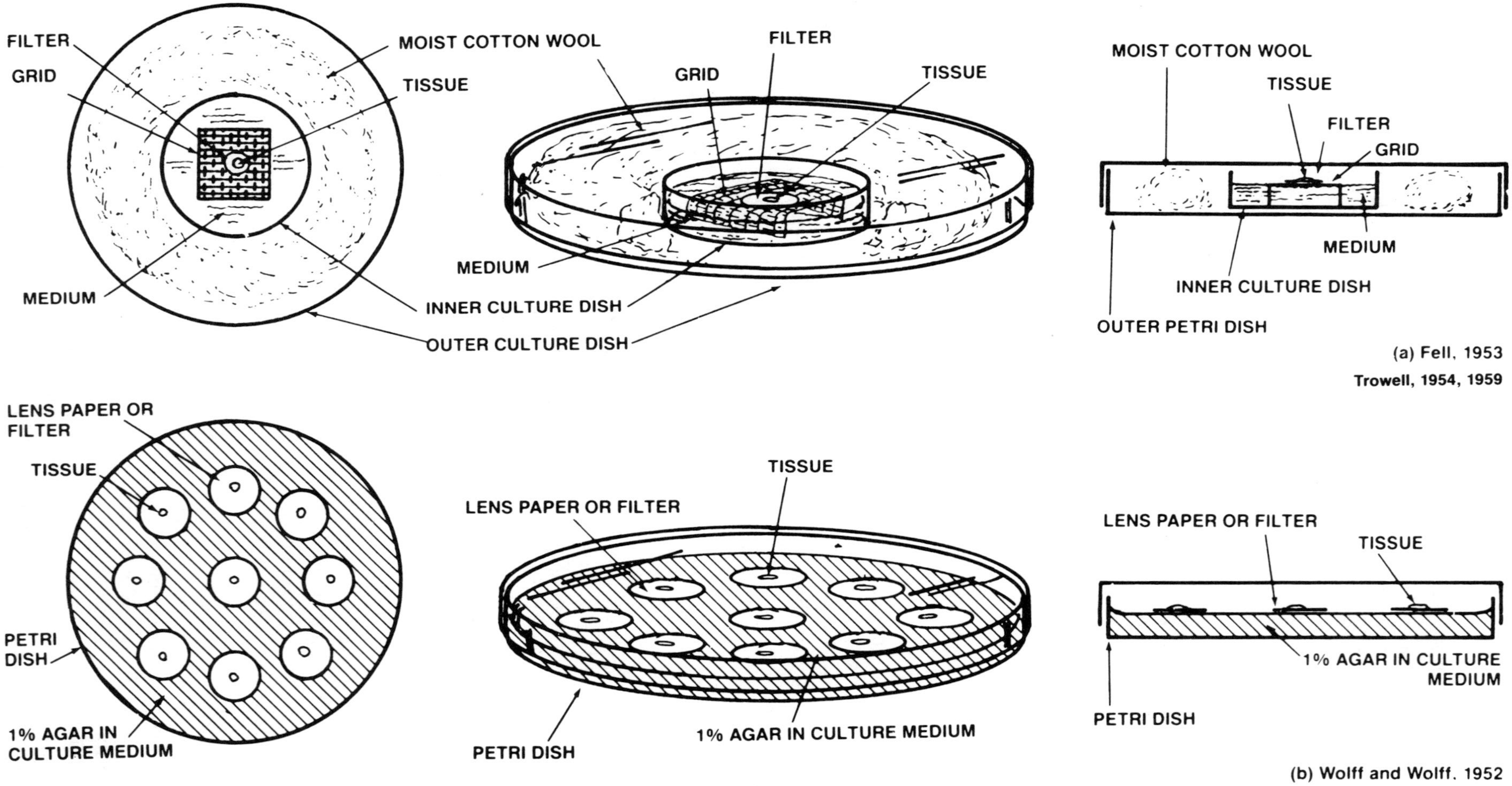

Fig. 22.1. *Types of organ culture. a. Wire gauze support in liquid medium [Fell, 1953; Trowell, 1959]. b. On filter or lens-paper rafts on agar medium [Wolff and Wolff, 1952]. c. Falcon organ culture dish (No. 3010) with stainless steel mesh support. d. Modification for embryonic induction studies [Grobstein, 1953a,b].*

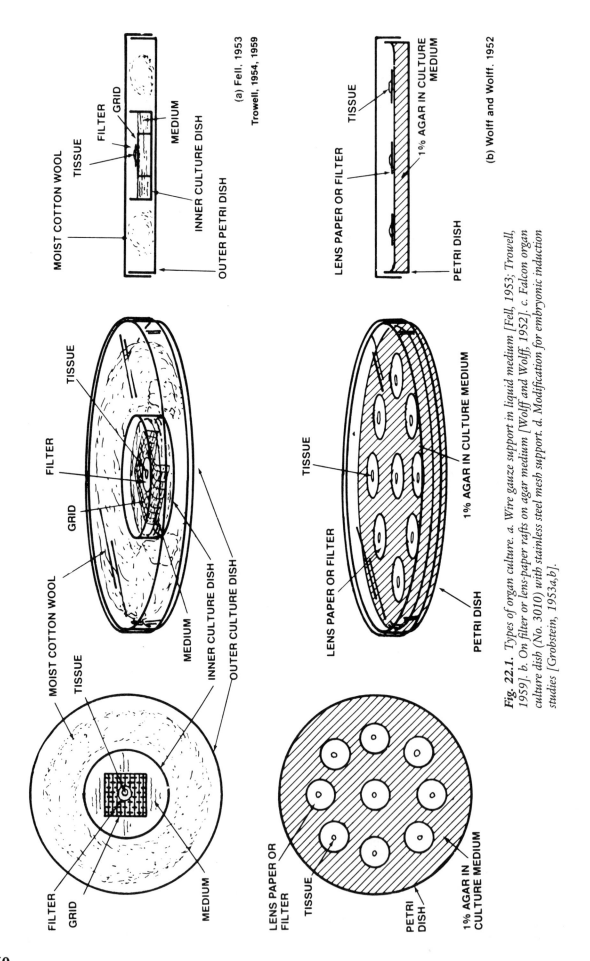

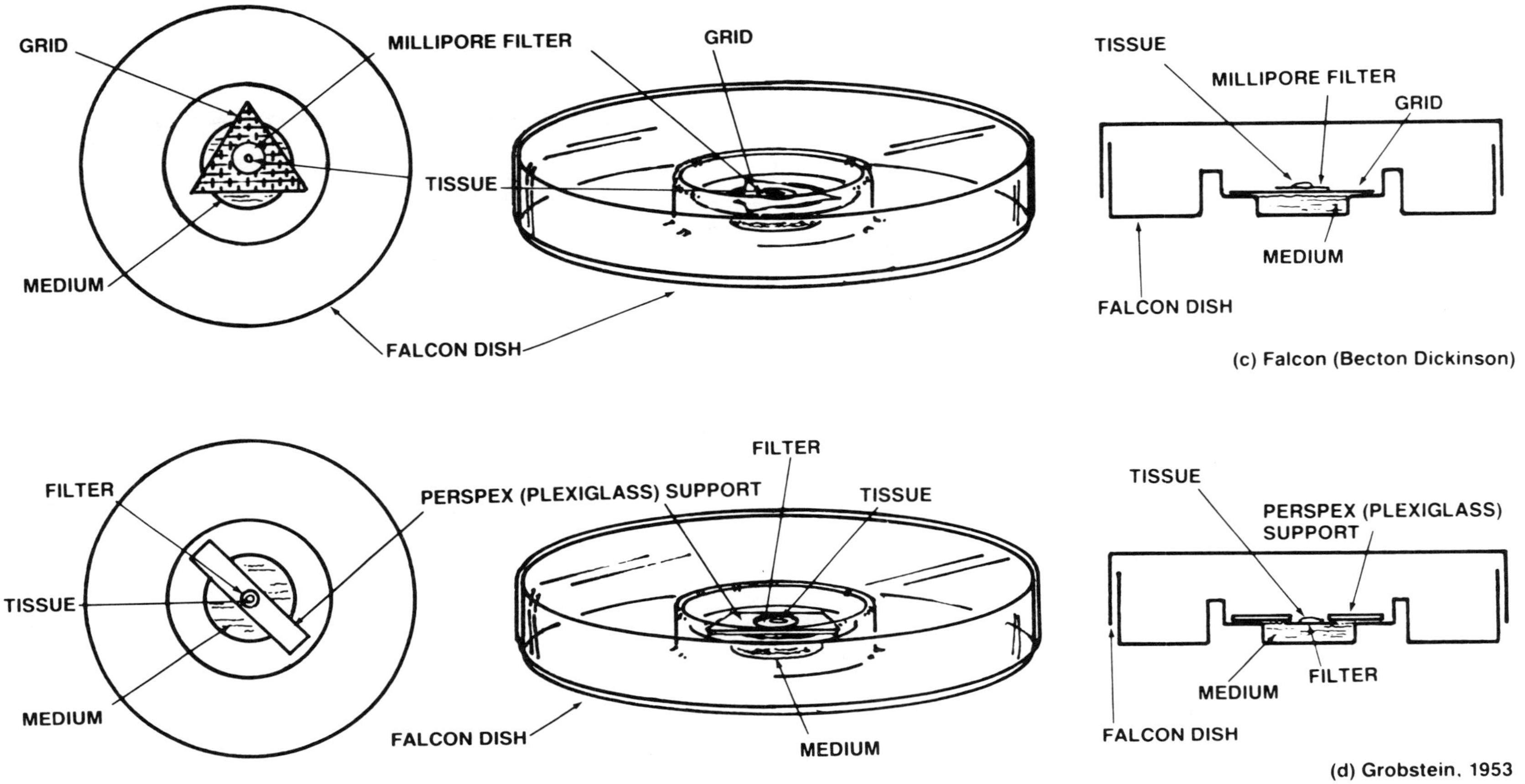

Fig. 22.1. (Continued)

phase [see Lechner, Chapter 20; Nicosia et al., 1983]. Stirred suspensions of tissue fragments tend to be restricted to embryonic or newborn tissue.

Several techniques have been described for gas–liquid interface culture, but one that has become popular because of the provision of suitable disposable plasticware is the grid technique derived from that of Trowell [1954, 1959].

Outline

Dissect out organ or tissue, reduce to 1 mm^3, or to thin membrane or rod, place on support at gas (air)/medium interface, and incubate in humid CO_2 incubator, changing the medium as required.

Materials (All sterile)

Instruments
Stainless steel grids
Sterile filters, 0.5-μm Nuclepore (sterilize by autoclaving)
Medium
Organ culture dishes (Falcon No. 3010; similar dish made by Costar for embryo culture)
Filter wells (see below) may also be used.

Protocol

1. Prepare grids (Falcon) with sterile filters (Nuclepore) in position (see Fig. 22.1) in dishes (Falcon), or use filter well of appropriate size (see below).
2. Add enough medium to wet the filter but not so much that it floats (~1.1 ml).
3. Place dishes in humid CO_2 incubator to equilibrate at 36.5°C.
4. Prepare tissue or dissect out whole embryonic organs, e.g., 8-day femur or tibiotarsus of chick embryo (see Chapter 9 for dissection). Tissue must not be more than 1 mm thick, preferably less, in one dimension, e.g., 8-day embryonic tibiotarsus is perhaps 5 mm long but only 0.5–0.8 mm in diameter. A fragment of skin might be 10 mm^2 but only 200 μm thick. Tissue like liver or kidney that must be chopped down to size should be no more than 1 mm^3.

   For short dissections (<1 hr), HBSS is sufficient, but for longer dissections, use 50% serum in HBSS buffered with HEPES to pH 7.4.
5. Take dishes from incubator and transfer tissue carefully to filters. A pipette is usually best and can be used to aspirate any surplus fluid transferred with the explant, although care should be taken to wet the inside of the pipette with medium before aspiration, to avoid fragments of tissue sticking to the pipette.
6. Check level of medium, making sure tissue is wetted, and return dishes to incubator.
7. Incubate for 1–3 wk, changing medium every 2 or 3 days.

Analysis

Usually by histology, autoradiography, or immunocytochemistry, but assay of total amounts of cellular constituents or enzyme activity is possible, although variation between replicates will be high.

Variations

Most variations are in:

(1) Type of medium: 199 or CMRL 1066 may be used with or without serum, and BJG [Biggers et al., 1961] for cartilage or bone.

(2) Type of support (see Fig. 22.1): The Grobstein technique has a number of advantages. Different types of tissue may be combined on the opposite sides of the filter to study their interaction (see above). Furthermore, the well formed on the top side of the filter assembly generates a meniscus of medium with a large surface area available for gas exchange. It is also possible to alter the configuration of the tissue by raising or lowering the level of medium in the dish, and thereby in the well; deeper medium gives a spherical explant and shallower medium flattens the explant. Although Grobstein filter assemblies are not commercially available, many of those currently available may be adaptable.

(3) O_2 tension: Embryonic cultures are usually best kept in air, but late-stage embryos, newborn, and adult tissue are better kept in 95% O_2 [Trowell, 1959; De Ridder and Mareel, 1978].

Organ cultures are useful in the demonstration of processes such as embryonic induction (see discussion of organogenesis [Auerbach & Grobstein, 1958, etc.] above), where the maintenance of the integrity of whole tissue is important. However, they are slow to prepare and present problems of reproducibility between samples. Growth is limited by diffusion (although growth is perhaps not necessary and may even be undesirable) and mitosis is nonrandomly distributed throughout the explant. Mitosis occurs around the periphery only, while the centers of explants frequently become necrotic (Fig. 22.2). It has been argued that this type of geometry makes organ cultures good models of tumor growth, where peripheral cell division is often accompanied by central necrosis.

HISTOTYPIC CULTURE

Various attempts have been made to regenerate tissue-like architecture from dispersed monolayer cultures.

Kruse and Miedema [1965] demonstrated that perfused monolayers could grow to more than 10 cells deep and organoid structures can develop in multilayered cultures if kept supplied with medium [Schneider et al., 1963; Bell et al., 1979]. Green & Thomas [1978] has shown that human epidermal keratinocytes will form dematoglyphs (friction ridges) if kept for several weeks without transfer, and Folkman and Haudenschild [1980] were able to demonstrate formation of capillary tubules in cultures of vascular endothelial cells cultured in the presence of endothelial growth factor and medium conditioned by tumor cells.

Sponge Techniques

Leighton first demonstrated that cells would penetrate cellulose sponge [Leighton et al., 1968]. Both normal and malignant cells can do this and it does not seem to reflect malignant behavior. Collagen coating of the sponge may facilitate occupation and Gelfoam (a gelatin sponge matrix used in reconstructive surgery) may be used in place of cellulose [Sorour et al., 1975]. These systems require histological analysis and are limited in dimensions, like organ cultures, by gaseous and nutrient diffusion.

Collagen gel (native collagen as distinct from denatured collagen coating) provides a matrix for the morphogenesis of primitive epithelial structures. Yang et al. [1979, 1980, 1981] and Flynn et al., [1982] have shown that breast epithelium forms rudimentary tubular and glandular structures grown in collagen. Analysis is again by histology.

Hollow Fibers

Since medium supply and gas exchange become limiting at high cell densities, Knazek et al. [1972; Gullino and Knazek, 1979] developed a perfusion chamber from a bed of plastic capillary fibers now available commercially (Amicon, Endotronics). The fibers are gas- and nutrient-permeable and support cell growth on their outer surfaces. Medium, saturated with 5% CO_2 in air, is pumped through the centers of the capillaries, and cells are added to the outer chamber surrounding the bundle (Fig. 22.3; see also Fig. 7.3). The cells attach and grow on the outside of the capillary fibers fed by diffusion from the perfusate and can reach tissue-like cell densities. There is an option between two types of plastic and different ultrafiltration properties giving molecular weight cut-off points at 10,000, 50,000, or 100,000 dalton, regulating the diffusion of macromolecules from the medium to the cells.

It is claimed that cells in this type of high-density culture behave as they would *in vivo*. Choriocarcinoma cells release more human chorionic gonadotrophin [Knazek et al., 1974] and colonic carcinoma cells produce elevated levels of CEA [Rutzky et al., 1979; Quarles et al., 1980]. There are considerable technical difficulties in setting up the chambers, however, and they are costly. Sampling cells from these chambers and determination of the cell concentration are difficult. However, they appear to present an ideal system for studying the synthesis and release of biopharmaceuticals, and are now being exploited on a semi-industrial scale (Endotronics) (see also Chapter 23).

Reaggregation and Spheroids

When dissociated cells are cultured in a gyratory shaker, they may reassociate into clusters. Dispersed cells from embryonic tissues will sort during reaggregation in a highly specific fashion [Linser and Moscona, 1980]; e.g., Muller cells of the chick embryo retina reaggregated with neuronal cells from the retina were inducible for glutamine synthetase, but those reaggregated with neurons from other parts of the brain were not. Cells in these heterotypic aggregates appear to be capable of sorting themselves into groups and forming tissue-like structures. This property is less easily demonstrated in adult cells, although some results suggest that it may be possible for adult cells to form organoid structures [Douglas et al., 1976, 1980; Bell et al., 1979].

Homotypic reaggregation also occurs fairly readily, and spheroids generated in gyratory shakers or by growth on agar have been used as models for chemotherapy *in vitro* [Twentyman, 1980] and for the characterization of malignant invasion [Mareel et al., 1980]. As with organ cultures, growth is limited by diffusion and a steady state may be reached where cell proliferation in the outer layers is balanced by central necrosis.

The following protocol for generating spheroid cultures has been contributed by Tom Wheldon, Institute of Radiotherapeutics and Oncology, Belvidere Hospital, Glasgow, Scotland.

Principle

Multicellular tumor spheroids provide a proliferating model for avascular micrometastases. The three-dimensional structure of spheroids allows the experimental study of aspects of drug penetration and resistance to radiation or chemotherapy dependent on intercellular contact. Human tumor spheroids are more easily developed from established cell lines or from xenografts than from primary tumors [Sutherland, 1988].

Outline

From single cell suspension (trypsinized monolayer or disaggregated tumor), cells are inoculated into flasks, previously base-coated with agar, and incubated to allow formation of small aggregates over 3–5 d. Aggregates may be transferred to fresh flasks, to

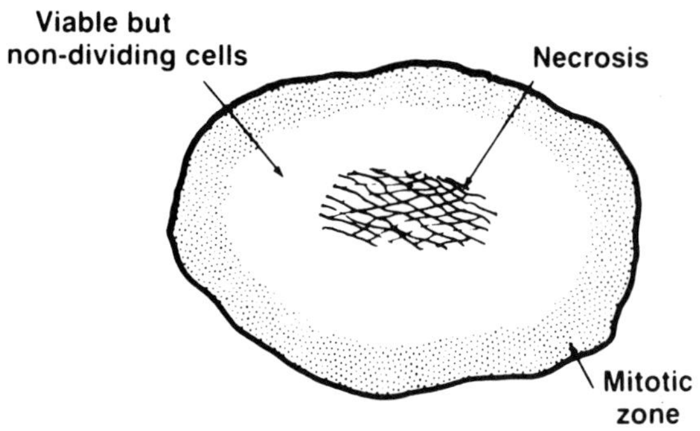

Fig. 22.2. Diagrammatic representation of the expected distribution of mitoses (stippled area) and necrosis (shaded central area) in an organ culture explant.

magnetic stirrer vessels, or to multiwell plates, where continued growth over 2 wk yields spheroids of maximum size, about 1,000 μm [Yuhas et al., 1977].

Materials

Sterile:
Noble agar (Difco)
Growth medium
Distilled water (sterile)
Trypsin 0.25% in PBS
25-cm² flasks or 24-well plates
9-cm petri dishes
Note. Where agar coating is used, flasks, plates, and dishes should be sterile but not necessarily tissue culture grade.

Protocol

1. Agar coating. In 25-cm² flasks: add 1 g Noble agar (Difco) to 20 ml distilled water in a 150-ml Erlenmeyer flask and boil gently until the agar has completely dissolved (about 10 min). Add contents immediately to 60 ml of growth medium, previously heated to 37°C, and put 5-ml aliquots into each flask. Ensure that the agar is free from bubbles. It will set (at room temperature) in about 5 min, giving a 1.25% agar-coated flask. In multiwell plates: add 0.5 g agar to 10 ml distilled water, heat as above, and then add 40 ml of distilled water. Place 0.5 ml of the resultant solution in each well of a 24-well plate (Corning 25820) to give a base coat of 1% agar. Accuracy and careful placement is important to ensure easy well-to-well focus of the microscope in subsequent viewing of spheroids.
2. Spheroid initiation. Trypsinize confluent monolayer (established lines) (see Chapter 10) or dis-

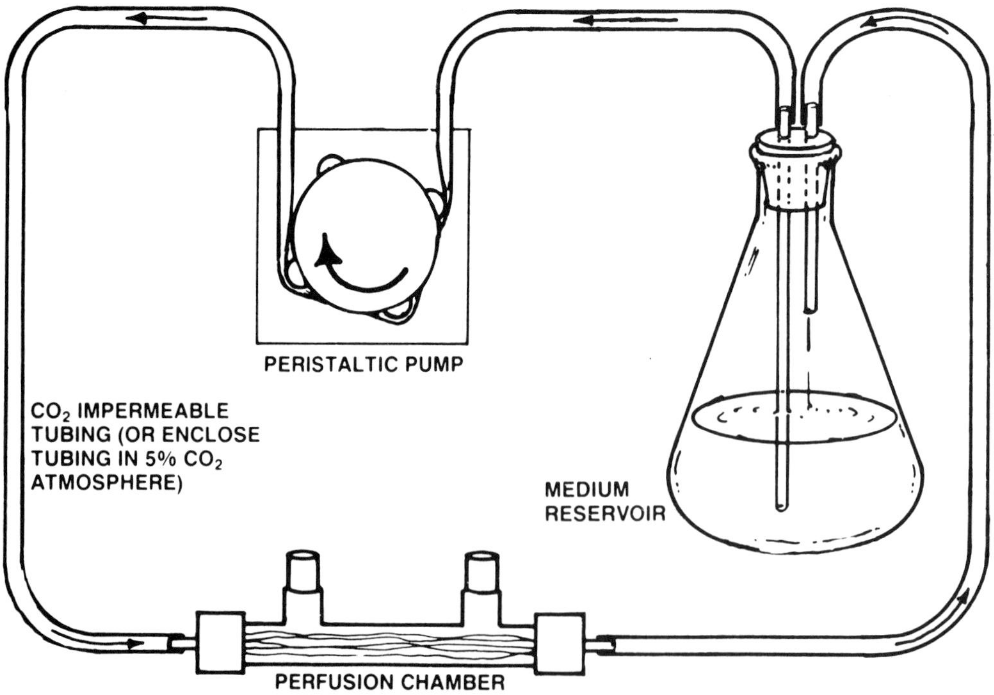

Fig. 22.3. Vitafiber (Amicon) apparatus for perfused culture on capillary bundles of permeable plastic. Medium is circulated from a reservoir to the culture chamber by a peristaltic pump. As silicone tubing is gas permeable, the apparatus should be enclosed in an atmosphere of 5% of CO_2. Alternatively, the length of tubing between the pump and the culture chamber may be enclosed in a polyethylene bag purged with CO_2.

aggregate (solid tumors) (see Chapter 9) to give single cell suspension. Neutralize trypsin with medium containing serum (if necessary). Count cells (Coulter counter or hemocytometer). Place 5×10^5 cells in each agar-coated 25-cm^2 flask in 5 ml of growth medium and incubate. If the cells are capable of spheroid formation, small aggregated clumps (about 100–300-μM diameter) will form spontaneously in 3–5 d.

3. For subsequent growth, spheroids should be transferred to new vessels. If growth is to be continued in 25-cm^2 flasks, the contents of the original flasks should be transferred to conical centrifuge tubes or universal containers, the spheroids allowed to settle, and single cells removed with the supernatant. Spheroids may then be resuspended in fresh medium and transferred to new agar-coated flasks, where growth will proceed by division of cells in the outer layer.
4. For growth in wells, decant the contents of a 25-cm^2 flask into a 25-cm^2 petri dish and, under laminar flow conditions, select individual spheroids of chosen dimensions under low-power magnification (×40). Using a Pasteur pipette and a Pi-Pump (Shuco International), transfer selected spheroids of similar diameter individually to agar-coated wells or a 24-well plate, each containing 0.5 ml medium. Place plates in CO_2 incubator. Replace medium once or twice weekly (exchanging 0.5 ml each time) or add 0.5 ml medium (without removal) once or twice weekly giving 2–4 wk for a 2-ml well.
5. Spheroid growth in wells or flasks may be quantified by regular measurement (e.g., two to three times weekly) of diameter using a microscope graticule or, better, by measurement of cross-sectional area using an image analysis scanner. The most accurate growth curves are obtained when spheroids are grown in wells and individually monitored.

Variations

(1) Some spheroids can be grown only when subjected to continuous agitation. Transfer single cells in suspension, or preformed aggregates, to siliconized glass culture vessels and agitate on a magnetic stirrer (Techne).

(2) As normal stromal cells are usually excluded from spheroids, spheroids formed from a disaggregated tumor cell suspension will usually contain only tumor cells. Monolayer cultures of tumor cells may then be established by placing spheroids in uncoated tissue culture petri dishes, to which they will attach. Monolayers will grow out from each attached spheroid and may be harvested by trypsinization [Bruland et al., 1985].

Applications

Spheroids have wide applications in assessment of cytotoxic treatment. Endpoints include treatment-induced growth delay, proportion of spheroids sterilized ("cured") by treatment, and colony formation in monolayer following disaggregation of treated spheroids [Freyer & Sutherland, 1980]. An important area of applications is in the use of spheroids to study penetration of cytotoxic drugs, antibodies, or other molecules used in targeted therapy [Sutherland, 1988; Carlsson and Nederman, 1989]. This category of application represents a special application of spheroids, as it is not possible in single cell suspensions or monolayer cultures.

FILTER WELLS

Several attempts have been made to generate dense populations of cells supported on a filter, either perfused from below by aerated medium [Dickson and Suzangar, 1976] or located at the gas–liquid interface or near to it. Mauchamp demonstrated the development of polarity and functional integrity in thyroid epithelium explanted on a collagen-coated filter in a specially constructed mount [Chambard et al., 1983] (see Chapter 14). Others have used the system to study invasion by granulocytes or malignant cells [McCall et al., 1981; Elvin et al., 1985].

One of the major advantages of the system is that it allows the recombination of cells at very high tissue-like densities, with ready access to medium and gas exchange, but in a multireplicate form. McCall et al. [1981] constructed chambers from disposable 2-ml syringes that utilized multiwell plates for support, and Millipore produces the Millicell-HA chambers in two sizes, 12-mm and 30-mm diameter (Fig. 22.4a), suitable for 24-well plates, 6-well plates, or larger dishes. These filters are available in HA-grade Millipore filters of different porosities. Similar filters are produced by Costar, Falcon, and Nunc (Anotec), already mounted in polystyrene (Transwells) in a wide range of sizes (Table 22.1) and filter materials.

Outline

Cells are seeded into filter chambers and cultured in excess medium in deep petri dishes or sample pots (Sterilin, Macom).

Materials (All sterile)

Approximately 0.5×10^6 cells per cm^2 of filter
Growth medium, 1–20 ml per filter (depending on vessel housing filter)

Filter inserts (Costar, Falcon, Millipore, Nunc)
Containers for filters: Corning 9-cm petri dishes, deep sample pots (Sterilin or Macom) if a high medium:cell ratio is required. Otherwise 6- or 24-well plates supplied with filter inserts will suffice.

Protocol

1. Place filter wells in dish or container.
2. Add medium, tilting dish to allow medium to occupy space below filter and displace air with minimum entrapment. Add medium until level with filter (15 ml in 9-cm petri dish, 2.5 ml in 6-well plate, 1.0 ml in 24-well plate).
3. Level dish and add 2×10^6 cells, 10^6/ml in 2 ml medium to top of filter for a 25-mm diameter filter, or 5×10^5 in 200 µl for an 8-mm diameter filter, taking care not to perforate filter.
4. Place in humid CO_2 incubator in protective box (see Chapter 11). It is critical to avoid shaking the box, and the cultures should not be moved in the incubator to avoid spillage and resultant contamination.
5. Monolayer (or multilayer) should become established in about 5 d. It forms before then but complete integrity (formation of matrix, cell contacts, polarity, etc.) takes several days.
6. Culture may be maintained indefinitely, replacing medium every 3–5 d or transferring to fresh dish. A second cell type may be added on top if it is desired to study interactions, or to the other side of the filter by inverting the filter holder.

Analysis

(1) Penetration of cells through filter. Trypsinize and count each side of filter in turn (trypsinized cells will not pass through even an 8-µm filter, as their spherical diameter in suspension exceeds this), or fix, embed, and section by electron microscope or conventional histology. Visualization is possible in whole mounts by mounting the fixed, stained (Giemsa) filter on a slide in DePeX under a coverslip under pressure to flatten it. Differential counting can then be performed by focusing on each plane alternately.

(2) Detachment of cells from filter to bottom of dish. Count by trypsinization or scanning.

(3) Partition above and below filter (chemotaxis or invasion) either by counting as in step 1 above or by prelabeling the cells with rhodamine or fluorescein isothiocyanate (5 µg/ml for 30 min in trypsinized suspension) and measuring the fluorescence of solubilized cells (0.1% SDS in 0.3 *N* NaOH for 30 min) trypsinized from either side of the filter.

(4) Invasion. Precoat filter with cell layer (normal fibroblasts, MDCK, etc.), ensure confluence is achieved by microscopic examination, and then seed EDTA-dissociated test cells on top of preformed layer (10^5–10^6 cells per filter). If test cells are RITC- or FITC-labeled, fluorescent measurements will reveal appearance of cells below filter.

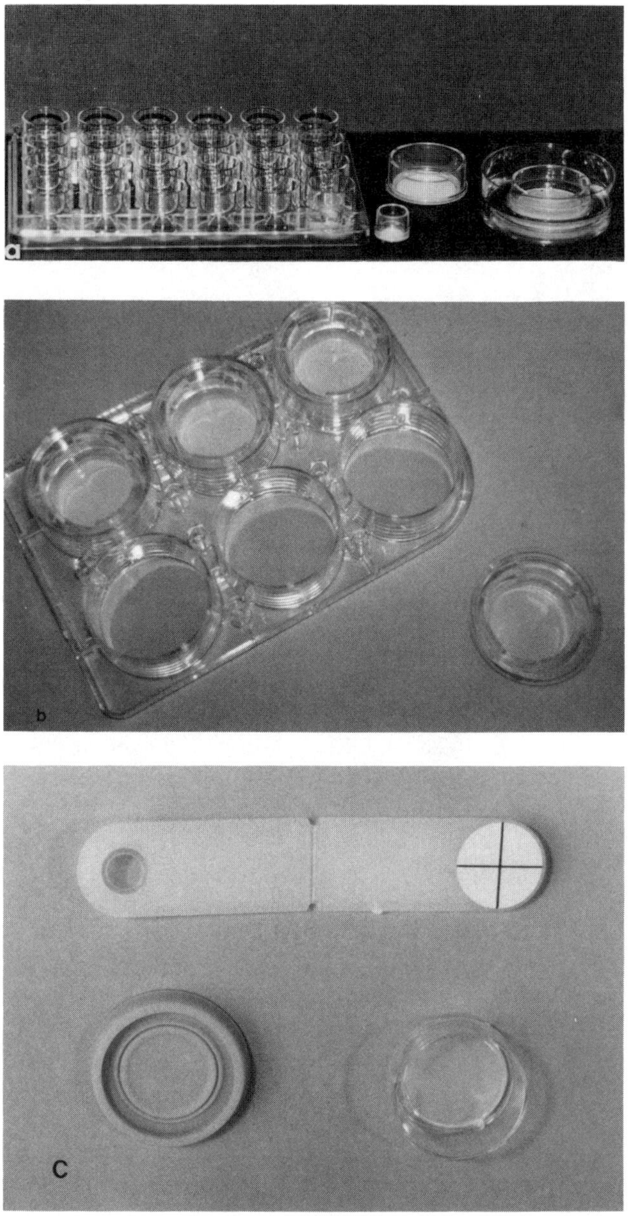

Fig. 22.4. Filter-well units for culture. a. Millipore, disposable. b. Three Costar "Transwells" in six-well plate, with Nuclepore filter well alongside. c. Ceramic filter (Nunc) with tool for removing filter from holder for cytology or extraction.

Several assays for invasion have been described using Matrigel or collagen coating of the filter [Repesh, 1989; Schlechte et al., 1990]. This has the advantage of being simpler to set up, and penetration is generally quicker, but there are potential problems of validity where no cellular barrier is being crossed.

TABLE 22.1. Types of Filter Wells and Applications

Make	Name	Material	Qualities	Transparency	Porosity (μm)
Millipore	Millicel	Nitrocell	Mesh	Opaque	0.45
		Polyolefin	Mesh	Transparent	0.45
Costar	Transwells	Polycarb	Absolute	Transparent	5–8
				Translucent[a]	0.45–1.0
Falcon	Inserts	Melinex	Absolute	Transparent	0.45–3
Nunclon	Anocel	Ceramic	Sieve	Transparent	0.01
Nuclepore	NuCell	Polycarb	Absolute	See Costar	0.45–8
Earl-Clay	Ultraclone	Collagen	Mesh	Transparent	

[a]The higher the pore frequency, the lower the transparency. Low-porosity filters have a high pore frequency and are, consequently, less transparent.

Variations

(1) Filter porosity. One-micrometer filters allow cell interaction and contact without transit of filter. Eight-micrometer filters allow live cells to cross. Filters of 0.2 μm probably do not allow cell contact. Low-porosity filters may be used to study cell interaction without intermingling.

(2) Transfilter combinations. Invert filter and load underside first with 0.5 ml, 2×10^6/ml of cell suspension. After 2 d invert filter and load well as above.

(3) Organotypic culture. The advent of filter-well technology, boosted by commercial availability, has produced a rapid expansion in the study of organotypic culture methods. Skin equivalents have been generated by co-culturing dermis with epidermis, with an intervening layer of collagen, or with dermal fibroblasts incorporated into the collagen [Fusenig, 1992], and models for paracrine control of growth and differentiation have been developed with lung [Speirs et al., 1991], prostate [Cunha et al., 1983], and breast [Van Roozendahl et al., 1992]. Several assays for malignant invasion have been described using Matrigel or collagen coating of the filter [Hendrix et al., 1987, 1990; Repesh, 1989; Schlechte et al., 1990].

CHAPTER 23

Specialized Techniques

It is intended in this chapter to provide detailed information on some specialized techniques referred to in the text but so far not described. In some cases, these are not tissue culture techniques *per se*, but rather techniques associated with tissue culture and that might be used in a number of different tissue culture-based experiments, e.g., autoradiography. Other techniques involving tissue culture directly but of a very specialized nature, e.g., monoclonal antibody production, are also described.

MASS CULTURE TECHNIQUES

Mass culture might be defined as encompassing from 10^9 cells to semi-industrial pilot plant 10^{11} or 10^{12} cells (1–1,000 l). The method employed depends on whether the cells proliferate in suspension or require to be anchored to the substrate.

Suspension Culture

Scale-up of suspension cultures is relatively simple since only the volume need be increased. Above about 5-mm-deep agitation of the medium is necessary, and above 10 cm sparging with CO_2 and air is required to maintain adequate gas exchange (Fig. 23.1). Stirring of such cultures is best done slowly with a large surface area paddle or large-diameter magnetic stirrer bar. The stirring speed should be between 30 and 100 rpm, sufficient to prevent cell sedimentation but not so fast as to grind or shear the cells. If a bar is used, it must be kept off the base of the culture vessel with a collar or be suspended from above. Antifoam (Dow Chemical Co.) must be included where the serum concentration is above 2%, particularly if the medium is sparged. In the absence of serum, it may be necessary to increase the viscosity of the medium with (1–2%) carboxymethyl cellulose (molecular weight $\sim 10^5$).

The procedure for setting up a 4-l culture of suspended cells is as follows:

Outline

Grow pilot culture of cells and add to prewarmed, pregassed aspirator of medium. Stir slowly with sparging until required cell concentration is reached and harvest.

Materials

Sterile:

Growth medium with antifoam (Dow Corning, BDH)

Pilot culture

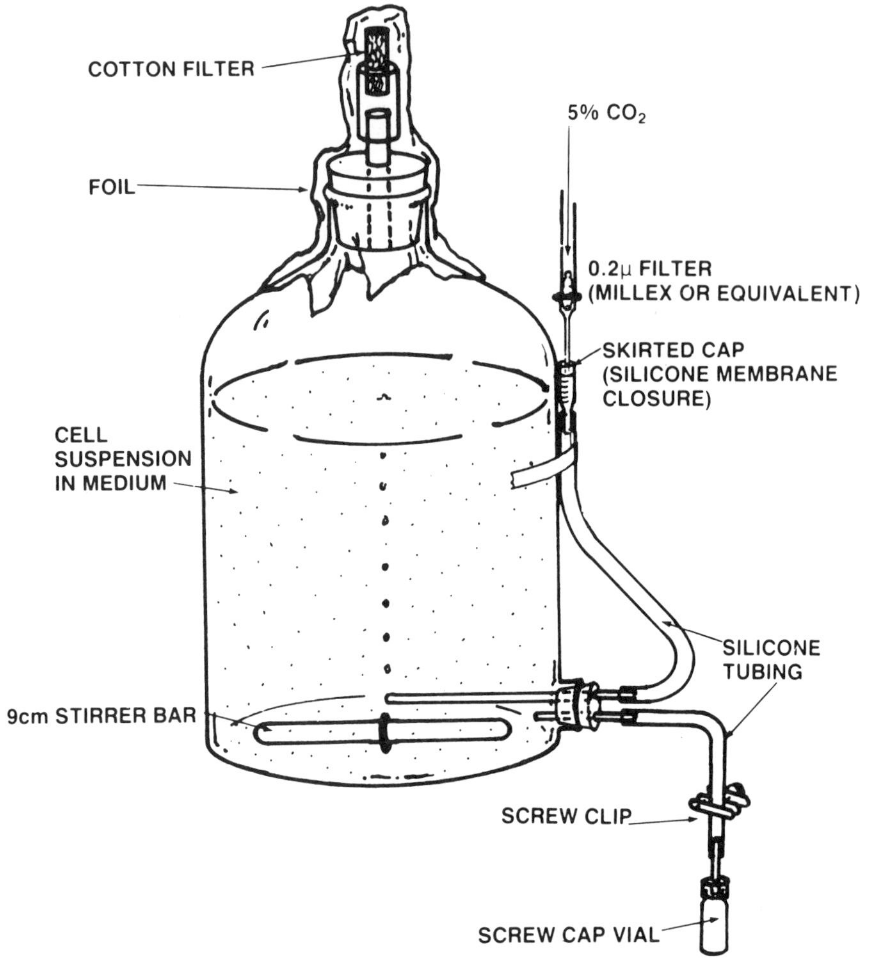

Fig. 23.1. *Bulk culture of cells in suspension. Standard 5- or 10-l aspirators may be modified as illustrated. Optimum mixing is achieved by using a large stirrer bar, with a central collar, and a slow stirring speed (~60 rpm). (Apparatus developed by the staff of the Beatson Institute for Cancer Research, Glasgow, Scotland.)*

Prepared aspirator or spinner culture vessel (Bellco, Techne) (Figs. 23.1, 23.10)
Nonsterile:
Magnetic stirrer and follower, or paddle
Supply of 5% CO_2
Bunsen burner or portable gas torch
Gas lighter
Counting fluid
Cell counter or hemocytometer

Protocol

1. Prepare a standard 5- or 10-l aspirator as in Figure 23.1, with a one-holed silicon rubber stopper at the top and a two-holed stopper at the bottom. The top stopper carries a glass tube with a cotton plug and the bottom stopper has (a) a glass tube and silicone rubber connection to an inlet port closed with a silicone membrane closure ("skirted cap") and (b) a screw cap fixed to a tube leading from the stopper with a screw-cap vial inserted in the cap. The hole for the bottom stopper should be about 15 mm above the base of the aspirator bottle and the stopper should fill it completely and leave no crevices for cells to lodge.

   A Teflon-coated bar magnet is placed in the aspirator. It should be as large as possible, while still able to turn freely in the bottom of the aspirator (~9–12 cm), and should have a central collar to raise it up from the base of the aspirator to avoid grinding cells below the bar. The collar may be made from silicone tubing if it is not already present.

   Alternatively, prepare a Techne or Bellco stirrer vessel, with the paddle in place and at least one of the vents closed with a permeable cap or cotton plug.

Sterilize by autoclaving 100 kPa (15 lb/in^2) for 20 min.

2. Set up "starter culture," using a standard screw-capped 1-l reagent bottle, or equivalent, with a Teflon-coated bar magnet in the bottom (see Fig. 10.2). Add 400 ml medium and seed with cells at $5 \times 10^4-10^5$/ml. Place on magnetic stirrer (see Chapter 4) rotating at 60 rpm and incubate until $5 \times 10^5-10^6$ cells/ml is reached.

   Small spinner vessels, 250–500 ml, are also availabe from Bellco and Techne (Fig. 23.10), suitable for use for starter cultures. Like the larger vessel, they should be autoclaved with one vent sealed with a permeable closure or cotton plug.
3. Add 4 l medium and 0.4 ml antifoam to sterile aspirator. The antifoam should be added directly to the aspirator using a disposable pipette or syringe. Place on magnetic stirrer in 36.5°C room or incubator. Connect 5% CO_2 air line via fresh, sterile, 25-mm, 0.2-μm Millex filter to sterile disposable hypodermic needle and insert needle into skirted cap or membrane seal. Turn on gas at a flow rate of approximately 10–15 ml/min and stir at 60 rpm. Incubate for about 2 hr to allow temperature and CO_2 tension to equilibrate.
4. Bring aspirator and starter culture back to laminar flow hood and remove top stopper (keeping aluminum foil in place) and starter bottle cap, taking care to keep stopper sterile. Flame neck of aspirator and starter culture bottle and pour starter culture into aspirator.
5. Replace stopper in aspirator and return to incubator or hot room. Reconnect 5% CO_2 line and restart stirrer at 60 rpm. Adjust gas flow to 10–15 ml/min.
6. Incubate for 4–7 d, sampling (see below) every day to check cell growth. When cell concentration reaches desired level, disconnect aspirator, run off cells into centrifuge bottles, and centrifuge at 100 g for 10 min.

Sampling. On aspirator, open screw clip and run 5–10 ml into vial to remove cells and medium that have been stagnant in the delivery line. Discard vial and contents and replace with fresh vial. Collect second 5–10 ml, perform a cell count, and check viability by dye exclusion.

Sampling from the Techne or Bellco stirrers can be performed by inserting a pipette into one of the side arms. This must be done under aseptic conditions. Alternatively, the stirrer vessel can be set up with a sampling tube already in place, taking care to avoid the stirrer paddle or pendulum.

Analysis. For best results, cells should not show a lag period of more than 24 hr and should still be in exponential growth when harvested. Plot cell counts daily and harvest at approximately 10^6 cells/ml.

Variations.

Continuous culture–"Biostat." If it is required that the cells be maintained at a set concentration, e.g., at mid-log phase, cells may be removed and medium added daily or cells may be run off and medium added continuously using the skirted cap entry port to add medium and the screw-cap outlet to collect into a larger reservoir (Fig. 23.2). The flow rate of medium may be calculated from the growth rate of the culture [Griffiths, 1992] but is better determined experimentally by serial cell counting at different flow rates of medium. Flow rate may be regulated by a variable peristaltic pump or a screw clip on the line above the drip feed, using gravity to produce the flow.

Production of cells in bulk, in the 1–20-l range, is best done by the "batch" method outlined first. The "steady-state" method is required for monitoring metabolic changes related to cell density but is more expensive in medium and is more likely to lead to contamination. However, if the operation is in the 50–1,000-l range, more investment and time is spent in generating the culture, and the batch method becomes more costly in time, materials, and "down time."

Suspension cultures can also be grown in bottles rotating on a special rack as for monolayer cultures (see below).

Monolayer cells. Anchorage-dependent cells cannot be grown in liquid suspension except on microcarriers (see below), but transformed cells, e.g., virally transformed or spontaneously transformed continuous cell lines, can. Because these cells are still capable of attachment, the culture vessels will need to be coated with a water-repellent silicone (e.g., Repelcote) and the calcium concentration may need to be reduced. MEMS medium is a variation of Eagle's MEM with no calcium in the formulation that has been used for the culture of HeLa-S_3 and other cells in suspension.

Monolayer Culture

For anchorage-dependent monolayer cultures, it is necessary to increase the surface area of the substrate in proportion to the cell number and volume of medium. This requirement has prompted a variety of different strategies, some simple, others complex.

Nunclon cell factory. The simplest system for scaling up monolayer cultures is the Nunclon Cell Factory (Figs. 23.3, 23.4) (see Table 7.1). This is made up of

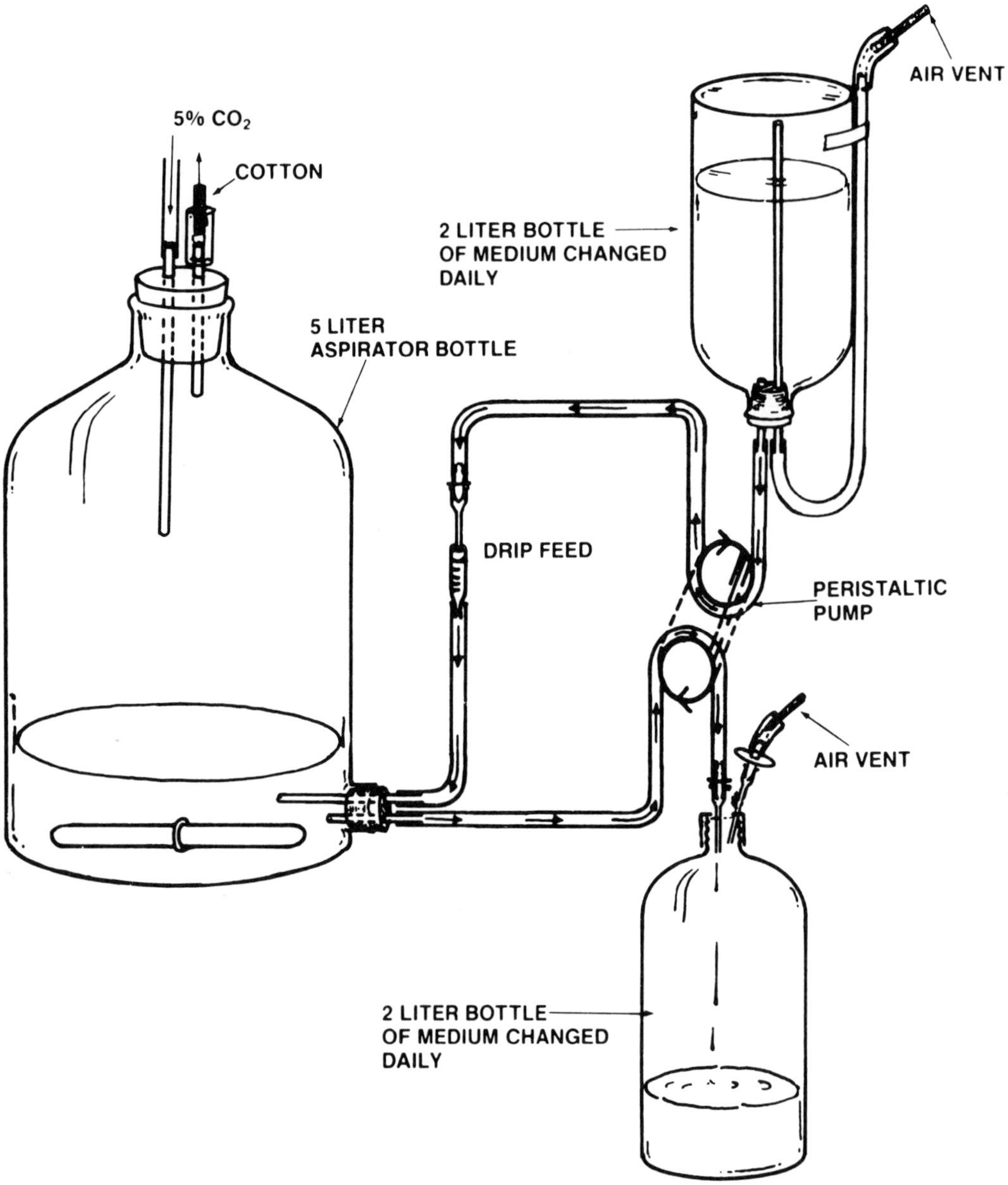

Fig. 23.2. "Biostat." A modification of the suspension culture vessel of Figure 23.1, with continuous matched input of fresh medium and output of cell suspension. (Based on a system developed by John Paul and George Lanyon.) The objective is to keep the culture conditions constant rather than to produce large numbers of cells. Bulk culture, per se, *is best performed in batches in the apparatus in Figure 23.1.*

rectangular petri dish-like units, giving a total surface area from 600–24,000 cm², interconnected at two adjacent corners by vertical tubes. Because of the positions of the apertures in the vertical tubes, medium can only flow between compartments when the unit is placed on end. When the unit is rotated and laid flat, the liquid in each compartments is isolated, although the apertures in the interconnecting tubes still allow connection of the gas phase. The cell factory has the advantage that it is not different in the geometry or the nature of its substrate from a conventional flask or petri dish. The recommended method of use is as follows:

Outline

Prepare a cell suspension in medium and run into the chambers of the unit. Lay the unit flat and gas with CO_2. Seal and incubate.

Materials (All sterile)

Monolayer cells
Growth medium

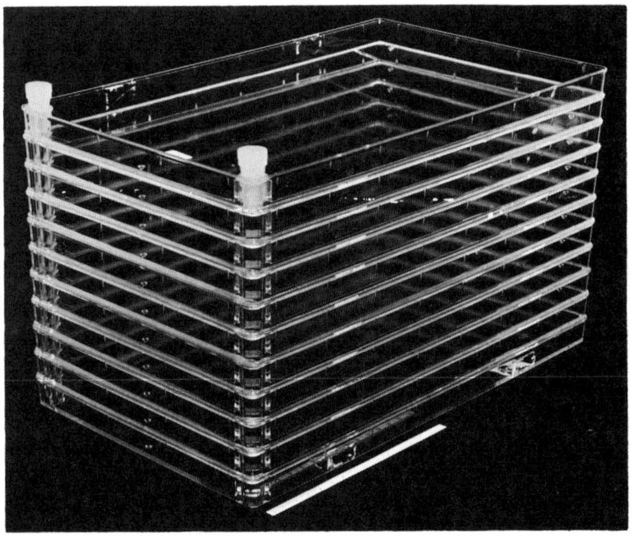

Fig. 23.3. *Nunclon "Cell Factory." Versions of this type of culture vessel are available both smaller (~1,500 cm²) and larger (~100,000 cm²), working on the same principle.*

0.25% crude trypsin
PBS
Hemocytometer or cell counter and counting fluid
Culture chamber
Silicone tubing and connectors

Protocol

1. Trypsinize cells (Chapter 10), resuspend, and dilute to 2 x 10^4 cells/ml in 1,500 ml medium.
2. Place chamber on long edge with supply tube to the bottom (see Fig. 23.4) and run cells and medium in through supply tube. Medium in all chambers will reach the same level.
3. Clamp off supply tube and disconnect from medium reservoir.
4. Rotate unit through 90° in the plane of the monolayer so that the unit lies on a short edge with the supply tube at the top.
5. Rotate unit through 90° perpendicular to the plane of the monolayer so that it now lies flat on its base with the culture surfaces horizontal. To transport to incubator, tip medium away from supply port.
6. If it is necessary to gas the culture, loosen clamp on supply line and purge unit with 5% CO_2 in air for 5 min, then clamp off both supply and outlet. The unit may be gassed continuously if desired.
7. To change medium (or collect medium), reverse step 5 and then 4, flame clamped line, open clamp, and drain off medium.
8. Replace medium as in steps 2–6.
9. To collect cells, remove medium as in step 7, add 500 ml PBS, and remove. Add 500 ml trypsin at 4°C and remove after 30 s. Incubate, add medium after 15 min, and shake to resuspend cells. Run off cells as in step 7.
10. The residue may be used to seed the next culture, although this does make it difficult to control the seeding density. It is better to discard the chamber and start fresh.

Analysis. Following growth in these chambers is difficult, so a single tray or chamber is supplied to act as a pilot culture. It is assumed the single tray will behave as the multichamber unit.

The supernatant medium can be collected repeatedly for virus or cell product purification. Collection of cells for analysis depends on the efficiency of trypsinization.

This technique has the advantage of simplicity but can be expensive if the unit is discarded each time cells are collected. It was designed primarily for harvesting supernatant medium but is a good method for producing large numbers of cells (3×10^8–3×10^9) for a pilot run, or on an intermittent basis.

Roller culture. If cells are seeded into a round bottle or tube that is then rolled around its long axis, the

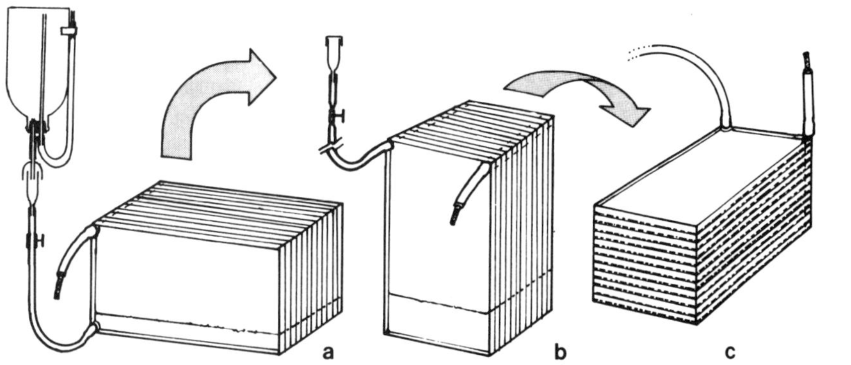

Fig. 23.4. *Filling Nunclon Cell Factory. a. Run medium in. b. Rotate onto short side away from inlet. c. Lay down flat, seal inlet, or connect to 5% CO_2 line.*

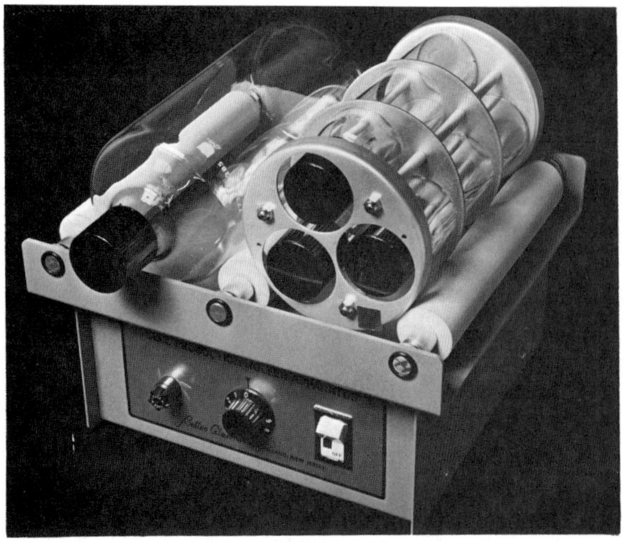

a

b

Fig. 23.5. *Roller culture bottles on racks. a. Small, bench-top rack (Bellco). b. Large bench-top or free-standing extendable rack. (Courtesy of New Brunswick Scientific.)*

medium carrying the cells runs around the inside of the bottle (Figs. 23.5, 23.6). If the cells are nonadhesive, they will be agitated by the rolling action but remain in the medium. If the cells are adhesive they will gradually attach to the inner surface of the bottle and grow to form a monolayer. This system has three major advantages over static monolayer culture: (1) the increase in surface area; (2) the constant, but gentle, agitation of the medium; and (3) the increased ratio of medium surface area to volume, allowing gas exchange to take place at an increased rate through the thin film of medium over cells not actually submerged in the deep part of the medium.

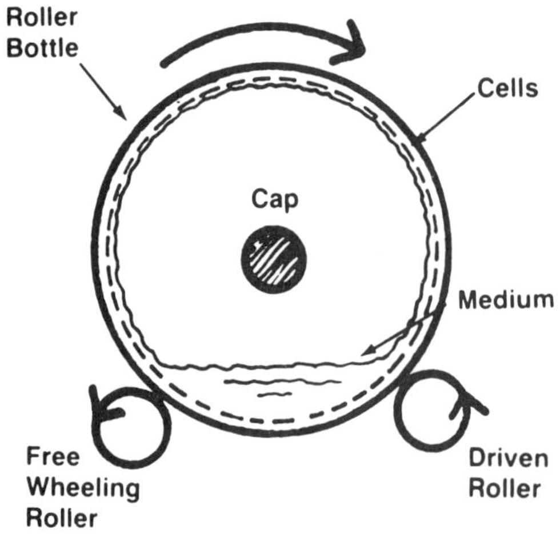

Fig. 23.6. *Roller bottle culture. Cell monolayer (dotted line) is constantly bathed in liquid but only submerged for about one-fourth of the cycle, enabling free gas exchange.*

Outline

Seed cell suspension in medium into round bottle and rotate slowly on a roller rack.

Materials

Sterile:
Medium and medium dispenser
PBSA
0.25% crude trypsin
Monolayer culture
Roller bottles
Nonsterile:
Hemocytometer, or cell counter and counting fluid
Supply of 5% CO_2
Roller apparatus

Protocol

1. Trypsinize cells and seed at usual density.
   Note. The gas phase is large in a roller bottle so it may be necessary to blow a little 5% CO_2 into the bottle (e.g., 2 s at 10 l/min). If medium is CO_2/HCO_3- buffered, then the gas phase should be purged with 5% CO_2 (30 s–1 min at 20 l/min, depending on the size of the bottle; see Chapter 10).
2. Rotate bottle slowly around its axis at 20 rev/hr until cells attach (24–48 hr).
3. Increase rotational speed to 60–80 rev/hr as cell density increases.
4. To feed or harvest medium, take bottles to sterile work area and draw off medium as usual and replace with fresh medium. A transfusion device (see Fig. 4.13) is useful for adding fresh medium,

provided that the volume is not critical. If the volume of medium is critical, it may be dispensed by pipette or metered by a peristaltic pump (Camlab, Jencons) (see Chapter 4).

5. To harvest cells, remove medium, rinse with 50–100 ml PBSA and discard PBSA, add 50–100 ml trypsin at 4°C, and roll for 15 s by hand or on rack at 20 rpm. Draw off trypsin, incubate the bottle for 5–15 min, add medium, shake, and/or wash off cells by pipetting.

Analysis. Monitoring cells in roller bottles can be difficult, but it is usually possible to see cells on an inverted microscope. With some microscopes, the condensor needs to be removed, and with others, the bottle may not fit on the stage. Choose a microscope with sufficient stage accommodation.

For repeated harvesting of large numbers of cells or for repeated collection of supernatant medium, the roller-bottle system is probably most economical, although it is labor-intensive and requires investment in a bottle-rolling unit (roller rack, Fig. 23.5).

Variations.

Aggregation. Some cells may tend to aggregate before they attach. This is difficult to overcome but may be improved by reducing the initial rotational speed to 5 or even 2 rev/hr or trying a different type or batch of serum.

Size. A range of bottles, both disposable and reusable, is available (see Table 7.1, Figs. 23.5, 23.7).

Volume. Medium volume may be varied. A low volume will give better gas exchange and may be better for untransformed cells. Transformed cells, which are more anaerobic and produce more lactic acid, may be better in a larger volume. The volumes given in Table 7.1 are mean values and may be halved or doubled as appropriate.

Mechanics. The system where bottles are supported on rollers (see Fig. 23.5) is now the most popular because it is most economical in space. Roller drums (see Fig. 23.8), once popular, demand more space. They are still used for smaller bottles or tubes.

Microcarriers. Monolayer cells may be grown on plastic microbeads of approximately 100-μm diameter, made of polystyrene (Biosilon, Nunc), Sephadex (Supabeads, Flow and Cytodex, Pharmacia), polyacrylamide (Biorad), collagen (Pharmacia), gelatin (KC Biologicals), or polyacrylamide (Biorad) [Griffiths, 1992]. Culturing monolayer cells on microbeads gives maximum ratio of surface area to medium volume and has the additional advantage that the cells may be treated as a suspension. While the Nunclon Cell Factory gives an increase in scale with conventional geometry, microcarriers require a significant departure from usual substrate design. This has relatively little effect at the microscopic level, as the cells are still growing on a smooth surface at the solid–liquid interface. The major difference created by microcarrier systems is in the mechanics of handling [Griffiths, 1992]. Efficient stirring without grinding the beads is essential and a pad-

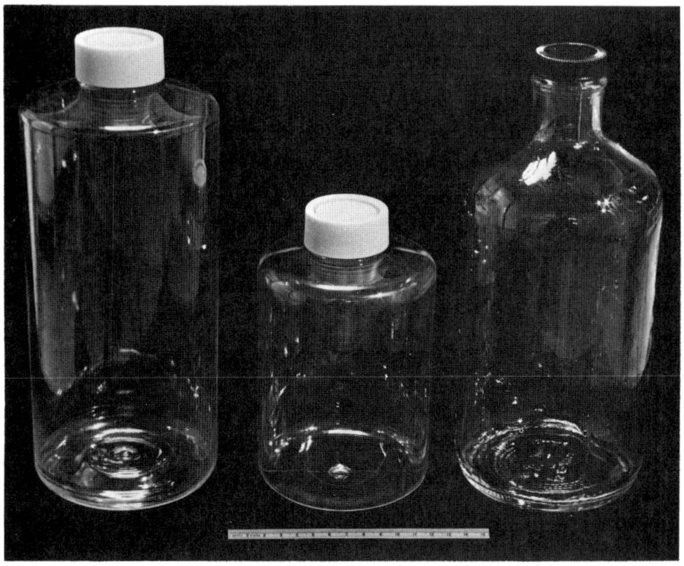

Fig. 23.7. *Examples of roller culture bottles. Center and left, disposable plastic (Falcon, Corning); right, glass.*

Fig. 23.8. *Roller drum apparatus for roller culture of large numbers of small bottles or tubes (New Brunswick Scientific).*

dle system (Fig. 23.9) rotating at 30 rpm is preferable to a plain cylindrical magnet. The rotating pendulum system (Techne) is becoming increasingly popular. Technical literature is available from microcarrier suppliers to assist in setting up satisfactory cultures.

Outline

Cells are seeded at a high cell and bead concentration, diluted, stirred, and sampled as required.

Materials

Sterile:

Microcarriers (Gelatin, KC Biologicals)
Growth medium
Donor culture
Culture vessel (Techne, Bellco)

Nonsterile:

Magnetic stirrer (Techne, BTL, Bellco)

Protocol

1. Suspend beads in one-third of the final volume of medium required.
2. Trypsinize cells, count, and seed at three to five times the normal seeding concentration into the bead suspension.
3. Stir at 10–25 rpm for 8 hr.
4. Increase stirring speed to approximately 60 rpm.
5. If pH falls, feed by switching off stirrer for 5 min, allowing beads to settle, and replacing one-half to two-thirds of medium.
6. To harvest cells, remove medium and wash by settling, digest beads with trypsin EDTA, spin down, and wash cells.

Analysis. Cell counting on beads can be difficult, so growth rate should be checked by determination of DNA (see Chapter 18), or protein if nonproteinaceous beads are used, or dehydrogenase activity using the MTT assay (see Chapter 19) on a sample of the beads.

Variation. Most variations on the method arise from the choice of bead or design of the culture vessel and stirrer [Griffiths, 1992].

Many other mass culture techniques exist [Speir and Griffiths, 1985–1990; Griffiths, 1992], but they cannot all be described in detail here. Costar has introduced Cellcube, a perfused, multisurface, single-use propagator with growth surface areas from 21,250 to 85,000 cm^2, with associated pumps, oxygenator, and system controller.

Many systems depend on filter membrane technology either in sheet or capillary bundle format. Membrane perfusion systems have been exploited by Millipore (MCCS), where the culture bed is a flat, permeable sheet, folded over into many layers, while Amicon produce a hollow-fiber reactor based on the Vitafiber system (see Chapter 22). Endotronics supply larger perfusion chambers in a similar style to the Vitafiber system, called Acusyst, and a flat-bed modular hollow-fiber system called Acumouse (also available as Tecnomouse from Tecnomara). Several other companies also produce hollow-fiber perfusion systems (see Trade Index). The potential of these systems lies in the re-creation of high tissue-like cell densities, matrix interactions, and the establishment of cell polarity, all of which may be important for posttranslational processing of proteins and for exocytosis.

A porous ceramic matrix has been used in the Opticell system, available at 4,250 cm^2 and units of 4.8 m^2, and provided with a perfusion control system. This system is a rigid, perfused matrix, but perfusion systems have also been developed with beds of glass beads or fluidized beds of natural substance beads such as collagen (Verax). These systems are described in greater detail by Speir and Griffiths [1985–1990] and Griffiths [1992].

There is a recurrent problem when cell cultures are scaled up, particularly in a fixed-bed bioreactor: that it is no longer possible to observe the cells directly, and monitoring the progress of a culture by cell counting becomes difficult. One approach to this problem (Kevin Brindle, personal communication) uses nuclear magnetic resonance (NMR) to assay the contents of the culture chamber. By placing the cells in a perfused hollow-fiber chamber within the magnetic field of an NMR spectroscope, a characteristic NMR spectrum is generated, enabling the identification and quantitation of specific metabolites (Fig. 23.10). NMR can also be used as an imaging device, producing a quasi-optical section through the chamber revealing the distribution of the cells, and it may even be able to distinguish between proliferating and nonproliferating zones of the culture.

Regulations regarding the use of transformed cells for production of biopharmaceuticals is now being relaxed, but the argument for anchorage-dependent cells remains. While suspension-growing cells are easier to manage, anchored cells, with their resultant potential for developing polarity, may yet represent a better model for the secretory cell, regardless of whether secretion is an intrinsic function of the cell or has been introduced via gene transfection technology.

LYMPHOCYTE PREPARATION

There is a variety of methods for the preparation of lymphocytes, but flotation on a combination of Ficoll and sodium metrizoate (e.g., Hypaque) is most widely

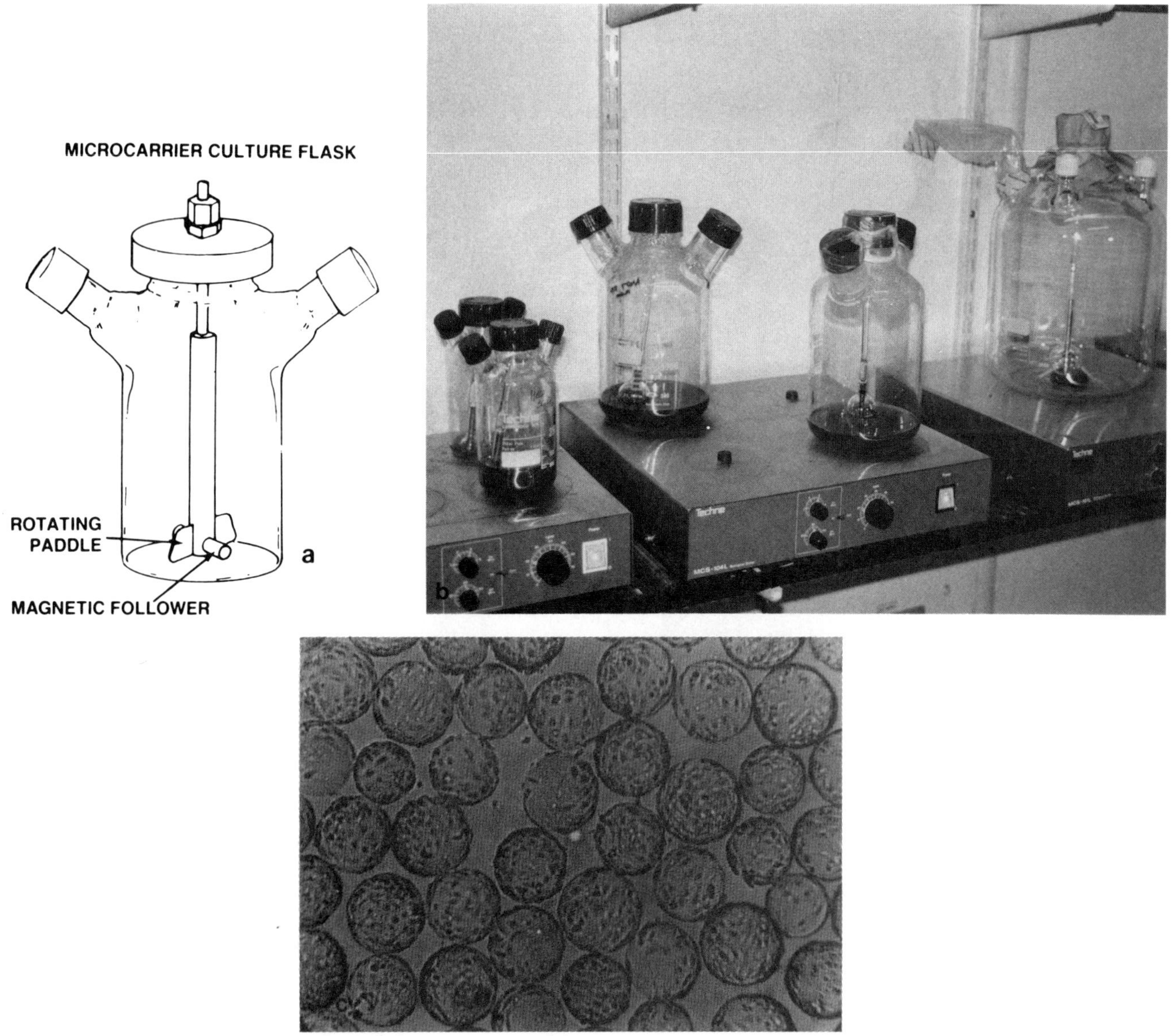

Fig. 23.9. *Microcarrier culture. a. The apparatus is similar to that used for suspension culture, as the principle is the same. Modifications include a larger surface area for stirring, with maximum liquid displacement, provided by a suspended paddle, the blades of which should increase more than in proportion to volume as the size of the vessel increases [see Griffiths, 1992], designed to minimize shearing and grinding of the cells (Bellco). b. A range of microcarrier culture vessels on multiplace stirrer racks (Techne). c. Vero cells growing on microcarriers. (Courtesy of ICN-Flow Laboratories, Irvine, Scotland.)*

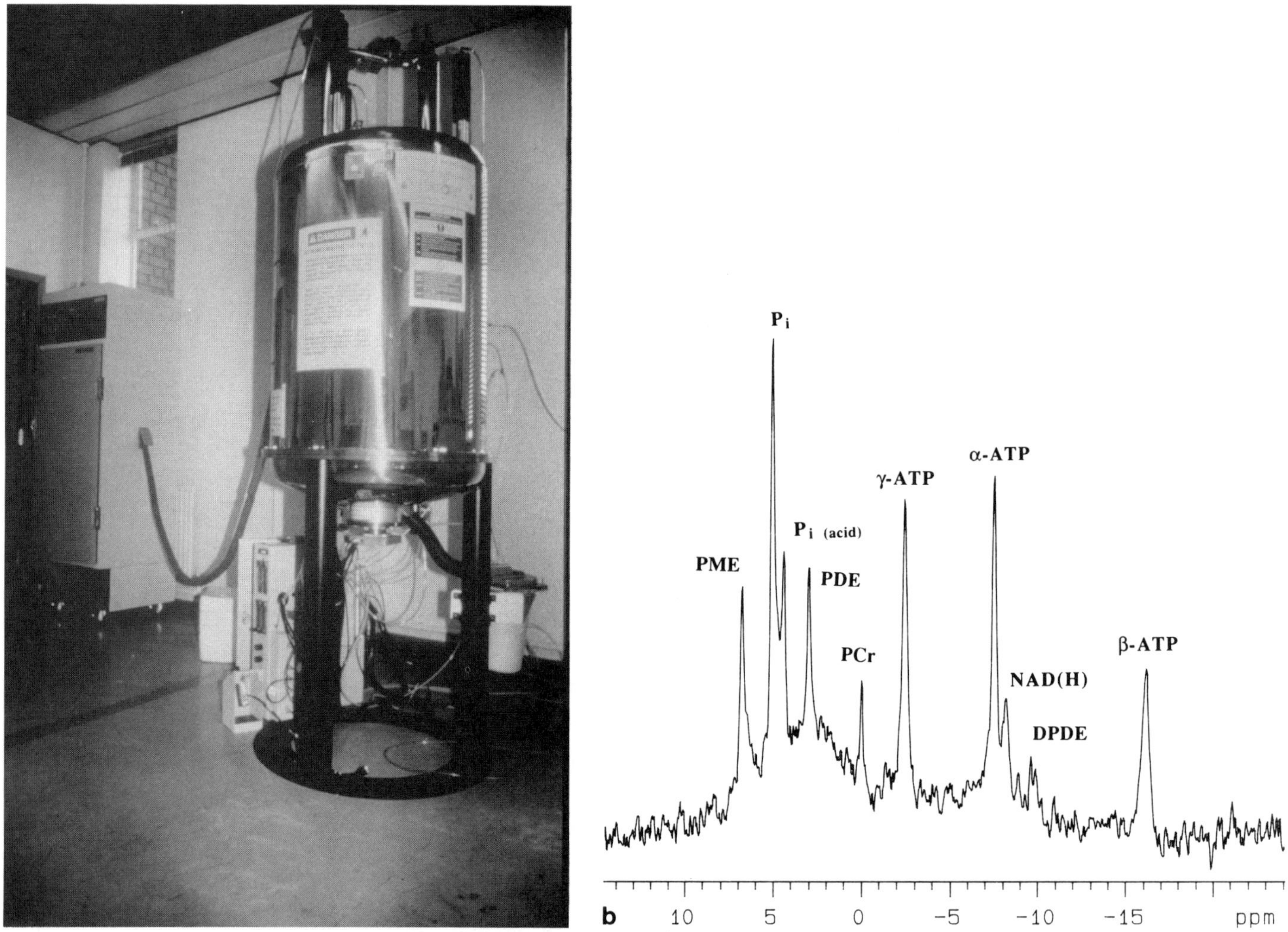

a b

Fig. 23.10. Cell growth in hollow fibers analyzed by NMR. a. Interfaced to a high-field NMR spectrometer. The environmental cabinet in the background contains buffer reservoirs, pumps, and gas-exchange equipment. This maintains the bulk of the apparatus at 37°C. Oxygenated growth medium is pumped at a flow rate of 50 ml/min through insulated tubing to the hollow-fiber cartridge, which is situated in the bore of the high-field magnet (foreground). b. ^{31}P NMR spectrum of CHO cells growing in a hollow- fiber reactor. The cells had been grown on macroporous beads in the extracapillary space of a specially constructed hollow-fiber cartridge that could be accommodated within a 25-mm diameter NMR probe. The cells were present at a density of approximately 7×10^7 cells/ml. Abbreviations: PME, phosphomonoesters including phosphocholine and phosphoethanolamine; Pi, extracellular inorganic phosphorus; $Pi^{(acid)}$, an acidic (pH 6.7) extracellular Pi pool within the bioreactor; PCr, phosphocreatine; γ-ATP, γ-phosphate of ATP; α-ATP, α-phosphate of ATP; DPDE, diphosphodiesters including UDP-glucose; β-ATP, β-phosphate of ATP. The chemical shift scale is referenced to phosphocreatine at 0.0 ppm. (Courtesy of Dr. Kevin Brindle.)

used [Boyum, 1968a,b; Perper et al., 1968]. Ready-to-use preparations are available from Nygaard, Pharmacia, and ICN-Flow.

Outline

Whole citrated blood or plasma, depleted in red cells by dextran-accelerated sedimentation, is layered on top of a dense layer of Ficoll and sodium metrizoate. After centrifugation, most of the lymphocytes are found at the interface between the Ficoll/metrizoate and the plasma.

Materials

Sterile:

Blood sample

Clear centrifuge tubes or universal containers
Dextraven 110 (Fisons)
PBSA
Lymphocyte preparation medium (ICN-Flow, Nygaard, Pharmacia) (Ficoll/metrizoate, adjusted to 1.077 g/cc)
Serum-free medium
Syringes or Pasteur pipettes
Nonsterile:
Hemocytometer or cell counter
Centrifuge

Protocol

1. Add Dextraven 110 to blood sample to final concentration of 10% and incubate at 36.5°C for 30 min to allow most of the erythrocytes to sediment.
2. Collect supernatant plasma, dilute 1:1 with PBSA, and layer 9 ml onto 6 ml Lymphoprep or other Ficoll/sodium metrizoate mixture. This should be done in a wide, transparent centrifuge tube with a cap such as the 25-ml Sterilin or Nunclon universal container, or the clear plastic Corning 50-ml tube, using double the above volumes.
3. Centrifuge for 15 min at 400 *g* (measured at center of interface).
4. Carefully remove plasma/PBSA without disturbing the interface.
5. Collect the interface with a syringe or Pasteur pipette and dilute to 20 ml in serum-free medium (e.g., RPMI 1640).
6. Centrifuge at 70 *g* for 10 min.
7. Discard supernatant fluid and resuspend pellet in 2 ml serum-free medium. If several washes are required, e.g., to remove serum factors, resuspend cells in 20-ml serum-free medium, and centrifuge two or three times more, and finally resuspend pellet in 2 ml.
8. Count cells on hemocytometer (count only nucleated cells) or on electronic counter.

Lymphocytes will be concentrated in the interface, along with some platelets and monocytes. Granulocytes will be found mostly in the Ficoll/metrizoate and in the 400 *g* pellet, and together with the erythrocytes in the pellet at the bottom of the tube. Removal of monocytes and residual granulocytes can be achieved by their adherence to glass (beads or flask surface) or to nylon mesh. If purer preparations are required, fractionation on density gradients of metrizamide (Nygaard) or Percoll (Pharmacia) or by centrifugal elutriation (see Chapter 12) may be attempted. Alternatively, specific subpopulations of lymphocytes may be purified on antibody- or lectin-bound affinity columns (Pharmacia) panning or flow cytometry (see Chapters 11 and 12).

Blast Transformation [Hume and Weidemann, 1980]

Lymphocytes in purified preparations, or in whole blood, may be stimulated with mitogens such as phytohemagglutinin (PHA), pokeweed mitogen (PWM), or antigen [Berger, 1979]. The resultant response may be used to quantify the immunocompetence of the cells. PHA stimulation is also used to produce mitosis for chromosomal analysis of peripheral blood [Rooney and Czepulkowski, 1986; Watt and Stephen, 1986] (see Chapter 13).

Materials

Medium + 10% FBS or autologous serum
Phytohemagglutinin (PHA), 50 μg/ml
Test tubes or universal containers
Microscope slides
Colcemid, 0.01 μg/ml in BSS
0.075 M KCl

Protocol

1. Using the washed interface fraction from step 7 above, incubate 2×10^6 cells/ml in medium, 1.5–2.0 cm deep, in HEPES or CO_2-buffered DMEM, CMRL 1066, or RPMI 1640 supplemented with 10% autologous serum or fetal bovine serum.
2. Add PHA, 5 μg/ml (final), to stimulate mitosis from 24 to 72 hr later.
3. Collect samples at 24, 36, 48, 60, and 72 hr and prepare smears or cytocentrifuge slides to determine optimum incubation time (peak mitotic index).
4. Add 0.001 μg/ml (final concentration) Colcemid for 2 hr when peak of mitosis is anticipated [Berger, 1979].
5. Centrifuge cells after Colcemid treatment, resuspend in 0.075 M KCl for hypotonic swelling, and proceed as for chromosome preparation in Chapter 13.

AUTORADIOGRAPHY

The following description is intended to cover autoradiography of any small molecular precursor into a cold acid-insoluble macromolecule such as DNA, RNA, or protein. Other variations may be derived from this or found in the literature [Rogers, 1979; Stein and Yanishevsky, 1979].

TABLE 23.1. Isotopes Suitable for Autoradiography

Isotope	Emission	Energy (mV) (mean)	T½
^{3}H	β^-	0.018	12.3 yr
^{55}Fe	x-rays	0.0065	2.6 yr
^{125}I	x-rays	0.035 0.033	60 d
^{14}C	β^-	0.155	5,570 yr
^{35}S	β^-	0.167	87 d
^{45}Ca	β^-	0.254	164 d

Isotopes suitable for autoradiography are listed in Table 23.1. A low-energy emitter, e.g., ^{3}H or ^{55}Fe, in combination with a thin emulsion gives high intracellular resolution. Slightly higher energy emitters, e.g., ^{14}C and ^{35}S, give localization at the cellular level. Still higher energy isotopes, e.g., ^{131}I, ^{59}Fe, and ^{32}P, give poor resolution at the microscopic level but are used for autoradiographs of chromatograms and electropherograms where absorption of low-energy emitters limits detection. Low concentrations of higher energy isotopes (^{14}C and above) used in conjunction with thick nuclear emulsions produce tracks useful in locating a few highly labeled particles, e.g., virus particles infecting a cell.

Tritium is used most frequently for autoradiography at the cellular level because the β-particles released have a mean range of about 1 μm, giving very good resolution. Tritium-labeled compounds are usually less expensive than the ^{14}C- or ^{35}S-labeled equivalents and have a long half-life. Because of the low energy of emission, however, it is important that the radiosensitive emulsion be positioned in close proximity to the specimen, with nothing between the cell and the emulsion. Even in this situation only the top 1 μm of the specimen will irradiate the emulsion.

β-particles entering the emulsion produce a latent image in the silver halide crystal lattice within the emulsion at the point where they stop and release their energy. The image may be visualized as metallic silver grains by treatment with an alkaline reducing agent (developer) with subsequent removal of the remaining unexposed silver halide by an acid fixer.

The latent image is more stable at low temperature and in anhydrous conditions, so sensitivity (signal versus background) may be improved by exposing in a refrigerator over desiccant. This will reduce background silver-grain formation by thermal activity.

Outline

Cultured cells are incubated with the appropriate isotopically labeled precursor (e.g., [^{3}H]-thymidine to label DNA), washed, fixed, and dried (Fig. 23.11). Any extractions necessary, e.g., to remove unincorporated precursors, are performed, and the specimen is coated with emulsion in the dark and left to expose. When subsequently developed in photographic developer, silver grains can be seen overlying areas where radioisotope was incorporated (Fig. 23.12).

Materials

Setting up culture:

Sterile:

Cells
PBSA
Trypsin
Growth medium
Coverslips or slides and petri dishes (may be nontissue culture grade if coverslips or slides are used) or plastic bottles

Nonsterile:

Hemocytometer, or cell counter and counting fluid

Labeling with isotope and setting up autoradiographs:

Sterile:

Isotope
HBSS

Nonsterile:

Protective gloves
Containers for disposal of radioactive pipettes
Container for radioactive liquid waste
Acetic methanol (1:3, ice-cold, freshly prepared)
DePeX
10% TCA
Emulsion (Kodak NTB2, Ilford G5, diluted 1 + 2 in distilled or deionized water)
Light-tight microscope slide boxes (Clay Adams, Raven)
Silica gel
Dark vinyl tape
Black paper or polyethylene

Processing:

D19 developer (Kodak)
Photographic fixer (Kodak, Ilford)
Hypoclearing agent (Kodak)
Coverslips (00)
Giemsa stain
0.01 M phosphate buffer (pH 6.5)

Protocol

1. Prepare culture. Monolayer cells may be grown on coverslips (Lux Thermanox or Polystyrene), slides (Nunc, Bellco), or in conventional plastic bottles or petri dishes.
2. Add isotope, usually in the range 0.1–10

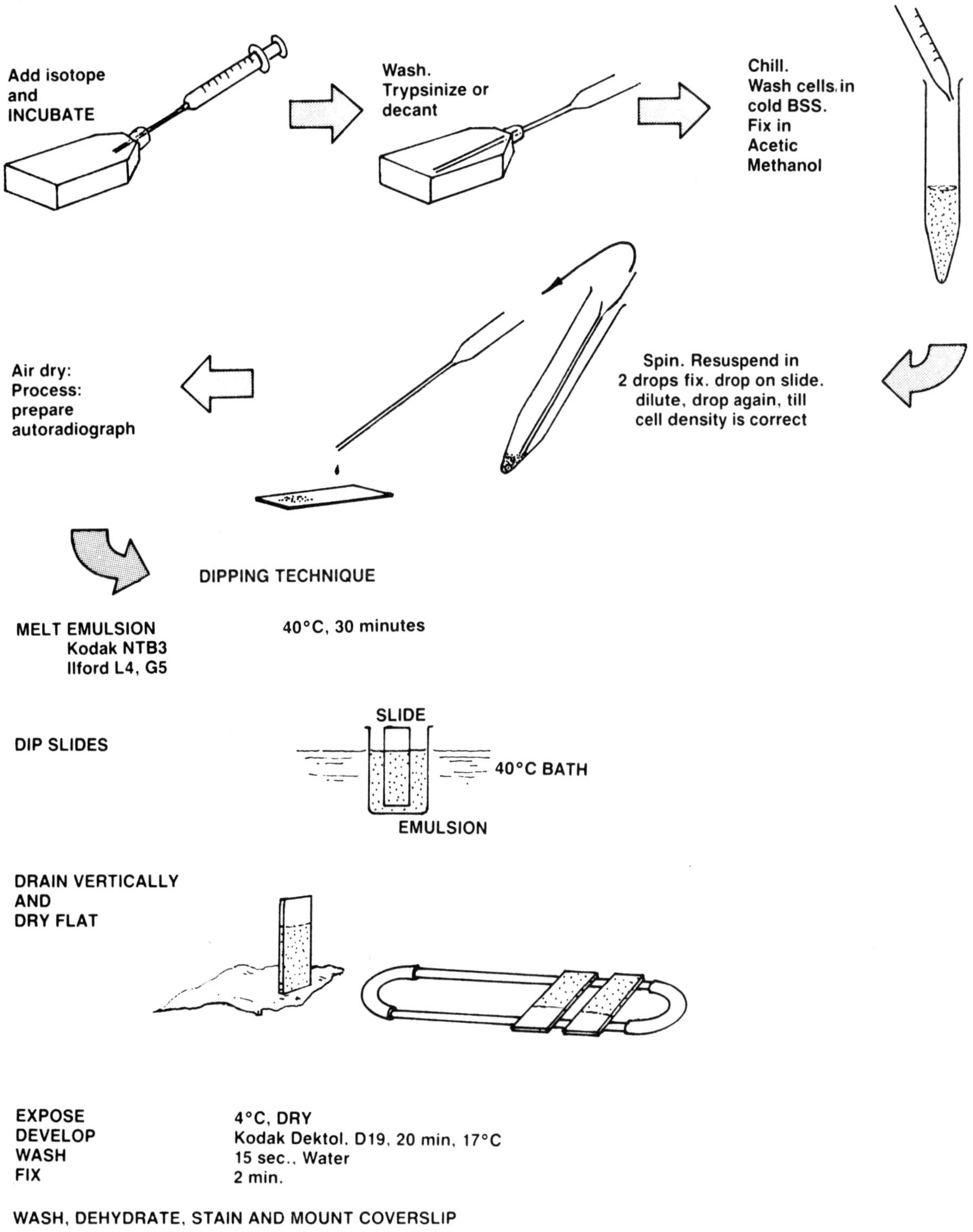

Fig. 23.11. *Steps in preparing an autoradiograph from a cell culture.*

μCi/ml (~4.0 KBq–0.4 MBq/ml), 100 Ci/mmol (~4 GBq/μmol) for 0.5–48 hr as appropriate.

◇ Follow local rules for handling radioisotopes. Since these vary, no special recommendation is made here, other than to wear gloves, do not work in horizontal laminar flow, use a shallow tray with an absorbant liner to contain any accidental spillage, incubate cultures in a box or tray labeled for radioactivity, and regulate the disposal of radioactive waste according to local limits.

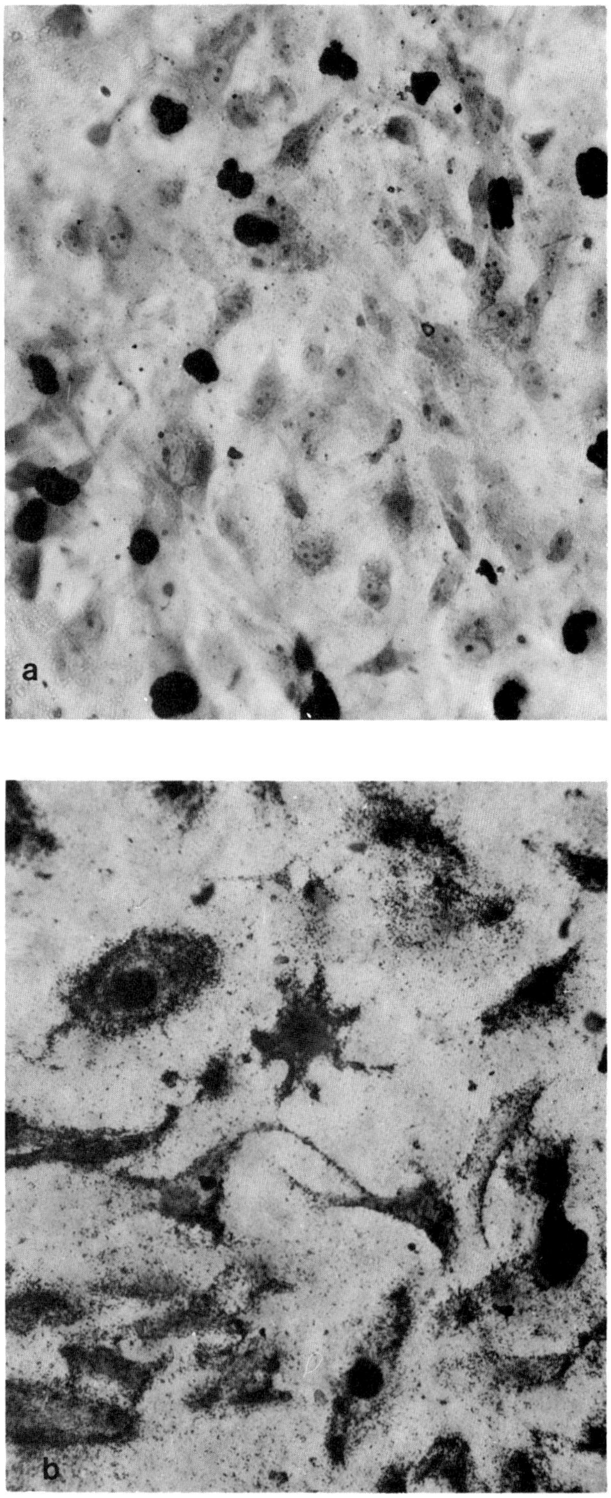

Fig. 23.12. Autoradiograph. This is an example of [^{3}H]-thymidine incorporation into a cell monolayer. Normal glial cells were incubated with 0.1 μCi/ml (3.7 KBq/ml), 200 Ci/mMol (7.4 GBq/μmol) [^{3}H]-thymidine for 24 hr, washed, and processed as in text. a. Typical densely labeled nuclei, suitable for determining labeling index (see Chapter 18). b. Similar culture infected with mycoplasma; cytoplasm is now labeled also.

3. Remove medium containing isotope, wash cells carefully in BSS, discarding medium and washes.

   ◇ Radioactive! ^{3}H-nucleosides are highly toxic due to their ultimate localization in DNA (see Chapter 6).

   All glassware must be carefully washed and free of isotopic contamination. Plastic coverslips should be used in preference to glass to minimize radioactive background. Be particularly careful with spillages; mop up right away. Wear gloves and change regularly, e.g., when you move from incubation (high level of isotope) to handling washed, fixed slides (low level of isotope).
4. Fix cells in ice-cold acetic methanol for 10 min. Coverslips should be mounted on a slide with DPX or Permount, cells uppermost. Cell suspensions may be centrifuged onto a slide (Cytospin or Cytobuckets) or drop preparations made (see Chapter 13). Prepare several extra control slides for use later to determine correct duration of exposure. All preparations will be referred to as "slides" from now on.
5. Extract acid-soluble precursors (when labeling DNA, RNA, or protein) with ice-cold 10% TCA (3 × 10 min), and perform any other control extractions, e.g., with lipid solvent or enzymatic digestion.
6. Wash slides in distilled water and methanol and dry.
7. Take to darkroom and, under dark red safelight, melt emulsion in water bath at 40°C and dilute 50:50 with deionized distilled water. It is convenient to place aliquots of the diluted emulsion in containers suitable for the number of slides to be handled at one time. If sealed in a dark box, these may be stored at 4°C until required.
8. Still under safelight, dip slides in emulsion, making sure that the cells are completely immersed, withdraw, blot the end of the slide, drain vertically for 5 s, then allow to dry flat on a tissue or filter paper.
9. When dry (~30 min), transfer to light-tight microscope slide boxes (Clay Adams, Raven Scientific) with a desiccant, such as silica gel, and seal with dark vinyl tape (e.g., electrical insulation tape).
10. Wrap in black paper and place in refrigerator. Make sure that this refrigerator is not used for storage of isotopes.
11. Leave at 4°C for 24 hr to 2 wk. The time required will depend on the activity of the specimen and can be determined by processing one of the extra slides at intervals.

12. To develop, return to darkroom (dark red safelight), unseal box, and allow slides to come to atmospheric temperature and humidity (~2 min).
13. Place slides in developer (e.g., Kodak D19) for 10 min with gentle intermittent agitation.
14. Wash briefly in distilled deionized water.
15. Transfer to photographic fixer for 3–5 min.
16. Rinse in deionized water and place in hypo-clearing agent (Kodak) for 1 min.
17. Wash in deionized water, five changes over 5 min.
18. Dry slides and examine on microscope. Phase contrast may be used by mounting a thin glass (00) coverslip in water. Remove coverslip when finished before water dries out or it will stick to the emulsion.
19. If staining is desired, immerse dry slide in neat Giemsa stain for 1 min then dilute 1:10 in 0.01 M, pH 6.5, phosphate buffer for 10 min. Rinse thoroughly under running tap water until color is removed from emulsion but not from cells. Staining solution should be removed by upward displacement with water. The slides should not be withdrawn or the stain poured off, or the scum that forms on top of the stain will adhere to the specimen.

    If a coverslip is used, it must be 00 with a minimum of mountant to allow sufficient working distance for a 100× objective.

Analysis.

Qualitative. Determine specific localization of grains, e.g., over nuclei only, or over one cell type rather than another.

Quantitative. (1) *Grain counting.* Count number of grains per cell, per nucleus, etc. This requires a low grain density, about 5–20 grains per nucleus, 10–50 grains per cell, no overlapping grains, and a low uniform background.

(2) *Labeling index.* Count number of labeled cells as a proportion of the total (see also Chapter 18). Grain density should be higher than in (1) to ease the recognition of labeled cells. If the grain density is high (e.g., ~100 grains per nucleus), set the lower threshold at, say, 10 grains per nucleus or per cell; but remember that low levels of labeling, significantly over background, may yet contain useful information.

Autoradiography is a useful tool for determining the distribution of isotope incorporation within a population, but it is less suited to total quantitation of isotope uptake or incorporation, when scintillation counting is preferable.

Variation. Autoradiographic localization of water-soluble precursors is possible with rapidly frozen specimens that have been freeze-dried or freeze-substituted to remove the water. These may be mounted dry, clamped to the radiosensitive film, or with a minimal amount of moisture (obtained by brief condensation on a cold slide) to promote adhesion of the emulsion [Hassbroek et al., 1962; Novak, 1962]. Both processes require Kodak AR-10 stripping film.

Isotopes of two different energies, e.g., ^3H and ^{14}C, may be localized in one preparation by coating the slide first with a thin layer of emulsion, coating that with gelatin alone, and finally coating the gelatin with a second layer of emulsion [Rogers, 1979]. The weaker β-emission from ^3H is stopped by the first emulsion and the gelatin overlay, while the higher energy β-emission from ^{14}C, having a longer mean path length of around 20 μm, will penetrate the upper emulsion.

Adams [1980] described a method for autoradiographic preparations from petri dishes or flasks where liquid emulsion is poured directly onto fixed preparations without the necessity for trypsinization.

Soft β-emitters may also be detected in electron microscope preparations using very thin films of emulsion or silver halide sublimed directly onto the section [Rogers, 1979].

Fluorescent and luminescent probes (Amersham International) are now being used in place of radioisotopically labeled probes (see FISH, this chapter). The resolution is often superior, quantitation is possible by confocal microscopy (Biorad) or a CCD camera (Dage-MTI), and disposal of reagents is environmentally friendly. Equipment for microscopic evaluation is, however, expensive ($50,000–200,000).

CULTURE OF CELLS FROM POIKILOTHERMS

The approach to the culture of cells from cold-blooded animals (poikilotherms) has been similar to that employed for warm-blooded animals largely because the bulk of present-day experience has been derived from birds and mammals. Thus, the dissociation techniques for primary culture employ proteolytic enzymes such as trypsin and EDTA as a chelating agent. Fetal bovine serum appears to substitute well for homologous serum or hemolymph (and is more readily available), but modified media formulations may improve growth. A number of these media are available through commercial suppliers (see Reagent Appendix and Trade Index) and the procedure is much the same as for mammalian cells–try those media and sera that are currently available, assessing for growth, plating efficiency, and specialized functions (see Chapter 7). Since the development of media for many invertebrate cell lines is in its infancy, it may prove necessary to develop new formulations if an untried class of invertebrates or type of tissue is examined. Most of the accumulated experience so far relates to insects and molluscs.

Two reviews cover some of the early developments of the field [Vago, 1971, 1972; Maramorosch, 1976], and some recent exploitation in biotechnology is described in a publication of the European Society for Animal Cell Technology [Jain et al., 1991; Klöppinger et al., 1991; Speir et al., 1991]. This relates to the use of the baculovirus vector in insect cell lines, which has many of the advantages of posttranslational modification, found in mammalian cells without the regulatory problems related to the isolation of biopharmaceutics from human and mammalian cells.

Culture of vertebrate cells other than birds and mammals has also followed procedures for warm-blooded vertebrates, and so far there has not been a major divergence in technique. Since this is a developmental area, certain basic parameters will still need to be considered to render culture conditions optimal, and if a new species is being investigated, optimal conditions for growth may need to be established, e.g., pH, osmolality (which will vary from species to species), nutrients, and mineral concentration. Temperature may be less vital but it should be fixed within the appropriate environmental range and regulated within ±0.5°C.

CELL SYNCHRONY

The percentage-labeled mitosis method for determining the duration of the stages of the cell cycle has been described in Chapter 18. In order to follow the progression of cells through the cell cycle, a number of techniques have been developed whereby a cell population may be fractionated or blocked metabolically so that on return to regular culture they will all be at the same phase.

Cell Separation

Techniques for cell separation have been described in Chapter 12. Sedimentation at unit gravity (Figs. 12.2, 12.3) is the simplest [Shall and McClelland, 1971; Shall, 1973], but centrifugal elutriation is preferable if a large number of cells ($>5 \times 10^7$) is required [Meistrich et al., 1977a,b] (see Figs. 12.5, 12.6). Fluorescence-activated cell sorting (see Figs. 12.13–12.15) can also be used in conjunction with a nontoxic, reversible DNA stain such as Hoechst 33342. The yield is lower than unit-gravity sedimentation ($\sim 10^7$ cells or less) but the purity of the fractions is higher.

One of the simplest techniques for separating syn-

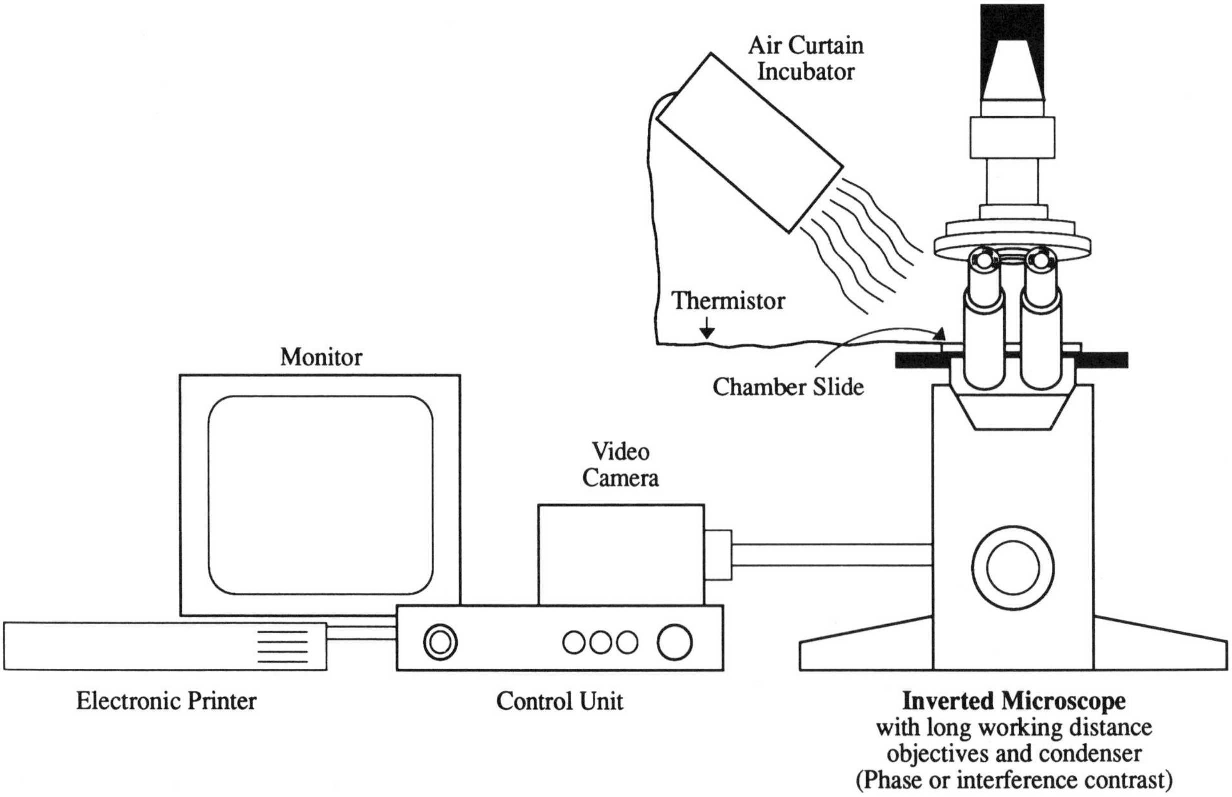

Fig. 23.13. *Suggested layout for time-lapse videomicrography. Microscope should be mechanically isolated from air curtain incubator.*

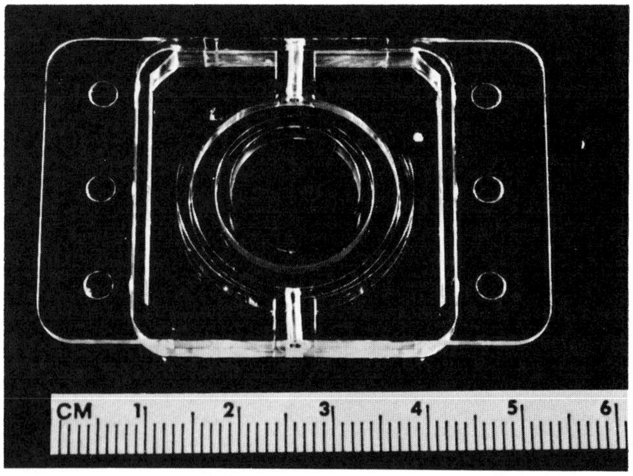

Fig. 23.14. Perfusion slide for use in cytological observation such as time-lapse videomicrography. Medium can be perfused in and out via the holes at top and bottom (Sterilin).

chronized cells is mitotic "shake-off." Mitotic cells tend to round up and detach when the flask is shaken. This works well with CHO cells [Tobey et al., 1967; Petersen et al., 1968] and some sublines of HeLa-S_3. Placing the cells at 4°C for 30 min to 1 hr a few hours previously enhances the yield at shake-off [Newton and Wildy, 1959; Lesser and Brent, 1970; Miller et al., 1972].

Blockade

Two types of blocking have been used:

DNA synthesis inhibition (S-phase). Thymidine, hydroxyurea, cytosine arabinoside, aminopterin, etc. [Stubblefield, 1968]: The effects of these agents are variable because many are toxic, as blocking cells in cycle at phases other than G_1 tends to lead to deterioration. Hence, the culture will contain nonviable cells, cells blocked in S but viable, and cells that have escaped the block.

Nutritional deprivation (G_1 phase). In these cases serum [Chang and Baserga, 1977] or isoleucine [Ley and Tobey, 1970] is removed from the medium for 24 hr and then restored, whereupon transit through cycle is resumed in synchrony.

A high degree of synchrony (e.g., >80%) is only achieved in the first cycle; by the second cycle it may be >60% and by the third cycle, close to random. Chemical blockade is often toxic to the cells and nutritional deprivation does not work well in many transformed cells. Physical fractionation techniques are probably most effective and do less harm to the cells.

TIME-LAPSE RECORDING

The following protocol for time-lapse recording has been contributed by John Lackie, Yamanouchi Research Institute, Littlemore Hospital, Oxford, OX4 4XN, England.

Principle

This is a technique whereby living cultures are filmed or videotaped in order that their behavior (movement, interactions with other cells, division) can be accelerated. With animal cells the requirement is nearly always to accelerate behavior, but basically similar techniques can be used to study rapid events. The major requirement is a good optical system, of whatever configuration (inverted or normal) is convenient for the particular system, and some form of camera and recording system (Fig. 23.13). Originally, most work was done with cine film, but this has largely been superseded by video techniques that have improved greatly, even since the previous edition of this book.

When first introduced, video systems offered convenience and immediacy at the expense of resolution, but modern video processing allows much greater resolution than was possible with direct-film methods. The key has been digital image storage, which makes it possible to clean up the image by subtracting background noise and, in certain circumstances, to enhance contrast by numerical processing of a

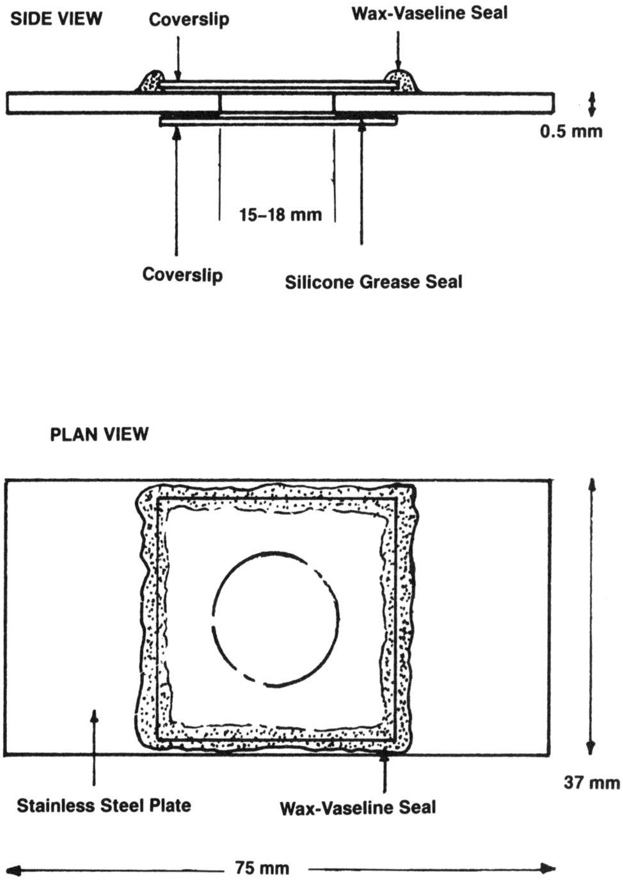

Fig. 23.15. Filming chamber.

digitized image. It is no longer worth giving serious consideration to film-based methods, especially now that it is possible to project videotape records to a large audience.

A further advantage of modern videocameras is that they work at low light levels and the problems that often arose because of local heating or light-induced damage are no longer serious.

The essential requirement is a good optical system; image processing is not a substitute for the use of good optics. Since cells are often followed for extended periods, it is essential to choose an optical set-up that allows cells to be kept in a sealed flask or sterile chamber at 37°C and to ensure that the mechanical system of the microscope does not allow the image to drift out of focus.

Phase-contrast or differential interference-contrast (DIC or Zeiss Nomarski) systems are the most commonly used, though neither is suitable for automated image analysis because of edge effects (with phase contrast) or apparent oblique illumination (with DIC). Possibly the only system that allows good image analysis is based on double-beam interference microscopy [Brown and Dunn, 1989], which may therefore enjoy a renaissance. The old problems of limited contrast with interference systems can be overcome now that it is possible to amplify a selected portion of the contrast range using digital image enhancement. For most laboratories, however, phase-contrast or DIC optics are suitable.

So many video-based systems are available, and so many new systems are being introduced, that it is difficult to give detailed advice. One system, used by Dr. J.P. Heath, involves the following (costs are approximate only).

Recording Equipment

Camera: Dage Instruments MTI CCD72 with HI Gain control box

Image averaging, background subtraction and integration: Videoscope LKH-9000 in series with Dage DSP 200

Video recorder: Panasonic AG 6720 S-VHS time-lapse

Video monitor: Sony PVM 1343

Images can be stored using a Truevision Targa Plus framestore in a PC; digital images can be printed from the framestore using a Sony UP-5000 color printer, obviating the need for a separate still-camera system.

A minimal system requires only a simple monochrome videocamera (color is irrelevant if phase contrast is being used), the time-lapse video recorder, and a monitor.

In general terms, the system must fulfill certain minimal requirements: It should work at relatively low light levels, there should be the facility to record at a range of speeds from real time down to 1/480 (so that an hour of real time can be replayed in a few seconds), and there should be some form of automatic time–date recording. It is also worth getting a system that is compatible with standard domestic video players.

Microscope and Incubation

The other considerations here concern the practicalities of setting up the cells for observation. Inverted optics are usually more convenient; because inverted microscopes are designed for looking at cells in culture flasks, they usually have long working-distance objectives and condensers, though this may be at the expense of some loss of resolution. Keeping the cells at 37°C can be achieved either using an air-curtain incubator (Sage) (a miniature-bead thermistor on the microscope stage connected to a proportional controller that regulates the current supply to a fan heater or hairdryer will serve) or by putting the microscope in an incubator housing or, even better, from the point of view of thermal stability, in a temperature-controlled room. Many solutions are possible, but take care, as ambient temperature in an open laboratory may change drastically overnight, or even significantly during normal regulation, and the temperature fluctuation may cause the image to go out of focus.

Observation Chamber

Normally, regular culture flasks or dishes may be used, but if very high resolution is required, then recourse may be had to a simple chamber that can be used on a normal microscope. It may be possible to obtain these commercially (e.g., Fig. 23.14) but the author prefers a chamber made from a stainless steel slide with a central hole (Fig. 23.15). A coverslip is sealed onto the lower surface and the chamber overfilled with medium; an upper coverslip with cells is then placed over the hole, excess medium is drawn off with tissue, and the upper coverslip sealed in place with paraffin wax–Vaseline (3:2, melts about 60°C) (petrolatum). These simple chambers can be thin, so high numerical aperture objectives and condensers can be used, and the optical path is very clean. Adherent cells, such as fibroblasts, will move quite happily over the ceiling of the chamber, and any debris will fall out of focus.

Note. Make sure that the video sequence includes a brief section of stage scale to allow calibration. Also, videotape builds up fast and a good system of recording the details of the sequence is essential. Bulk archival storage may require that certain sequences be transferred to optical disc.

Movement Analysis
There is an extensive literature on the problems of analyzing cell motility [Noble and Levine, 1986, *inter alia*], and the problem is nontrivial. Automated tracking has been used in a few special cases, where the recognition of cells was straightforward [Dow et al., 1987].

CONFOCAL MICROSCOPY

Time-lapse records, or real-time records made on videotape, can be made via the confocal microscope with a conventional high-resolution videocamera or CCD. These systems allow the recording of events within the cell depicted by the distribution, relocation, and changes in staining intensity of fluorescent probes. This can be used to localize cell organelles, such as the nucleus or Golgi complex [Lippincott-Schwartz et al., 1990], measure fluctuations in intracellular calcium [Cobbold and Rink, 1987], or follow the penetration and movement within the cell of a drug [Neyfakh, 1987; Bucana et al., 1990]. Measurements can be made in three dimensions, as the excitation and detection system is capable of "cutting" optical sections through the cell, and over very short time periods.

An application of this system is described under the protocol for chromosome painting (see below).

CULTURE OF AMNIOCYTES

The human fetal karyotype can be determined by culturing amniotic fluid cells obtained by amniocentesis (Fig. 23.16). Amniocentesis can now be performed from 11 weeks of gestation onwards, but for practical clinical reasons few amniocenteses are done after the 18th week of gestation.

Now inborn errors of metabolism and other sex-linked or autosomal recessive and dominant conditions are mainly diagnosed on placental tissues obtained by chorionic villus sampling and using direct techniques (uncultured material). Chromosome analysis from amniotic fluid does, however, require the culture of cells.

This protocol has been submitted by Marie Ferguson-Smith, East Anglian Regional Genetics Service, Cytogenetics Department, Level 2, Addenbrooke's Hospital, Hills Road, Cambridge, CB2 2QQ, England.

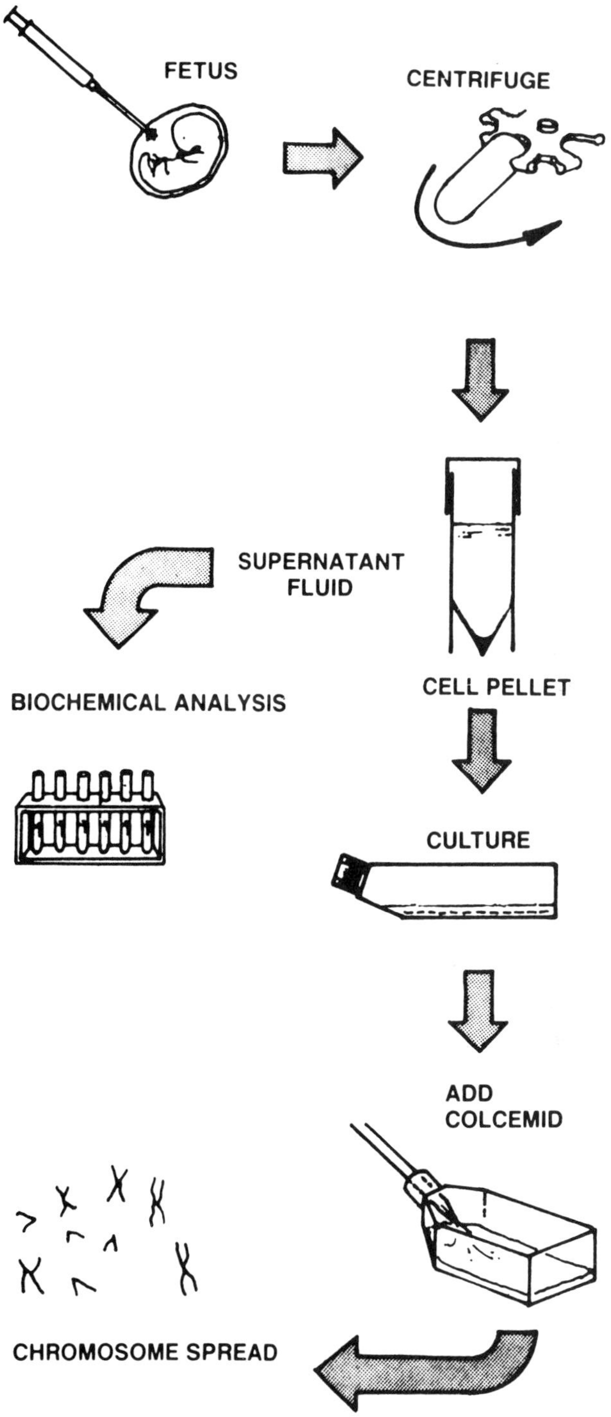

Fig. 23.16. *Culture of amniocentesis samples for detection of chromosomal abnormalities.*

Principle
The principle of culturing amniotic fluid cells is based on separation of cells from the supernatant by centrifugation and setting up the cell suspension using the *in situ*/coverslip method or a flask method. Good

and quick results can be obtained using the open/humid culture system.

Outline

In open system, the cell culture is incubated in 5% CO_2, 95% humidity incubator on: (1) coverslip in a petri dish (*in situ* method) or in a (2) Leighton tube without coverslip and with lightly screwed on cap (for suspension harvest method).

Materials

Sterile

Complete medium Ham's F10 or TC 199
L-Glutamine (200 m*M*), working concentration 4 M/ml
Penicillin 10,000 U/ml, streptomycin 10 mg/ml, use at 100 U/ml and 10 μg/ml, respectively
Fetal bovine serum, working concentration 10–20%
Ultra ser G (serum substitute–LKB), working concentration 1–2%
Colcemid (GIBCO or Seralab) 10 μg/ml; final working concentration 0.1 μg/ml
Amniotic fluid sample 10–20 ml (between 11 and 18 weeks' gestation)
Trypsin, Bacto (Difco) 0.25% in 0.85% NaCl: follow manufacturer's instructions for reconstitution
T/V 0.3 ml reconstituted Bacto Trypsin (Difco) in 20 ml versene 1.0 m*M* (ICN-Flow)
Plastic disposable pipettes or teat pipettes for setting up of cultures
Syringes and needles or quills for feeding the cultures
35-mm plastic petri dishes
22-mm glass coverslips (dry-heat sterilized)
Leighton tubes
Universal (conical base) containers for collection of amniotic fluid samples and centrifugation
Sterile forceps

Nonsterile:

Hypotonic solutions 0.05 M KCl in distilled water
Fixative: 3 parts methanol to 1 part glacial acetic acid
Trypsin Bacto (Difco): follow manufacturer's instructions for reconstitution. Keep reconstituted stock solution sterile in refrigerator.
Phosphate buffer, 0.01 M, pH 6.8 (see Giemsa phosphate buffer, Reagent Appendix; adjust to pH 6.8 with 1 M NaOH)
Leishman stain (Gurr), 1 part plus 3 parts 0.15 M NaCl
5% CO_2 incubator
Inverted microscope to inspect the cultures
Centrifuge (bench top)
Plastic disposable pipettes and rubber teats or teat pipettes for harvesting

Protocol: Setting Up and Cell Culture

1. Collect amniotic fluid in one or two conical-based universal containers.
2. Centrifuge sample in universal containers at 150 g for 10 min.
3. Remove supernatant, leaving 1 ml of fluid above the cell pellet.
4. Label the appropriate type and number of culture vessels to be used.
5. Resuspend the cell pellet(s) in an appropriate amount of medium for the number of culture vessels to be set up (see below).
   Leighton tube–1 to 1.5 ml of medium to 0.5 ml of cell suspension, petri dish (35 mm)–2 to 2.5 ml of medium to 0.5 ml of cell suspension.
6. Incubate cultures at 37°C in 5% CO_2/humid incubator (Leighton tubes should have slack caps).
7. After 5–7 d the cultures should be inspected for cell growth and media changed (replace the removed medium with the same volume of fresh medium).
8. Inspect cultures and change medium twice a week. Cells are ready for harvest when actively growing colonies are observed, usually 10 to 14 d after initiating culture.

In Situ Harvesting of Coverslips

1. Remove coverslip into a fresh petri dish with medium 24 hr before intending to harvest (you can keep the original dish as a source of extra cells).
2. Add colcemid to a final concentration of 0.1 μg/ml for 2–3 hr.
3. Remove medium gently and replace very gently with 2 ml 0.05 M KCl, prewarmed to 37°C. Place culture in incubator for 8–10 min (time depends on size and activity of colonies).
4. Add six drops of fixative gently down the side of the dish. Allow fixative to disperse in KCl.
5. Remove supernatant from petri dish and replace with fresh fixative, adding it gently down the side of the dish. Leave for 10–20 min.
6. Remove coverslip, standing it against the dish on absorbent paper, allowing excess fixative to drain and the coverslip to air dry.

Suspension Harvesting from Leighton Tubes

1. Feed culture 24 hr prior to harvest.
2. Add 0.1 μg/ml of colcemid for 2–3 hr and incubate at 37°C.

3. Check the vessel for rounded up cells (dividing cells).
4. Remove medium and wash culture with 1 ml T/V.
5. Remove solution and add 1 ml T/V; after about 2 min cells should be rounding off and detaching from surface.
6. Add 1.5 ml of distilled water (prewarmed to 37°C) to tube (this will stop the trypsin action) and leave for 2 min.
7. Centrifuge at 150 *g* for a further 7 min.
8. Remove supernatant and gently add 2 ml cold fixative, tapping the tube to ensure cell pellet is resuspended.
9. Centrifuge at 150 *g* for 7 min.
10. Repeat steps 8 and 9, then remove supernatant and add a few drops of fresh fixative.
11. Cell suspension is ready for slide making or storage in −20°C. Make slides by dropping cell suspension onto cleaned slides.

G-Banding by Trypsin

1. Age slides or coverslips by placing them in hot oven (60°C) overnight or on a hot plate (75°C) for 2 hr.
2. Make up trypsin solution by mixing 0.4 ml of reconstituted Difco trypsin in 50 ml of saline.
3. Place slides or coverslip into solution and agitate gently for a few seconds (times vary according to age of slides and method of slide making).
4. Rinse slide thoroughly with pH 6.8 0.01 M phosphate buffer.
5. Stain in diluted Leishman stain for 1½–3 min (time depends on age of stain).
6. Rinse slide or coverslip in pH 6.8 buffer and blot dry, taking care not to rub off cells.
7. Slides or coverslips can be examined for quality of bands before mounting.
8. Mount slides and coverslips in DePeX.

SOMATIC CELL FUSION

For many years mammalian, and particularly human, genetic analysis was hampered by the limitations of the duration of the breeding cycle and difficulties in performing breeding experiments. The discovery by Barski et al. [1961] and Sorieul and Ephrussi [1961] that somatic cells would fuse in the presence of Sendai virus led to a burst of activity that has developed into the field of somatic cell genetics.

Briefly, somatic cells fuse if cultured with inactivated Sendai virus, or with polyethylene glycol (PEG) [Pontecorvo, 1975; Milstein, 1979]. A proportion of the cells that fuse will progress to nuclear fusion, and a proportion of these will progress through mitosis such that both sets of chromosomes replicate together and a hybrid is formed. In some interspecific hybrids, e.g., human–mouse, one set of chromosomes (the human) is gradually lost [Weiss and Green, 1967]. Thus, genetic recombination is possible *in vitro* and, in some cases, segregation as well.

Since the proportion of viable hybrids is low, selective media are required to favor the survival of the hybrids at the expense of the parental cells. TK^- and $HGPRT^-$ mutants (see below) of the two parental cell types are used, and the selection is carried out in HAT medium (*h*ypoxanthine, *a*minopterin, and *t*hymidine) (Fig. 23.17) [Littlefield, 1964a]. Only cells formed by the fusion of two different parental cells (heterokaryons) survive, as the parental cells, and fusion products of the same parental cell type (homokaryons) are deficient in either thymidine kinase or hypoxanthine guanine phosphoribosyl transferase. They cannot, therefore, utilize thymidine or hypoxanthine from the medium, and since aminopterin blocks endogenous synthesis of purines and pyrimidines, they are unable to synthesize DNA.

The following protocol for somatic cell fusion has been contributed by Ivor Hickey, Division of Genetic Engineering, School of Biology and Biochemistry, Queen's University, Medical Biology Centre, 97 Lisburn Road, Belfast BT9 7BL, Northern Ireland.

Principle

Although many cell lines will undergo spontaneous fusion, the frequency of such events is very low. In order to produce hybrids in significant numbers, cells are treated with either inactivated Sendai virus [Harris and Watkins, 1965] or, more commonly, the chemical fusogen polyethylene glycol (PEG) [Pontecorvo, 1975]. Selection systems that kill parental cells but not hybrids are then used to isolate clones of hybrid cells.

Outline

Cells to be fused are brought into close contact either in suspension or in monolayers before treatment with PEG, which is brief to reduce cell killing. Usually cells are given a 24-hr period to recover before selection for hybrids is exercised.

Materials

Sterile:
PEG 1,000 (BDH)
Complete growth medium
Serum-free growth medium
1.0 M NaOH
50-mm tissue culture dishes
Plastic universal containers (Sterilin)

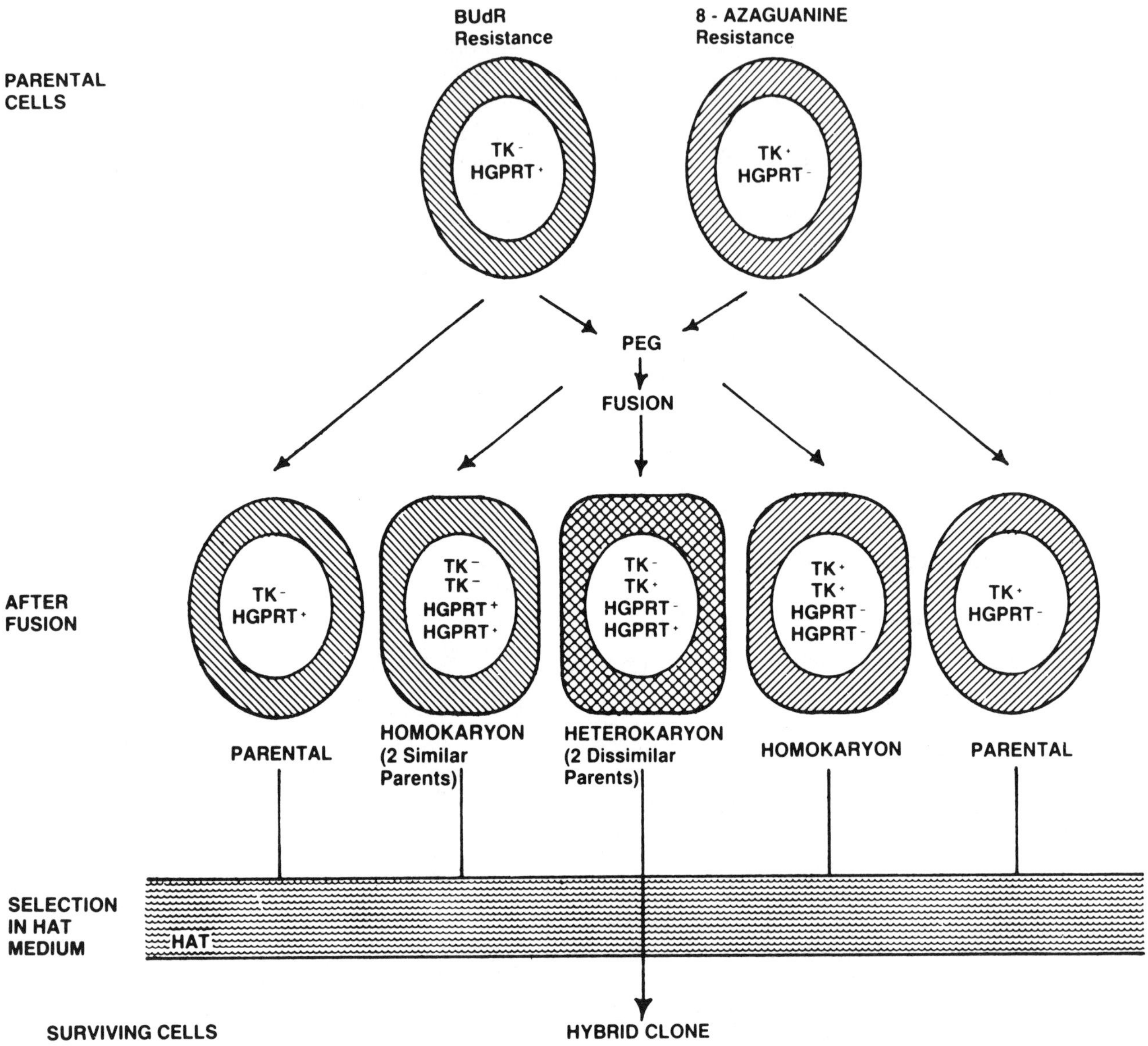

Fig. 23.17. Somatic cell hybridization. Selection of hybrid cells after fusion (see text).

Protocol: (a) Monolayer Fusion

1. Autoclave PEG 1,000. This both liquifies and sterilizes the PEG. Allow it to cool to 37°C and mix with an equal volume of serum-free medium prewarmed to the same temperature. Adjust the pH to approximately 7.6–7.9 with 1.0 M NaOH. This solution can be stored at 4°C for up to 2 wk.
2. Inoculate equal numbers of the cells to be fused into 50-mm tissue culture dishes. Between 2.5×10^5 and 10^6 of each parental cell line per dish is usually sufficient. The mixed culture is incubated overnight.
3. Warm the PEG solution to 37°C. It may be necessary at this point to readjust the pH with NaOH.
4. Remove the medium thoroughly from the cultures and wash once with serum-free medium. Add 3.0 ml of the PEG solution and spread over the monolayer of cells.
5. Remove the PEG after exactly 1.0 min, and rinse the monolayer three times with 10.0-ml volumes of serum-free medium before returning to complete medium.
6. Culture the cells overnight before adding selection medium.

(b) Suspension Fusion

1. Prepare PEG as above.
2. Centrifuge a mixture of 4×10^6 cells of each of the two parental cell lines at 150 g for 5 min at room temperature. Centrifugation and subse-

quent fusion are conveniently carried out in 30-ml plastic universal containers or centrifuge tubes.

3. Resuspend the pellet in 15.0 ml of serum-free medium and repeat the centrifugation.
4. Aspirate off all of the medium and resuspend the cells in 1.0 ml PEG solution by gentle pipetting.
5. After 1.0 min, dilute with 9.0 ml of serum-free medium and transfer half of the suspension to each of two universal containers or centrifuge tubes containing a further 15.0 ml of serum-free medium.
6. Centrifuge as before, remove the supernatant, and resuspend in complete medium.
7. After overnight incubation, clone the cells in selection medium.

Variations. A large number of variations of the PEG fusion technique have been reported. While the procedures described here work well with a range of mouse, hamster, and human cells in interspecific and intraspecific fusions, they are unlikely to be optimal for all cell lines. Inclusion of 10% DMSO in the PEG solution has the advantage of reducing its viscosity, and has been reported to improve fusion [Norwood et al., 1976]. The molecular weight of the PEG used need not be 1,000 D. Preparations with molecular weights from 400 to 6,000 D have been successfully used to produce hybrids. Although now largely superseded by PEG as a fusogen, for reasons of convenience, Sendai virus fusion remains a reliable method. If a source of virus is available, the method of Harris and Watkins [1965] can be used.

Selection of Hybrid Clones

The method of selection used in any particular instance depends on the species of origin of the two parental cell lines, their growth properties, and whether selectable genetic markers are present in either or both cell lines. Hybrids are most frequently selected using the HAT system: 10^{-4} M hypoxanthine, 6 × 10^{-7} M aminopterin, and 1.6 × 10^{-6} M thymidine [Littlefield, 1964a]. This can be used to isolate hybrids made between pairs of mutant cell lines deficient in the enzymes thymidine kinase (TK^-) and hypoxanthine guanosine phosphoribosyl transferase ($HGPRT^-$), respectively. TK^- cells are selected by exposure to BUdR and $HGPRT^-$ by exposure to thioguanine by the procedures described by Biedler in Chapter 11. Where only one parent carries such a mutation, HAT selection can still be applied if the other cell line does not grow, or grows poorly in culture, e.g., lymphocytes or senescing primary cultures.

Differential sensitivity to the cardiac glycoside ouabain is an important factor in the selection of hybrids between rodent cells and cells from a number of other species, including human. Rodent cells are resistant to concentrations of this antimetabolite up to 2.0 m*M*, while human cells are killed at 10^{-5} M ouabain. The hybrids are much more resistant to ouabain than the human parental cells. If a rodent cell line that is HGPRT deficient is fused to unmarked human cells, then the hybrids can be selected in medium containing HAT and low concentrations of ouabain.

Although many other selection systems have been reported, only complementation of auxotrophy [Kao et al., 1969] has been widely used.

It must be stressed that whatever method is used to isolate clones of putative hybrid cells, confirmation of the hybrid nature of the cells must be obtained. This is usually done using cytogenetic (see Chapter 13) or biochemical techniques. In certain cases comparison of numbers of hybrids with frequencies of revertants may be the only way of doing this.

GENE TRANSFER

In addition to fusion of whole cells, fusion of isolated nuclei, individual chromosomes, and even purified genes or gene fragments with whole cells or enucleated cytoplasts is now possible (Fig. 23.18) [Shows and Sakaguchi, 1980]. Enucleation is performed by centrifuging cytochalasin-B-treated cells such that the nuclei detach from an anchored monolayer and pellet at the bottom of the tube. This gives cytoplasmic residues without nuclei (cytoplasts) and nuclei with only some residual plasma membrane surrounding them (karyoplasts). Incubation of karyoplasts with cytoplasts, or whole cells, in the presence of polyethylene glycol results in fusion.

Chromosomes may be isolated from metaphase cells by hypotonic lysis and incubation of these with whole cells after co-precipitation with calcium phosphate results in their incorporation into the nucleus. The chromosomes may be fractionated by density centrifugation or flow cytophotometry and individual chromosome pairs inserted into recipient cells.

In order to study the function of individual genes, the sequence of interest can be cloned and then transferred into host cells by a variety of techniques, such as transfection, lipofection, or retroviral infection. Cloned DNA is often conveniently maintained as part of a bacterial plasmid. Many plasmids can attain high copy number during bacterial growth, thus ensuring a plentiful stock of DNA for experimentation. Plasmid DNA is purified from the bacteria prior to use. Once the sequence of interest has been cloned, it can be manipulated further to isolate subclones containing,

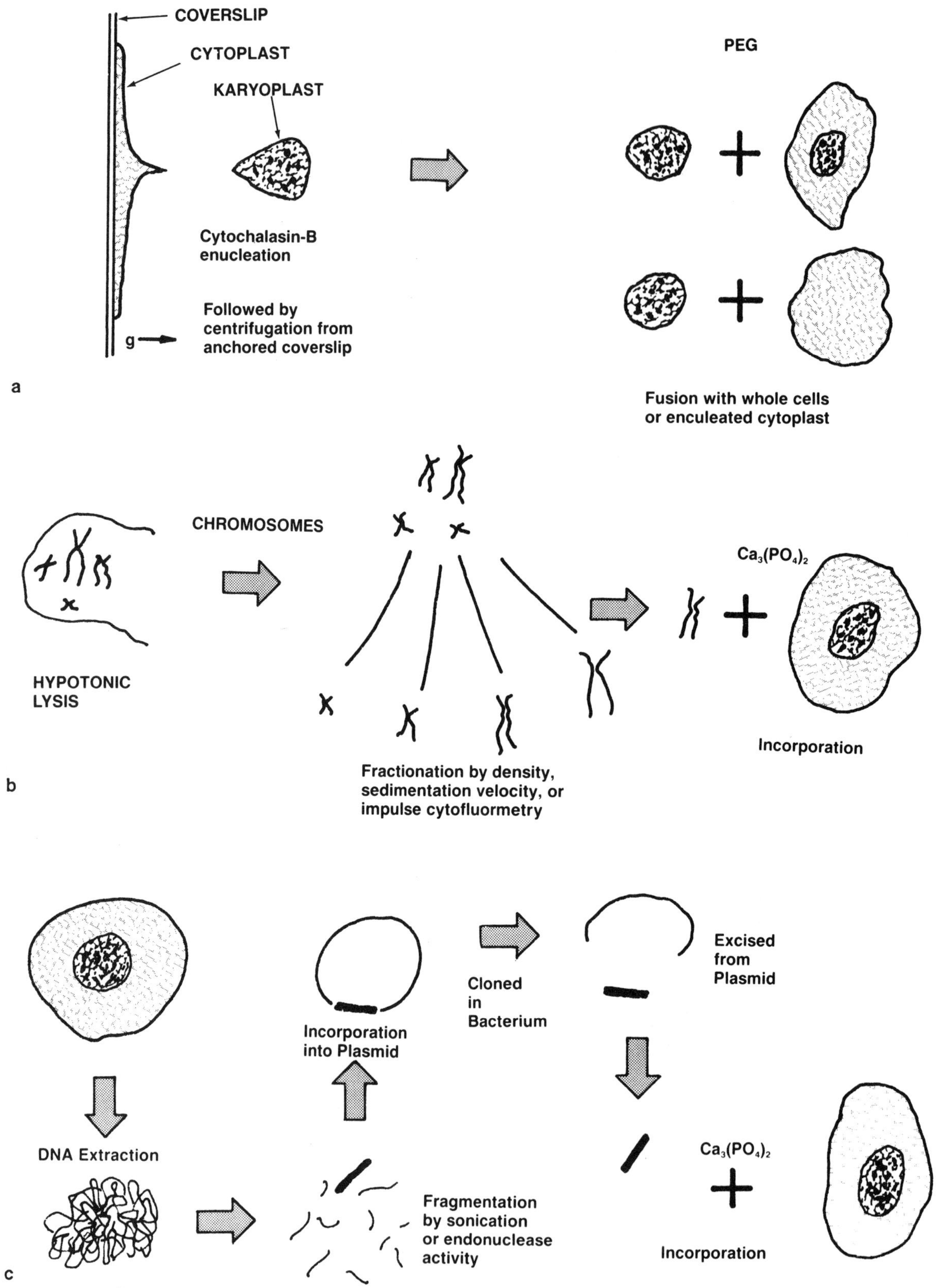

Fig. 23.18. *Gene transfer techniques. a. Whole nuclei extruded by treatment with cytochalasin B hybridized with whole cells or enucleated cytoplast. b. Chromosomes isolated from cells in mitotic arrest and fractionated by density gradient centrifugation or flow cytophotometry added to whole cells. c. Isolated DNA endonuclease fragments, amplified by gene cloning techniques, added to whole cells and incorporated by treatment with lipofection, electroporation, or co-preciptated with* $Ca_3(PO_4)_2$.

for example, promoter sequences. By genetic manipulation promoter sequences can be linked to a reporter gene (e.g., β-*gal*) whose product (β-galactosidase) can be readily assayed subsequent to transfection. In this way, tissue-specific gene expression can be analyzed in detail. Oncogenes and tumor suppressor gene function can also be analyzed by similar manipulations.

The following protocol for gene transfer into mammalian cells by the calcium phosphate technique has been contributed by Demetrios A. Spandidos, National Hellenic Research Foundation, Institute of Biological Research and Biotechnology, 48 Vas. Constantinou Ave., Athens 11635, Greece.

Principle

The calcium phosphate technique for introducing genes into mammalian cells was first described by Graham and Van der Eb [1973] and is still widely used.

Exogenous DNA is mixed with calcium chloride and is then added to a solution containing phosphate ions. A calcium phosphate–DNA coprecipitate is formed, which is taken up by mammalian cells in culture, resulting in expression of the exogenous gene. This method can be used to introduce any DNA into mammalian cells for transient expression assays or long-term transformation.

Outline

Transfect cells with the appropriate DNA carrying a selectable marker, i.e., the aminoglycoside phosphotransferase (*aph*) gene, and apply selection to eliminate cells that have not taken up and expressed the exogenous gene.

Materials

Geneticin (GIBCO)
Methocel MC4000 CP (Fluka)
Plasmids carrying the *aph* gene [see Spandidos and Wilkie, 1984a, for a variety of *aph* recombinant plasmids]
Carrier salmon sperm DNA (Sigma)
Growth medium, SF12: Ham's F12 (ICN-Flow), supplemented with Eagle's MEM amino acids (ICN-Flow), with 15% fetal bovine serum (Sterile Systems Inc.)
Liquid medium selection procedure:
2 × HEPES-buffered saline (2 × HBS):
1.63 g NaCl (Analar, BDH)
1.19 g HEPES (Sigma)
0.023 g Na_2HPO_4-$2H_2O$ (Analar, BDH)
Distilled water to 100 ml
Adjust the pH to 7.1 with 0.5 M NaOH. Filter sterilize the solution and store it at 4°C. The final composition of 2 × HBS is 0.28 M NaCl, 1.5 m*M* Na_2HPO_4, 50 m*M* HEPES, pH 7.1.
2.5 M $CaCl_2$: dissolve 10.8 g of $CaCl_2.6H_2O$ in 20 ml (final volume) of distilled water. Filter sterilize the solution and store it at 4°C.
Tris-EDTA buffer (TEB): 0.1 m*M* EDTA, 1.0 m*M* Tris-HCl, pH 8.0. Mix 50 μl of 0.2 M EDTA, pH 8.0 and 100 μl of 1.0 M Tris-HCl, pH 8.0, with distilled water to 100-ml final volume. Filter sterilize the solution and store it at 4°C.
Methocel selection procedure:
Methocel medium:
Mix 3 g of Methocel with 200 ml of distilled water and autoclave. The Methocel dissolves to yield a clear solution that can be stored at 4°C for at least 6 months (see Reagent Appendix). Just before use, warm the medium to 37°C and add:
22.0 ml 10 × Ham's F12 medium
4.0 ml 50 × MEM essential amino acids
4.0 ml 0.1 M sodium pyruvate
2.5 ml 0.2 M glutamine
5.0 ml 7.5% sodium bicarbonate
Next add 100 ml of serum and the appropriate concentration of the drugs to be used for selection. Thus, for Methocel medium containing geneticin (200 μg/ml), add 3.4 ml of 20 mg geneticin/ml water (stock solution, filter sterilized).
The final concentration of Methocel and serum are 0.9% and 30%, respectively. Note that the composition of the Methocel medium in step 2 depends upon the cell line being used, while the nature of the components in step 3 depends upon the selection marker being used.

Protocol: Liquid Medium Selection Procedure for Attached Cells

1. Harvest exponentially growing cells (i.e., mouse LATK⁻ cells) by trypsinization (see Chapter 10).
2. Replate the cells at a density of 5×10^5 per flask (25-cm^2 growth area) containing 5 ml of SF12 medium containing 15% fetal bovine serum. Incubate at 37°C for 24 hr.
3. Into a plastic bijoux vial, place 0.5 ml of donor DNA plus carrier DNA at a concentration of 80 μg/ml in TEB. Add 0.4 ml TEB and 0.1 ml of 2.5 M $CaCl_2$. Mix.
4. Add this DNA solution slowly (about 30 s) with continuous mixing to 1.0 ml of 2 × HBS already in a second bijoux vial. Mix immediately by vortexing and leave the solution at room temperature for 30 min. The DNA concentration at this stage is 20 μg/ml.
5. After the incubation, a fine precipitate will have formed. Add 0.5 ml of this DNA–calcium phos-

phate suspension to each flask containing cells in 5 ml of growth medium.

6. Incubate the flasks at 37°C for 24 hr to allow absorption of the DNA–calcium phosphate coprecipitate by the cells.
7. *Preselection expression stage.* Replace the medium with fresh, prewarmed medium and incubate at 37°C for a further 24 hr to allow expression of the transferred gene(s) to occur.
8. *Selection stage.* Replace the medium with an appropriate selection medium, in this case SF12 medium containing 15% serum and 200 μg/ml geneticin. (Cultured cell lines differ in their sensitivity to geneticin and the most suitable concentration of geneticin to use must be empirically determined.)
9. Renew the selection medium every 2–3 d for up to 2–3 wk when colonies are routinely counted.

Analysis

To pick individual colonies, remove the growth medium and then cut off the top of the flask using a heated scalpel, or use flask sealed with flexible, removable film (Nunc). Place a stainless steel cloning ring (diameter 0.4–0.8 cm) over each colony and detach the cells by incubation for 2–3 min at room temperature with 0.1 ml of 0.25% trypsin (see Chapter 11). After the cells have detached, use a sterile Pasteur pipette or micropipette to transfer them to 25-cm² flasks containing fresh prewarmed growth medium. If a permanent record is desired, the cells in some of the flasks can be fixed in ice-cold methanol for 15 min, then stained with Giemsa (see Chapter 13).

Methocel Selection Procedure for Anchorage-Independent Cells

1. Start transformation as described in the protocol for liquid medium selection procedure, steps 1–8. After allowing for preselection expression (step 8), trypsinize and then count the cells.
2. Mix 0.2 ml of cells with 20 ml of Methocel medium in a plastic universal container and plate on bacteriological plates (9-cm diameter). Up to 2 $\times 10^6$ cells per plate can be plated, the choice of plating density depending on the transformation frequency expected.
3. Incubate the plates at 37°C for 7–10 d, depending on the doubling time of the recipient cell line.

Analysis

Count the colonies using an inverted microscope. If required, pick individual colonies using a Pasteur pipette and grow these in a suitable growth medium (5 ml per flask of 25-cm² growth area or 6-cm diameter dish).

Variations. DNA can also be introduced into cells by electroporation, where a high cell concentration is briefly exposed to a high-voltage electric field in the presence of the DNA to be transfected [Chu et al., 1987]. Small holes are generated transiently in the cell membrane and DNA is allowed to enter the cell, and, in some of the cells, becomes incorporated into the genome. Equipment for electroporation is available commercially (BioRad).

It has been known for some time that substances packaged in liposomes (synthetic, hollow, lipid microspheres) can gain facilitated entry into viable cells. This principle is used to promote uptake of DNA in transfection [Felgner et al., 1987] using Lipofectin (Life Technologies).

Retroviruses are becoming increasingly popular as vectors for introducing foreign DNA by infection, as they are able to incorporate larger DNA fragments than plasmids and will infect host cells spontaneously. Further details of this and other gene transfer techniques are to be found in Sambrook et al. [1989] and Frederick et al. [1993].

PRODUCTION OF MONOCLONAL ANTIBODIES

One of the most exciting developments of somatic cell fusion arises from the demonstration that sensitized plasma cells from the spleen of an immunized mouse can be fused to continuous lines of mouse myeloma cells. Some of the resultant fusion products are capable of synthesizing immunoglobulins; if cloned, each clone produces a single specific monoclonal antibody [Milstein et al., 1979]. The steps in the technique are outlined in Figure 23.19. A mouse is first immunized with crude antigen, and later the spleen is removed. It is then minced and placed in culture with TK^- mouse myeloma cells in the presence of polyethylene glycol. The cells fuse, and the fusion products are cloned in HAT medium. The parental cells or homokaryons will not grow in HAT medium because the myeloma is TK^- and the spleen cells are unable to grow *in vitro*. The clones are then grown and their specificity tested by immunoassay of their supernatant medium with appropriate target cells or antigen, anchored to microtitration plates. Where a specific antigen–antibody complex forms, it can be detected autoradiographically with ^{131}I-labeled or enzyme-conjugated antimouse immunoglobulin.

The required clones are recloned, retested, and then frozen. For subsequent antibody production, the cells

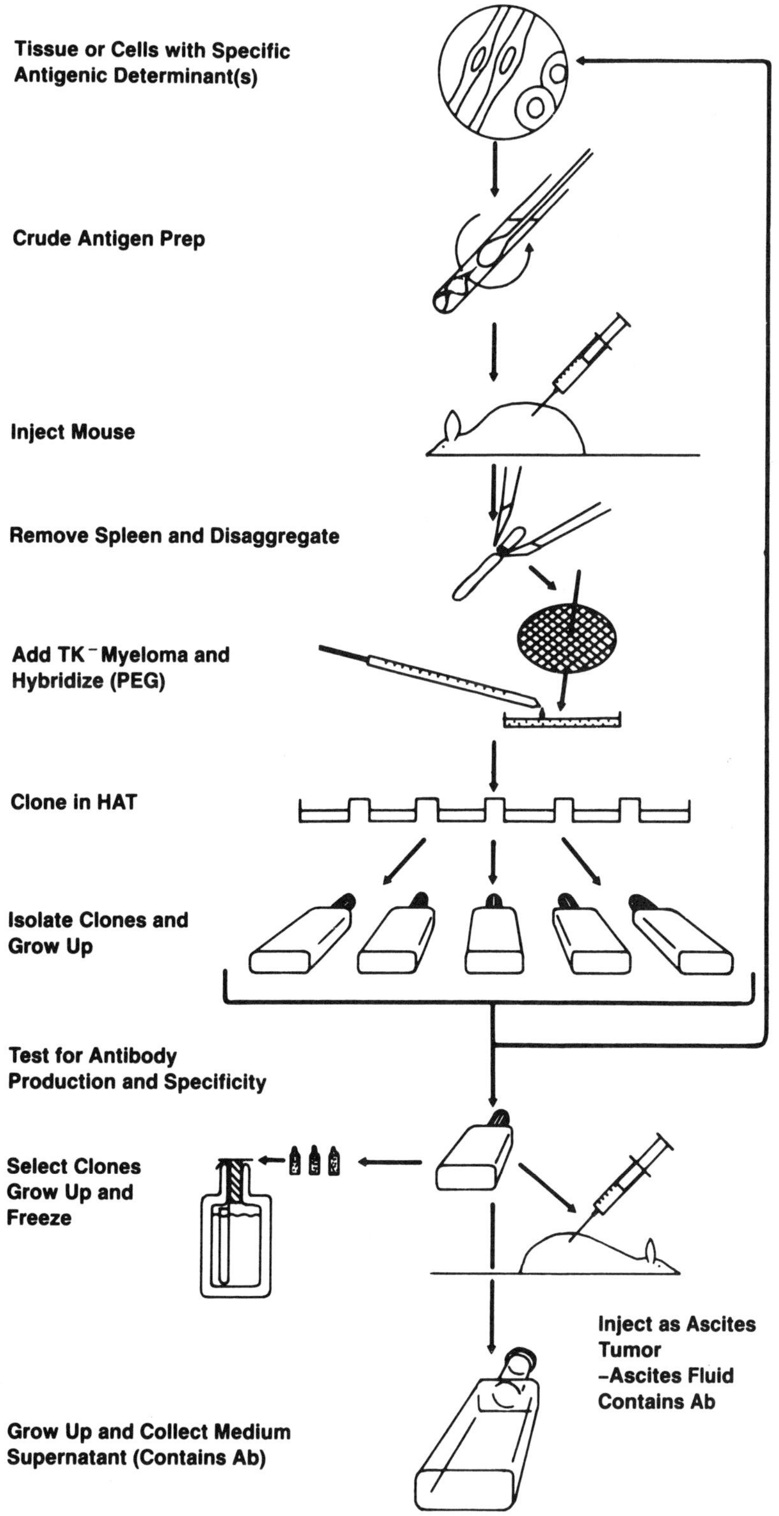
Tissue or Cells with Specific Antigenic Determinant(s)
Crude Antigen Prep
Inject Mouse
Remove Spleen and Disaggregate
Add TK⁻ Myeloma and Hybridize (PEG)
Clone in HAT
Isolate Clones and Grow Up
Test for Antibody Production and Specificity
Select Clones Grow Up and Freeze
Inject as Ascites Tumor
–Ascites Fluid Contains Ab
Grow Up and Collect Medium Supernatant (Contains Ab)

Fig. 23.19. *Schematic diagram of the production of hybridoma clones capable of secreting monoclonal antibodies. (Drawing by David Tallach.)*

may be propagated in suspension culture and the supernatant medium used, or they may be passaged in mice as an ascites tumor. Ascitic fluid gives very high yields of antibody although contaminated by globulins from the host animal.

The potential of this technique is enormous for the study of the immune system, production of cell type–specific markers, performance of structural analysis of proteins, in the production of vaccines, and in purification of proteins by immunoaffinity chromatography.

The following protocol for the preparation of hybridomas has been contributed by Chris Morris, European Collection of Animal Cell Cultures, PHLS, CAMR, Porton Down, Salisbury, England.

Principle

The production of heterokaryons is a result of fusion between myeloma cells and immunocompetent spleen cells. The resulting hybrids are selected for in HAT medium and subsequently screened for the production of a specific (monoclonal) antibody. Stable secreting hybrids are established by recloning the antibody-producing cells.

Outline

Immunize rats or mice with the desired antigen from whole or partially purified cells. It may be necessary to conjugate the antigen to increase antigenicity, e.g., with hemocyanin. Spleen cells are fused with myelomas using polyethylene glycol (PEG) and stable hybrids are selected by diluting the cells into 24- or 96-well plates in medium containing HAT medium.

Culture supernatants are assayed for specific antibody activity between 7–15 d after fusion. Cells from positive wells are cloned by dilution in 96-well trays and rescreened prior to expansion. The positive cells should be cloned twice more; frozen stock should be made from amplified cultures at each cloning and stored in liquid nitrogen.

Materials

Sterile:

24- and 96-well tissue culture trays
Screened NBCS and FCS
DMEM (4.5 g/l glucose) or RPMI 1640
0.34 M sucrose
15-ml and 50-ml sterile conical tubes (Falcon, Corning)
Sterile coverslips (20 × 20 mm)
Myelomas, e.g., NS/1, Sp2/0-Ag14 (mouse); y/o (rat) (European Culture Collection or ATCC)
Pristane (2,6,10,14-tetramethyl pentadecane) (Sigma)
HT stock (100×)
Hypoxanthine (6-hydroxypurine, Sigma No. 9377) 136.1 mg (10^{-2} M) and thymidine (Sigma No. T5018) 37.8 mg (1.6×10^{-3} M) dissolved in 100 ml dH_2O. Heat to 50°C until dissolved. Filter at 0.2 μm and aliquot. Store at −20°C. Keep working stocks at 4°C.

A Stock (100×)

Aminopterin (4-aminofolic acid; 4-aminopter-oyl-glutamic acid, Sigma No. 1784) 1.76 mg (4×10^{-5} M is added to 90 ml d H_2O. Add 0.5 ml IN NaOH. When dissolved add 0.5 ml IN HCl. Make up to 100 ml, 0.2-μm filter, and aliquot. Store at −20°C. Protect working stock (4°C) from light.

Myeloma Growth Medium

Dulbecco's MEM (4.5 g/l glucose) or RPMI 1640 with 10% serum (FBS, or a mixture of FBS + NBCS, e.g., 5 + 5) with 10 ml of HT per l, 10 ml of 200 m*M* L-glutamine per l. Prescreen sera by cloning the myelomas in 96-well trays, e.g., at 3, 1, and 0.3 cells/well concentrations, and selecting the one giving the highest number of wells with growing cells. All serum used should be *held* at 56°C for 40 min prior to use.

Hybridoma Growth Medium

Prepare as for myeloma medium, with an additional 5% serum, e.g., 10% FBS + 5% NBCS. Aminopterin is added as required.

PEG Solution (50% v/v)

Two molecular weights have been used successfully by the author, i.e., 1,500 and 4,000. It may be necessary to try several brands for the best results, e.g., BDH, Merck, Roth.

Melt 2 × 10 g by autoclaving, and cool to 50°C. With a glass pipette heated in a flame, measure accurately the PEG volume of one (this PEG is then discarded). Add an equivalent volume of DMEM (no serum), heated to 50°C, to the other container of PEG and mix rapidly. If any of the PEG solidifies, discard and start again with fresh PEG. The solution is kept at 4°C and should go alkaline (purple color) after 2–3 d. Use only alkaline solutions.

Nonsterile:

Fluorescence microscope
Mountant suitable for UV illumination
Hemocytometer (improved Neubauer)
Trypan blue solution (see Chapter 19)
Antibody screening reagents, e.g., ELISA, RIA, immunofluorescence
Mice (balb/c) or rats (Lou)

Protocol

A fusion should not be attempted until a reliable screening method for specific antibodies has been devised. The essentials of a successful screening method are speed, ease, and reliability, e.g., ELISA, radioimmunoassay.

1. Culture myelomas in HT medium and maintain between 2–6 × 10^5 cells/ml for 7–10 d prior to a fusion. The cells should be from a reliable source (known to give *stable* fusion products) and must be screened for mycoplasma contamination.
2. One to three days prior to the fusion, aseptically prepare peritoneal macrophages from 4–8-week-old normal healthy mice by injecting 5 ml of ice-cold 0.34 M sucrose. Count the cells and seed at 10^5/well for 24-well trays (1 ml) and 2 × 10^4/well for 96-well trays (0.2 ml) in HAT medium.
3. After the final antigen injection, screen, using your chosen system, the immunized animals' blood for specific antibody production (tail or eye bleeding). Successfully immunized animals are sacrificed by neck dislocation or CO_2 asphyxiation. Aseptically remove the spleen. Transfer to a laminar flow cabinet and place in a petri dish with 5 ml of growth medium (HT). Tease spleen apart with surgical needes and mix with a pipette. Transfer the dispersed cells in medium to a 15-ml tube and allow 1–2 min for large debris to settle. Transfer supernatant to another tube and count cells. Centrifuge (80 *g* for 5 min).
4. Aliquot enough myeloma cells into 50-ml tubes to give a final ratio of one myeloma cell for every two to three spleen cells, e.g., 5 × 10^7 myelomas to 10^8 spleen cells. Centrifuge (80 *g* for 5 min) and resuspend both myeloma and spleen cell pellets in 20–30 ml of serum-free medium in one 50-ml tube. Centrifuge (80 *g* for 5 min). Pour off supernatant and drain pellet by inverting the tube onto sterile tissue. The cells must be fully drained to avoid diluting the PEG.
5. Hold the tube in water at 37°C and add *dropwise* 1 ml of PEG solution onto the wall of the tube just above the pellet, over a 5–10-s period. Gently vibrate the base of tube with finger or thumb while adding PEG, but do not break up the pellet. Hold the tube for a further 50 s at 37°C and mix the contents gently every 10 s with a pipette tip. Remove the tube from 37°C and immediately start to add dropwise, with gentle swirling, 5 ml of serum-free medium at the rate of 1 ml per min. A further 5 ml is added at 2 ml per min. Stand the tube for 2–3 min (capped). Centrifuge gently (60–70 *g* for 5 min).
6. Decant the medium. Resuspend the cells by gently adding HAT medium to give a final myeloma concentration of 1.5 × 10^6/ml for 24-well or 5 × 10^5/ml for 96-well plates. Add 0.1 ml per well to macrophage plates. Transfer to 7–8% CO_2 chambers at 37°C.
7. Start to examine wells for colonies after 3–4 d. Maintain expanding cultures in HAT for at least 10–14 d and medium change (50%) growing wells at least once prior to screening. Screen wells when more than 50% of the base is covered, expand positives to at least 3 × 10^6 cells, then clone in 96-well trays (HAT medium). Rescreen them using HT medium for subsequent steps, expand, and reclone. Repeat once more to establish stable hybrids. Expand selected positive hybrids at each cloning step and freeze cells at 5–6 × 10^6 per ampule. Store in liquid nitrogen.
8. *Ascites production*: Hybridomas grown in peritoneal cavity will grow to very high densities, thus producing high concentrations of antibody (up to 100 times higher than in cell culture). To facilitate *in vivo* growth, the mouse is given an intraperitoneal injection of 0.5 ml Pristane, 7–10 d before injecting hybrids (minimum 3 × 10^6/mouse). Tumors usually take 7–14 d to grow. Remove ascites fluid from swollen mice with a syringe and centrifuge. If taken aseptically, the cells can be cultured *in vitro* or, alternatively, reinjected into a new mouse. If second or third drainings of ascites are taken, the fluid usually becomes contaminated with blood. It is considered humane at this point to kill the mouse.
9. *Cryopreservation* (see Chapter 17).

IN SITU MOLECULAR HYBRIDIZATION

Nucleic acid hybridization is used routinely for the detection of specific nucleotide sequences in DNA (Southern blotting) or RNA (Northern blotting). This technique can be applied to fixed cells to detect nucleotide sequences *in situ* [Conkie et al., 1974; Harrison et al., 1974](Fig. 23.20).

The following protocol for *in situ* hybridization has been contributed by David Conkie, the Beatson Institute for Cancer Research, Garscube Estate, Switchback Road, Bearsden, Glasgow, Scotland.

Principle

Molecular hybridization *in situ* is the annealing of nucleic acid probes, single-stranded and highly radioactive, to complementary regions of nucleic acid fixed in conventional cytological preparations. Specifically, the technique described permits localization of RNA transcripts within individual cells and is readily

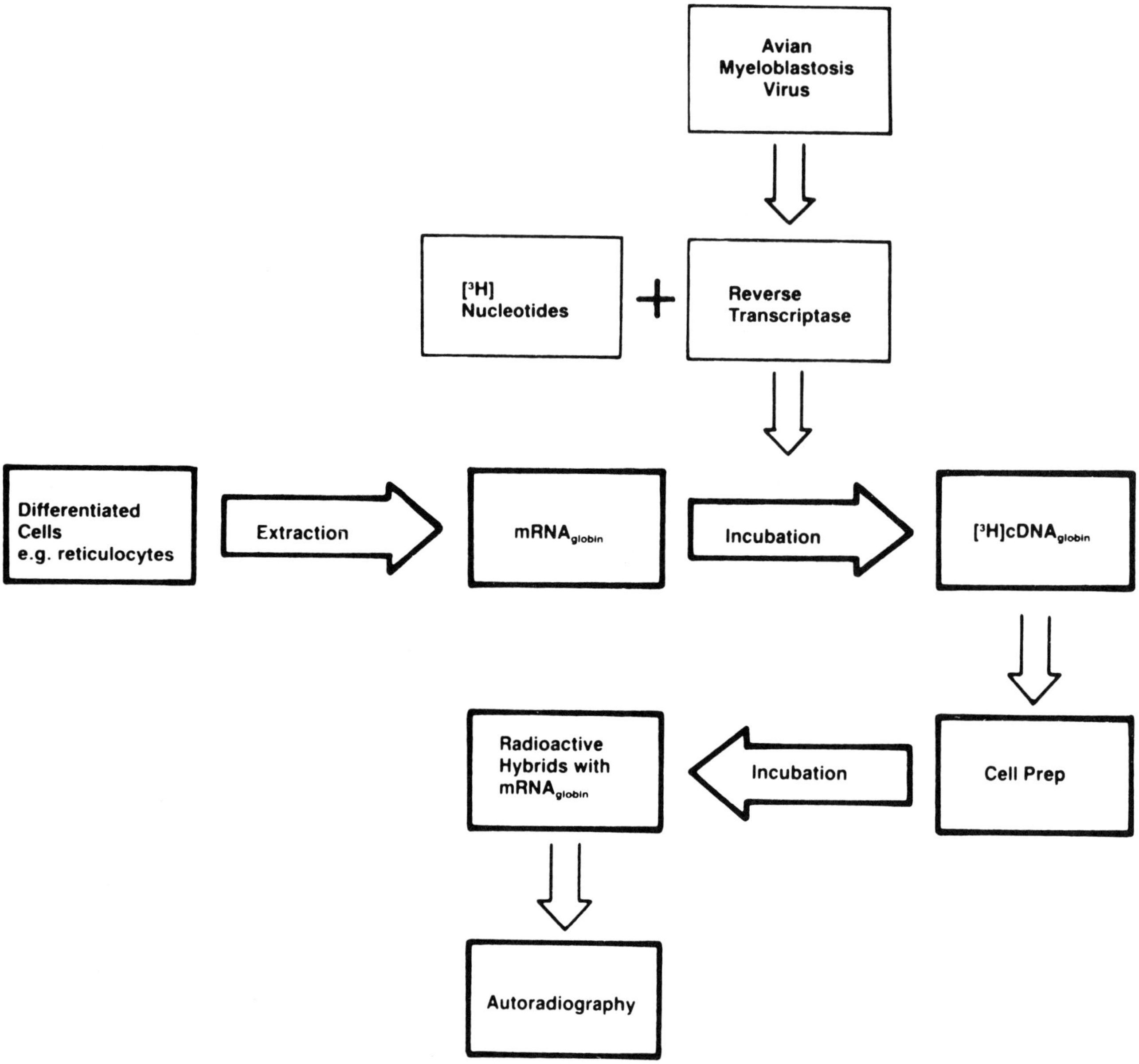

Fig. 23.20. In situ *hybridization of radioactive complementary DNA, synthesized on a messenger RNA template by the action of reverse transcriptase, with intracellular messenger RNA.*

accomplished without the necessity for high specific activity probes or signal enhancement.

Outline

Apply the radiolabeled probe directly onto fixed cells at 43°C for 18 hr. Wash in 2× SSC at 55°C and then at 18°C. Detect molecular hybrids by autoradiography; for cytological discrimination, stain the target tissue.

Materials

Water bath at 43°C and then at 55°C
Glass slide carriers with lid
Methanol
10% trichloroacetic acid
70% ethanol
Gelatin chrome alum (gelatin 5 g, chrome alum 0.5 g, distilled water 1 l).
Gelatin-coated glass slides (immerse slides in chromic acid, wash thoroughly in water, dip into gelatin chrome alum freshly prepared, and allow to drain and dry dust-free; store at 4°C)
16-mm diameter 00 glass coverslips (immerse in chromic acid, then wash thoroughly in water; store in 70% ethanol)
2× SSC (2× standard saline citrate; sodium

chloride 17.5g, sodium citrate 8.8 g, distilled water 1 l, adjust to pH 7.0)
3x SSC (sodium chloride 26.25 g, sodium citrate 13.2 g, distilled water 1 l, adjust to pH 7.0)
60–80° petroleum ether
Rubber solution (Cow Proofings Ltd.; reduce viscosity by adding 1/10th volume of petroleum ether)
Hybridization buffer (3x SSC containing 40% formamide [Fluka] adjusted to pH 6.5)
Hybridization probe: radioactive nucleic acid probes may be prepared by reverse transcriptase or nick translation in the presence of ^{35}S- or ^{3}H-labeled precursors; incorporation of a single ^{35}S-nucleotide precursor results in a specific activity of up to 10^9 dpm/μg using reverse transcriptase or 3×10^8 dpm/μg by nick translation (kits containing essential reagents and protocol for reverse transcription and nick translation are available from Amersham or NEN Division of DuPont); dissolve the labeled nucleic acid probe in the hybridization buffer at about 5×10^7dpm/ml; if double stranded, denature at 70°C for 10 min immediately before use.
Autoradiography materials as in "Autoradiography" section above
DPX mountant

Protocol

1. Centrifuge suspension cultures onto gelatin-coated glass slides using a cytocentrifuge to give 10^6 cells/cm^2. Alternatively, prepare thin cryostat sections, tissue imprints, or monolayer cultures grown on glass or Thermanox (ICN-Flow).
2. Fix the cytological preparations in methanol for 5 min followed by three immersions in 10% trichloracetic acid at 4°C.
3. Dehydrate in 70% ethanol and air dry. At this stage coverslip cultures should be fixed to a glass slide with cells uppermost, using a drop of DPX. Fixed preparations may be stored at 4°C.
4. Warm the fixed cells to 43°C by placing the slide on a metal tray floating in a water bath at 43°C. Place 5 μl of the hybridization probe directly onto the fixed cells.
5. Cover with a 16-mm diameter coverslip.
6. Apply rubber solution to seal the edges of the coverslip.
7. Incubate the slides for 18 hr on the flat metal tray at 43°C floating in a water bath with lid. The high humidity prevents dehydration of the hybridization buffer if the seal is incomplete.
8. After incubation, cool the slides on ice and then peel the rubber seal from the coverslip using fine forceps.
9. Dislodge the coverslip either under the surface of 2× SSC or by using a jet of 2× SSC.
10. Immerse the slides in a glass-slide carrier containing 2× SSC.
11. Seal the lid with tape and partially immerse the slide carrier in a 55°C water bath for 30 min.
12. Wash the slides four times in 2× SSC at 18°C.
13. Dehydrate in 70% ethanol and air dry.
14. Detect molecular hybrids by autoradiography using stripping film or liquid emulsion as in the "Autoradiography" section in this chapter.
15. Stain the cells through the autoradiographic emulsion (see above).

Analysis. Microscopy reveals the localization of silver grains over specific cells that contain the target RNA transcripts.

Variations. Alternative nonradioactive techniques for detecting molecular hybrids in cytological preparations have been devised (see below). These methods provide a faster detection time with enhanced resolution and signal-to-noise ratio.

Uridine triphosphates containing a biotin molecule attached through "linker" allylamine groups can be incorporated into DNA probes by nick translation [Langer et al., 1981]. These probes have been used in the *in situ* hybridization technique by Langer-Safer et al. [1982] and are detected by either: (1) immunological methods using antibiotin antibodies and a second antibody tagged with a fluorescent, enzymatic, or electron-dense reagent or (2) affinity labeling with avidin conjugated to rhodamine or biotinylated derivatives of peroxidase or alkaline phosphatase.

The following protocol for fluorescence *in situ* hybridization (FISH) has been provided by Nicol Keith, CRC Department of Medical Oncology, University of Glasgow, Garscube Estate, Bearsden, Glasgow G61 1BD, Scotland.

Fluorescence *In Situ* Hybridization (FISH) in the Analysis of Genes and Chromosomes

In situ hybridization is the most direct way of determining the linear order of genes on chromosomes. Using chromosome- and gene-specific probes, numerical and structural aberrations can also be analyzed within individual cells. These techniques have wide applications in the diagnosis of genetic disease, identification of gene deletions, translocations, and gene amplification during cancer development [Bar-Am et al., 1992; Kallioniemi et al., 1992; Matsumura et al., 1992; Ried et al., 1992; Telenius et al., 1992].

Principle

Nucleic acid probes are labeled nonisotopically by incorporation of nucleotides modified with molecules including biotin or digoxigenin. After hybridization of the labeled probes to the chromosomes, detection of the hybridized sequences is achieved by forming antibody complexes that recognize the biotin or digoxigenin within the probe. The hybridization is visualized by using antibodies conjugated to fluorochromes. The fluorescent signal can be detected in a number of ways. If the signal is strong enough, standard fluorescent microscopy can be used. However, data analysis and storage can be improved considerably by the use of digital imaging systems such as confocal laser scanning microscopy or cooled CCD camera. The major advantages of fluorescence *in situ* hybridization (FISH) are that it is nonisotopic, rapid, good for data storage and manipulation, sensitive, shows accurate signal localization, allows simultaneous analysis of two or more fluorochromes, and provides a quantitative and spatial distribution of the signal.

Protocol for FISH Using Single Copy Genomic Probes and Chromosome Painting

Outline

Biotinylated or digoxigenin-labeled probes are hybridized to denatured chromosomes and detected by double-antibody fluorescent staining.

Materials

Labeled probe:

Large cosmid clones are most suitable for probes to detect single copy genes. However, cDNA probes can be used. DNA is labeled by nick translation using commercial kits. The nick translation kit marketed by Boehringer Mannheim can be used to incorporate either biotin or digoxigenin and the manufacturer's instructions are followed.

Precipitation of probe containing repetitive sequences:

Large cosmid probes often contain repetitive sequences that if not suppressed prior to hybridization would result in high levels of nonspecific hybridization. The repeat sequences can be suppressed by competition with unlabeled human Cot1 DNA sequences that are enriched for repeat sequences. The Cot1 DNA can be included at the precipitation step.

(a) For a 20-μl nick translation reaction, add 1 μl 0.5 M EDTA, 2.5 μl 4.0 M LiCl, 1 μl glycogen, 100- to 1,000-fold excess human Cot1 DNA (Life Technologies), 100 μl ethanol.

(b) Precipitate on dry ice for 30 min or at −20°C overnight.

(c) Spin in microfuge for 15 min to pellet DNA.

(d) Wash pellet in 70% ethanol.

(e) Spin to repellet DNA and dry.

(f) Resuspend DNA at 2–10 ng/μl in hybridization buffer.

Precipitation of probes without repetitive sequences:

Follow protocol above but leave out Cot1 DNA from precipitation.

Hybridization mix:

50% formamide
5% dextran sulfate
2× SSC
500 μg/ml salmon sperm DNA
SSC, 1×: 0.15 M NaCl, 0.015 M sodium citrate

Probe denaturation:

(a) For probes containing repetitive sequences that need to be suppressed using Cot1 DNA. Heat probe mix to 70°C for 10 min. Transfer probe mix to 37°C for 1 hr prior to application to slide.

(b) For probes without repetitive sequences. Heat probe mix to 70°C for 10 min. Chill on ice for 10 min.

Fixative:

Methanol:acetic acid, 3:1

TBST:

0.05% Tween 20 in 0.1 M Tris, 0.15 M NaCl, pH 7.5

Antibodies:

First antibody: e.g., sheep polyclonal antiserum to digoxigenin or biotin (titration recommended by supplier [Boehringer Mannheim]). Dilute antibodies in 3% BSA in TBST.

Second antibody: e.g., FITC-conjugated donkey anti-sheep IgG (Jackson Immunoresearch Inc.)

Mountant:

Vectashield; contains inhibitor of photo-bleaching (Vector Labs)

Protocol: (a) Chromosome Preparation and Denaturation

1. Prepare metaphase-arrested cells by standard technique (see above) and drop onto slides. Mark area of spreads with diamond pencil or objective slide marker (Nikon).
2. Fix slides for 1 hr in fresh methanol:acetic acid, 3:1, and air dry.
3. Rinse in 2 × SSC for 2 min at room temperature.
4. Incubate in 100 μg/ml RNAase in 2 × SSC at 37°C for 1 hr.
5. Rinse in PBS.
6. (Optional step) Fix in freshly prepared 1% paraformaldehyde in PBS 10 min, room temperature.
7. Rinse in PBS.

8. Dehydrate: 2 × 2 min 70% ethanol, 2 × 2 min 100% ethanol, air dry.
9. Denature chromosomes in 70% formamide in 2 × SSC at 70°C for 2–4 min (determine experimentally, starting at 2 min). Make sure 70% formamide is at 70°C before use.
10. Wash slides in several changes of ice-cold 70% ethanol.
11. Dehydrate as in step 8 and air dry.
    Chromosomes are now ready for hybridization.

(b) Hybridization

1. Apply 10 μl of denatured probe over area of spreads and cover with 22 × 22 mm coverslip.
2. Seal coverslip with rubber latex adhesive.
3. Place slides in humidified box at 37°C overnight.

(c) Probe Detection

1. Remove coverslips by emersion in 2 × SSC (room temperature) and peel off adhesive. Place slides in Coplin jar.
2. Soak slides, 2 × 10 min, in 50% formamide, 1 × SSC at 42°C.
3. Wash slides, 2 × 10 min, in 2 × SSC, 42°C.
4. Rinse in TBST.
5. Block with 3% bovine serum albumin (BSA) in TBST for 30–60 min at 37°C.
6. Add first antibody to slides. Use 100 μl/slide, cover with Parafilm coverslip.
7. Incubate 1 hr at 37°C.
8. Wash in 500 ml TBST for 10 min at room temperature.
9. Add second fluorochrome-conjugated antibody, in 3% BSA/TBST, to slides at a titration recommended by supplier or determined by experiment.
    Note. Be sure to use correct antibody combinations; for example, sheep polyclonal antiserum to digoxigenin as a first antibody, followed by FITC-conjugated donkey anti-sheep IgG.
10. Incubate for 30 min at 37°C.
11. Wash for 30 min in 500 ml TBST at room temperature.
12. Counterstain with 0.8 μg/ml DAPI and/or 0.4 μg/ml propidium iodide in TBST for 10 min. Mount slides in antifade mountant.
13. View slides using fluorescence microscopy using appropriate filter combinations.

(d) Chromosome Painting (see below)

Chromosome paints are available commercially from a number of sources, including Life Technologies, Cambio, and Oncor. It is therefore no longer necessary to prepare your own paints. The hybridization and detection protocols vary with each commercial source. However, in general the protocol from (a), above, on chromosome preparation and denaturation, can be used prior to hybridization. Hybridization and detection can then be carried out according to the supplier's instructions. If the paint is labeled with biotin such as the Cambio paints, the detection protocol, step (c), can be followed using the appropriate antibody combinations.

Applications and Recent Advances

Chromosome paints are a collection of probes specific for any chromosome. Thus, complex translocation events can be studied. Chromosome paints are now available commercially for most chromosomes and are most valuable in situations where good metaphase spreads can be achieved, such as hematological malignancies or cell lines. A new application of chromosome painting is termed reverse painting and involves flow sorting a chromosome of unknown origin, generating a paint for this chromosome, and painting normal lymphocytes with these sequences. In this way the normal chromosomes from which the unknown marker is derived will be visualized [Telenius et al., 1992].

FISH has many exciting applications in molecular cytogenetics. In particular it is proving to be useful in the study of the molecular events underlying solid tumor development. By hybridizing with gene- and chromosome-specific probes, which have been labeled with different fluorochromes, molecular detail on aneuploidy, allele loss, and gene amplifications can be achieved [Kallioniemi et al., 1992; Matsumura et al., 1992]. Through detection of probes labeled with different haptens, multiple genetic events in individual cells can be visualized. By varying the ratios of hapten, up to seven probes have been visualized simultaneously to date [Ledbetter, 1992; Ried et al., 1992].

Recent progress in image analysis and probe labeling has made it possible for nucleotides conjugated directly to fluorochromes to be used for FISH. This approach is less time-consuming, as it does not involve any immunological detection and reduces background, thus aiding accurate probe localization [Bajalica et al., 1992]. Another new approach is primed *in situ* labeling (PRINS), which may replace the need for the standard hybridization step [Koch et al., 1992]. PRINS requires short oligonucleotide primers specific for the sequence under study to be annealed to the chromosomes. The oligos serve as primers for the synthesis of DNA using the chromosome target as the template. Either hapten-modified bases or bases modified by conjugation to fluorochromes are included to allow visualization of the reaction. PRINS has the potential to be very versatile and may reduce the time for *in situ* hybridizations even further and negate the requirement for cloning large genomic fragments for FISH.

VIRUS PREPARATION AND ASSAY

Many of the mass propagation methods described above were developed to produce large quantities of virus for analytical and preparative purposes. The current status of our understanding of viruses and their use in the manufacture of vaccines owes much to the development of tissue culture [Habel and Salzman, 1969; Kuchler, 1974, 1977; Petricciani, 1979]. The corollary is also true, that the development of large-scale culture techniques and serum-free media was prompted in part by the requirements of virology.

Viral assays are of two main types: (1) cytopathic and (2) transforming [Weiss et al., 1982,1985]. Cytopathic viruses may be assayed by their antimetabolic effects in microtitration plates or by the formation of characteristic plaques in monolayers of the appropriate host cell. A viral suspension is serially diluted and added to monolayer culture plates. The number of plaques forming at the limiting dilution is taken as equivalent to the number of infectious particles in the supernatant medium, allowing the concentration of virus in the initial sample to be calculated. Characterization of the virus may be performed with specific antisera, measuring inhibition of the cytopathic effect of the virus, or by radioimmunoassay.

Transforming viruses may be assayed by the selective growth of transformed clones in suspension [Macpherson and Montagnier, 1964], or by looking for transformation foci in monolayer cultures [Temin and Rubin, 1958] (see Chapter 15).

The following protocol for virus preparation and assay has been contributed by Joan C.M. Macnab, Medical Research Council Institute of Virology, University of Glasgow, Scotland.

Principle

Viruses require a susceptible host-cell system to replicate. Virus particles adsorb to cultured cells at physiological temperatures. The virus penetrates, uncoats, replicates its nucleic acid in the nucleus or cytoplasm, and after transcription and translation assembles progeny virions that egress either by cell lysis or budding. The number of progeny virus particles released from one cell is referred to as the burst size. Virus is titrated by adsorbing serial 10-fold dilutions to cells and counting the resultant plaques after appropriate incubation at the permissive temperature.

Outline

Herpes simplex virus (HSV) is used as the example.

Permissive cell cultures are infected with an aliquot of virus (1 hr), incubated at 37°C for 2 d, and the resultant virus harvested by (1) sonicating and extracting the cells and (2) pelleting virus from the supernatant. Virus is titrated by adsorbing dilutions of virus to cells (1 hr) and counting resultant plaques (37°C for 2 d).

Materials

Sterile:

Stock virus of known titer–refer to virus assay (see below) if titer not known
80-oz (2-l) autoclaved round glass bottles or plastic (Falcon) bottles suitable for roller culture or Nunclon cell factory or large plastic flasks (Nunclon, Falcon)
50-mm plastic culture dishes (Nunclon, Falcon)
Growth medium with 5% serum for culturing virus
Optional–human serum (pooled human serum with HSV antibodies) instead of calf serum to prevent virus spread and secondary plaques
Bijoux bottles for dilutions
Black cup vial or ampules
PBSA
1-, 5-, and 10-ml glass pipettes
Eppendorf pipettes
Micropipette tips (Eppendorf, Gilson)

Nonsterile:

Suitable gassed incubator or warm room
Roller culture facilities if available
Bath sonicator, e.g., Cole Palmer
Ice
Storage space at −70°C or −186°C
Scraper suitable for removing virus-infected cells (Costar, Corning)

Protocol: (a) Virus Preparation

1. Remove medium from cells.
2. Infect cells with 0.001 plaque-forming units per ml (pfu/ml) virus in enough medium (without serum) to just cover cell monolayer.
3. Incubate for 60–90 min at 37°C with gassing if necessary.
4. Add medium with 5% serum, 17 ml for large culture flask and 20–30 ml for roller bottle.
5. Incubate 37°C for 2 d.
6. Check for virus cytopathic effect (cpe). Cells should be rounded, giant, syncytial, and coming off substrate into medium. Shake or scrape remaining cells into medium.
7. Reserve supernatant fluid, pellet cells by low-speed centrifugation (50–100 g, 5–10 min).

Two virus stocks, (1) supernatant virus and (2) cell-associated virus, are now prepared:

1. Remove all but 20–30 ml supernatant to sterilized centrifuge bottles. Pellet virus at 12,000 *g* for 2–3 hr. Resuspend virus pellet in 10 ml supernatant fluid, sonicate for 10 min on ice in 60-s bursts, aliquot to ampules (0.5 ml each), and freeze at −70°C. Call this preparation (1) supernatant.
2. Add 10 ml supernatant fluid to cells pelleted at low speed.
3. Sonicate 10 min on ice in 60-s bursts until cells are obviously disrupted.
4. Pellet cell debris (100 *g*, 10 min).
5. Collect supernatant and reserve as cell-associated virus extraction (2a).
6. Aliquot and freeze. Re-extract cell pellet as in steps 2–6. Call this cell-associated virus extraction (2b).

Virus Assay

Titrate virus preparations separately (the titers will be different).

1. To sterile bijoux add 0.9 ml PBSA + 5% calf serum (to improve viability of virus).
2. Add 0.1 ml and, using fresh Eppendorf tip or pipette for each dilution, dilute virus in serial 10-fold steps.
3. Remove medium from 50-mm dishes. (Each preparation is usually titered in duplicate.)
4. Carefully add 0.1 ml virus dilution at rim of plate and gently run this over all the surface of the dish. (Do not pipette roughly onto cell sheet or cells will detach.)
5. Incubate at 37°C for 1–1½ hr in gassed incubator.
6. Overlay with 4.0 ml medium with 5% calf serum or 5% human serum (if desired). Human serum contains HSV antibodies and prevents secondary plaques.
7. Incubate at 37°C in gassed incubator 2 d.
8. Fix monolayer after careful removal of medium.
9. Stain with Giemsa.
10. Count number of plaques (these usually appear as holes in the cell monolayer) with characteristic virus cpe, i.e., cell rounding, syncytia, or giant cells.

Analysis

When countable plaques are visualized, then note dilution. An average of eight plaques at 10^{-7} dilution represents the titer of 0.1 ml virus stock. Stock titer is therefore 8×10^8 pfu/ml (assuming that one plaque arises from one infectious particle).

Variations. If poor titers of virus are obtained, this may be due to poor yield of infectious particles per cell, i.e., burst size. Defective noninfectious virus may be present in high amounts. This may be calculated by determining the total number of virus particles by the electron microscope (EM) and calculating the number of those that are infectious particles. This gives the particle:infectivity ratio and should preferably not exceed 50:1. High ratios of particles (seen in EM) to infectious particles frequently interfere with replication of infectious virions. Some viruses release few particles into the supernatant, e.g., HSV-2, and most virus is cell associated.

Virus stock should be checked for contamination with bacteria, fungi, and mycoplasma by plating on suitable medium or infecting cells.

Most viruses are titrated by using an overlay of 0.6% purified agarose, Noble (Difco) agar, or methyl cellulose to prevent spread of virus to secondary plaques. However, HSV is very sensitive to impurities in agarose and this step is generally omitted unless plaque purified stocks are being made.

Herpes simplex virus replicates in a wide variety of human and animal cell types, whereas other viruses, e.g., human cytomegalovirus (CMV), replicate only in human fibroblast cells [Oram et al., 1982]. Some Bunyaviridae replicate in amphibian cell lines [Watret et al., 1985]. The tumor virus SV40 grows in monkey (BSC-1) cells, whereas polyoma virus grows in mouse embryo cells. The best culture system for a specific virus may be found by consulting reference textbooks such as *Comprehensive Virology* [Fraenkel-Conrat and Wagner, 1979] or *Molecular Biology of the Tumour Viruses* [Weiss et al., 1982, 1985].

Note. Some important viruses cannot, as yet, be cultured in tissue culture systems *in vitro*, e.g., papilloma virus and the AIDS virus (HIV).

IN CONCLUSION

Tissue culture is not limited to mammalian and avian systems, though much of modern technology and our current understanding of cellular and molecular biology have derived from these cultures. Using appropriate culture conditions, it is possible to culture cells from cold-blooded vertebrates such as reptiles, amphibia, and fish and many cell lines are now available (commercial catalogues, e.g., Flow, GIBCO, American Type Culture Collection [ATCC], and the European Collection for Animal Cell Cultures [ECACC]; see Trade Index and Chapter 17, Tables 17.3, 17.4).

To cover these and many other fascinating aspects of this field would take many volumes and defeat the objective of this book. It has been more my intention to provide sufficient information to set up a laboratory and prepare the necessary materials with which to perform basic tissue culture and to develop some of the more important techniques required for the characterization and understanding of your cell lines. This book may not be sufficient on its own, but with help and advice from colleagues and other laboratories, it may make your introduction to tissue culture easier and more profitable than it otherwise might have been.

Reagent Appendix

Acetic/Methanol
1 part glacial acetic acid
3 parts methanol
Make up fresh each time used and keep at 4°C.

Agar 2.5%
2.5 g agar
100 ml water
Boil to dissolve.
Sterilize by autoclaving.
Store at room temperature.

Amido Black
See "Naphthalene black"

Amino Acids—Essential
See Eagle's MEM: "Amino Acids," Chapter 7.
(Available as 50× concentrate in 0.1 *N* HCl from commercial suppliers such as ICN/Flow and GIBCO/Life Technologies.)
Make up tyrosine and tryptophan together at 50× in 0.1 *N* HCl and remaining amino acids at 100× in ultrapure water (working strength concentrations are given in Chapter 7).
Dilute for use as in Chapter 8.
Sterilize by filtration.
Store at 4°C in the dark.

Amino Acids—Nonessential

Ingredient	*g/l (100×)*
L-alanine	0.89
L-asparagine H_2O	1.50
L-aspartic acid	1.33
Glycine	0.75
L-glutamic acid	1.47
L-proline	1.15
L-serine	1.05
Water	1,000 ml

Sterilize by filtration.
Store at 4°C.
Use 1:100.

Antibiotics
See under specific headings, e.g., penicillin, streptomycin, kanamycin, gentamycin, mycostatin.

Antifoam
(e.g., RD emulsion 9964.40; *see* Antifoam, Trade Index, Sources of Materials)
Aliquot and autoclave to sterilize.
Store at room temperature.
Dilute 0.1 ml/l liter, i.e., 1:10,000.

Bactopeptone, 5%
5 g Difco bactopeptone dissolved in 100 ml Hanks' BSS
Stir to dissolve.
Dispense in aliquots appropriate to a 1:50 dilution, and autoclave.
Store at room temperature.
Dilute 1/10 for use.

Balanced Salt Solutions (BSS)

See Table 7.4.
Dissolve each constituent separately, adding $CaCl_2$ last, and make up to 1 l. Adjust to pH 6.5.
Sterilize by autoclaving or filtration. If autoclaved, pH must be kept below pH 6.5 to avoid precipitation of phosphate; alternatively, calcium may be omitted and added later. If glucose is included, solution should be filtered to avoid carmelization of the glucose, or the glucose may be autoclaved separately (see below) at a higher concentration (e.g., 20%) and added later.
If autoclaving is used, mark liquid level before autoclaving. Store at room temperature and make up to mark with sterile ultrapure water before use, if evaporation has occurred. If borosilicate glass is used, the bottle may be autoclaved sealed and no evaporation will occur.
Hanks' BSS
Without phenol red: as regular recipe but omit phenol red.
Dissection BSS
Hanks' BSS without bicarbonate; previously sterilized by autoclaving.
250 U/ml penicillin
250 μg/ml streptomycin
100 μg/ml kanamycin or 50 μg/ml gentamycin
2.5 μg/ml Amphotericin B
(All preparations sterile; see below)
Store at −20°C.

Broths

See manufacturers' instructions (Difco, Microbiological Associates, GIBCO) for preparation. *See also* bactopeptone and tryptose phosphate broth.
Sterilize by autoclaving.

Carboxymethylcellulose (CMC)

Weigh out 4 g CMC and place in beaker.
Add 90 ml Hanks' BSS, and bring to boil to "wet" the CMC.
Allow to stand overnight at 4°C to clear.
Make volume to 100 ml with Hanks' BSS.
Sterilize by autoclaving. The CMC will solidify again during autoclaving but will dissolve at 4°C.
For use, e.g., to increase viscosity of medium in suspension cultures, use 3 ml per 100 ml growth medium.

Chick Embryo Extract [Paul, 1975]

1. Remove embryos from eggs as described in Chapter 9 and place in 9-cm petri dishes.
2. Take out eyes using two pairs of sterile forceps.
3. Transfer embryos to flat- or round-bottomed 50-ml containers, two embryos to each container.
4. Add an equal volume of Hanks' BSS to each.
5. Using a sterile glass rod that has been previously heated and flattened at one end, mash the embryos in the BSS until they have broken up.
6. Stand for 30 min at room temperature.
7. Centrifuge for 15 min at 2,000 *g*.
8. Remove supernatant and after keeping a sample to check sterility (see Chapter 8), aliquot and store at −20°C.

Extracts of chick and other tissues may also be prepared by homogenization in a Potter homogenizer or Waring blender [Coon and Cahn, 1966].

1. Homogenize chopped embryos with an equal volume of Hanks' BSS.
2. Transfer homogenate to centrifuge tubes and spin at 1,000 *g* for 10 min.
3. Transfer the supernatant to fresh tubes and centrifuge for a further 20 min at 10,000 *g*.
4. Check sample for sterility (see Chapter 8), aliquot remainder, and store at −20°C.

0.1 M Citric Acid/0.1% Crystal Violet

21.0 g citric acid
1.0 g crystal violet
Make up to 1,000 ml with deionized water.
Stir to dissolve. To clarify, filter through Whatman No. 1 filter paper.

CMC

See "Carboxymethylcellulose"

Colcemid, 100× Concentrate

100 mg colcemid
100 ml Hanks' BSS
Stir to dissolve.
Sterilize by filtration.
Aliquot and store at −20°C.
◇ Toxic; handle with care; weigh in fume cupboard and wear gloves.

Collagenase

2,000 U/ml in Hanks' BSS
100,000 units Worthington CLS grade collagenase, or equivalent (specific activity = 1,500–2,000 U/mg)
50 ml Hanks' BSS
To dissolve: stir at 36.5°C for 2 hr or at 4°C overnight.
Sterilize by filtration as for serum (see Chapter 8).
Divide into aliquots suitable for 1–2 wk use.
Store at −20°.

Collagenase-Trypsin-Chicken Serum (CTC) [Coon and Cahn, 1966]

	Volume	*Final Concentration*
Calcium-and magnesium-free saline [Moscona, 1952], sterile	85 ml	
Trypsin stock, 2.5%, sterile	4 ml	0.1%
Collagenase stock, 1%, sterile	10 ml	0.1%
Chick serum	1 ml	1.0%

Aliquot and store at −20°C.

Collection Medium (for tissue biopsies)

Growth medium	500 ml
Penicillin	125,000 units
Streptomycin	125 mg
Kanamycin or	50 mg
Gentamycin	25 mg
Amphotericin B	1.25 mg

Store up to 3 wk at 4°C or at −20°C for longer periods.

Counting Fluid × 10 [Paul, 1975]

Sodium chloride	700 g
Trisodium citrate ($2H_2O$)	170 g

Distilled water up to 10 l.
Stir ingredients in water until dissolved.
Filter through Whatman Filter Paper No. 1, then dispense into Pyrex bottles.
Adjust pH to 7.2
Autoclave and store at room temperature.
Dilute 1:10 to use, when osmolality should be 290 mOsm/Kg.

Crystal Violet 0.1% in Water

Crystal violet	100 mg
Water	100 ml

Available ready-made from BDH.

Developer, photographic

See manufacturer's instructions

D19 (Kodak)

See manufacturer's instructions

Dexamethasone (Merck)

1 mg/ml (100×)
This comes already sterile in glass vials. To dissolve, add 5 ml water by syringe to vial, remove, and dilute to give a concentration of 1 mg/ml.
Aliquot and store at −20°C.
1 mg/ml is approximately 2.5 m*M*. Dilute for use to give 10–50 n*M* (physiological concentration range), 0.1–1.0 μ*M* (pharmacological dose range) or 25–100 μ*M* (high-dose range).
β-methasone (Glaxo) and methylprednisolone (Sigma) may be prepared in the same way.

Dissection BSS

See "Balanced Salt Solutions"

EDTA (Versene)

Prepare as 10 m*M* concentrate, 0.374 g/l in PBSA.
Sterilize by autoclaving or filtration.
Dilute 1:10, or 1:5 for use at 1.0–2.0 m*M*, or, exceptionally, 1:2 for use at 10 m*M*, diluted in PBSA or trypsin in PBSA (see below).

EGTA

As for EDTA, but may be used at higher concentrations due to its lower toxicity.

Ficoll, 20%

Sprinkle 20 g Ficoll (Pharmacia) on the surface of 80 ml ultrapure water and leave overnight to settle and dissolve.
Make up to 100 ml in ultrapure water.
Sterilize by autoclaving.
Store at room temperature.

Fixative, photographic (Kodak, Agfa, Fugi or Ilford)

See manufacturer's instructions

Fixative for Tissue Culture

See "Acetic/Methanol"
Alternatively, use pure anhydrous ethanol or methanol (see Chapter 13), 10% formalin, 1% glutaraldehyde, or 5% paraformaldehyde.

Gentamycin (ICN/Flow, GIBCO/Life Technologies, Schering)

Dilute to 50 μg/ml for use.

Gey's Balanced Salt Solution

	g
NaCl	7.00
KCl	0.37
$CaCl_2$	0.17
$MgCl_2 \cdot 6H_2O$	0.21
$MgSO_4 \cdot 7H_2O$	0.07
$Na_2HPO_4 \cdot 12H_2O$	0.30
KH_2PO_4	0.03
$NaHCO_3$	2.27
Glucose	1.00
Water to	1,000 ml
CO_2	5%

Giemsa Stain

Buffer:

$NaH_2PO_4 \cdot 2H_2O$	0.01 M	1.38 g/l
$Na_2HPO_4 \cdot 7H_2O$	0.01 M	2.68 g/l

Combine to give pH 6.5.
Dilute prepared Giemsa concentrate (Gurr, BDH, Fisher) 1:10 in 100 ml buffer. Filter through Whatman No. 1 filter paper to clarify. Use fresh each time (precipitates on storage).
Alternatively, apply undiluted Giemsa stain to anhydrous, fixed preparation for 1–2 min, dilute 10-fold with water, leave 5–10 min, and rinse by upward displacement (see Chapter 13).

Glucose, 20%

Glucose 20 g
Dissolve in Hanks' BSS and make up to 100 ml.
Sterilize by autoclaving.
Store at room temperature.

Glutamine 200 m*M*

L-glutamine 29.2 g
Hanks' BSS 1,000 ml
Dissolve glutamine in BSS and sterilize by filtration (see Chapter 8).
Aliquot and store at −20°C.

Glutathione

Make 100× stock, i.e., 0.10 M in BSS or PBSA, and dilute to 1 m*M* for use.
Sterilize by filtration.
Aliquot and store at −20°C.

Growth Medium

That medium normally used to maintain the cells in exponential growth with the appropriate serum added if specified. Varies with individual cell type and conditions required (see Chapters 7 and 8).

Ham's F12
See Chapter 7.

Hanks' BSS
See above and "Balanced Salt Solutions," Chapter 7.

HAT Medium

Drug	*Concentration*	*Dissolve in*	*Molarity (100× final)*
Hypoxanthine	136 mg/100 ml	0.05 *N* HCl	1×10^{-2} M
Aminopterin	1.76 mg/100 ml	0.1 *N* NaOH	4×10^{-5} M
Thymidine	38.7 mg/100 ml	BSS	1.6×10^{-3} M

For use in the HAT selective medium, mix equal volumes, sterilize by filtration, and add mixture to medium at 3% V/V. Store H and T at 4°C, A at −20°C.

HB Medium
Add the following to CMRL 1066 medium:

Insulin	5 μg/ml
Hydrocortisone	0.36 μg/ml
β-retinyl acetate	0.1 μg/ml
Glutamine	1.17 m*M*
Penicillin	50 U/ml
Streptomycin	50 μg/ml
Gentamycin	50 μg/ml
Fungisone	1.0 μg/ml
Fetal bovine serum	1%

HBSS
See "Hanks' BSS," Chapter 7.

Hoechst 33258 [Chen, 1977]
2-2(4-hydroxyphenol)-6-benzimidazolyl-6-(1-methyl-4-pierpazyl)-benzimidalol-trihydrochloride)
Make up 1 mg/ml stock in BSS without phenol red and store at −20°C. For use dilute 1:20,000 (1.0μl → 20 ml) in BSS without phenol red at pH 7.0.
◇ This substance may be carcinogenic. Handle with extreme care. Weigh in fume cupboard and wear gloves.

Kanamycin Sulfate "Kannasyn" (10 mg/ml)
4 × 1 g vials
Hanks' BSS 400 ml
Add 5 ml BSS from a 400-ml bottle of BSS to each vial.
Leave for a few minutes to dissolve.
Remove the BSS and kanamycin from the vials and add back to the BSS bottle.
Add another 5 ml BSS to each vial to rinse and return to BSS bottle. Mix well.
Dispense 20 ml into sterile containers and store at −20°C.
Test sterility: add 2 ml to 10 ml sterile medium, free of all other antibiotics, and incubate at 36.5°C for 72 hr.
Use at 100 μg/ml.

Lactalbumin Hydrolysate 5% (10×)

Lactalbumin hydrolysate	5 g
Hanks' BSS	100 ml

Heat to dissolve.
Sterilize by autoclaving.
Use at 0.5%.

McIlvaines Buffer pH 5.5

		To make 20 ml	*100 ml*
0.2 M Na_2HPO_4	(28.4 g/l)	11.37 ml	56.85 ml
0.1 M Citric acid	(21.0 g/l)	8.63 ml	43.15 ml

Check pH and adjust to 5.5.

Media
The constituents of some media in common use are listed in Chapter 7 with the recommended procedure for their preparation. For those not described, see Morton [1970], manufacturers' catalogues (see under "Media" in Trade Index, Source of Materials), or the original literature references.

MEM
See "Eagle's MEM," Chapter 7.

2-Mercaptoethanol (M.W. 78)
Stock solution 5×10^{-3} M (100×)
3.9 mg in 10 ml HBSS
Sterilize by filtration in fume cupboard.
Store at −20°C or make fresh each time.

Methocel
See "Methylcellulose"

Methylcellulose (1.6% in medium)
1. Add 8.0 g methylcellulose (4,000 centipose) to a 500-ml medium storage bottle containing 250 ml distilled water at 80–100°C and a magnetic stirrer bar.
2. Shake by hand until the methylcellulose is wetted.
◇ Wear gloves!
3. Place on a magnetic stirrer and stir until cooled to room temperature.
4. Remove to cold room and continue to stir overnight.
5. Autoclave the Methocel, when it will form an opaque solid.
6. Allow to cool to room temperature and then cool to 4°C.
7. Add 250 ml of 2× strength medium and stir overnight at 4°C.
8. Dispense into sterile 100-ml bottles and store at −20°C.
Thawed bottles may be kept at 4°C for 1 month.
For use, dilute methylcellulose medium with appropriate volume of serum and add cell suspension in sufficient growth medium to give final concentration of 0.8% methylcellulose.

Mitomycin C (Stock solution 10 μg/ml [50×])
2-mg vial
Measure 20 ml HBSS into a sterile container.
Remove 2 ml by syringe and add to vial of mitomycin.
Allow to dissolve, withdraw, and add back to container.
Store for 1 wk only at 4°C in the dark. (Cover container with aluminum foil.)
For longer periods store at −20°C.
Dilute to 2 μg/10^6 cells, 0.2 μg/ml, for use, but check for

effective concentration when using each new cell line as a feeder layer.
◇ Toxic; weigh in fume cupboard and wear gloves.

Mycostatin (Nystatin) (2 mg/ml, 100×)

Mycostatin	200 mg
Hanks' BSS	100 ml

Make up by same method as kanamycin.
Final concentration 20 μg/ml.

MTT 3-(4,5-dimethylthiazol-2-yl)-2,5-diphenyltetrazolium bromide (MTT, Sigma)

50 mg/ml in PBS
Sterilize by filtration.

Napthalene Black 1% in Hanks' BSS

Napthalene black	1 g
Hanks' BSS	100 ml

Dissolve as much as possible of the stain, then filter through Whatman No. 1 filter paper.

PBS

See "Phosphate-buffered saline"

PBSA

See "Phosphate-buffered saline"

Penicillin (e.g., Crystapen Benzylpenicillin [sodium]) 1,000,000 units per vial

Use 4 vials and 400 ml Hanks' BSS.
Make up as for kanamycin, stock concentration 10,000 U/ml.
Use at 50–100 U/ml.

Percoll (Pharmacia)

Made up and sterile and should be diluted with medium or HBSS until correct density is achieved.
Check osmolality and adjust to 290 mOsm/Kg by adding water or less medium.

Phosphate-Buffered Saline PBS (Dulbecco "A") (See Table 7.4, Chapter 7)

Oxoid tablets, Code BR 14a, 1 tablet per 100 ml distilled water. Dispense, then autoclave. Store at room temperature, pH 7.3, osmolality 280 mOsm/Kg.
PBSB contains the calcium and magnesium and should be made up and sterilized separately. Mix with PBSA, if required, immediately before use.
PBS may also be made up from pure single reagents (see Chapter 7) or premixed powder (see Sources of Materials, below). In each case the Ca^{2+} and Mg^{2+} salts should be made up separately and added just before use, if required. In general, dissociation solutions, and simple isotonic salt solutions for rinsing, should be made up without Ca^{2+} and Mg^{2+} salts, while incubation solutions should be complete, and generally will contain glucose (see Chapter 7).

Phosphate-Buffered Saline/EDTA (PBS/EDTA) 10 m*M* (PE)

Make PBSA as above.
Add EDTA, 3.72 g/l, and stir.
Dispense, autoclave, and store at room temperature.
Use 1:10 at 1 m*M* for normal applications or 1:2 (5 m*M*) for high chelating conditions, e.g., trypsinization of CaCo-2 cells.

Phytohemagglutinin

Stock 500 μg/ml (100×).
Lyophilized (Sigma)
Dissolve powder by adding HBSS by syringe to ampule. Aliquot and store at −20°C.
Dilute 1:100 for use.

SF12

Ham's F12 (see Chapter 7) with 2× Eagle's MEM essential amino acids, and 1× nonessential amino acids and lacking thymidine and with 10× folic acid concentration (devised for Friend erythroleukemia cells and L5178Y lymphoma).

Sodium Citrate/Sodium Chloride (SSC)

	0.03 M	*0.09 M*
Trisodium citrate (dihydrate)	8.82 g	26.46 g
NaCl	7.53 g	52.60 g
Water	1.0 l	1.0 l

SSC

See "Sodium Citrate/Sodium Chloride"

Streptomycin Sulfate

1 g per vial
(Similar method to Kanamycin)
Take 2 ml from a bottle containing 100 ml sterile Hanks' BSS and add to 1 g vial of streptomycin.
When solution complete, return 2 ml to 98 ml Hanks'.
Dilute 1/200 for use.
Final concentration, 50 μg/ml.

Trypsin Diluent, Buffered

	g
Sodium chloride	6.0
Trisodium citrate	2.9
Tricine (N-Tris (hydroxymethyl)methyl glycine)	1.79
Phenol red	0.005
Ultrapure water to	1,000 ml

Stir ingredients until dissolved, adjust pH to 7.8.
Filter through Whatman No. 1 filter paper.
Dispense and autoclave.
Check osmolality and adjust to 290 mOsm/Kg.

Trypsin Stock

2.5% in 0.85% (0.14 M) NaCl
Trypsin solutions can be bought commercially. Alternatively, to make up a 2.5% solution in 0.85% NaCl, stir for 1 hr at room temperature or 10 hr at 4°C. If trypsin does not dissolve completely, clarify by filtration through Whatman No. 1 filter paper.
Sterilize by filtration, aliquot, and store at −20°C.

Note. Trypsin is available as crude (e.g., Difco 1:250) or purified (e.g., Worthington or Sigma 3× recrystallized) preparations. Crude preparations contain several other proteases that may be important in cell dissociation but may also be harmful to more sensitive cells. The usual practice is to use crude trypsin unless cell damage reduces viability or reduced growth is observed, when purified trypsin may be used. Pure trypsin has a higher specific activity and should therefore be used at a proportionally lower concentration, e.g., 0.01 or 0.05%. Check for mycoplasma if preparing for raw trypsin.

Trypsin, Versene, Phosphate (TVP)

Trypsin (Difco 1:250)	25 mg (or 1 ml 2.5%)
Phosphate-buffered saline	98 ml
Disodium EDTA ($2H_2O$)	37 mg
Chick serum (Flow)	1 ml

Mix PBS and EDTA, autoclave, and store at room temperature.
Add chick serum and trypsin. If using powdered trypsin, sterilize by filtration before adding chick serum.
Check for mycoplasma and aliquot and store at −20°C.

Tryptose Phosphate Broth

10% in Hanks' BSS

Tryptose phosphate (Difco)	100 g
Hanks' BSS	1,000 ml

Stir until dissolved.
Aliquot and sterilize in the autoclave.
Store at room temperature.
Use 1:100 (final concentration 0.1%).

Tyrodes's Solution

	g
NaCl	8.00
KCl	0.20
$CaCl_2$	0.20
$Mg_2Cl_2 \cdot 6H_2O$	0.10
$NaH_2PO_4 \cdot H_20$	0.05
Glucose	1.00
Ultrapure water to	1,000 ml
Gas phase	Air

Versene

See "EDTA"

Viability Stain

See "Naphthalene black"

Vitamins

See media recipes (Chapter 7)
Make up at 100× concentrated.
Sterilize by filtration.
Store at −20°C in the dark.

Trade Index

SOURCES OF MATERIALS

Agar	GIBCO
	Difco
Air-curtain incubator	Sage Instruments
Amino acids	Sigma
Aminopterin	Sigma
Ampules, glass	Wheaton
Ampules, plastic	Nunc (*see* GIBCO in UK)
Ampule sealer	Kahlenberg-Globe
Anemometers	*See* Laminar flow cabinets
Antibiotics	*See* individual antibiotics
Antibodies	Amersham
	Biopool
	Boehringer
	DAKO
	GIBCO
	ICN
	Serotec
	Sigma
	Upstate Biotechnology
	Vector
Antifoam RD emulsion 9964.40	Dow-Corning
	BDH/Merck
	Miles
Automatic glassware washing machines	Betterbuilt (John Burge, UK)
	Lancer
	Miele
	Vernitron (USA)
Automatic pipettes	Alpha
	Becton Dickinson
	Boehringer
	Costar
	Gilson
	ICN
	Jencons
	Labsystems
	Rainin
	Tecnomara
	Zinsser
Automatic pipette plugger	Bellco (A.R. Horwell, UK)
	Volac, *see* Camlab
Bactopeptone	Difco
Benchcote	Johnson & Johnson
Biochemicals	Aldrich
	Boehringer
	BDH/Merck
	Calbiochem
	Fluka
	Sigma
	U.S. Biochemical Co.
Bottles	Bellco
	Corning
	Schott

	Elga
	Fisons
Densitometers	Beckman Instruments
	Gelman
	Gilford Instruments Ltd.
	Helena
	Joyce-Loebl
	Molecular Dynamics
Density meter	Paar
DePeX, permount	*See* Stains
Detergents	Alconox
	Decon
	Diversey
	ICN
	Pierce
Dexamethasone	Sigma
	Merck
Dextran	Calbiochem
	Sigma
Dextraven 100	Fisons
Diacetyl fluorescein	Fisher
	Sigma
Dialysis tubing	Laboratory suppliers
Disc filter assembly for sterilization	*See* Filters
Disinfectants	Bio Medical Products
	Day-Impex
	ICN
	Polysciences
	Sigma
	Thomas
	Whatman
Dispase	Boehringer Mannheim
Dispensers, liquids	Accuramatic
	Boehringer
	Denley
	Gilson
	Labs Systems
	Zinsser
DMSO	Merck
DNase	Sigma
	Worthington
Ecoscint	National Diagnosis
EDTA	*See* Chemicals
Ehrlenmeyer flasks	*See* Glassware
Electronic cell counter for cell sizing	Coulter
Electronic thermometer	Comark
	Fisher
	Grant
	Omega
	Rustrak
Electrophoresis	Bio-Rad
	GIBCO/BRL
	Haake
	Innovative Chemistry
	LKB
	Pharmacia
	Shandon
Emulsion	*See* Photographic developers
Epidermal growth factor	*See* Growth factors
Ethidium bromide	Sigma
Fibroblast growth factor	*See* Growth factors
Fibronectin	*See* Growth factors
Fermentors	Biotech Instruments
	Braun
	Chemap
	Invitron
	L.H. Fermentation
	New Brunswick
Ficoll	Pharmacia
Filter holders	*See* Filters
Filter wells	Becton Dickinson
	Costar
	Earl Clay
	Millipore
	Nuclepore
	Nunc
Filters	Becton Dickinson
	Gelman
	Millipore
	Nalge
	Nuclepore
	Sartorius
	Pall (Europe) Ltd.
Filtration equipment	*See* Filters
Flow cytometers	Becton Dickinson
	Coulter
	Ortho Instruments
Fluorescence-activated cell sorter (FACS)	Becton Dickinson
Fluorescence fade retardant	Vector
Freezers, −20°C	Fisher
	Fisons
	New Brunswick Scientific
	Local discount warehouses
Freezers, −70°C	Fisher
	Fisons
	Revco
	Kelvinator
Freezers, liquid nitrogen	Cryo-med
	L'Aire Liquide
	MVE
	Planer
	Taylor Wharton
	Thermolyne
Fungicide (Mycostatin)	GIBCO
Gas-permeable caps	Becton Dickinson
	Corning
	Nunc
Gentamicin	Schering

	Bio-Rad
	Canberra-Packard
	Dynatech
	Molecular Dynamics
Plate sealers	ICN/Flow
	Conway
Pleated cartridge filter	Pall
	Gelman
Pluronic-F68	Serva
Polaroid camera	Polaroid
Poly-D-lysine	Sigma
Polystyrene flasks, petri dishes	*See* Tissue culture flasks
Polyvinyl pyrrolidone	Calbiochem
	Sigma
Pressure cooker, bench-top autoclave	Astell Hearson
	Harvard/LTE
	Heinicke
	Napco
	Valley Forge
Projection table	Bellco (Horwell, UK)
Pronase	Sigma
Pumps, peristaltic	Autoclude
	Cole Parmer
	Gilson
	Horwell
	Millipore
	Pharmacia
	Tecnomara
	Watson-Marlowe
	Zinsser
Pumps, vacuum	Millipore
	Gelman
Quinacrine dihydrochloride	Sigma
Recording thermometer	Cambridge Instruments
	Fisher
Refractometer	Beckman
	Cole Parmer
	Fisher
	Fisons
Refrigerators	Local discount warehouses
Reusable in-line filter assembly	*See* Filters
Reverse osmosis	Elga
	Fisons
	Millipore
Roller bottles, glass	Bellco
Roller bottles, plastic	*See* Tissue culture flasks
Roller bottle rack	Bellco
	New Brunswick Scientific
	Tecnomara
Rubber pipette bulb	Bellco
	Bibby
	Fisher
	Fisons
	Jencons Scientific Ltd.
Scalpels	*See* Instruments, dissecting
Scintillation fluid	Econoscint
	National Diagnostics
	New England Nuclear (DuPont)
	Packard
Scintillation vials, minivials	Canberra Packard
Semipermeable nylon film	*See* Packaging
Serum	GIBCO
	Globepharm
	ICN
	JRH Biologicals/Seralab
	Microbiological Associates
Serum substitutes	
Serxtend	NEN (DuPont)
Ventrex	Ventrex
Ultroser	IBF
Nutricyte	Brooks Labs
Nu-Serum	Collaborative Research
CLEX	Dextran Products Ltd.
Serum-free media	Biological Industries
	Boehringer
	Clonetics
	GIBCO
	Hyclone
	IBF
	ICN
	Irvine
	JRH Biologicals
	Promacell
	Sigma
	Whittaker
Sieves	Endecotts
	Sigma
Silica gel	Fisher
	BDH/Merck
Silicon tubing	ESCO
Silicone	Fisher
Slide boxes, light tight	Clay Adams
	Raven Scientific
Slide containers for emulsion (Cyto-Mailer)	Lab-Tek, *See* Fisher
Stains	BDH/Merck
	Fisher
Stainless steel mesh	Falcon, *See* Becton Dickinson
Sterility indicators, Thermalog	Browne
	Bennet
	Popper
Sterilization bags	*See* Patapar paper
Sterilization film	*See* Portex, Cedanco
Sterilizing and drying oven	Astell Hearson
	Cole Parmer
	Camlab Ltd.

	Fisher
	Fisons
	Forma
	Harvard/LTE
	Horwell
	LEEC
	New Brunswick Scientific
	Precision Scientific Co.
Sterilizing tape (indicator)	Fisher
	Shamrock
Stills	Corning
	Fisons
	Jencons
Stirrers	*See* Magnetic stirrers
Streptomycin	Glaxo
Supplements, e.g., tryptose (lactalbumin hydrolysate, bactopepone)	Difco
Syringes	Local laboratory suppliers
Temperature controllers	*See* Thermostats
Temperature recorders	*See* Recording thermometer
Thermalog	Bennet (UK)
Thermostats, proportional controllers	Controls & Automation, Ltd.
	Fisher
	Jumo
	Napco
α-Thioglycerol	Sigma
Thymidine	Sigma
Time-lapse video	*See* Videocamera, Video recorder
Tissue culture flasks, etc.	Nunc (GIBCO)
	Becton Dickinson
	Bellco
	Bibby
	Corning
	Costar
	Linbro, Lux (ICN)
Tissue culture media	*See* Media
Trypan blue	*See* Stains
Trypsin	ICN
	GIBCO
	Sigma
	Worthington
Trypsin inhibitor (soya bean)	Sigma
Tryptophan	*See* Amino acids
Tryptose phosphate broth	Difco
	Oxoid
Tyrosine	*See* Amino acids
Universal containers	Sterilin
	Nunc (GIBCO)
Vacuum pump	*See* Pumps
Video camera	Dage-MTI
	Hamamatsu
	See also Microscopes
Video recorder	Panasonic
Vinblastin	Sigma
Vinyl tape	3M (Lab suppliers)
Vitafiber	Amicon
Vitamins, solid	Sigma
Vitamins, solution	*See* Tissue culture media
Vortex mixer	Gallenkamp
	Fisher
	General laboratory suppliers
Washing machine	Vernitron/Betterbuilt (Burge, UK)
X-ray film	Eastman Kodak Company
	Fuji

ADDRESSES OF COMMERCIAL SUPPLIERS

Accuramatic
Watington King's Lynn, Norfolk PE33 0JB, England.
0553 810282
Automatic dispenser.

Adelphi Manufacturing Co.
207 Duncan Terrace, London NI8, England.

Advanced Instruments Inc.
2 Technology Way, Norwood, MA 02062.
(617) 320-9000; Fax: (617) 320-8181

Advanced Protein Products Ltd.
Unit 18H, Premier Partnership Estate, Leys Road, Brockmoor, Brierley Hill, West Midlands DY5 3UP, England.
0384 263862
Serum: bovine; fetal calf; horse; porcine; sheep; rabbit; chicken; mouse; albumin; transferrin; endothelial growth factor; PDGF; TGF-Beta; fibronectin; fibrinogen; hemoglobin; transferrin; growth factors; hemoglobin, appro-sera; platelets; serum substitutes.

Advanced Tissue Sciences
10933 North Torrey Pines Rd., La Jolla, CA 92037-1005.
(619) 450-5730; Fax: (619) 450-5703

Janssen Biotech
Lammerdries 55, B-2250 Olen, Belgium.
014 22 40 15; Fax: 014 23 15 33
Cytotoxicity assay kits; Filter well assays of cytotoxicty and inflammation; "Skin2."

Air Products & Chemicals Inc.
7201 Hamilton Blvd., Allentown, PA 18195.
(215) 481-8257; Fax: (215) 481-5036
Gases; nitrogen; oxygen; CO_2.

Albert Browne Ltd.
Chancery Street, Leicester, LE1 5NA, England.
Sterilization indicator tubes.

Alconox, Inc.
9 E. 40th St., New York, NY 10016.
(212) 532-4040; Fax: (212) 532-0172
Detergents.

Aldrich Chemical Co.
1001 W. St. Paul Ave., P.O. Box 355, Milwaukee, WI 53201.
(414) 273-3850; (800) 558-9160; Fax: (800) 962-9591
Chemicals; biochemicals.

Alpha Laboratories
40 Parham Drive, Eastleigh Hants, S05 4NU England.
07093 610911
Microtitration; pipettors and dispensers; micropipettes; plate readers; multiwell plates; microplates; monoclonal antibodies; disposables; leucocyte surface antigens; immunoglobulins; APAAP; microplate readers; microplate software.

American Type Culture Collection (ATCC)
12301 Parklawn Drive, Rockville, MD 20852-1776.
(301) 881-2600; (800) 638-6597; Fax: (301) 231-5826
Cell lines; hybridomas; plasmids.

Amersham Corporation
2636 S. Clearbrook Drive, Arlington Heights, IL 60005.
(708) 593-6300 Fax: (708) 437-1640
Amersham International
Amersham Place, Little Chalfont, Amersham, Bucks., HP7 9NA, England.
0404 544000
Radioisoptopes; antibodies; luminescence probes; avidin-biotin.

Amicon Inc.
72 Cherry Hill Drive, Beverly, MA 01915.
(508) 777-3622; Fax: (508) 777-6204
Amicon Ltd.
Amicon House, 2 Kingsway, Woking, Surrey, GU21 1UR, England.
Molecular filtration; Vitafiber; perfused capillary bundles.

Analytical Measuring Systems
Shirehill Industrial Estate, Shirehill, Saffron Walden, Essex CB11 3AQ, England.
0799 24080
Image analysis; colony counters.

Anotec Separations Ltd.
Wildmere Road, Banbury, Oxon OX16 7JU, England.
0295 272323
Filtration; filter wells.

Applied Immune Sciences Inc.
200 Constitution Drive, Menlo Park, CA 94025.
(800) 247-8881; (415) 326-7302; Fax: (415) 326-0923
TechGen International Ltd.
Suite 8, 50 Sulivan Rd., London, SW6 3DX, England.
071 371 5922; Fax: 071 371 0496
Panning; treated flasks for cell separation by panning.

Applied Scientific
2405 35th Ave., San Francisco, CA 94116.
(415) 759-9851
Plastics; tips; tubes; micropipettes.

Aquatron Water Stills
J Bibby Science Products Ltd., Stone, Staffs, England.
Water purification; stills.

Artek Systems Corp.
14340 Sullyfield Circus, Chantilly, VA 22021.
(703) 631-7800; Fax: (703) 631-7816
Image analysis; colony counters.

Aseptic-Thermo Indicator Co.
See PyMaH Corporation
Sterilization indicators.

Assab Ltd.
32 St. Andrews Road, Glasgow G41 1ST, Scotland.
041 429 8181; Fax: 041 636 1076
Incubators; ovens; CO_2 incubators.

Astell Hearson
Equipment Division, Marlow House, Hailey Rd., Erith, Kent DA18 4AL, England.
01 310 1313
Ovens; autoclaves.

Autoclude Peristatic Pumps
Arisdale Ave., S. Ockendon, Essex, RM15 5DP, England.
0708 856125/6; Fax: 0708 857366
Peristaltic pumps.

Baker Co., Inc.
Sanford Airport, P.O. Drawer E, Sanford, ME 04073.
(800) 992-2537; (207) 324-8773; Fax: (207) 324-3869
Laminar flow cabinets; biohazard cabinets; safety cabinets.

Barnstead Thermolyne Corporation
P.O. Box 797, Dubuque, IA 52004-0797.
(319) 589-0538; Fax: (319) 589-0530
Labmart (Cambridge) Ltd.
1 Pembroke Ave., Waterbeach, Cambridge CB5 9QR, England.
223 861665; Fax: 223 861990
Water purification; nitrogen feezers; distillation; reverse osmosis; sterilizers; ovens; $C0_2$ incubators; stirrer; hot plates.

Baxter International Inc.
1 Baxter Parkway, Deerfield, IL 60015.
(708) 948-2000; Fax: (708) 948-3948
Culture bags; biosensors; DNA probes; immunoassays; antibodies.

Bayer UK Ltd.
Computer Division, Bayer House, Strawberry Hill, Newbury, Berks. RG13 1JA, England.
0635 39000
Gases; gas detectors

BDH *See* Merck

Beckman Instruments, Inc.
2500 Harbor Blvd., Fullerton, CA 92634.
(714) 871-4848; (800) 742-2345; Fax: (714) 773-8898
Beckman UK
Progress Road, Sands Industrial Estate, High Wycombe, Bucks HP12 4LJ, England.
0494 41181
Centrifuges; scintillation counters; spectrophotometers; pH meters; microtitration; plate reader.

Becton Dickinson Inc.
1 Becton Drive, Franklin Lakes, NJ 07417-1886.
(800) 235-5953; (201) 847-4200; Fax: (201) 847-4975
Becton Dickinson (UK) Ltd.
Between Towns Road, Cowley, Oxford OX4 3LY, England.
0865 748844
Plastic flasks; petri dishes; filters; pipettes; syringes; filters; filter wells; multiwell plates; organ culture dishes; growth factors; Matrigel; fibronectin.

Bellco Biotechnology
P.O. Box B, 340 Elrudo Road, Vineland, NJ 08360.
(800) 257-7043; (609) 691-1075; Fax: (609) 691-3247
See also Arnold R. Horwell Ltd.
Stirrers; roller racks; glassware; flasks.

Bennet & Co. Ltd.
Field House, Newton, Tony, Salisbury Wilts, SP4 0HF, England.
0980 64488
Sterilizers; indicators; Thermalog.
Bethesda Research Laboratories: *see* GIBCO/BRL
Bibby, J. Science Products Ltd.
Stone, Staffordshire ST15 0SA, England.
0785 812121
Tissue culture flasks; dishes; pipette aids; tubes; roller flasks; video.
Bio Medical Products Corp.
59 White Meadow Road, Rockaway, NJ 07866.
(201) 627-6010; Fax: (201) 627-7632
Laboratory and hospital supplies; disinfectants.
Bio-Rad Microscience Div.
19 Blackstone St., Cambridge, MA 02139.
(617) 864-5820; (800) 444-1422; Fax: (617) 864-9328
Bio-Rad Laboratories (UK) Ltd.
Mayland Avenue, Hemel Hempstead, Herts., HP2 7TD, England.
0442 232552; Fax: 0442 259118
Bio-Rad Labs
1414 Harbour Way, South Richmond, CA 94804.
Molecular biology; electrophoresis; isoelectric focusing; Rotofor; blotting; immunology; chromatography; HPLC; high-performance electrophoresis; confocal microscopy.
Bio-Tek Instruments, Inc.
Highland Park, P.O. Box 998, Winooski, VT 05404.
(802) 655-4040; (800) 451-5172; Fax: (802) 655-7941
Microtitration; plate readers.
Biofluids, Inc.
1146 Taft Stree, Rockville, MD 20850.
(301) 424-4140; Fax: (301) 424-3619
Media; supplements; growth factors.
Biological Industries
Kibbutz Bet Haemek, 25115, Israel.
972 4 960595; Fax: 972 4 968896
Biological Industries Ltd.
Media House, Dunswood Rd., Cumbernauld, G67 3EN, Scotland.
0236 7287000
Medium; serum; serum-free medium; extracellular matrix-coated plates.
Biomedical Technologies, Inc.
22 Thorndike Street, Cambridge, MA 02141.
(617) 344-9942; Fax: (617) 344-1451
Bionique Laboratories, Inc.
Bloomingdale Road, Saranac Lake, NY 12983.
(518) 891-2356; Fax: (518) 891-5753
Culture chamber for microscopic observation.
Biopool UK Ltd.
Wesley St., Old Glossop, Derbyshire SK13 9RY, England.
0457 868921
Biopool International Inc.
Box 792, Grand Island, NY 14072-0792.
(800) 668-8717
Antibodies; UPA; TPA; plasminogen.
Bioprocessing Ltd.
Consett No. 1, Industrial Estate, Medomsley Road, Consett Co., Durham DH8 6TJ, England.
0207 590565
Growth factors; PDGF; filtration.
Biotech Instruments, Ltd.
183 Camford Way, Luton, Beds., LU3 3AN, England.
0582 502388; Fax: 0582 597091
Fermentors.
BioTech Trade & Service GmbH
Franz-Antoni-Str 22, D-6837, St Leon-Rot 1, Germany.
06227 51308; Fax: 06227 53694
Promocell serum-free medium; serum; cell cultures: fibroblasts; melanocytes; keratinocytes; endothelium.
BOC Ltd.
The Priestley Centre, 10 Priestley Road, The Surrey Research Park, Guildford, Surrey, GU2 5XY, England.
0483 5798577; Fax: 0483 32115
Gases; 5% CO_2 in air; oxygen; nitrogen.
Boehringer Mannheim GmbH
Sandhofer Str. 116, P.O. Box 310120, D6800 Mannheim 31, Germany.
062 17591; Fax: 0621 7594004
Boehringer Mannheim (UK) Ltd.
Boehringer Mannheim House, Bell Lane, Lewes, East Sussex BN7 1LG, England.
0273 480444
Boehringer Mannheim Biochemicals
P.O. Box 50414, Indianapolis, IN 46250-0414.
(800) 262-1640; Fax: (317) 576-2754
Biochemicals; enzymes; mycoplasma; micropipettes; media; sera; serum-free media supplements; growth factors; cytokines; attachment factors; antibodies; BM Cyline (mycoplasma inhibtor); DAPI; collagenase; dispase.
Boots Pharmaceuticals
North Higham, Nottingham, England.
Antibiotics; Crystapen.
Boro Labs Ltd.
Aldermaston, Berkshire RG7 4QU, England.
0734 811731
CO_2 incubator.
Boyden Data Papers Ltd.
Parkhouse Street, Camberwell, London.
Melinex film.
Braun, B., Biotech International
Schwarzenberger Weg 73-79, D-3508 Melsungen, Germany.
05661 71 3704; Fax: 05661 713702
B. Braun Biotech
999 Postal Road, Allentown, PA 18103.
(215) 266-6262; (800) 258-9000; Fax: (215) 266-9319
Fermentors.
British Bio-Technology Ltd.
4–10 The Quadrant, Barton Lane, Abingdon, Oxon OX14 3YS, England.
0235 529449; Fax 0235 533420
Antibodies; probes; endothelial CAMS; cell adhesion.
BS & S (Scotland) Ltd.
5/7 West Telferton, Portobello Industrial Estate, Edinburgh EH7 6UL, Scotland.
031 669 2282
ECOSCINT.

Buck Scientific, Inc.
58 Fort Point St., East Norwalk, CT 06855-1097.
(203) 853-9444; (800) 562-5566; Fax: (203) 853-0569
Sterilization packaging paper; semipermeable nylon film.

Calbiochem Corp.
10937 N. Torrey Pines Rd., La Jolla, CA 92037.
(619) 450-9600; Fax: (619) 453-3552
Biochemicals; water-driven magnetic stirrer.

Cambridge BioScience
25 Signet Court, Stourbridge Common Business Centre, Swann's Road, Cambridge CB5 8LA, England.
0223 316855; Fax: 0223 60732
Calbiochem; Serva; biochemicals; growth factors; antibodies; enzymes; collagenase peptides; signal transduction inhibitors.

Cambridge Instruments
Kent Industrial Instruments, Howard Road, Eaton-Socon, Huntingdon, Cambs. PE19 3EU, England.
0480 735321
Temperature recorders.

Cambridge Instruments Ltd.
Viking Way, Bar Hill, Cambridge CB3 8EL, England.
0954 82020
Photozoom; inverted microscope.

Camlab Limited
Nuffield Road, Cambridge CB4 1TH, England.
0223 424222
Magnetic stirrers; Variomag.

Canberra Packard
Brook House, 14 Station Road, Pangbourne, Berks. RG8 7DT, England.
0734 844981
Scintillation counters; scintillation chemicals; microtitration radioactivity and luminescence detectors.

CD Medical, Inc.
P.O. Box 9308, Miami Lakes, FL 33014.
Perfused capillary culture.

Cedanco
P.O. Box 42, Wellesley, MA 02181.
Sterilization packaging paper.

Cell Enterprise Inc.
15719 Crabbs Brancy Way, Derwood, MD 20855.

Cellon
22 rue Dernier, Sol L-2543 Luxembourg.
352495975
Serum; serum replacement; fermentors.

CellPro Incorporated
Suite 100, 22322–20th Ave, Southeast Bothwell, WA 98021.
(206) 485-7644
Cell separation; immunoaffinity; avidin biotin; bone marrow stem cells.

Chemap AG
Holzliwisenstrasse 5, CH-8604, Volketswil, Switzerland.
01 947 22 22; Fax: 01 945 39 30

Chemap, Inc., Alfa Laval Group
901 Hadley Road, South Plainfield, NJ 07080.
(201) 757-7006; (800) 323-8831; Fax: (201) 757-7006
Fermentors; biotechnology.

Chemtech International Ltd.
Merritress House, Church Lane, Aebridge Romsey, Hants. SO51 0HN, England.
0794 41244
Serum substitutes; Clex.

CIBA Pharmaceuticals Ltd.
Horsham, Sussex, England.

CIS (UK) Ltd.
Unit 5, Lincoln Park Business Centre, Lincoln Road, High Wycombe, Bucks. HP12 3RD, England.
0494 35922
Serum-free medium; supplements; antibodies; Ultradoma; anti-P-glycoprotein.

Clandon Scientific Ltd.
Lysons Avenue, Ahs Vale, Aldershot, Hants. GU12 5QF, England.
0252 514711
Osomometer; Gonotec.

Clay Adams, Division of Becton Dickinson & Co.
299 Webro Road, Parsippany, NJ 07054.
Laboratory equipment; microscope and histology; slide dishes; autoradiography; light-tight slide boxes.

Clonetics Corporation
9620 Chesapeake Dr., San Diego, CA 92123.
(619) 541-2726; (619) 541-0823
Serum-free media; keratinocyte cultures; keratinocyte medium; endothelial cells and medium; melanocytes and medium; mammary cell cultures and medium.

Cole-Parmer Instrument Co.
7425 North Oak Park Avenue, Niles, IL 60648.
(708) 647-7600; Fax: (708) 647-9660
Labortory equipment; pumps.

Collaborative Research Inc.
2 Oak Park, Bedford, MA 01730.
(617) 275-0004; Fax: (617) 275-0043
(*See also* Becton Dickinson.)
Growth factors; matrix products; FGF; EGF; Matrigel; fibronectin.

Collagen Corporation
2455 Faber Place, Palo Alto, CA 94303.
(415) 856-0200; Fax: (415) 856-2238
Collagen.

Comark Electronics Ltd.
Martenfield, Welwyn, Garden City, Herts. AL7 1JP, England.
0707 375550; Fax: 0707 331202
Recorders; temperature.

Connaught Laboratories Ltd.
1755 Steeles Avenue W., Willowdale, Ontario M2R 3T4, Canada.
(416) 667-2701; Fax: (416) 667-2900
Media; cell lines.

Contamination Control, Inc.
P.O. Box 316, Kulpsville, PA 19443.
Laminar flow cabinets; safety cabinets.

Controls & Automation Ltd.
Bury Mead Road, Hitchin, Herts. SG5 1RT, England.
0462 36161
Temperature controllers.

Conway Laboratory Industries Ltd.
Hazeley House, Hartley Wintney, HANTS., RG27 8LT, England

0252 844580; Fax 0252 842991
Microtitration plate sealers
Cooper Biomedical
Malvern, PA.
Corning, Inc.
HP-AB-03, Corning, NY 14831.
(607) 974-4667; Fax: (607) 974-7919
Glassware; plastics; bottles; water distillation; culture flasks; petri dishes; pipettes; filters.
Costar Corp.
1 Alewife Center, Cambridge, MA 02140.
(617) 868-6200; (800) 492-1110; Fax: (617) 868-2076
Costar Europe Ltd.
Sloterweg 305a 1171 VC, Badhoevedorp, The Netherlands.
020 659 6051; Fax: 020 659 6051
Plastic flasks; petri dishes; multiwell plates; filters; filter wells; organ culture dishes; IVF culture dish; pipettes; tips.
Coulter Corp.
601 W. 20th St., West Hialeah, FL 33012.
(305) 885-0131; Fax: (305) 883-6820
Coulter Electronics Ltd.
Northwell Drive, Luton, Beds., LU3 3RH, England.
0582 491414; Fax: 0582 490390
Cell counters; flow cytometer.
Cow Proofings Ltd.
Slough, Berks., England.
Cow gum.
Cryodiffusion
49 Rue de Verdun, F27690 Lery, France.
Cryomed
51529 Birch Street, New Baltimore, MI 48047.
(313) 725-4614; (800) 322-COLD; Fax: (313) 725-7501
Life Science Laboratories
Sedgewick Road, Luton, Beds., LU4 9DT, England.
0582 597676
Nitrogen freezers; narrow-necked tray systems.
Cryoservice Ltd.
Blackpole Trading Estate
Liquid nitrogen; CO_2; nitrogen freezers.
Cryotechnics Refrigeration
20/21 New Broompark Industrial Estate, Granton Park Ave., Edinburgh, EH5 1RS, Scotland.
031 552 0755; Fax: 031-551 1238
Nitrogen freezers; MVE.
Dage-MTI, Inc.,
701 N Roeske Ave., Michigan City, IN 46360.
(219) 872-5514; Fax: (219) 872-5559.
CCD cameras
Dako Corp.
6392 Via Real, Carpenteria, CA 93013.
(805) 566-6655; Fax: (805) 566-6688
Dako Ltd.
16 Manor Courtyard, Hughenden Avenue, High Wycombe, Bucks. HP13 5RE, England.
0494 452016
Antibodies; A571/alpha-1-fetoprotein/4000 13000; A297; calcitonin; alpha-2-macroglobulin; choronic gonadotropin (hCG); desmin; glial fibrillary acidic protein (GFA); plasminogen; myelin basic protein; neuron-specific enolase; S-100; somatostain; transferrin.
Damon/IEC (UK) Ltd.
Unit 7, Lawrence Way, Brewers Hill Road, Dunstable, Bedfordshire LU6 1BD, England.
Centrifuges; Hotpak incubators; B22 high-speed refrigerated centrifuge; rotor; CO_2 incubators; ovens; dryers; glassware washers; roller culture apparatus; sterilizers; microtitration plate carriers.
Day-Impex
Station Works, Earls Colne, Colchester, Essex C06 2ER, England.
0787 223232; Fax: 0787 224171
Nitrogen freezers; Vircon disinfectant.
Decon Laboratories Ltd.
Freepost, Conway Street, Hove, East Sussex BN3 2ZZ, England.
Detergents; microbiology and tissue culture; LEEC CO_2 incubators; Bellco; bioreactor; alginate beads; monoclonal antibodies; culture media; plant cell culture; plant tissue culture; water purification; Astell Scientific; reverse osmosis.
Denley Instruments Ltd.
Natts Lane, Billingshurst, West Sussex RH14 9EY, England.
0403 813441; 0403 784796; Fax: 0403 784796
Microtitration; freezer racks; trays; plate reader; washer; dispenser; replicate sampler.
Devro Ltd.
7872261 ext. 220
Collagen.
Dextran Products Ltd.
P.O. Box 1360, Princeton, NJ 08542.
Dextran.
Difco Laboratories
P.O. Box 1058, Detroit, MI 48232.
(800) 521-0851; Fax: (313) 462-8517
Difco Laboratories
P.O. Box 14B, Central Avenue, East Molesley, Kent KT8 0SE, England.
Media; blood products; agar; natural media; bactopeptone; tryptose phosphate broth; trypsin.
Distillers Co2. Ltd.
Cedar House, 39 London Road, Reigate, Sussex RH2 9QE, England.
0737 241133; Fax: 0737 241842
Liquid CO_2.
Diversey Ltd.
Weston Favell Centre, Northampton, NN3 4PD, England.
Detergents.
Dow Corning Corp.
Dow Corning Center, Box 0994, Midland, MI 48686-0994.
(517) 496-4000
BDH Laboratory Supplies
Merck House, Poole, BH15 1TD, England.
0202 669700; Fax: 0202 666536
Silicones; adhesives; antifoam; water repellant.

DuPont Company, Biotechnology Systems
Barley Mill Plaza, P-24, Wilmington, DE 19898.
(800) 551-2121; Fax: (617) 542-8468
Du Pont (UK) Ltd.
Wedgwood Way, Stevenage, Hertfordshire SG1 4QN, England.
0438 734015; Fax: 0438 734621
Sorvall centrifuges; rotors; molecular probes; NGF; bFGF; TGF-alpha; TGF-beta; antibodies; transgenic mice; oncogenes; radioisotopes, NEN.
(Incorporating New England Nuclear).

Dynal Inc.
475 Northern Blvd., Great Neck, NY 11021.
(516) 829-0039; Fax: (516) 829-0045
Dynal (UK) Ltd.
Station House, 26 Grove Street, Wirral, Merseyside L62 2AB, England.
051 644 6555
Dynal A.S.
P.O. Box 158, Skoyen N-0212, Oslo 2, Norway.
472 52 90 00
Antibodies; Dynabeads: magnetized beads; immunoaffinity; cell separation.

Dynatech Laboratories Inc.
14340 Sullyfield Circus, Chantilly, VA 22021.
(703) 631-7800; (800) 336-4543; Fax: (703) 631-7816
Dynatech Laboratories Ltd.
Daux Rd., Billingshurst, West Sussex RH14 9SJ, England.
0403 813381/2; 0403 814565/7; Fax: 0403 814397
Microtitration; plate readers; auto samplers; diluters; dispensers; mixers; washers.

Earl-Clay Laboratories, Inc.
890 Lamont Ave., Novato, CA 94947.
Medium; growth chambers; collagen filter wells.

Eastman Kodak Co.
Eastman Fine Chemicals, 343 State St., B-701, Rochester, NY 14652-3512.
(800) 225-5352; Fax: (716) 722-3179
Eastman Kodak, Laboratory & Research Products Div.
P.O. Box 3263, CH-6300 Zug 3, Switzerland.
042 232525; Fax: 042 211252
High-grade organic chemicals; thermal video printers; photographic films and chemicals.

EDT Instruments Ltd.
Lorne Road, Dover, Kent CT16 2AA, England.
0304 213555; Fax: 0304 204297
Medium; serum free.

EDT Analytical
14 Trading Estate Road, London NW10 7LU, England.
01 961 9210
Serum; microcarriers; bulk cell culture; fermentors; bioreactors; Hyclone.

Elga Ltd.
Lane End, High Wycombe, Bucks. HP14 3JH, England.
0494 881393; Fax: 0494 881007
Water purification.

Endotronics
8500 Evergreen Blvd., Coon Rapids, MN 55433.
(612) 786-0302
Acusyst-s; hollow fiber; microcapillary perfusion culture; large-scale culture.

Envair Ltd.
York Ave., Haslingden, Rossendale, Lancashire BB4 4HX, England.
0706 228416; Fax: 0706 831957
Laminar flow; biohazard; safety cabinets.

Eppendorf-Netheler-Hinz GmbH
Postfach 650670, D-2000 Hamburg 65, Germany.
040 53801 688; Fax: 040 53801 556
Eppendorf North America, Inc.
545 Science Drive, Madison, WI 53711.
(608) 231-1188; Fax: (608) 231-1339
Micropipettes; microcentrifuges; tubes.

ESCO Rubber
Sterilin Limited, Lampton House, Lampton Road, Hounslow, Middlesex TW4 3EE, England.
01 572 2468
Silicone rubber; membranes; silicone tubing; rubber products; bottle-cap liners; Silescol sheets; Silescol bungs (stoppers); butyl rubber.

European Collection of Animal Cell Cultures
CAMR, Porton Down, Salisbury, SP4 0JG, England.
0980 610391; Fax: 0980 611315
Cell lines; hybridomas; DNA fingerprinting.

Expanded Metal Co.
Hartlepool, England.
Stainless steel mesh.

Falcon Plastics
See Becton Dickinson.

Fenwall Laboratories, Division of Travenol Labs.
See Baxter International.
Plastic culture bags.

Fisher Scientific
711 Forbes Avenue, Pittsburgh, PA 15219.
(412) 787-6322; Fax: (412) 562-8313
Chemicals; laboratory equipment; magnetic stirrers.

Fisons Scientific Equipment
Bishop Meadow Road, Loughborough, Leics. LE11 0RG, England.
0509 231166; Fax: 0509 231893
General laboratory suppliers; MSE centrifuges; chemicals; Fistream stills.

Flexcell Corp.
Box 890, McKeesport, PA 15132.
(412) 664-FLEX
Flexible cell culture supports; Flexercell strain unit.

Flow Laboratories
See ICN Biomedicals.

Fluka Chemie AG.
Industriestrasse 25, CH-9470 Buchs, Switzerland.
081 755 25 11; Fax: 081 756 54 49
Fluka Chemical Corp.
980 S. Second St., Ronkonkoma, NY 11779.
(516) 467-0890; (800) FLUKAUS; Fax: (516) 467-0663
Chemicals; biochemicals.

Fluorochem Ltd.
Wesley Street, Old Glossop, Derbyshire SK13 9RY, England.
Fluorochemicals; FC43 high-density medium.

FMC BioProducts
191 Thomaston St., Rockland, ME 04841.
(800) 341-1574; (207) 594-3941; Fax: (207) 594-3941
FMC BioProducts, Europe
Risingevej 1 DK-2665, Vallensbaek Strand, Denmark.
45 42 73 11 22; Fax: 45 42 73 56 92
Forma Scientific, Inc.
P.O. Box 649, Marietta, OH 45750.
(800) 848-3080; Fax: (614) 373-6770
Incubators; CO_2 incubators.
Fuji Photo Film
350 Fifth Avenue, New York, NY 10001.
Photographic film and chemicals.
Gelman Sciences Inc.
600 S. Wagner Road, Ann Arbor, MI 48106.
(313) 665-0651; (800) 521-1520; Fax: (313) 761-1208
Gelman Sciences Ltd.
10 Harrowden Road, Brackmills, Northampton, NN4 0EZ, England.
0604 765141
Filters; laminar flow cabinets; biohazard cabinets; safety cabinets.
Gen-Probe Inc.
9880 Campus Point Dr., San Diego, CA 92121.
(619) 546-8000; Fax: (619) 452-5848
Lab Impex Ltd.
111–113 Waldegrave Rd., Teddington, Middlesex, England.
081 977 3266; Fax: 081 977 0170
Gen-Probe mycoplasma detection.
Genzyme Corporation
1 Kendall Square, Cambridge, MA 02139-1562
(617) 252-7500; Fax: (617) 252-7600.
New Brunswick Scientific (UK) Ltd.
Edison House, 163 Dixons Hill Road, North Mimms, Hatfield, Herts AL9 7JE, England
0707 275733; Fax: 0707 267859
Growth factors; antibodies.
Germfree Laboratories
7435 NW 41st St., Miami, FL 33166.
(305) 592-1780; (800) 922-1780; Fax: (305) 591-7280
Laminar flow; safety cabinets.
GIBCO BRL, Div. Life Technologies, Inc.
P.O. Box 9418, Gaithersburg, MD 20898.
(301) 840-8000; Fax: (301) 670-8539
GIBCO/BRL/Life Technologies
P.O. Box 35, Trident House, Renfrew Road, Paisley PA3 4EF, Scotland.
041 889 6100; Fax: 041 887 1167
Media; sera; plastics; filters; growth factors; peel-apart flasks; inserts; cell scrapers; cryotubes; filter wells; mycoplasma detection kits; keratinocyte medium; serum-free medium; insect; chromosome painting; BRL; molecular probes; Nunc plastics.
Gilford Systems, CIBA-Corning Diagnostics Corp.
132 Artino St., Oberlin, OH 44074.
(216) 774-1041; (800) 445-3673; Fax: (216) 774-3939
Spectrophotometers.
Gilson Medical Electronics, Inc.
3000 W. Beltline Hwy., P.O. Box 27, Middleton, WI 53562.
(608) 836-1551; (800) 445-7661; Fax: (608) 831-4451
Gilson Medical Electronics (France) S.A.
72 Rue Gambetta, B.P. 45, F-95400 Villiers le Bel, France.
1 34 29 50 00; Fax: 1 34 29 50 80
Micropipettes; dispensers.
Glaxo Laboratories Ltd.
Greenford, Essex, England.
Globepharm Ltd.
P.O. Box 89C, Esher, Kent, KT10 9NA, England.
0372 465307; Fax: 0372 468818
Serum.
Gonotec, GmbH
Eisenacher Strasse 56, D-1000 Berlin 62, Germany.
(030) 784 60 27; Fax: 030 788 1201
Osmometer.
Gordon Keeble Ltd.
Petersfield House, St. Peter's St., Duxford, Cambridge, CB2 4RP, England.
Plastic pipettes and containers; pipette tips.
Grant Instruments (Cambridge) Ltd.
Barrington, Cambridge CB2 5QZ, England.
0763 260811; Fax: 0763 262410
Water baths; incubators; temperature recorders; CO_2 incubators.
Greiner GmbH
Maybachstrasse 2, Postfach 1162, D-7443 Frickenhausen, Germany.
07022 501 0; Fax: 07022 501 514
Culture flasks; plates; dishes.
Hamamatsu Photonics K.K., Systems Division
812 Joko-cho, Hamamatsu City, 431-32, Japan.
0534 35 1562; Fax: 0534 35 1574
Photonic Microscopy Inc.
2625 Butterfield Rd., Suite 204-S, Oak Brook, IL 60521.
(312) 571-1244; Fax: (312) 571-1244
Video cameras; time-lapse video recording.
Hana Media, Inc.
626 Bancroft Way, Berkeley, CA 94710.
(800) 772-HANA
See DuPont.
Medium; serum; serum-free media; Mycotrim mycoplasma detection.
Hanna Instruments
Eden Way, Pages Industrial Park, Leighton Buzzard, Beds., LU7 8TZ, England.
0525 850855; Fax: 0525 853668
Harvard/LTE
Greenfield, Oldham OL3 7BR, England.
0457 876221; Fax: 0457 870131
Harvard/LTE Inc.
22 Pleasant Street, South Natick, MA 01760.
(508) 650-3939; Fax: (508) 655-6029
Drying cabinets; ovens; autoclaves; sterilizers; bench-top autoclave; hot plates.
Hazelton Research Products
See JRH Biosciences.
Heinicke Instruments
Postfach 1203, Friedrich-Ebert-Strasse 10, D-8223, Trostberg/Alz, Germany.
Heinicke Instruments Co.
3000 Taft St., Hollywood, FL 33021.
Autoclaves; incubators; bench-top autoclave.

Helena Laboratories
P.O. Box 752, Beaumont, TX 77704-0752.
(800) 231-5663; (409) 842-3714; Fax: (409) 842-6241
Scanning densitometers.

Hepaire Manufacturing Ltd.
Aire Cool House, Spring Gardens, London Road, Romford, Essex RM7 9LY, England.
Laminar flow cabinets; biohazard cabinets, class II.

Heraeus Instruments GmbH
Heraeusstr. 12–14, P.O. Box 15 63, D6450 Hanau 1, Germany.
061 81 35 465, Fax: 061 81 35 749
Heraeus Equipment Ltd.
Unit 9, Wates Way, Ongar Road, Brentwood, Essex, England.
Heraeus Equipment
111A Corporate Blvd., S. Plainfield, NJ 07080.
(201) 755-4800; (800) 441-2554; (201) 754-9494
Incubators; ovens; centrifuges; CO_2 incubators; Mikro-12 reusable dishes; Petriperm dishes.

Hoechst (UK) Ltd.
Hoechst House, Salisbury Road, Hounslow, Middx. TW4 6HJ, England.
Pharmaceuticals; stains; reagents; mycoplasma stain 33285.

Horiba Ltd.
1 Harrowden Road, Brackmills, Northampton NN4, England.
6504 765171
Cell counters; cell sizing.

Horwell, Arnold R. Ltd.
Laboratory & Clinical Supplies, 73 Maygrobe Road, West Hampstead, London NW6 2BP, England.
081 328 1551
Labgard; autoflow; CO_2 incubators; laminar flow; biohazard cabinets.

Horwell A. R., Ltd.
Laboratory & Clinical Supplies, 73 Maygrove Road, West Hampstead, London NW6 2BP, England.
071 328 1551; Fax: 071 372 5259
Laboratory equipment; ovens; incubators; pipettors; Pi-pump; bulb pipette; Vibratome microtome.

HyClone Laboratories Inc.
1725 S. State Highway 89-91, Logan, UT 84321.
(801) 492-5663
Tissue culture; serum; medium.

IBF Biotechnics
35 Ave Jean-Jaures, F-92390 Villeneuve la Garenne, France.
1 47 98 83 53; Fax: 1 47 92 26 55
IBF Biotechnics, Inc.
7151 Columbia Gateway Dr., Columbia, MD 21046.
(301) 290-1505; Fax: (301) 290-1509
Media supplements; Ultroser G.

ICN Biomedicals
P.O. Box 28050, Cleveland, OH 44128.
(216) 831-3000; (800) 321-6842; Fax: (216) 831-2569
ICN Biomedicals
Eagle House, Peregrine Business Park, High Wycombe, Bucks. HP13 7DL, England.
0494 443826; Fax: 0494 473162
Media; sera; plastic flasks; petri dishes; growth factors; radioisotopes; antibodies; equipment; incubators; Lab Tek slides; Thermanox coverslips; CO_2 incubators; mycoplasma removal agent (MRA).

Ilford Ltd.
Mobberley, Knutsford, Cheshire, England.
0565 650000; Fax: 0565 872 734.
Ilford
West 70 Century Road, Paramus, NJ 07652.
Photographic film; developers; fixers; nuclear emulsions for autoradiography; CIBAchrome processing.

Imcera Bioproducts, Inc.
P.O. Box 207, Terre Haute, IN 47808.

Imperial Laboratories (Europe) Ltd.
West Portway, Andover, Hampshire SP10 3LF, England.
0264 33 33 11
Medium; serum: fetal bovine; horse; lamb; porcine; rabbit.

Innovative Chemistry, Inc.
P.O. Box 90, Marshfield, MA 02050.
(617) 837-6709; Fax: (617) 834-7325
Isoenzyme electrophoresis kits.

Interlab; Intermed
See Nunc.

International Bio-Technologies Ltd.
Kiryat Hadassah, P.O. Box 12000, Jerusalem 91120, Israel.
02 431214
Extracellular matrix-coated plates.

International Equipment Co.
300 Second Ave., Needham Heights, MA 02194.
(617) 449-8060; (800) 843-1113; Fax: (617) 444-6743
Damon/IEC (UK) Ltd.
Unit 7, Lawrence Way, Brewers Hill Rd., Dunstable, Beds. LU6 1BD, England.
0582 604669; Fax: 0582 609257
Centrifuges; sterilizers; roller culture apparatus; Hotpak CO_2 incubators; ovens; cytocentrifuge slide carriers.

Invitron Corp.
4649 LeBourget Dr., St. Louis, MO 63134.
(314) 426-5000; Fax: (314) 426-6331
Fermentors; large-scale culture systems; fixed-bed bioreactors.

Irvine Scientific
2511 Daimler St., Santa Ana, CA 92705.
(800) 437-5706; (714) 261-7800; Fax: (714) 261-6522
Medium; serum; serum-free medium.

J.R.H. Biosciences
P.O. Box 14848, Lenaxa, KS 66215.
(800) 255-6032; Fax: (913) 469-5584
Sera-Lab Ltd.
Hophurst Lane, Crawley Down, Sussex, RH10 4FF, England.
0342 716366; Fax: 0342 717351
Medium; serum; antibodies.

J. Staniar & Co.
34 Stanley Road, Whitefield, Manchester M25, England.
061 767 9026; Fax: 061 767 9033
Nylon mesh; gauze.

Jencons (Scientific) Ltd.
Cherrycourt Way, Industrial Estate, Stanbridge Road, Leighton Buzzard, Beds. LU7 8UA, England.
0525 372010; Fax: 0525 379547

Laboratory equipment; glassware; roller racks; stirrers; distillation apparatus; nitrogen freezers; Labsystems; micropipettes.

John Burge Equipment Ltd.
35 Furze Platt Road, Maidenhead, Berks., England.
Glassware washing machines; autoclaves.

John Poulten Ltd.
77-93 Tanner Street, Barking, Essex 1G11 8QD, England.
01 594 4256
Pipette plugging.

Johnson & Johnson Medical Ltd.
Coronation Road, Ascot, Berks., SL5 9EY, England
0344 872626; Fax 0344 872599
Benchcote

Jouan Ltd.
130 Western Road, Tring, Herts. HP23 4BU, England.
0442 89020; Fax: 0442 826880
Jouan Inc.
110B Industrial Dr., Winchester, VA 22602.
(800) 662-7477; Fax: (703) 869-8626
Incubators; ovens; centrifuges; peristaltic pump; automatic dispenser.

Joyce-Loebl Ltd.
Dukesway, Team Valley, Gateshead, NE11 0PZ, England.
091 482 2111; Fax: 091 482 5249
Image analysis; densitometry; Magiscan image analyzer; microdensitometer; Chromoscan digital densitometer.

JRH Biosciences
13804 W. 107th St., Lenexa, KS 66215.
(800) 255-6032; Fax: (913) 469-5584
Sera-Lab Ltd.
Hophurst Lane, Crawley Down, Sussex RH10 4FF, England.
0342 716366; Fax: 0342 717351
Media; sera; supplements.

Jumo Instrument Co.
Hysol, Harlow, Essex CM18 6QZ, England.
0279 24606
Temperature controllers.

KabiVitrum Ltd.
KabiVitrum House, Riverside Way, Uxbridge, Middlesex UB8 2YF, England.
0895 51144
Plasminogen activator inhibitor; t-PA; S-2251 chromogenic substrate; plasminogen; protease assay; endotoxin; ENA processor.

Kahlenberg-Globe Equipment Co.
Sarasota, FL.
Ampule sealer.

Kelvinator Scientific
707 Robins Street, Conway, AR 72032.
(501) 450-3700; Fax: (501) 327-0663
Freezers; −80°C freezers.

Kent Industrial Instruments
Howard Rd., Eaton Socon, Huntingdon, Cambridgeshire PE19 3EU, England.
0480 75321
Circular chart temperature recorder.

Kodak Ltd.
P.O. Box 66, Kodak House, Station Road, Hemel Hempstead, Herts. HP1 1UJ, England.
Photographic film and chemicals; x-ray film.

Kor Biochemicals
69 Rue de la Petrusse, L-8084 Betrange, Luxembourg.

L.H. Fermentation Ltd.
Unit 10, Nimrod Industrial Estate, Elgar Road, S. Reading, Berks. RG2 0EB, England.
0734 753300; Fax: 0734 755147
L.H. Fermentation
3942 Trust Way, Hayward, CA 94545.
Biological fermentors; bioreactors.

L'Air Liquide
57 av Carnot, BP13 94503 Champigny Cedex, France.
1 49 83 5555
L'Aire Liquide, Cryogenic Equip. Div.
P.O. Box 395, Allentown, PA 18105.
(215) 439-1993; Fax: (215) 459-9216
L'Aire Liquide (UK) Ltd.
44 Hertford St., London W1Y 7TF, England.
Nitrogen freezers; dewars.

Lab-Impex Research
111-113 Waldegrave Rd., Teddington, Middlesex TW11 8LL, England.
081 977 3266; Fax: 081 977 0170
Laboratory equipment; laboratory furniture; freezer racking.

Lab-Line Instruments, Inc.
15th & Bloomingdale Ave., Melrose Park, IL 60160-1491.
(708) 450-2600; (800) LAB-LINE; Fax: (708) 450-0943.
Incubators; CO_2 incubators.

Labsystems (UK) Ltd.
12 Redford Way, Uxbridge, Middlesex UB8 1SZ, England.
0895 38421
Microtitration; plate readers; pipettors,; dispensers; diluters; micropipettes; antibodies; diagnostic kits; cytokeratin; GFAP; desmin; vimentin; neurofilaments.

Lancer UK Ltd.
1 Pembroke Avenue, Waterbeach, Cambridge CB5 9QR, England.
0223 861665
Lancer (USA) Inc.
715-JW. State Road 434, Longwood, FL 32750.
(407) 332-1855; Fax: (407) 332-0040
Glassware washing machines.

Laser Laboratory Systems Ltd.
P.O. Box 166, Southampton S09 7LP, England.
0702 260487
Pipetting systems; pipette aid; micropipettes; tips.

LEEC Ltd.
Private Road, No. 7, Colwick Estate, Nottingham NG4 2AJ, England.
0602 616222; Fax: 0602 616680
Incubators, ovens; CO_2 incubators; bottle brush.

Leica Inc.
P.O. Box 123, Buffalo, NY 14240-0123.
(716) 891-3000; Fax: (716) 891-3155
Leica Mikroscopie und Systeme GmbH
Ernst Leitz Strasse, P.O. Box 20 40, W 6330 Wetzlar 1, Germany.
064 41 290: Fax: 064 41 29 33 99
Leica UK Ltd.
Davy Avenue, Knowhill, Milton Keynes, MK5 8BR,

England.
0908 666 663
Microscopes: dissecting; inverted; confocal.

Life Technologies Ltd.
See GIBCO.

LiquiPure Europe Limited
Wedgewood Road, Industrial Estate, Bicester, Oxon OX6 7UL, England.
0869 249919
Water purification; reverse osmosis; ultrafiltration; deionization.

M.A. Bioproducts
Building 100, Biggs Ford Road, Walkersville, MD 21793.
MA Bioproducts
11841 Mississippi Ave., Los Angeles, CA 90025.
Medium; serum; glassware.

Malvern Instruments Ltd.
Spring Lane, South Malvern, Worcestershire WR14 1AQ, England.
0684 892456; Fax: 0684 892789
Cell sizing; particle sizing; laser scanner.

MDH Ltd.
Walworth Road, Andover, Hampshire SP10 5AA, England.
0264 62111; Fax: 0264 356452
Air sterilization; laminar flow/safety cabinets.

Media-Cult a/s
Symbion Science Park, Haraldsgade 68, DK2100 Copenhagen, Denmark.
Cytotoxicity testing.

Medical Air Technology
Canto House, Wilton St., Denton, Manchester M34 3LZ, England.
061 320 0652; Fax: 061 335 0313
Laminar flow cabinets.

Merck Sharpe & Dohme
4545 Oleatha Avenue, St. Louis, MO 63116.
(800) 325-9034; Fax: (314) 353-3754
Pharmaceuticals; chemicals; dexamethasone.

Merck, E.
Frankfurter Strasse 250, Postfach 4119, D-6100 Darmstadt, Germany.
061 51720; Fax: 061 5172 2000
BDH
Broom Road, Poole, Dorset, BH12 4NN, England.
(0202) 745520; Fax: (0202) 738299.
Chemicals.

Microbiological Associates
Life Sciences Center, 9900 Blackwell Road, Rockville, MD 20850.
(301) 738-1000; Fax: (301) 738-1036
Medium; serum; glassware; mycoplasma testing; cytotoxicity testing; virus screening.

Microgon Inc.
23152 Verdugo Drive, Laguna Hills, CA 92653.
(714) 581-3880; Fax: (714) 855-6120
Filtration; hollow-fiber perfusion; filter sterilization; capillary perfusion.

Miele Laboratory Technology
P.O. Box 24 00, 4830 Guetersloh, Germany.
05241 890; Fax: 05241 89 1500
Miele Co. Ltd.
Abingdon, Oxon OX14 1TW, England.
0235 554455; Fax: 0235 554477
Miele Appliances Inc.
22D Worlds Fair Dr., Somerset, NJ 08873.
(908) 560-0899; (800) 843-7231; Fax: (908) 560-9649
Glassware washing machines.

Miles Inc., Diagnostics Div.
195 W. Birch St., Kankakee, IL 60901.
(815) 937-8270; (800) 227-9412; Fax: (815) 937-8285
Miles Inc., Diagnostics Div.
P.O. Box 70, Elkhart, IN 46515.
(800) 284-2637; Fax: (219) 262-6704
Miles Ltd.
Stoke Court, Stoke Poges Slough SL2 4LY, England.
02814 5151; Fax: 02814 3993
Bayer Diagnostics
Stoke Court, Stoke Poges Slough SL2 4LY, England.
0753 645151; Fax: 0753 643893
Diagnostic reagents; cytospin slide holders; biochemicals; Ex-cyte; lipids; Pentex; media supplements; albumin.

Millipore Corp.
80 Ashby Rd., Bedford, MA 01730.
(617) 275-9200; Fax: (617) 275-5550
Millipore Ltd.
The Boulevard, Blackmoor Lane, Watford, Herts. WD1 8YW, England.
0923 816375
Chromatography; filtration; CONSEP; protein purification; Millex filters; filter sterilization; water purification.

Minnesota Valley Engineering
407 7th St., NW, New Prague, MN 56071.
(612) 758-4400; (800) 247-4446; Fax: (612) 758-8252
See Planer Biomed.
Liquid nitrogen freezers; storage tanks; dewars.

Molecular Dynamics
880 E. Arques Ave., Sunnyvale, CA 94086.
(800) 333-5703; (408) 733-1222; Fax: (408) 773-8343
Microtitration plate readers; analytical densitometers; computerized densitometry; confocal microscopy.

Molecular Devices Corporation
Menlo Oaks Corporate Center, 4700 Bohannon Drive, Menlo Park, CA 94025.
Microtitration; plate readers.

Morgan Sheet Metal Co.
See Kahlenberg-Globe.

Nalgene Labware Dept., Nalge Co.
P.O. Box 20365, Rochester, NY 14602.
(716) 264-3898; Fax: (716) 586-3294
See Labmart, Cambridge.
Plasticware; sterilization filters; storage bottles; laboratory apparatus.

NAPCO
10855 S.W. Greeburg Road, Portland, OR 97223.
Incubators; CO_2 incubators.

National Diagnostics
1013-1017 Kennedy Blvd., Manville, NJ 08835.
(201) 722-8600
ECOSCINT.

New England Nuclear
See DuPont.
New Brunswick Scientific Co., Inc.
44 Talmadge Rd., Box 4005, Edison, NJ 08818-4005.
(800) 631-5417; (908) 287-1200; Fax: (908) 287-4222
New Brunswick Scientific (UK) Ltd.
Edison House, 163 Dixons Hill Rd., North Mymms, Hatfield, Herts. AL9 7JE, England.
07072 75733; 07072 75707; Fax: 07072 67859
Incubators; freezers; centrifuges; rotary shaker incubators.
Nikon Corporation
Fuji Building, 2-3 Marunouchi 3-chome, Chiyoda-ku, Tokyo 100, Japan.
81 3 3216 1039; Fax: 81 3 3201 5856
Nikon Inc.
1300 Walt Whitman Road, Melville, NY 11747-3064.
(516) 547-8500; Fax: (516) 547-0306
Nikon Europe B.V., Instrument Dept.
P.O. Box 222, NL-1170 AE, Badhoevedorp, The Netherlands.
31 20 659 4406; Fax: 31 20 659 8335
Nikon (UK) Ltd., Instrument Division
Haybrook, Halesfield 9 Telford, Shropshire TF7 4EW, England.
0952 587444; Fax: 0952 588009
Microscopes: dissecting; stereo zoom; photo stereo; inverted.
Novabiochem (UK) Ltd.
3 Heathcoat Building, Highfields Science Park, University Boulevard, Nottingham NG7 1BR, England.
0602 430840
Peptides; growth factors; cytokines; biologically active peptides; antipeptide antibodies; serum proteins; Calbiochem.
Novo Biolabs
Novo Industri A/S Novo Alle, DK-2880 Bagsvaerd, Denmark.
Monoclonal antibodies; diagnostic kits.
Nuaire, Inc.
2100 Fernbrook Lane, Plymouth, MN 55447.
(612) 553-1270; (800) 328-3352; Fax: (612) 553-0459
Jencons (Scientific) Ltd.
Cherrycourt Way Industrial Estate, Stanbridge Rd., Leighton Buzzard, Beds. LU7 8UA, England.
0525 372010; Fax: 0525 379547
Safety cabinets; water-jacketed CO_2 incubator
Nuclepore Corp.
7035 Commerce Circus, Pleasantón, CA 94566-3294.
(415) 463-2530; Fax: (415) 463-2029
Nuclepore Corp.
63 Charlwood Drive, Oxshott, Surrey KT22 0HB, England.
0372 844370; Fax: 0372 844248
Filters: polycarbonate filters; filter wells.
Nunc A/s
P.O. Box 280, Kamstrup, DK-4000, Roskilde, Denmark.
45 42 359065
(*See* GIBCO/BRL/Life Technologies in UK.)
Nunc Inc.
2000 N. Aurora Rd., Naperville, IL 60563-1796.
(800) 288-6862; Fax: (708) 416-2556
Plastic flasks; petri dishes; multiwell plates; microtitration plates; chamber slides; slide flasks.
Nycomed Pharma.
Box 4284, Torshov, N-0401 Oslo 4, Norway.
472 226 350; Fax: 472 712535
Nycomed (UK) Ltd.
Nycomed House, 2111 Coventry Rd., Sheldon, Birmingham B26 3EA, England.
021 742 2444; Fax: 021 722 2190
Iodinated density media; Nycodenz; Metrizamide; cell separation.
Olympus Optical Co., Ltd.
2-43-2, Hatagaya, Shibuya-ku, Tokyo, Japan.
Olympus Corp., Precision Instrument Div.
4 Nevada Drive, Lake Success, NY 11042-1179.
(516) 488-3880; (800) 446-5967; Fax: (516) 222-7920
Olympus Optical Co., UK, Ltd.
2–8 Honduras St., London EC1Y 0TX, England.
071 253 2772; Fax; 071 251 6330
Microscopes: inverted; zoom stereo.
Omega
Omega Drive, Box 4047, Stamford, CT 06907.
(203) 359-7700
Temperature indicator; electronic thermometer.
Organogenesis, Inc.
83 Rogers St., Cambridge, MA 02142.
(617) 864-0640; (800) 776-7546; Fax: (617) 876-6811
Cytotoxicity screening/testing; living skin models; "Testskin"; inflammation.
Ortho Pharmaceuticals Inc.
410 University Avenue, Westwood, MA 02090.
Ortho Pharmaceuticals Ltd.
Enterprise House, Station Road, Loudwater, High Wycombe, Bucks. England.
Pharmaceuticals; flow cytometer.
Oxoid USA Inc.
Wade Road, Basingstoke, Hants. RG24 0PW, England.
Oxoid USA Inc.
P.O. Box 691, Ogdensburg, NY 13669, USA.
(800) 567-8378
Blood products; agar plates; PBS tablets.
Paar Scientific Instruments
594 Kingston Rd., London SW20 8DN, England.
081 540 8553; Fax: 081 543 8737
Anton Paar USA Inc.
1030A Wilmer Ave., Richmond, VA 23227.
(800) 722-7556; (804) 264-1097: Fax: (804) 262-0805
Density meters.
Paesel & Lorei GmbH & Co.
Borsigallee 6, P.O. Box 630 347, D-6000 Frankfurt am Main 63, Germany.
069 42 20 95-99
Biochemicals; tissue culture media; growth factors; antibodies.
Pall Ultrafine Filtration Co.
2200 Northern Blvd., East Hills, NY 11548.
(516) 484-5400; Fax: (516) 621-3976
Pall Process Filtration Ltd.
Europa House, Havant Street, Portsmouth P01 3PD, England.
0705 753545; Fax: 0705 831324
Filter sterilization; filter testing.

Panasonic
See local video suppliers or general scientific suppliers
Particle Data Inc.
236 N. York, P.O. Box 265, Elmhurst, IL 60126.
(708) 832-5653; (800) 323-6140; Fax: (708) 832-5686
Particle sizing and counting; cell counters and cell sizers.
Pfeifer & Langen Dormagen
Frankenstrasse 25 D-4047 Dormagen, Germany
02106 52-1
pH, conductivity, temperature, oxygen meters.
Pharmacia LKB Biotechnology
Bjorkgatan 30, S-751 82, Uppsala, Sweden.
46 181 65000; Fax: 46 181 43820
Pharmacia LKB Biotechnology
800 Centennial Avenue, P.O. Box 1327, Piscataway, NJ 08855-1327.
(908) 457-8000; Fax: (908) 457-0557
Pharmacia Biosystems Ltd.
Davy Avenue, Knowhill, Milton Keynes, MK5 8PH, England.
Fax: 0908 690 091
Electrophoresis; chromatography; Cytodex microcarriers; Percoll density medium; cell separation.
Pierce Chemical Co.
P.O. Box 117, 3747 N. Meridian Rd., Rockford, IL 61105.
(800) 874-3723; (815) 968-0747; Fax: (815) 968-7316
Pierce & Warriner (UK)
44 Upper Northgate St., Chester, Cheshire CH1 4EF, England.
0244 382525
Pierce Europe BV
European Corporate Headquarters, Int Antwoordnummer CCRI Numbero 364 3260 WB, Oud-Beijerland
31 1860 19277; Fax: 31 1860 19179
Life Science Laboratories Ltd.
Sedgewick Road, Luton LU4 9DT, England.
0582 597676
Chemicals; biochemicals; protein assay; detergents; surfactant; protein purification; dialysis; microdialysis.
Planer Biomed
Windmill Road, Sunbury-on-Thames, Middlesex TW16 7HD, England.
0932 786262
Liquid nitrogen storage; nitrogen freezers; MVE; controlled-rate coolers.
Platon, G.A.
Platon Park, Viables, Baskingstoke, Hants. RG22 4PS, England.
0256 460122
Gases; flow meters.
Polaroid (UK) Ltd.
575 Technology Square 9-P, Cambridge, MA 02139.
(617) 577-2000; (800) 225-1618
Polaroid UK Ltd.
Ashely Road, St. Albans, AL1 5PR, England.
0727 59191; Fax: 0727 869 335
Instant photographic film; copy cameras; slide makers.
Polysciences Inc.
400 Valley Road, Warrington, PA 18976.
(215) 343-6484; Fax: (215) 343-0214
Polyfiltronics Ltd.
2 Waterside Hamm Moor Lane, Weybridge, Surrey KT15 2SN, England.
0932 858457; Fax: 0932 842959
Microtitration; filters; manifolds; tubes; laboratory suppliers; disinfectants.
Popper & Sons, Inc.
300 Denton Ave., New Hyde Park, NY 11040.
(516) 248-0300; Fax: (516) 747-1188
Syringes and needles; blunt canulae.
Portland Plastics
Portex Ltd., The Reachfields, Hythe, Kent, England.
Autoclavable nylon film.
Precision Scientific, Inc.
3737 W. Cortland St., Chicago, IL 60647.
(800) 621-8820; (312) 227-2660; Fax: (312) 227-1828
Prior Clave Limited
129 Nathan Way, Woolwich Industrial Estate, London SE28 0AB, England.
01 316 6620
Autoclaves; sterililzers.
Promega Corporation
Delta House, Chilworth Research Centre, Southampton SO1 7NS, England.
0800 181037
Growth factors; cytokines.
PromoCell
Leimer Str., 2 D-6900 Heidelberg, Germany.
06227 315369; Fax: 06221 315769
BioTech Trade & Service, Laborbedarf GmbH
Franz Antoni Strasse 22, D-6837 St. Leon-Rot 1, Germany.
06227 51308; Fax: 06227 53694
Serum-free medium; keratinocytes; neutral red assay; cytotoxicity.
Protein Polymer Technologies Inc.
10655 Sorrento Valley Road, 1st Floor, San Diego, CA 92121.
(800) 755-0407; Fax: (619) 558-6477
Matrix; Pronectin.
Purite Ltd.
Unit E, Bandet Way, Thame Industrial Estate, Chinnor Road, Thame, Oxon OX9 3SJ, England.
084 421 7141; Fax: 084 421 8098
Water purification; reverse osmosis; deionization; distillation.
PyMaH Corporation
89 South Route 206, Somerville NJ 08876.
908 526 1222; Fax: 908 526 9358
Queue Systems, Inc.
P.O. Box 1901, Parkersburg, WV 26102.
(800) 222-6902; Fax: (304) 464-4229
Incubators; CO_2 incubators.
Rainin Instrument Co., Inc.
Mack Road, Woburn, MA 01888-4026.
Anachem
Charles St., Luton, Beds. LU2 0EB, England.
0582 456666
Pipettes: micropipettes; motorized; automatic.

Revco Scientific, Inc.
275 Aiken Road, Asheville, NC 28804.
(704) 658-2711; (800) 252-7100; Fax: (704) 645-3368
Revco Scientific International
P.O. Box 321, NL-8600 AH Sneek, The Netherlands.
31 5157 5105; Fax: 31 5157 4659
Freezers; ultra-deep freezers.

Rustrak Instruments
The Hyde, Brighton BN2 4JU, England.
273 606271; Fax: 273 609990
Rustrak Instruments
Rt. 12 & Middle Rd., East Greenwich, RI 02818.
(401) 884-6800; Fax: (401) 884-4872
Recorders; electronic thermometers.

Safetech
Enterprise House, Plassey Technological Park, Limerick, Ireland.
353 61 338177
Laminar flow; mini-cleanroom; Cleansphere.

Sage Instruments, Division of Orion Research, Inc.
529 Main St., The Schrafft Center, Boston, MA 02129.
(617) 242-3900; (800) 225-1480; Fax: (617) 242-8594
Air curtain incubator.

Sanyo Gallenkamp plc
Park House, Meridian East, Meridian Business Park, Leicester, Leicestershire LE3 2UZ, England.
0533 630530; Fax: 0533 630353
Water purification; distillation; centrifuges; incubators; freezers.

Sartorius AG
P.O. Box 32 43, Weender Landstrasse 94–108, 3400 Goettingen, Germany.
551 3080; Fax: 551 308 289
Sartorius Ltd.
Blenheim Road, Longmead Industrial Estate, Epsom, Surrey KT19 9BR, England.
03727 45811; Fax: 03727 20799
Sartorius Corp.
140 Wilbur Place, Bohemia, NY 11716.
(516) 563-5120; Fax: (516) 563-5065
Sterilization filters; molecular filtration; balances.

Saxon Micro
P.O. Box 28, Newmarket, Suffolk CB8 8NY, England.
0638 665120
Micromanipulation; cloning; plate reading; computerized stage; Quixell 42.

Schering Chemicals Ltd.
Pharmaceutical Division, Burgess Hill, Sussex, RH15 9NE, England.

Schering Corporation
2000 Galloping Hill Road, Kenilworth, NJ 07033-1310.
Pharmaceuticals.

Schott Glaswerke, Chemical Div., Laboratory Product Group
Hattenbergstr. 10, Postfach 2480, D-6500 Mainz 1, Germany.
06131 633462; Fax: 06131 622006
Schott America
3 Odell Plaza, Yonkers, NY 10701.
(914) 968-8900; Fax: (914) 968-4422
Glassware; media bottles.

Schuco Scientific Ltd.
Woodhouse Rd., London N12 0NE, England.
081 368 1642; Fax: 081 361 3761
Pi-Pump pipettor; Securatab tubing sealers and tool.

Secomak
Stanmore, Middlesex HA7 1BR, England.
Thermostatic controls; time-lapse incubator.

Seescan Analytical Services Ltd.
Unit 9, 25 Gwydir St., Cambridge CB1 2LG, England.
223 460004; Fax: 223 460116
Image analysis.

Septracor Bioprocessing Europe
35 Avenue Jean-Jaures, 92395 Villeneuve la Garenne-Cedex, France.
331 46859200; Fax: 331 47922655
Growth factors; protein purification; chromatography; filtration.

Sera-Lab Ltd.
Crawley Down, Sussex RH10 4FF, England.
0342 716366
See also JRH Biosciences.
Serum; anti-sera; antibodies.

Serotec
22 Bankside Station Approach, Kidlington, Oxford OX5 1JE, England.
08675 79941; Fax: 08675 3899
Antibodies; interferon alpha; magnetizable beads; TNF; integrin; fibronectin.

Serva Feinbiochemica GmBH
Carl-Benz-Srasse 7, D-6900 Heidelberg, Germany
06221 502 0; Fax: 06221 502 113
Biochemicals, growth factors

Shamrock Scientific Speciality Systems, Inc.
34 Davis Dr., Bellwood, IL 60104.
(312) 992-1187; (800) 323-0294
High/Low temperature-indicating tape and labels; sterilizer tapes.

Shandon Scientific Ltd.
Chadwich Road, Astmoor, Runcorn, Cheshire WA7 1PR, England.
0928 566611; Fax: 0928 565845
Shandon Southern Instruments, Inc.
515 Broad Street, Sewickley, PA 15143.
Laboratory equipment; cytocentrifuge.

Sigma Chemical Co.
3050 Spruce Street, St. Louis, MO 63178.
(800) 325-3010; (314) 771-5750; Fax: (800) 325-5052
Sigma Chemical Co. Ltd.
Fancy Road, Poole, Dorset BH17 7NH, England.
0202 733114
Biochemicals; media; sera; antibodies; aidin; alkaline phosphatase; FITC:TRITC; growth factors; hormones; steroids; cyclodextrins; collagenase; trypsin; protease; anti-protease.

Signal Instrument Co. Ltd.
9 Krooner Road, Camberley, Surrey GU15 2QP, England.
0276 66833
Gas mixers/blenders.

Squibb & Sons Ltd.
Regal House, Twickenham, Middx., TW1 3QT, England.
Pharmaceuticals; Mycostatin; fungicide.

Staniar, John, & Co.
Sherborne Street, Manchester M3 1FD, England.
Nylon mesh filters; Nybolt.
Taylor Wharton Cryogenics
P.O. Box 568, Theodore, AL 36590.
(205) 443-8680; Fax: (205) 653-2209
Taylor-Wharton Cryogenics
Oxford Street, Bilston, West Midlands WV14 7EG, England.
0902 494353; Fax: 0902 354717
Nitrogen freezers (were Union Carbide).
Techne (Cambridge) Limited
Duxford, Cambridge CB2 4PZ, England.
0223 832401
Laboratory equipment; biological stirrers; suspension culture.
Tecnomara AG
Industriestrasse, 44 CH-8304 Wallisellen, Zurich, Switzerland.
01 830 2277
Pipette aids; roller racks; peristaltic dosing pump; autoclaves.
Tekmar Co.
P.O. Box 429576, Cincinnati, OH 45242-9576.
(513) 247-7000; (800) 543-4461; Fax: (513) 247-7050
Nylon gauze/mesh.
Thermolyne
2555 Kerper Blvd., Dubuque, IA 52001.
(319) 556-2241
Labmart (Cambridge) Ltd.
1 Pembroke Ave., Waterbeach, Cambridge CB5 9QR, England.
0223 861665
Nitrogen freezers; Thermolyne locator; Nalgene storage boxes; rack storage.
Thermolyne
See Barnstead Thermolyne Corp.
Thomas Scientific.
99 High Hill Road, Swedesboro, NJ 08085-9904.
(800) 345-2100; (609) 467-2000; Fax: (609) 467-3087
Laboratory equipment; magnetic stirrers; disinfectants.
Tissue Culture Services Ltd.
Botolph, Claydon, Buckingham MK18 2LR, England.
029 671 4071/4072; Fax: 029 671 4806
Antibodies: mycoplasma; cytokeratin; CEA; alpha fetoprotein; AFP; serum; agar plates; broth.
Treff Lab
Precision Laboratory Products, CH-9113, Degersheim, Switzerland.
071 54 22 42 – 54 54 54
Pipette tips; storage.
U.S. Biochemical Corp.
P.O. Box 22400, Cleveland, OH 44122.
(800) 321-9322; (216) 765-5000; Fax: (216) 464-5075, (800) 535-0898
Cambridge Biosciences
Growth factors; antibodies; oligonucleotides.
Union Carbide
See Taylor Wharton Cryogenics.
Universal Biologicals Ltd.
12-14 St Ann's Crescent, London SW18 2LS, England.
01 870 8753
Antibodies/insulin-like growth factor I (IGF-I), human.
Universal Biologicals Ltd.
30 Merton Road, London SW18 1QY, England.
081 870 8753; Fax: 081 874 2563
Collagen; growth factors.
Upjohn Corp.
7000 Portage Road, Kalamazoo, MI 49001.
(616) 323-4000; Fax: (616) 323-4077
Upjohn
Puurs, Lichtevstraat, B-2670 Belgium.
Pharmaceuticals; β-methasone.
Upstate Biotechnology Inc.
89 Saranac Place, Lake Placid, NY 12946.
(800) 233-3991 (orders); (800) 548-7853 (technical service); Fax: (518) 523-1336
Antibodies; matrix products; growth factors.
Valley Forge Instrument Co., Inc.
55 Buckwalter Road, Phoenixville, PA 19460.
(215) 933-1806
See John Burge Equipment Ltd.
Glassware washing machines; sterilizers.
Vector Laboratories, Inc.
30 Ingold Road, Burlingame, CA 94010.
(415) 697-3600; Fax: (415) 697-0339
Vector Laboratories
16 Wulfric Square, Bretton, Peterborough PE3 8RF, England.
0733 265530; Fax: 0733 263048
Antibodies; lectins; biotin/avidin; glucose oxidase; alkaline phosphatase; UV bleaching inhibitor; Vectastain; enzyme immunoassays and hybridoma screening.
Ventrex Laboratories Inc.
217 Read Street, Portland, ME 04103.
Serum substitute.
Vernitron/Betterbuilt
S. Empire Blvd., Carlstadt, NJ 07072.
See John Burge Equipment Ltd.
Glassware washing machines; autoclaves; sterilizers.
Vindon Scientific Ltd.
Ceramyl Works, Diggle, Oldham, Lancs. OL3 5JY, England.
0457 876616; Fax: 0457 871355
Incubators; ovens; CO_2 *incubators.*
Volac
77-93 Tanner Street, Barking, Essex IG11 8QD, England.
Glassware; pipettes; pipette plugger.
Wallac Oy
P.O. Box 10, SF-20 101 Turku, Finland.
358 678 111
Microtitration; scintillation counters.
Watson-Marlowe Ltd.
Falmouth, Cornwall TR11 4RU, England.
Pumps; peristaltic pumps.
0326 373461; Fax: 0326 376009
Wescor
ChemLab Scientific Products Ltd.
Construction House, Grenfell Avenue, Hornchurch, Essex RM12 4EH, England.

04024 76162
Osmometer.
Whatman LabSales
P.O. Box 1359, Hillsboro, OR 97123.
(503) 648-0762; (800) 942-8626; Fax: (503) 648-8118
Anachem Scientific, Ltd.
Whatman House, St. Leonard's Road, 20/20 Maidstone, Kent ME16 0LS, England.
Laboratory supplies; filters; chromatography; autoclave tape; disinfectants.
Wheaton Scientific
1000 N. Tenth Street, Millville, NJ 08332.
(609) 825-1100; Fax: (609) 825-1368
Jencons Scientific Ltd.
Cherrycourt Way Industrial Estate, Stanbridge Road, Leighton Buzzard, Beds. LU7 8UA, England.
0525 372010
Roller culture racks; glassware; culture flasks; bottles; slides.
Whittaker Bioproducts
8830 Biggs Ford Road, Walkersville, MD 21793-0127.
Media; serum-free media; supplements; growth factors.
Worthington Biochemical Corporation
Halls Mill Road, Freehold, NJ 07728.
(800) 445-9603; Fax: (908) 308-4453
Enzymes; trypsin; collagenase.
Zeiss, Carl, Germany
P.O. Box 1380, D-7082 Oberkochen, Germany.
07364 200; Fax: 07364 6808
Carl Zeiss Oberkochen Ltd.
Woodfield Road, Welwyn Garden City, Herts. AL7 1LU, England.
0707 331144; Fax: 0707 330237
Carl Zeiss, Inc.
Microscope Division, 1 Zeiss Drive, Thornwood, NY 10594.
(800) 233-2343; Fax: (914) 681-7446
Microscopes; inverted microscope; research microscopes.
Zinsser Analytic (UK) Ltd.
Howarth Road, Stafferton Way, Maidenhead, Berks. SL6 1AP, England.
0628 773202
See Wheaton and Jencons.
Peristaltic pump dispensers.

SCIENTIFIC SOCIETIES WITH INTERESTS IN TISSUE CULTURE

American Tissue Culture Association
Executive Director: 48815 Centre Park Drive, Suite 210, Columbia, MD 21045.
European Tissue Culture Society (ETCS)
Secretary: Alan Doyle, ECACC, CAMR, Porton Down, Salisbury, SP4 0JG, England. ETCS can also supply information regarding National Tissue Culture Societies in Europe.
European Society for Animal Cell Culture Technology (ESACT)
Secretary: Caroline MacDonald, Dept. of Biological Sciences, Paisley University, Paisley, Scotland.
American Society for Cell Biology
Executive Officer: 9650 Rockville Pike, Bethesda, MD 20814.
European Cell Biology Organisation
(Federation: Membership via National Cell Biology Societies in Europe) Secretary General: Dr. Michael Balls, Department of Human Morphology, University of Nottingham, England.
British Society for Cell Biology
Secretary: R.T. Johnson, Department of Zoology, University of Cambridge, Cambridge CB2 3EJ, England.

Glossary

[Modified after Schaeffer, 1990]

Adaptation. Induction or repression of synthesis of a macromolecule (usually a protein) in response to a stimulus, e.g., enzyme adaptation—an alteration in enzyme activity brought about by an inducer or repressor and involving an altered rate of enzyme synthesis or degradation.

Allograft. See *Homograft.*

Amniocentesis. Prenatal sampling of the amniotic cavity.

Anchorage-dependent. Requiring attachment to a solid substrate for survival or growth.

Anemometer. An instrument for measuring airflow rate.

Aneuploid. Not an exact multiple of the haploid chromosome number (haploid = that number present in germ cells after meiosis; i.e., each chromosome represented once).

Aseptic. Free of microbial infection.

Balanced salt solution. An isotonic solution of inorganic salts present in approximately the correct physiological concentrations. May also contain glucose but usually free of other organic nutrients.

Bioreactor. Culture vessel for large-scale production of cells, either anchored to substrate or propagated in suspension.

Biostat. Culture vessel where physical, physicochemical, and physiological conditions, and cell concentration, are kept constant, usually by perfusion and monitoring.

Carcinoma. A tumor derived from epithelium, usually from endodermally or ectodermally derived cells.

Cell culture. Growth of cells dissociated from the parent tissue by spontaneous migration or mechanical or enzymatic dispersal.

Cell fusion. Formation of single cell body by fusion of two other cells; either spontaneously or, more often, by induced fusion with inactivated Sendai virus or polyethylene glycol.

Cell hybridization. See *Hybrid cell.*

Cell line. A propagated culture after the first subculture.

Cell strain. A characterized cell line derived by selection or cloning.

Chemically defined. Used of medium to imply that it is made entirely from pure defined constituents. Distinct from "serum-free," where other poorly characterized constituents may be used to replace serum.

Clone. A population of cells derived from one cell.

Confluent. Where all the cells are in contact all around their periphery with other cells, and no available substrate is left uncovered.

Contact inhibition. Inhibition of cell membrane ruffling and cell motility when cells are in complete contact with other adjacent cells, as in a confluent culture. Often precedes cessation of cell proliferation but not necessarily causally related.

Continuous cell line or cell strain. One having the capacity for infinite survival. Previously known as "established" and often referred to as "immortal."

Cyclic growth. Growth from a low cell density to a high cell density with a regular subculture interval. A regular repetition of the growth cycle for maintenance purposes.

Deadaptation. Reversible loss of a specific property due to the absence of the appropriate inducer (not always defined).

Dedifferentiation. A term implying irreversible loss of the specialized properties that a cell would have expressed *in vivo*. As evidence accumulates that cultures "dedifferentiate" by a combination of selection of undifferentiated cells or stromal cells and deadaptation resulting from the absence of the appropriate inducers, this term is going out of favor. It is still correctly applied to progressive loss of differentiated morphology in histological observations of, for example, tumor tissue.

Density limitation of growth. Mitotic inhibition correlated with an increase in cell density.

Diploid. Each chromosome represented as a pair, identical in the autosomes and female sex chromosomes and nonidentical in male sex chromosomes, and corresponding to the chromosome number and morphology of most somatic cells of the species from which the cells were derived.

Ectoderm. The outer germ layer of the embryo giving rise to the epithelium of the skin.

Embryonic induction. The interaction of cells from two different germ layers, promoting differentiation, often reciprocal.

Endoderm. The innermost germ layer of the embryo giving rise to the epithelial component of organs such as the gut, liver, and lungs.

Endothelium. An epithelial-like cell layer lining spaces within mesodermally derived tissues, such as blood vessels, and derived from the mesoderm of the embryo.

Enzyme induction. An increase in synthesis of an enzyme produced by, for example, hormonal stimulation.

Epithelial. Used of a culture to imply cells derived from epithelium but often used more loosely to describe any cells of a polygonal shape with clear, sharp boundaries between cells. *Pavement-like.* More correctly this should be termed "epithelioid" or "epithelial-like."

Epithelium. A covering or lining of cells, as in the surface of the skin or lining of the gut, usually derived from the embryonic endoderm or ectoderm but exceptionally derived from mesoderm, as with kidney tubules and mesothelium lining body cavities.

Euploid. Exact multiples of the haploid chromosome set. The correct morphology characteristic of each chromosome pair in the species from which the cells were derived is not implicit in the definition but is usually assumed to be the case. Otherwise it should be stated as "euploid but with some chromosomal aberrations."

Explant. A fragment of tissue transplanted from its original site and maintained in an artificial medium.

Fermentor. Large-scale culture vessel, usually applied to cells in suspension. Derived from term applied to microbiological culture.

Fibroblast. A proliferating precursor cell of the mature differentiated fibrocyte.

Fibroblastic. Resembling fibroblasts, i.e., spindle shaped (bipolar) or stellate (multipolar); usually arranged in parallel arrays at confluence if contact inhibited. Often used indiscriminately for undifferentiated mesodermal cells regardless of their relationship to the fibrocyte lineage. Implies a migratory type or cell with processes exceeding the nuclear diameter by threefold or more.

Finite cell line. A culture that has been propagated by subculture but is only capable of a limited number of cell generations *in vitro* before dying out.

Generation number. The number of population doublings (estimated from dilution at subculture) that a culture has undergone since explantation. Necessarily contains an approximation of the number of generations in primary culture.

Generation time. The interval from one point in the cell division cycle to the same point in the cycle, one division later. Distinct from doubling time or population doubling time, which is derived from the total cell count of a population and therefore averages different generation times, including the effect of nongrowing cells.

Genotype. The total genetic characteristics of a cell.

Glycocalyx. Glycosylated peptides, proteins, and lipids, and glycosaminoglycans attached to the surface of the cell.

Growth curve. A semilog plot of cell number on a log scale against time on a linear scale in a proliferating cell culture. Usually divided into lag phase, before growth is initiated, log phase, the period of exponential growth, and plateau, a stable cell count achieved when the culture stops growing at a high cell density.

Growth cycle. Growth interval from subculture to the top of the log phase, ready for a further subculture.

Haploid. That chromosome number where each chromosome is represented once. In most higher animals it is the number present in the gametes and half of the number found in most somatic cells.

Heterokaryon. Genetically different nuclei in a common cytoplasm, usually derived by cell fusion.

Heteroploid. A term used to describe a culture (not a cell) where the cells comprising the culture have chromosome numbers other than diploid.

Histotypic. A culture resembling tissue-like morphology *in vivo*. It is usually implied that this is a three-dimensional culture recreated from dispersed cell culture that attempts to retain, by cell proliferation and multilayering or by reaggregation, the tissue-like structure. Organ cultures cannot be propagated, whereas histotypic cultures can.

Homiothermic. Able to maintain a constant body temperature in spite of environmental fluctuation.

Homograft. (Allograft). A graft derived from a genetically different donor of the same species as the recipient.

Homokaryon. Genetically identical nuclei in a common cytoplasm, usually a product of cell fusion.

Hybrid cell. Mononucleate cell that results from the fusion of two different cells, leading to the formation of a synkaryon.

Ideogram. The arrangement of (in the case of genetic analysis of a cell) the chromosomes in order by size and morphology so that the karyotype may be studied.

Immortalization. The acquisition of an infinite life-span. May be induced in finite cell lines by transfection with oncogenes, or the large T-region of the SV40 genome, or infection with SV40 (whole virus) or Epstein-Barr virus (EBV). Does not necessarily imply malignant transformation, although it may be a component.

Induction. An increase in effect produced by a given stimulus.

Infection. (Other than the commonplace definition.) Transfer of genomic DNA by infection with a retroviral construct containing the DNA sequence under investigation, usually packaged with a promoter sequence and a reporter gene, such as β-galactosidase, where the product may be detected by staining with a chromogenic substrate.

Isograft. (Syngraft) A graft derived from a genetically identical or nearly identical donor of the same species as the recipient.

Karyotype. The distinctive chromosomal complement of a cell.

Laminar flow. The flow of a fluid that closely follows the shape of a streamlined surface without turbulence. Used in connection with laminar airflow cabinets to imply a stable flow of air over the work area such as to minimize turbulence.

Laminar flow cabinet or hood. A work station with filtered air flowing in a laminar nonturbulent flow parallel to (horizontal laminar flow) or perpendicular to (vertical laminar flow) the work surface, to maintain the sterility of the work.

Leukemia. Malignant disease of the hemopoietic system, evident as circulating blast cells.

Lipofection. Transfection of DNA by fusion with lipid encapsulated DNA.

Lymphoma. A solid tumor of lymphoid cells.

Log phase. See *Growth curve.*

Malignant. A term to describe a tumor that has become invasive or metastatic (i.e., colonizing other tissues). Usually progressive, leading to destruction of host cells and ultimately death of the host.

Malignant transformation. The development of the ability to invade normal tissue without regulation in space or time. May also lead to metastatic growth (colonization of a distant site with subsequent unregulated invasive growth).

Manometer. A "U"-shaped tube containing liquid, the levels of which in each limb of the "U" reflect the pressure difference between the ends.

Medium. A mixture of inorganic salts and other nutrients capable of sustaining cell survival *in vitro* for 24 hours. *Growth medium.* That medium which is used in routine culture such that the cell number increases with time. *Maintenance medium.* A medium that will retain cell survival without growth (cell proliferation), e.g., a low-serum or serum-free medium used with serum-dependent cells to maintain cell survival without cell proliferation. Plural *Media.*

Mesenchyme. Loose, often migratory, embryonic tissue derived from the mesoderm, giving rise to connective tissue, cartilage, muscle, hemopoietic cells, etc., in the adult.

Mesoderm. A germ layer in the embryo arising between the ectoderm and endoderm and giving rise to mesenchyme, which, in turn, gives rise to connective tissue, etc. (see *Mesenchyme,* above).

Monoclonal. Derived from a single clone of cells. *Monoclonal antibody.* Antibody produced by a clone of lymphoid cells either *in vivo* or *in vitro*. *In vitro* the clone is usually derived from a hybrid of a sensitized spleen cell and continuously growing myeloma cell.

Morphogenesis. The development of form and structure of an organism.

Myeloma. A tumor derived from myeloid cells. Used in monoclonal antibody production when the myeloma cell can produce immunoglobulin.

Neoplastic. A new, unnecessary proliferation of cells giving rise to a tumor.

Neoplastic transformation. The conversion of a nontumorigenic cell into a tumorigenic cell.

Oncogene. A gene that, when transfected or infected into normal cells, induces a malignant tumor. Usually genes coding for growth factors, receptors, signal transducers, or nuclear regulators. Positive acting.

Organ culture. The maintenance or growth of organ primordia or the whole or parts of an organ *in vitro* in a way that may allow differentiation and preservation of the architecture and/or function.

Organogenesis. The development of organs.

Organotypic. Histotypic culture involving more than one cell type to create a model of the cellular interactions characteristic of the organ *in vivo*. A reconstruction from dissociated cells or fragments of tissue is implied, as distinct from organ culture, where the structural integrity of the explanted tissue is retained.

Passage. The transfer or subculture of cells from one culture vessel to another. Usually, but not necessarily, implies subdivision of a proliferating cell population enabling propagation of a cell line or cell strain. *Passage number.* The number of times a culture has been subcultured.

Phenotype. The aggregate of all the expressed properties of a cell, being the product of the interaction of the genotype with the regulatory environment.

Plateau. See *Growth curve.*

Plating efficiency. The percentage of cells seeded at subculture giving rise to colonies. If each colony can be said to be derived from one cell, this is synonymous with cloning efficiency. Sometimes used loosely to describe the number of cells surviving after subculture, but this is better termed the "seeding efficiency."

Poikilothermic. Body temperature close to that of the environment and not regulated by metabolism.

Population density. The number of monolayer cells per unit area of substrate. For cells growing in suspension, this term is identical to the cell concentration.

Population doubling time. The interval required for a cell population to double at the middle of the logarithmic phase of growth.

Primary culture. A culture started from cells, tissue, or organs taken directly from an organism and before the first subculture.

Pseudodiploid. Numerically diploid chromosome number but with chromosomal aberrations.

Quasidiploid. See *Pseudodiploid.*

Sarcoma. A tumor derived from mesodermally derived cells, e.g., connective tissue, muscle (myosarcoma), or bone (osteosarcoma).

Saturation density. Maximum number of cells attainable per cm^2 (monolayer culture) or per ml (suspension culture) under specified culture conditions.

Seeding efficiency. The percentage of the inoculum that attaches to the substrate within a stated period of time (implying viability, or survival, but not necessarily proliferative capacity).

Somatic cell genetics. The study of cell genetics by recombination and segregation of genes in somatic cells. Usually by cell fusion.

Split ratio. The divisor of the dilution ratio of a cell culture at subculture, e.g., one flask divided into four or 100 ml up to 400 ml would be a split ratio of 4.

Subconfluent. Less than confluent. All of the available substrate is not covered.

Subculture. See *Passage.*

Substrate. The matrix or solid underlay upon which a monolayer culture grows.

Superconfluent. When a monolayer culture progresses beyond the state where all the cells are attached to the substrate and multilayering occurs.

Suppressor gene. A gene that exhibits the transformed (malignant) phenotype, usually associated with dominant negative regulation of cell proliferation or cell migration. Often mutated or deleted in transformed cells and cancer.

Suspension culture. Where cells will multiply suspended in medium.

Synkaryon. A hybrid cell that results from the fusion of the nuclei it carries.

Tetraploid. Twice the diploid (four times the haploid) number of chromosomes.

Tissue culture. Properly, the maintenance of fragments of tissue *in vitro* but now commonly applied as a generic term to include tissue explant culture, organ culture, and dispersed cell culture, including the culture of propagated cell lines and cell strains.

Transfection. The transfer, by artificial means, of genetic material from one cell to another. Implies transfer of less than the whole nucleus of the donor cell and is usually achieved by using isolated chromosomes, DNA, or cloned genes.

Transformation. A permanent alteration of the cell phenotype presumed to occur via an irreversible genetic change. May be spontaneous as in the development of rapidly growing continuous cell lines from slow-growing early passage rodent cell lines or induced by chemical or viral action. Usually produces cell lines that have an increased growth rate, an infinite life-span, a higher plating efficiency, and are often (but not necessarily) tumorigenic.

Variant. A cell line expressing a stable phenotype that is different from the parental culture from which it was derived.

Viral transformation. A permanent phenotypic change induced by the genetic and heritable effects of a transforming virus.

Xenograft. Transplantation of tissue to a different species from which it was derived. Often used to describe implantation of human tumors in athymic (nude), immune-deprived, or immune-suppressed mice.

References

Aaronson, S.A., Todaro, G.J. (1968) Development of 3T3-like lines from Balb/c mouse embryo cultures: Transformation susceptibility to SV40. J Cell Physiol 72:141–148.

Aaronson, S.A., Todaro, G.J.. Freeman, A.E. (1970) Human sarcoma cells in culture: Identification by colony-forming ability on monolayers of normal cells. Exp Cell Res 61:1–5.

Aaronson, S.A., Bottaro, D.P., Miki, T., Ron, D., Finch, P.W., Fleming, T.P., Ahn, J., Taylor, W.G., Rubin, J.S. (1991) Keratinocyte growth factor. A fibroblast growth factor family member with unusual target cell specificity. Ann NY Acad Sci 638:62–77.

Abaza, N.A., Leighton, J., Schultz, S.G. (1974) Effects of ouabain on the function and structure of a cell line (MDCK) derived from canine kidney. I. Light microscopic observations of monolayer growth. In Vitro 10:172–183.

Abercrombie, M., Heaysman, J.E.M. (1954) Observations on the social behaviour of cells in tissue culture, II. "Monolayering" of fibroblasts. Exp Cell Res 6:293–306.

Adams, D.O. (1979) Macrophages. In Jakoby, W.B., Pastan, I.H. (eds): "Methods of Enzymology, Vol. 57, Cell Culture." New York, Academic Press, pp. 494–506.

Adams, R.L.P. (1980) In Work, T.S., Burdon, R.H. (eds): "Laboratory Techniques in Biochemistry and Molecular Biology. Cell Culture for Biochemists." Amsterdam, Elsevier/North Holland Biomedical Press.

Adolphe, M. (1984) Multiplication and type II collagen production by rabbit articular chondrocytes cultivated in a defined medium. Exp Cell Res 155:527–536.

Advisory Committee on Dangerous Pathogens (ACDP) (UK) (1984) Health and Safety Commission, HMSO Publications, P.O. Box 276, London, SW8 5DT, England.

Aghamohammadi, S.Z., Savage, J.R. (1989) A pulse BrdU method for SCE. Mutation Res 216:259–266.

Aitken, M.L., Villalon, M., Verdugo, P., Nameroff, M. (1991) Enrichment of subpopulations of respiratory epithelial cells using flow cytometry. Am J Resp Cell Mol Biol 4:174–178.

Alberts, B., Bray, D., Lewis, J., Raff, M., Roberts, K., Watson, J.D. (1989) "Molecular Biology of the Cell," 2nd ed. New York, Garland Publishing, pp. 703–705.

Albrecht, A.M., Biedler, J.L., Hutchison, D.J. (1972) Two different species of dihydrotolate reductase in mammalian cells differentially resistant to amethopterin and methasquin. Cancer Res 32:1539–1546.

Aldhous, P. (1991) AIDS viruses. Spectre of contamination [news]. Nature 349:359.

Aletsee-Ufrecht, M.C., Langley, K., Rotsch, M., Havemann, K., Gratzl, M. (1990) NCAM: A surface marker for human small cell lung cancer cells. FEBS Lett 267:295–300.

Allen, R.D., Allen, N.S. (1983) Video-enhanced microscopy with a computer frame memory. J Microsc 129:3-17.

Alley, M.C., Scudiero, D.A., Monks, A., Hursey, M.L., Czerwiniski, M.J., Fine, D.L., Abbot, B.J., Mayo, J.G., Shoemaker, R.H., Boyd, M.R. (1988) Feasibility of drug screening with panels of human tumour cell lines using a microculture tetrazolium assay. Cancer Res 48:589–601.

Ambrose, E.J., Dudgeon, J.A., Easty, D.M., Easty, C.C. (1961) The inhibition of tumor growth by enzymes in tissue culture. Exp Cell Res 24:220-227.

Ames, B.N. (1980) ldentifying environmental chemicals causing mutations and cancer. Science 204:587-593.

Amsterdam, A., Zauberman, A., Meir, G., Pinhasi-Kimhi, O., Suh, B.S., Oren, M. (1988) Cotransfection of granulosa cells with simian virus 40 and Ha-RAS oncogene generates stable lines capable of induced steroidogenesis. Proc Natl Acad Sci USA 85:7582-7586.

Andersson, L.C., Nilsson, K., Gahmberg, C.G. (1979a) K562-a human erythroleukemic cell line. Int J Cancer 23:143-147.

Andersson, L.C., Jokinen, M., Klein, G., Nilsson, K. (1979b) Presence of erythrocytic components in the K562 cell line. Int J Cancer 24:5-14.

Andersson, L.C., Jokinen, M., Gahmberg, C.G. (1979c) Induction of erythroid differentiation in the human leukaemia cell line K562. Nature 278:364-365.

Andreoli, S.P., McAteer, J.A. (1990) Reactive oxygen molecule-mediated injury in endothelial and renal tubular epithelial cells in vitro. Kidney Int 38:785-794.

Antoniades, H.N., Scher, C.D., Stiles, C.D. (1979) Purification of human platelet-derived growth factor. Proc Natl Acad Sci USA 76:1809.

Arlett, C.F., Smith, D.M., Green, M.H.L., McGregor, D.B., Clarke, G.M., Cole, J., Asquith, J.C. (1989) Mammalian gene mutation assays based upon colony formation. In Kirkland, D.J. (ed): "Statistical Evaluation of Mutagenicity Test Data." New York, Cambridge University Press, pp. 66-101.

Armati, P.J., Bonner, J. (1990) A technique for promoting Schwann cell growth from fresh and frozen biopsy nerve utilizing D-valine medium. In Vitro Cell Dev Biol 26:1116-1118.

Arrighi, F.E., Hsu, T.C. (1974) Staining constitutive heterochromatin and Giemsa crossbands of mammalian chromosomes. In Yunis, J. (ed): "Human Chromosome Methodology, 2nd Ed." New York, Academic Press.

Askanas, V., Bornemann, A., Engel, W.K. (1990) Immunocytochemical localization of desmin at human neuromuscular junctions. Neurology 40:949-953.

Au, A.M.-J., Varon, S. (1979) Neural cell sequestration on immunoaffinity columns. Exp Cell Res 120:269.

Auerbach, R., Grobstein, C. (1958) Inductive interaction of embryonic tissues after dissociation and reaggregation. Exp Cell Res 15:384-397.

Augeron, C., Laboisse, C.L. (1984) Emergence of permanently differentiated cell clones in a human colonic cancer cell line in culture after treatment with sodium butyrate. Cancer Res 44:3961-3969.

Augusti-Tocco, G., Sato, G. (1969) Establishment of functional clonal lines of neurons from mouse neuroblastoma. Proc Natl Acad Sci USA 64:311-315.

Avrameas, S. (1970) Immunoenzyme techniques: Enzymes as markers for the localization of antigens and antibodies. In Bourne, G.H., Danielli, J.F. (eds): "International Review of Cytology." New York, Academic Press, pp. 349-385.

Balin, A.K., Goodman, B.P., Rasmussen, H., Cristofalo, V.J. (1976) The effect of oxygen tension on the growth and metabolism of WI-38 cells. J Cell Physiol 89:235-250.

Ballard, P.L. (1979) Glucocorticoids and differentiation. Glucocorticoid Horm Action 12:439-517.

Ballard, P.L., Tomkins, G.M. (1969) Dexamethasone and cell adhesion. Nature 244:344-345.

Balmforth, A.J., Ball, S.G., Freshney, R.I., Graham, D.I., McNamee, B., Vaughan, P.F.T. (1986) D-1 dopaminergic and beta-adrenergic stimulation of adenylate cyclase in a clone derived from the human astrocytoma cell line G-CCM. J Neurochem 47:715-719.

Bard, D.R., Dickens, M.J., Smith, A.U., Sarck, J.M. (1972) Isolation of living cells from mature mammalian bone. Nature 236:314-315.

Barde, Y.A., Lindsay, R.M., Monard, D., Thoenen, H. (1978) New factor released by cultured cells supporting survival and growth of sensory neurones. Nature 274:818.

Barile, M.F. (1977) In Acton, R.T., Lynn, J.D. (eds): "Cell Culture and Its Applications." New York, Academic Press, p. 291.

Barkley, W.E. (1979) Safety considerations in the cell culture laboratory. In Jakoby, W.B., Pastan, I. (eds.): "Methods of Enzymology, vol. 58." New York, Academic Press, pp. 36-43.

Barnes, D., Sato, G. (1980) Methods for growth of cultured cells in serum-free medium. Anal Biochem 102:255-270.

Barnes, W.D., Sirbasku, D.A., Sato, G.H. (eds) (1984a) "Cell Culture Methods for Molecular and Cell Biology, Vol. 1. Methods for Preparation of Media, Supplements, and Substrata for Serum-Free Animal Cell Culture." New York, Alan R. Liss.

Barnes, W.D., Sirbasku, D.A., Sato, G.H. (eds) (1984b) "Cell Culture Methods for Molecular and Cell Biology, Vol. 2. Methods for Serum-Free Culture of Cells of the Endocrine System." New York, Alan R. Liss.

Barnes, W.D., Sirbasku, D.A., Sato, G.H. (eds) (1984c) "Cell Culture Methods for Molecular and Cell Biology, Vol. 3. Methods for Serum-Free Culture of Epithelial and Fibroblastic Cells." New York, Alan R. Liss.

Barnes, W.D., Sirbasku, D.A., Sato, G.H. (eds) (1984d) "Cell Culture Methods for Molecular and Cell Biology, Vol. 4. Methods for Serum-Free Culture of Neuronal and Lymphoid Cells." New York, Alan R. Liss.

Barnett, S.C. (1993) Purification of rat olfactory nerve ensheathing cells from the olfactory bulb by fluorescence activated cell sorting. In Griffiths, J.B., Doyle, A., Newell, D.G. (eds): "Cell and Tissue Culture: Laboratory Procedures." (In press.)

Barnett, S.C., Hutchins, A-M., Noble, M. (1993) Purification of olfactory nerve ensheathing cells of the olfactory bulb. Dev Biol 155:337-350.

Barnstable, C. (1980) Monoclonal antibodies which recognize different cell types in the rat retina. Nature 286:231-234.

Barski, G., Sorieul, S., Cornefert, F. (1961) Production dans les cultures *in vitro* de deux souches cellulaires en association de cellules de caractère "hybride." CR Acad Sci [D] Paris 251:1825.

Bateman, A.E., Peckham, M.J., Steel, G.G. (1979) Assays of drug sensitivity for cells from human tumours: *In vitro* and *in vivo* tests on a xenografted tumour. Br J Cancer 40:81.

Battye, F.L., Shortman, K. (1991) Flow cytometry and cell-separation procedures. Curr Opin Immunol 3:238-241.

Beddington, R. (1992) Transgenic mutagenesis in the mouse. Trends Genetics 8:10.

Bell, E., Ivarsson, B., Merrill, C. (1979) Production of a tissue-like structure by contraction of collagen lattices by human fibroblasts of different proliferative potential *in vitro*. Proc Natl Acad Sci USA 76:1274–1279.

Benda, P., Lightbody, J., Sato, G., Levine, L., Sweet, W. (1968) Differentiated rat glial cell strain in tissue culture. Science 161:370.

Benders, A.A.G.M., van Kuppevelt, T.H.M.S.M., Oosterhof, A., Veerkamp, J.H. (1991) The biochemical and structural maturation of human skeletal muscle cells in culture: The effect of serum substitute Ultroser G. Exp Cell Res 195:284–294.

Benya, P.D., Padilla, S.R., Nimni, M.E. (1978) Independent regulation of collagen types by chondrocytes during the loss of differentiated function in culture. Cell 15:1313–1321.

Berenbaum, M.C. (1985) The expected effects of a combination of agents: The general solution. J Theor Biol 114:413–432.

Berger, S.L. (1979) Lymphocytes as resting cells. In Jakoby, W.B., Pastan, I.H. (eds): "Methods in Enzymology, Vol. 57, Cell Culture." New York, Academic Press, pp. 486–494.

Bergerat, J.P., Barlogie, B., Drewinko, B. (1979) Effects of cisdi-chloro-diammineplatinum (II) on human colon carcinoma cells *in vitro*. Cancer Res 39:1334.

Berky, J.J., Sherrod, P.C. (eds) (1977) "Short Term *in vitro* Testing for Carcinogenesis, Mutagenesis and Toxicity." Philadelphia, Franklin Institute Press.

Bernier, S.M., Desjardins, J., Sullivan, A.K. (1990) Establishment of an osseous cell line from fetal rat calvaria using an immunocytolytic method of cell selection: Characterisation of the cell line and of derived clones. J Cell Physiol 145:274–285.

Bernstein, A. (1975) Differentiation of clonal lines of teratocarcinoma cells: Formation of embryoid bodies in vitro. Proc Natl Acad Sci USA 72:1441–1445.

Berry, M.N., Friend, D.S. (1969) High yield preparation of isolated rat liver parenchymal cells. A biochemical and fine structural study. J Cell Biol 43:506–520.

Bertoncello, I., Bradley, T.R., Watt, S.M. (1991) An improved negative immunomagnetic selection strategy for the purification of primitive hemopoietic cells from normal bone marrow. Exp Hematol 19:95–100.

Bettger, W.J., Boyce, S.T., Walthall, B.J., Ham, R.G. (1981) Rapid clonal growth and serial passage of human diploid fibroblasts in a lipid-enriched synthetic medium supplemented with EGF, insulin and dexamethasone. Proc Natl Acad Sci USA 78:5588–5592.

Bhargava, M., Joseph, A., Knesel, J., Halaban, R., Li, Y., Pang, S., Golberg, I., Setter, E., Donovan, M.A., Zarnegar, R., Faletto, D., Rosen, E.M. (1992) Scatter factor and hepatocyte growth factor activities, properties, and mechanism. Cell Growth Differentiation 3:11–20.

Biedler, J.L. (1976) Chromosome abnormalities in human tumour cells in culture. In Fogh, J. (ed): "Human Tumour Cells *in vitro*." New York, Academic Press.

Biedler, J.L., Albrecht, A.M., Hutchinson, D.J., Spengler, B.A. (1972) Drug response dihydrofolate reductase, and cytogenetics of amethopterin-resistant Chinese hamster cells *in vitro*. Cancer Res 32:151–161.

Biggers, J.D., Gwatkin, R.B.C., Heyner, S. (1961) Growth of embryonic avian and mammalian tibiae on a relatively simple chemically defined medium. Exp Cell Res 25:41.

Bignami, A., Dahl, D., Rueger, D.G. (1980) Glial fibrillary acidic GFA protein in normal neural cells and in pathological conditions. In Federoff, S., Hertz, L. (eds): "Advances in Cellular Neurobiology, Vol. 1." New York, Academic Press.

Biosafety in Microbiological and Biomedical Laboratories (1984) Division of Safety, BG31, ICO2, NIH, Bethesda, MD.

Birch, J.R., Pirt, S.J. (1970) Improvements in a chemically-defined medium for the growth of mouse cells (strain LS) in suspension. J Cell Sci 7:661–670.

Birch, J.R., Pirt, S.J. (1971) The quantitative glucose and mineral nutrient requirements of mouse LS (suspension) cells in chemically-defined medium. J Cell Sci 8:693–700.

Birnie, G.D., Simons, P.J. (1967) The incorporation of ^{3}H-thymidine and ^{3}H-uridine into chick and mouse embryo cells cultured on stainless steel. Exp Cell Res 46:355–366.

Bishop, J.M. (1991) Molecular themes in oncogenesis. Cell 64:235–248.

Bissell, D.M., Arenson, D.M., Maher, J.J., Roll, F.J. (1987) Support of cultured hepatocytes by a laminin-rich gel. Evidence for a functionally significant subendothelial matrix in normal liver. J Clin Invest 79:801–812.

Bjerkvig, R., Laerum, O.D., Mella, O. (1986a) Glioma cell interactions with fetal rat brain aggregates *in vitro*, and with brain tissue*in vivo*. Cancer Res 46:4071–4079.

Bjerkvig, R., Steinsvag, S.K., Laerum, O.D. (1986b) Reaggregation of fetal rat brain cells in a stationary culture system I.: Methodology and cell identification. In Vitro 22:180–192.

Blaker, G.J., Birch, J.R., Pirt, S.J. (1971) The glucose, insulin and glutamine requirements of suspension cultures of HeLa cells in a defined culture medium. J Cell Sci 9:529–537.

Blum, J.L., Wicha, M.S. (1988) Role of the cytoskeleton in laminin induced mammary expression. J Cell Physiol 135:13–22.

Bobrow, M., Madan, J., Pearson, P.L. (1972) Staining of some specific regions on human chromosomes, particularly the secondary constriction of no. 9. Nature 238:122–124.

Bochaton-Piallat, M.L., Gabbiani, F., Ropraz, P., Gabbiani, G. (1992) Cultured aortic smooth muscle cells from newborn and adult rats show distinct cytoskeletal features. Differentiation 49:175–185.

Bögler, O., Wren, D., Barnett, S.C., Land, H., Noble, M. (1990) Cooperation between two growth factors promotes extended self-renewal and inhibits differentiation of oligodendrocyte-type-2 astrocyte (O-2A) progenitor cells. Proc Natl Acad Sci USA 87:6368–6372.

Bohnert, A., Hornung, J., Mackenzie, I.C., Fusenig, N.E. (1986) Epithelial–mesenchymal interactions control basement membrane production and differentiation in cultured and transplanted mouse keratinocytes. Cell Tissue Res 244:413–429.

Booyse, F.M., Sedlak, B.J., Rafelson, M.E. (1975) Culture of arterial endothelial cells. Characterization and growth of bovine aortic cells. Thromb Diathes Ahemorrh 34:825–839.

Bornstein, M.B., Murray, M.R. (1958) Serial observations on patterns of growth, myelin formation, maintenance and degeneration in cultures of newborn rat and kitten cerebellum. J Biophys Biochem Cytol 4:499.

Boukamp, P., Petrusevska, R.T., Breitkreutz, D., Hornung, J., Markham, A. (1988) Normal keratinisation in a spontaneously immortalised, aneuploid human keratinocyte cell line. J Cell Biol 106:761–771.

Bowman, P.D., Betz, A.L., Ar, D., Wolinsky, J.S., Penney, J.B., Shivers, R.R., Goldstein, G. (1981) Primary culture of capillary endothelium from rat brain. In Vitro 17:353–362.

Boyce, S.T., Ham, R.G. (1983) Calcium-regulated differentiation of normal human epidermal keratinocytes in chemically defined clonal culture and serum-free serial culture. J Invest Dermatol 81:33–40s.

Boyce, S.T., Hansbrough, J.F. (1988) Biologic attachment, growth, and differentiation of cultured human epidermal keratinocytes on a graftable collagen and chondroitin-6-sulfate substrate. Surgery 103:421–431.

Boyd, M.R. (1989) Status of the NCI preclinical antitumor drug discovery screen. Prin Prac Oncol 10:1–12.

Boyes, B.G., Rogers, C.G., Karpinsky, K., Stapley, R. (1990) A statistical evaluation of the reproducibility of micronucleus, sister-chromatid exchange, thioguanine-resistance and ouabain-resistance assays in V79 cells exposed to ethyl methanesulfonate and 7,12-dimethylbenz[a]anthracene. Mutation Res 234:81–89.

Boyum, A. (1968a) Isolation of leucocytes from human blood. A two-phase system for removal of red cells with methylcellulose as erythrocyte aggregative agent. Scand J Clin Lab Invest (Suppl 97) 21:9–29.

Boyum, A. (1968b) Isolation of leucocytes from human blood. Further observations. Methylcellulose, dextran and Ficoll as erythrocyte aggregating agents. Scand J Clin Lab Invest (Suppl 97) 31:50.

Braa, S.S., Triglia, D. (1991) Predicting ocular irritation using 3-dimensional human fibroblast cultures. Cosmetics Toiletries 106:55–58.

Braaten, J.T., Lee, M.J., Schewk, A., Mintz, D.H. (1974) Removal of fibroblastoid cells from primary monolayer cultures of rat neonatal endocrine pancreas by dosium ethylmercurithiosalicylate. Biochem Biophys Res Comm 61:476–482.

Bradford, M. (1976) A rapid and sensitive method for the quantitation of microgram quantities of protein using the principle of protein-dye binding. Anal Biochem 72:248–254.

Bradley, N.J., Bloom, H.J.G., Davies, A.J.S., Swift, S.M. (1978) Growth of human gliomas in immune-deficient mice: A possible model for pre-clinical therapy studies. Br J Cancer 38:263.

Bradley, T.R., Metcalf, D. (1966) The growth of mouse bone marrow cells *in vitro.* Aust J Biol Med 44:287–300.

Brattain, M.G., Marks, M.E., McCombs, J., Finely, W., Brattain, D.E. (1983) Characterization of human colonic carcinoma cell lines isolated from a single primary tumor. Br J Cancer 47:373–381.

Breen, G.A.M., De Vellis, J. (1974) Regulation of glycerol phosphate dehydrogenase by hydrocortisone in dissociated rat cerebral cell cultures. Dev Biol 41:255–266.

Breitman, T.R., Selonick, S.E., Collins, S.J. (1980) Induction of differentiation of the human promyelocytic leukaemia cell line (HL-60) by retinoic acid. Proc Natl Acad Sci USA 71:2936–2940.

Breitman, T.R., Kene, B.R., Hemmi, H. (1984) Studies of growth and differentiation of human myelomonocytic leukaemia cell lines in serum-free medium. In Barnes, D.W., Sirbasku, D.A., Sato, G.H. (eds): "Methods for Serum-Free Culture of Neuronal and Lymphoid Cells." New York, Alan R. Liss, pp. 215–236.

Bretzel, R.G., Bonath, K., Federlin, K. (1990) The evaluation of neutral density separation utilizing Ficoll-sodium diatrizoate and Nycodenz and centrifugal elutriation in the purification of bovine and canine islet preparations. Hormone Metab Res (Suppl) 25:57–63.

Bridges, M.A., Davidson, A.G.F., Walker, D.C. (1991) Cystic fibrosis and control nasal epithelial cells harvested by a brushing procedure. In Vitro Cell Dev Biol 27A:684–686.

Brockes, J.P., Fields, K.L., Raff, M.C. (1979) Studies on cultured rat Schwann cells. I. Establishment of purified populations from cultures of peripheral nerve. Brain Res 165:105.

Brouty-Boyé, D. Gresser, I., Baldwin, C. (1979) Reversion of the transformed phenotype to the parental phenotype by subcultivation of x-ray transformed $C_3H/1OT½$ at low cell density. Int J Cancer 2:253–260.

Brouty-Boyé, D., Tucker, R.W., Folkman, J. (1980) Transformed and neoplastic phenotype: Reversibility during culture by cell density and cell shape. Int J Cancer 26:501–507.

Brouty-Boyé, D., Kolonias, D., Savaraj, N., Lampidis, T.J. (1992) Alpha-smooth muscle actin expression in cultured cardiac fibroblasts of newborn rat. In Vitro Cell Dev Biol 28A:293–296.

Brower, M., Carney, D.N., Oie, H.K., Gazdar, A.F., Minna, J.D. (1986) Growth of cell lines and clinical specimens of human nonsmall cell lung cancer in a serum-free defined medium. Cancer Res 46:798–806.

Brown, A.F., Dunn, G.A. (1989) Microinterferometry of the movement of dry matter in fibroblasts. J Cell Sci 92:379–389.

Broxmeier, H. (1994) In Freshney, R.I., Pragnell, I.B., Freshney, M.G. (eds): "Culture of Haemopoietic Cells." New York, Wiley-Liss (in press).

Bruland, Ø., Fodstad, Ø., Pihl, A. (1985) The use of multicellular spheroids in establishing human sarcoma cell lines in vitro. Int J Cancer 35:793–798.

Brunk, C.F., Jones, K.C., James, T.W. (1979) Assay for nanogram quantities of DNA in cellular homogenates. Anal Biochem 92:497–500.

Bruynell, E.A., Debray, H., de Mets, M., Mareel, M.M., Montreuil, J. (1990) Altered glycosylation in Madin-Darby canine kidney (MDCK) cells after transformation by murine sarcoma virus. Clin Expl Metastasis 8:241–253.

Brysk, M.M., Snider, J.M., Smith, E.B. (1981) Separation of newborn rat epidermal cells on discontinuous isokinetic gradients of percoll. J Invest Dermatol 77:205–209.

Bucana, C.D., et al. (1990) Retention of vital dyes correlates inversely with the multidrug-resistant phenotype of

adriamycin-selected murine fibrosarcoma variants. Exp Cell Res 190:69.
Buckingham, M. (1992) Making muscle in mammals. Trends Genetics 8:144–149.
Buehring, G.C. (1972) Culture of human mammary epithelial cells. Keeping abreast of a new method. J Natl Cancer Inst 49:1433–1434.
Buick, R.N., Stanisic, T.H., Fry, S.E., Salmon, S.E., Trent, J.M., Krosovich, P. (1979) Development of an agar-methyl cellulose clonogenic assay for cells of transitional cell carcinoma of the human bladder. Cancer Res 39:5051–5056.
Buonassisi, V., Sato, G., Cohen, A.I. (1962) Hormone-producing cultures of adrenal and pituitary tumor origin. Proc Natl Acad Sci USA 48:1184–1190.
Burchell, J., Taylor-Papadimitriou, J. (1989) Antibodies to human milk fat globule molecules. Cancer Invest 17:53–61.
Burchell, J., Durbin, H., Taylor-Papadimitriou, J. (1983) Complexity of expression of antigenic determinants recognised by monoclonal antibodies HMFG 1 and HMFG 2 in normal and malignant human mammary epithelial cells. J Immunol 131:508–513.
Burchell, J., Gendler, S., Taylor-Papadimitriou, J., Girling, A., Lewis, A., Millis, R., Lamport, D. (1987) Development and characterisation of breast cancer reactive monoclonal antibodies directed to the core protein of the human milk mucin. Cancer Res 47:5476–5482.
Burgess, A.W., Metcalf, D. (1980) The nature and action of granulocyte-macrophage colony stimulating factors. Blood 56:947–958.
Burgess, A.W., Nicola, N.A. (1983) Growth factors and stem cells. London: Academic Press, 355 pp.
Burgess, W.H., Maciag, T. (1989) The heparin-binding fibroblast growth factor family of proteins. Ann Rev Biochem 58:575–606.
Burke, J.M., Ross, R. (1977) Collagen synthesis by monkey arterial smooth muscle cells during proliferation and quiescence in culture. Exp Cell Res 107:387–395.
Burt, A.M., Pallett, C.D., Sloane, J.P., O'Hare, M.J., Schafler, K.F., Yardeni, P., Eldad, A., Clarke, J.A., Gusterson, B.A. (1989) Survival of cultured allografts in patients with burns assessed with probe specific for Y chromosome. Br Med J 298:915–917.
Burwen, S.J., Pitelka, D.R. (1980) Secretory function of lactating moose mammary epithelial cells cultured on collagen gels. Exp Cell Res 126:249–262.
Buset, M., Winawer, S., Friedman, E. (1987) Defining conditions to promote the attachment of adult human colonic epithelial cells. In Vitro 23:403–412.
Cahn, R.D., Lasher, R. (1967) Simultaneous synthesis of DNA and specialized cellular products by differentiating cartilage cells *in vitro*. Proc Natl Acad Sci USA 58:1131–1138.
Cahn, R.D., Cooh, H.G., Cahn, M.B. (1967) In Wilt, F.H., Wessells, N.K. (eds): "Methods in Developmental Biology." New York, Thomas Y. Crowell, 493 pp.
Campion, D.G. (1984) The muscle satellite cell–a review. Int Rev Cytol 87:225–251.
Carlsson, J., Nederman, T. (1989) Tumour spheroid technology in cancer therapy research. Eur J Clin Oncol 25:1127–1133.
Carmichael, J., DeGraff, W.G., Gazdar, A.F., Minna, D.J., Mitchell, J.B. (1987a) Evaluation of a tetrazolium-based semiautomated colorimetric assay: Assessment of chemosensitivity testing. Cancer Res 47:936–942.
Carmichael, J., DeGraff, W.G., Gazdar, A.F., Minna, J.D., Mitchell, J.B. (1987b) Evaluation of a tetrazolium-based semiautomated colorimetric assay: Assessment of radiosensitivity. Cancer Res 47:943–946.
Carney, D.N., Bunn, P.A., Gazdar, A.F., Pagan, J.A., Minna, J.D. (1981) Selective growth in serum-free hormone-supplemented medium of tumor cells obtained by biopsy from patients with small cell carcinoma of lung. Proc Natl Acad Sci USA 78:3185–3189.
Carpenter, G., Cohen, S. (1977) Epidermal growth factor. In Acton, R.T., Lynn, J.D. (eds): "Cell Culture and Its Application." New York, Academic Press, pp. 83–105.
Carrel, A. (1912) On the permanent life of tissues outside the organism. J Exp Med 15:516–528.
Casillas, F.L., Cheifetz, S., Doody, J., Andres, J.L., Lane, W.S., Massague, J. (1991) Structure and expression of the membrane proteoglycan betaglycan, a component of the TGF-β receptor system. Cell 67:785–795.
Caspersson, T., Farber, S., Foley, C.E., Kudynowski, J., Modest, E.J., Simonsson, E., Wagh, U., Zech, L. (1968) Chemical differentiation along metaphase chromosomes. Exp Cell Res 49:219–222.
Catsimpoolas, N., Griffith, A.L., Skrabut, E.M., Valeri, C.R. (1978) An alternate method for the preparative velocity sedimentation of cells at unit gravity. Anal Biochem 87:243–248.
Center for Disease Control, Office of Biosafety, Atlanta, GA 00333, USA (1985) "Proposed Biosafety Guidelines for Microbiological and Bacteriological Laboratories." Publications Dept., DHHS, Public Health Service.
Ceriani, R.L., Taylor-Papadimitriou, J., Peterson, J.A., Brown, P. (1979) Characterization of cells cultured from early lactation milks. In Vitro 15:356–362.
Chambard, M., Vemer, B., Gabrion, J., Mauchamp, J., Bugeia, J.C., Pelassy, C., Mercier, B. (1983) Polarization of thyroid cells in culture; evidence for the basolateral localization of the iodide "pump" and of the thyroid-stimulating hormone receptor-adenyl cyclase complex. J Cell Biol 96:1172–1177.
Chambard, M., Mauchamp, J., Chaband, O. (1987) Synthesis and apical and basolateral secretion of thyroglobulin by thyroid cell monolayers on permeable substrate: Modulation by thyrotropin. J Cell Physiol 133:37–45.
Chang, H., Baserga, R. (1977) Time of replication of genes responsible for a temperature sensitive function in a cell cycle specific to mutant from a hamster cell line. J Cell Physiol 92:333–343.
Chang, R.S. (1954) Continuous subcultivation of epithelial-like cells from normal human tissues. Proc Soc Exp Biol Med 87:440-443.
Chang, S.E., Taylor-Papadimitriou, J. (1983) Modulation of phenotype in cultures of human milk epithelial cells and its relation to the expression of a membrane antigen. Cell Diff 12:143–154.
Chang, S.E., Keen, J., Lane, E.B., Taylor-Papadimitriou, J. (1982) Establishment and characterisation of SV40-

transformed human breast epithelial cell lines. Cancer Res 42:2040–2053.

Chaproniere, D.M., McKeehan, W.L. (1986) Serial culture of adult human prostatic epithelial cells in serum-free medium containing low calcium and a new growth factor from bovine brain. Cancer Res 46:819–824.

Chen, T.C., Curthoys, N.P., Lagenaur, C.F., Puschett, J.B. (1989) Characterization of primary cell cultures derived from rat renal proximal tubules. In Vitro Cell Dev Biol 25:714–722.

Chen, T.R. (1977) In situ detection of mycoplasm contamination in cell cultures by fluorescent Hoechst 33258 stain. Exp Cell Res 104:255.

Choi, K.W., Bloom, A.D. (1970) Cloning human lymphocytes *in vitro*. Nature 227:171–173.

Christensen, B., Kieler, J., Villien, M., Don, P., Wang, C.Y., Wolf, H. (1984) A classification of human urothelial cells propagated in vitro. Anticancer Res 4:319–338.

Chu, G., Hayakawa, H., Berg, P. (1987) Electroporation for the efficient transfection of mammalian cells with DNA. Nucleic Acids Res 15:1311–1326.

Chung, Y.S., Song, I.S., Erickson, R.H., Sleisinger, M.H., Kim, Y.S. (1985) Effect of growth and sodium butyrate on brush border membrane-associated hydrolases in human colorectal cancer cell lines. Cancer Res 45:2976–2982.

Cioni, C., Filoni, S., Aquila, C., Bernardini, S., Bosco, L. (1986) Transdifferentiation of eye tissue in anuran amphibians: Analysis of the transdifferentiation capacity of the iris of *Xenopus laevis* larvae. Differentiation 32:215–220.

Clark, J.M., Pateman, J.A. (1978) Long-term culture of Chinese hamster Kupffer cell lines isolated by a primary cloning step. Exp Cell Res 112:207–217.

Clarke, G.D., Ryan, P.J. (1980) Tranquilizers can block mitogenesis in 3T3 cells and induce differentiation in Friend cells. Nature 287:160–161.

Clayton, R.M., Bower, D.J., Clayton, P.R., Patek, C.E., Randall, F.E., Sime, C., Wainwright, N.R., Zehir, A. (1980) Cell culture in the investigation of normal and abnormal differentiation of eye tissues. In Richards, R.J., Rajan, K.T. (eds): "Tissue Culture in Medical Research (II)." Oxford, Pergamon Press.

Cobbold, P.H., Rink, T.J. (1987) Fluorescence and bioluminescence measurement of cytoplasmic free Ca^{2+}. Biochem J 248:313.

Cohen, J., Balazs, R., Hojos, F., Currie, D.N., Dutton, G.R. (1978) Separation of cell types from the developing cerebellum. Brain Res 148:313–331.

Cohen, S. (1962) Isolation of a mouse submaxillary gland protein accelerating incisor eruption and eyelid opening in the new-born animal. J Biol Chem 237:1555–1562.

Cole, J., Arlett, C.F. (1984) The detection of gene mutations in cultured mammalian cells. In Vennitt, S., Parry, J.M. (eds): "Mutagenicity Testing–A Practical Approach." Oxford, IRL Press, pp. 233–273.

Cole, J., Fox, M., Garner, R.C., McGregor, D.B., Thacker, J. (1990) Gene mutation assays in cultured mammalian cells. In Kirkland, D.J. (ed): "Basic Mutagenicity Tests: UKEMS Recommended Procedures, Part 1 (revised)." Cambridge, Cambridge University Press, pp. 87-114.

Cole, J., Richmond, F.R., Bridges, B.A. (1991) The mutagenicity of 2-amino-N6-hydroxyadenine to L5178Y tk+/-3.7.2c mouse lymphoma cells: Measurement of mutation to ouabain, 6-thioguanine and trifluorothymidione and the induction of micronuclei. Mutation Res 253:55–62.

Cole, R.J., Paul, J. (1966) The effects of erythropoietin on haem synthesis in mouse yolk sac and cultured foetal liver cells. J Embryol Exp Morphol 15:245–260.

Cole, S.P.C. (1986) Rapid chemosensitivity testing of human lung tumour cells using the MTT assay. Cancer Chemother Pharmacol 17:259–263.

Collins, S.J., Gallo, R.C., Gallagher, R.E. (1977) Continuous growth and differentiation of human myeloid leukaemic cells in suspension culture. Nature 270:347–349.

Committee for a Standardized Karyotype of *Rattus norvegicus*. (1973) Standard karyotype of the Norway rat, *Rattus norvegicus*. Cytogenet Cell Genet 12:199–205.

Committee on Standardized Genetic Nomenclature for Mice. (1972) Standard karyotype of the mouse *Mus musculis*. J Hered 63:69.

Conkie, D. (1992) Separation of viable cells by centrifugal elutriation. In Freshney, R.I. (ed): "Animal Cell Culture, a Practical Approach." Oxford, IRL at Oxford University Press, pp. 149–164.

Conkie, D., Affara, N., Harrison, P.R., Paul, J., Jones, K., (1974) *In situ* localization of globin messenger RNA formation. II. After treatment of Friend virus-transformed mouse cells with dimethyl sulphoxide. J Cell Biol 63:414–419.

Control of Substances Hazardous to Health Regulations (1988) Health and Safety Commission, HMSO Publications, P.O. Box 276, London, SW8 5DT, England.

Coon, H.D. (1968) Clonal cultures of differentiated rat liver cells. J Cell Biol 39:29a.

Coon, H.G., Cahn, R.D. (1966) Differentiation *in vitro:* Effects of sephadex fractions of chick embryo extract. Science 153:1116–1119.

Coons, A.H., Kaplan, M.M. (1950) Localization of antigen in tissue cells. II. Improvements in a method for the detection of antigen by means of fluorescent antibody. J Exp Med 91:1–13.

Cooper, G.W. (1965) Induction of somite chondrogenesis by cartilage and notochord: A correlation between inductive activity and specific stages of cytodifferentiation. Dev Biol 12:185–212.

Cooper, P.D., Burt, A.M., Wilson, J.N. (1958) Critical effect of oxygen tension on rate of growth of animal cells in continuous suspended culture. Nature 182:1508–1509.

Cou, J.Y. (1978) Establishment of clonal human placental cells synthesizing human choriogonadotropin. Proc Natl Acad Sci USA 75:1854–1858.

Cour, I., Maxwell, G., Hay, R.J. (1979) Tests for bacterial and fungal contaminants in cell cultures as applied at the ATCC. In Evans, V.J., Perry, V.P., Vincent, M.M. (eds): "Manual of the American Tissue Culture Association." 5:1157–1160.

Courtenay, V.D., Selby, P.I., Smith, I.E., Mills, J., Peckham, M.J. (1978) Growth of human tumor cell colonies from biopsies using two soft-agar techniques. Br J Cancer 38:77–81.

Coutinho, L.H., Gilleece, M.H., de Wynter, E.A., Will, A., Testa, N.G. (1992) Clonal and long-term cultures using human bone marrow. In Testa, N.G., Molineux, G. (eds): "Haemopoiesis: A Practical Approach." Oxford, IRL Press at Oxford University Press, pp. 75-106.

Crabb, I.W., Armes, L.G., Johnson, C.M., McKeehan, W.L. (1986) Characterization of multiple forms of prostatropin (prostate epithelial growth factor) from bovine brain. Biochem Biophys Res Commun 136:1155-1161.

Creasey, A.A., Smith, H.S., Hackett, A.I., Fukuyama, K., Epstein, W.L., Madin, S.H. (1979) Biological properties of human melanoma cells in culture. In Vitro 15:342.

Croce, C.M. (1991) Genetic approaches to the study of the molecular basis of human cancer. Cancer Res (Suppl) 51:5015s-5018s.

Crouch, E.C., Stone, K.R., Bloch, M., McDivitt, R.W. (1987) Heterogeneity in the production of collagens and fibronectin by morphologically distinct clones of a human tumor cell line: Evidence for intratumoral diversity in matrix protein biosynthesis. Cancer Res 47(22):6086-6092.

Crowe, R., Ozer, H. Rifkin, D. (1978) "Experiments with Normal and Transformed Cells." Cold Spring Harbor, NY, Cold Spring Harbor Laboratory.

Cunha, G.R. (1984) Androgenic effects upon prostatic epithelium are mediated via tropic influences from stroma. In "New Approaches to the Study of Benign Prostatic Hyperplasia." New York, Alan R. Liss, pp. 81-102.

Cunha, G.R., Young, P. (1992) Role of stroma in oestrogen-induced epithelial proliferation. Epith Cell Biol 1:18-31.

Cuttitta, F., Carney, D.N., Mulshine, J., Moody, T.W., Fedorko, J., Fischler, A., Minna, J.D. (1985) Bombesin-like peptides can function as autocrine growth factors in human small-cell lung cancer. Nature 316:823.

Damon, D.H., Lobb, R.R., D'Amore, P.A., Wagner, J.A. (1989) Heparin potentiates the action of acidic fibroblast growth factor by prolonging its biological half-life. J Cell Physiol 138:221-226.

Das, S.K., Stanley, E.R. (1982) Structure-function studies of a colony stimulating factor (CSF-1). J Biol Chem 257:13679-13684.

Davison, P.M., Bensch, R., Karasek, M.A. (1983) Isolation and long-term serial cultivation of endothelial cells from microvessels of the adult human dermis. In Vitro 19:937-945.

Dawe, C.J., Potter, M. (1957) Morphologic and biologic progression of a lymphoid neoplasm of the mouse in vivo and in vitro. Am J Pathol 33:603.

De Leij, L., Poppema, S., Nulend, I.K., Haar, A.T., Schwander, E., Ebbens, F., Postmus, P.E., Hauw The, T. (1985) Neuroendocrine differentiation antigen on huam lung carcinoma and Kulchitski cells. Cancer Res 45:2192-2200.

De Ridder, L., Calliauw, L. (1990) Invasion of human brain tumors in vitro: Relationship to clinical evolution. J Neurosurg 72:589-593.

De Ridder, L., Mareel, M. (1978) Morphology and ^{125}I-concentration of embryonic chick thyroids cultured in an atmosphere of oxygen. Cell Biol Int Rep 2:189-194.

De Vitry, F., Camier, M., Czernichow, P., Benda, P., Cohen, P., Tixier-Vidal, A. (1974) Establishment of a clone of mouse hypothalamic neurosecretory cells synthesizing neurophysin and vasopressin. Proc Natl Acad Sci USA 71:3575-3579.

De Vonne, T.L., Mouray, H. (1978) Human α_2-macroglobulin and its antitrypsin and antithrombin activities in serum and plasma. Clin Chim Acta 90:83-85.

Dean, B.J., Danford, N. (1984) Assays for the detection of chemically induced chromosome damage in cultured mammalian cells. In Venitt, S., Parry, J.M. (eds): "Mutgenicity Testing, A Practical Approach." Oxford, IRL Press.

Defendi, V. (ed) (1964) "Retention of Functional Differentiation in Cultured Cells." Philadelphia, Wistar Institute Press.

Del Vecchio, P., Smith, J.R. (1981) Expression of angiotensin converting enzyme activity in cultured pulmonary artery endothelial cells. J Cell Physiol 108:337-345.

DeLarco, J.E., Todaro, G.J. (1978) Epithelioid and fibroblastoid rat kidney cell clones: Epidermal growth factor receptors and the effect of mouse sarcoma virus transformation. J Cell Physiol 94:335-342.

DeMars, R. (1957) The inhibition of glutamine of glutamyl transferase formation in cultures of human cells. Biochim Biophys Acta 27:435-436.

Dendy, P.P., Hill, B.T. (1983) Human tumour sensitivity testing *in vitro*: Techniques and clinical applications. London, Academic Press.

Dennis, L.W. (1992) Tissue-cultured skin grafts. J Burn Care Rehab 13:93-94.

Detrisac, C.J., Sens, M.A., Garvin, A.J., Spicer, S.S., Sens, D.A. (1984) Tissue culture of human kidney epithelial cells of proximal tubule origin. Kidney Int 25:383-390.

Dexter, D.L., Barbosa, J.A., Calabresi, P. (1979) N,N-dimethylformamide-induced alteration of cell culture characteristics and loss of tumorigenicity in cultured human colon carcinoma cells. Cancer Res 39:1020-1025.

Dexter, T.M., Allen, T.D., Lajtha, L.G. (1977) Conditions controlling the proliferation of haemopoietic stem cells in vitro. J Cell Physiol 91:335-345.

Dexter, T.M., Allen, T.D., Scott, D., Teich, N.M. (1979) Isolation and characterisation of a bipotential haematopoietic cell line. Nature 277:417-474.

Dexter, T.M., Testa, N.G., Allen, T.D., Rutherford, S., Scolnick, E. (1981) Molecular and cell biological aspects of erythropoiesis in long-term bone marrow cultures. Blood 58:699-707.

Dexter, T.M., Spooncer, E., Simmons, P., Allen, T.D. (1984) Long-term marrow culture: An overview of technique and experience. In Wright, D.G., Greenberger, J.S. (eds): "Long-Term Bone Marrow Culture," Kroc Foundation Series 18. New York, Alan R. Liss, pp. 57-96.

Dickson, J.A., Suzangar, M. (1976) The in vitro response of human tumours to cytotoxic drugs and hyperthermia (42°) and its relevance to clinical oncology. In Balls, M., Monnickendam, M. (eds): "Organ Culture and Biomedical Research." Cambridge, Cambridge University Press, pp. 417-446.

Dickson, J.D., Flanigan, T.P., Kemshead, J.T. (1983) Monoclonal antibodies reacting specifically with the cell surface of human astrocytes in culture. Biochem Soc Trans 11:208.

DiPaolo, J.A. (1965) In vitro test systems for cancer chemotherapy. III. Preliminary studies of spontaneous mammary tumors in mice. Cancer Chemother Rep 44:19-24.

Dodson, M.V., Mathison, B.A., Mathison, B.D. (1990) Effects

of medium and substratum on ovine satellite cell attachment, proliferation and differentiation in vitro. Cell Diff Dev 29(1):59–66.

Doherty, P., Ashton, S.V., Moore, S.E., Walsh, F.S. (1991) Morphoregulatory activities of NCAM and N-cadherin can be accounted for by G protein-dependent activation of L- and N-type neuronal Ca^{2+} channels. Cell 67:21–33.

Donato, M.T., Gomez-Lechon, M.J., Castell, J.V. (1990) Drug metabolizing in rat hepatocytes co-cultured with cell lines. In Vitro Cell Dev Biol 26:1057–1062.

Dotto, G.P., Parada, L.F., Weinberg, R.A. (1985) Specific growth response of *ras*-transformed embryo fibroblasts to tumour promoters. Nature 318:472–475.

Douglas, W.H.J., Moorman, G.W., Teel, R.W. (1976) The formation of histotypic structures from monodispersed rat lung cells cultured on a three-dimensional substrate. In Vitro 12:373–381.

Douglas, W.H.J., McAteer, J.A., Dell'Orco, R.T., Phelps, D. (1980) Visualization of cellular aggregates cultured on a three-dimensional collagen sponge matrix. In Vitro 16:306–312.

Dow, J.A.T., Lackie, J.M., Crocket, K.V. (1987) A simple microcomputer-based system for real-time analysis of cell behaviour. J Cell Sci 87:171–182.

Drejer, J., Larsson, O.M., Schousboe, A. (1983) Characterization of uptake and release processes for D- and L-aspartate in primary cultures of astrocytes and cerebellar granule cells. Neurochem Res 8:231–243.

Duffy, M.J., Reilly, D., O'Sullivan, C., O'Higgins, N., Fennelly, J.J., Andreasen, P. (1990) Urokinase-plasminogen activator, a new and independent prognostic marker in breast cancer. Cancer Res 50:6827–6829.

Duksin, D., Maoz, A., Fuchs, S. (1975) Differential cytotoxic activity of anticollagen serum on rat osteoblasts and fibroblasts in tissue culture. Cell 5:83–86.

Dulak, N.C., Temin, H.M. (1973a) A partially purified rat liver cell conditioned medium with multiplication-stimulating activity for embryo fibroblasts. J Cell Physiol 81:153–160.

Dulak, N.C., Temin, H.M. (1973b) Multiplication-stimulating activity for chicken embryo fibroblasts from rat liver cell conditioned medium: A family of small peptides. J Cell Physiol 81:161–170.

Dulbecco, R., Elkington, J. (1973) Conditions limiting multiplication of fibroblastic and epithelial cells in dense cultures. Nature 246:197–199.

Dulbecco, R., Freeman, G. (1959) Plaque formation by the polyoma virus. Virology 8:396–397.

Dulbecco, R., Vogt, M. (1954) Plaque formation and isolation of pure cell lines with poliomyelitis viruses. J Exp Med 199:167–182.

Dunham, L.J., Stewart, H.L. (1953) A survey of transplantable and transmissible animal tumors. J Natl Cancer Inst 13:1299–1377.

Eagle, H. (1955) The specific amino acid requirements of mammalian cells (stain L) in tissue culture. J Biol Chem 214:839.

Eagle, H. (1959) Amino acid metabolism in mammalian cell cultures. Science 130:432.

Eagle, H. (1973) The effect of environmental pH on the growth of normal and malignant cells. J Cell Physiol 82:1–8.

Eagle, H., Foley, G.E., Koprowski, H., Lazarus, H., Levine, E.M., Adams, R.A. (1970) Growth characteristics of virus-transformed cells. J Exp Med 131:863–879.

Earle, W.R., Schilling, E.L., Stark, T.H., Straus, N.P., Brown, M.F., Shelton, E. (1943) Production of malignancy in vitro. IV. The mouse fibroblast cultures and changes seen in the living cells. J Natl Cancer Inst 4:165–212.

Easton, T.G., Valinsky, J.E., Reich, E. (1978) M540 as a fluorescent probe of membranes: Staining of electrically excitable cells. Cell 13:476–486.

Easty, D.M., Easty, G.C. (1974) Measurement of the ability of cells to infiltrate normal tissues in vitro. Br J Cancer 29:36–49.

Ebendal, T. (1976) The relative roles of contact inhibition and contact guidance in orientation of axons extending on aligned collagen fibrils in vitro. Exp Cell Res 98:159–169.

Ebendal, T. (1979) Stage-dependent stimulation of neurite outgrowth exerted by nerve growth factor and chick heart in cultured embryonic ganglia. Dev Biol 72:276.

Ebendal, T., Jacobson, C.O. (1977) Tissue explants affecting extension and orientation of axons in cultured chick embryo ganglia. Exp Cell Res 105:379–387.

Edelman, G.M. (1973) Nonenzymatic dissociations. B. Specific cell fractionation of chemically derivatized surfaces. In Kruse, P.F., Jr., Pattcrson, M.K., Jr. (eds): "Tissue Culture Methods and Applications." New York, Academic Press, pp. 29–36.

Edelman, G.M. (1986) Cell adhesion molecules in the regulation of animal form and tissue pattern. Annu Rev Cell Biol 2:81–116.

Edelman, G.M. (1988) Morphoregulatory molecules. Biochemistry 27:3533–3543.

Edwards, P.A.W., Easty, D.M., Foster, C.S. (1980) Selective culture of epithelioid cells from a human squamous carcinoma using a monoclonal antibody to kill fibroblasts. Cell Biol Int Rep 4:917–922.

Eisen, H., Bach, R., Emery, R. (1977) Induction of spectrin in Friend erythroleukaemic cells. Proc Natl Acad Sci USA 74:3898–4002.

Eisenbarth, G.S., Walsh, F.S., Nirenberg, M. (1979) Monoclonal antibody to a plasma membrane antibody of neurons. Proc Natl Acad Sci USA 76:4913–4917.

Eisinger, M., Lee, J.S., Hefton, J.M., Darzykiewicz, A., Chiao, J.W., Deharven, E. (1979) Human epidermal cell cultures–growth and differentiation in the absence of dermal components or medium supplements. Proc Natl Acad Sci USA 76:5340.

Elliget, K.A., Lechner, J.F. (1992) Normal human bronchial epithelial cell cultures. In Freshney, R.I. (ed): "Culture of Epithelial Cells." New York, Wiley-Liss, pp. 181–196.

Elsdale, T., Bard, J. (1972) Collagen substrata for studies on cell behaviour. J Cell Biol 54:626–637.

Elvin, P., Wong, V., Evans, C.W. (1985) A study of the adhesive, locomotory and invasive behaviour of Walker 256 carcinosarcoma cells. Exp Cell Biol 53:9–18.

Eng, L.F., Bigbee, J.W. (1979) Immunochemistry of nervous-system specific antigens. In Aprison (ed): "Advances in Neurochemistry." New York, Plenum Press, pp. 43–98.

Engel, L.W., Young, N.A., Tralka, T.S., Lippman, M.E., O'Brien,, S.J., Joyce, M.J. (1978) Establishment and characterization of three new continuous cell lines derived from breast carcinomas. Cancer Res 38:3352–3364.

Epstein, M.A., Barr, Y.M. (1964) Cultivation in vitro of human lymphoblasts from Burkitt's malignant lymphoma. Lancet 1:252.

Espmark, J.A., Ahlqvist-Roth, L. (1978) Tissue typing of cells in cultures. I. Distinction between cell lines by the various patterns produced in mixed haemabsorption with selected multiparous sera. J Immunol Methods 24:141–153.

Evans, V.J., Bryant, J.C. (1965) Advances in tissue culture at the National Cancer Institute in the United States of America. In Ramakrishnan, C.V. (ed): "Tissue Culture." The Hague, W. Junk, pp. 145–167.

Evans, V.J., Bryant, J.C., Fioramonti, M.C., McQuilkin, W.T., Sanford, K.K., Earle, W.R. (1956) Studies of nutrient media for tissue C cells in vitro. I. A protein-free chemically defined medium for cultivation of strain L cells. Cancer Res 16:77.

Fantini, J., Galons, J.P., Abadie, B., Canioni, P., Cozzone, P.J., Marvali, J., Tirard, A. (1987) Growth in serum-free medium of human colonic adenocarcinoma cell lines on microcarriers: A two-step method allowing optimal cell spreading and growth. In Vitro 23:641–646.

Federoff, S. (1975) In Evans, V.J., Perry, V.P., Vincent, M.M. (eds): "Manual of the Tissue Culture Association" 1:53–57.

Felgner, P.L., Gadek, T.R., Holm, M., Roman, R., Chan, H.W., Wenz, M., Northrop, J.P., Ringold, G.M., Danielsen, M. (1987) Lipofection–a highly efficient, lipid-mediated DNA-transfection procedure. Proc Natl Acad Sci USA 84:7413–7417.

Fell, H.B. (1953) Recent advances in organ culture. Sci Prog 162:212.

Fell, H.B., Robison, R. (1929) The growth, development and phophatase activity of embryonic avian femora and limb buds cultivated in vitro. Biochem J 23:767–784.

Fergusson, R.J., Carmichael, J., Smyth, J.F. (1980) Human tumour xenografts growing in immunodeficient mice: A useful model for assessing chemotherapeutic agents in bronchial carcinoma. Thorax 41:376–380.

Finbow, M.E., Pitts, J.D. (1981) Permeability of junctions between animal cells. Exp Cell Res 131:1–13.

Finlay, C.A., Hinds, P.W., Levine, A.J. (1989) The p53 protooncogene can act as a suppressor of transformation. Cell 57:1083–1093.

Fisher, H.W., Puck, T.T., Sato, G. (1958) Molecular growth requirements of single mammalian cells: The action of fetuin in promoting cell attachment of glass. Proc Natl Acad Sci USA 44:4–10.

Flynn, D., Yang, J., Nandi, S. (1982) Growth and differentiation of primary cultures of mouse mammary epithelium embedded in collagen gel. Differentiation 22:191.

Fogh, J. (1973) "Contamination in Tissue Culture." New York, Academic Press.

Fogh, J. (1977) Absence of HeLa cell contamination in 169 cell lines derived from human tumors. J Natl Cancer Inst 58:209–214.

Fogh, J., Trempe, G. (1975) In Fogh, J. (ed): "Human Tumor Cells In Vitro." New York, Academic Press, pp. 115–159.

Foley, J.F., Aftonomos, B.T. (1973) Pronase. In Kruse, P.F., Jr., Patterson, M.K. Jr. (eds): "Tissue Culture Methods and Applications." New York, Academic Press, pp. 185–188.

Folkman, I., Moscona, A. (1978) Role of cell shape in growth control. Nature 273:345–349.

Folkman, I., Haudenschild, C.C., Zetter, B.R. (1979) Long-term culture of capillary endothelial cells. Proc Natl Acad Sci USA 76:5217–5221.

Folkman, J. (1992) Angiogenesis and cancer. Sem Cancer Biol

Folkman, J., Haudenschild, C. (1980) Angiogenesis *in vitro*. Nature 288:551–556.

Fontana, A., Hengarner, H., de Tribolet, N., Weber, E. (1984) Glioblastoma cells release interleukin 1 and factors inhibiting interleukin 2-mediated effects. J Immunol 132(4): 1837–1844.

Foreman, J., Pegg, D.E. (1979) Cell preservation in a programmed cooling machine: The effect of variations in supercooling. Cryobiology 16:315–321.

Foster R., Martin, G.S. (1992) A mutation in the catalytic domain of pp60v-src is responsible for the host- and temperature-dependent phenotype of the Rous sarcoma virus mutant tsLA33-1. Virology 187:145–155.

Fraenkel-Conrat, H., Wagner, R.R. (eds) (1979) "Comprehensive Virology." New York and London, Plenum Press.

Frame, M., Freshney, R.I., Shaw, R., Graham, D.I. (1980) Markers of differentiation in glial cells. Cell Biol Int Rep 4:732.

Franks, L.M. (1980) Primary cultures of human prostate. Methods Cell Biol 21:153–169.

Fraser, C.M., Venter, J.C. (1980) The synthesis of β-adrenergic receptors in cultured human lung cells: Induction by glucocorticoids. Biochem Biophys Res Commun 94:390–398.

Frazier, J.M. (1992) "In Vitro Toxicity Testing." New York, Marcel Dekker.

Frederick, M.A., Brent, R., Kingston, R.E., Moore, R.E., Seidman, J.G., Smith, J.A., Struhl, K. (1993) "Current Protocols in Molecular Biology, Vol. I and II." New York, John Wiley & Sons.

Fredin, B.L., Seiffert, S.C., Gelehrter, T.D. (1979) Dexamethasone-induced adhesion in hepatoma cells: The role of plasminogen activator. Nature 277:312–313.

Freedman, V.H., Shin, S. (1974) Cellular tumorigenicity in nude mice: Correlation with cell growth in semi-solid medium. Cell 3:355–359.

Freeman, A.E., Igel, H.J., Herrman, B.J., Kleinfeld, K.L. (1976) Growth and characterisation of human skin epithelial cultures. In Vitro 12:352–362.

Freshney, R.I. (1972) Tumour cells disaggregated in collagenase. Lancet 2:488–489.

Freshney, R.I. (1976) Separation of cultured cells by isopycnic centrifugation in metrizamide gradients. In Rickwood, D. (ed): "Biological Separations." London and Washington, Information Retrieval, pp. 123–130.

Freshney, R.I. (1978) Use of tissue culture in predictive testing of drug sensitivity. Cancer Topics 1:5–7.

Freshney, R.I. (1980) Culture of glioma of the brain. In Thomas, D.G.T., Graham, D.I. (eds): "Brain Tumours, Scientific Basic, Clinical Investigation and Current Therapy." London, Butterworths, pp. 21–50.

Freshney, R.I. (1985) Induction of differentiation in neoplastic cells. Anticancer Res 5:111–130.

Freshney, R.I. (ed) (1992) "Culture of Epithelial Cells." New York, Wiley-Liss.

Freshney, R.I., Hart, E. (1982) Clonogenicity of human glia in suspension. Br J Cancer 46:463.

Freshney, R.I., Paul, J., Kane, I.M. (1975) Assay of anti-cancer drugs in tissue culture: Conditions affecting their ability to incorporate 3H-leucine after drug treatment. Br J Cancer 31:89–99.

Freshney, R.I., Morgan, D., Hassanzadah, M., Shaw, R., Frame, M. (1980a) Glucocorticoids, proliferation and the cell surface. In Richards, R.J., Rajan, K.T. (eds): "Tissue Culture in Medical Research (II)." Oxford, Pergamon Press, pp. 125–132.

Freshney, R.I., Sherry, A., Hassanzadah, M., Freshney, M., Crilly, P., Morgan, D. (1980b) Control of cell proliferation in human glioma by glucocorticoids. Br J Cancer 41:857–866.

Freshney, R.I., Celik, F., Morgan, D. (1982a) Analysis of cytotoxic and cytostatic effects. In Davis, W., Malvoni, C., Tanneberger, St. (eds): "The Control of Tumor Growth and Its Biological Base." Fortschritte in der Onkologie, Band 10. Berlin, Akademie-Verlag, pp. 349–358.

Freshney, R.I., Hart, E., Russell, J.M. (1982b) Isolation and purification of cell cultures from human tumours. In Reid, E., Cook, G.M.W., Morre, D.J. (eds): "Cancer Cell Organelles. Methodological Surveys (B): Biochemistry, Vol. 2." Chichester, England, Horwood, pp. 97–110.

Freshney, R.I., Pragnell, I.B., Freshney, M.G. (1994) "Culture of Haemopoietic Cells." New York, Wiley-Liss (in press).

Freyer, J.P., Sutherland, R.M. (1980) Selective dissociation and characterization of cells from different regions of multicell tumour spheroids. Cancer Res 40:3956–3965.

Friedenstein, A.J., Luria, E., Molineux, G. (1992) Assays of the haemopoietic microenvironment. In Testa, N.G., Molineux, G. (eds): "Haemopoiesis: A Practical Approach". Oxford, IRL Press at Oxford University Press, pp. 189–200.

Friend, C., Patuleia, M.C., Nelson, J.B. (1966) Antibiotic effect of tylosine on a mycoplasma contaminant in a tissue culture leukemia cell line. Proc Soc Exp Biol Med 121:1009.

Friend, C., Scher, W., Holland, J.G., Sato, T. (1971) Hemoglobin synthesis in murine virus-induced leukemic cells *in vitro.* 2. Stimulation of erythroid differentiation by dimethyl sulfoxide. Proc Natl Acad Sci USA 68:378–382.

Friend, K.K., Dorman, B.P., Kucherlapati, R.S., Ruddle, F.H. (1976) Detection of interspecific translocations in mouse–human hybrids by alkaline Giemsa staining. Exp Cell Res 99:31–36.

Fry, J., Bridges, J.W. (1979) The effect of phenobarbitone on adult rat liver cells and primary cell lines. Toxicol Lett 4:295–301.

Fuller, B.B., Meyskens, F.L. (1981) Endocrine responsiveness in human melanocytes and melanoma cells in culture. J Natl Cancer Inst 66:799–802.

Fusenig, N.E. (1986) Mammalian epidermal cells in culture. In Bereiter-Hahn, J., Matoltsy, A.G., Richards, K.S. (eds): "Biology of the Integument, Vol. 2, Vertebrates." New York, Springer Verlag, pp. 409–442.

Fusenig, N.E. (1992) Cell intercation and epithelial differentiation. In Freshney, R.I. (ed): "Culture of Epithelial Cells." New York, Wiley-Liss, pp. 26–57.

Fusenig, N.E., Worst, P.K.M. (1975) Mouse epidermal cell cultures. II. Isolation, characterization and cultivation of epidermal cells from perinatal mouse skin. Exp Cell Res 93:443–457.

Gabrilove, J.L., Welte, K., Harris, P., Platzer, E., Lu, L., et al. (1986) Pluripoietin alpha: A second hematopoietic colony stimulating factor produced by the human bladder carcinoma cell line 5637. Proc Natl Acad Sci USA 83:2478–2482.

Gallico, G.G. III (1990) Biologic skin substitutes. Clin Plas Surg 17:519–526.

Gartler, S.M. (1967) Genetic markers as tracers in cell culture. Second Bicennial Review Conference on Cell, Tissue and Organ Culture. NCI Monographs, pp. 167–195.

Gartner, S., Kaplan, H.S. (1980) Long-term culture of human bone marrow cells. Proc Natl Acad Sci USA 77:4756–4759.

Gaudernack, T., Leivestad, T., Ugelstad, J., Thorsby, E. (1986) Isolation of pure functionally active CD8+ T cells. Positive selection with monoclonal antibodies directly conjugated to monosized magnetic microspheres. J Immunol Meth 90:179–187.

Gaush, C.R., Hard, W.L., Smith, T.F. (1966) Characterization of an established line of canine kidney cells (MDCK). Proc Soc Exp Biol Med 122:931–933.

Gazdar, A.F., Carney, D.N., Minna, J.D. (1983) The biology of nonsmall cell lung cancer. Sem Oncol 10:3–19.

Geppert, E.F., Williams, M.C., Mason, R.J. (1980) Primary culture of rat alveolar type II cells on floating collagen membranes. Exp Cell Res 128:363–374.

Gey, G.O., Coffman, W.D., Kubicek, M.T. (1952) Tissue culture studies of the proliferative capacity of cervical carcinoma and normal epithelium. Cancer Res 12:364–365.

Ghosh, D., Danielson, K.C., Alston, J.T., Heyner, S. (1991) Functional differential of mouse uterine epithelial cells grown on collagen gels or reconstituted basement membranes. In Vitro Cell Dev Biol 27A:713–719.

Giard, D.J., Aaronson, S.A., Todaro, G.J., Arnstein, P., Kersey, J.H., Dosik, K., Parks, W.P. (1972) *In vitro* cultivation of human tumors: Establishment of cell lines derived from a series of solid tumors. J Natl Cancer Inst 51:1417.

Gignac, S.M., Uphof, C.C., MacLeod, R.A.F., Steube, K., Voges, M., Drexler, H.G. (1992) Treatment of mycoplasma contaminated continuous cell lines with mycoplasma removal agent (MRA). Leukaemia Res 16:815–822.

Gilbert, S.F., Migeon, B.R. (1975) D-Valine as a selective agent for normal human and rodent epithelial cells in culture. Cell 5:11–17.

Gilbert, S.F., Migeon, B.R. (1977) Renal enzymes in kidney cells selected by D-Valine medium. J Cell Physiol 92:161–168.

Gilchrest, B.A., Nemore, R.E., Maciag, T. (1980) Growth of human keratinocytes on fibronectin-coated plates. Cell Biol Int Rep 4:1009–1016.

Gilchrest, B.A., Albert, L.S., Karassik, R.L., Yaar, M. (1985) Substrate influences human epidermal melanocyte attachment and spreading in vitro. In Vitro 21:114.

Gilden, D.H., Wroblewska, Z., Eng, L.F., Rorke, L.B. (1976) Human brain in tissue culture. Part 5. Identification of glial cells by immunofluorescence. J Neurol Sci 29:177–184.

Gillis, S., Watson, J. (1981) Interleukin-2 dependent culture of cytolytic T cell lines. Immunol Rev 54:81-109.
Gimbrone, M.A., Jr., Cotran, R.S., Folkman, J. (1974) Human vascular endothelial cells in culture, growth and DNA synthesis. J Cell Biol 60:673-684.
Giovanella, B.C., Stehlin, J.S., Williams, L.J. (1974) Heterotransplantation of human malignant tumors in "nude" mice. II. Malignant tumors induced by injection of cell cultures derived from human solid tumors. J Natl Cancer Inst 52:921.
Gjerset, R., Yu, A., Haas, M. (1990) Establishment of continuous cultures of T-cell acute lymphoblastic leukemia cells at diagnosis. Cancer Res 50:10-14.
Goldberg, B. (1977) Collagen synthesis as a marker for cell type in mouse 3T3 lines. Cell 11:169-172.
Golde, D.W. (1984) "Hematopoiesis." New York, Churchill Livingstone, p. 361.
Golde, D.W., Cline, M.J. (1973) Cultivation of normal and neoplastic human bone marrow leucocytes in liquid suspension. In "Proceedings of the 7th Leucocyte Culture Conference." New York, Academic Press.
Goldman, B.I., Wurzel, J. (1992) Effects of subcultivation and culture medium on differentiation of human fetal cardiac myocytes. In Vitro Cell Dev Biol 28A:109-119.
Goldwasser, E. (1975) Erythropoietin and the differentiation of red blood cells. Fed Proc 34:2285-2292.
Good, N.E., Winget, G.D., Winter, W., Connolly, T.N., Izawa, S., Singh, R.M.M. (1966) Hydrogen ion buffers and biological research. Biochemistry 5:467-477.
Goodwin, G., Shaper, J.H. Abezoff, M.D., Mendelsohn, G., Baylin, S.B. (1983) Analysis of cell surface proteins delineates a differentiation pathway linking endocrine and nonendocrine human lung cancers. Proc Natl Acad Sci 80:3807-3811.
Gospodarowicz, D. (1974) Localization of fibroblast growth factor and its effect alone and with hydrocortisone on 3T3 cell growth. Nature 249:123-127.
Gospodarowicz, D., Mescher, A.L. (1977) A comparison of the responses of cultured myoblasts and chondrocytes to fibroblast and epidermal growth factors. J Cell Physiol 93:117-128.
Gospodarowicz, D., Moran, J. (1974) Growth factors in mammalian cell cultures. Annu Rev Biochem 45:531-558.
Gospodarowicz, D., Moran, J., Braun, D., Birdwell, C. (1976) Clonal growth of bovine vascular endothelial cells: Fibroblast growth factor as a survival agent. Proc Natl Acad Sci USA 73:4120-4124.
Gospodarowicz, D., Moran, J.S., Braun, D.L. (1977) Control of proliferation of bovine vascular endothelial cells. J Cell Physiol 91:377-386.
Gospodarowicz, D., Greenburg, G., Bialecki, H., Zetter, B.R. (1978a) Factors involved in the modulation of cell proliferation *in vivo* and *in vitro*: The role of fibroblast and epidermal growth factors in the proliferative response of mammalian cells. In Vitro 14:85-118.
Gospodarowicz, D., Greenburg, G., Birdwell, C.R. (1978b) Determination of cell shape by the extracellular matrix and its correlation with the control of cellular growth. Cancer Res 38:4155-4171.
Gospodarowicz, D., Delgado, D., Vlodavsky, I. (1980) Permissive effect of the extracellular matrix on cell proliferation *in vitro*. Proc Natl Acad Sci USA 77:4094-4098.
Gough, N.M., Gough, J., Metcalf, D., Kelso, A., Grail, A., et al. (1984) Molecular cloning of cDNA encoding a murine haemopoietic growth regulator granulocyte/macrophage colony stimulating factor. Nature 309:763-767.
Graham, F.L., Van der Eb, A.J. (1973) A new technique for the assay of infectivity of human adenovirus 5 DNA. Virology 52:456-461.
Granner, D.K., Hayashi, S., Thompson, E.B., Tornkins, G.M. (1968) Stimulation of tyrosine aminotransferase synthesis by dexamethasone phosphate in cell culture. J Mol Biol 35:291-301.
Green, A.E., Athreya, B., Lehr, H.B., Coriell, L.L. (1967) Viability of cell cultures following extended preservation in liquid nitrogen. Proc Soc Exp Biol Med 124:1302-1307.
Green, C.L., Pretlow, T.P., Tucker, K.A., Bradley, E.L., Jr., Cook, W.J., Pitts, A.M., Pretlow II, T.G. (1980) Large-capacity separation of malignant cells and lymphocytes from the Furth mast cell tumor in a reorienting zonal rotor. Cancer Res 40:1791-1796.
Green H. (1977) Terminal differentiation of cultured human epidermal cells. Cell 11:405-416.
Green, H., Kehinde, O. (1974) Sublines of mouse 3T3 cells that accumulate lipid. Cell 1:113-116.
Green, H., Thomas, J. (1978) Pattern formation by cultured human epidermal cells: Development of curved ridges resembling dermatoglyphs. Science 200:1385-1388.
Green, H., Kehinde, O., Thomas. J. (1979) Growth of cultured human epidermal cells into multiple epithelia suitable for grafting. Proc Natl Acad Sci USA 76:5665-5668.
Greenberger, J.S. (1980) Self-renewal of factor-dependent haemopoietic progenitor cell lines derived from long-term bone marrow cultures demonstrate significant mouse strain genotypic variation. J Supramol Struct 13:501-511.
Greenleaf, R.D., Mason, R.J., Williams, M.C. (1979) Isolation of alveolar type II cells by centrifugal elutriation. In Vitro 15:673.
Griffiths, J.B. (1992) Scaling up of animal cell cultures. In Freshney, R.I. (ed): "Animal Cell Culture, a Practical Approach." Oxford, IRL Press at Oxford University Press, pp. 47-93.
Griffiths, J.B., Pirt, G.J. (1967) The uptake of amino acids by mouse cells (Strain LS) during growth in batch culture and chemostat culture: The influence of cell growth rate. Proc R Soc Biol 168:421-438.
Grobstein, C. (1953) Epithelio-mesenchymal specificity in the morphogenesis of mouse submandibular rudiments *in vitro*. J Exp Zool 124:383.
Gross, M., Goldwasser, E. (1971) On the mechanism of erythropoietin-induced differentiation. J Biol Chem 246:2480-2486.
Guguen-Guillouzo, C. (1992) Isolation and culture of animal and human hepatocytes. In Freshney, R.I. (ed): "Culture of Epithelial Cells." New York, Wiley-Liss, pp. 198-223.
Guguen-Guillouzo, C., Guillouzo, A. (eds) (1986) Methods for preparation of adult and fetal hepatocytes. In "Isolated and Culture Hepatocytes." Paris, Les Editions INSERM, John Libbey Eurotext, pp. 1-12.
Guguen-Guillouzo, C., Campion, J.P., Brissot, P., et al. (1982)

High yield preparation of isolated human adult hepatocytes by enzymatic profusion of the liver. Cell Biol Int Rep 6:625–628.

Guguen-Guillouzo, C., Clement, B., Baffet, G., et al. (1983) Maintenance and reversibility of active albumin secretion by adult rat hepatocytes co-cultured with another liver epithelial cell type. Exp Cell Res 143:47–54.

Guilbert, L.I., Iscove, N.N. (1976) Partial replacement of serum by selenite, transferrin, albumin and lecithin in haemopoietic cell cultures. Nature 263:594–595.

Guillouzo, A., Guguen-Guillouzo, C., Bourel, M. (1981) Hepatocytes in culture: Expression of differentiated functions and their application to the study of metaboiism. Triangle (Sandoz J Med Sci) 20:121–128.

Guillouzo, A.M.A. (1989) Methodes in vitro en pharmacotoxiocologie. Les Edition INSERM 170:200.

Gullino, P.M. (1985) Angiogenesis, tumor vascularization, and potential interference with tumor growth. In Mihich, E. (ed): "Biological Responses in Cancer." New York, Plenum.

Gullino, P.M., Knazek, R.A. (1979) Tissue culture on artificial capillaries. In Jakoby, W.B., Pastan, I. (eds): "Methods in Enzymology, Vol. 58. Cell Culture." New York, Academic Press, pp. 178–184.

Gumbiner, B. (1992) Epithelial morphogenesis. Cell 69:385–387.

Guner, M., Freshney, R.I., Morgan, D., Freshney, M.G., Thomas, D.G.T., Graham, D.I. (1977) Effects of dexamethasone and betamethasone on *in vitro*: Cultures from human astrocytoma. Br J Cancer 35:439–47.

Gwatkin, R.B.L. (1973) Pronase. In Kruse, P.F., Jr., Patterson, M.K., Jr. (eds): "Tissue Culture Methods and Applications." New York, Academic Press, pp. 3–5.

Habel, K., Salzman, N.P. (eds) (1969) "Fundamental Techniques in Virology." New York, Academic Press.

Halton, D.M., Peterson, W.D., Jr., Hukku, B. (1983) Cell culture quality control by rapid isoenzymatic characteristics. In Vitro 19:16–24.

Ham, R.G. (1963) An improved nutrient solution for diploid Chinese hamster and human cell lines. Exp Cell Res 29:515.

Ham, R.G. (1965) Clonal growth of mammalian cells in a chemically defined synthetic medium. Proc Natl Acad Sci USA 53:288.

Ham, R.G. (1984) Growth of human fibroblasts in serum-free media. In Barnes, D.W., Sirbasku, D.A., Sato, G.H. (eds): "Cell Culture Methods for Molecular and Cell Biology, Vol. 3." New York, Alan R. Liss, pp. 249–264.

Ham, R.G., McKeehan, W.L. (1978) Development of improved media and culture conditions for clonal growth of normal diploid cells. In Vitro 14:11–22.

Ham, R.G., McKeehan, W.L. (1979) Media and growth requirements. In Jakoby, W.B., Pastan, I.H. (eds): "Methods in Enzymology, Vol. 58, Cell Culture." New York, Academic Press, pp. 44–93.

Hamburger, A.W., Salmon, S.E. (1977) Primary bioassay of human tumor stem cells. Science 197:461–463.

Hamburger, A.W., Salmon, S.E., Kim, M.B., Trent, J.M., Soehnlen, B., Alberts, D.S., Schmidt, H.J. (1978) Direct cloning of human ovarian carcinoma cells in agar. Cancer Res 38:3438-3444.

Hames, B.D., Glover, D.M. (1991) "Oncogenes." Oxford, IRL Press at Oxford University Press.

Hamilton, W.G., Ham, R.G. (1977) Clonal growth of Chinese hamster cell lines in protein-free media. In Vitro 13:537–547.

Hammond, S.L., Ham, R.G., Stampfer, M.R. (1984) Serum free growth of human mammary epithelial cells: Rapid clonal growth in defined medium and extended serial passage with pituitary extract. Proc Natl Acad Sci USA 81:5435–5439.

Hanks, J.H., Wallace, R.E. (1949) Relation of oxygen and temperature in the preservation of tissues by refrigeration. Proc Exp Biol Med 71:196.

Hapel, A.J., Fung, M.C., Johnson, R.M., Young, I.G., Johnson, G., Metcalf, D. (1985) Biologic properties of molecularly cloned and expressed murine Interleukin-3. Blood 65: 1453–1459.

Harrington, W.N., Godman, G.C. (1980) A selective inhibitor of cell proliferation from normal serum. Proc Natl Acad Sci USA Biol Sci 77:423–427.

Harris, H., Hopkinson, D.A. (1976) "Handbook of Enzyme Electrophoresis in Human Genetics." New York, American Elsevier.

Harris, H., Watkins, J.F. (1965) Hybrid cells from mouse and man: Artificial heterokaryons of mammalian cells from different species. Nature 205:640–646.

Harris, L.W., Griffiths, J.B. (1977) Relative effects of cooling and warming rates on mammalian cells during the freeze-thaw cycle. Cryobiology 14:662–669.

Harrison, P.R., Conkie, D., Affara, N., Paul, J. (1974) *In situ* localization of globin messenger RNA formation. I. During mouse foetal liver development. J Cell Biol 63:402–413.

Harrison, R.G. (1907) Observations on the living developing nerve fiber. Proc Soc Exp Biol Med 4:140–143.

Hart, I.R., Fidler, I.J. (1978) An *in vitro* quantitative assay for tumor cell invasion. Cancer Res 38:3218–3224.

Hartley, R.S., Yablonka-Reuveni, Z. (1990) Long-term maintenance of primary myogenic cultures on a reconstituted basement membrane. In Vitro Cell Dev Biol 26:955–961.

Hassbroek, F.J., Neggle, J.C., Fleming, A.L. (1962) High-resolution auto-radiography without loss of water-soluble ions. Nature 195:615–616.

Haudenschild, C.C., Zahniser, D., Folkman, J., Klagsbrun, M. (1976) Human vascular endothelial cells in culture. Exp Cell Res 98:175–183.

Hauschka, S.D., Konigsberg, I.R. (1966) The influence of collagen on the development of muscle clones. Proc Natl Acad Sci USA 55:119–126.

Hay, E.D. (ed.) (1991) "Cell Biology of Extracellular Matrix." New York, Plenum Press.

Hay, R.J. (1979) Identification, separation and culture of mammalian tissue cells. In Reid, E. (ed): "Methodological Surveys in Biochemistry. Vol. 8, Cell Populations." London, Ellis Horwood, pp. 143–160.

Hay, R.J. (1992) Cell line preservation and characterization. In Freshney, R.I. (ed): "Culture of Animal Cells, a Practical Approach." Oxford, IRL Press at Oxford University Press, pp. 95–148.

Hay, R.J., Strehler, B.L. (1967) The limited growth span of cell strains isolated from the chick embryo. Exp Gerontol 2:123.

Hay, R.J., Kern, J., Caputo, J. (1979) Testing for the presence of viruses in cultured cell lines. In "Manual of American Tissue Culture Association," 5:1127–1130.

Hay, R.J., Caputo, J., Macy, M.L. (1992) Establishing or verifying cell line identity. In "ATCC Quality Control Methods for Cell Lines," 2nd ed. Rockville, MD, American Type Culture Collection, pp. 52–66.

Hayashi, I., Sato, G.H. (1976) Replacement of serum by hormones permits growth of cells in a defined medium. Nature 259:132–134.

Hayflick, L., Moorhead, P.S. (1961) The serial cultivation of human diploid cell strains. Exp Cell Res 25:585–621.

Heald, K.A., Hail, C.A., Downing, R. (1991) Isolation of islets of Langerhans from the weanling pig. Diabetes Res 17:7–12.

Heffelfinger, S.C., Hawkins, H.H., Barrish, J., Taylor, L., Darlington, G. (1992) SK HEP-1: A human cell line of endothelial origin. In Vitro Cell Dev Biol 28A:136–142.

Heldin, C.H., Westermark, B., Wasteson, A. (1979) Platelet-derived growth factor: Purification and partial charactenzation. Proc Natl Acad Sci USA 76:3722–3726.

Hemstreet, G.P., Enoch, P.G., Pretlow, T.G. (1980) Tissue disaggregation of human renal cell carcinoma with further isopyknic and isokinetic gradient purification. Cancer Res 40:1043–1049.

Hendrix, M.J., Wood, W.R., Seftor, E.A., Lotan, D., Nakajina, M., Misiorowski, R.L., Seftor, R.E., Stetler-Stevenson, W.G., Bevacqua, S.J., Liotta, L.A., et al. (1990) Retinoic acid inhibition of human melanoma cell invasion through a reconstituted basement membrane and its relation to decreases in the expression of protealytic enzymes and motility factor receptor. Cancer Res 50:4121–4130.

Hendrix, M.J.C., Seftor, E.A., Seftor, E.B., Fidler, I.J. (1987) A simple quantitative assay for studying the invasive potential of high and low human metastatic variants. Cancer Lett 38:137–147.

Hennings, H., Michael, D., Cheng, C., Steinert, P., Holbrook, K., Yuspa, S.H. (1980) Calcium regulation of growth and differentiation of mouse epidermal cells in culture. Cell 19:245–254.

Hering, B.J., Muench, K.P., Schelz, J., Amelang, D., Heitfeld, M., Bretzel, R.G., Bonath, K., Federlin, K. (1990) The evaluation of neutral density separation utilizing Ficoll-sodium diatrizoate and Nycodenz and centrifugal elutriation in the purification of bovine and canine islet preparations. Horm Metab Res (Suppl) 25:57–63.

Herzenberg, L.A., Sweet, R.G., Herzenberg, L.A. (1976) Fluorescence-activated cell sorting. Sci Am 234:108–117.

Heyderman, E., Steele, K., Ormerod, M.G. (1979) A new antigen on the epithelial membrane: Its immunoperoxidase localisation in normal and neoplastic tissue. J Clin Pathol 32:35–39.

Heyworth, C.M., Spooncer, E. (1992) In vitro clonal assays for murine multipotential and lineage restricted myeloid progenitor cells. In Testa, N.G., Molineux, G. (eds): "Haemopoiesis: A Practical Approach." Oxford, IRL Press at Oxford University Press, pp. 37–54.

Higuchi, K. (1977) Cultivation of mammalian cell lines in serum-free chemically defined medium. Methods Cell Biol 14:131.

Hill, B.T. (1983) An overview of clonogenic assays for human tumour biopsies. In Dendy, P.P., Hill, B.T. (eds): "Human Tumour Drug Sensitivity Testing In Vitro." New York, Academic Press, pp. 91–102.

Hilwig, I., Gropp, A. (1972) Staining of constitutive heterochromatin in mammalian chromosomes with a new fluorochrome. Exp Cell Res 75:122–126.

Hince, T.A., Roscoe, J.P. (1980) Differences in pattern and level of plasminogen activator production between a cloned cell line from an ethylnitrosourea-induced glioma and one from normal adult rat brain. J Cell Physiol 104:199–207.

Hirai, Y., Takebe, K., Takashina, M., Kobayashi, S., Takeichi, M. (1992) Epimorphin: A mesenchymal protein essential for epithelial morphogenesis. Cell 69:471–481.

Holbrock, K.A., Hennings, H. (1983) Phenotypic expression of epidermal cells in vitro: A review. J Invest Dermatol 81:11s–24s.

Holden, H.T., Lichter, W., Sigel, M.M. (1973) Quantitative methods for measuring cell growth and death. In Kruse, P.F., Jr., Patterson, M.K., Jr. (eds): "Tissue Culture Methods and Applications." New York, Academic Press, pp. 408–412.

Hollenberg, M.D., Cuatrecasas, P. (1973) Epidermal growth factor: Receptors in human fibroblasts and modulation of action by cholera toxin. Proc Natl Acad Sci USA 70:2964–2968.

Holley, R.W., Armour, R., Baldwin, J.H. (1978) Density-dependent regulation of growth of BSC-1 cells in cell culture: Growth inhibitors formed by the cells. Proc Natl Acad Sci USA 75:1864–1866.

Honn, K.V., Singley, J.A., Chavin, W. (1975) Fetal bovine serum: A multivariate standard (38805). Proc Soc Exp Biol Med 149:344–347.

Hopps, H., Bernheim, B.C., Nisalak, A., Tjio, J.H., Smadel, J.E. (1963) Biologic characteristics of a continuous cell line derived from the African green monkey. J Immunol 91:416–424.

Horibata, K., Harris, A.W. (1970) Mouse myelomas and lymphomas in culture. Exp. Cell Res 60:61–77.

Horita, A., Weber, L.J. (1964) Skin penetrating property of drugs dissolved in dimethylsulfoxide (DMSO) and other vehicles. Life Sci 3:1389–1395.

Horst, J., Kluge, F., Beyreuther, K., Gerok, W. (1975) Gene transfer to human cells: Transducing phage λ*plac* gene expression in Gm gsangliosidosis fibroblasts. Proc Natl Acad Sci USA 72:3531–3535.

Horster, M. (1979) Primary culture of mammalian nephron epithelia: Requirements for cell outgrowth and proliferation from defined explanted nephron segments. Pflugers Arch 382:209–215.

Howard, B.V., Macarak, E.J., Gunson, D., Kefalides, N.A. (1976) Characterization of the collagen synthesized by endothelial cells in culture. Proc Natl Acad Sci USA 73:2361–2364.

Howard, M., Kessler, S., Chused, T., Paul, W.E. (1981) Long term culture of normal mouse B lymphocytes. Proc Natl Acad Sci USA 78:5788–5792.

Howie Report (1978) "Code of Practice for Prevention of Infection in Clinical Laboratories and Post-Mortem Rooms." London, H.M. Stationery Office.

Hoyer, L.W., de los Santos, R.P., Hoyer, J.R. (1973) Antihemophilic factor antigen: Localization in endothelial cells by immunofluorescence microscopy. J Clin Invest 52:2737-2744.

Hozier, J., Applegate, M., Moore, M.M. (1992) In vitro mammalian mutagenesis as a model for genetic lesions in human cancer. Mutation Res 270:201-209.

Hsu, T.C., Benirschke, K. (1967) "Atlas of Mammalian Chromosomes, Vols. 1-4." New York, Springer.

Huang, H.J.S., Yee, J.K., Shew, J.Y., Chen, P.L., Bookstein, R., Friedman, T., Lee, H.P., Lee, W.H. (1988) Suppression of the neoplastic phenotype by replacement of the Rb gene product in human cancer cells. Science 242:1563-1566.

Hull, R.N., Cherry, W.R., Weaver, G.W. (1976) The origin and characteristics of a pig kidney cell strain, LLC-PKI. In Vitro 12:670-677.

Hume, D.A., Weidemann, M.J. (1980) "Mitogenic Lymphocyte Transformation." Amsterdam, Elsevier/North Holland Biomedical Press.

Human Cytogenetic Nomenclature: *see* International System for Human Cytogenetic Nomenclature.

Hynes, R.O. (1973) Alteration of cell-surface proteins by viral transformation and by proteolysis. Proc Natl Acad Sci USA 70:3170-3174.

Hynes, R.O. (1974) Role of cell surface alterations in cell transformation. The importance of proteases and cell surface proteins. Cell 1:147-156.

Hynes, R.O. (1976) Cell surface proteins and malignant transformation. Biochim Biophys Acta 458:73-107.

Hynes, R.O. (1992) Integrins: Versatility, modulation, and signaling in cell adhesion. Cell 69:11-25.

Hyvonen T., Alakuijala L., Andersson L., Khomutov A.R., Khomutov R.M., Eloranta T.O. (1988) 1-Amino-oxy-3-aminopropane reversibly prevents the proliferation of cultured baby hamster kidney cells by interfering with polyamine synthesis. J Biol Chem 263:1138-1144.

Ilsie, A.W., Puck, T.T. (1971) Morphological transformation of Chinese hamster cells by dibutyryl adenosine cycline 3':5'-monophosphate and testosterone. Proc Natl Acad Sci USA 2:358-361.

International System for Human Cytogenetic Nomenclature, An. (1978) Report of the Standing Committee on Human Cytogenetic Nomenclature. Washington, DC, The National Foundation of the March of Dimes.

Imagawa, S., Smith, B.R., Palmer-Crocker, R., Bunn, H.F. (1989) The effect of recombinant erythropoietin on intracellular free calcium in erythropoietin-responsive cells. Blood 73:1452-1457.

Ireland, G.W., Dopping-Hepenstal, P.J., Jordan, P.W., O'Neill, C.H. (1989) Limitation of substratum size alters cytoskeletal organization and behaviour of Swiss 3T3 fibroblasts. Cell Biol Int Rep 13:781-790.

Iscove, N., Melchers, F. (1978) Complete replacement of serum by albumin, transferrin and soybean lipid in cultures of lipopolysaccharide-reactive B lymphocytes. J Exp Med 147:923-933.

Iscove, N.N. (1984) Culture of lymphocytes and hemopoietic cells in serum-free medium. In Barnes, D.W., Sirbasku, D.A., Sato, G.H. (eds): "Methods for Serum-Free Culture of Neuronal and Lymphoid Cells." New York, Alan R. Liss, pp. 169-186.

Iscove, N.N., Guilbert, L.W., Weyman, C. (1980) Complete replacement of serum in primary cultures of erythropoitin-dependent red cell precursors (CFU-E) by albumin, transferrin, iron, unsaturated fatty acid, lecithin and cholesterol. Exp Cell Res 126:121-126.

Itagaki, A., Kimura, G. (1974) TES and HEPES buffers in mammalial cell cultures and viral studies: Problems of carbon dioxide requirements. Exp Cell Res 83:351-360.

Jacobs, K., Shoemaker, C., Rudersdorf, R., Neill, S.D., Kaufman, R.J., et al. (1985) Isolation and characterisation of genomic and cDNA clones of human erythropoietin. Nature 313:806-810.

Jacobs, J.P. (1970) Characteristics of a human diploid cell designated MRC-5. Nature 227:168-170.

Jaffe, E.A., Nachman, R.L., Becker, G.C., Minick, C.R. (1973) Culture of human endothelial cells derived from umbilical veins. J Clin Invest 52:2745-2744.

Jain, D., Ramasubramamanyan, K., Gould, S., Lenny, A., Candelore, M., Tota, M., Strader, C., Alves, K., Cuca, C., Tung, J.S., Hunt, G., Junker, B., Buckland, B.C., Silberklang, M. (1991) In Speir, R.E., Griffiths, J.B., Meignier, B. (eds): "Production of Biologicals from Animal Cells in Culture." Oxford, Butterworth-Heinemann, pp. 345-351.

Jakoby, W.B., Pastan, I.H. (1979) Cell culture. In Colowick, S.P., Kaplan, N.D. (eds): "Methods in Enzymology, Vol. 58." New York, Academic Press.

Jeffreys, A.J., Wilson, V., Thein S.L. (1985) Individual specific "fingerprints of human DNA." Nature 316:76-79.

Jenkins, N. (ed.) (1992) "Growth Factors, a Practical Approach." Oxford, IRL Press at Oxford University Press.

Jenssen, D. (1984) A quantitative test for mutagenicity in V79 Chinese hamster cells. In Kilbey, B.J., Legator, M., Nichols, W., Ramel, C. (eds): "Handbook of Mutagenicity Test Procedures." Amsterdam, Elsevier Science Publishers BV, pp. 269-290.

Jessell, T.M., Melton, D.A. (1992) Diffusible factors in vertebrate embryonic induction. Cell 68:257-270.

Jetten, A.M., Smets, H. (1985) Regulation and differentiation of tracheal epithelial cells by retinoids. In "Retinoids: Differentiation and Disease." Ciba Foundation Symposium. London, Pitman, pp. 61-76.

Jones, T.L., Haskill, J.S. (1973) Polyacrylamide: An improved surface for cloning of primary tumors containing fibroblasts. J Natl Cancer Inst 51:1575-1580.

Jones, T.L., Haskill, J.S. (1976) Use of polyacrylamide for cloning of primary tumors. Methods Cell Biol 14:195.

Joyce, C. (1987) The race to map the human genome. New Scientist 1550 (Mar 5):35-39.

Kahn, P., Shin, S.-L. (1979) Cellular tumorigenicity in nude mice. Test of association among loss of cell-surface fibronectin, anchorage independence, and tumor-forming ability. J Cell Biol 82:1.

Kaltenbach, J.P., Kaltenbach, M.H., Lyons, W.B. (1958) Nigrosin as a dye for differentiating live and dead ascites cells. Exp Cell Res 15:112-117.

Kaminska, B., Kaczmarek, L., Grzelakowska-Sztabert, B. (1990) The regulation of G0-S transition in mouse T lymphocytes by polyamines. Exp Cell Res 191:239-245.

Kao, F.-T., Puck, T.T. (1968) Genetics of somatic mammalian cells. VII. Induction and isolation of nutritional mutants in Chinese hamster cells. Proc Natl Acad Sci USA 60:1275-1281.

Kao, F.T., Chasin, L., Puck, T.T. (1969) Genetics of somatic mammalian cells, X. Complementation analysis of glycine-requiring auxotrophs. Proc Natl Acad Sci USA 64:1284-1291.

Kao, W-Y., Prockop, D.I. (1977) Proline analogue removes fibroblasts from cultured mixed cell populations. Nature 266:63-64.

Karasek, M.A. (1983) Culture of human keratinocytes in liquid medium. J Invest Dermatol 81:21s-28s.

Kawamura, A. (ed) (1969) "Fluorescent Antibody Techniques and Their Application." Tokyo, University of Tokyo Press.

Kawasaki, E.S., Ladner, M.B., Wang, A.M., van Arsdell, J., Waffen, M.K., et. al. (1985) Molecular cloning of a complementary DNA encoding human macrophage-specific colony-stimulating factor (SCF-I). Science 230:291-296.

Kédinger, M., Simon-Assmann, P., Alexandre, E., Haffen, K. (1987) Importance of a fibroblastic support for *in vitro* differentiation of intestinal endodermal cells and for their response to glucocorticoids. Cell Diff 20:171-182.

Kelley, D.S., Becker, J.E., Potter, V.R. (1978) Effect of insulin, dexamethasone, and glucagon on the amino acid transport ability of four rat hepatoma cell lines and rat hepatocytes in culture. Cancer Res 38:4591-4601.

Kempson, S.A., McAteer, J.A., Al-Mahrouq, H.A., Dousa, T.P., Dougherty, G.S., Evan, A.P. (1989) Proximal tubule characteristics of cultured human renal cortex epithelium. J Lab Clin Med 113:285-296.

Kenworthy, P., Dowrick, P., Baillie-Johnson, H,. McCann, B., Tsubouchi, H., Arakaki, N., Daikuhara, Y., Warn, R.M. (1992) The presence of scatter factor in patients with metastatic spread to the pleura. Br J Cancer 66:243-247.

Kern, P.A., Knedler, A., Eckel, R.H. (1983) Isolation and culture of microvascular endothelium from human adipose tissue. J Clin Invest 71:1822-1829.

Kibbey, M.C., Royce, L.S., Dym, M., Baum, B.J., Kleinman, H.K. (1992) Glandular-like morphogenesis of the human submandibular tumour cell line A253 on basement membrane components. Exp Cell Res 198:343-351.

Kim, Y.S., Whitehead, J.S., Perdomo, J. (1979) Glycoproteins of cultured epithelial cells from human colonic adenocarcinoma and fetal intestine. Eur J Cancer 15:725-735.

Kimhi, Y.H., Palfrey, C., Spector, I. (1976) Maturation of neuroblastoma cells in the presence of dimethyl suphoxide. Proc Natl Acad Sci USA 73:462-466.

Kinard, F., De Clercq, L., Billen, B., Amory, B., Hoet, J-J., Remacle, C. (1990) Culture of endocrine pancreatic cells in protein-free chemically defined media. In Vitro Cell Dev Biol 26:1004-1010.

Kindler, V., Thorens, B., de Kossodo, S., Allet, B., Eliason, J.F. (1986) Stimulation of hematopoiesis *in vivo* by recombinant bacterial murine interleukin 3. Proc Natl Acad Sci USA 83:1001-1005.

Kingsbury, A., Gallo, V., Woodhams, P.L., Balazs, R. (1985) Survival, morphology and adhesion properties of cerebellar interneurons cultured in chemically defmed and serum-supplemented medium. Dev Brain Res 17:17-25.

Kinsella, J.L., Grant, D.S., Weeks, B.S., Kleinman, H.K. (1992) Protein kinase C regulates endothelial cell tube formation on basement membrane matrix, Matrigel. Exp Cell Res 199:56-62.

Kirkland, S.C., Bailey, I.G. (1986) Establishment and characterisation of six human colorectal adenocarcinoma cell lines. Br J Cancer 53:779-785.

Kissane, J.M., Robbins, E. (1958) The fluorometric measurement of deoxyribonucleic acid in animal tissues with specific reference to the central nervous system. J Biol Chem 233:184-188.

Kitos, P.A., Sinclair, R., Waymouth, C. (1962) Glutamine metabolism by animal cells growing in a synthetic medium. Exp Cell Res 27:307-316.

Klagsbrun, M., Baird, A. (1991) A dual receptor system is required for basic fibroblast growth factor activity. Cell 67:229-231.

Klann, R.C., Marchok, A.C. (1982) Effects of retinoic acid on cell proliferation and cell differentiation in a rat tracheal epithelial cell line. Cell Tissue Kinet 15:473-482.

Klein, B., Pastink, A., Odijk, H., Westerveld, A., van der Eb, A.J. (1990) Transformation and immortalization of diploid xeroderma pigmentosum fibroblasts. Exp Cell Res 191:256-262.

Kleinman, H.K., McGoodwin, E.B., Rennard, S.I., Martin, G.R. (1981) Preparation of collagen substrates for cell attachment: Effect of collagen concentration and phosphate buffer. Anal Biochem 94:308.

Klevjer-Anderson, P., Buehring, G.C. (1980) Effect of hormones on growth rates of malignant and nonmalignant human mammary epithelia in cell culture. In Vitro 16:491-501.

Klöppinger, M., Fertig, G., Fraune, E., Miltenburger, H.G. (1991) High cell density perfusion culture of insect cells for production of baculovirus and recombinant protein. In Speir, R.E., Griffiths, J.B., Meignier, B. (eds): "Production of Biologicals from Animal Cells in Culture." Oxford, Butterworth-Heinemann, pp. 470-474.

Knazek, R.A., Gullino, P., Kohler, P.O., Dedrick, R. (1972) Cell culture on artificial capillaries. An approach to tissue growth *in vitro*. Science 178:65-67.

Knazek, R.A., Kohler, P.O., Gullino, P.M. (1974) Hormone production by cells grown *in vitro* on artificial capillaries. Exp Cell Res 84:251.

Knedler, A., Ham, R.G. (1987) Optimized medium for clonal growth of human microvascular endothelial cells with minimal serum. In Vitro 23(7):481-491.

Kohler, G., Milstein, C. (1975) Continuous cultures of fused cells secreting antibody of predefined specificity. Nature 256:495-497.

Kohlhepp, E.A., Condon, M.E., Hamburger, A.W. (1987) Recombinant human interferon-α enhancement of retinoic acid induced differentiation of HL-60 cells. Exp Hematol 15:414-418.

Konigsberg, I.R. (1979) Skeletal myoblasts in culture. In Jakoby, W.B., Pastan, I.H. (eds): "Methods in Enzymology, Vol. 57, Cell Culture." New York, Academic Press, pp. 511-527.

Koren, H.S., Handwerger, B.S., Wunderlich, J.R. (1975) Identification of macrophage-like characteristics in a murine tumor cell line. J Immunol 114:894-897.

Kosher, R.A., Church, R.L. (1975) Stimulation of *in vitro*

somite chondrogenesis by procollagen and collagen. Nature 258:327–330.

Kralovanszky, J., Harrington, F., Greenwell, A. (1990) Isolation of viable intestinal epithelial cells and their use for *in vitro* toxicity studies. In Vivo 4:201–204.

Kreisberg, J.L., Sachs, G., Pretlow, T.G.E., McGuire, R.A. (1977) Separation of proximal tubule cells from suspensions of rat kidney cell by free-flow electrophoresis. J Cell Physiol 93:169–172.

Kreth, W., Herzenberg, L.A. (1974) Fluorescence-activated cell soning of human T and B lymphocytes. Cell Immunol 12:396–406.

Krog, H.H. (1976) Identification of inbred strains of mice, *Mus musculus.* I. Genetic control of mice using starch gel electrophoresis. Biochem Genet 14:319–326.

Kruse, P., Patterson, M.K. (1973) "Tissue Culture Techniques and Applications." New York, Academic Press.

Kruse, P.F., Jr., Miedema, E. (1965) Production and characterization of multiple-layered populations of animal cells. J Cell Biol 27:273.

Kruse, P.F., Jr., Keen, L.N., Whittle, W.L. (1970) Some distinctive characteristics of high density perfusion cultures of diverse cell types. In Vitro 6:75–78.

Kuchler, R.J. (ed) (1974) "Animal Cell Culture and Virology." Stroudsburg, PA, Dowden, Hutchinson & Ross.

Kuchler, R.J. (1977) "Biochemical Methods in Cell Culture and Virology." New York, Academic Press.

Kuriharcuch, W., Green, H. (1978) Adipose conversion of 3T3 cells depends on a serum factor. Proc Natl Acad Sci USA 75:6107–6110.

Kurtz, J.W., Wells, W.W. (1979) Automated fluorometric analysis of DNA, protein, and enzyme activities: Application of methods in cell culture. Anal Biochem 94:166.

Labarca, C., Paigen, K. (1980) A simple, rapid, and sensitive DNA assay procedure. Anal Biochem 102:344–352.

Laferte S., Loh, L.C. (1992) Characterization of a family of structurally related glycoproteins expressing beta 1-6-branched asparagine-linked oligosaccharides in human colon carcinoma cells. Biochem J 283:193–201.

Lammers, R., Gray, A., Schlessinger, J., Ullrich, A. (1989) Differential signalling potential of insulin- and IGF-1-receptor cytoplasmic domains. EMBO J 8:1369–1375.

Lan, S., Smith, H.S., Stampfer, M.R. (1981) Clonal growth of normal and malignant human breast epithelia. J Surg Oncol 18:317–322.

Lane, E.B. (1982) Monoclonal antibodies provide specific intramolecular markers for the study of tonofilament organisation. J Cell Biol 92:665–673.

Lange, W., Brugger, W., Rosenthal, F.M., Kanz, L., Lindemann, A. (1991) The role of cytokines in oncology. Int J Cell Clon 9:252–273.

Langer, P.R., Waldrop, A.A., Ward, D.C. (1981) Enzymatic synthesis of biotin-labelled polynucleotides: Novel nucleic acid affinity probes. Proc Natl Acad Sci USA 78:6633.

Langer-Safer, P.R., Levine, M., Ward, D.C. (1982) Immunological methods for mapping genes on *Drosophila* polytene chromosomes. Proc Natl Acad Sci USA 79:4381.

Lasfargues, E.Y. (1973) Human mammary tumors. In Kruse, P., Patterson, M.K. (eds): "Tissue Culture Methods and Applications." New York, Academic Press, pp. 45–50.

Lasnitzki, I. (1992) Organ culture. In Freshney, R.I. (ed): "Animal Cell Culture, a Practical Approach." Oxford, IRL Press at Oxford University Press, pp. 213–261.

Lasnitzki, I., Mizuno, T. (1979) Role of mesenchyme in the induction of the rat prostate gland by androgens in organ culture. J Endocrinol 82:171.

Laug, W.E., Tokes, Z.A., Benedict, W.F., Sorgente, N. (1980) Anchorage independent growth and plasminogen activator production by bovine endothelial cells. J Cell Biol 84:281–293.

Law, L.W., Dunn, T.B., Boyle, P.J., Miller, J.H. (1949) Observations on the effect of a folic acid antagonist on transplantable lymphoid leukemia in mice. J Nat Cancer Inst 10:179–192.

Le Roith, D., Raizada, M.K. (eds) (1989) "Molecular and Cellular Biology of Insulin-like Growth Factors and Their Receptors." New York, Plenum.

Lebeau, M.M., Rowley, J.D. (1984) Heritable fragile sites in cancer. Nature 308:607–608.

Lechner, J.F., LaVeck, M.A. (1985) A serum free method for culturing normal human bronchial epithelial cells at clonal density. J Tissue Cult Methods 9:43–48.

Lechner, J.F., Haugen, A., Autrup, H., McClendon, I.A., Trump, B.F., Harris, C.C. (1981) Clonal growth of epithelial cells from normal adult human bronchus. Cancer Res 41:2294–2304.

Leder, A., Leder, P. (1975) Butyric acid, a potent inducer of erythroid differentiation in cultured erythroleukemic cells. Cell 5:319–322.

Legrand, A., Greenspan, P., Nagpal, M.L., Nachtigal, S.A., Nachtigal, M. (1991) Characterization of human vascular smooth muscle cells transformed by the early genetic region of SV40 virus. Am J Pathol 139:629–640.

Leibo, S.P., Mazur, P. (1971) The role of cooling rates in low-temperature preservation. Cryobiology 8:447–452.

Leibovitz, A. (1963) The growth and maintenance of tissue cell cultures in free gas exchange with the atmosphere. Am J Hyg 78:173–183.

Leighton, J. (1951) A sponge matrix method for tissue culture. Formation or organized aggregates of cells *in vitro.* J Natl Cancer Inst 12:545–561.

Leighton, J. (1991) Radial histophysiologic gradient culture chamber rationale and preparation. In Vitro Cell Dev Biol 27A:786–790.

Leighton, J., Mark, R., Rush, G. (1968) Patterns of three-dimensional growth in collagen coated cellulose sponge: Carcinomas and embryonic tissues. Cancer Res 28:286–296.

Lesser, B., Brent, T.P. (1970) Cold storage as a method for accumulating mitotic HeLa cells without impairing subsequent synchronous growth. Exp Cell Res 62:470–473.

Lever, J. (1986) Expression of differentiated functions in kidney epithelial cell lines. Min Elec Metab 12:14–19.

Levi-Montalcini, R. (1964) Growth control of nerve cells by a protein factor and its antiserum. Science 143:105–110.

Levi-Montalcini, R.C.P. (1979) The nerve-growth factor. Sci Am 240:68.

Levine, E.M., Becker, B.G. (1977) Biochemical methods for detecting mycoplasma contamination. In McGarrity, G.T., Murphy, D.G., Nichols, W.W. (eds): "Mycoplasma Infection of Cell Cultures." New York, Plenum Press, pp. 87–104.

Ley, K.D., Tobey, R.A. (1970) Regulation of initiation of DNA synthesis in Chinese hamster cells. II. Induction of DNA synthesis and cell division by isoleucine and glutamine in G_1-arrested cells in suspension culture. J Cell Biol 47:453–459.

Li, A.P., Carver, J.H., Choy, W.N., Gupta, R.S., Loveday, K.S., O'Neill, J.P., Riddle, J.C. Stankowski, L.F., Yang, L.L. (1987) A guide for the performance of the Chinese hamster ovary cell/hypoxanthine guanine phosphoribosyltransferase gene mutation assay. Mutation Res 189:135–141.

Li, A.P., Roque, M.M., Beck, D.J., Kaminski, D.L. (1992) Isolation and culturing of hepatocytes from human livers. J Tissue Cult Methods 14:139–146.

Li, Y., Joseph, A., Bhargava, M.M., Rosen, E.M., Nakamura, T., Golberg, I. (1992) Effect of scatter factor and hepatocyte growth factor on motility and morphology of MDCK cells. In Vitro Cell Dev Biol 28A:364–368.

Liber, H.L., Thilly, W.G. (1982) Mutation assays at the thymidine kinase locus in diploid human lymphoblasts. Mutation Res 94:467–485.

Lieber, M., Mazzetta, J., Nelson-Rees, W., Kaplan, M., Todaro, G. (1975) establishment of a continuous tumor-cell line (PANC-1) from a human carcinoma of the exocrine pancreas. Int J Cancer 15:741–747.

Liebermann, D., Sachs, L. (1978) Nuclear control of neurite induction in neuroblastoma cells. Exp Cell Res 113:383–390.

Lillie, I.H., MacCallum, D.K., Jepsen, A. (1980) Fine structure of subcultivated stratified squamous epithelium grown on collagen rafts. Exp Cell Res 125:153–165.

Lim, R., Mitsunobu, K. (1975) Partial purification of morphological transforming factor from pig brain. Biochem Biophys Acta 400:200–207.

Limat, A., Hunziker, T., Boillat, C., Bayreuther, K.J., Noser, F. (1989) Post-mitotic dermal fibroblasts efficiently support the growth of human follicular keratinocytes. J Invest Dermatol 92:758–762.

Lin, C.C., Uchida, I.A. (1973) Fluorescent banding of chromosomes (Q-bands). In Kruse, P.F., Patterson, M.K. (eds): "Tissue Culture Methods and Applications." New York, Academic Press, pp. 778–781.

Lin, M.A., Latt, S.A., Davidson, R.L. (1974) Identification of human and mouse chromosomes in human-mouse hybrids by centromere fluorescence. Exp Cell Res 87:429–433.

Lindgren, A., Westermark, B., Ponten, J. (1975) Serum stimulation of stationary human glia and glioma cells in culture. Exp Cell Res 95:311–319.

Lindsay, R.M. (1979) Adult rat brain astrocytes support survival of both NGF-dependent and NGF-insensitive neurones. Nature 282:80.

Linser, P., Moscona, A.A. (1980) Induction of glutamine synthetase in embryonic neural retina-localization in Muller fibers and dependence on cell interaction. Proc Natl Acad Sci USA 76:6476–6481.

Liotta, L. (1987) The role of cellular proteases and their inhibitors in invasion and metastasis. Introductionary overview. Cancer Metastasis Rev 9:285–287.

Lippincott-Schwartz, J., et al. (1990) Forskolin inhibits and reverses the effects of Brefeldin A on Golgi morphology by a cAMP-independent mechanism. J Cell Biol 112:567.

Littauer, U.Z., Giovanni, M.Y., Glick, M.C. (1979) Differentiation of human neuroblastoma cells in culture. Biochem Biophys Res Commun 88:933–939.

Littlefield, J.W. (1964a) Selection of hybrids from matings of fibroblasts in vitro and their presumed recombinants. Science 145:709–710.

Littlefield, J.W. (1964b) Three degrees of guanylic acid pyrophosphorylase deficiency in mouse fibroblasts. Nature 203:1142–1144.

Litwin, J. (1973) Titanium disks. In Kruse, P.F., Patterson, M.K. (eds): "Tissue Culture Methods and Applications." New York, Academic Press, pp. 383–387.

Lloyd, K.O., Travassos, L.R., Takahashi, T., Old, L.J. (1979) Cell surface glycoproteins of human tumor cell lines: Unusual characteristics of malignant melanoma. J Natl Cancer Inst 63:623.

Lord, B.I., Molineux, G., Testa, N.G., Kelly, M., Spooncer, E., Dexter, T.M. (1986) The kinetic response of haemopoietic precursor cells, *in vivo,* to highly purified recombinant interleukin-3. Lymphokine Res 5:97–104.

Lotan, R., Lotan, D. (1980) Simulation of melanogenesis in a human melanoma cell line by retinoids. Cancer Res 40:33–45.

Lovelock, J.E., Bishop, M.W.H. (1959) Prevention of freezing damage to living cells by dimethyl sulphoxide. Nature 183:1394–1395.

Lowry, O.N., Rosebrough, N.J., Farr, A.L., Randall, R.J. (1951) Protein measurement with the folin phenol reagent. J Biol Chem 193:265–275.

Luikart, S.D., Maniglia, C.A., Furcht, L.T., McCarthy, J.B., Oegama, T.R. (1990) A heparan sulphate-containing fraction of bone marrow stroma induces maturation of HL-60 cell *in vitro.* Cancer Res 50:3781–3785.

Lutz, M.P., Gaedicke, G., Hartmann, W. (1992) Large-scale cell separation by centrifugal elutriation. Anal Biochem 200:376–380.

MacDonald, C.M., Freshney, R.I., Hart, E., Graham, D.I. (1985) Selective control of human glioma cell proliferation by specific cell interaction. Exp Cell Biol 53:130–137.

Maciag, T., Cerondolo, J., Ilsley, S., Kelley, P.R., Forand, R. (1979) Endothelial cell growth factor from bovine hypothalamus-identification and partial characterization. Proc Natl Acad Sci USA 76:5674–5678.

Macieira-Coelho, A. (1973) Cell cycle analysis. A. Mammalian cells. In Kruse, P.F., Patterson, M.K. (eds): "Tissue Culture Methods and Applications." New York, Academic Press, pp. 412–422.

Macklis, I.D., Sidman, R.L., Shine, H.D. (1985) Cross-linked collagen surface for cell culture that is stable, uniform, and optically superior to conventional surfaces. In Vitro 21:189–194.

Macpherson, I. (1973) Soft agar technique. In Kruse, P.F., Patterson, M.K. (eds): "Tissue Culture Methods and Applications." New York, Academic Press, pp. 276–280.

Macpherson, I., Bryden, A. (1971) Mitomycin C treated cells as feeders. Exp Cell Res 69:240–241.

Macpherson, I., Montagnier, L. (1964) Agar suspension culture for the selective assay of cells transformed by polyoma virus. Virology 23:291–294.

Macpherson, I., Stoker, M. (1962) Polyoma transformation of hamster cell clones–an investigation of genetic factors affecting cell competence. Virology 16:147.

Macy, M. (1978) Identification of cell line species by isoenzyme analysis. Manual Am Tissue Cult Assoc 4:833–836.

Madsen, O.D., Nielsen, J.H., Michelsen, B., Westermark, P., Betsholtz, C., Nishi, M., Steiner, D.F. (1991) Islet amyloid polypeptide and insulin expression are controlled differently in primary and transformed islet cells. Mol Endocrinol 5:143–148.

Mahdavi, V., Hynes, R.O. (1979) Proteolytic enzymes in normal and transformed cells. Biochim Biophys Acta 583:167–178.

Malan-Shibley, L., Iype, P.T. (1981) The influence of culture conditions on cell morphology and tyrosine aminotransferase levels in rat liver epithelial cell lines. Exp Cell Res 131:363–371.

Maltese, W.A., Volpe, I.J. (1979) Induction of an oligodendroglial enzyme in C-6 glioma cells maintained at high density or in serum-free medium. J Cell Physiol 101:459–470.

Management of Health and Safety at Work Regulations (1992) Health and Safety Executive, Broad Lane, Sheffield S3 7HQ, England.

Maniatis, T., Hardison, R.C., Lacy, E., Lauer, J., O'Connell, C., Quon, D., Sim, G.K., Efstradiadis, A. (1978) The isolation of structural genes from libraries of eukaryotic DNA. Cell 15:687–701.

Maramorosch, K. (1976) "Invertebrate Tissue Culture." New York, Academic Press.

Marcus, M., Lavi, U., Nattenberg, A., Ruttem, S., Markowitz, O. (1980) Selective killing of mycoplasmas from contaminated cells in cell cultures. Nature 285:659–660.

Mardh, P.H. (1975) Elimination of mycoplasmas from cell cultures with sodium polyanethol sulphonate. Nature 254:515–516.

Mareel, M., Kint, J., Meyvisch, C. (1979) Methods of study of the invasion of malignant C3H-mouse fibroblasts into embryonic chick heart *in vitro*. Virchows Arch B Cell Pathol 30:95–111.

Mareel, M.M., Bruynell, E., Storme, G. (1980) Attachment of mouse fibrosarcoma cells to precultured fragments of embryonic chick heart. Virchows Arch B Cell Pathol 34:85–97.

Mark, J. (1971) Chromosomal characteristics of neurogenic tumours in adults. Hereditas 68:61–100.

Markus, G., Takita, H., Camiolo, S.M., Corsanti, J., Evers, J.L., Hobika, J.H. (1980) Content and characterization of plasminogen activators in human lung tumours and normal lung tissue. Cancer Res 40:841–848.

Marshall, C.J. (1991) Tumor suppressor genes. Cell 64:313–326.

Martin, G.R. (1975) Teratocarcinomas as a model system for the study of embryogenesis and neoplasia. Cell 5:229–243.

Martin, G.R. (1978) Advantages and limitations of teratocarcinoma stem cells as models of development. In Johnson, M.H. (ed): "Development in Mammals, Vol. 3." Amsterdam, North-Holland Publishing, p. 225.

Martin, G.R., Evans, M.J. (1974) The morphology and growth of a pluripotent teratocarcinoma cell line and its derivatives in tissue culture. Cell 2:163–172.

Massague, J. (1990) The transforming growth factor-β family. Annu Rev Cell Biol 6:597–641.

Massague, J., et al. (1992) Transforming growth factor-β. Cancer Surveys 12:81–103.

Masui, T., Wakefield, L.M., Lechner, J.F., LaVeck, M.A., Sporn, M.B., Harris, C.C. (1986a) Type beta transforming growth factor is the primary differentiation-inducing serum factor for normal human bronchial epithelial cells. Proc Natl Acad Sci USA 83(8):2438–2442.

Masui, T., Lechner, J.F., Yoakum, G.H., Willey, J.C., Harris, C.C. (1986b) Growth and differentiation of norrnal and transformed human bronchial epithelial cells. J Cell Physiol (Suppl) 4:73–81.

Mather, J. (1979) Testicular cells in defined medium. In Jakoby, W.B., Pastan, I.H. (eds): "Methods in Enzymology, Vol. 57, Cell Culture." New York, Academic Press, p. 103.

Mather, J.P., Sato, G.H. (1979a) The growth of mouse melanoma cells in hormone supplemented, serum-free medium. Exp Cell Res 120:191.

Mather, J.P., Sato, G.H. (1979b) The use of hormone supplemented serum free media in primary cultures. Exp Cell Res 124:215.

Matsamura, T., Nitta, K., Yoshikawa, M., Takaoka, T., Katsuta, H. (1975) Action of bacterial protease on the dispersion of mammalian cells in tissue culture. Jpn J Exp Med 45:383–392.

Matsui, A., Zsebo K., Hogan, B.L.M. (1992) Derivation of pluripotential embryonic stem cells from murine primordial germ cells in culture. Cell 70:841–847.

Maurer, H.R. (1988) Cell culture techniques for testing of biologically active peptides and drugs: Clonogenic assays using agar-containing glass capillaries. In "Modern Methods in Protein Chemistry," 3:335–357. Berlin, Walter de Gruyter & Co.

Maurer, R. (1992) Towards chemically defined serum-free media for mammalian cell culture. In Freshney, R.I. (ed): "Animal Cell Culture, a Practical Approach," 2nd ed. Oxford, IRL Press at Oxford University Press, pp. 15–46.

Mazur, P., Leibo, S.P., Farrant, J., Chu, E.H.Y., Hanna, M.G., Jr., Smith, C.H. (1970) Interactions of cooling rate, warming rate and protective additive on the survival of frozen mammalian cells. In Wolstenholme, G.E.W., O'Conor, M. (eds): "The Frozen Cell." CIBA Foundation Symposium. London, J.A. Churchill, pp. 69–85.

McAteer, J.A., Kempson, S.A., Evan, A.P. (1991) Culture of human renal cortex epithelial cells. J Tissue Cult Methods 13:143–148.

McCall, E., Povey, J., Dumonde, D.C. (1981) The culture of vascular endothelial cells on microporous membranes. Thromb Res 24:417–431.

McCool, D., Miller, R.J., Painter, R.H., Bruch, W.R. (1970) Erythropoietin sensitivity of rat bone marrow cells separated by velocity sedimentation. Cell Tissue Kinet 3:55–66.

McGarrity, G.J. (1982) Detection of mycoplasmic infection of cell cultures. In Maramorosch, K. (ed): "Advances in Cell Culture, Vol. 2." New York, Academic Press, pp. 99–131.

McGowan, J.A. (1986) Hepatocyte proliferation in culture. In

Guillouzo, A., Guguen-Guillouzo, C. (eds): "Isolated and Cultured Hepatocytes." Paris, Les Editions Inserm, John Libbey Eurotext, pp. 13–38.

McGregor, D.B., Edwards, I., Riach, C.J., Cattenach, P., Martin, R., Mitchell, A., Caspary, W.J. (1988) Studies of an S9 based metabolic activation system used in the mouse lymphoma L51768Y cell mutation assay. Mutagenesis 3:485–490.

McKay, I., Taylor-Papadimitriou, J. (1981) Junctional communication pattern of cells cultured from human milk. Exp Cell Res 134:465–470.

McKeehan, W.L. (1977) The effect of temperature during trypsin treatment on viability and multiplication potential of single normal human and chicken fibroblasts. Cell Biol Int Rep 1:335–343.

McKeehan, W.L., Ham, R.G. (1976) Stimulation of clonal growth of normal fibroblasts with substrata coated with basic polymers. J Cell Biol 71:727–734.

McKeehan, W.L., McKeehan, K.A. (1979) Oxocarboxylic acids, pyridine nucleotide-linked oxidoreductases and serum factors in regulation of cell proliferation. J Cell Physiol 101:9–16.

McKeehan, W.L., Hamilton, W.G., Ham, R.G. (1976) Selenium is an essential trace nutrient for growth of WI-38 diploid human fibroblasts. Proc Natl Acad Sci USA 73:2023–2027.

McKeehan, W.L., McKeehan, K.A., Hammond, S.L., Ham, R.G. (1977) Improved medium for clonal growth of human diploid cells at low concentrations of serum protein. In Vitro 13:399–416.

McKeehan, W.L., Adams, P.S., Rosser, M.P. (1982) Modified nutrient medium MCDB 151 (WJAC401), defined growth factors, cholera toxin, pituitary factors, and horse serum support epithelial cell and suppress fibroblast proliferation in primary cultures of rat ventral prostate cells. In Vitro 18:87–91.

McKeehan, W.L., Adams, P.S., Rosser, M.P. (1984) Direct mitogenic effects of insulin, epidermal growth factor, cholera toxin, unknown pituitary factors and possibly prolactin, but not androgen, on normal rat prostate epithelial cells in serum-free primary cell culture. Cancer Res 44:1998–2010.

McLean, J.S., Frame, M.C., Freshney, R.I., Vaughan, P.F.T., Mackie, A.E. (1986) Phenotypic modification of human glioma and non-small cell lung carcinoma by glucocorticoids and other agents. Anticancer Res 6:1101–1106.

Meera Khan, P. (1971) Enzyme electrophoresis on cellulose acetate gel: Zymogram patterns in man-mouse and man-Chinese hamster somatic cell hybrids. Arch Biochem Biophys 145:470–483.

Mege, R.M., Matsuzaki, F., Gallin, W.J., Goldberg, J.I., Cunningham, B.A., Edelman, G.M. (1989) Construction of epithelioid sheets by transfection of mouse sarcoma cells with cDNAs for chicken cell adhesion molecules. Proc Natl Acad Sci USA 85:7274–7278.

Meier, S., Hay, E.D. (1974) Control of corneal differentiation by extracellular materials. Collagen as a promoter and stabilizer of epithelial stroma production. Dev Biol 38:249–270.

Meier, S., Hay, E.D. (1975) Stimulation of corneal differentiation by interaction between cell surface and extracellular matrix. J Cell Biol 66:275–291.

Meistrich, M.L., Meyn, R.E., Barlogie, B. (1977a) Synchronization of mouse L-P59 cells by centrifugal elutriation separation. Exp Cell Res 105:169.

Meistrich, M.L., Gordina, D.J., Meyn, R.E., Barlogie, B. (1977b) Separation of cells from mouse solid tumors by centrifugal elutriation. Cancer Res 37:4291–4296.

Melera, P.W., Wolgemuth, D., Biedler, J.L., Hession, C. (1980) Antifolate-resistant Chinese hamster cells. Evidence from independently derived sublines for the overproduction of two dihydrofolate reductases encoded by different mRNAs. J Biol Chem 255:319–322.

Merril, C.R. (1971) Bacterial virus gene expression in human cells. Nature 233:398–400.

Messer, A. (1977) The maintenance and identification of mouse cerebellar granule cells in monolayer culture. Brain Res 130:1–12.

Messing, E.M., Fahey, I.L., deKernion, I.B., Bhuta, S.M., Bubbers, I.E. (1982) Serum-free medium for the in vitro growth of normal and malignant urinary bladder epithelial cells. Cancer Res 42:2392–2397.

Metcalf, D. (1985a) Haemopoietic colonies *in vitro.* In "Recent Results in Cancer Research 61." Berlin, Springer-Verlag, p. 227.

Metcalf, D. (1985b) The granulocyte-macrophage colony stimulating factors. Science 229:16–22.

Metcalf, D. (1986) The molecular biology and functions of the granulocyte macrophage colony-stimulating factors. Blood 67:257–267.

Metcalf, D., Begley, C.G., Nicola, N.A., Lopez, A.F., Williamson, D.J. (1986) Effects of purified bacterially synthesized murine multi-CSF (IL-3) on hematopoiesis in normal adult mice. Blood 68:46–57.

Meyskens, F.L., Fuller, B.B. (1980) Characterization of the effects of different retinoids on the growth and differentiation of a human melanoma cell line and selected subclones. Cancer Res 40:2194–2196.

Meyskens, F.L., Berglund, E.B., Saxe, D.F., Fuller, B.B., Pacelli, L.Z., Hall, J.D., Ray, C.G. (1980) Biological and biochemical properties of a human uveal melanocyte-derived cell line. In Vitro 16:775–780.

Michalopoulos, G., Pitot, H.C. (1975) Primary culture of parenchymal liver cells on collagen membranes. Fed Proc 34:826.

Michler-Stuke, A., Bottenstein, J. (1982) Proliferation of glial-derived cells in defined media. J Neurosci Res 7:215–228.

Miller, D.R., Hamby, K.M., Allison, D.P., Fischer, S.M., Slaga, T.J. (1980) The maintenance of a differentiated state in cultured mouse epidermal cells. Exp Cell Res 129:63–71.

Miller, G.G., Walker, G.W.R., Giblack, R.E. (1972) A rapid method to determine the mammalian cell cycle. Exp Cell Res 72:533–538.

Miller, R.G., Phillips, R.A. (1969) Separation of cells by velocity sedimentation. J Cell Physiol 73:191–201.

Milo, G.E., Ackerman, G.A., Noyes, I. (1980) Growth and ultrastructural characterization of proliferating human keratinocytes in *vitro* without added extrinsic factors. In Vitro 16:20–30.

Milstein, C., Galfre, G., Secher, D.S., Springer, T. (1979) Mini review, monoclonal antibodies and cells surface antigens. Cell Biol Int Rep 3:1–16.

Minna, I., Gilman, A. (1973) Genetic analysis of the nervous system using somatic cell hybrids. In Davidson, R.L., de la Cruz, F.F. (eds): "Somatic Cell Hybridization." New York, Raven Press, pp. 191–196.

Minna, I., Glazer, D., Nirenberg, M. (1972) Genetic dissection of neural properties using somatic cells hybrids. Nature New Biol 235:225–231.

Minna, I.D., Carney, D.N., Cuttitta, F., Gazdar, A.F. (1983) The biology of lung cancer. In Chabner, B. (ed): "Rational Basis for Chemotherapy." New York, Alan R. Liss.

Mitaka, T., Sattler, C.A., Sattler, G.L., Sargent, L.M., Pitot, H.C. (1991) Multiple cell cycles occur in rat hepatocytes cultured in the presence of nicotinamide and epidermal growth factor. Hepatology 13:21–30.

Moll, R., Franke, W.W., Schiller, D.L. (1982) The catalog of human cytokeratins: Patterns of expression in normal epithelia, tumours and cultured cells. Cell 31:11–24.

Montagnier, L. (1968) Corrélation entre la transformation des cellule BHK21 et leur résistance aux polysaccharides acides en milieu gélifié. CR Acad Sci D 267:921–924.

Montes de Oca, F., Macy, M.L., Shannon, J.E. (1969) Isoenzyme characterization of animal cell cultures. Proc Soc Exp Biol Med 132:462–469.

Montesano, R., Matsumonto, K., Nakamura, T., Orci, L. (1991) Identification of a fibroblast-derived epithelial morphogen as hepatocyte growth factor. Cell 67:901–908.

Moore, A.E., Sabachewsky, L., Toolan, H.W. (1955) Cancer Res 15:598.

Moore, G.E., Gerner, R.E., Franklin, H.A. (1967) Culture of normal human leukocytes. J Am Med Assoc 199:519–524.

Moore, M.M., Clive, D., Hozier, J., Howard, B.E., Batson, A.G., Turner, N.T., Sawyer, J. (1985) Analysis of trifluorothymidine-resistant (TFT^r) mutants of L5178Y mouse lymphoma cells. Mutation Res 151:147–159.

Morgan, J.E., Moore, S.E., Walsh, F.S., Partridge, T.A. (1992) Formation of skeletal muscle *in vivo* from the mouse C2 cell line. J Cell Sci 102:779–787.

Morgan, J.G., Morton, H.J., Parker, R.C. (1950) Nutrition of animal cells in tissue culture. I. Initial studies on a synthetic medium. Proc Soc Exp Biol Med 73:1.

Morton, H.J. (1970) A survey of commercially available tissue culture media. In Vitro 6:89–108.

Moscona, A.A. (1952) Cell suspension from organ rudiments of chick embryos. Exp Cell Res 3:535.

Moscona, A.A., Piddington, R. (1966) Stimulation by hydrocortisone of premature changes in the developmental pattern of glutamine synthetase in embryonic retina. Biochim Biophys Acta 121:409–411.

Mosmann, T. (1983). Rapid colorimetric assay for cellular growth and survival: Application to proliferation and cytotoxicity assays. J Immunol Methods 65:55–63.

Moss, P.S., Strohman, R.C. (1976) Myosin synthesis by fusion-arrested chick embryo myoblasts in cell culture. Dev Biol 48:431–437.

Moss, P.S., Spector, D.H., Glass, C.A., Strohman, R.C. (1984) Streptomycin retards the phenotypic maturation of chick myogenic cells. In Vitro 20:473–478.

Mowles, J. (1988) The use of ciprofloxacin for the elimination of mycoplasma from naturally infected cell lines. Cytotechnology 1:355–358.

Muirhead, E.E., Rightsel, W.A., Pitcock, J.A., Inagami, T. (1990) Isolation and culture of juxtaglomerular and renomedullary interstitial cells. Methods Enzymol 191:152–167.

Munthe-Kaas, A.C., Seglen, P.O. (1974) The use of metrizamide as a gradient medium for isopycnic separation of rat liver cells. FEBS Lett 43:252–256.

Murakami, H. (1984) Serum-free cultivation of plasmacytomas and hybridomas. In Barnes, D.W., Sirbasku, D.A., Sato, G.H. (eds): "Methods for Serum-Free Culture of Neuronal and Lymphoid Cells." New York, Alan R. Liss, pp. 197–206.

Murakami, H., Masui, H. (1980) Hormonal control of human colon carcinoma cell growth in serum-free medium. Proc Natl Acad Sci USA 77:3464–3468.

Murao, S., Gemmell, M.A., Callaghan, M.F., et al. (1983) Control of macrophage cell differentiation in human promyelocytic leukemia by 1,25-dihydroxyvitamin D3 and phorbol-12-myristate-13-acetate. Cancer Res 43:4989–4996.

Murphy, S.J., Watt, D.J., Jones, G.E. (1992) An evaluation of cell separation techniques in a model mixed cell population. J Cell Sci 102:789–798.

Naeyaert, J.M., Eller, M., Gordon, P.R., Park, H-Y., Gilchrest, B.A. (1991) Pigment content of cultured human melanocytes does not correlate with tyrosine message level. Br J Dermatol 125:297–303.

Nagaoka, S., Tanzawa, H., Suzuki, J. (1990) Cell proliferation of hydrogels. In Vitro Cell Dev Biol 26:51–61.

Nagy, B., Ban, K., Bradar, B. (1977) Fibrinolysis associated with human neoplasia: Production of plasminogen activator by human tumors. Int J Cancer 19:614–620.

Nakamura, T., Nawa, K., Ichihara, A., Kaise, N., Nishino, T., (1987) Purification and subunit structure of hepatocyte growth factor from rat platelets. FEBS Lett 224:311–316.

Nakamura, T., Nishizawa, T., Hagiya, M., Seki, T., Shimonishi, M., Sugimura, A., Tashiro, K., Shimizu, S. (1989) Molecular cloning and expression of human hepatocyte growth factor. Nature 342:440–443.

Nardone, R.M., Todd, G., Gonzalezx, P., Gaffney, E.V. (1965) Nucleoside incorporation into strain L cells: Inhibition by pleuropneumonia-like organisms. Science 149:1100–1101.

Nelson, P.G., Lieberman, M. (1981) "Excitable Cells in Tissue Culture." New York, Plenum.

Nelson-Rees, W., Flandermeyer, R.R. (1977) Inter- and intraspecies contamination of human breast tumor cell lines HBC and BrCa5 and other cell cultures. Science 195:1343–1344.

Neugut, A.I., Weinstein, I.B. (1979) Use of agarose in the determination of anchorage-independent growth. In Vitro 15:351.

Newton, A.A., Wildy, P. (1959) Parasynchronous division of HeLa cells. Exp Cell Res 16:624–635.

Neyfakh, A.A. (1987) Use of fluorescent dyes as molecular probes for the study of multidrug resistance. Exp Cell Res 174:168.

Nichols, E.A., Ruddle, F.H. (1973) A review of enzyme polymorphism, linkage and electrophoretic conditions for mouse and somatic cell hybrids in starch gels. J Histochem Cytochem 21:1066–1081.

Nichols, W.W., Murphy, D.G., Christofalo, V.J. (1977) Characterization of a new diploid human cell strain, IMR-90. Science 196:60–63.

Nicola, N.A. (1987) Hemopoietic growth factors and their interactions with specific receptors. J Cell Physiol (Suppl) 5:9–14.

Nicola, N.A., Metcalf, D., Matsumoto, M., Johnson, G.R. (1983) Purification of a factor inducing differentiation in murine myelomonocytic leukaemic cells. J Biol Chem 258:9017–9023.

Nicola, N.A., Begley, C.A., Metcalf, D. (1985) Identification of the human analogue of a regulator that induces differentiation in murine leukamic cells. Nature 314:625–628.

Nicolson, G.L. (1976) Trans-membrane control of the receptors on normal and tumor cells. II. Surface changes associated with transformation and malignancy. Biochim Biophys Acta 458:1–72.

Nicosia, R.F., Leighton, J. (1981) Angiogenesis *in vitro:* Light microscopic, radioautographic and ultrastructural studies of rat aorta in histophysiological gradient cluture. In Vitro 17:204.

Nicosia, R.F., Tchao, R., Leighton, J. (1983) Angiogenesis-dependent tumor spread in reinforced fibrin clot culture. Cancer Res 43:2159–2166.

Nielsen, V. (1989) Vibration patterns in tissue culture vessels. Nunc Bulletin 2 (May 1986, rev March 1989). Roskilde, Denmark, A/S Nunc.

Nilos, R.M., Makarski, J.S. (1978) Control of melanogenesis in mouse melanoma cells of varying metastatic potential. J Natl Cancer Inst 61:523–526.

Noble, M., Murray, K. (1984) Purified astrocytes promote the *in vitro* division of a bipotential glial progenitor cell. EMBO J 3:2243–2247.

Noble, P.B., Levine, M.D. (1986) "Computer-Assisted Analyses of Cell Locomotion and Chemotaxis." Boca Raton, Florida, CRC Press.

Noguchi, P., Wallace, R., Johnson, J., Earley, E.M., O'Brien, S., Ferrone, S., Pellegrino, M.A., Milstein, J., Needy, C., Browne, W., Petricciaru, J. (1979) Characterization of WiDr: A human colon carcinoma cell line. In Vitro 15:401.

Norwood, T.H., Zeigler, C.J., Martin, G.M. (1976) Dimethyl sulphoxide enhances polyethylene glycol-mediated somatic cell fusion. Somatic Cell Genet 2:263–270.

Novak, J. (1962) A high-resolution autoradiographic apposition method for water-soluble tracers and tissue constituents. Int J Appl Radiat Isot 13:187–190.

O'Brien, S.J., Kleiner, G., Olson, R., Shannon, J.E. (1977) Enzyme polymorphisms as genetic signatures in human cells cultures. Science 195:1345–1348.

O'Brien, S.J., Shannon, J.E., Gail, M.H. (1980) Molecular approach to the identification and individualization of human and animal cells in culture: Isozyme and allozyme genetic signatures. In Vitro 16:119–135.

Oda, D., Watson, E. (1990) Human oral epithelial cell culture I. Improved conditions for reproducible culture in serum-free medium. In Vitro Cell Dev Biol 26:589–595.

O'Farrell, P.H. (1975) High resolution two-dimensional electrophoresis of proteins. J Biol Chem 250:4007–4021.

O'Garra, A., Warren, D.J., Holman, M., Popham, A.M., Sanderson, C.J., et al. (1986) Interleukin 4 (B-Cell growth factor II/eosinophil differentiation factor) is a mitogen and differentiation factor for preactivated murine B lymphocytes. Proc Natl Acad Sci USA 83:5228–5232.

O'Hare, M.J., Ellison, M.L., Neville, A.M. (1978) Tissue culture in endocrine research: Perspectives, pitfalls, and potentials. Curr Top Exp Endocrinol 3:1–56.

Ohno, T., Saijo-Kurita, K., Miyamoto-Eimori, N., Kurose, T., Aoki Y, Yosimura, S. (1991) A simple method for in situ freezing of anchorage-dependent cells including rat liver parenchymal cells. Cytotechnology 5:273–277.

Olmsted, C.A. (1967) A physico-chemical study of fetal calf sera used as tissue culture nutrient correlated with biological tests for toxicity. Cell Res 48:283–299.

Olsson, I., Ologsson, T. (1981) Induction of differentiation in a human promyelocytic leukemic cell line (HL-60). Exp Cell Res 131:225–230.

O'Neill, J.P., McGinniss, M.J., Berman, J.K., Sullivan, L.M., Nichlas, J.A., Albertini, R.J. (1987) Refinement of a T-lymphocyte cloning assay to quantify the *in vivo* thioguanine resistant mutant frequency in humans. Mutagenesis 2:87–94.

Oram, J.D., Downing, R.G., Akrigg, A., Dollery, A.A., Duggleby, C.J., Wilkinson, G.W.G., Greenaway, P.J. (1982) Use of recombinant plasmids to investigate the structure of the human cytomegalovirus genome. J Gen Virol 59:111–129.

Orly, J., Sato, G., Erickson, G.F. (1980) Serum suppresses the expression of hormonally induced function in cultured granulosa cells. Cell 20:817–827.

Osborne, C.K., Hamilton, B., Tisus, G., Livingston, R.B. (1980) Epidermal growth factor stimulation of human breast cancer cells in culture. Cancer Res 40:2361–2366.

Osborne, H.B., Bakke, A.C., Yu, J. (1982) Effect of dexamethasone on HMBA-induced Friend cell erythrodifferentiation. Cancer Res 42:513–518.

Ostertag, W., Pragnell, I.B. (1978) Changes in genome composition of the Friend virus complex in erythroleukaemia cells during the course of differentiation induced by DMSO. Proc Natl Acad Sci USA 75:3278–3282.

Ostertag, W., Pragnell, I.B. (1981) Differentiation and viral involvement in differentiation of transformed mouse and rat erythroid cells. Curr Topics Microbiol Immunol 94/95:143–208.

Owens, R.B., Smith, H.S., Hackett, A.J. (1974) Epithelial cell culture from normal glandular tissue of mice. Mouse epithelial cultures enriched by selective trypsinisation. J Natl Cancer Inst 53:261–269.

Oyama, V.I., Eagle, H. (1956) Measurement of cell growth in tissue culture with a phenol reagent (Folin-Ciocalteau). Proc Soc Exp Biol Med 91:305–307.

Pahlman, S., Ljungstedt-Pahlman, A., Sanderson, P.J., Ward, G.A., Hermon-Taylor, J. (1979) Isolation of plasma-membrane components from cultured human pancreatic cancer cells by immunoaffinity chromatography of anti-βM Sepharose 6MB. Br J Cancer 40:701.

Paraskeva, C., Williams, A.C. (1992) The colon. In Freshney, R.I. (ed): "Culture of Epithelial Cells." New York, Wiley-Liss, pp. 82–105.

Paraskeva, C., Buckle, B.G., Sheer, D., Wigley, C.B. (1984) The isolation and characterisation of colorectal epithelial cell lines at different stages in malignant transfromation from familial polyposis coli patients. Int J Cancer 34:49–56.

Paraskeva, C., Buckle, B.G., Thorpe, P.E. (1985) Selective killing of contaminating human fibroblasts in epithelial cultures derived from colorectal tumors using an anti-Thy-1 antibody-ricin conjugate. Br J Cancer 51:131–134.

Pardee, A.B., Cherington, P.V., Medrano, E.E. (1984) On deciding which factors regulate cell growth. In Barnes, D.W., Sirbasku, D.A., Sato, G.H. (eds): "Methods for Serum-Free Culture of Epithelial and Fibroblastic Cells." New York, Alan R. Liss, pp. 157–166.

Parenjpe, M.S., Boone, C.W., Ande Eaton, S. (1975) Selective growth of malignant cells by *in vitro* incubation on Teflon. Exp Cell Res 93:508–512.

Paris Conference (1971, Suppl 1975): Standardization in human cytogenetics. Cytogenet Cell Genet 15:201–238.

Parker, R.C., Castor, L.N., McCulloch, E.A. (1957) Altered cell strains in continuous culture. Special publications, NY Acad Sci 5:303–313.

Parkinson, E.K., Yeudall, W.A. (1992) The epidermis. In Freshney, R.I. (ed): "Culture of Epithelial Cells." New York, Wiley-Liss, pp. 59–80.

Parks, W.M., Gingrich, R.D., Dahle, C.E., Hoak, J.C. (1985) Identification and characterization of an endothelial, cell-specific antigen with a monoclonal antibody. Blood 66:816–823.

Patel, K., Moore, S.E., Dickinson, G., Rossell, R.J., Beverley, P.C,, Kemshead, J.T., Walsh, F.S. (1989) Neural cell adhesion molecule (NCAM) is the antigen recognised by monoclonal antibodies of similar specificity in small-cell lung carcinoma and neuroblastoma. Int J Cancer 44:573–578.

Patueleia, M.C., Friend, C. (1967) Tissue culture studies on murine virus-induced leukemia cells: Isolation of single cells in agar-liquid medium. Cancer Res 27:726–730.

Paul, J. (1975) "Cell and Tissue Culture." Edinburgh, Churchill Livingstone, pp. 172–184.

Paul, J., Fottrell, P.F. (1961) Molecular variation in similar enzymes from different species. Ann NY Acad Sci 94:668–677.

Paul, J., Conkie, D., Freshney, R.I. (1969) Erythropoietic cell population changes during the hepatic phase of erythropoiesis in the foetal mouse. Cell Tissue Kinet 2:283–294.

Paul, W.E., Sredni, B., Schwartz, R.H. (1981) Long-term growth and cloning of non-transformed lymphocytes. Nature 294:697–699.

Pearse, A.G.E. (1968) "Histochemistry,·Theoretical and Applied." Boston, Little, Brown, pp. 255–264.

Peat, N., Gendler, S.J., Lalani, N., Duhig, T., Taylor-Papadimitriou J., (1992) Tissue-specific expression of a human polymorphic epithelial mucin (MUC1) in transgenic mice. Cancer Res 52:1954–1960.

Peehl, D.M., Ham, R.G. (1980) Clonal growth of human keratinocytes with small amounts of dialysed serum. In Vitro 16:526–540.

Pegolo, G., Askanas, V., Engel, W.K. (1990) Expression of muscle-specific isozymes of phosphorylase and creatine kinase in human muscle fibers cultured aneurally in serum-free, hormonally/chemically enriched medium. Int J Dev Neurosci 8:299–308.

Pereira, M.E.A., Kabat, E.A. (1979) A versatile immunoadsorbent capable of binding lectins of various specificities and its use for the separation of cell populations. J Cell Biol 82:185–194.

Pereira-Smith, O., Smith, J. (1988) Genetic analysis of indefinite division in human cells: Identification of four complementation groups. Proc Natl Acad Sci USA 85:6042–6046.

Perper, R.J., Zee, T.W., Mickelson, M.M. (1968) Purification of lymphocytes and platelets by gradient centrifugation. J Lab Clin Med 72:842–868.

Pertoft, H., Laurent, T.C. (1977) Isopycnic separation of cells and cell organelles by centrifugation in modified colloidal silica gradients. In Catsimpoolas, N. (ed): "Methods of Cell Separation." New York, Plenum Press.

Petersen, D.F., Anderson, E.C., Tobey, R.A. (1968) Mitotic cells as a source of synchronized cultures. In Prescott, D.M. (ed): "Methods in Cell Physiology." New York, Academic Press, pp. 347–370.

Peterson, E.A., Evans, W.H. (1967) Separation of bone marrow cells by sedimentation at unit gravity. Nature 214:824–825.

Petricciani, J.C., Hoops, H.E., Chapple, P.J. (eds) (1979) "Cell Substrates: Their Use in the Production of Vaccines and Other Biologicals." New York, Plenum Press.

Pfeffer, L.M., Eisenkraft, B.L. (1991) The antiproliferative and antitumour effects of human alpha interferon on cultured renal carcinomas correlate with the expression of a kidney-associated differentiation antigen. Interferons Cytokines 17:30–31.

Phillips, P., Steward, J.K., Kumar, S. (1976) Tumor angiogenesis factor (TAF) in human and animal tumors. Int J Cancer 17:549–558.

Phillips, P., Kumar, P., Kumar, S., Waghe, M. (1979) Isolation and characterization of endothelial cells from adult rat brain white matter. J Anat 129:261.

Phillips, P.D., Cristofalo, V.J. (1988) Classification system based on the functional equivalency of mitogens that regulate WI-38 cell proliferation. Exp Cell Res 175:396–403.

Pignatelli, M., Bodmer, W.F. (1988) Genetics and biochemistry of collagen binding-trigered glandular differentiation in a human colon carcinoma cell line. Proc Natl Acad Sci USA 85:5561–5565.

Pines, J. (1992) Cell proliferation and control. Curr Opin Cell Biol 4:144–148.

Pitot, H., Periano, C., Morse, P., Potter, V.R. (1964) Hepatomas in tissue culture compared with adapting liver *in vitro.* Natl Cancer Inst Monogr 13:229–245.

Pizzonia, J.H., Gesek, F.A., Kennedy, S.M., Coutermarach, B.A., Bacskal, B.J., Friedman, P.A. (1991) Immunomagnetic separation, primary culture, and characterisation of cortical thick ascending limb plus distal convluted tubule cells from mouse kidney. In Vitro Cell Dev Biol 27A:409–416.

Platsoucas, C.D., Good, R.A., Gupta, S. (1979) Separation of human lymphocyte-T subpopulations (T-mu, T-gamma) by density gradient electrophoresis. Proc Natl Acad Sci USA 76:1972.

Plumb, J.A., Milroy, R., Kaye, S.B. (1989) Effects of the pH dependence of 3-(4,5-dimethylthiazol-2-yl)-2,5-diphenyltetra-zolium bromide-formazan absorption on chemosensitivity determined by a novel tetrazolium-based assay. Cancer Res 49:4435–4440.

Polinger, I.S. (1970) Separation of cell types in embryonic heart cell cultures. Exp Cell Res 63:78–82.

Pollack, M.S., Heagney, S.D., Livingston, P.O., Fogh, J. (1981) HLA-A, B, C & DR alloantigen expression on forty-six cultured human tumor cell lines. J Natl Cancer Inst 66:1003–1012.

Pollard, J.W., Walker, J.M. (eds) (1990) "Animal Cell Culture. Methods in Molecular Biology," 5. Clifton, NJ, Humana Press, pp. 83–97.

Pollock, M.F., Kenny, G.E. (1963) Mammalian cell cultures contaminated with pleuro-pneumonia-like organisms. III. Elimination of pleuro-pneumonia-like organisms with specific antiserum. Proc Soc Exp Biol Med 112:176–181.

Pontecorvo, G. (1975) Production of mammalian somatic cell hybrids by means of polyethylene glycol treatment. Somat Cell Genet 1:397–400.

Pontén, J. (1975) Neoplastic human glia cells in culture. In Pontén, J., Macintyre, E. (1968) Interaction between normal and transformed bovine fibroblasts in culture. II. Cells transformed by polyoma virus. J Cell Sci 3:603–668.

Pontén, J., Westermark, B. (1980) Cell generation and aging of nontransformed glial cells from adult humans. In Fedorof, S., Hertz, L. (eds): "Advances in Cellular Neurobiology, Vol. 1." New York, Academic Press, pp. 209–227.

Post, M., Floros, J., Smith, B.T. (1984) Inhibition of lung maturation by monoclonal antibodies against fibroblast-pneumocyte factor. Nature 308:284–286.

Povey, S., Hopkinson, D.A., Harris, H., Franks, L.M. (1976) Characterization of human cell lines and differentiation from HeLa by enzyme typing. Nature 264:60–63.

Prasad, K.N., Edwards-Prasad, J., et al. (1980) Vitamin E increases the growth inhibitory and differentiating effects of tumour therapeutic agents on neuroblastoma and glioma cells in culture. Proc Soc Exp Biol Med 164:158–163.

Pretlow, T.G. (1971) Estimation of experimental conditions that permit cell separations by velocity sedimentation on isokinetic gradients of Ficoll in tissue culture medium. Anal Biochem 41:248–255.

Pretlow, T.G., Pretlow, T.P. (1989) Cell separation by gradient centrifugation methods. Methods Enzymol 171:462–482.

Pretlow, T.G., Delmoro, C.M., Dilley, G.G., Spadafora, C.G., Pretlow, T.P. (1991) Transplantation of human prostatic carcinoma into nude mice in Matrigel. Cancer Res 51:3814–3817.

Pretlow, T.P., Stinson, A.I., Pretlow, T.G., Glover, G.L. (1978) Cytologic appearance of cells dissociated from rat colon and their separation by isokinetic and isopyknic sedimentation in gradients of Ficoll. J Natl Cancer Inst 61:1431–1437.

Prince, G.A., Jenson, A.B., Billups, L.C., Notkins, A.L. (1978) Infection of human pancreatic beta cell cultures with mumps virus. Nature 271:158–161.

Prop, F.J.A., Wiepjes, G.J. (1973) Sequential enzyme treatment of mouse mammary gland. In Kruse, P.F., Patterson, M.K. (eds): "Tissue Culture Methods and Applications." New York, Academic Press, pp. 21–24.

Provision and Use of Work Equipment (1992) Health and Safety Executive, Broad Lane, Sheffield S3 7HQ, England.

Prunieras, M., Regnier, M., Woodley, D. (1983) Methods for cultivation of keratinocytes with an air–liquid interface. J Invest Dermatol 81:28s–33s.

Puck, T.T., Marcus, P.I. (1955) A rapid method for viable cell titration and clone production with HeLa cells in tissue culture: The use of X-irradiated cells to supply conditioning factors. Proc Natl Acad Sci USA 41:432–437.

Puck, T.T., Cieciura, S.J., Robinson, A. (1958) Genetics of somatic mammalian cells: III. Long term cultivation of euploid cells from human and animal subjects. J Exp Med 108:945–956.

Quarles, J.M., Morris, N.G., Leibovitz, A. (1980) Carcinoembryonic antigen production by human colorectal adenocarcinoma cells in matrix-perfusion culture. In Vitro 16:113–118.

Quastler, H. (1963) The analysis of cell population kinetics. In Lamerton, L.F., Fry, R.J.M. (eds): "Cell Proliferation." Philadelphia, Davis, pp. 18–34.

Quax, P.H., Frisdal, E., Pedersen, N., Bonavaud, S., Thibert, P., Martelly, I., Verheijen, J.H., Blasi, F., Barlovatz-Meimon, G. (1992) Modulation of activities and RNA level of the components of the plasminogen activation system during fusion of human myogenic satellite cells in vitro. Dev Biol 151:166–175.

Quintanilla, M., Brown, K., Ramsden, M., Balmain, A. (1986) Carcinogen specific mutation and amplification of Ha-ras during mouse skin carcinogenesis. Nature 322:78–79.

Rabito, C.A., Tchao, R., Valentich, J., Leighton, J. (1980) Effect of cell substratum interaction of hemicyst formation by MDCK cells. In Vitro 16:461–468.

Raff, M.C. (1990) Glial cell diversification in the rat optic nerve. Science 243:1450–1455.

Raff, M.C., Abney, E., Brockes, J.P., Hornby-Smith, A. (1978) Schwann cell growth factors. Cell 15:813–822.

Raff, M.C., Fields, K.L., Hakomori, S.L., Minsky, R., Pruss, R.M., Winter, J. (1979) Cell-type-specific markers for distinguishing and studying neurons and the major classes of glial cells in culture. Brain Res 174:283–309.

Raff, M.C., Miller, R.H., Noble, M. (1983) A glial progenitor cell that develops *in vitro* into an astrocyte or an oligodendrocyte depending on the culture medium. Nature 303:390–396.

Raff, M.C., Lillien, L.E., Richardson, W.D., Burne, J.F., Noble, M. (1988) Platelet-derived growth factor from astrocytes drives the clock that times oligodendrocyte differentiation in culture. Nature 333:562–565.

Ramaekers, F.C.S., Puts, J.J.G., Kant, A., Moesker, O., Jap, P.H.K., Vooijs, G.P. (1982) Use of antibodies to intermediate filaments in the characterization of human tumors. Cold Spring Harbor Symp Quant Biol 46:331-339.

Ranscht, B., Clapshaw, P.A., Price, J., Noble, M., Seifert, W., (1982) Development of oligodendrocytes and Schwann cells studied with monoclonal antibody against galactocerebroside. Proc Natl Acad Sci USA 79:2709–2713.

Raz, A. (1982) B16 melanoma cell variants: Irreversible inhibition of growth and induction of morphologic differentiation by anthracycline antibiotics. J Natl Cancer Inst 68:629–638.

Reddy, J.K., Rao, M.S., Warren, J.R., Minnick, O.T. (1979) Concanavalin A agglutinability and surface microvilli of dissociated normal and neoplastic pancreatic acinar cells of the rat. Exp Cell Res 120:55–61.

Reel, J.R., Kenney, F.T. (1968) "Superinduction" of tyrosine transaminase in hepatoma cell cultures: Differential inhibition of synthesis and turnover by actinomycin D. Proc Natl Acad Sci USA 61:200–206.

Reeves, M.E. (1992) A metastatic tumour cell line has greatly reduced levels of a specific homotypic cell adhesion molecule activity. Cancer Res 52:1546–1552.

Reid, L.M. (1990) Stem cell biology, hormone/matrix synergies and liver differentiation. Curr Opin Cell Biol 2:121–130.

Reid, L.M., Rojkind, M. (1979) New techniques for culturing differentiated cells: Reconstituted basement membrane rafts. In Jakoby, W.B., Pastan, I.H. (eds.): "Methods in Enzymology, Vol. 57, Cell Culture." New York, Academic Press, pp. 263–278.

Reitzer, L.J., Wice, B.M., Kennel, D. (1979) Evidence that glutamine, not sugar, is the major energy source for cultured HeLa cells. J Biol Chem 254:2669–2677.

Repesh, L.A. (1989) A new *in vitro* assay for quantitating tumor cell invasion. Invas Metast 9:192–208.

Rheinwald, J.G., Beckett, M.A. (1981) Tumorigenic keratinocyte lines requiring anchorage and fibroblast support cultured from human squamous cell carcinomas. Cancer Res 41:1657–1663.

Rheinwald, J.G., Green, H. (1975) Serial cultivation of strains of human epidermal keratinocytes: The formation of keratinizing colonies from single cells. Cell 6:331–344.

Rheinwald, J.G., Green, H. (1977) Epidermal growth factor and the multiplication of cultured human keratinocytes. Nature 265:421–424.

Rhim, J.S., Trimmer, R., Arnstein, O., Huebner, R.J. (1981) Neoplastic transformation of chimpanzee cells induced by adenovirus type 12 simian virus 40 hybrid virus (Ad12-SV40). Proc Natl Acad Sci USA 78:13–17.

Richardson, W.D., Raff, M., Noble, M. (1990) The oligodendrocyte-type-2 astrocyte lineage. Semin Neurosci 2:445–454.

Richler, C., Yaffe, D. (1970) The in vitro cultivation and differentiation capacities of myogenic cell lines. Dev Biol 23:1–22.

Richmond, A., Lawson, D.H., Nixon, D.W., Chawla, R.K. (1985) Characterization of autostimulatory and transforming growth factors from human melanoma cells. Cancer Res 45:6390–6394.

Rickwood, D., Birnie, G.D. (1975) Metrizamide, a new density gradient medium. FEBS Lett 50:102–110.

Rifkin, D.B., Loeb, J.N., Moore, G., Reich, E. (1974) Properties of plasminogen activators formed by neoplastic human cell cultures. J Exp Med 139:1317–1328.

Rindler, M.J., Chuman, L.M., Shaffer, L., Saier, M.H., Jr. (1979) Retention of differentiated propenies in an established dog kidney epithelial cell line (MDCK). J Cell Biol 81:635–648.

Robinson, W.D., Green, M.H.L., Cole, J., Garner, R.C., Healy, M.J.R., Gatehouse, D. (1989) Statistical evaluation of bacterial/mammalian fluctuation tests. In Kirkland, D.J. (ed): "Statistical Evaluation of Mutatgenicity Test Data." Cambridge, Cambridge University Press, pp. 102–140.

Rockwell, G.A., Sato, G.H., McClure, D.B. (1980) The growth requirements of SV40 virus transformed Balb/c-3T3 cells in serum-free monolayer culture. J Cell Physiol 103:323–331.

Rofstad, E.K. (1991) Spheroids and xenografts. In Masters, J.R.W. (ed): "Human Cancer in Primary Culture." London, Kluwer Academic Publishers, pp. 81–102.

Rogers, A.W. (1979) "Techniques of Autoradiography," 3rd ed. Amsterdam, Elsevier/North-Holland Biomedical Press.

Rojkind, M., Gatmaitan, Z., Mackensen, S., Giambrone, M.A., Ponce, P., Reid, L.M. (1980) Connective tissue biomatrix: Its isolation and utilization for long term cultures of normal rat hepatocytes. J Cell Biol 87:255–263.

Rooney, D.E., Czepulkowski, B.H. (eds) (1986) "Human Cytogenetics, a Practical Approach." Oxford, IRL Press at Oxford University Press.

Rosenfeld, M.A., Yoshimura, K., Trapnell, B.C., Yoneyama, K., Rosenthal, E.R., Dalemenas, W., Fukayama, M., Bargon, J., Stier, L.E., Stratford-Perricaudet, L., Perricaudet, M., Guggino, W.B., Pavirani, A., Lecocq, J-P., Crystal, R.G. (1992) *In vivo* transfer of the human cystic fibrosis transmembrane conductance regulator gene to the airway epithelium. Cell 68:143–155.

Rosenman, S.J., Gallatin, W.M. (1991) Cel surface glycoconjugates in intercellular and cell-substratum interactions. Sem Cancer Biol 2:357–366.

Ross, R. (1971) The smooth muscle cell. II. Growth of smooth muscle in culture and formation of elastic fibers. J Cell Biol 50:172–186.

Rossi, G.B., Friend, C. (1967) Erythrocytic maturation of (Friend) virus-induced leukemic cells in spleen clones. Proc Natl Acad Sci USA 58:1373–1380.

Rothfels, K.H., Siminovitch, L. (1958) An air drying technique for flattening chromosomes in mammalian cells growth in vitro. Stain Technol 33:73–77.

Rotman, B., Papermaster, B.W. (1966) Membrane properties of living mammalian cells as studied by enzymatic hydrolysis of fluorogenic esters. Proc Natl Acad Sci USA 55:134–141.

Rovera, G., O'Brien, T.G., Diamond, L. (1979) Induction of differentiation in human promyelocytic leukemia cells by tumor promoters. Science 204:868–870.

Royal College of Physicians of London (1990) "Guidelines on the Practice of Ethics Committees in Medical Research Involving Human Subjects," 2nd ed. ISBN 0 900596 902.

Rudland, P.S., Davies, A.T., Warburton, M.J. (1982) Prostaglandin induced differentiation or dimethyl sulphoxide induced differentiation: Reduction of the neoplastic potential of a rat mammary tumour stem cell line. J Natl Cancer Inst 69:1083–1093.

Ruoff, N.M., Hay, R.J. (1979) Metabolic and temporal studies on pancreatic exocrine cells in culture. Cell Tissue Res 204:243–252.

Rutzky, L.P., Tomita, J.T., Calenoff, M.A., Kahan, B.D. (1979) Human colon adenocarcinoma cells. III. In vitro organoid expression and carcino-enbryonic antigen kinetics in hollow fiber culture. J Natl Cancer Inst 63:893.

Rygaard, K., Moller, C., Bock, E., Spang-Thomsen, M. (1992) Expression of cadherin and NCAM in human small cell lung cancer cell lines and xenografts. Br J Cancer 65:573–577.

Sachs, L. (1978) Control of normal cell differentiation and the phenotypic reversion of malignancy in myeloid leukaemia. Nature 274:535–539.

Sachs, L. (1982) Normal development programmes in myeloid leukaemia: Regulatory proteins is the control of growth and differentiation. Cancer Surv 1:321–342.

Safe Working and the Prevention of Infection in Clinical Laboratories (1991) Health and Safety Commission, HMSO Publications, P.O. Box 276, London, SW8 5DT, England.

Sager, R. (1992) Tumor suppressor genes in the cell cycle. Curr Opin Cell Biol 4:155–160.

Saier, M.H. (1984) Hormonally defined, serum free medium for a proximal tubular kidney epithelial cell line, LLC-PKI. In Barnes, W.D. (ed): "Methods for Serum Free Culture of Epithelial and Fibroblastic Cells." New York, Alan R. Liss, pp. 25–31.

Sambrook, J., Fritsch, E.F., Maniatis, T. (1989) "Molecular Cloning. A Laboratory Manual," 2nd ed. Cold Spring Harbor, NY, Cold Spring Harbor Laboratory Press, 3 vols.

Sandberg, A.A. (1982) Chromosomal changes in human cancers: Specificity and heterogeneity. In Owens, A.H., Coffey, D.S., Baylin, S.B. (eds): "Tumour Cell Heterogeneity." New York, Academic Press, pp. 367–397.

Sandström, B. (1965) Studies on cells from liver tissue cultivated *in vitro*. I. Influence of the culture method on cell morphology and growth pattern. Exp Cell Res 37:552–568.

Sanford, K.K., Earle, W.R., Likely G.D. (1948) The growth *in vitro* of single isolated tissue cells. J Natl Cancer Inst 9:229.

Sanford, K.K., Earle, W.R., Evans, V.J., Waltz, H.K., Shannon, I.E. (1951) The measurement of proliferation in tissue cultures by enumeration of cell nuclei. J Natl Cancer Inst 11:773.

Sato, G. (1979) The growth of cells in serum-free hormone-supplemented medium. In Jakoby, W.B., Pastan, I.H. (eds): "Methods in Enzymology." New York, Academic Press, pp. 94–109.

Sato, G. (ed) (1981) "Functionally Differentiated Cell Lines." New York, Alan R. Liss.

Sato, G.H., Yasumura, Y. (1966) Retention of differentiated function in dispersed cell culture. Trans NY Acad Sci 28:1063–1079.

Sattler, G.A., Michalopoulos, G., Sattler, G.L., Pitot, H.C. (1978) Ultrastructure of adult rat hepatocytes cultured on floating collagen membranes. Cancer Res 38:1539–1549.

Savage, C.R., Jr., Bonney, R.I. (1978) Extended expression of differentiated function in primary cultures of adult liver parenchymal cells maintained on nitrocellulose filters. I. Induction of phosphoenol pyruvate carboxykinase and tyrosine aminotransferase. Exp Cell Res 114:307–315.

Schaeffer, W.I. (1990) Terminology associated with cell, tissue and organ culture, molecular biology and molecular genetics. In Vitro Cell Dev Biol 26:97–101.

Schengrund, C.L., Repman, M.A. (1979) Differential enrichment of cells from embryonic rat cerebra by centrifugal elutriation. I. Neurochemistry 33:283.

Scher, W., Holland, J.G., Friend, C. (1971) Hemoglobin synthesis in murine virus-induced leukemic cells *in vitro*. I. Partial purification and identification of hemoglobins. Blood 37:428–437.

Schimmelpfeng, L., Langenberg, U., Peters, I.M. (1968) Macrophages overcome mycoplasma infections of cells *in vitro*. Nature 285:661.

Schlechte, W., Brattain, M., Boyd, D. (1990) Invasion of extracellular matrix by cultured colon cancer cells: Dependence on urokinase receptor display. Cancer Comm 2:173–179.

Schmidt, R., Reichert, U., Michel, S., Shrott, B., Boullier, M. (1985) Plasma membrane transglutaminase and cornified envelope competence in cultured human keratinocytes. FEBS Lett 186:204.

Schneider, E.L., Stanbridge, E.I. (1975) A simple biochemical technique for the detection of mycoplasma contamination of cultured cells. Methods Cell Biol 10:278–290.

Schneider, H., Muirhead, E.E., Zydeck, F.A. (1963) Some unusual observations of organoid tissues and blood elements in monolayer cultures. Exp Cell Res 30:449–459.

Schnook, L.B., Otz, U., Lazary, S., De Week, A.L., Minowada, J., Odavic, R., Kniep, E.M., Edy, V. (1981) Lymphokine and monokine activities in supernatants from human lymphoid and myeloid cell lines. Lymphokines 2:1–19.

Schoenlein, P.V., Shen, D.-W., Barrett, J.T., Pastan, I.T., Gottesman, M.M. (1992) Double minute chromosomes carrying the human multidrug resistance 1 and 2 genes are generated from the dimerization of submicroscopic circular DNAs in colchicine-selected KB carcinoma cells. Mol Biol Cell 3:507–520.

Schousboe, A., Thorbek, P., Hertz, L., Krogsgaard-Larsen, P. (1979) Effects of GABA analogues of restricted conformation on GABA transport in astrocytes and brain cortex slices and on GABA receptor binding. J Neurochem 33:181.

Schulman, H.M. (1968) The fractionation of rabbit reticulocytes in dextran density gradients. Biochim Biophys Acta 148:251–255.

Schwartz, S.M. (1978) Selection and characterization of bovine aortic endothelial cells. In Vitro 14:966.

Schwarz, M.A., Mitchell, M., Emerson, D.L. (1990) Reconstituted basement membrane enhances neurite outgrowth in PC12 cells induced by nerve growth factor. Cell Growth Diff 1:13–18.

Scotto, K.W., Biedler, I.L., Melera, P.W. (1986) Amplification and expression of genes associated with multidrug resistance in mammalian cells. Science 232:751–755.

Seeds, N.W. (1971) Biochemical differentiation in reaggregating brain cell culture. Proc Natl Acad Sci USA 68:1858–1861.

Segal, S. (1964) Hormones, amino-acid transport and protein synthesis. Nature 203:17–19.

Seglen, P.O. (1975) Preparation of isolated rat liver cells. Methods Cell Biol 13:29–83.

Seifert, W., Müller, H.W. (1984) Neuron-glia interaction in mammalian brain: Preparation and quantitative bioassay of a neurotropic factor (NTF) from primary astrocytes. In Barnes, D.W., Sirbasku, D.A., Sato, G.H. (eds): "Methods for Serum-Free Culture of Neuronal and Lymphoid Cells." New York, Alan R. Liss, pp. 67–78.

Selby, P.J., Thomas, M.J. Monaghan, P., Sloane, J., Peckham, M.J. (1980) Human tumour xenografts established and serially transplanted in mice immunologically deprived by thymectomy, cytosine arabinoside and whole-body irradiation. Br J Cancer 41:52.

Shall, S. (1973) Sedimentation in sucrose and Ficoll gradients of cells grown in suspension culture. In Kruse, P.F., Patterson, M.K. (eds): "Tissue Culture Methods and Applications." New York, Academic Press, pp. 198–204.

Shall, S., McClelland, A.J. (1971) Synchronization of mouse fibroblast LS cells grown in suspension culture. Nature New Biol 229:59–61.

Shapiro, H.M. (1988) "Practical Flow Cytometry, Second Edition." New York, Wiley-Liss.

Sharpe, P.T. (1988) "Methods of Cell Separation." Amsterdam, Elsevier.

Shiba Y., Kanno Y. (1989) Modulation of survival and proliferation of BSC-1 cells through changes in spreading behavior caused by the tumor-promoting phorbol ester TPA. Cell Struct Funct 14:685–696.

Shows, T.B., Sakaguchi, A.Y. (1980) Gene transfer and gene mapping in mammalian cells in culture. In Vitro 16:55–76.

Silvestri, F.F., Banavale, S.D., Hulette, B.C., Civin, C.I., Preisler, H.D. (1991) Isolation and characterization of the $CD34^+$ hematopoietic progenitor cells from the peripheral blood of patients with chronic myeloid leukemia. Int J Cell Cloning 9:474–490.

Simon-Assmann, P., Kedinger, M., Haffen, K. (1986) Immunocytochemical localization of extracellular matrix proteins in relation to rat intestinal morphogenesis. Differentiation 32:59–66.

Simonian, M.H., White, M.L., Foggia, D.A. (1987) Clonal growth and culture life span of bovine adrenocortical cells in a serum-free medium. In Vitro 23(4):247–256.

Sinha, M.K., Buchanan, C., Raineri-Maldonado, C., Khazanie, P., Atkinson, S., DiMarchi, R., Caro, J.F. (1990) IGF-II receptors and IGF-II-stimulated glucose transport in human fat cells. Am J Physiol 258:E534–542.

Sirica, A.E., Hwand, C.G., Sattler, G.L., Pitot, H.C. (1980) Use of primary cultures of adult rat hepatocytes on collagen gel-nylon mesh to evaluate carcinogen-induced unscheduled DNA synthesis. Cancer Res 40:3259–3267.

Skehan, P., Storeng, R., Scudiero, N., et al. (1989) Evaluation of colorimetric protein and biomass stains for assaying *in vitro* drug effects upon human tumor cell lines. Proc Am Soc Cancer Res 30:612.

Sladek, N.E. (1973) Bioassay and relative cytotoxic potency of cyclophosphamide metabolites generated *in vitro* and *in vivo*. Cancer Res 33:1150–1158.

Smith, H.S., Owens, R.B., Hiller, A.J., Nelson-Rees, W.A., Johnston, J.O. (1976) The biology of human cells in tissue culture. I. Characterization of cells derived from osteogenic sarcomas. Int J Cancer 17:219–234.

Smith, H.S., Lan, S., Ceriani, R., Hackett, A.J., Stampfer, M.R. (1981) Clonal proliferation of cultured non-malignant and malignant human breast epithelia. Cancer Res 41:4637–4643.

Smith, S.M., Schroedl, N.A. (1992) Heme containing compounds replace chick embryo extract and enhance differentiation in avian muscle cell culture. In Vitro Cell Dev Biol 28A:387–390.

Smith, W.L., Garcia-Perez, A. (1985) Immunodissection: Use of monoclonal antibodies to isolate specific types of renal cells. Am J Physiol 248:F1–F7.

Snyder, E.Y., Deitcher, D.L., Walsh, C., Arnold-Aldea, S., Hartweig, E.A., Cepko, C.L. (1992) Multipotent cell lines can engraft and participate in development of mouse cerebellum. Cell 68:33–51.

Sordillo, L.M., Oliver, S.P., Akers, R.M. (1988) Culture of bovine mammary epithelial cells in D-valine modified medium: Selective removal of contaminating fibroblasts. Cell Biol Int Rep 12:355.

Sorieul, S., Ephrussi, B. (1961) Karylogical demonstration of hybridization of mammalian cells *in vitro*. Nature 190:653–654.

Sorour, O., Raafat, M., El-Bolkainy, N., Mohamad, R. (1975) Infiltrative potentiality of brain tumors in organ culture. J Neurosurg 43:742–749.

Soule, H.D., Vasquez, J., Long, A., Albert, S.. Brennan, M. (1973) A human cell line from a pleural effusion derived from a breast carcinoma. J Natl Cancer Inst 51:1409–1416.

Souza, L.M., Boone, T.C., Gabrilove, I., Lai, P.H., Zsebo, K.M., et al. (1986) Recombinant human granulocyte colony-stimulating factor: Effects on normal and leukemic myeloid cells. Science 232:61–65.

Spandidos, D.A., Wilkie, N.M. (1984a) Malignant transformation of early passage rodent cells by a single mutated human oncogene. Nature 310:469–475.

Spandidos, D.A., Wilkie, N.M. (1984b) Expression of exogenous DNA in mammalian cells. In Hames, B.D., Higgins, S.J. (eds): "In Vitro Transcription and Translation–A Practical Approach." Oxford, IRL Press, pp. 1–48.

Speir, R., Griffiths, J.B. (1985–1990) "Animal Cell Biotechnology." London, Academic Press, 4 vols.

Speir, R.E., Griffiths, J.B., Meignier, B. (eds) (1991) "Production of Biologicals from Animal Cells in Culture." Oxford, Butterworth-Heinemann.

Speirs, V., Ray, K.P., Freshney, R.I. (1991) Paracrine control of differentiation in the alveolar carcinoma, A549, by human foetal lung fibroblasts. Br J Cancer 64:693–699.

Spinelli, W., Sonnenfeld, K.H., Ishii, N. (1982) Effects of phorbol ester tumor promoters and nerve growth factor on neurite outgrowth in cultured human neuroblastoma cells. Cancer Res 42:5067–5073.

Splinter, T.A.W., Beudeker, M., Beek, A.V. (1978) Changes in cell density induced by isopaque. Exp Cell Res 111:245–251.

Spooncer, E., Eliason, J., Dexter, T.M. (1992) Long-term mouse bone marrow cultures. In Testa, N.G., Molineux, G. (eds): "Haemopoiesis: A Practical Approach." Oxford, IRL Press at Oxford University Press, pp. 55–74.

Spremulli, E.N., Dexter, D.L. (1984) Polar solents: A novel class of antineoplastic agents. J Clin Oncol 2:227–241.

Sredni, B., Sieckmann, D.G., Kumagai, S.H., Green, I., Paul, W.E. (1981) Long term culture and cloning of non-transformed human B-lymphocytes. J Exp Med 154:1500–1516.

Stacey, G.N., Bolton, B.J., Morgan, D., Clark, S.A., Doyle, A. (1992) Multilocus DNA fingerprint analysis of cellbanks: Stability studies and culture identification in human B-lymphoblastoid and mammalian cell lines. Cytotechnology 8:13–20.

Stampfer, M., Halcones, R.G., Hackett, A.J. (1980) Growth of normal human mammmary cells in culture. In Vitro 16:415–425.

Stanbridge, E.J., Doersen, C.-J. (1978) Some effects that mycoplasmas have upon their injected host. In McGarrity, G.J., Murphy, D.G., Nichols, W.W. (eds): "Mycoplasma Infection of Cell Cultures." New York, Plenum Press, pp. 119–134.

Stanley, E.R., Guilbert, I. (1981) Methods for the purification, assay, characterisation and target cell binding of a colony stimulating factor (CSF-I). J Immunol Methods 45:253–289.

Stanley, M.A., Parkinson, E. (1979) Growth requirements of human cervical epithelial cells in culture. Int J Cancer 24:407–414.

Stanners, C.P., Eliceri, G.L., Green, H. (1971) Two types of ribosome in mouse-hamster hybrid cells. Nature New Biol 230:52–54.

Stanton, B.A., Biemesderfer, D., Wade, J.B., Giebisch, G. (1981) Structural and functional study of the rat distal nephron: Effects of potassium adaptation and depletion. Kidney Int 19:36–48.

States, B., Foreman, J., Lee, J., Segal, S. (1986) Characteristics of cultured human renal cortical epithelia. Biochem Med Metab Biol 36:151–161.

Steel, G.G. (1979) Terminology in the description of drug-radiation interactions. Int J Radiat Oncol Biol Phys 5:1145–1150.

Steele, M.P., Levine, R.A., Joyce-Brady, M., Brody, J.S. (1992) A rat alveolar type II cell line developed by adenovirus 12SE1A gene, transfer. Am J Resp Cell Mol Biol 6:50–56.

Steele, V.E., Marchok, A.C., Nettesheim, P. (1978) Establishment of epithelial cell lines following exposure of culture tracheal epithelium to 12-0-tetradecanoylphorbol- 13-acetate. Cancer Res 38:3563–3565.

Stein, H.G., Yanishevsky, R. (1979) Autoradiography. In Jakoby, W.B., Pastan, I.H. (eds): "Methods in Enzymology, Vol. 57. Cell Culture." New York, Academic Press, pp. 279–292.

Stephenson, J.R., Axelrad, A.A., McLeod, D.I., Schreeve, M.M. (1971) Induction of colonies of hemoglobin-synthesizing cells by erythropoietin in vitro. Proc Natl Acad Sci USA 68:1542–1546.

Stockdale, F.E., Topper, Y.J. (1966) The role of DNA synthesis and mitosis in hormone dependent differentiation. Proc Natl Acad Sci USA 56:1283–1289.

Stoker, M., O'Neill, C., Berryman, S., Waxman, B. (1968) Anchorage and growth regulation in normal and virus transformed cells. Int J Cancer 3:683–693.

Stoker, M., Perryman, M., Eeles, R. (1982) Clonal analysis of morphological phenotype in cultured mammary epithelial cells from human milk. Proc R Soc Lond Ser B 215:231–240.

Stoker, M.G.P. (1973) Role of diffusion boundary layer in contact inhibition of growth. Nature 246:200–203.

Stoker, M.G.P., Rubin, H. (1967) Density dependent inhibition of cell growth in culture. Nature 215:171–172.

Stolwijk, J.A.M., Prop, F.J.A., Eijenstein, L., Karten, F., Peters, K., Polak, M., Spies, M., Souw, L. (1987) Representativity of human mammary tumour cell cultures: DNA-cytophotometry as a method for checking tumour cell characteristics. Eur J Cancer Clin Oncol 23:187–193.

Stoner, G.D., Harris, C.C., Myers, G.A., Trump, B.F., Connor, R.D. (1980) Putrescine stimulates growth of human bronchial epithelial cells in primary culture. In Vitro 16:399–406.

Stoner, G.D., Katoh, Y., Foidart, J-M., Trump, B.F., Steinert, P., Harris, C.C. (1981) Cultured human bronchial epithelial cells: Blood group antigens, keratin, collagens and fibronectin. In Vitro 17:577–587.

Strickland, S., Beers, W.H. (1976) Studies on the role of plasminogen activator in ovulation. *In vitro* response of granulosa cells to goandotropins, cyclic nucleotides, and prostaglandins. J Biol Chem 251:5694–5702.

Stubblefield, E. (1968) Synchronization methods for mammalian cell cultures. In Prescott, D.M. (ed): "Methods in Cell Physiology." New York, Academic Press, pp. 25–43.

Styles, J.A. (1977) A method for detecting carcinogenic organic chemicals using mammalian cells in culture. Br J Cancer 36:558.

Su, H.Y., Bos, T.J., Monteclaro, F.S., Vogt, P.K. (1991) Jun inhibits myogenic differentiation. Oncogene 6:1759–1766.

Su, R.T., Chang, Y.C. (1989) Transformation of human epidermal cells by transfection plasmid containing simian virus 40 DNA linked to a neomycin gene in a defined medium. Exp Cell Res 180:117–133.

Sundqvist, K., Liu, Y., Arvidson, K., Ormstad, K., Nilsson, L., Toftgård, R., Grafström, R.C. (1991) Growth regulation of serum-free cultures of epithelial cells from normal human buccal mucosa. In Vitro Cell Dev Biol 27A:562–568.

Sutherland, R.M. (1988) Cell and micro environment interactions in tumour microregions: The multicell spheroid model. Science 240:117–184.

Sykes, J.A., Whitescarver, J., Briggs, L., Anson, J.H. (1970) Separation of tumor cells from fibroblasts with use of discontinuous density gradients. J Natl Cancer Inst 44:855–864.

Taderera, J.V. (1967) Control of lung differentiation in vitro. Dev Biol 16:489–512.

Takahashi, K., Okada, T.S. (1970) Analysis of the effect of "conditioned medium" upon the cell culture at low density. Dev Growth Diff 12:65–77.

Takeda, K., Minowada, J., Bloch, A. (1982) Kinetics of appearance of differentiation-associated characteristics in ML-1, a line of human myeloblastic leukaemia cells, after treatment with TPA, DMSO, or Ara-C. Cancer Res 42:5152–5158.

Tarella, C., Ferrero, D., Gallo, E., Luyca Pagliardi, G., Ruscetti, F.W. (1982) Induction of differentiation of HL-60 cells by dimethylsulphoxide: Evidence for a stochastic model not linked to the cell division cycle. Cancer Res 42:445–449.

Tashjian, A.H., Jr. (1979) Clonal strains of hormone-producing pituitary cells. In Jakoby, W.B., Pastan, I.H. (eds): "Methods in Enzymology, Vol. 57, Cell Culture." New York, Academic Press, pp. 527–535.

Tashjian, A.H., Yasamura, Y., Levine, L., Sato, G.H., Parker, M. (1968) Establishment of clonal strains of rat pituitary tumor cells that secrete growth hormone. Endocrinology 82:342–352.

Taub, M. (1984) Growth of primary and established kidney

cell cultures in serm-free media. In Barnes, D.W., Sirbasku, D.A., Sato, G.H. (eds): "Methods for Serum-Free Culture of Epithelial and Fibroblastic Cells." New York, Alan R. Liss, pp. 3–24.

Taub, M.L., Yang, S.I., Wang, Y. (1989) Primary rabbit proximal tubule cell cultures maintain differentiated functions when cultured in a hormonally defined serum-free medium. In Vitro Cell Dev Biol 25:770–775.

Taylor, C.R. (1978) Immunoperoxidase techniques. Arch Pathol Lab Med 102:113–121.

Taylor-Papadimitriou, J., Shearer, M., Stoker, M.G.P. (1977) Growth requirement of human mammary epithelial cells in culture. Int J Cancer 20:903–908.

Taylor-Papadimitriou, J., Purkiss, P., Fentiman, I.S. (1980) Choleratoxin and analogues of cyclic AMP stimulate the growth of cultured human epithelial cells. J Cell Physiol 102:317–322.

Taylor-Robinson, D. (1978) Cultural and serologic procedures for mycoplasmas in tissue culture. In McGarrity, G., Murphy, D.G., Nichols, W.W. (eds): "Mycoplasma Infection of Cell Cultures." New York, Plenum Press, pp. 47–56.

Temin, H.M. (1966) Studies on carcinogenesis by avian sarcoma viruses. III. The differential effect of serum and polyanions on multiplication of uninfected and converted cells. J Natl Cancer Inst 37:167–175.

Temin, H.M., Rubin H. (1958) Characteristics of an assay for Rous sarcoma virus and Rous sarcoma cells in tissue culture. Virology 6:669–688.

Terasaki, T., Kameya, T., Nakajima, T., Tsumuraya, M., Shimosato, Y., Kato, K., Ichinose, H., Nagatsu, T., Hasegawa, T. (1984) Interconversion of biological *characteristics* of small cell lung cancer cells depending on the culture conditions. Gann 75:1689–1699.

Testa, N.G. (1985) Clonal assays for haemopoietic and lymphoid cells in vitro. In Potten, C.S., Hendry, J.H. (eds): "Cell Clones." Edinburgh, Churchill Livingstone, pp. 27–43.

Thilly, W.G., Levine, D.W. (1979) Microcarrier culture: A homogenous environment for studies of cellular biochemistry. In Jakoby, W.B., Pastan, I.H. (eds): "Methods in Enzymology, Vol. 57, Cell Culture." New York, Academic Press, pp. 184–194.

Thomas, D.G.T., Darling, J.L., Paul, E.A., Mott, T.C., Godlee, J.N., Tobias, J.S., Capra, L.G., Collins, C.D., Mooney, C., Bozek, T., Finn, G.P., Arigbabu, S.O., Bullard, D.E., Shannon, N., Freshney, R.I. (1985) Assay of anti-cancer drugs in tissue culture: Relationship of relapse free interval (RFI) and in vitro chemosensitivity in patients with malignant cerebral glioma. Br J Cancer 51:525–532.

Thompson, L.H., Baker, R.M. (1973) Isolation of mutants of cultured mammalian cells. In Prescott, D. (ed): "Methods in Cell Biology, Vol. 6." New York, Academic Press, pp. 209–281.

Thomson, A.W. (ed) (1991) "The Cytokine Handbook." London, Academic Press.

Thornton, S.C., Mueller, S.N., Levine, E.M. (1983) Human endothelial cells: Use of heparin in cloning and long-term serial cultivation. Science 222:623–625.

Till, J.E., McCulloch, E.A. (1961) A direct measurement of the radiation sensitivity of normal mouse bone marrow cells. Radiation Res 14:213–222.

Tobey, R.A., Anderson, E.C., Petersen, D.F. (1967) Effect of thymidine on duration of G1 in chinese hamster cells. J Cell Biol 35:53–67.

Todaro, G.J., DeLarco, I.E. (1978) Growth factors produced by sarcoma virus-transformed cells. Cancer Res 38:4147–4154.

Todaro, G.J., Green, H. (1963) Quantitative studies of the growth of mouse embryo cells in culture and their development into established lines. J Cell Biol 17:299–313.

Tom, B.H., Rutzky, L.P., Iakstys, M.M., Oyasu, R., Kaye, C.I., Kahan, B.D. (1976) Human colonic adenocarcinoma cells. I. Establishment and description of a new line. In Vitro 12:180.

Tozer, B.T., Pirt, S.J. (1964) Suspension culture of mammalian cells and macromolecular growth promoting fractions of calf serum. Nature 201:375–378.

Traganos, F., Darzynkiewicz, Z., Sharpless, T., Melamed, M.R. (1977) Nucleic acid content and cell cycle distribution of five human bladder cell lines analyzed by flow cytofluorometry. Int J Cancer 20:30–36.

Trapp, B.D., Honegger, P., Richelson, E., Webster, H. de F. (1981) Morphological differentiation of mechanically dissociated fetal rat brain in aggregating cell cultures. Brain Res 160:235–252.

Traxinger, R.R., Marshall, S. (1989) Role of amino acids in modulating glucose-induced desensitization of the glucose transport system. J Biol Chem 264:20910–20916.

Trickett, A.E., Ford, D.J., Lam-Po Tang, P.R.L., Vowels, M.R. (1990) Comparison of magnetic particles for immunomagnetic bone marrow purging using an acute lymphoblastic leukaemia model. Transpl Proc 22:2177–2178.

Triglia, D., Braa, S.S., Yonan, C., Naughton, G.K. (1991) Cytotoxicity testing using neutral red and MTT assays on a three-dimensional human skin substrate. Toxic In Vitro 5:573–578.

Trowell, O.A. (1954) A modified technique for organ culture in vitro. Exp Cell Res 6:246.

Trowell, O.A. (1959) The culture of mature organs in a synthetic medium. Exp Cell Res 16:118–147.

Troyer, D.A., Kreisberg, J.I., (1990) Isolation and study of glomerular cells. Methods Enzymol 191:141–152.

Tsao, M.C., Walthall, B.I., Ham, R.G. (1982) Clonal growth of normal human epidermal keratinocytes in a defined medium. J Cell Physiol 110:219–229.

Tsuji, K., Hayata, Y., Sato, M., et al. (1976) Neuronal differentiation of OAT cell carcinoma in vitro by dibutyryl cyclic 3′5′-monophosphate. Cancer Lett 1:311–318.

Turner, N.T., Batson, A.G., Clive, D. (1984) Procedures for the L5178Y $tk^{+/-}$–$TK^{-/-}$ mouse lymphoma cell mutagenicity assay. In Kilbey, B.J., Legator, M., Nichols, W., Ramel, C. (eds): "Handbook of Mutagenicity Test Procedures". Amsterdam, Elsevier Science Publishers BV, pp. 269–290.

Turner, R.W.A., Siminovitch, L., McCulloch, E.A., Till, J.E. (1967) Density gradient centrifugation of hemopoietic colony-forming cells. J Cell Physiol 69:73–81.

Tveit, K.M., Pihl, A. (1981) Do cells lines in vitro reflect the properties of the tumours of origin? A study of lines derived from human melanoma xenografts. Br J Cancer 44:775–786.

Twentyman, P.R. (1980) Response to chemotherapy of EMT6 spheroids as measured by growth delay and cell survival. Eur J Cancer 42:297–304.

Uchida, I.A., Lin, C.C. (1974) Quinacrine fluorescent patterns. In Yunis, J. (ed): "Human Chromosome Methodology," 2nd ed. New York, Academic Press, pp. 47–58.

Unkless, I., Dano, K., Kellerman, G., Reich, E. (1974) Fibrinolysis associated with oncogenic transformation. Partial purification and characterization of cell factor, a plasminogen activator. J Biol Chem 249:4295–4305.

Ure, J.M. (1992) A rapid and efficient method for freezing and recovering clones of embryonic stem cells. Trends Genet 8:6.

Vago, C. (ed) (1971) "Invertebrate Tissue Culture, Vol. 1." New York, Academic Press.

Vago, C. (ed) (1972) "Invertebrate Tissue Culture, Vol. 2." New York, Academic Press.

Vaheri, A., Ruoslahti, E., Westermark, B., Ponten, J. (1976) A common cell-type specific surface antigen in cultured human glial cells and fibroblasts: Loss in malignant cells. J Exp Med 143:64–72.

Van Beek, W.P., Glimelius, B., Nilson, K., Emmelot, P. (1978) Changed cell surface glycoproteins in human glioma and osteosarcoma cells. Cancer Lett 5:311–317.

Van der Bosch, J., Masui, H., Sato, G. (1981) Growth characteristics of primary tissue cultures from heterotransplanted human colorectal carcinomas in serum-free medium. Cancer Res 41:611–618.

Van Roozendahl, C.E.P., van Ooijen, B., Klijn, J.G.M., Claasen, C., Eggermont, A.M.M., Henzen-Logmans, S.C., Foekens, J.A. (1992) Stromal influences on breast cancer cell growth. Br J Cancer 65:77–81.

Van Someren, H., Van Hemegowyen, H.B., Los, W., Wurzer-Figurelli, E., Doppert, B., Yerylolt, M. Meera Khan, P. (1974) Enzyme electrophoresis on cellulose acetate gel II zymogram patterns in man-Chinese hamster cell hybrids. Humangenetik 25:189–201.

VanDiggelen, O., Shin, S., Phillips, D. (1977) Reduction in cellular tumorigenicity after mycoplasma infection and elimination of mycoplasma from infected cultures by passage in nude mice. Cancer Res 37:2680–2687.

Van't Hof, J. (1968) In Prescott, D.M. (ed): "Methods in Cell Physiology." New York, Academic Press, p. 95.

Van't Hof, J. (1973) Cell cycle analysis B. In Kruse, P., Patterson, M.K. (eds): "Tissue Culture Techniques and Application." New York, Academic Press, pp. 423–428.

Varner, H.H., Hewitt, A.T., Martin, G.R. (1984) Isolation of chondronectin. In Barnes, D.W., Sirbasku, D.A., Sato, G.H. (eds): "Cell Culture Methods for Molecular and Cell Biology, Vol. 1." New York, Alan R. Liss, pp. 239–244.

Varon, S., Manthorpe, M. (1980) Separation of neurons and glial cells by affinity methods. In Fedoroff, S., Hertz, L. (eds): "Advances in Cellular Neurobiology, Vol. 1." New York, Academic Press, pp. 405–442.

Venitt, S. (1984) "Mutagenicity Testing. A Practical Approach." Oxford, IRL Press.

Venitt, S., Parry, J.M. (eds) (1984) Background to mutgenicity testing. In "Mutgenicity Testing, A Practical Approach." Oxford, IRL Press.

Venitt, S., Crofton-Sleigh, C., Forster, R. (1984) Bacterial mutation assays using reverse mutation. In Venitt, S., Parry, J.M. (eds): "Mutgenicity Testing, A Practical Approach." Oxford, IRL Press.

Visser, J.W., De Vries, P. (1990) Identification and purification of murine hematopoietic stem cells by flow cytometry. Methods Cell Biol 33:451–468.

Vistica, D.T., Skehan, P., Scudiero, D., Monks, A., Pittman, A., Boyd, M.R. (1991) Tetrazolium-based assays for cellular viability: A critical examination of selected parameters affecting formazan production. Cancer Res 51:2515–2520.

Vlodavsky, I., Lui, G.M., Gospodarowicz, D. (1980) Morphological appearance, growth behavior and migratory activity of human tumor cells maintained on extracellular matrix versus plastic. Cell 19:607–617.

Von Hoff, D.D., Clark, G.M., Weis, G.R., Marshall, M.H., Buchok, J.B., Knight, W.A., Lemaistre, C.F. (1986) Use of in vitro dose rsponse effecst to select antineoplastics for high dose or regional administration regimens. J Clin Oncol 4:18–27.

Voyta, J.C., Via, D.P., Butterfield, C.E., Zetter, B.R. (1984) Identification and isolation of endothelial cells based on their increased uptake of acetylated-low density lipoprotein. J Cell Biol 99:2034–2040.

Vries, J.E., Benthem, M., Rumke, P. (1973) Separation of viable from nonviable tumor cells by flotation on a Ficoll-triosil mixture. Transplantation 15:409–410.

Walker, C.R., Bandman, E., Strohman, R.C. (1979) Diazepam induces relaxation of chick embryo muscle fibers *in vitro* and inhibits myosin synthesis. Exp Cell Res 123:285–291.

Wallace, D.H., Hegre, O.D. (1979) Development *in vitro* of epithelial-cell monolayers derived from fetal rat pancreas. In Vitro 15:270.

Walter, H. (1975) Partition of cells in two-polymer aqueous phases: A method for separating cells and for obtaining information on their surface properties. In Prescott, D.M. (ed): "Methods in Cell Biology." New York, Academic Press, pp. 25–50.

Walter, H. (1977) Partition of cells in two-polymer aqueous phases: A surface affinity method for cell separation. In Catsimpoolas, N. (ed): "Methods of Cell Separation." New York, Plenum Press, pp. 307–354.

Wang, H.C., Fedoroff, S. (1972) Banding in human chromosomes treated with trypsin. Nature New Biol 235:52–53.

Wang, H.C., Fedoroff, S. (1973) Karyology of cells in culture E. Trypsin technique to reveal G-bands. In Kruse, P.F., Patterson, M.J. (eds): "Tissue Culture Methods and Applications." New York, Academic Press, pp. 782–787.

Wang, R.I. (1976) Effect of room fluorescent light on the deterioration of tissue culture medium. In Vitro 12:19–22.

Warnock, M. (1985) "A Question of Life. The Warnock Report on Human Fertilisation and Embryology." Oxford, Basil Blackwell.

Warren, L., Buck, C.A., Tuszynski, G.P. (1978) Glycopeptide changes and malignant transformation. A possible role for carbohydrate in malignant behavior. Biochim Biophys Acta 516:97.

Watret, G.E, Pringle, C.R., Elliott, R.M. (1985) Synthesis of

Bunyavirus-specific proteins in a continuous cell line (XTC-2) derived from *Xenopus laevis*. J Gen Virol 66:473-482.

Watson, J.V. (1991) Introduction to flow cytometry. Laboratory Handbooks & Techniques. Cambridge, Cambridge University Press.

Watson, J.V., Erba, E. (1992) Flow Cytometry. In Freshney, R.I. (ed): "Animal Cell Culture, a Practical Approach." Oxford, IRL Press at Oxford University Press, pp. 165-212.

Watt, F. (1991) Annual Meeting of European Tissue Culture Society, Krackow, Poland.

Watt, J.L., Stephen, G.S. (1986) Lymphocyte culture for chromosome analysis. In Rooney, D.E., Czepulkowski, B.H. (eds): "Human Cytogenetics, a Practical Approach." Oxford, IRL Press at Oxford University Press, pp. 39-56.

Waymouth, C. (1959) Rapid proliferation of sublines of NCTC clone 929 (Strain L) mouse cells in a simple chemically defined medium (MB752/1). J Natl Cancer Inst 22:1003.

Waymouth, C. (1970) Osmolality of mammalian blood and of media for culture of mammalian cells. In Vitro 6:109-127.

Waymouth, C. (1974) To disaggregate or not to disaggregate. Injury and cell disaggregation, transient or permanent? In Vitro 10:97-111.

Waymouth, C. (1977) In Evans, V.I., Perry, V., Vincent, M.M. (eds): "Manual of American Tissue Culture Association," 3:521.

Waymouth, C. (1979) Autoclavable medium AM 77B. J Cell Physiol 100:548-550.

Waymouth, C. (1984) Preparation and use of serum-free culture media. In Barnes, W.D. Sirbasku, D.A., Sato, G.H. (eds): "Cell Culture Methods for Molecular and Cell Biology, Vol. 1. Methods for Preparation of Media, Supplements, and Substrata for Serum-Free Animal Cell Culture." New York, Alan R. Liss, pp. 23-68.

Weber, M.M., Stonington, O.G., Poche, P.A. (1974) Epithelial outgrowth from suspension cultures of human prostatic tissue. In Vitro 10:196-205.

Weibel, E.R., Palade, G.E. (1964) New cytoplasmic components in arterial endothelia. J Cell Biol 23:101-102.

Weichselbaum, R., Epstein, I., Little, J.B. (1976) A technique for developing established cell lines from human osteosarcomas. In Vitro 12:833-836.

Weinberg, R.A. (ed) (1989) "Oncogenes and the Molecular Origins of Cancer." Cold Spring Harbor, NY, Cold Spring Harbor Laboratory Press.

Weiss, M.C., Green, H. (1967) Human-mouse hybrid cell lines containing partial complements of human chromosomes and functioning human genes. Proc Natl Acad Sci USA 58:1104-1111.

Weiss, R., Teich, N., Varmus, H., Coffin, C. (eds) (1982, 1985) "Molecular and Biology of the Tumour Viruses." Cold Spring Harbor, NY, Cold Spring Harbor Laboratory Press.

Wells, D.L., Lipper, S.L., Hilliard, J.K., Stewart, J.A., Holmes, G.P., Herrmann, K.L., Kiley, M.P., Schonberger, L.B. (1989) Herpes virus simiae contamination of primary rhesus monkey kidney cell cultures. CDC recommendations to minimize risks to laboratory personnel. Diagn Microbiol Infect Dis 12:333-335.

Wessells, N.K. (1977) "Tissue Interactions and Development." Menlo Park, CA, W.A. Benjamin.

Westermark, B. (1974) The deficient density-dependent growth control of human malignant glioma cells and virus-transformed glialike cells in culture. Int J Cancer 12:438-451.

Westermark, B. (1978) Growth control in miniclones of human glial cells. Exp Cell Res 111:295-299.

Westermark, B., Wasteson, A. (1975) The response of cultured human normal glial cells to growth factors. In Luft and Hall (eds): "Advances in Metabolic Disorders, Vol. 8." New York, Academic Press, pp. 85-100.

Westermark, B., Ponten, J., Hugosson, R. (1973) Determinants for the establishment of permanent tissue culture lines from human gliomas. Acta Pathol Microbiol Scand A 81:791-805.

Whetton, A.D., Dexter, T.M. (1986) Haemopoietic growth factors. Trends Biochem Sci 11:207-211.

White, R., Lalouel, J-M. (1987) Chromosome mapping with DNA markers. Sci Am 258:20-28.

Whitlock, C.A., Robertson, D., Witte, O.N. (1984) Murine B cell lymphopoiesis in long term culture. J Immunol Methods 67:353-369.

Whittle, W.L., Kruse, P.F. (1973) Replicate roller bottles. In Kruse, P.F., Patterson, M.K. (eds): "Tissue Culture Methods and Applications." New York, Academic Press, pp. 327-331.

Whur, P., Magudia, M., Boston, I., Lockwood, J., Williams, D.C. (1980) Plasminogen activator in cultured Lewis lung carcinoma cells measured by chromogenic substrate assay. Br J Cancer 42:305-312.

Wiepjes, G.J., Prop, F.J.A. (1970) Improved method for preparation of single-cell suspensions from mammary glands of adult virgin mouse. Exp Cell Res 61:451-454.

Wigler, M., Sweet, R., Sim, G.K., Wold, B., Pellicer, A., Lacy, E., Maniatis, T., Silverstein, S., Axel, R. (1979) Transformation of mammalian cells with genes from procaryotes and eucaryotes. Cell 16:777-785.

Wilkins, L., Gilchrest, B.A., Szabo, G., Weinstein, R., Maciag, T. (1985) The stimulation of normal human melanocyte proliferation in vitro by melanocyte growth factor from bovine brain. J Cell Physiol 122:350.

Willey, J.C., Moser, C.E., Jr., Lechner, J.F., Harris, C.C. (1984) Differential effects of 12-0-tetradecanoylphorbol-13-acetate on cultured normal and neoplastic human bronchial epithelial cells. Cancer Res 44:5124-5126.

Willingham, M.C., Pastan, I. (1975) Cyclic AMP modulates microvillus formation and agglutinability in transformed and normal mouse fibroblasts. Proc Natl Acad Sci USA 72:1263-1267.

Wilson, J.K.V., Bittner, G.N., Oberley, T.D., Meisner L.F., Weese, J.L. (1987) Cell culture of human colon adenomas and carcinomas. Cancer Res 47:2704-2713.

Wilson, P.D., Dillingham, M.A., Breckon, R., Anderson, R.J. (1985) Defined human renal tubular epithelia in culture: Growth, characterization, and hormonal response. Am J Physiol 248:F436-F443.

Wilson, P.D., Schrier, R.W., Breckon, R.D., Gabow, P.A.

(1986) A new method for studying human polycystic kidney disease epithelia in culture. Kidney Int 30:371–378.

Winterton, A. (1989) "Review of the Guidance on the Research Use of Fetuses and Fetal Material." London, Her Majesty's Stationery Office, Cm 762.

Witkowski, J.A. (1990) The inherited character of cancer–an historial survey. Cancer Cells 2:229–257.

Wolff, D.A., Pertoft, H. (1972) Separation of HeLa cells by colloidal silica density gradient centrifugation. J Cell Biol 55:579.

Wolff, E., Wolff, E. (1952) La determination de la differentiation sexuelle de la syrinx du canard cultivé *in vitro*. Bull Biol 86:325.

Wolff, E.T., Haffen, K. (1952) Sur une méthode de culture d'organes embryonnaires in vitro. Tex Rep Biol Med 10:463-472.

Wright, J.E., Dendy, P.P. (1976) Identification of abnormal cells in short-term monolayer cultures of human tumor specimens. Acta Cytol (Baltimore) 20:328–334.

Wright, W.C., Daniels, W.P., Fogh, J. (1981) Distinction of seventy-one cultured human tumor cell lines by polymorphic enzyme analysis. J Natl Cancer Inst 66:239–248.

Wu, R., Wu, M.M.J. (1986) Effects of retinoids on human bronchial epithelial cells: Differential regulation of hyaluronate synthesis and keratin protein synthesis. J Cell Physiol 127:73–82.

Wu, Y.J., Parker, L.M., Binder, N.E., Beckett, M.A., Sinard, J.H., Griffiths, C.T., Rheinwald, J.G. (1982) The mesothelial keratins: A new family of cytoskeletal proteins identified in cultured mesothelial cells and nonkeratinizing epithelia. Cell 31:693–703.

Wuarin, L., Verity, M.A., Sidell, N. (1991) Effects of interferon-gamma and its interaction with retinoic acid on human neuroblastoma differentiation. Int J Cancer 48:136–141.

Wurster-Hill, D., Cannizzaro, L.A., Pettengill, O.S., Sorenson, G.D., Cate, C.C., Maurer, L.H. (1984) Cytogenetics of small cell carcinoma of the lung. Cancer Genet Cytogenet 13:303–330.

Wyllie, F.S., Bond, J.A., Dawson, T., White, D., Davies, R., Wynford-Thomas, D. (1992) A phenotypically and karyotypically stable human thyroid epithelial line conditionally immortalized by SV40 large T antigen. Cancer Res 52:2938–2945.

Wysocki, L.J., Sata, V.L. (1978) "Panning" for lymphocytes: A method for cell selection. Proc Natl Acad Sci USA 75:2844–2848.

Yaffe, D. (1968) Retention of differentiation potentialities during prolonged cultivation of myogenic cells. Proc Natl Acad Sci USA 61:477–483.

Yaffe, D. (1971) Developmental changes preceding cell fusion during muscle cell differentiation in vitro. Exp Cell Res 66:33–48.

Yamada, K.M. (1991) Fibronectin and other cell interactive glycoproteins. In Hay, E.D. (ed): "Cell Biology of Extracellular Matrix", 2nd ed. New York, Plenum Press, pp. 111–148.

Yan, G., Fukabori, Y., Nikolaropoulos, S., Wang, F., McKeehan, W.L. (1993) Heparin-binding keratinocyte growth factor is a candidate stromal to epithelial cell andromedin. Mol Endocrinol (in press).

Yanai, N., Suzuki, M., Obinata, M. (1991) Hepatocyte cell lines established from transgenic mice harboring temperature-sensitive simian virus 40 large T-antigen gene. Exp Cell Res 197:50–56.

Yang, J., Richards, J., Bowman, P., Guzman, R., Enami, J., McCormick, K., Hamamoto, S., Pitelka, D.. Nandi, S. (1979) Sustained growth and 3-dimensional organization of primary mammary tumor epithelial cells embedded in collagen gels. Proc Natl Acad Sci USA 76:3401.

Yang, J., Richards, J. Guzman, R., Imagawa, W. Nandi, S. (1980) Sustained growth in primary cultures of normal mammary epithelial cells embedded in collagen gels. Proc Natl Acad Sci USA 77:2088–2092.

Yang, J., Elias, J.J., Petrakis. N.L., Wellings, S.R., Nandi, S. (1981) Effects of hormones and growth factors on human mammary epithelial cells in collagen gel culture. Cancer Res 41:1021–1027.

Yang, Y-C., Ciarletta, A.B., Temple, P.A., Chung, M.P., Kovacic, S., et al. (1986) Human IL3 (Multi-CSF): Identification by expression cloning of a novel hematopoietic growth factor related to murine IL-3. Cell 47:3–10.

Yasin, R., Kundu, D., Thomson, E.J. (1981) Growth of adult human cells in culture at clonal densities. Cell Diff 10:131–137.

Yasumura, Y., Tashijian, A.H., Sato, G. (1966) Establishment of four functional clonal strains of animal cells in culture. Science 154:1186–1189.

Yavin, Z., Yavin, E. (1980) Survival and maturation of cerebral neurons on poly(L-lysine) surfaces in the absence of serum. Dev Biol 75:454–460.

Yeoh, G.C.T., Hilliard, C., Fletcher, S., Douglas, A. (1990) Gene expression in clonally derived cell lines produced by in vitro transformation of rat fetal hepatocytes: Isolation of cell lines which retain liver-specific markers. Cancer Res 50:75–93.

Yerganian, G., Leonard, M.J. (1961) Maintenance of normal in situ chromosomal features in long-term tissue cultures. Science 133:1600–1601.

Yoshida, Y., Hilborn, V., Hassett, C., Mezfi, P., Byers, M.J., Freeman, A.G. (1980) Characterization of mouse fetal lung cells cultured on a pigskin substrate. In Vitro 16:433–445.

Yuhas, J.M., Li, A.P., Martinex, A.O., Ladman, A.J. (1977) A simplified method for production and growth of multicellular tumour spheroids (MTS). Cancer Res 37:3639–3643.

Yuspa, S.H., Hawley-Nelson, P., Stanley, I.R., Hennings, H. (1980) Epidermal cell culture. Transplant Proc 12:114–122.

Yuspa, S.H., Koehler, B., Kulesz-Martin, M., Hennings, H. (1981) Clonal growth of mouse epidermal cells in medium with reduced calcium concentration. J Invest Dermatol 76:144–146.

Yusufi, A.N.K., Szczepanska-Konkel, M., Kempson, S.A., McAteer, J.A., Dousa, T.P. (1986) Inhibition of human renal epithelial Na^+/Pi cotransport by phosphonoformic acid. Biochem Biophys Res Commun 139:679–686.

Zaroff, L., Sato, G.H., Mills, S.E. (1961) Single-cell platings from freshly isolated mammalian tissue. Exp Cell Res 23:565–575.

Zawydiwski, R., Duncan, G.R. (1978) Spontaneous ^{51}Cr release by isolated rat hepatocytes: An indicator of membrane damage. In Vitro 14:707–714.

Zetter, B.R. (1981) The endothelial cells of large and small blood vessels. Diabetes 30(suppl 2):24–28.

Zwain, I.H., Morris, P.L., Cheng, C.Y. (1991) Identification of an inhibitory factor from a Sertoli clonal cell line (TM4) that modulates adult rat Leydig cell steroidogenesis. Mol Cell Endocrinol 80:115–126.

GENERAL TEXTBOOKS FOR FURTHER READING

Adolphe, M., Barlovatz-Meimon, G. (1985) "Culture de Cellules Animales; Methodologies, Applications." Paris, Editions INSERM. *Collection of techniques-oriented chapters on basic and advanced aspects of tissue culture.*

Alberts, B., Bray, D., Lewis, J., Raff, M., Roberts, K., Watson, J.D. (1989) "Molecular Biology of the Cell, 2nd edition." New York, Garland Publishing, Inc. *Probably the best general textbook on cell and molecular biology.*

Barnes, D.W., Sirbasku, D.A., Sato, G.H. (eds) (1984) "Cell Culture Methods for Molecular and Cell Biology." 4 vols. New York, Alan R. Liss.

Butler, M. (ed) (1991) "Mammalian Cell Biotechnology, a Practical Approach." Oxford, IRL Press at Oxford University Press. *Useful introduction to basic biotechnology.*

Crowe, R., Ozer, H., Rifkin, D. (1978) "Experiments with Normal and Transformed Cells." Cold Spring Harbor, NY, Cold Spring Harbor Laboratory. *Laboratory exercises for senior undergraduate and graduate students.*

Dealtry, G.B., Rickwood, D. (1992) "Cell Biology LabFax." Oxford, Bios Scientific Publishers. *Useful collection of data on microscopy, cell structure, oncogenes, growth factors, inhibitors, and radioisotopes in biology.*

Dixon, R.A. (1985) "Plant Cell Culture, a Practical Approach." Oxford, IRL Press.

Doyle, A., Griffiths, J.B., Newell, D.G. (eds) (1993) "Cell and Tissue Culture: Laboratory Procedures." Chichester, England, John Wiley & Sons. *Loose-leaf compendium of general and specialized techniques with regular updates. Very expensive but very good source for a wide variety of techniques.*

Doyle, A., Hay, R., Kirsop, B.E. (eds) (1990) "Living Resources for Biotechnology." Cambridge, Cambridge University Press. *Useful information on databases and quality control.*

Fogh, J. (1975) "Human Tumor Cells In Vitro." New York, Plenum Press. *Some useful listings of human tumor types in culture.*

Freshney, R.I. (1992) "Animal Cell Culture, a Practical Approach," 2nd ed. Oxford, IRL Press.

Freshney, R.I. (ed) (1992) "Culture of Epithelial Cells." New York, Wiley-Liss. *Invited chapters on specialized culture of epithelium; technique oriented.*

Freshney, R.I., Pragnell, I.B., Freshney, M.G. (eds) (1994) "Culture of Haemopoietic Cells." New York, Wiley-Liss (In press). *Second in the series "Culture of Specialized Cells." Invited chapters on specialized techniques.*

Harris, C.C., Trump, B.F., Stoner, G.D. (1981) "Normal Human Tissue and Cell Culture." In Prescott, D.M. (series ed): "Methods in Cell Biology." New York, Academic Press.

Jakoby, W.B., Pastan, I.H. (eds) (1979): "Methods in Enzymology, Vol. 57, Cell Culture." New York, Academic Press. *Good for specialized techniques, matrix, serum-free media but some sections a bit dated.*

Kruse, P.F., Patterson, M.K. (eds) (1973): "Tissue Culture Methods and Applications." New York, Academic Press. *Some useful specialized techniques but a little out of date in places.*

Maramorosch, K. (1976) "Invertebrate Tissue Culture." New York, Academic Press.

Masters, J.R.W. (1991) "Human Cancer in Primary Culture." London, Kluwer. *Product of an European Tissue Culture Society workshop.*

Paul, J. (1975) "Cell and Tissue Culture." Edinburgh, Churchill Livingstone. *Though dated, still a good basic textbook.*

Pollack, R. (ed) (1981) "Reading in Mammalian Cell Culture," 2nd ed. Cold Spring Harbor, NY, Cold Spring Harbor Laboratory. *Very good compilation of key papers in the field. Good tutorial and general interest. Good for teaching.*

Reinert, J., Yeoman, M.M. (1982) "Plant Cell and Tissue Culture. A Laboratory Manual." Berlin, Heidelberg, New York, Springer-Verlag.

Sato, G., Pardee, A.B., Sirbasku, D.A. (1982) "Growth of Cells in Hormonally Defined Media." Cold Spring Harbor Conferences on Cell Proliferation, Vol. 9. Cold Spring Harbor, NY, Cold Spring Harbor Laboratory. *Good review of serum-free culture.*

Shahar, A., de Vellis, J., Vernadakis, A., Haber, B. (1989) "A Dissection and Tissue Culture Manual of the Nervous System." New York, Wiley-Liss. *Useful short protocols; well illustrated.*

Spier, R.E., Griffiths, J.B. (eds) (1985) "Animal Cell Biotechnology," 4 vols. New York, Academic Press. *Invited chapters covering a wide range of biotechnological applications of cell culture.*

USEFUL JOURNALS

Tissue Culture Techniques Oriented

Cytotechnology
In Vitro, Cell and Developmental Biology
Journal of Tissue Culture Methods

Cell Biology

Cell
Cell Biology, International Reports
Cellular Biology
Current Opinion in Cell Biology
European Journal of Cell Biology
European Journal of Cancer and Clinical Oncology
Experimental Cell Biology
Experimental Cell Research
Journal of Cell Biology
Journal of Cellular Physiology
Journal of Cell Science

Cancer

British Journal of Cancer
Cancer Research
Cell Growth & Differentiation
International Journal of Cancer
Journal of the National Cancer Institute

Index

The numbers in bold type indicate primary references